HAZARDOUS WASTE REGULATIONS

HAZARDOUS WASTE REGULATIONS

AN INTERPRETIVE GUIDE

ALEX MALLOW

 VAN NOSTRAND REINHOLD COMPANY
NEW YORK   CINCINNATI   ATLANTA   DALLAS   SAN FRANCISCO
LONDON   TORONTO   MELBOURNE

Van Nostrand Reinhold Company Regional Offices:
New York   Cincinnati   Atlanta   Dallas   San Francisco

Van Nostrand Reinhold Company International Offices:
London   Toronto   Melbourne

Copyright © 1981 by Van Nostrand Reinhold Company

Library of Congress Catalog Card Number: 81-402
ISBN: 0-442-21935-0

Manufactured in the United States of America

Published by Van Nostrand Reinhold Company
135 West 50th Street, New York, N.Y. 10020

Published simultaneously in Canada by Van Nostrand Reinhold Ltd.

15   14   13   12   11   10   9   8   7   6   5   4   3   2   1

Library of Congress Cataloging in Publication Data

Mallow, Alex.
    Hazardous waste regulations.

    Includes index.
    1. Hazardous wastes—Law and legislation—United
States. I. Title.
KF3958.M34   344.73'0462   81-402
ISBN 0-442-21935-0   AACR1

PREFACE

The purpose of this interpretive guide is to reformat the almost endless flow of words covering some 150 pages of 3-column Federal Register hazardous waste regulations into small, distinct, and digestible pieces—with descriptive titles. This further permits a check-list approach for easy identification of applicable pieces of the regulatory puzzle by the regulated community. The great utility of this approach is readily demonstrated in Chapter 10, wherein hard regulatory requirements, along with regulations of somewhat lesser concern and interest, are offered in a check-list format for five different scenarios. Make no mistake about it, the regulations are lengthy and complex, and will require much time and effort to understand the full meaning, consequences, and impact of what is destined to be the Environmental Protection Agency's (EPA) largest regulatory program. Hopefully, this book will help to ease the task.

The regulation content of this book parallels the related sections of the hazardous waste law (Subtitle C of the Resource Conservation and Recovery Act or RCRA). Chapter 1 provides an introduction to the hazardous waste law and regulations. Chapter 2 deals with general provisions applicable to several of the regulatory parts. Identification and listing of hazardous waste is covered in Chapter 3. Chapters 4 and 5 address standards applicable to generators and transporters, respectively, of hazardous waste. Chapter 6 covers final and interim standards applicable to owners and operators of hazardous waste treatment, storage, and disposal facilities. The seventh chapter covers permitting for treatment, storage, and disposal facilities. Chapter 8 provides guidelines for authorized State hazardous waste programs. Preliminary notification of hazardous waste activity is covered in Chapter 9. The tenth and final chapter provides a check-list approach to regulatory requirements for five different scenarios.

Appendix A provides definitions of words and phrases that appear in the regulations. It is essential to learn many of these definitions. As one EPA official recently counseled, "To understand the regulations you must become a definition-person."

Appendix B contains a complete copy of the Resource Conservation and Recovery Act (RCRA), as amended by the Quiet Communities Act of 1978.

All recently promulgated regulations relating to hazardous waste management are included as Appendix C.

Appendix D contains proposed hazardous waste regulations, which must go through a normal public review process and will therefore likely undergo some changes.

Finally, Appendix E contains the contents tables only of Department of Transportation (DOT) regulations of concern for those handling hazardous wastes.

ACKNOWLEDGMENTS

Appreciation is extended to Roy Fagin, M. D., director of Grumman Aerospace Corporation's Environmental Planning and Control Program. He urged and directed me to achieve an in depth involvement and understanding of hazardous waste regulations, and was later greatly supportive of my efforts in writing this book. To my wife Marilyn Mallow, and Susan Heuler, I offer much thanks and appreciation for having typed and proofread the major portion of the manuscript.

A. Mallow
Commack, New York

CONTENTS

HAZARDOUS WASTE REGULATIONS

1
INTRODUCTION

Improper management of hazardous waste is probably the most serious environmental problem in the United States today. The Environmental Protection Agency (EPA) estimates that in 1979 this country generated almost 60 million metric tons of hazardous waste, but only 10 percent was managed in an environmentally sound manner. The remainder—more than 50 million tons—was transported, treated, stored, or disposed of in a manner which potentially threatens human health and the environment.

EPA estimates that there are 760,000 large and small generators of hazardous wastes producing the 60 million tons of hazardous waste a year. The greatest amount of these wastes come from very large generators, typically large manufacturing facilities. Just over 5 percent, or 40,000, of the total number of generators each produce more than 5,000 kilograms per month of hazardous wastes. Yet, these large generators account for 97.7 percent of the total quantity of hazardous waste. Roughly 91 percent, or 695,000 of the generators produce less than 1,000 kilograms per month (regulatory exemption breakpoint), yet contribute only 1 percent or 600,000 tons per year of the total hazardous waste generated. At levels of generation below 100 kilograms per month, 74 percent or 563,000 of the generators produce only 0.23 percent or 138,000 tons per year of hazardous waste.

From the above, it is seen that EPA expects about 65,000 generators in the 1,000-kilogram-per-month class to be initially regulated. With an additional 10,000 transporters and 30,000 treatment, storage, and disposal facilities coming under the domain of the Resource Conservation and Recovery Act, it is easy to see why this program is destined to be EPA's largest. Even with an initial 100,000 or more comprising the regulated community, this is merely the start. The agency will initiate rulemaking within 2-5 years to phase in expanded coverage of small generators down to those generating more than 100 kilograms per month. This will bring in an additional 140,000 generators, raising the regulated community to nearly one-quarter million—a monumental regulatory task.

OVERVIEW OF REGULATIONS AND ASSOCIATED RCRA SECTIONS

Subtitle C of the Solid Waste Disposal Act, as amended by the Resource Conservation and Recovery Act (RCRA) of 1976, as amended, directs the Environmental Protection Agency to promulgate comprehensive regulations to protect human health and the environment from the improper management of hazardous waste. When fully implemented, this program will provide "cradle-to-grave" regulation of hazardous wastes.

Table 1-1 relates the part number in Title 40 of the Code of Federal Regulation (CFR) with its associated RCRA Section number. The commonly used format relating Title with Part number is illustrated by the following example: 40 CFR 260. The "40" represents the title number of the Code of Federal Regulation (CFR). The "260" represents the Part number for a particular group of regulations. Part number is then broken down into decimal divisions or sections, representing a more detailed tuning of the many regulatory pieces.

Part 260 sets forth definitions of words and phrases which appear in Parts 261-265, corresponding to RCRA Sections 3001-3004. That Part also contains provisions which are generally applicable to all those regulations.

Section 3001 of Subtitle C directs EPA to identify the characteristics of and to list those hazardous wastes which are subject to regulation under Subtitle C. The regulations are codified in Part 261, as seen from Table 1-1.

Parts 262 and 263, corresponding to Sections 3002 and 3003 of RCRA, require EPA to establish standards for generators and transporters of hazardous waste. These rules will ensure proper recordkeeping and reporting, the use of a manifest system to track hazardous waste, the use of proper labels and containers, and the delivery of the waste to properly permitted facilities.

To ensure that these facilities are designed, constructed, and operated in a manner which protects human health and the environment, Section 3004 of RCRA directs EPA to promulgate technical, administrative, monitoring, and financial standards. Parts 264 and 265 reflect the final and interim facility standards. Part 266 will be promulgated in the future covering waste-specific facility standards.

The facility standards will be used by EPA to issue permits to owners and operators of facilities under RCRA Section 3005 (Parts 122 and 124).

For those States interested in administering the RCRA program instead of EPA, Section 3006 (Part 123) requires the Agency to issue guidelines under which States may seek authorization to carry out the program.

Table 1-1

RCRA Section No.	40 CFR Part No.	Descriptive Title
—	260	(Definitions used in other parts corresponding to Sections 3001–3004 rules, and general provisions applicable to these Parts)
3001	261	Identification and Listing of Hazardous Waste
3002	262	Standards Applicable to Generators of Hazardous Waste
3003	263	Standards Applicable to Transporters of Hazardous Waste
3004	264	Standards Applicable to Owners and Operators of Hazardous Waste Treatment, Storage, and Disposal Facilities
3004	265	Interim Status Standards Applicable to Owners and Operators of Hazardous Waste Treatment, Storage, and Disposal Facilities[1]
3005	122 and 124	Permits for Treatment, Storage, and Disposal Facilities
3006	123	Guidelines For Authorized State Hazardous Waste Programs
3010	[2]	Preliminary Notification of Hazardous Waste Activity

[1] Waste-specific facility standards will be promulgated in the future under Part 266.
[2] Federal Register (FR) notice of 2/26/80 (See Chapter 9 and Appendix C of this volume).

Finally, under Section 3010, all person engaging in activities subject to control under Sections 3002 through 3004 must notify EPA or States having authorized RCRA hazardous waste programs.

THE REGULATORY OUTLOOK

While the regulations that have recently been promulgated represent the bulk of the hazardous waste regulation package, changes can be expected over the next decade as EPA gains experience in managing this huge and complex program. In the nearer term, EPA expects to amend Part 261 bringing additional wastes into the hazardous waste management system in late 1980 or early 1981. The same is true for facility standards under Part 264, which will also affect some portions of the permitting requirements of Part 122.

Some proposed hazardous waste regulations are included in Appendix C of this volume. These will have to pass through a normal public review process and will thus likely undergo changes.

SUMMARY

The bulk of the contents of Chapters 2-9 are tables that provide detailed hazardous waste regulatory requirements, along with the associated Section number of the regulations. This affords the reader the opportunity to refer back to the exact regulations of concern for further study and clarification as may be necessary. Detailed regulatory requirements for five different scenarios are provided in the last chapter.

2
GENERAL PROVISIONS APPLICABLE TO PARTS 260-265: PART 260

The purpose of Part 260 is to consolidate in one place several provisions that apply to Parts 260-265.

Rules concerning the designation and handling of confidential information are covered, as are rules of grammatical construction intended to assist EPA in eliminating awkward phrases in preparation of the rules. General procedures are included which EPA will use in acting on petitions to amend Parts 260-265. Also included are specific procedures applicable to petitions for approval of equivalent testing and analytical methods, as well as petitions for delisting wastes at a particular facility.

Definitions of key words and phrases used in Parts 261-265 form another element of Part 260. These definitions, along with additional ones included in RCRA and Parts 122, 123, and 124, have been consolidated in Appendix A of this book.

Table 2-1 addresses the details of Part 260 (other than definitions), and contains the following 7 areas:

    I. Applicability
   II. Availability of Information to Public
  III. Use of Number and Gender
  IV. Definitions
   V. General Approach and Requirements for Rulemaking Petitions
  VI. Additional Requirements for Petitions for Equivalent Testing or Analytical Methods
 VII. Additional Requirements for Petitions to Amend Part 261 to Exclude a Waste Produced at a Particular Facility (Delisting)

Table 2-1. General Provisions Applicable To Parts 260-265.

Section No.	Description
	I. Applicability
260.1(a)	A. General Part 260 provides definitions of terms, general standards, and overview information applicable to Parts 260-265.
260 Authority	B. Federal Statutes/Authority Sections 1006, 2002(a), 3001-3007, 3010, and 7004 of the Solid Waste Disposal Act, as amended by the Resource Conservation and Recovery Act of 1976, as amended (42 U.S.C. 6905, 6912(a), 6921-6927, 6930, and 6974).
	II. Availability of Information to Public

260.2(a)	A. Nonconfidential Information
	Any nonconfidential information provided to EPA under Parts 260-265 is available to the public.
	(1) Authorities Describing Extent and Manner of Information Availability
	   (a) Freedom of Information Act, 5 U.S.C. Section 552
	   (b) Section 3007(b) of RCRA
	   (c) EPA regulations (40 CFR 2—Public Information) implementing the Freedom of Information Act and Section 3007(b).

260.2(b)	B. Confidential Information (1) Claim of Business Confidentiality Such a claim may be asserted by any person submitting information to EPA in accordance with Parts 260-265, covering part or all of the information.

Table 2-1. General Provisions Applicable To Parts 260–265. (Continued)

Section No.	Description

	(2) Authority Setting Forth Procedures to Follow
	Procedures come under 40 CFR 2.203(b) regulations.
	(3) Extent of Information Disclosed by EPA
	    (a) If Confidentiality Claim Is Filed
	        Disclose in accordance with 40 CFR 2, Subpart B regulations.
	    (b) If No Confidentiality Claim Is Filed
	        All information received by EPA is available to public without notice to person submitting it.

260.3	III. Use of Number and Gender A. General Section 260.3 establishes simple rules of grammatical construction concerning number and gender, in order to allow EPA to simplify the drafting of its final 260 through 265 regulations. It eliminates the need for such awkward phrases as "he/she/it" or "the owner (or in the event there is more than one owner, the owners)."
260.3(a)	B. Word in the Masculine Gender Also includes the feminine and neuter gender.
260.3(b)	C. Words in the Singular Also includes plural.
260.3(c)	D. Words in the Plural Also includes the singular.
Subpart B	IV. Definitions
260.10	A. General Section 260.10 contains definitions and phrases which are used in Parts 260–265, and has been consolidated in Appendix A (Definitions) of this book, along with additional definitions included in RCRA and Parts 122, 123, and 124.
260.20	V. General Approach and Requirements for Rulemaking Petitions
260.20(a)	A. Who May Petition? Any person may petition Administrator to modify or revoke any provisions in Parts 260 through 265.
260.20(b) 260.20(b)(1) 260.20(b)(2) 260.20(b)(3) 260.20(b)(4)	B. Petition Submittal Requirements (1) Petition must be sent to Administrator by certified mail. (2) Name and address of petitioner. (3) Statement of petitioner's interest in proposed action. (4) Description of proposed action, with suggested regulatory language (where appropriate). (5) Statement of need and justification for proposed action. (6) Supporting tests, studies, or other information.

260.20(c)	C. Tentative Administrator Decision to Grant or Deny Petition
	(1) Publication In Federal Register Required
	    (a) Advance notice of proposed rulemaking; or
	    (b) Proposed rule; or
	    (c) Tentative determination to deny petition.
	(2) *Federal Register* notice offers opportunity for written public comment.

260.20(d)	D. Informal Public Hearing on Administrator's Tentative Decision
	(1) At Request of Interested Person
	    (a) Written request must be made to Administrator for his approval.
	    (b) Statement of issues to be raised is required.
	    (c) Explanation why written comments not adequate to communicate person's views also required.
	(2) Administrator may decide on his own motion to hold an informal hearing.

260.20(e)	E. Final Administrator Decision on Petition After evaluating public comments the Administrator will make a final decision by publishing in the *Federal Register* a regulatory amendment or a denial of the petition.

Table 2-1. General Provisions Applicable To Parts 260–265. (Continued)

Section No.	Description
260.21	VI. Additional Requirements for Petitions for Equivalent Testing or Analytical Methods
260.21(a)	A. General Any person seeking to add a testing or analytical method to Parts 261, 264, or 265 may petition for a regulatory amendment under this section and Section 260.20 (general petition requirements).
260.21(a)	B. Basic Requirements For Successful Petition Person must demonstrate to Administrator that proposed method is equal to or superior to corresponding method prescribed in Parts 261, 264, or 265 in terms of: (1) Sensitivity (2) Accuracy (3) Precision (e.g., reproducibility).
260.21(b)	C. Petition Submittal Requirements (1) Information required by Section 260.20(b)
260.21(b)(1)	(2) Full Description of Proposed Method Includes all procedural steps and equipment used in the method.
260.21(b)(2)	(3) Description of types of wastes or waste matrices for which proposed method may be used.
260.21(b)(3)	(4) Comparative Results Proposed method versus corresponding methods of Parts 261, 264, and 265.
260.21(b)(4)	(5) Assessment of factors which may interfere with or limit use of proposed method.
260.21(b)(5)	(6) Quality Control Procedures Description of such procedures necessary to ensure sensitivity, accuracy and precision of proposed method.
260.21(c)	D. Administrator Request for Additional Information after Receiving Petition Such a request on the proposed method may be made in order to adequately evaluate it.
260.21(d)	E. Upon Amendment of Regulations Permitting Use of New Testing Method Method will be incorporated in test manual: "Test Methods For The Evaluation Of Solid Waste: Physical/Chemical Methods," SW-846.
260.21(d) Note	(1) Availability of Test Manual It will be provided to any person on request for inspection or copying at EPA headquarters or any EPA Regional Office.
260.22	VII. Additional Requirements for Petitions to Amend Part 261 to Exclude a Waste Produced at a Particular Facility (Delisting)
260.22(a)	A. General Any person seeking to exclude a waste at a particular generating facility from the lists in Subpart D of Part 261 may petition for a regulatory amendment under this Section and Section 260.20 (general petition requirements).
260.22(a)	B. Basic Demonstration Requirements for a Successful Delisting Petition Petitioner must demonstrate to Administrator that waste produced by a particular generating facility: (1) Does not meet any of the criteria under which the waste was listed as a hazardous waste. (2) Does not meet the criterion of Section 261.11(a)(3) for an acutely hazardous waste listed under Section 261.11(a)(2). (3) Does not exhibit hazardous waste characteristics defined in Subpart C of Part 261 (e.g., ignitable, corrosive, reactive, or toxic).
260.22(b)	C. Procedures to Be Used to Petition Administrator for Regulatory Amendment to Exclude Waste under Sections 261.3(a)(2)(ii) or (c) (1) General May use procedures of this Section (260.22) and Section 260.20. (2) Waste may be either: (a) Listed in Subpart D of Part 261 (b) Contain a waste listed in Subpart D (c) Derived from waste listed in Subpart D. (3) Exclusions may only be issued for a particular generator, storage, treatment, or disposal facility.

Table 2-1. General Provisions Applicable To Parts 260–265. (Continued)

Section No.	Description

	(4) Petitioner must make same demonstration as required in Section 260.22(a). Exception:
	 (a) Where waste is:
	    (i) Mixture of solid waste and one or more listed wastes; or
	    (ii) Is derived from one or more hazardous wastes,
	 the demonstration may be made with respect to each constituent listed waste or the waste mixture as a whole.
	
	 Note: A waste so excluded may still be a hazardous waste by operation of Subpart C (Characteristics of Hazardous Waste) of Part 261.

260.22(c)	D. If Waste Is Listed With Codes I, C, R, or E in Subpart D of Part 261 Petitioner must show that demonstration samples of waste do not exhibit characteristics defined below using applicable test methods prescribed therein: (1) Section 261.21–Ignitability (2) Section 261.22–Corrosivity (3) Section 261.23–Reactivity (4) Section 261.24–EP Toxicity.
260.22(d) 260.22(d)(1) 260.22(d)(2)	E. If Waste Is Listed with Code T (Toxic Waste) in Subpart D of Part 261 Petitioner must demonstrate that: (1) Demonstration samples of waste do not contain the constituent as defined in Appendix VII (Basis for Listing Hazardous Waste) of Part 261, that caused Administrator to list the waste. (a) Appropriate test method(s) of Appendix III (Chemical Analysis Test Methods) of Part 261 must be used. (2) Waste does not meet criterion of Section 261.11(a)(3) when considering the eleven factors of that Section.
260.22(e) 260.22(e)(1) 260.22(e)(2)	F. If Waste Is Listed With Code H (Acute Hazardous Waste) In Subpart D of Part 261 Petitioner must demonstrate that waste does not meet both criteria that follow: (1) The criterion of Section 261.11(a)(2) (Acute hazardous waste). (2) The criterion of Section 261.11(a)(3) when considering the eleven factors cited therein.
260.22(f)	G. This Section reserved for listed radioactive wastes.
260.22(g)	H. This Section reserved for listed infectious wastes.

260.22(h)	I. Demonstration Sample Requirements
	 (1) Quantity
	    Must consist of enough representative samples, but in no case less than four samples.
	 (2) Timing
	    Samples must be taken over a period of time sufficient to represent the variability or uniformity of the waste.

260.22(i)
 	
 260.22(i)(1)
 260.22(i)(2)
 260.22(i)(3)
 260.22(i)(4)
 260.22(i)(5)
 	
 260.22(i)(6)
 	
 260.22(i)(7)
 	
 260.22(i)(8)
 260.22(i)(9)
	J. Petition Requirements
	 In addition to information necessary in general requirements of Section 260.20(b) each, petition must include following:
	 (1) Name and address of laboratory performing sampling or tests of waste.
	 (2) Names and qualifications of persons sampling and testing waste.
	 (3) Dates of sampling and testing.
	 (4) Location of generating facility.
	 (5) Description of manufacturing process or other operations and feed materials producing waste.
	 (6) Assessment of whether such processes, operations or feed materials might produce waste not covered by demonstration.
	 (7) Description of waste.
	 (8) Estimate of average and maximum monthly and annual quantities of waste covered by demonstration.
	 (9) Where Demonstration Based on (Eleven) Factors of Section 261.11(a)(3)
	    Include pertinent data on and discussion of factors delineated in respective criterion for listing a hazardous waste.
	 (10) Description of methodologies and equipment used to obtain the representative samples.
	 (11) Description of Sample Handling and Preparation Techniques
	    Includes techniques used for extraction, containerization, and preservation of samples.

Table 2-1. General Provisions Applicable To Parts 260-265. (Continued)

Section No.	Description
260.22(i)(10)	(12) Description of Tests Performed Includes results.
260.22(i)(11)	(13) Names and model numbers of instruments used in performing tests.
260.22(i)(12)	(14) Following statement signed by generator of waste or his authorized representative is required: "I certify under penalty of law that I have personally examined and am familiar with the information submitted in this demonstration and all attached documents, and that, based on my inquiry of those individuals immediately responsible for obtaining the information, I believe that the submitted information is true, accurate, and complete. I am aware that there are significant penalties for submitting false information, including the possibility of fine and imprisonment."
260.22(j)	K. Administrator Request for Additional Information May be made in order for him to adequately evaluate petition.
260.22(k)	L. An Exclusion Will Only Apply to Waste Generated at Individual Facility Covered by Demonstration It will not apply to waste from any other facility.
260.22(l)	M. Partial Exclusion of Waste Administrator may exclude only part of waste for which the demonstration is submitted where he believes variability of waste justifies partial exclusion.
260.22(m)	N. Temporary Exclusion by Administrator Such an exclusion prior to final decision may be granted if he feels great likelihood that exclusion will be finally granted. (1) Administrator will publish notice of temporary exclusion in *Federal Register.*

3
IDENTIFICATION AND LISTING OF HAZARDOUS WASTE: PART 261

The keystone of Subtitle C of RCRA, covering hazardous waste management, is Section 3001, which requires EPA to identify the characteristics of and to list those solid wastes requiring management as hazardous wastes.

Part 261 regulations is the first phase of EPA's implementation of Section 3001. Four characteristics of hazardous waste are identified and must be used by persons handling solid wastes to determine if the waste is hazardous. Further, the standard lists 103 process wastes as hazardous wastes, and about 400 chemicals.

Persons who generate, transport, treat, store or dispose of hazardous wastes identified or listed in Part 261 must comply with all applicable requirements of Parts 262–265 and Parts 122 and 124, and notification requirements of Section 3010 of RCRA.

In addition to identifying and listing hazardous wastes, this regulation also sets forth the criteria used by EPA to identify characteristics of hazardous wastes and to list hazardous wastes.

Table 3-1 addresses the detailed requirements of Part 261 and contains the following nine areas:

    I. Purpose, Scope, and Applicability
    II. Definitions
    III. Exclusions
    IV. Special Requirements for Hazardous Waste Generated by Small Quantity Generators
    V. Special Requirements for Hazardous Waste Which Is Used, Reused, Recycled, or Reclaimed
    VI. Criteria for Identifying the Characteristics of Hazardous Waste
    VII. Criteria for Listing Hazardous Waste
    VIII. Characteristics of Hazardous Waste
    IX. Lists of Hazardous Wastes

Item IX includes a proposed listing of eleven waste streams as well as listings of 17 generic and 11 process wastes expected to be promulgated in the near future. Other areas that EPA will eventually address in Part 261 include:

- PCB integration
- Regulation of wastes which are used, reused, recovered, or reclaimed
- Radioactive wastes
- Infectious wastes
- Other listed wastes
- Lowering 1000 kilogram per month exemption to 100 kilograms per month (within 2–5 years).

Figures 3-1 through 3-6 contain various groupings of listed wastes, and have been included in this chapter for ready reference. (See pp. 17–22.)

Table 3-1. Identification and Listing of Hazardous Waste.

Section No.	Description
	I. Purpose, Scope, and Applicability
261.1(a)	A. General Part 261 identifies those solid wastes subject to regulation as hazardous wastes under the following Parts: (1) Part 262–Standards Applicable to Generators of Hazardous Waste (2) Part 263–Standards Applicable to Transporters of Hazardous Waste (3) Parts 264 and 265–Final and Interim Status Standards Applicable to Owners and Operators of Hazardous Waste Treatment, Storage, and Disposal Facilities (4) Parts 122 and 124–Permit Requirements (5) Part 123–State Program Requirements (6) Notification Requirement of Section 3010 of RCRA.
261.1(a)(1)	B. Subpart Organization (1) Subpart A (a) Defines term "solid waste" (Section 261.2). (b) Defines term "hazardous waste" (Section 261.3). (c) Identifies those wastes excluded from regulation under Parts 262 through 265 and 122–124 (Section 261.4). (d) Establishes special management requirements for hazardous waste produced by small quantity generators (Section 261.5). (e) Establishes special requirements for hazardous waste which is used, reused, recycled, or reclaimed (Section 261.6).
261.1(a)(2)	(2) Subpart B Sets forth criteria used by EPA to identify characteristics of hazardous waste and to list particular hazardous wastes.
261.1(a)(3)	(3) Subpart C Identifies four characteristics of hazardous wastes: (a) Ignitability (Section 261.21) (b) Corrosivity (Section 261.22) (c) Reactivity (Section 261.23) (d) EP toxicity (Section 261.24).
261.1(a)(4)	(4) Subpart D Lists particular hazardous wastes.
261.1(b) 261.1(b)(2)	C. Basis of Substance Being Considered a Hazardous Waste Yet Not So Designated in Part 261 (1) Imminent Hazard Per Section 7003 of RCRA The Administrator may institute civil actions to abate imminent hazards caused by any "hazardous waste."
261.1(b)(1)	(2) Obtaining Information Per Section 3007 of RCRA EPA is authorized to obtain information on any "hazardous wastes" in order to develop regulations or enforce RCRA.
261 Authority	D. Federal Statutes/Authority Section 1006, 202(a), 3001, and 3002 of the Solid Waste Disposal Act, as amended by the Resource Conservation and Recovery Act of 1976, as amended (42 U.S.C. 6905, 6912, 6921, and 6922).
	II. Definitions
261.2 261.2(a)(1)	A. Definition of Solid Waste (1) Any garbage, refuse, sludge, or "other waste material" not excluded under Section 261.4(a) (Exclusions)
261.2(b)	(2) "Other Waste Material" Any solid, liquid, semisolid, or contained gaseous material which:
261.2(b)(1)	(a) Is "discarded" or being accumulated, stored, or physically, chemically, or biologically treated prior to being "discarded;" or
261.2(b)(2)	(b) Has served its original intended use and sometimes is "discarded;" or (c) Is a manufacturing or mining byproduct and sometimes is "discarded."
261.2(c)	B. Definition of "Discarded" Material A material is considered "discarded" if: (1) It is abandoned by being:
261.2(c)(1)	(a) "Disposed of;" or

Table 3-1. Identification and Listing of Hazardous Waste. (Continued)

Section No.	Description
261.2(c)(2)	(b) Burned or incinerated (but not burned as a fuel); or
261.2(c)(3)	(c) Physically, chemically or biologically treated (other than burned or incinerated) instead of or prior to being "disposed of;" and
	(2) Is not used, reused, reclaimed or recycled.
261.2(d)	C. Definition of "Disposed of" Material
	A material is "disposed of" if it is discharged, deposited, injected, dumped, spilled, leaked, or placed into or on any land or water so that such material or any constituent thereof may enter the environment or be emitted into the air or discharged into ground or surface waters.
261.2(e)	D. Definition of "Manufacturing or Mining Byproduct" Material
	(1) Such a material is not a primary product of manufacturing or mining operation.
	(2) It is a secondary and incident product of a particular operation.
	(3) It would not be solely or separately mined or manufactured.
	(4) The term does not include an intermediate manufacturing or mining product.
261.3	E. Definition of Hazardous Waste
261.3(a)	(1) A solid waste is a hazardous waste if:
261.3(a)(1)	(a) It is not excluded from regulation as a hazardous waste under Section 261.4(b) (e.g., household waste, fly ash waste, etc.); and
261.3(a)(2)	(b) It meets any of following criteria:
261.3(a)(2)(i)	(i) It is listed in Subpart D (Lists of Hazardous Waste) of Part 261 and has not been delisted via petitioning provisions under Sections 260.20 and 260.22.
261.3(a)(2)(ii)	(ii) It is a mixture of solid waste and one or more hazardous wastes listed in Subpart D of Part 261 and has not been delisted via Sections 260.20 and 260.22.
261.3(a)(2)(iii)	(iii) It exhibits any of the characteristics of hazardous waste identified in Subpart C (Characteristics of Hazardous Waste) of Part 261.
261.3(c)(2)	(c) It is generated from the treatment, storage, or disposal of a hazardous waste.
	(i) Includes:
	• Sludge
	• Spill residue
	• Ash
	• Emission control dust
	• Leachate.
	(ii) Does not include precipitation runoff.
261.3(b)	(2) Conditions When Solid Waste Becomes a Hazardous Waste
	(a) Solid waste is not excluded from regulation under Section 261.4(a)(1); and
261.3(b)(1)	(b) When solid waste first meets listing description set forth in Subpart D of Part 261 (Lists of Hazardous Wastes); or
261.3(b)(2)	(c) Case of Mixture of Solid Waste and One or More Listed Hazardous Wastes
	As soon as a listed hazardous waste of Subpart D is first added to the solid waste; or
261.3(b)(3)	(d) When solid waste exhibits any characteristics identified in Subpart C or Part 261 (Characteristics of Hazardous Waste).
261.3(d)	(3) When Waste Ceases to be Hazardous Waste
261.3(d)(1)	(a) It does not exhibit any characteristics of hazardous waste identified in Subpart C of Part 261.
261.3(d)(2)	(b) Waste Delisted under Sections 260.20 and 260.22
	(i) For listed waste under Subpart D of Part 261.
	(ii) For waste containing a listed waste under Subpart D.
	(iii) For waste derived from a waste listed in Subpart D.
261.4	III. Exclusions
261.4(a)	A. Materials Which Are Not Solid Wastes
	Following materials are not solid wastes for purposes of this Part.
261.4(a)(1)(i)	(1) Domestic Sewage
261.4(a)(1)(ii)	That is, untreated sanitary wastes that pass through a sewer system.
	(2) Any mixture of domestic sewage and other wastes that passes through a sewer system to publicly owned treatment works (POTW).
261.4(a)(2)	(3) Industrial Wastewater Discharges
	(a) Includes such discharges that are point source discharges subject to regulation under Section 402 of Clean Water Act, as amended.

Table 3-1. Identification and Listing of Hazardous Waste. (Continued)

Section No.	Description

261.4(a)(2) Comment	(b) Does not include:
	  (i) Industrial wastewaters while being collected, stored, or treated before discharge.
	  (ii) Sludges that are generated by industrial wastewater treatment.

261.4(a)(3)	(4) Irrigation return flows.
261.4(a)(4)	(5) Source, Special Nuclear or Byproduct Material As defined by the Atomic Energy Act of 1954, as amended, 42 U.S.C. 2011 et seq.
261.4(a)(5)	(6) Materials subject to in-situ mining techniques which are removed from the ground as part of the extraction process.
261.4(b)	B. Solid Wastes Which Are Not Hazardous Wastes Following solid wastes are not hazardous wastes.
261.4(b)(1)	(1) Household Waste That is, any waste material including garbage, trash, and sanitary wastes in septic tanks derived from households (including single and multiple residences, hotels, and motels). (a) Includes household waste that has been collected, transported, stored, treated, disposed, recovered (e.g., refused-derived fuel), or reused.
261.4(b)(2)	(2) Agricultural Wastes Solid wastes generated by any of following activities and which are returned to soils as fertilizers:
261.4(b)(2)(i)	(a) Growing and harvesting of agricultural crops.
261.4(b)(2)(ii)	(b) Raising of animals, including animal manures.
261.4(b)(3)	(3) Mining overburden returned to mine site.

261.4(b)(4)	(4) Waste Generated Primarily from Combustion of Coal or Other Fossil Fuels
	  (a) Fly ash waste
	  (b) Bottom ash waste
	  (c) Slag waste
	  (d) Flue gas emission control waste.

261.4(b)(5)	(5) Wastes Associated With Exploration, Development, or Production of Crude Oil, Natural Gas, or Geothermal Energy
	  (a) Drilling fluids
	  (b) Produced waters
	  (c) Other wastes.

261.5	IV. Special Requirements for Hazardous Waste Generated by Small Quantity Generators
261.5(a)	A. Generators of Less Than 1000 Kilograms of Hazardous Wastes in Calendar Month Wastes not subject to RCRA regulations (e.g., Parts 262 through 265 and Parts 122 and 124 and notification requirement of Section 3010 of RCRA), except as otherwise specified below.
261.5(b)	B. Generator of Less than 1000 Kilograms of Hazardous Waste per Calendar Month Who Accumulates More than 1000 Kilograms of Such Waste Accumulated wastes subject to RCRA regulations.
261.5(c)	C. Other Categories of Small Quantity Generators and Accumulators of Hazardous Waste Wastes are subject to RCRA regulations if quantities shown below are exceeded for either monthly calendar generation or total accumulation.
261.5(c)(1) and (2)	(1) One Kilogram of On- or Off-Specification Commercial Product or Manufacturing Chemical Intermediate with Generic Name Listed in Section 261.33(e) (see Fig. 3-1) Exclusion applies to listed substance in pure or near-pure (e.g., off-specification) form, and not process wastes containing an acutely hazardous substance of Fig. 3-1.
261.5(c)(5)	(2) 100 Kilograms of Spill Residue Containing Acutely Hazardous Waste Listed in Fig. 3-1 Can be any residue, contaminated soil, water, or other debris resulting from cleanup of spill, into or on any land or water.
261.5(c)(3) and (4)	D. Containers or Liners Used to Hold "Pure" Acutely Hazardous Wastes Listed in Fig. 3-1 Containers or liners that were used to hold such wastes are subject to RCRA regulations if capacity exceeds values shown below:
261.5(c)(3)	(1) 20 liter container capacity
261.5(c)(4)	(2) 10 kilogram inner liner capacity.
261.5(d)	E. Small Quantity Hazardous Waste Exclusion Not Unqualified Following compliance by generator is necessary: (1) Must comply with Section 261.11 (Hazardous Waste Determination)

Table 3-1. Identification and Listing of Hazardous Waste. (Continued)

Section No.	Description
	(2) Must treat or dispose of waste in on-site facility or ensure delivery to off-site treatment, storage, or disposal facility, either of which is:
261.5(d)(1)	(a) Permitted by EPA under Part 122 or by a State with a hazardous waste management program authorized under Part 123; or
261.5(d)(2)	(b) In interim status under Parts 122 and 265; or
261.5(d)(3)	(c) Permitted, licensed, or registered by a State to manage municipal or industrial solid waste.
261.5(e)	F. Mixture of "Small Quantity" Hazardous Waste with Nonhazardous Waste
	(1) Mixture remains subject to reduced regulatory requirements although resultant mixture exceeds quantity limitations identified in this Section (261.5).
	(2) If mixture meets any characteristics of hazardous waste identified in Subpart C, wastes are subject to RCRA regulations.
261.6	V. Special Requirements for Hazardous Waste Which Is Used, Reused, Recycled, or Reclaimed
261.6(a)	A. Hazardous Waste Which Is Used, Reused, Recycled, or Reclaimed
	Except as otherwise provided in Section 261.6(b), waste is not subject to RCRA regulations if:
261.6(a)(1)	(1) It is being beneficially used or reused or legitimately recycled or reclaimed; or
261.6(a)(2)	(2) It is being accumulated, stored, or physically, chemically, or biologically treated prior to beneficial use or reuse or legitimate recycling or reclamation.
261.6(b)	B. Hazardous Waste Which Is Transported or Stored Prior to Being Used, Reused, Recycled, or Reclaimed
	If the hazardous waste is a sludge, or is listed in Subpart D, or contains one or more hazardous wastes listed in Subpart D, the following requirements relative to transportation or storage must be met:
261.6(b)(1)	(1) Notification requirements under Section 3010 of RCRA.
261.6(b)(2)	(2) Part 262.
261.6(b)(3)	(3) Part 263.
261.6(b)(4)	(4) Subparts A, B, C, D, and E of Part 264.
261.6(b)(5)	(5) Subparts A, B, C, D, E, G, H, I, J, and L of Part 265.
261.6(b)(6)	(6) Parts 122 and 124 with respect to storage facilities.
261.10	VI. Criteria for Identifying the Characteristics of Hazardous Waste
261.10(a)	A. Criteria Used by Administrator to Identify and Define a Characteristic of Hazardous Waste (in Subpart C)
261.10(1)	(1) Health and Environmental Criteria
	The characteristic exhibited may:
261.10(1)(i)	(a) Cause or significantly contribute to:
	(i) An increase in mortality; or
	(ii) An increase in serious irreversible or incapacitating reversible illness; or
261.10(1)(ii)	(b) Pose a substantial present or potential hazard to human health or the environment when it is improperly treated, stored, transported, disposed of, or otherwise managed.
	(2) Test Measurement or Analytical Criteria
	The characteristic can be:
261.10(2)(i)	(a) Measurable by standardized and available testing protocols and is reasonably within the capability of waste generators or private sector laboratories which are available to the generators; or
261.10(2)(ii)	(b) Reasonably detected by generators through knowledge of their waste.
261.11	VII. Criteria for Listing Hazardous Waste
261.11(a)	A. Criteria Used by Administrator to List a Solid Waste as a Hazardous Waste
261.11(a)(1)	(1) It Exhibits Any of the Characteristics of Hazardous Waste Identified in Subpart C of Part 261; or
261.11(a)(2)	(2) It Exhibits Acutely Hazardous Characteristics
	(a) Substance has been found to be fatal to humans in low doses; or
	(b) Substance has shown following toxicity characteristics in studies with animals:
	(i) Oral LD 50 toxicity of less than 50 milligrams per kilogram for rats; or
	(ii) Inhalation LD 50 toxicity of less than 2 milligrams per liter for rats; or
	(iii) Dermal LD 50 toxicity of less than 200 milligrams per kilogram for rabbits; or
	(c) Substance is otherwise capable of causing or significantly contributing to an increase in serious irreversible or incapacitating reversible illness; or

Table 3-1. Identification and Listing of Hazardous Waste. (Continued)

Section No.	Description
261.11(a)(3)	(3) It Contains Any of the Toxic Constituents Listed in Appendix VIII of Part 261. (See Appendix C of this book) Nonetheless, the Administrator may conclude that the waste is not hazardous to human health or the environment upon consideration of the following factors:
261.11(a)(3)(i)	(a) Nature of toxicity presented by constituent.
261.11(a)(3)(ii)	(b) Concentration of constituent in waste.
261.11(a)(3)(iii)	(c) Potential of constituent or toxic degradation product of constituent to migrate from waste into environment with improper management considered in Section 261.11(a)(3)(vii).
261.11(a)(3)(iv)	(d) Persistence of constituent or any toxic degradation product of constituent.
261.11(a)(3)(v)	(e) Potential for constituent or any toxic degradation product of constituent to degrade into nonharmful constituents and rate of degradation.
261.11(a)(3)(vi)	(f) Degree to which constituent or any degradation product of constituent bioaccumulates in ecosystems.
261.11(a)(3)(vii)	(g) Plausible types of improper management to which waste could be subjected.
261.11(a)(3)(viii)	(h) Quantities of waste generated at individual generation sites or on a regional or national basis.
261.11(a)(3)(ix)	(i) Nature and severity of human health and environmental damage that has occurred as a result of improper management of wastes containing constituent.
261.11(a)(3)(x)	(j) Action taken by other governmental agencies or regulatory programs based on health or environmental hazard posed by waste or waste constituent.
261.11(a)(3)(xi)	(k) Other appropriate factors.
	Notes: (1) Substances will be listed in Appendix VIII only if shown in scientific studies to have toxic, carcinogenic, mutagenic, or teratogenic effects on humans or other life forms. (2) Appendix VIII is for EPA use only, and does not affect regulated community.
261.11(b)	B. Listing Classes or Types of Solid Waste as Hazardous Waste (1) This may be done by Administrator if he has reason to believe that individual wastes, within the class or type of waste, typically or frequently are hazardous under the definition of hazardous waste found in Section 1004(5) of RCRA. (2) Definition of Hazardous Waste Per Section 1004(5) of RCRA "The term 'hazardous waste' means a solid waste, or combination of solid wastes, which because of its quantity, concentration, or physical, chemical, or infectious characteristics may — (a) Cause, or significantly contribute to an increase in mortality or an increase in serious irreversible, or incapacitating reversible, illness; or (b) Pose a substantial present or potential hazard to human health or the environment when improperly treated, stored, transported, or disposed of, or otherwise managed."
261.11(c)	C. Exclusion Limits for Acute Hazardous Wastes The Administrator will use the criteria for listing specified in this Section (261.11) to establish exclusion limits for acute hazardous wastes referred to in Section 261.5(c).
Subpart C	VIII. Characteristics of Hazardous Waste
261.20	A. General
261.20(a)	(1) Solid Waste Is Hazardous If It Exhibits Any Characteristic Identified in This Subpart Following conditions must also apply: (a) Meets Section 261.2 definition of solid waste. (b) It is not excluded from regulation as a hazardous waste under Section 261.4(b).
261.20(a) Comment	(2) Generator Is Responsible for Determining If Wastes Exhibit One or More Characteristics Identified in This Subpart Under Section 262.11 (Hazardous Waste Determination).
261.20(b)	(3) EPA Hazardous Waste Number (a) For Hazardous Waste Identified by Characteristic in this Subpart But Not Listed as Hazardous Waste: The waste is assigned an EPA Hazardous Waste Number as follows: (i) Ignitable characteristic—D001. (ii) Corrosive characteristic—D002. (iii) Reactive characteristic—D003. (iv) EP Toxic characteristic—As specified in Table 1 of Part 261 (see Appendix C of this book). (b) When Number Must be Used In complying with: (i) Notification requirements of Section 3010 of RCRA.

Table 3-1. Identification and Listing of Hazardous Waste. (Continued)

Section No.	Description
	(ii) Certain recordkeeping and reporting requirements under Parts 262 through 265 and Part 122.
261.20(c)	(4) Sampling Methods
	(a) Representative Sample
	Any waste sample obtained using sampling methods of Appendix I of Part 261 (see Appendix C of this book) will be considered a "representative sample" as defined in Part 260.
261.20(c) Comment	(b) Use of Alternative Sampling Methods
	Since Appendix I sampling methods are not formally adopted by Administrator, use of an alternative method does not require demonstration of equivalency set forth in Sections 260.20 and 260.21.
261.21	B. Characteristic Of Ignitability
261.21(a)	(1) How Solid Waste Exhibits Characteristic of Ignitability
	Representative sample has any of following properties:
261.21(a)(1)	(a) It Is a Liquid
	(i) But not an aqueous solution containing less than 24% alcohol by volume.
	(ii) Has flash point less than 60°C (140°F) as determined by:
	• A Pensky–Martens closed cup tester using test method specified in ASTM standard D-93-79; or
	• A Setaflash closed cup tester using test method specified in ASTM standard D-3278-78; or
	• An equivalent test method approved by EPA under Sections 260.20 and 260.21.
261.21(a)(2)	(b) It Is Not a Liquid
	(i) Is capable under standard temperature and pressure of causing fire through friction, absorption of moisture, or spontaneous chemical changes.
	(ii) When ignited, it burns so vigorously and persistently that it creates a hazard.
261.21(a)(3)	(c) It Is an Ignitable Compressed Gas
	(i) As defined in 49 CFR 173.300 (see Appendix E of this book for DOT regulation contents only); and
	(ii) As determined by test methods described in that regulation or equivalent test methods approved under Sections 260.20 and 260.21.
261.21(a)(4)	(d) It Is an Oxidizer
	As defined in 49 CFR 173.151 (see Appendix E of this book for DOT regulation contents only).
261.21(b)	(2) EPA Hazardous Waste Number
	A solid waste exhibiting the characteristic of ignitability, but not listed as a hazardous waste in Subpart D has EPA Hazardous Waste Number D001.
261.22	C. Characteristic of Corrosivity
261.22(a)	(1) How Solid Waste Exhibits Characteristic of Corrosivity
	Representative sample has any of following properties:
261.22(a)(1)	(a) It Is Aqueous
	Has a pH less than or equal to 2 or greater than or equal to 12.5 as determined by:
	• A pH meter using either the test method specified in "Test Methods for the Evaluation of Solid Waste, Physical/Chemical Methods;" or
	• An equivalent test method approved by EPA under Sections 260.20 and 260.21.
261.22(a)(2)	(b) It Is a Liquid
	Corrodes steel (SAE 1020) at a rate greater than 6.35 mm (0.250 inch) per year at a test temperature of 55°C (130°F) as determined by:
	• Test method specified in NACE (National Association of Corrosion Engineers) Standard TM-01-69 as standardized in "Test Methods for the Evaluation of Solid Waste, Physical/Chemical Methods;" or
	• An equivalent test method approved by EPA under Sections 260.20 and 260.21.
261.22(b)	(2) EPA Hazardous Waste Number
	A solid waste exhibiting the characteristic of corrosivity, but not listed as a hazardous waste in Subpart D has the EPA Hazardous Waste Number D002.
261.23	D. Characteristic of Reactivity
261.23(a)	(1) How Solid Waste Exhibits Characteristic of Reactivity
	Representative sample has any of following properties:
261.23(a)(1)	(a) It is normally unstable and readily undergoes violent change without detonating.
261.23(a)(2)	(b) It reacts violently with water.
261.23(a)(3)	(c) It forms potentially explosive mixtures with water.
261.23(a)(4)	(d) When mixed with water, it generates toxic gases, vapors, or fumes in a quantity sufficient to present a danger to human health or the environment

Table 3-1. Identification and Listing of Hazardous Waste. (Continued)

Section No.	Description
261.23(a)(5)	(e) It Is a Cyanide or Sulfide Bearing Waste When exposed to pH conditions between 2 and 12.5 it can generate toxic gases, vapors, or fumes in a quantity sufficient to present a danger to human health or the environment.
261.23(a)(6)	(f) It is capable of detonation or explosive reaction if subjected to a strong initiating source or if heated under confinement.
261.23(a)(7)	(g) It is readily capable of detonation or explosive decomposition or reaction at standard temperature and pressure.
261.23(a)(8)	(h) It is a forbidden explosive per 49 CFR 173.51 (see Appendix E of this book for DOT regulation content only); or (i) It is a Class A explosive per 49 CFR 173.53 (see Appendix E of this book for DOT regulation content only); or (j) It is a Class B explosive per 49 CFR 173.88 (see Appendix E of this book for DOT regulation content only).
261.23(b)	(2) EPA Hazardous Waste Number A solid waste exhibiting the characteristic of reactivity, but not listed as a hazardous waste in Subpart D has the EPA Hazardous Waste Number D003.
261.24 261.24(a)	F. Characteristic of EP Toxicity (1) How Solid Waste Exhibits Characteristics of EP Toxicity The extract from a representative sample of waste contains any of the contaminants listed in Table I (see Part 261 of Appendix C of this book) at a concentration equal to or greater than respective values given in Table 1 using: (a) Test methods described in Appendix II (see Part 261 of Appendix C of this book—EP Toxicity Test Procedures); or (b) Equivalent methods approved by EPA under Sections 260.20 and 260.21. (2) Where Waste Contains Less than 5% Filterable Solids The waste itself, after filtering, is considered to be the extract for the purposes of this Section.
261.24(b)	(3) EPA Hazardous Waste Number A solid waste exhibiting the characteristic of EP toxicity, but not listed as a hazardous waste in Subpart D, has the EPA Hazardous Waste Number specified in Table I which corresponds to the toxic contaminant causing it to be hazardous.
Subpart D	IX. Lists of Hazardous Wastes
261.30 261.30(a)	A. General (1) Solid Waste Is Hazardous If Listed in This Subpart Unless excluded under Sections 260.20 and 260.22.
261.30(b)	(2) EPA Hazard Codes Indicate Basis for Listing Classes or Types of Wastes Listed in This Subpart Hazard codes follow and more than one may be used: (a) Ignitable Waste (I) (b) Corrosive Waste (C) (c) Reactive Waste (R) (d) EP Toxic Waste (E) (e) Acute Hazardous Waste (H) (f) Toxic Waste (T).
261.30(c)	(3) EPA Hazardous Waste Number (a) Such a number is assigned to each listed waste in this Subpart (b) When Number Must Be Used In compliance with: (i) Notification requirements of Section 3010 of RCRA. (ii) Certain recordkeeping and reporting requirements under Parts 262–265 and Part 122.
261.30(b) 261.31	B. Listed Wastes Of Sections 261.31 and 261.32 (1) Section 261.31 (see Figs. 3-2 and 3-5) Eighteen (18) hazardous wastes from nonspecific sources.
261.32	(2) Section 261.32 (see Figs. 3-3 and 3-5) Eighty-five (85) hazardous wastes from specific sources.
261.30(b)	(3) Appendix VII (see Part 261 of Appendix C of this book) Identifies constituent that caused EPA to designate waste of Sections 261.31 and 261.32 as EP Toxic Waste (E) or Toxic Waste (T).
261.33	C. Discarded Commercial Chemical Products, Off-Specification Species, Containers, and Spill Residues Thereof

Table 3-1. Identification and Listing of Hazardous Waste. (Continued)

Section No.	Description
261.33(d) Comment	(1) Clarification of Phrase "Commercial Chemical Product or Manufacturing Chemical Intermediate Having the Generic Name Listed in . . ."
	(a) Phrase refers to a listed chemical substance, in its pure or near pure (e.g., "off-specification") form, which is manufactured or formulated for commercial or manufacturing use. Applicable listed chemical groupings follow:
261.33(e)	(i) Acutely hazardous wastes listed in Fig. 3-1 (Section 261.33(e))
	These substances are subject to small quantity exclusion defined in Section 261.5(c).
261.33(f)	(ii) Toxic Wastes Listed in Fig. 3-4 (Section 261.33(f))
	These substances are subject to small quantity exclusion defined in Section 261.5(a) and (b).
261.33(d) Comment	(b) Phrase does not refer to a material, such as a manufacturing process waste, that contains any of the substances listed in Sections 261.33(e) and (f) (Figs. 3-1 and 3-4, respectively).
	(c) For Manufacturing Process Waste Containing Listed Wastes of Sections 261.33(e) or (f) (Figs. 3-1 and 3-4, respectively):
	(i) Such waste is of itself not necessarily a hazardous waste under these regulations.
	(ii) It is deemed to be a hazardous waste if:
	• It is listed in Sections 261.31 or 261.32 (Figs. 3-2 and 3-3, respectively); or
	• It is identified as a hazardous waste by the characteristics set forth in Subpart C of this Part.
261.33	(2) Materials or Items Considered as Hazardous Wastes If/When Discarded or Intended for Discarding
261.33(a)	(a) Commercial chemical product or manufacturing chemical intermediate with generic name listed in Sections 261.33(e) or (f) (Figs. 3-1 and 3-4, respectively).
261.33(b)	(b) Off-specification commercial chemical product or manufacturing intermediate with generic name listed in Sections 261.33(e) or (f) (Figs. 3-1 and 3-4, respectively).
261.33(c)	(c) Container or Liner Used to Hold Commercial Chemical Product or Manufacturing Chemical Intermediate with Generic Name Listed in Section 261.33(e) (Fig. 3-1) Exceptions follow:
261.33(c)(1)	(i) Container or inner liner triple rinsed using solvent capable of removing commercial chemical product or manufacturing chemical intermediate.
261.33(c)(2)	(ii) Container or inner liner cleaned by another method shown in scientific literature or by tests conducted by generator to achieve equivalent removal.
261.33(c)(3)	(iii) Container exempted if removed inner liner prevented contact between Fig. 3-1 substance and container.
261.33(d)	(d) Spill Residues
	Any residue or contaminated soil, water or other debris resulting from cleanup of a spill into or on any land or water, of commercial chemical product or manufacturing chemical intermediate listed in Sections 261.33(e) or (f) (Figs. 3-1 and 3-4, respectively).
FR pp. 47833–47834, 7/16/80	D. July 16, 1980 Promulgated Hazardous Wastes
	(1) See Fig. 3-5 for two generic hazardous wastes under Section 261.31.
	(2) See Fig. 3-5 for 16 hazardous wastes from specific sources under Section 261.32.
FR pp. 33118–33119, 5/19/80	E. Listed Hazardous Wastes Scheduled for Fall (1980) Promulgation
	See Fig. 3-6 for 17 generic wastes and 11 process wastes.
	F. Proposed Amendments to Section 261.32: List of Hazardous Waste from Specific Sources
FR p. 33137, 5/19/80	(1) From May 19, 1980 *Federal Register*
	See Appendix D of this book for proposed list of 11 waste streams to Section 261.32 (Fig. 3-3).
	(2) From July 16, 1980 *Federal Register*
FR p. 47836, 7/16/80	See Appendix D of this book for proposed list of 7 waste streams to Section 261.32 (Fig. 3-3).

Hazardous waste No.	Substance [1]
	1080 see P058
	1081 see P057
	(Acetato)phenylmercury see P092
	Acetone cyanohydrin see P069
P001	3-(alpha-Acetonylbenzyl)-4-hydroxycoumarin and salts
P002	1-Acetyl-2-thiourea
P003	Acrolein
	Agarin see P007
	Agrosan GN 5 see P092
	Aldicarb see P069
	Aldifen see P048
P004	Aldrin
	Algimycin see P092
P005	Allyl alcohol
P006	Aluminum phosphide (R)
	ALVIT see P037
	Aminoethylene see P054
P007	5-(Aminomethyl)-3-isoxazolol
P008	4-Aminopyridine
	Ammonium metavanadate see P119
P009	Ammonium picrate (R)
	ANTIMUCIN WDR see P092
	ANTURAT see P073
	AQUATHOL see P088
	ARETIT see P020
P010	Arsenic acid
P011	Arsenic pentoxide
P012	Arsenic trioxide
	Athrombin see P001
	AVITROL see P008
	Aziridene see P054
	AZOFOS see P061
	Azophos see P061
	BANTU see P072
P013	Barium cyanide
	BASENITE see P020
	BCME see P016
P014	Benzenethiol
	Benzoepin see P050
P015	Beryllium dust
P016	Bis(chloromethyl) ether
	BLADAN-M see P071
P017	Bromoacetone
P018	Brucine
P019	2-Butanone peroxide
	BUFEN see P092
	Butaphene see P020
P020	2-sec-Butyl-4,6-dinitrophenol
P021	Calcium cyanide
	CALDON see P020
P022	Carbon disulfide
	CERESAN see P092
	CERESAN UNIVERSAL see P092
	CHEMOX GENERAL see P020
	CHEMOX P.E. see P020
	CHEM-TOL see P090
P023	Chloroacetaldehyde
P024	p-Chloroaniline
P025	1-(p-Chlorobenzoyl)-5-methoxy-2-methylindole-3-acetic acid
P026	1-(o-Chlorophenyl)thiourea
P027	3-Chloropropionitrile
P028	alpha-Chlorotoluene
P029	Copper cyanide
	CRETOX see P108
	Coumadin see P001
	Coumafen see P001
P030	Cyanides
P031	Cyanogen
P032	Cyanogen bromide
P033	Cyanogen chloride
	Cyclodan see P050
P034	2-Cyclohexyl-4,6-dinitrophenol
	D-CON see P001
	DETHMOR see P001
	DETHNEL see P001
	DFP see P043
P035	2,4-Dichlorophenoxyacetic acid (2,4-D)
P036	Dichlorophenylarsine
	Dicyanogen see P031
P037	Dieldrin
	DIELDREX see P037
P038	Diethylarsine
P039	0,0-Diethyl-S-(2-(ethylthio)ethyl)ester of phosphorothioic acid
P040	0,0-Diethyl-0-(2-pyrazinyl)phosphorothioate
P041	0,0-Diethyl phosphoric acid, 0-p-nitrophenyl ester
P042	3,4-Dihydroxy-alpha-(methylamino)-methyl benzyl alcohol
P043	Di-isopropylfluorophosphate
	DIMETATE see P044

Hazardous Waste No.	Substance [1]
	1,4:5,8-Dimethanonaphthalene, 1,2,3,4,10,10-hexachloro-1,4,4a,5,8,8a-hexahydro endo, endo see P060
P044	Dimethoate
P045	3,3-Dimethyl-1-(methylthio)-2-butanone-O-[(methylamino)carbonyl] oxime
P046	alpha,alpha-Dimethylphenethylamine
	Dinitrocyclohexylphenol see P034
P047	4,6-Dinitro-o-cresol and salts
P048	2,4-Dinitrophenol
	DINOSEB see P020
	DINOSEBE sde P020
	Disulfoton see P039
P049	2,4-Dithiobiuret
	DNBP see P020
	DOLCO MOUSE CEREAL see P108
	DOW GENERAL see P020
	DOW GENERAL WEED KILLER see P020
	DOW SELECTIVE WEED KILLER see P020
	DOWICIDE G see P090
	DYANACIDE see P092
	EASTERN STATES DUOCIDE see P001
	ELGETOL see P020
P050	Endosulfan
P051	Endrin
	Epinephrine see P042
P052	Ethylcyanide
P053	Ethylenediamine
P054	Ethyleneimine
	FASCO FASCRAT POWDER see P001
	FEMMA see P091
P055	Ferric cyanide
P056	Fluorine
P057	2-Fluoroacetamide
P058	Fluoroacetic acid, sodium salt
	FOLODOL-80 see P071
	FOLODOL M see P071
	FOSFERNO M 50 see P071
	FRATOL see P058
	Fulminate of mercury see P065
	FUNGITOX OR see P092
	FUSSOF see P057
	GALLOTOX see P092
	GEARPHOS see P071
	GERUTOX see P020
P059	Heptachlor
P060	1,2,3,4,10,10-Hexachloro-1,4,4a,5,8,8a-hexahydro-1,4:5,8-endo, endo-dimethanonaphthalene
	1,4,5,6,7,7-Hexachloro-cyclic-5-norbornene-2,3-dimethanol sulfite see P050
P061	Hexachloropropene
P062	Hexaethyl tetraphosphate
	HOSTAQUICK see P092
	HOSTAQUIK see P092
	Hydrazomethane see P068
P063	Hydrocyanic acid
	ILLOXOL see P037
	INDOCI see P025
	Indomethacin see P025
	INSECTOPHENE see P050
	Isodrin see P060
P064	Isocyanic acid, methyl ester
	KILOSEB see P020
	KOP-THIODAN see P050
	KWIK-KIL see P108
	KWIKSAN see P092
	KUMADER see P001
	KYPFARIN see P001
	LEYTOSAN see P092
	LIQUIPHENE see P092
	MALIK see P050
	MAREVAN see P001
	MAR-FRIN see P001
	MARTIN'D MAR-FRIN see P001
	MAVERAN see P001
	MEGATOX see P005
P065	Mercury fulminate
	MERSOLITE see P092
	METACID 50 see P071
	METAFOS see P071
	METAPHOR see P071
	METAPHOS see P071
	METASOL 30 see P092
P066	Methomyl
P067	2-Methylaziridine
	METHYL-E 605 see P071
P068	Methyl hydrazine
	Methyl isocyanate see P064
P069	2-Methyllactonitrile
P070	2-Methyl-2-(methylthio)propionaldehyde-o-(methylcarbonyl) oxime

Hazardous Waste No.	Substance [1]
	METHYL NIRON see P042
P071	Methyl parathion
	METRON see P071
	MOLE DEATH see P108
	MOUSE-NOTS see P108
	MOUSE-RID see P108
	MOUSE-TOX see P108
	MUSCIMOL see P007
P072	1-Naphthyl-2-thiourea
P073	Nickel carbonyl
P074	Nickel cyanide
P075	Nicotine and salts
P076	Nitric oxide
P077	p-Nitroaniline
P078	Nitrogen dioxide
P079	Nitrogen peroxide
P080	Nitrogen tetroxide
P081	Nitroglycerine (R)
P082	N-Nitrosodimethylamine
P083	N-Nitrosodiphenylamine
P084	N-Nitrosomethylvinylamine
	NYLMERATE see P092
	OCTALOX see P037
P085	Octamethylpyrophosphoramide
	OCTAN see P092
P086	Oleyl alcohol condensed with 2 moles ethylene oxide
	OMPA see P085
	OMPACIDE see P085
	OMPAX see P085
P087	Osmium tetroxide
P088	7-Oxabicyclo[2.2.1]heptane-2,3-dicarboxylic acid
	PANIVARFIN see P001
	PANORAM D-31 see P037
	PANTHERINE see P007
	PANWARFIN see P001
P089	Parathion
	PCP see P090
	PENNCAP-M see P071
	PENOXYL CARBON N see P048
P090	Pentachlorophenol
	Pentachlorophenate see P090
	PENTA-KILL see P090
	PENTASOL see P090
	PENWAR see P090
	PERMICIDE see P090
	PERMAGUARD see P090
	PERMATOX see P090
	PERMITE see P090
	PERTOX see P090
	PESTOX III see P085
	PHENMAD see P092
	PHENOTAN see P020
P091	Phenyl dichloroarsine
	Phenyl mercaptan see P014
P092	Phenylmercury acetate
P093	N-Phenylthiourea
	PHILIPS 1861 see P008
	PHIX see P092
P094	Phorate
P095	Phosgene
P096	Phosphine
P097	Phosphorothioic acid, 0,0-dimethyl ester, 0-ester with N,N-dimethyl benzene sulfonamide
	Phosphorothioic acid 0,0-dimethyl-0-(p-nitrophenyl) ester see P071
	PIED PIPER MOUSE SEED see P108
P098	Potassium cyanide
P099	Potassium silver cyanide
	PREMERGE see P020
P100	1,2-Propanediol
	Propargyl alcohol see P102
P101	Propionitrile
P102	2-Propyn-1-o1
	PROTHROMADIN See P001
	QUICKSAM see P092
	QUINTOX see P037
	RAT AND MICE BAIT see P001
	RAT-A-WAY see P001
	RAT-B-GON see P001
	RAT-O-CIDE #2 see P001
	RAT-GUARD see P001
	RAT-KILL see P001
	RAT-MIX see P001
	RATS-NO-MORE see P001
	RAT-OLA see P001
	RATOREX see P001
	RATTUNAL see P001
	RAT-TROL see P001
	RO-DETH see P001
	RO-DEX see P108
	ROSEX see P001

Fig. 3-1. Acute hazardous wastes (Code H). From Federal Register, **Vol. 45**; No. 98, May 19, 1980.

Hazardous Waste No.	Substance[1]	Hazardous Waste No.	Substance[1]	Hazardous Waste No.	Substance[1]
	ROUGH & READY MOUSE MIX see P001		TAG FUNGICIDE see P092	P117..........	Thiuram
	SANASEED see P108		TEKWAISA see P071		THOMPSON'S WOOD FIX see P090
	SANTOBRITE see P090		TEMIC see P070		TIOVEL see P050
	SANTOPHEN see P090		TEMIK see P070	P118..........	Trichloromethanethiol
	SANTOPHEN 20 see P090		TERM-I-TROL see P090		TWIN LIGHT RAT AWAY see P001
	SCHRADAN see P085	P109..........	Tetraethyldithiopyrophosphate		USAF RH–8 see P069
P103..........	Selenourea	P110..........	Tetraethyl lead		USAF EK–4890 see P002
P104..........	Silver Cyanide	P111..........	Tetraethylpyrophosphate	P119..........	Vanadic acid, ammonium salt
	SMITE see P105	P112..........	Tetranitromethane	P120..........	Vanadium pentoxide
	SPARIC see P020		Tetraphosphoric acid, hexaethyl ester see P062		VOFATOX see P071
	SPOR-KIL see P092		TETROSULFUR BLACK PB see P048		WANADU see P120
	SPRAY-TROL BRAND RODEN-TROL see P001		TETROSULPHUR PBR see P048		WARCOUMIN see P001
	SPURGE see P020	P113..........	Thallic oxide		WARFARIN SODIUM see P001
P105..........	Sodium azide		Thallium peroxide see P113		WARFICIDE see P001
	Sodium coumadin see P001	P114..........	Thallium selenite		WOFOTOX see P072
P106..........	Sodium cyanide	P115..........	Thallium (I) sulfate		YANOCK see P057
	Sodium fluoroacetate see P056		THIFOR see P092		YASOKNOCK see P058
	SODIUM WARFARIN see P001		THIMUL see P092		ZIARNIK see P092
	SOLFARIN see P001		THIODAN see P050	P121..........	Zinc cyanide
	SOLFOBLACK BB see P048		THIOFOR see P050	P122..........	Zinc phosphide (R,T)
	SOLFOBLACK SB see P048		THIOMUL see P050		ZOOCOUMARIN see P001
P107..........	Strontium sulfide		THIONEX see P050		
P108..........	Strychnine and salts		THIOPHENIT see P071		
	SUBTEX see P020	P116..........	Thiosemicarbazide		
	SYSTAM see P085		Thiosulfan tionel see P050		

[1]The Agency included those trade names of which it was aware; an omission of a trade name does not imply that the omitted material is not hazardous. The material is hazardous if it is listed under its generic name.

Fig. 3-1. Acute hazardous wastes (Code H). From Federal Register, **Vol. 45**; No. 98, May 19, 1980. (Continued)

Industry and EPA hazardous waste No.	Hazardous waste	Hazard code
Generic:		
F001	The spent halogenated solvents used in degreasing, tetrachloroethylene, trichloroethylene, methylene chloride, 1,1,1-trichloroethane, carbon tetrachloride, and the chlorinated fluorocarbons; and sludges from the recovery of these solvents in degreasing operations.	(T)
F002	The spent halogenated solvents, tetrachloroethylene, methylene chloride, trichloroethylene, 1,1,1-trichloroethane, chlorobenzene, 1,1,2-trichloro-1,2,2-trifluoroethane, o-dichlorobenzene, trichlorofluoromethane and the still bottoms from the recovery of these solvents.	(T)
F003	The spent non-halogenated solvents, xylene, acetone, ethyl acetate, ethyl benzene, ethyl ether, n-butyl alcohol, cyclohexanone, and the still bottoms from the recovery of these solvents.	(I)
F004	The spent non-halogenated solvents, cresols and cresylic acid, nitrobenzene, and the still bottoms from the recovery of these solvents	(T)
F005	The spent non-halogenated solvents, methanol, toluene, methyl ethyl ketone, methyl isobutyl ketone, carbon disulfide, isobutanol, pyridine and the still bottoms from the recovery of these solvents.	(I, T)
F006	Wastewater treatment sludges from electroplating operations...	(T)
F007	Spent plating bath solutions from electroplating operations..	(R, T)
F008	Plating bath sludges from the bottom of plating baths from electroplating operations	(R, T)
F009	Spent stripping and cleaning bath solutions from electroplating operations..	(R, T)
F010	Quenching bath sludge from oil baths from metal heat treating operations..	(R, T)
F011	Spent solutions from salt bath pot cleaning from metal heat treating operations	(R, T)
F012	Quenching wastewater treatment sludges from metal heat treating operations...	(T)
F013	Flotation tailings from selective flotation from mineral metals recovery operations...	(T)
F014	Cyanidation wastewater treatment tailing pond sediment from mineral metals recovery operations..	(T)
F015	Spent cyanide bath solutions from mineral metals recovery operations..	(R, T)
F016	Dewatered air pollution control scrubber sludges from coke ovens and blast furnaces..	(T)

Fig. 3-2. Hazardous waste from nonspecific sources. From Federal Register, **Vol. 45**; No. 98, May 19, 1980.

Industry and EPA hazardous waste No.	Hazardous waste	Hazard code
Wood Preservation: K001...............	Bottom sediment sludge from the treatment of wastewaters from wood preserving processes that use creosote and/or pentachlorophenol	(T)
Inorganic Pigments:		
K002................................	Wastewater treatment sludge from the production of chrome yellow and orange pigments	(T)
K003................................	Wastewater treatment sludge from the production of molybdate orange pigments	(T)
K004................................	Wastewater treatment sludge from the production of zinc yellow pigments	(T)
K005................................	Wastewater treatment sludge from the production of chrome green pigments	(T)
K006................................	Wastewater treatment sludge from the production of chrome oxide green pigments (anhydrous and hydrated)	(T)
K007................................	Wastewater treatment sludge from the production of iron blue pigments	(T)
K008................................	Oven residue from the production of chrome oxide green pigments...	(T)
Organic Chemicals:		
K009................................	Distillation bottoms from the production of acetaldehyde from ethylene..	(T)
K010................................	Distillation side cuts from the production of acetaldehyde from ethylene...	(T)
K011................................	Bottom stream from the wastewater stripper in the production of acrylonitrile...	(R, T)
K012................................	Still bottoms from the final purification of acrylonitrile in the production of acrylonitrile	(T)
K013................................	Bottom stream from the acetonitrile column in the production of acrylonitrile	(R, T)
K014................................	Bottoms from the acetonitrile purification column in the production of acrylonitrile	(T)
K015................................	Still bottoms from the distillation of benzyl chloride..	(T)
K016................................	Heavy ends or distillation residues from the production of carbon tetrachloride	(T)
K017................................	Heavy ends (still bottoms) from the purification column in the production of epichlorohydrin	(T)
K018................................	Heavy ends from fractionation in ethyl chloride production..	(T)
K019................................	Heavy ends from the distillation of ethylene dichloride in ethylene dichloride production...	(T)
K020................................	Heavy ends from the distillation of vinyl chloride in vinyl chloride monomer production...	(T)
K021................................	Aqueous spent antimony catalyst waste from fluoromethanes production..	(T)
K022................................	Distillation bottom tars from the production of phenol/acetone from cumene...	(T)

Fig. 3-3. Hazardous waste from specific sources. From Federal Register, **Vol. 45**; No. 98, May 19, 1980.

Industry and EPA hazardous waste No.	Hazardous waste	Hazard code
K023	Distillation light ends from the production of phthalic anhydride from naphthalene	(T)
K024	Distillation bottoms from the production of phthalic anhydride from naphthalene	(T)
K025	Distillation bottoms from the production of nitrobenzene by the nitration of benzene	(T)
K026	Stripping still tails from the production of methyl ethyl pyridines	(T)
K027	Centrifuge residue from toluene diisocyanate production	(R, T)
K028	Spent catalyst from the hydrochlorinator reactor in the production of 1,1,1-trichloroethane	(T)
K029	Waste from the product stream stripper in the production of 1,1,1-trichloroethane	(T)
K030	Column bottoms or heavy ends from the combined production of trichloroethylene and perchloroethylene	(T)
Pesticides:		
K031	By-products salts generated in the production of MSMA and cacodylic acid	(T)
K032	Wastewater treatment sludge from the production of chlordane	(T)
K033	Wastewater and scrub water from the chlorination of cyclopentadiene in the production of chlordane	(T)
K034	Filter solids from the filtration of hexachlorocyclopentadiene in the production of chlordane	(T)
K035	Wastewater treatment sludges generated in the production of creosote	(T)
K036	Still bottoms from toluene reclamation distillation in the production of disulfoton	(T)
K037	Wastewater treatment sludges from the production of disulfoton	(T)
K038	Wastewater from the washing and stripping of phorate production	(T)
K039	Filter cake from the filtration of diethylphosphorodithoric acid in the production of phorate	(T)
K040	Wastewater treatment sludge from the production of phorate	(T)
K041	Wastewater treatment sludge from the production of toxaphene	(T)
K042	Heavy ends or distillation residues from the distillation of tetrachlorobenzene in the production of 2,4,5-T	(T)
K043	2,6-Dichlorophenol waste from the production of 2,4-D	(T)
Explosives:		
K044	Wastewater treatment sludges from the manufacturing and processing of explosives	(R)
K045	Spent carbon from the treatment of wastewater containing explosives	(R)
K046	Wastewater treatment sludges from the manufacturing, formulation and loading of lead-based initiating compounds	(T)
K047	Pink/red water from TNT operations	(R)
Petroleum Refining:		
K048	Dissolved air flotation (DAF) float from the petroleum refining Industry	(T)
K049	Slop oil emulsion solids from the petroleum refining industry	(T)
K050	Heat exchanger bundle cleaning sludge from the petroleum refining industry	(T)
K051	API separator sludge from the petroleum refining industry	(T)
K052	Tank bottoms (leaded) from the petroleum refining industry	(T)
Leather Tanning Finishing:		
K053	Chrome (blue) trimmings generated by the following subcategories of the leather tanning and finishing industry: hair pulp/chrome tan/retan/wet finish; hair save/chrome tan/retan/wet finish; retan/wet finish; no beamhouse; through-the-blue; and shearling.	(T)
K054	Chrome (blue) shavings generated by the following subcategories of the leather tanning and finishing industry: hair pulp/chrome tan/retan/wet finish; hair save/chrome tan/retan/wet finish; retan/wet finish; no beamhouse; through-the-blue; and shearling.	(T)
K055	Buffing dust generated by the following subcategories of the leather tanning and finishing industry: hair pulp/chrome tan/retan/wet finish; hair save/chrome tan/retan/wet finish; retan/wet finish; no beamhouse; and through-the-blue.	(T)
K056	Sewer screenings generated by the following subcategories of the leather tanning and finishing industry: hair pulp/chrome tan/retan/wet finish; hair save/chrome tan/retan/wet finish; retan/wet finish; no beamhouse; through-the-blue; and shearling.	(T)
K057	Wastewater treatment sludges generated by the following subcategories of the leather tanning and finishing industry: hair pulp/chrome tan/retan/wet finish; hair save/chrome tan/retan/wet finish; retan/wet finish; no beamhouse; through-the-blue and shearling.	(T)
K058	Wastewater treatment sludges generated by the following subcategories of the leather tanning and finishing industry: hair pulp/chrome tan/retan/wet finish; hair save/chrome tan/retan/wet finish; and through-the-blue.	(R, T)
K059	Wastewater treatment sludges generated by the following subcategory of the leather tanning and finishing industry: hair save/non-chrome tan/retan/wet finish.	(R)
Iron and Steel:		
K060	Ammonia still lime sludge from coking operations	(T)
K061	Emission control dust/sludge from the electric furnace production of steel	(T)
K062	Spent pickle liquor from steel finishing operations	(C, T)
K063	Sludge from lime treatment of spent pickle liquor from steel finishing operations	(T)
Primary Copper: K064	Acid plant blowdown slurry/sludge resulting from the thickening of blowdown slurry from primary copper production	(T)
Primary Lead: K065	Surface impoundment solids contained In and dredged from surface impoundments at primary lead smelting facilities	(T)
Primary Zinc:		
K066	Sludge from treatment of process wastewater and/or acid plant blowdown from primary zinc production	(T)
K067	Electrolytic anode slimes/sludges from primary zinc production	(T)
K068	Cadmium plant leach residue (iron oxide) from primary zinc production	(T)
Secondary Lead: K069	Emission control dust/sludge from secondary lead smelting	(T)

Fig. 3-3. Hazardous waste from specific sources. From Federal Register, **Vol. 45**; No. 98, May 19, 1980. (Continued)

(f) The commercial chemical products or manufacturing chemical intermediates, referred to in paragraphs (a), (b) and (d) of this section, are identified as toxic wastes (T) unless otherwise designated and are subject to the small quantity exclusion defined in § 261.5 (a) and (b). These wastes and their corresponding EPA Hazardous Waste Numbers are:

Hazardous Waste No.	Substance [1]
	AAF see U005
U001	Acetaldehyde
U002	Acetone (I)
U003	Acetonitrile (I,T)
U004	Acetophenone
U005	2-Acetylaminoflourene
U006	Acetyl chloride (C,T)
U007	Acrylamide
	Acetylene tetrachloride see U209
	Acetylene trichloride see U228
U008	Acrylic acid (I)
U009	Acrylonitrile
	AEROTHENE TT see U226
	3-Amino-5-(p-acetamidophenyl)-1H-1,2,4-triazole, hydrate see U011
U010	6-Amino-1,1a,2,8,8a,8b-hexahydro-8-(hydroxymethyl)8-methoxy-5-methylcarbamate azirino(2',3':3,4) pyrrolo(1,2-a) indole-4, 7-dione (ester)
U011	Amitrole
U012	Aniline (I)
U013	Asbestos
U014	Auramine
U015	Azaserine
U016	Benz[c]acridine
U017	Benzal chloride
U018	Benz[a]anthracene
U019	Benzene
U020	Benzenesulfonyl chloride (C,R)
U021	Benzidine
	1,2-Benzisothiazolin-3-one, 1,1-dioxide see U202
	Benzo[a]anthracene see U018
U022	Benzo[a]pyrene
U023	Benzotrichloride (C,R,T)
U024	Bis(2-chloroethoxy)methane
U025	Bis(2-chloroethyl) ether
U026	N,N-Bis(2-chloroethyl)-2-naphthylamine
U027	Bis(2-chloroisopropyl) ether
U028	Bis(2-ethylhexyl) phthalate
U029	Bromomethane
U030	4-Bromophenyl phenyl ether
U031	n-Butyl alcohol (I)
U032	Calcium chromate
	Carbolic acid see U188
	Carbon tetrachloride see U211
U033	Carbonyl fluoride
U034	Chloral

Fig. 3-4. Toxic wastes (Code T). From Federal Register, **Vol. 45**; No. 98, May 19, 1980.

Hazardous Waste No.	Substance[1]
U035	Chlorambucil
U036	Chlordane
U037	Chlorobenzene
U038	Chlorobenzilate
U039	p-Chloro-m-cresol
U040	Chlorodibromomethane
U041	1-Chloro-2,3-epoxypropane
	CHLOROETHENE NU see U226
U042	Chloroethyl vinyl ether
U043	Chloroethene
U044	Chloroform (I,T)
U045	Chloromethane (I,T)
U046	Chloromethyl methyl ether
U047	2-Chloronaphthalene
U048	2-Chlorophenol
U049	4-Chloro-o-toluidine hydrochloride
U050	Chrysene
	C.I. 23060 see U073
U051	Cresote
U052	Cresols
U053	Crotonaldehyde
U054	Cresylic acid
U055	Cumene
	Cyanomethane see U003
U056	Cyclohexane (I)
U057	Cyclohexanone (I)
U058	Cyclophosphamide
U059	Daunomycin
U060	DDD
U061	DDT
U062	Diallate
U063	Dibenz[a,h]anthracene
	Dibenzo[a,h]anthracene see U063
U064	Dibenzo[a,i]pyrene
U065	Dibromochloromethane
U066	1,2-Dibromo-3-chloropropane
U067	1,2-Dibromoethane
U068	Dibromomethane
U069	Di-n-butyl phthalate
U070	1,2-Dichlorobenzene
U071	1,3-Dichlorobenzene
U072	1,4-Dichlorobenzene
U073	3,3'-Dichlorobenzidine
U074	1,4-Dichloro-2-butene
	3,3'-Dichloro-4,4'-diaminobiphenyl see U073
U075	Dichlorodifluoromethane
U076	1,1-Dichloroethane
U077	1,2-Dichloroethane
U078	1,1-Dichloroethylene
U079	1,2-trans-dichloroethylene
U080	Dichloromethane
	Dichloromethylbenzene see U017
U081	2,4-Dichlorophenol
U082	2,6-Dichlorophenol
U083	1,2-Dichloropropane
U084	1,3-Dichloropropene
U085	Diepoxybutane (I,T)
U086	1,2-Diethylhydrazine
U087	O,O-Diethyl-S-methyl ester of phosphorodithioic acid
U088	Diethyl phthalate
U089	Diethylstilbestrol
U090	Dihydrosafrole
U091	3,3'-Dimethoxybenzidine
U092	Dimethylamine (I)
U093	p-Dimethylaminoazobenzene
U094	7,12-Dimethylbenz[a]anthracene
U095	3,3'-Dimethylbenzidine
U096	alpha,alpha-Dimethylbenzylhydroperoxide (R)
U097	Dimethylcarbamoyl chloride
U098	1,1-Dimethylhydrazine
U099	1,2-Dimethylhydrazine
U100	Dimethylnitrosoamine
U101	2,4-Dimethylphenol
U102	Dimethyl phthalate
U103	Dimethyl sulfate
U104	2,4-Dinitrophenol
U105	2,4-Dinitrotoluene
U106	2,6-Dinitrotoluene
U107	Di-n-octyl phthalate
U108	1,4-Dioxane

Hazardous Waste No.	Substance[1]
U109	1,2-Diphenylhydrazine
U110	Dipropylamine (I)
U111	Di-n-propylnitrosamine
	EBDC see U114
	1,4-Epoxybutane see U213
U112	Ethyl acetate (I)
U113	Ethyl acrylate (I)
U114	Ethylenebisdithiocarbamate
U115	Ethylene oxide (I,T)
U116	Ethylene thiourea
U117	Ethyl ether (I,T)
U118	Ethylmethacrylate
U119	Ethyl methanesulfonate
	Ethylnitrile see U003
	Firemaster T23P see U235
U120	Fluoranthene
U121	Fluorotrichloromethane
U122	Formaldehyde
U123	Formic acid (C,T)
U124	Furan (I)
U125	Furfural (I)
U126	Glycidylaldehyde
U127	Hexachlorobenzene
U128	Hexachlorobutadiene
U129	Hexachlorocyclohexane
U130	Hexachlorocyclopentadiene
U131	Hexachloroethane
U132	Hexachlorophene
U133	Hydrazine (R,T)
U134	Hydrofluoric acid (C,T)
U135	Hydrogen sulfide
	Hydroxybenzene see U188
U136	Hydroxydimethyl arsine oxide
	4,4'-(Imidocarbonyl)bis(N,N-dimethyl)aniline see U014
U137	Indeno(1,2,3-cd)pyrene
U138	Iodomethane
U139	Iron Dextran
U140	Isobutyl alcohol
U141	Isosafrole
U142	Kepone
U143	Lasiocarpine
U144	Lead acetate
U145	Lead phosphate
U146	Lead subacetate
U147	Maleic anhydride
U148	Maleic hydrazide
U149	Malononitrile
	MEK Peroxide see U160
U150	Melphalan
U151	Mercury
U152	Methacrylonitrile
U153	Methanethiol
U154	Methanol
U155	Methapyrilene
	Methyl alcohol see U154
U156	Methyl chlorocarbonate
	Methyl chloroform see U226
U157	3-Methylcholanthrene
	Methyl chloroformate see U156
U158	4,4'-Methylene-bis-(2-chloroaniline)
U159	Methyl ethyl ketone (MEK) (I,T)
U160	Methyl ethyl ketone peroxide (R)
	Methyl iodide see U138
U161	Methyl isobutyl ketore
U162	Methyl methacrylate (R,T)
U163	N-Methyl-N'-nitro-N-nitrosoguanidine
U164	Methylthiouracil
	Mitomycin C see U010
U165	Naphthalene
U166	1,4-Naphthoquinone
U167	1-Naphthylamine
U168	2-Naphthylamine
U169	Nitrobenzene (I,T)
	Nitrobenzol see U169
U170	4-Nitrophenol
U171	2-Nitropropane (I)
U172	N-Nitrosodi-n-butylamine
U173	N-Nitrosodiethanolamine
U174	N-Nitrosodiethylamine

Hazardous Waste No.	Substance[1]
U175	N-Nitrosodi-n-propylamine
U176	N-Nitroso-n-ethylurea
U177	N-Nitroso-n-methylurea
U178	N-Nitroso-n-methylurethane
U179	N-Nitrosopiperidine
U180	N-Nitrosopyrrolidine
U181	5-Nitro-o-toluidine
U182	Paraldehyde
	PCNB see U185
U183	Pentachlorobenzene
U184	Pentachloroethane
U185	Pentachloronitrobenzene
U186	1,3-Pentadiene (I)
	Perc see U210
	Perchlorethylene see U210
U187	Phenacetin
U188	Phenol
U189	Phosphorous sulfide (R)
U190	Phthalic anhydride
U191	2-Picoline
U192	Pronamide
U193	1,3-Propane sultone
U194	n-Propylamine (I)
U196	Pyridine
U197	Quinones
U200	Reserpine
U201	Resorcinol
U202	Saccharin
U203	Safrole
U204	Selenious acid
U205	Selenium sulfide (R,T)
	Silvex see U233
U206	Streptozotocin
	2,4,5-T see U232
U207	1,2,4,5-Tetrachlorobenzene
U208	1,1,1,2-Tetrachloroethane
U209	1,1,2,2-Tetrachloroethane
U210	Tetrachloroethene
	Tetrachloroethylene see U210
U211	Tetrachloromethane
U212	2,3,4,6-Tetrachlorophenol
U213	Tetrahydrofuran (I)
U214	Thallium (I) acetate
U215	Thallium (I) carbonate
U216	Thallium (I) chloride
U217	Thallium (I) nitrate
U218	Thioacetamide
U219	Thiourea
U220	Toluene
U221	Toluenediamine
U222	o-Toluidine hydrochloride
U223	Toluene diisocyanate
U224	Toxaphene
	2,4,5-TP see U233
U225	Tribromomethane
U226	1,1,1-Trichloroethane
U227	1,1,2-Trichloroethane
U228	Trichloroethene
	Trichloroethylene see U228
U229	Trichlorofluoromethane
U230	2,4,5-Trichlorophenol
U231	2,4,6-Trichlorophenol
U232	2,4,5-Trichlorophenoxyacetic acid
U233	2,4,5-Trichlorophenoxypropionic acid alpha, alpha, alpha- Trichlorotoluene see U023
	TRI-CLENE see U228
U234	Trinitrobenzene (R,T)
U235	Tris(2,3-dibromopropyl) phosphate
U236	Trypan blue
U237	Uracil mustard
U238	Urethane
	Vinyl chloride see U043
	Vinylidene chloride see U078
U239	Xylene

[1] The Agency included those trade names of which it was aware; an omission of a trade name does not imply that it is not hazardous. The material is hazardous if it is listed under its generic name.

Fig. 3-4. Toxic wastes (Code T). From Federal Register, **Vol. 45**; No. 98, May 19, 1980. (Continued)

Title 40 of the Code of Federal
Regulations is amended as follows:
1. In § 261.31, add the following waste
streams:

§ 261.31   Hazardous waste from nonspecific sources.

Industry	EPA hazardous waste No.	Hazardous waste	Hazard code
Generic	F017	Paint residues or sludges from industrial painting in the mechanical and electrical products industry.	(T)
	F018	Wastewater treatment sludge from industrial painting in the mechanical and electrical products industry.	(T)

2. In § 261.32, add the following waste streams:

§ 261.32   Hazardous waste from specific sources.

Industry	EPA hazardous waste No.	Hazardous waste	Hazard code
Inorganic Chemicals........................	K071	Brine purification muds from the mercury cell process in chlorine production, where separately prepurified brine is not used.	(T)
	K073	Chlorinated hydrocarbon wastes from the purification step of the diaphragm cell process using graphite anodes in chlorine production.	(T)
	K074	Wastewater treatment sludges from the production of TiO₂ pigment using chromium bearing ores by the chloride process.	(T)
Paint Manufacturing	K078	Solvent cleaning wastes from equipment and tank cleaning from paint manufacturing.	(I, T)
	K079	Water or caustic cleaning wastes from equipment and tank cleaning from paint manufacturing.	(T)
	K081	Wastewater treatment sludges from paint manufacturing........	(T)
	K082	Emission control dust or sludge from paint manufacturing......	(T)
Organic Chemicals	K083	Distillation bottoms from aniline production.............................	(T)
	K085	Distillation or fractionating column bottoms from the production of chlorobenzenes.	(T)
Ink Formulation..................................	K086	Solvent washes and sludges, caustic washes and sludges, or water washes and sludges from cleaning tubs and equipment used in the formulation of ink from pigments, driers, soaps, and stabilizers containing chromium and lead.	(T)
Veterinary Pharmaceuticals.............	K084	Wastewater treatment sludges generated during the production of veterinary pharmaceuticals from arsenic or organoarsenic compounds.	(T)
Coking..	K087	Decanter tank tar sludge from coking operations...................	(T)
Primary Aluminum.............................	K088	Spent potliners from primary aluminum reduction....................	(T)
Ferroalloys.......................................	K090	Emission control dust or sludge from ferrochromium-silicon production.	(T)
	K091	Emission control dust or sludge from ferrochromium production	(T)
	K092	Emission control dust or sludge from ferromanganese production.	(T)

Fig. 3-5. List of hazardous wastes promulgated July 16, 1980. From Federal Register, **Vol. 45**; No. 138, July 16, 1980.

Appendix B*—Scheduled Fall Promulgation

Generic

1. Reactor clean-up wastes from the chlorination, dehydrochlorination, or oxychlorination of aliphatic hydrocarbons
2. Fractionation bottoms from the separation of chlorination hydrocarbons
3. Distillation bottoms from the separation of chlorinated aliphatic hydrocarbons
4. Washer wastes from the production of chlorinated aliphatic hydrocarbons
5. Spent catalyst from the production of chlorinated aliphatic hydrocarbons
6. Reactor clean-up wastes from the chlorination of cyclic aliphatic hydrocarbons
7. Fractionation bottoms from the separation of chlorinated cyclic aliphatic hydrocarbons
8. Distillation bottoms from the separation of chlorinated cyclic aliphatic hydrocarbons
9. Washer wastes from the production of chlorinated cyclic aliphatic hydrocarbons
10. Spent catalyst from the production of chlorinated cyclic aliphatic hydrocarbons
11. Batch residues from the batch production of chlorinated polymers
12. Solution residues from the production of chlorinated polymers
13. Reactor clean-up wastes from the chlorination of aromatic hydrocarbons
14. Fractionation bottoms from the separation of chlorinated aromatic hydrocarbons
15. Distillation bottoms from the separation of chlorinated aromatic hydrocarbons
16. Washer wastes from the production of chlorinated aromatic hydrocarbons
17. Waste Oil [Comment: This listing description was originally proposed on December 18, 1978 (43 FR 58957) as Waste lubricating oil and Waste hydraulic or cutting oil.]
18. Polychlorinated biphenyls (PCB) and PCB items as defined in 40 CFR Part 761 [Comment: The Agency indicated in the preamble to the Section 3004

Fig. 3-6. List of hazardous wastes scheduled for promulgation in Fall 1980. From Federal Register, **Vol. 45**; No. 98, May 19, 1980.

regulations (43 FR 58993), their intention to integrate the TSCA regulations for the disposal of PCB's with the RCRA hazardous waste regulations.]

Process Wastes

1. Sub-ore from underground and surface mining of uranium, overburden from surface mining of uranium and waste rock from underground mining of uranium with a radium-226 activity in excess of 5pCi/gm [Comment: This listing description was originally proposed on December 18, 1978 (43 FR 58958) as: Waste rock and overburden from uranium mining.]
2. Leach zone overburden and discarded phosphate ore from phosphate surface mining and slimes from phosphate ore beneficiation [Comment: This listing description was originally proposed on December 18, 1978 (43 FR 58958) as: Overburden and slimes from phosphate surface mining.]
3. Waste gypsum from processing phosphate ore to produce phosphoric acid [Comment: This listing description was originally proposed on December 18, 1978 (43 FR 58958) as: Waste gypsum from phosphoric acid production.]
4. Slag and fluid bed prills from processing phosphate ore to produce elemental phosphorous [Comment: This listing description was originally proposed on December 18, 1978 (43 FR 58958) as: Slag and fluid bed prills from elemental phosphorous production.]
5. Washwater/sludges from ink printing equipment clean-up [Comment: This listing description includes three wastes which were originally proposed on August 22, 1979 (44 FR 49403 and 49404) as: Waste from equipment cleaning from flexoprinting in the manufacture of paperboard boxes; Waste from press clean-up in newspaper printing and Wash water from printing ink equipment cleaning.]
6. Wastes from photographic processing [Comment: This listing was originally proposed on August 22, 1979 (44 FR 49404) as: Waste Ferricyanide bleach, dichromate bleach, color developer (Agfa), bleach fix (Agfa) and acid solution from photographic processing.]
7. Lead acid storage battery production wastewater treatment sludges
8. Lead acid storage battery production clean-up wastes from cathode and anode paste production
9. Nickel cadmium battery production wastewater treatment sludges
10. Lead slag from lead alkyl production
11. Emission control dust/sludge from reverberatory furnace and converters from primary copper production [Comment: This listing description was included in the listing description originally proposed on December 18, 1978 (43 FR 58959) as: Primary copper smelting and refining electric furnace slag, converter dust, acid plant sludge and reverberatory dust.]

Fig. 3-6. List of hazardous wastes scheduled for promulgation in Fall 1980. From Federal Register, **Vol. 45**; No. 98, May 19, 1980. (Continued)

4
STANDARDS APPLICABLE TO GENERATORS OF HAZARDOUS WASTE: PART 262

Part 262 establishes standards for generators of hazardous waste. The regulation requires a generator of solid waste to determine if his waste is a hazardous waste. Generators must prepare manifests for all shipments of hazardous wastes that are sent to off-site treatment, storage, or disposal facilities. For these shipments, generators must also package, label, mark, and placard the waste according to EPA/DOT regulations.

Additionally, generators are required to keep records, report those shipments which do not reach the facility designated on the manifest, and submit an annual summary of their activities. This Part also establishes special requirements for international shipment of hazardous waste and provides for the temporary (90 day) accumulation of hazardous wastes without a permit.

Table 4-1 addresses the detailed requirements of Part 262 and contains the following nine areas:
  I. Applicability
  II. Hazardous Waste Determination
  III. EPA Identification Numbers
  IV. Manifest Requirements
  V. Pretransport Requirements
  VI. Accumulation Time
  VII. Recordkeeping Requirements
  VIII. Reporting Requirements
  IX. International Shipments

Figure 4-1 is a copy of EPA's Forms 8700-13 and 8700-13A, including instructions, which is to be used by generators for annual reporting of off-site hazardous waste shipments. (See pp. 29–32.)

Table 4-1. Standards Applicable To Generators Of Hazardous Waste.

Section No.	Description
	I. Applicability
262.10(a)	A. General Part 262 regulations establish standards for generators of hazardous waste.
262.10(b)	B. Part 262 Applicability for On-Site Generators Who Treat, Store, or Dispose of Hazardous Wastes (1) Section 262.11—For determining whether or not generator has a hazardous waste. (2) Section 262.12—For obtaining an EPA identification number. (3) Section 262.40(c) and (d)—For recordkeeping. (4) Section 262.43—For additional reporting. (5) Section 262.51—For farmers, if applicable.
262.10 Note	C. Other Title 40 Applicability for On-Site Generators That Treat, Store, or Dispose of Hazardous Wastes (1) Part 122—Permit requirements. (2) Parts 264 and 265—Final and interim standards applicable to owners and operators of hazardous waste treatment, storage, and disposal facilities. (3) Part 266—Waste-specific facility standards (not yet promulgated).
262.10(c)	D. Importers of Hazardous Waste into United States Such persons become hazardous waste generators and must comply with standards applicable to generators established in this Part (262).
262.10(d) and 262.51	E. Exemptions (1) Farmers Disposing of Hazardous Waste Pesticides from Own Use

Table 4-1. Standards Applicable To Generators Of Hazardous Waste. (Continued)

Section No.	Description
	Farmers not required to comply with standards of this Part or Parts 122, 264, or 265 provided: (a) Each emptied pesticide container is triple-rinsed per Section 261.33(c). (b) Disposal of pesticide(s) on own farm is consistent with instructions on pesticide label.
262 Authority	F. Federal Statutes/Authority Sections 2002(a), 3001, 3002, 3003, 3004, and 3005, Solid Waste Disposal Act, as amended by Resource Conservation and Recovery Act of 1976 and as amended by the Quiet Communities Act of 1978 (42 U.S.C. 6912(a), 6921, 6922, 6923, 6924, and 6925).
262.10(e)	G. Violations/Penalties Penalties are prescribed in Section 3008 of RCRA for violations of this Part, for persons generating hazardous wastes as defined in 40 CFR 261.
	II. Hazardous Waste Determination
262.11	A. Requirement for Determining If Waste Is Hazardous Such a determination must be made by a person generating solid waste, as defined in 40 CFR 261.2 (Definition of Solid Waste).
262.11(a)	B. Methods and Sequence For Determining If Waste Is Hazardous (1) Check Exclusion of Waste From Regulation (a) Under Section 261.4 – Exclusions. (b) Under Section 261.5 – Special Requirements for Hazardous Waste Generated by Small Quantity Generators.
262.11(b)	(2) Determine If Waste Is Listed as Hazardous Under Subpart D (Lists of Hazardous Wastes) of Part 261
262.11(c)	(3) If Not Listed Determine If Waste Is Ignitable, Corrosive, Reactive, or Toxic (under Subpart C of Part 261) Alternate methods follow:
262.11(c)(1)	(a) Testing wastes per methods of Subpart C of Part 261.
262.11(c)(1)	(b) Testing wastes per equivalent method approved by Administrator (Section 260.21 – Petitions for Equivalent Testing or Analytical Methods).
262.11(c)(2)	(c) Applying knowledge of hazard characteristic in light of materials or processes used.
262.11(b) Note	C. Demonstration by Generator That Listed Waste from Particular Facility or Operation Is Not Hazardous Such a demonstration to the Administrator is allowed under Section 260.22 (Petitions to Amend Part 261 to Exclude a Waste Produced at a Particular Facility).
	III. EPA Identification Numbers
262.12(a)	A. Prohibitions on Generator without EPA Identification Number Such a generator must not: (1) Treat (2) Store (3) Dispose of (4) Transport (5) Offer for transport hazardous wastes.
262.12(c)	B. Prohibitions on Generator Using Outside Hazardous Waste Service without EPA Identification Number Generators must not offer hazardous wastes to such: (1) Transporters (2) Treatment facilities (3) Storage facilities (4) Disposal facilities.
262.12(b)	C. Obtaining an EPA Identification Number If such a number has not been received by the generator, apply for one from the Administrator using EPA Form 8700-12 (see Fig. 9-1).

Table 4-1. Standards Applicable To Generators Of Hazardous Waste. (Continued)

Section No.	Description
Subpart B	IV. Manifest Requirements
262.20	A. General Requirements
262.20(a)	(1) Preparation of Manifest Prior to Transporting Hazardous Waste Off-Site Such is required of generator who transports or offers for transportation hazardous waste for off-site treatment, storage, or disposal.
262.20(b)	(2) Designation of Single Facility with Permit to Handle Waste Generator must so designate on manifest.
262.20(c)	(3) Designation of Single Alternative Facility with Permit to Handle Waste Generator may so designate on manifest, for emergency that prevents delivery of waste to primary facility
262.20(d)	(4) If Hazardous Waste Cannot Be Delivered to Designated or Alternate Facility The generator must: (a) Designate another facility; or (b) Instruct transporter to return waste.
262.21	B. Content of Manifest
262.21(a)(1)	(1) Manifest document number.
262.21(a)(2)	(2) Generator's name, mailing address, telephone number, and EPA identification number
262.21(a)(3)	(3) Name and EPA identification number of each transporter.
262.21(a)(4)	(4) Name, address, and EPA identification number of designated facility and alternate facility, if any.
262.21(a)(5)	(5) Description of waste(s) (e.g., proper shipping name, etc.) required by Department of Transportation regulations in: (a) 49 CFR 172.101 (see Appendix E of this book for regulation contents only). (b) 49 CFR 172.202 (see Appendix E of this book for regulation contents only). (c) 49 CFR 172.203 (see Appendix E of this book for regulation contents only).
262.21(a)(6)	(6) Total quantity of each hazardous waste loaded into or onto the transport vehicle. (a) By units of weight or volume. (b) Type and number of containers.
262.21(b)	(7) Following certification must be on manifest: "This is to verify that the above named materials are properly classified, described, packaged, marked, and labeled and are in proper condition for transportation according to the applicable regulations of the Department of Transportation and EPA."
262.22	C. Number of Copies Required Manifest copies will be needed as shown: (1) Generator—initial plus returned copy. (2) Each transporter. (3) Owner or operator of designated facility.
	D. Manifest Handling Requirements
262.23(a)	The generator must:
262.23(a)(1)	(1) Sign manifest certification by hand.
262.23(a)(2)	(2) Obtain handwritten signature of initial transporter and date of acceptance on manifest.
262.23(a)(3)	(3) Retain one copy in accordance with the recordkeeping requirements of Section 262.40(a).
262.23(b)	(4) Give transporter remaining copies of manifest.
262.23(c)	(5) For Shipment of Hazardous Waste within United States Solely by Railroad or Water (Bulk Shipment Only) (a) Send three copies of manifest to owner/operator of designated facility. (b) Date and sign manifest copies. (c) Copy of manifest not required for each transporter.
Subpart C	V. Pretransport Requirements
262.30	A. Packaging Requirements for Off-Site Transportation The generator must package the hazardous waste per applicable Department of Transportation regulations on packaging under: (1) 49CFR173 (see Appendix E of this book for regulation contents only). (2) 49CFR178 (see Appendix E of this book for regulation contents only). (3) 49CFR179 (see Appendix E of this book for regulation contents only).

Table 4-1. Standards Applicable To Generators Of Hazardous Waste. (Continued)

Section No.	Description
262.31	B. Labeling Requirements for Off-Site Transportation The generator must label each package of hazardous waste per applicable Department of Transportation regulations on hazardous materials under 49CFR172, Subpart E (see Appendix E of this book for regulation contents only).

262.32
262.32(a)
	C. Marking Requirements for Off-Site Transportation
	   (1) General
	       The generator must mark each package of hazardous waste per applicable Department of Transportation regulations on hazardous materials under 49 CFR 172, Subpart D (see Appendix E of this book for regulation contents only).

262.32(b)	(2) Markings for 110 Gallon or Less Containers
	       The generator must mark the container holding the hazardous waste with the following words and information displayed per Department of Transportation regulations under 49 CFR 172.304:
	
	       "Hazardous Waste–Federal Law Prohibits Improper Disposal. If found, contact the nearest police or public safety authority or the U.S. Environmental Protection Agency.
	       Generator's Name and Address _____
	       Manifest Document Number _____."

262.33	D. Placarding Requirements for Off-Site Transportation The generator must placard or offer the initial transporter the appropriate placards per Department of Transportation regulations for hazardous materials under 49 CFR 172, Subpart F.
	VI. Accumulation Time
262.34(a)	A. On-Site Accumulation of Hazardous Wastes for 90 Days or Less The generator does not require a permit provided that:
262.34(a)(1)	(1) Waste is shipped off-site in 90 days or less
262.34(a)(2)	(2) Waste Placed in Contiainers
262.34(a)(2) and (4)	(a) Packaging, labeling and marking standards per Sections 262.30, 262.31, and 262.32, respectively, are met.
262.34(a)(2)	(b) Inspection and special requirements for ignitable or reactive waste standards per Sections 265.174 and 265.176, respectively, are met.
262.34(a)(2)	(3) Waste Placed in Tanks Requirements of Subpart J (e.g., Tanks) of Part 265 are met, except for Section 265.193 (Waste analysis and trial tests).
262.34(a)(3)	(4) Tank or container is clearly marked and visible for inspection, with date upon which each period of accumulation begins.

262.34(a)(5)	(5) Preparedness and prevention requirements of Subpart C of Part 265 are met.
	   (6) Contingency plan and emergency procedures of Subpart D of Part 265 are met.
	   (7) Personnel training requirements of Section 265.16 are met.

262.34(b)	B. On-Site Accumulation of Hazardous Wastes for More Than 90 Days The generator is an operator of a storage facility and is subject to the requirements of 40 CFR 264 and 265 and the permit requirements of 40 CFR 122.
	VII. Recordkeeping Requirements

262.40(a)	A. Manifests
	Generator must keep signed and dated copy of each manifest for three years. Keep either:
	   (1) Copy from initial transporter; or
	   (2) Copy from designated facility.

262.40(b)	B. Annual Report Generator must keep copy of each annual report for at least three years from due date of report (March 1).
	C. Exception Report Generator must keep copy of each exception report for at least three years.
262.40(c)	D. Records of Test Results, Water Analyses, or Other Determinations under Section 262.11 (Hazardous Waste Determination)

Table 4-1. Standards Applicable To Generators Of Hazardous Waste. (Continued)

Section No.	Description
	Generator must keep such records for at least three years from date that waste was last sent to on-site or off-site treatment, storage, or disposal facility.
262.40(d)	E. Retention Requirements during Unresolved Enforcement Action The periods cited above are extended automatically or as requested by Administrator.
	VIII. Reporting Requirements
262.41 262.41(a) 262.41(a)(1) 262.41(a)(2) 262.41(a)(3) 262.41(b)	A. Annual Reporting (1) For Generators Who Ship Hazardous Waste Off-Site Must submit annual reports: (a) On EPA Forms 8700-13 and 8700-13A (see Fig. 4-1 for forms and instructions). (b) To the generator's Regional Administrator (c) No later than 1 March for the preceeding calendar year. (2) For Generators Who Treat, Store, or Dispose of Hazardous Wastes On-Site Must submit annual report in accordance with provisions of: (a) Parts 264 and 265—Final and interim standards applicable to owners and operators of hazardous waste treatment, storage, and disposal facilities. (b) Part 266—Waste-specific facility standards (not yet promulgated). (c) Part 122—Permit requirements.
262.42 262.42(a) 262.42(b) 262.42(b)(1) 262.42(b)(2)	B. Exception Reporting (1) Waste Status Inquiry for Nonreceipt of Signed Manifest Copy from Designated Facility Such is required of generator, who must contact transporter and/or owner/operator of facility, within 35 days of acceptance of waste by initial transporter (2) Exception Report for Nonreceipt of Signed Manifest Copy from Designated Facility Such is required of generator to his Regional Administrator, within 45 days of acceptance of waste by initial transporter. Content of exception report follows: (a) Legible copy of manifest for which generator does not have confirmation of delivery. (b) Cover letter signed by generator or authorized representative explaining: (i) Efforts taken to locate hazardous wastes. (ii) Results of these efforts.
262.43	C. Additional Reporting Required by Administrator (1) Authority Sections 2002(a) and 3002(b) of the Act. (2) Content The Administrator may require generators to furnish additional reports concerning the quantities and disposition of wastes identified or listed in Part 261.
262.50	IX. International Shipments
262.50(a)	A. General Any person who exports hazardous waste to a foreign country or imports hazardous waste from a foreign country into the United States must comply with: (1) Requirements of this Part (262); and (2) Special requirements of this Section (262.50).
262.50(b) 262.50(b)(1) 262.50(b)(1)(i) 262.50(b)(1)(ii) 262.50(b)(1)(iii)	B. Exporters of Hazardous Waste Outside of United States The generator must: (1) Notify Administrator in Writing (a) Within four weeks prior to initial shipment of hazardous waste to each country in each calendar year. (b) Properly Identify Waste (i) By its EPA hazardous waste identification number. (ii) By its DOT shipping description. (c) Include name and address of foreign consignee. (d) Send notice to: Hazardous Waste Export Division for Oceans and Regulatory Affairs (A-107) United States Environmental Protection Agency Washington, DC 20460

Table 4-1. Standards Applicable To Generators Of Hazardous Waste. (Continued)

Section No.	Description
262.50(b)(1)(iii) Note	(e) This requirement will not be delegated from the Administrator to States under 40 CFR 123 (State Program Requirements).
262.50(b)(2)	(2) Have foreign Consignee Confirm Delivery of Waste Copy of manifest signed by consignee may be used.
262.50(b)(3)	(3) Meet Manifest Content Requirements of Section 262.21 Exceptions:
262.50(b)(3)(i)	(a) Name and address of foreign consignee, instead of name and address and EPA identification number of designated facility.
262.50(b)(3)(ii)	(b) Identify point of departure from United States through which waste must travel before entering a foreign country.
262.50(c)	(4) File An Exception Report
262.50(c)(1)	(a) For Nonreceipt of Manifest Signed by Transporter and Stating Date and Place of Departure from United States Such is required of generator within 45 days of acceptance of waste by initial transporter; or
262.50(c)(2)	(b) For Nonreceipt of Written Confirmation from Foreign Consignee that Hazardous Waste Was Received Such is required of generator within 90 days of acceptance of waste by initial transporter
262.50(d)	C. Importers of Hazardous Waste within the United States Must meet manifest content requirements of Section 262.21. Exceptions:
262.50(d)(1)	(1) Name and address of foreign generator and importer's name and address and EPA identification number, instead of generator's name, address, and EPA identification number.
262.50(d)(2)	(2) U.S. importer or agent must sign and date the certification and obtain the signature of the initial transporter, instead of the generator's signature on the certification statement.

Please print or type with ELITE type *(12 characters per inch)*.

GSA No. 12345-XX
Form Approved OMB No. 158-R00XX

♻EPA	U.S. ENVIRONMENTAL PROTECTION AGENCY HAZARDOUS WASTE REPORT	I. TYPE OF HAZARDOUS WASTE REPORT

PART A: GENERATOR ANNUAL REPORT

THIS REPORT IS FOR THE YEAR ENDING DEC. 31, | 1 | 9 | | |

PART B: FACILITY ANNUAL REPORT

THIS REPORT FOR YEAR ENDING DEC. 31, | 1 | 9 | | |

PART C: UNMANIFESTED WASTE REPORT

THIS REPORT IS FOR A WASTE RECEIVED *(day, mo., & yr.)* | | – | | – | 1 | 9 | |

PLEASE PLACE LABEL IN THIS SPACE

INSTRUCTIONS: You may have received a preprinted label attached to the front of this pamphlet; affix it in the designated space above—left. If any of the information on the label is incorrect, draw a line through it and supply the correct information in the appropriate section below. If the label is complete and correct, leave Sections II, III, and IV below blank. If you did not receive a preprinted label, complete all sections. "Installation" means a single site where hazardous waste is generated, treated, stored, or disposed of. Please refer to the specific instructions for generators or facilities before completing this form. The information requested herein is required by law *(Section 3002/3004 of the Resource Conservation and Recovery Act)*.

II. INSTALLATION'S EPA I.D. NUMBER

F | | | | | | | | | | | | 1

III. NAME OF INSTALLATION

IV. INSTALLATION MAILING ADDRESS

STREET OR P.O. BOX

3 | | | | | | | | | | | | | | | | | | | | | | | | | | | | | |

CITY OR TOWN | ST. | ZIP CODE

4 | | | | | | | | | | | | | | | | | | | | | | |

V. LOCATION OF INSTALLATION

STREET OR ROUTE NUMBER

5 | | | | | | | | | | | | | | | | | | | | | | | | | | | |

CITY OR TOWN | ST. | ZIP CODE

6 | | | | | | | | | | | | | | | | | | | | | | |

VI. INSTALLATION CONTACT

NAME *(last and first)* | PHONE NO. *(area code & no.)*

2 | | | | | | | | | | | | | | | | | | | | | | | – | | | – | |

VII. TRANSPORTATION SERVICES USED *(for Part A reports only)*

List the EPA Identification Numbers for those transporters whose services were used during the reporting year represented by this report.

VIII. COST ESTIMATES FOR FACILITIES *(for Part B reports only)*

A. COST ESTIMATE FOR FACILITY CLOSURE	B. COST ESTIMATE FOR POST CLOSURE MONITORING AND MAINTENANCE *(disposal facilities only)*
G $	$

IX. CERTIFICATION

I certify under penalty of law that I have personally examined and am familiar with the information submitted in this and all attached documents, and that based on my inquiry of those individuals immediately responsible for obtaining the information, I believe that the submitted information is true, accurate, and complete. I am aware that there are significant penalties for submitting false information, including the possibility of fine and imprisonment.

A. PRINT OR TYPE NAME	B. SIGNATURE	C. DATE SIGNED

EPA Form 8700-13 (5-80)

PAGE ____1____ OF _____

Fig. 4-1. EPA Form 8700-13, Generator's Annual Reporting Of Off-Site Hazardous Waste Shipments.

Please print or type with ELITE type (12 characters/inch).

GSA No. 12345-XX
Form Approved OMB No. 158-R00XX

⊕EPA

U.S. ENVIRONMENTAL PROTECTION AGENCY
GENERATOR ANNUAL REPORT – PART A
(Collected under the authority of Section 3002 of RCRA.)

FOR OFFICIAL USE ONLY (Items 1 and 2)	► 1. DATE RECEIVED	—	— 1 9	X. GENERATOR'S EPA I.D. NO.	T/A C 1
	► 2. TYPE OF REPORT		G		

XI. FACILITY'S EPA I.D. NO.

16   27

XIII. FACILITY ADDRESS (street or P.O. box, city, state, & zip code)

XII. FACILITY NAME (specify)

XIV. WASTE IDENTIFICATION

LINE NUMBER	A. DESCRIPTION OF WASTE	B. DOT HAZARD CLASS	C. EPA HAZARDOUS WASTE NUMBER (see instructions)	D. AMOUNT OF WASTE	E. UNIT OF MEASURE (enter code)
1			30 - 33 34 - 37		
		28 29	38 - 41 42 - 45 46	54	55
2					
3					
4					
5					
6					
7					
8					
9					
10					
11					
12			30 - 33 34 - 37		
		28 29	38 - 41 42 - 45 46 - 54		55

XV. COMMENTS (enter information by line number — see instructions)

EPA Form 8700-13A (5-80)

BILLING CODE 6560-01-C

PAGE _____ OF _____

Fig. 4-1. EPA Form 8700-13A, Generator's Annual Reporting Of Off-Site Hazardous Waste Shipments. (Continued)

General Instructions, Hazardous Waste Report (EPA Form 8700–13)

Important: READ ALL INSTRUCTIONS BEFORE COMPLETING THIS FORM.

Section I. Type of Hazardous Waste Report

Part A: Generator Annual Report—For generators who ship their waste off-site to facilities which they do not own or operate, fill in the reporting year for this report (e.g., 1982).

Note.—Generators who ship hazardous waste off-site to a facility which they own or operate must complete the facility (Part B) report instead of the Part A report.

Part B: Facility Annual Report—For owners or operators of on-site or off-site facilities that treat, store, or dispose of hazardous waste, fill in the reporting year for this report (e.g., 1982).

Part C: Unmanifested Waste Report—For facility owners or operators who accept for treatment, storage, or disposal any hazardous waste from an off-site source without an accompanying manifest, fill in the date the waste was received at the facility (e.g. 04–12–1982).

Section II thru Section IV. Installation I.D. Number, Name of Installation, and Installation Mailing Address

If you received a preprinted label from EPA, attach it in the space provided and leave Sections II through IV blank. If there is an error or omission on the label, cross out the incorrect information and fill in the appropriate item(s). If you did not receive a preprinted label, complete Section II through Section IV.

Section V. Location of Installation

If your installation location address is different than the mailing address, enter the location address of your installation.

Section VI. Installation Contact

Enter the name (last and first) and telephone number of the person who may be contacted regarding information contained in this report.

Section VII. Transportation Services Used (For Part A Reports ONLY)

List the EPA Identification Number for each transporter whose services you used during the reporting year.

Section VIII. Cost Estimates for Facilities (For Part B Reports ONLY)

A. Enter the most recent cost estimate for facility closure in dollars. See Subpart H of 40 CFR Parts 264 or 265 for more detail.

B. For disposal facilities only, enter the most recent cost estimate for post closure monitoring and maintenance. See Subpart H of 40 CFR Parts 264 or 265 for more detail.

Section IX. Certification

The generator or his authorized representative (Part A reports) or the owner or operator of the facility or his authorized representative (Parts B and C reports) must sign and date the certification where indicated. The printed or typed name of the person signing the report must also be included where indicated.

Note.—Since more than one page is required for each report, enter the page number of each sheet in the lower right corner as well as the total number of pages.

Generator Annual Report, Part A Instructions (EPA Form 8700–13A)

Generator Annual Report for generators who ship their hazardous waste off-site to facilities which they do not own or operate.

Important: READ ALL INSTRUCTIONS BEFORE COMPLETING THIS REPORT.

Section X. Generator's Identification Number

Enter your EPA identification number.

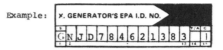

Section XI. Facility's Identification Number

Enter the EPA identification number of the facility to which you sent the waste described below in Section XIV (a separate sheet must be used for each facility to which you sent hazardous waste.)

Section XII. Facility Name

Enter the name of the facility corresponding to the facility's EPA identification number in Section XI.

Section XIII. Facility Address

Enter the address of the facility corresponding to the facility's EPA identification number in Section XI.

Section XIV. Waste Identification

All information in this section must be entered by line number. Each line entry will describe the total annual amount of each waste shipped to the facility identified in Section XI, above.

Section XIV–A. Description of Waste

For hazardous wastes that are listed under 40 CFR Part 261, Subpart D, enter the EPA listed name, abbreviated if necessary. Where

mixtures of listed wastes were shipped, enter the description which you believe best describes the waste.

For unlisted hazardous waste identified under 40 CFR Part 261, Subpart C, enter the description which you believe best describes the waste. Include the specific manufacturing or other process generating the waste (e.g., green sludge from widget manufacturing) and, if known, the chemical or generic chemical name of the waste.

Section XIV–B. DOT Hazard Class

Enter the two digit code from Table 1 which corresponds to the DOT hazard class of the waste described. (If the waste described has been shipped under more than one DOT hazard class, use a separate line for each DOT hazard class.)

Table 1

DOT hazard class	Code
Combustible	01
Corrosive	02
Etiologic agent	03
Explosive A	04
Explosive B	05
Flammable gas	06
Flammable liquid	07
Flammable solid	08
Irritating agent	09
Nonflammable gas	10
Organic peroxide	11
ORM–E	12
Oxidizer	13
Poison A	14
Poison B	15
Radioactive	16

Section XIV–C. EPA Hazardous Waste Number

For listed wastes, enter the EPA Hazardous Waste Number from 40 CFR Part 261, Subpart D, which identifies the waste.

For a mixture of more than one listed waste, enter each of the applicable EPA Hazardous Waste Numbers. Four spaces are provided. If more space is needed, continue on the next line(s) and leave all other information on that line blank.

XIV. WASTE IDENTIFICATION					
LINE NUMBER	A. DESCRIPTION OF WASTE	B. DOT HAZARD CLASS	C. EPA HAZARDOUS WASTE NUMBER (see instructions)	D. AMOUNT OF WASTE	UNIT OF MEASURE (enter code)
1	Steel finishing sludge	0 2	K 0 6 0 K 0 6 1 / K 0 6 2 K 0 6 3	4 1 2 4 6	T
2			K 0 6 4		

For unlisted hazardous wastes, enter the EPA Hazardous Waste Numbers from 40 CFR Part 261, Subparts C, applicable to the waste. If more than four spaces are required, follow the procedure described above.

Section XIV–D. Amount of Waste

Enter the amount of this waste you shipped to the facility identified in Section XI and include the weight of containers if left at the treatment, storage, or disposal facility.

Fig. 4-1. EPA Form 8700-13, Generator's Annual Reporting Of Off-Site Hazardous Waste Shipments. (Continued)

Section XIV–E. Unit of Measure

Enter the unit of measure code for the quantity of waste described on this line. Units of measure which must be used in this report and the appropriate codes are:

Units of measure	Code
Pounds	P
Short tons (2,000 lbs)	T
Kilograms	K
Tonnes (1,000 kg)	M

Units of volume may not be used for reporting but must be converted into one of the above units of weight taking into account the appropriate density or specific gravity of the waste.

Section XV. Comments

This space may be used to explain or clarify any entry. If used, enter a cross reference to the appropriate Section number.

Note.—Since more than one page is required for each report, enter the page number of each sheet in the lower right corner as well as the total number of pages.

[FR Doc. 80–14665 Filed 5–16–80; 8:45 am]

BILLING CODE 6560-01-M

Fig. 4-1. EPA Form 8700-13, Generator's Annual Reporting Of Off-Site Hazardous Waste Shipments. (Continued)

5
STANDARDS APPLICABLE TO TRANSPORTERS OF HAZARDOUS WASTE: PART 263

Part 263 contains regulations that apply to persons transporting hazardous waste requiring a manifest (per 40 CFR 262). The standards require a transporter to obtain an EPA identification number, comply with the manifest system, deliver the wastes to the designated treatment, storage, or disposal facility, and to keep records.

Also, a transporter must take prescribed measures if a hazardous waste discharge occurs during transit.

Table 5-1 that follows contains five major areas and addresses the detailed requirements of Part 263:

    I. Applicability
   II. EPA Identification Numbers
  III. Manifest System Operation for Transporters
  IV. Recordkeeping Requirements
   V. Hazardous Waste Discharges

Table 5-1. Standards Applicable to Transporters of Hazardous Waste.

Section No.	Description
	I. Applicability
263.10(a)	A. General Part 263 regulations establish standards for transporters of hazardous waste within the United States, if its transportation requires a manifest under 40 CFR 262 (Standards Applicable To Generators Of Hazardous Waste).

263.10(a) Note	B. EPA Adoption of Certain DOT Regulations Governing Transportation Of Hazardous Materials
	(1) Main Content of Adopted DOT Regulations
	   (a) Labeling requirements.
	   (b) Marking requirements.
	   (c) Placarding requirements.
	   (d) Using proper containers.
	   (e) Reporting discharges.
	(2) Purposes of Adoption Of DOT Regulations
	   (a) To satisfy statutory obligation to promulgate regulations necessary to protect human health and environment while transporting hazardous waste.
	   (b) To ensure consistency with DOT requirements and thus avoid duplicative or conflicting requirements.
	(3) Interstate and Intrastate Transportation of Hazardous Waste
	EPA-adopted DOT regulations are applicable.
	(4) Caution to Transporters of Hazardous Waste
	DOT regulations are fully applicable to their activities and are enforceable by DOT.
	(5) EPA Retention of Its Authority to Enforce These Regulations
	This is so regardless of DOT action.
	(6) EPA/DOT Joint Development of Standards for Transporters of Hazardous Waste
	This is done to avoid conflicting requirements.

263.10(b)	C. Exemptions (1) On-site transportation of hazardous waste by generators. (2) On-site transportation of hazardous waste by owners or operators of permitted waste management facilities.

Table 5-1. Standards Applicable to Transporters of Hazardous Waste. (Continued)

Section No.	Description
263.10(c)(1)	D. Transporters of Hazardous Waste into the United States Such transporters become generators and must comply with 40 CFR 262 (Standards Applicable To Generators Of Hazardous Waste).
263.10(c)(2)	E. Transporters Who Mix Hazardous Waste of Different Shipping Descriptions into Single Container Such transporters become generators and must comply with 40 CFR 262.
263.10(c) Note	F. Transporters Who Store Hazardous Waste Such transporters must comply with: (1) Storage standards of 40 CFR 264 and 265. (2) Permit requirements of 40 CFR 122.
263 Authority	G. Federal Statutes/Authority Sections 2002(a), 3002, 3003, 3004, and 3005 of the Solid Waste Disposal Act, as amended by the Resource Conservation and Recovery Act of 1976 and as amended by the Quiet Communities Act of 1978 (42 U.S.C. 6912, 6922, 6923, 6924, and 6925).
263.11	II. EPA Identification Numbers
263.11(a)	A. Prohibition Transporters must not move hazardous wastes without an EPA identification number.
263.11(b)	B. Obtaining an EPA Identification Number If such a number has not been received by transporter, apply for one from Administrator using EPA Form 8700-12 (see Fig. 9-1).
263.20	III. Manifest System Operation for Transporters
263.20(a)	A. Acceptance of Hazardous Waste from Generator Transporter requires signed manifest from generator, per 40 CFR 262.
263.20(b)	B. Requirements Prior to Transporting Hazardous Waste The transporter must: (1) Sign and date manifest, acknowledging acceptance of waste from generator. (2) Return signed copy to generator before leaving his property.
263.20(c)	C. Accompaniment of Manifest with Hazardous Waste Shipment This is responsibility of transporter.
263.20(d) 263.20(d)(1) 263.20(d)(2) 263.20(d)(3)	D. Delivery of Hazardous Waste to Another Transporter or Designated Facility Transporter must: (1) Obtain date of delivery and handwritten signature on manifest. (2) Retain copy of manifest per recordkeeping requirements of Section 263.22. (3) Give remaining copies of manifest to accepting transporter or designated facility.

263.20(e)
 	
263.20(e)(1)
263.20(e)(2)
 	
263.20(e)(3)
 	
263.20(e)(4)
 	
263.20(e)(5)
	E. Bulk Shipment by Rail or Water
	Requirements of Sections 263.20(c) and 263.20(d) do not apply if following conditions met:
	(1) Hazardous waste delivered to designated facility by rail or water (bulk shipment).
	(2) Shipping paper accompanies waste and has all manifest information except:
	    (a) EPA identification numbers.
	    (b) Generator certification.
	    (c) Generator signature.
	(3) Delivering transporter gets delivery date and signature of owner/operator of designated facility on either:
	    (a) Manifest, or
	    (b) Shipping paper.
	(4) Person delivering hazardous waste to initial rail or water (bulk shipment) transporter:
	    (a) Obtains date of delivery and signature of that transporter on the manifest.
	    (b) Forwards signed manifest to designated facility.
	(5) Copy of shipping paper or manifest retained by each rail or water (bulk shipment) transporter, per record-keeping requirements of Section 263.22.

263.20(f) 263.20(f)(1)	F. Transporters Exporting Hazardous Waste Out of United States Such transporters must: (1) Indicate on manifest date waste left U.S.A.

Table 5-1. Standards Applicable to Transporters of Hazardous Waste. (Continued)

Section No.	Description
263.20(f)(2)	(2) Sign manifest.
	(3) Retain one copy of manifest per recordkeeping requirements of Section 263.22(c).
263.20(f)(3)	(4) Return signed copy of manifest to generator.
263.21	G. Compliance with Manifest
263.21(a)	(1) Transporter must deliver entire quantity of hazardous waste accepted from generator or other transporter to:
263.21(a)(1)	(a) Designated facility listed on manifest; or
263.21(a)(2)	(b) Alternate designated facility, if emergency prevents delivery; or
263.21(a)(3)	(c) Next designated transporter; or
263.21(a)(4)	(d) Place outside U.S.A. designated by generator.
263.21(b)	(2) If Hazardous Waste Cannot Be Delivered Per Section 263.21(a) Transporter must:
	(a) Contact generator for further instructions.
	(b) Revise manifest according to the generator's instructions.
263.22	IV. Recordkeeping Requirements
263.22(a)	A. Transporters of Hazardous Waste Such transporters must retain copy of manifest:
	(1) Signed by generator, himself, and next designated transporter or owner/operator of designated facility.
	(2) For three years from date waste was accepted by initial transporter.
263.22(b)	B. Transporters of Hazardous Waste by Rail or Water (Bulk Shipment) Such transporters must retain a copy of shipping papers per Section 263.20(e)(2) for three years from date hazardous waste was accepted by initial transporter.
263.22(c)	C. Transporters Exporting Hazardous Waste Out of United States Such transporters must keep a manifest copy indicating that waste left U.S.A., for period of three years from date hazardous waste was accepted by initial transporter.
263.22(d)	D. Retention Requirements during Unresolved Enforcement Action The periods cited above are extended automatically or as requested by the Administrator.
Subpart C	V. Hazardous Waste Discharges
263.30	A. Immediate Actions Required in Event of Hazardous Waste Discharge During Transit
263.30(a)	(1) Transporter must take appropriate action to protect human health and environment. For example:
	(a) Notify local authorities.
	(b) Dike the discharge area.
263.30(b)	(2) Where Responsible Federal, State, or Local Official Requires Immediate Waste Removal to Protect Human Health or Environment Official may authorize removal of waste by transporters:
	(a) Not having an EPA identification number.
	(b) Without preparation of a manifest.
263.30(c)	(3) Response by Air, Rail, Highway, or Water Transporters
263.30(c)(1)	(a) Give notice of discharged hazardous waste, if required by 49 CFR 171.15 (see Appendix E of this book for regulation contents only), to National Response Center (800-424-8802 or 202-426-2675).
263.30(c)(2)	(b) Report in writing per 49 CFR 171.16 to: Director, Office of Hazardous Materials Regulations Materials Transportation Bureau Department of Transportation Washington, DC 20590
263.30(d)	(4) Response by Water (Bulk Shipment) Transporters Give same notice as required by 33 CFR 153.203 for oil and hazardous substances.
263.31	B. Discharge Clean Up Requirements The transporter must:
	(1) Clean up any hazardous waste discharge that occurs during transportation, or
	(2) Take such action as may be required or approved by Federal, State or local officials that eliminates any hazard to human health or environment.

6

STANDARDS APPLICABLE TO OWNERS AND OPERATORS OF HAZARDOUS WASTE TREATMENT, STORAGE, AND DISPOSAL FACILITIES: PARTS 264 AND 265

The existing Parts 264 and 265 regulations are the first phase of EPA's requirements under Section 3004 of RCRA. They apply to owners and operators of facilities that treat, store, and dispose of wastes which are identified or listed as hazardous under Part 261.

Part 265 regulations establish requirements applicable during the interim status period—that is, the period after an owner or operator has applied for a permit, but prior to final disposition of the application. Requirements include preparedness for and prevention of hazards, contingency planning and emergency procedures, the manifest system, recordkeeping and reporting, groundwater monitoring, facility closure and post-closure care, financial requirements, and the use and management of containers. Also covered is the design and operation of tanks; surface impoundments; waste piles; land treatment facilities; landfills; incinerators; thermal, physical, chemical, and biological treatment units; and injection wells. Additionally, there are general requirements respecting identification numbers, required notices, waste analysis, security at facilities, inspection of facilities, and personnel training.

Part 264 regulations include the first phase of the standards which will be used to issue permits for hazardous waste treatment, storage, and disposal facilities. Requirements include preparedness for and prevention of hazards, contingency planning and emergency procedures, the manifest system, and recordkeeping and reporting. Also included are general requirements for identification numbers, required notices, waste analysis, security at facilities, inspection of facilities, and personnel training. Additional Part 264 regulations will be promulgated in late 1980 or early 1981.

Table 6-1 that follows addresses the detailed requirements of Parts 264 and 265 and contains the following 27 areas:

    I. Applicability
   II. General Facility Requirements
  III. Waste Analysis Requirements
  IV. Security Requirements
   V. General Inspection Requirements
  VI. Personnel Training Requirements

  VII. General Requirements for Ignitable, Reactive, or Incompatible Wastes
 VIII. Preparedness and Prevention
  IX. Contingency Plan
   X. Emergency Coordinator and Emergency Procedures
  XI. Manifest System
  XII. Recordkeeping Requirements
 XIII. Reporting Requirements
 XIV. Ground-Water Monitoring
  XV. General Closure Requirements
 XVI. General Post-Closure Requirements
 XVII. Financial Requirements
XVIII. Use and Management of Containers
 XIX. Tanks
  XX. Surface Impoundments
 XXI. Waste Piles
 XXII. Land Treatment
XXIII. Landfills
XXIV. Incinerators
 XXV. Thermal Treatment
XXVI. Chemical, Physical, and Biological Treatment
XXVII. Underground Injection

Figure 6-1 is a copy of EPA's Forms 8700-13 and 8700-13B, including instructions, which is to be used by facility owners or operators for annual reports for on-site and off-site treatment, storage or disposal of hazardous waste, and for unmanifested waste reporting. (See pp. 79–84.)

Also included in this chapter for ready reference (rather than in Appendix C—Hazardous Waste Regulations), are the following common 264 and 265 appendices and tables:

- Appendix I—Recordkeeping Instructions and associated Tables 1 and 2.
- Appendix III—EPA Interim Primary Drinking Water Standards.
- Appendix IV—Tests for Significance.
- Appendix V—Examples of Potentially Incompatible Waste.

Table 6-1. Final (i.e., Part 264) and Interim (i.e., Part 265) Status Standards for Owners and Operators of Hazardous Waste Treatment, Storage, and Disposal Facilities.

Section No.	Description
	I. Applicability
	A. Purpose of Parts 264 And 265
	Establish minimum national standards which define acceptable management of hazardous waste:
265.1(a)	(1) During period of interim status (Part 265).
264.1(a)	(2) For final permitting (Part 264)
	B. Applicability of Standards of Parts 264 and 265
264.1(b) and 265.1(b)	(1) Standards Apply to Owners and Operators of Hazardous Waste Treatment, Storage, and Disposal Facilities
265.1(b)	(a) Existing Non-Permitted Facilities–Part 265
	Must have qualified for interim status by:
	(i) Filing notification to EPA by August 18, 1980.
	(ii) Applying for permit by November 18, 1980.
264.1(b)	(b) Finally Permitted Facilities–Part 264
264.1(b)	(2) Exceptions
and 265.1(b)	(a) As per Part 261
	(b) As specifically provided otherwise in Parts 264 and 265.
265.1(b) Comment	(3) Impact on Operation of Existing Facility Without Interim Status
	Operation is prohibited after November 18, 1980 unless a (final) permit is issued, which may take several years to receive. Therefore, it is essential to file notificaiton and apply for a permit by the specified schedule dates, so as to achieve interim status.
	C. Specific Exemptions and/or Conditions for Applicability
	(1) Ocean Disposal of Hazardous Waste
	(a) Disposal with Permit under Marine Protection, Research, and Sancturaries Act
265.1(c)(1)	(i) Part 265 requirements do not apply.
264.1(c)	(ii) Part 264 requirements apply only as included in Section 122.26(a) (Permits by Rule for ocean disposal barges or vessels).
264.1(c) Comment and 265.1(c)(1) Comment	(b) Treatment or Storage of Hazardous Waste before Loading onto Ocean Vessel Part 264 or 265 regulations apply.
	(2) Underground Injection Disposal of Hazardous Waste
	(a) Disposal with Permit under Injection Control Program of Safe Drinking Water Act
265.1(c)(2)	(i) Part 265 requirements do not apply.
264.1(d)	(ii) Part 264 requirements apply only to extent included in Section 122.45 (Requirements for Wells Injecting Hazardous Waste).
264.1(d) Comment and 265.1(c)(2) Comment	(b) Above-Ground Treatment or Storage of Hazardous Waste before Injecting It Underground Part 264 and 265 regulations apply.
265.1(c)(2) Comment	(c) Disposal of Hazardous Waste by Underground Injection Without a RCRA or UIC Permit but Having Interim Status Part 265 regulations apply until RCRA or UIC permit decision.
	(3) Publicly Owned Treatment Works (POTW) Handling Hazardous Waste
	(a) Treatment, Storage, or Disposal at POTW
265.1(c)(3)	(i) Part 265 requirements do not apply.
265.1(e)	(ii) Part 264 requirements apply only as included in Section 122.26(c) (Permits by Rule for POTW).
	(4) Authorized State RCRA Programs
	(a) Treatment, Storage, or Disposal of Hazardous Waste
264.1(f) and 265.1(c)(4)	(i) In States with Full Authorization Per Subparts A and B of Part 123 Neither Parts 264 nor 265 requirements apply, and owner/operator must use "substantially equivalent" State standards.
264.1(f) and 265.1(c)(4)	(ii) In States With Interim Authorization Per Subpart F of Part 123 For Phase II Interim Authorization neither Parts 264 nor 265 requirements apply and "substantially equivalent" State standards must be used.
	(b) Underground Injection Disposal of Hazardous Waste If State RCRA Program Does Not Cover Such Disposal Requirements Exemptions of Section 264.1(d) or 265.1(c)(2) apply.
264.1(g)(1) and 265.1(c)(5)	(5) Small Quantity Exclusion Part 264 or 265 requirements do not apply to owner/operator of State-permitted, licensed, or registered facility handling municipal or industrial solid waste if only hazardous waste treated, stored, or disposed of is excluded from regulation under small generator regulations of Section 261.5.

Table 6-1. Final (i.e., Part 264) and Interim (i.e., Part 265) Status Standards for Owners and Operators of Hazardous Waste Treatment, Storage, and Disposal Facilities. (Continued)

Section No.	Description
264.1(g)(2) and 265.1(c)(6)	(6) Recycled or Reused Hazardous Waste Exclusion Part 264 or 265 requirements do not apply to owner/operator of facility which recycles or reuses hazardous waste as specified in Section 261.6(a), except to extent that Section 261.6(b) provides otherwise.
264.1(g)(3) and 265.1(c)(7)	(7) Exclusion for Generators Who Store Hazardous Waste On-Site for Less than 90 Days Part 264 and 265 requirements do not apply except for certain Part 265 requirements specified in Section 262.34.
265.1(g)(4) and 265.1(c)(8)	(8) Exclusion for Farmers Disposing of Waste Pesticides Part 264 and 265 requirements do not apply if farmer disposes of waste pesticides from his own use in compliance with provisions of Section 262.51 (Farmers).
264.1(g)(5) and 265.1(c)(9)	(9) Totally Enclosed Treatment Facility Part 264 and 265 requirements do not apply to owner or operator of a totally enclosed treatment facility, as defined in Section 260.10.
264.3	D. Relationship of Part 264 Standards to Interim Standards of Part 265 Until final disposition of facility owner's/operator's permit application is made, which is based on Part 264 standards, compliance with Part 265 requirements is necessary. *Note:* Facility operator must have fully complied with requirements for interim status as defined in Section 3005(e) of RCRA and Section 122.23 (Interim Status), otherwise treatment, storage, or disposal of hazardous waste is prohibited after 11/18/80, except in accordance with a permit, which can take years to obtain.
264.4 and 265.4	E. Imminent Hazard Action Notwithstanding any other provisions of these regulations, enforcement actions may always be brought pursuant to Section 7003 of RCRA (Imminent Hazard).
	II. General Facility Requirements
264.11 and 265.11	A. Identification Number Facility owner/operator must apply to EPA for EPA identification number per notification procedures.
264.12 and 265.12 264.12(a) and 265.12(a)	B. Required Notices (1) Facility Receiving Foreign Hazardous Wastes Owner/operator must notify Regional Administrator in writing at least 4 weeks prior to expected arrival date.
264.12(b)	(2) Notice to Generator from Facility Owner/Operator (a) Must inform generator in writing that he has appropriate permit for, and will accept, waste being sent. (b) Copy of written notice must be kept by owner/operator.
264.12(c) and 265.12(b)	(3) Notification to New Owner or Operator of Facility Regarding Permit (Part 122) and Facility Standards (Part 264/265) Requirements This must be done in writing prior to transferring ownership or operation of a: (a) Facility during its operating life; or (b) Disposal facility during the post-closure care period.
264.12(c) Comment and 265.12(b) Comment	*Note:* Failure to notify new owner/operator of requirements does not relieve him of compliance requirements.
264.13 and 265.13	III. Waste Analysis Requirements
264.13(a) and 265.13(a) 264.13(a)(1) and 265.13(a)(1)	A. Detailed Chemical and Physical Analysis (1) General Requirement Prior to owner/operator treating, storing, or disposing of any hazardous waste, detailed chemical and physical analysis of a representative waste sample is required. (2) Minimum Analysis Requirements (a) Must contain all information needed to treat, store or dispose of waste per requirements of: (i) Part 264 and 265; and/or
264.13(a)(1)	(ii) Conditions of permit issued under Parts 122, Subparts A and B and Part 124.

Table 6-1. Final (i.e., Part 264) and Interim (i.e., Part 265) Status Standards for Owners and Operators of Hazardous Waste Treatment, Storage, and Disposal Facilities. (Continued)

Section No.	Description
264.13(a)(2) and 265.13(a)(2)	(b) May include data developed under Part 261. (c) May include published or documented data on the hazardous waste or hazardous waste generated from similar processes.
264.13(a)(2) Comment and 265.13(a)(2) Comment	(i) Facility's records of analysis on waste prior to effective date of regulations is example. (d) Owner/operator of off-site facility may arrange for generator to supply all or part of detailed chemical and physical analysis.
	(3) Responsibility If owner/operator accepts generator's waste he is ultimately responsible for obtaining waste analysis—whether provided by generator or not.
264.13(a)(3) and 265.13(a)(3)	(4) Repeat Analyses (a) General Analysis must be repeated as necessary to ensure it is accurate and up to date. (b) Minimum Requirements for Repeat Analysis
264.13(a)(3)(i) and 265.13(a)(3)(i)	(i) When owner/operator is notified or believes that process or operation generating hazardous waste has changed.
264.13(a)(3)(ii) and 265.13(a)(3)(ii)	(ii) When inspection per Section 264.13(a)(4) or 265.13(a)(4) shows that received waste differs from description on manifest or shipping paper.
264.13(a)(4) and 265.13(a)(4)	(5) Inspection of Hazardous Waste Inspection and/or analysis of received waste coming to facility is required to determine if it matches description on manifest or shipping paper.
264.13(b) and 265.13(b)	B. Waste Analysis Plan (1) General Requirements (a) Owner/operator must develop and follow written waste analysis plan which describes procedures for complying with analysis requirements of Section 264.13(a) or 265.13(a). (b) Plan must be kept at facility. (2) Detailed Requirements
264.13(b)(1) and 265.13(b)(1)	(a) Parameters by which each hazardous waste will be analyzed. (b) Rationale for Selection of Parameters Description of how analysis of parameters permit compliance with Section 264.13(a) or 265.13(a).
264.13(b)(2) and 265.13(b)(2)	(c) Test methods to be used to test the parameters.
264.13(b)(3) and 265.13(b)(3)	(d) Sampling method to be used to obtain representative sample of waste to be analyzed.
264.13(b)(3)(i) and 265.13(b)(3)(i)	(i) Use one of sampling methods described in Appendix I of Part 261; or
264.13(b)(3)(ii) and 265.13(b)(3)(ii)	(ii) Equivalent sampling method obtained via petitioning provisions of Part 260.
264.13(b)(4) and 265.13(b)(4)	(e) Frequency of Repeat Analysis So as to ensure that analysis is accurate and up to date.
264.13(b)(5) and 265.13(b)(5)	(f) Waste analysis that hazardous waste generators agreed to supply.
265.13(b)(6)	(g) Additional Interim Status Waste Analysis Requirements (i) Section 265.193—Waste Analysis and Trial Tests (for tanks). (ii) Section 265.225—Waste Analysis and Trial Tests (for surface impoundments). (iii) Section 265.252—Waste Analysis (for waste piles). (iv) Section 265.273—Waste Analysis (for land treatment). (v) Section 265.345—Waste Analysis (for incinerators). (vi) Section 265.375—Waste Analysis (for thermal treatment). (vii) Section 265.402—Waste Analysis and Trial Tests (for chemical, physical and biological treatment).
264.13(c) and 265.13(c)	(h) For Off-Site Facilities (i) Hazardous Waste Inspection Procedures Procedures to be used must be specified, and/or analysis, for received wastes, to determine if it matches description on manifest or shipping paper.
264.13(c)(1) and 265.13(c)(1)	(ii) Procedures to be used to determine identity of each movement of waste at facility.
264.13(c)(2) and 265.13(c)(2)	(iii) Sampling method to be used to obtain representative sample of waste to be identified, if identification method includes sampling.
264.13(c)(2) Comment	(3) Submittal of Waste Analysis Plan with Part B of Permit Application Required per Part 122 Subpart B.

Table 6-1. Final (i.e., Part 264) and Interim (i.e., Part 265) Status Standards for Owners and Operators of Hazardous Waste Treatment, Storage, and Disposal Facilities. (Continued)

Section No.	Description
264.14 and 265.14	IV. Security Requirements
264.14(a) and 265.14(a)	A. General Requirement Owner/operator of active portion of facility must assure that security system: (1) Prevents unknowing entry of people. (2) Minimizes possibility of unauthorized entry of persons or livestock.
	B. Variance (1) Conditions to Be Met
264.14(a)(1) and 265.14(a)(1) 264.14(a)(2) and 265.14(a)(2)	(a) Physical contact with waste structures or equipment within active portion of facility will not injure unknowing or unauthorized persons or livestock entering; and (b) Disturbance of waste or equipment by unknowing or unauthorized persons or livestock will not cause violation of Part 264 or 265 requirements. *Note:* Because these two conditions are rarely concurrently satisfied, EPA does not expect many sites to be exempt from security requirements.
264.14(a)(2) Comment	(2) Demonstration that Conditions of Variance Can Be Met Part 122 (Subpart B) requires owner/operator seeking variance to demonstrate in Part B of permit application that conditions of variance can be met.
264.14(b)(1) and 265.14(b)(1)	C. Surveillance and Controlled Entry Unless variance conditions are met, a 24-hour surveillance system is required for monitoring and controlling entry onto active portion of facility. (1) Types of Surveillance Systems (a) Television monitoring. (b) Guards or facility personnel.
264.14(b)(2)(ii) and 265.14(b)(2)(ii)	(2) Means to Control Entry Controlled entry through gates or other entrances to active portion of facility is required. (a) Possible Approaches to Control (i) Attendant. (ii) Television monitors. (iii) Locked entrance. (iv) Controlled roadway access.
264.14(b)(2) Comment and 265.14(b)(2) Comment	(3) Where Facility or Plant Is Within Larger Complex Having Suitable Surveillance System and Controlled Entry Surveillance and controlled entry requirements of this Section are satisfied.
264.14(b)(2)(i) and 265.14(b)(2)(i)	D. Artificial and/or Natural Barrier Such a barrier completely surrounding active portion of facility is required. (1) Types of Barriers (a) Fence in good repair. (b) Fence combined with a cliff.
264.14(b)(2) Comment and 265.14(b)(2) Comment	(2) Where Facility or Plant Is Within Larger Complex Having Barrier Barrier requirements of this Section are satisfied.
264.14(c) and 265.14(c)	E. Warning Signs Unless variance conditions are met such signs are required. (1) Requirements for Warning Signs (a) Legend (i) "Danger–Unauthorized Personnel Keep Out" (ii) Existing signs with other legend may be used if it indicates that only authorized personnel are allowed to enter and that entry can be dangerous. (iii) Written in English and other language predominant in area. (iv) Must be legible from 25 feet. (b) Posting Requirements (i) Signs at each entrance to active portion of facility. (ii) At other locations. (iii) Sufficient numbers to be seen from any approach to active facility.
265.14(c) Comment	F. Security Requirements at Disposal Facilities During Post-Closure Period See Section 265.117(b) for special requirements.

Table 6-1. Final (i.e., Part 264) and Interim (i.e., Part 265) Status Standards for Owners and Operators of Hazardous Waste Treatment, Storage, and Disposal Facilities. (Continued)

Section No.	Description
264.15 and 265.15	V. General Inspection Requirements
264.15(a) and 265.15	A. General Requirements (1) Owner/operator must inspect facility for malfunctions, deterioration, operator error, and discharges that may be causing, or may lead to:
264.15(a)(1) and 265.15(a)(1)	(a) Release of hazardous wastes into environment; or
264.15(a)(2) and 265.15(a)(2)	(b) Threat to human health.
264.15(a) and 265.15(a)	(2) Frequency of Inspections Often enough to identify problems in time to correct them before they harm human health or environment.
264.15(b)(1) and 265.15(b)(1)	B. Inspection Plan/Schedule (1) Develop and Follow Written Inspection Schedule This is required of owner/operator to prevent, detect or respond to environmental or human health hazards. Equipments to be inspected follow: (a) Monitoring equipment. (b) Safety equipment. (c) Emergency equipment. (d) Operating and structural equipment (such as dikes and sump pumps). (e) Security devices.
264.15(b)(2) and 265.15(b)(2)	(2) Schedule must be kept at facility.
264.15(b)(3) and 265.15(b)(3)	(3) Schedule Must Identify Types of Problems to Be Looked for During Inspection Would include following types of malfunctions or deteriorations: (a) Inoperative sump pump. (b) Leaking fitting. (c) Eroding dike.
264.15(b)(4) and 265.15(b)(4)	(4) Frequency of Inspections (a) May vary for different items on schedule. (b) Base on rate of deterioration of equipment and probability of environmental or human health incident if deterioration or malfunction or operator error goes undetected between inspections. (c) Areas Subject to Spills Must be inspected daily when in use (e.g., loading and unloading areas).
265.15(b)(4)	(d) Additional Interim Status Inspection Requirements As called for in: (i) Section 265.174–Inspections (for containers). (ii) Section 265.194–Inspections (for tanks). (iii) Section 265.226–Inspections (for surface impoundments). (iv) Section 265.347–Monitoring and Inspections (for incinerators). (v) Section 265.377–Monitoring and Inspections (for thermal treatment). (vi) Section 265.403–Inspections (for chemical, physical and biological treatment).
265.15(b)(4) Comment	(5) Inspection Schedule and Permit Application (a) Part 122 (Subpart B) requires inspection schedule submission with Part B of permit application. (b) EPA will evaluate schedule to ensure that it adequately protects human health and environment. (c) EPA may modify or amend schedule as necessary.
264.15(c) and 265.15(c)	C. Remedy for Any Deterioration or Malfunction of Equipment or Structures Revealed by Inspection Required of owner/operator: (1) Schedule that ensures problem does not lead to environmental or human health hazard. (2) Immediate remedial actions where hazard is imminent or has already occurred.
264.15(d) and 265.15(d)	D. Inspection Log (1) Owner/operator must record inspections in inspection log or summary. (2) Retention Time Records must be kept for three years from date of inspection. (3) Minimum Content Requirements (a) Date and time of inspection.

Table 6-1. Final (i.e., Part 264) and Interim (i.e., Part 265) Status Standards for Owners and Operators of Hazardous Waste Treatment, Storage, and Disposal Facilities. (Continued)

Section No.	Description
	(b) Name of inspector.
	(c) Notation of observations made.
	(d) Date and nature of any repairs or other remedial actions.
264.16 and 265.16	VI. Personnel Training Requirements
264.16(a)(1) and 265.16(a)(1)	A. Classroom Instruction or on-the-Job Training Facility personnel must successfully complete such a program teaching them to perform duties in manner ensuring compliance with Parts 264 and 265. (1) Owner/operator must ensure that program includes all elements described in document of Section 264.16(d)(3) or 265.16(d)(3).
264.16(a)(2) and 265.16(a)(2)	(2) Program must be directed by person trained in hazardous waste management procedures. (3) Program must include instruction which teaches facility personnel hazardous waste management procedures (including contingency plan implementation) relevant to position they have.
264.16(a)(3) and 265.16(a)(3)	(4) Minimum Training Program (a) Must be designed to ensure that facility personnel are able to respond effectively to emergencies by familiarizing them with emergency procedures, equipment and systems.
264.16(a)(3)(i) and 265.16(a)(3)(i)	(b) Procedures for using, inspecting, repairing, and replacing facility emergency and monitoring equipment.
264.16(a)(3)(ii) and 265.16(a)(3)(ii)	(c) Key parameters for automatic waste feed cut-off systems.
264.16(a)(3)(iii) and 265.16(a)(3)(iii)	(d) Communications or alarm systems.
264.16(a)(3)(iv) and 265.16(a)(3)(iv)	(e) Response to fires or explosions.
264.16(a)(3)(v) and 265.16(a)(3)(v)	(f) Response to ground-water contamination incidents.
264.16(a)(3)(vi) and 265.16(a)(3)(vi)	(g) Shutdown of operations.
264.16(b) and 265.16(b)	B. Timing Facility personnel must successfully complete program of Section 264.16(a) or 265.16(a) within: (1) Six months after effective date of these regulations; or (2) Six months after date of their employment or assignment to a facility or to a new position at a facility, whichever is later.
264.16(b) and 265.16(b)	C. Employees Hired after Effective Date of Regulations Must not work in unsupervised position until completion of training requirements of Section 264.16(a) or 265.16(a).
264.16(c) and 265.16(c)	D. Annual Review of Initial Training Required of facility personnel.
264.16(d) and 265.16(d)	E. Records Owner/operator must maintain following documents and records at the facility:
264.16(d)(1) and 265.16(d)(1)	(1) Job title for each position at facility related to hazardous waste management.
264.16(d)(2) and 265.16(d)(2)	(2) Name of employee filling each job. (3) Written Job Description For each position listed in Section 264.16(d)(1) or 265.16(d)(1) above. (a) Description may be consistent with description for similar positions in same company location or bargaining unit. (b) Content Required (i) Requisite skill. (ii) Education. (iii) Other qualifications. (iv) Duties of employees assigned to each position.
264.16(d)(3) and 265.16(d)(3)	(4) Written description of type and amount of both introductory and continuing training to be given to each person listed in Sections 264.16(d)(1) or 265.16(d)(1).
264.16(d)(4) and 265.16(d)(4)	(5) Documentation showing that training or job experience required under Sections 264.16(a)(b) and (c) or 265.16(a)(b) and (c) has been given to and completed by facility personnel.
264.16(e) and 265.16(e)	(6) Retention Time (a) Current Personnel Training records must be kept until closure of facility.

Table 6-1. Final (i.e., Part 264) and Interim (i.e., Part 265) Status Standards for Owners and Operators of Hazardous Waste Treatment, Storage, and Disposal Facilities. (Continued)

Section No.	Description
	(b) Former Employees Training records must be kept for at least three years from date employee last worked at facility.
	(c) Personnel Transferred within Same Company Training records may accompany personnel.
265.17	VII. General Requirements for Ignitable, Reactive, or Incompatible Wastes
265.17(a)	A. General Precautions Owner/operator must take precautions to prevent accidental ignition or reaction of ignitable or reactive waste.
	B. Separation/Protection of Ignitable and Reactive Wastes from Sources of Ignition or Reaction Such sources follow: (1) Open flames. (2) Smoking. (3) Cutting and welding. (4) Hot surfaces. (5) Frictional heat. (6) Sparks (static, electrical or mechanical). (7) Spontaneous ignition (e.g., from heat-producing chemical reactions). (8) Radiant heat.
	C. Limitations on Smoking and Open Flames While Handling Ignitable or Reactive Waste Owner/operator must: (1) Confine smoking and open flame to specially designated locations. (2) Place "No Smoking" signs conspicuously wherever there is hazard from ignitable or reactive waste.
265.17(b)	D. Treatment, Storage, or Disposal of Ignitable or Reactive Waste and Mixing of Incompatible Wastes or Incompatible Wastes and Materials Must be conducted so that it does not:
265.17(b)(1)	(1) Generate extreme heat or pressure, fire or explosion, or violent reaction.
265.17(b)(2)	(2) Produce uncontrolled toxic mists, fumes, dusts, or gases in sufficient quantities to threaten human health.
265.17(b)(3)	(3) Produce uncontrolled flammable fumes or gases in sufficient quantities to pose risk of fire or explosions.
265.17(b)(4)	(4) Damage structural integrity of device or facility containing waste.
265.17(b)(5)	(5) Through other like means threaten human health or environment.
Subpart C	VIII. Preparedness and Prevention
264.31 and 265.31 264.31 265.31 264.31 and 265.31	A. General Facility Design, Construction, Maintenance, and Operation Requirements With regard to: (1) Design, construction, maintenance and operation of new facilities, and (2) Maintenance and operation of existing facilities, possibility must be minimized of a fire, explosion, or any unplanned sudden or non-sudden release of hazardous waste or hazardous waste constituents to air, soil, or surface water which could threaten human health or the environment.
264.32 and 265.32	B. Alarm/Communication System and Control Equipment (1) Equipment Requirements for Facility
264.32(a) and 265.32(a)	(a) Internal Communications or Alarm System Must be capable of providing immediate instruction (voice or signal) to facility personnel.
264.32(b) and 265.32(b)	(b) Telephone or Hand-Held Two-Way Radio at Scene of Operations Must be capable of summoning emergency assistance from local police and/or fire departments, or State or local emergency response teams.
264.32(c) and 265.32(c)	(c) Portable Fire Extinguishers and Fire Control Equipment Including special extinguishing equipment such as that using foam, inert gas or dry chemicals. (d) Spill control and decontamination equipment.

Table 6-1. Final (i.e., Part 264) and Interim (i.e., Part 265) Status Standards for Owners and Operators of Hazardous Waste Treatment, Storage, and Disposal Facilities. (Continued)

Section No.	Description
264.32(d) and 265.32(d)	(e) Water at Adequate Volume and Pressure For supplying water hose streams, or foam producing equipment, or automatic sprinklers, or water spray systems.
	(2) Exemptions
265.32	(a) During Interim Status While little or no interaction with facility owner/operator is EPA's goal during interim status, the owner/operator should be prepared to demonstrate–if need be–that none of the hazards posed by waste handled at the facility could require the kind of equipment specified in this Section (265.32), should an exemption be sought.
264.32 and 264.32(d) Comment	(b) For Final Permitting Instead of preparation for a demonstration, as in (a) above, the facility owner/operator wanting exemption must demonstrate lack of hazard and need for equipment, in Part B of the permit application.
264.33 and 265.33	(3) Testing and Maintenance of Equipment All of following facility equipment that may be required must be tested and maintained as necessary to ensure its proper operation in time of emergency: (a) Communications or alarm systems. (b) Fire protection equipment. (c) Spill control equipment. (d) Decontamination equipment.
264.34 and 265.34	(4) Access to Communications or Alarm System
264.34(a) and 265.34(a)	(a) Whenever Hazardous Waste Is Being Poured, Mixed, Spread, or Otherwise Handled All personnel involved in the operation must have immediate access to an internal alarm or emergency communication device, either directly or through visual or voice contact with another employee. (i) Exemptions
265.34(a)	• During Interim Status If such a device is not required under Section 265.32.
264.34(a)	• For Final Permitting If Regional Administrator has ruled that such a device is not required under Section 264.32.
264.34(b) and 265.34(b)	(b) With Just One Employee on Premises While Facility Is Operating He must have immediate access to a device, such as a telephone or a hand-held two-way radio, capable of summoning external emergency assistance. (i) Exemptions
265.34(b)	• During Interim Status If such a device is not required under Section 265.32.
264.34(b)	• For Final Permitting If Regional Administrator has ruled that such a device is not required under Section 264.32.
264.35 and 265.35	C. Required Aisle Space for Emergency Situation Owner/operator must maintain sufficient aisle space to allow unobstructed movement of personnel, fire protection, spill control, and decontamination equipments to any area of facility operation in an emergency. (1) Exemptions
265.35	(a) During Interim Status Owner/operator should be prepared to demonstrate–if need be–that aisle space is not needed for movement of personnel and equipment in emergency situations, should an exemption be sought.
264.35 and 264.35 Comment	(b) For Final Permitting Owner/operator who wants an exemption must demonstrate in Part B of permit application that aisle space is not needed for movement of personnel and equipment in emergency situations.
264.36	D. Special Handling for Ignitable or Reactive Waste for Final Permitting Owner/operator must take precautions to prevent accidental ignition or reaction of ignitable or reactive waste. (1) Waste Must Be Separated and Protected from Sources of Ignition or Reaction Includes, but not limited to following: (a) Open flames. (b) Smoking.

Table 6-1. Final (i.e., Part 264) and Interim (i.e., Part 265) Status Standards for Owners of Hazardous Waste Treatment, Storage, and Disposal Facilities. (Continued)

Section No.	Description
	(c) Cutting and welding.
	(d) Hot surfaces.
	(e) Frictional heat.
	(f) Sparks (static, electrical, or mechanical).
	(g) Spontaneous ignition (e.g., from heat-producing chemical reactions).
	(h) Radiant heat.
	(2) While Handling Ignitable or Reactive Wastes Owner/operator must confine smoking and open flames to specially designated locations.
	(3) "No Smoking" Signs Must be conspicuously placed whenever there is a hazard from ignitable or reactive waste.
264.37 and 265.37	E. Arrangements with State or Local Authorities
264.37(a) and 265.37(a)	(1) Owner/Operator Must Attempt Arrangements with Authorities as Appropriate for Type Wastes Handled and Potential Need for Services
264.37(a)(1) and 265.37(a)(1)	(a) Arrangements for Familiarization of Police, Fire Departments, and Emergency Response Teams with Facility (i) Layout of facility. (ii) Properties of hazardous waste handled at facility. (iii) Associated hazards. (iv) Places where facility personnel would normally be working. (v) Entrances to and roads inside facility. (vi) Possible evacuation routes.
264.37(a)(2) and 265.37(a)(2)	(b) Where More than One Police and Fire Department Might Respond to Emergency (i) Agreements designating primary emergency authority to specific police and fire department. (ii) Agreements with others to provide support to primary emergency authority.
264.37(a)(3) and 265.37(a)(3)	(c) Agreements with State emergency response teams. (d) Agreements with emergency response contractors and equipment suppliers.
264.37(a)(4) and 265.37(a)(4)	(e) Arrangements for Familiarization of Local Hospitals with Facility (i) Properties of hazardous wastes handled at facility. (ii) Types of injuries or illnesses which could result from fires, explosions, or releases at facility.
264.37(b) and 265.37(b)	(2) Where State or Local Authorities Decline to Enter into Arrangements Owner/operator must document refusal in their operating record.
	IX. Contingency Plan
	A. General
264.51(a) and 265.51(a)	(1) Requirement for Contingency Plan Each owner/operator must have a contingency plan for his facility. (2) Purpose of Contingency Plan It must be designed to minimize hazards to human health or the environment from fires, explosions, or any unplanned sudden or non-sudden release of hazardous waste or hazardous waste constituents to air, soil, or surface water.
264.51(b) and 265.51(b)	(3) Implementation of Contingency Plan Provisions of plan must be carried out immediately whenever there is a fire, explosion, or release of hazardous waste or hazardous waste constituents which could threaten human health or the environment.
264.52 and 265.52	B. Content of Contingency Plan
264.52(a) and 265.52(a)	(1) Description of Actions Facility Personnel Must Take For compliance with Sections 264.51/265.51 above and Sections 264.56/265.56 (Emergency Procedures) in response to fires, explosions, unplanned sudden or non-sudden release of hazardous waste constituents to air, soil, or surface water at the facility.
264.52(b) and 265.52(b)	(2) Preparation of Spill Prevention Control and Countermeasures (SPCC) Plan If owner/operator prepared SPCC Plan per 40 CFR 112 or 151, or some other emergency or contingency plan, he need only amend it to incorporate appropriate hazardous waste management provisions.
264.52(c) and 265.52(c)	(3) Description of Arrangements with State and Local Authorities Per Sections 264.37 and 265.37.

Table 6-1. Final (i.e., Part 264) and Interim (i.e., Part 265) Status Standards for Owners and Operators of Hazardous Waste Treatment, Storage, and Disposal Facilities. (Continued)

Section No.	Description
264.52(d) and 265.52(d)	(4) Emergency Coordinator(s) (a) List of names, addresses, phone numbers (office and home) of all persons qualified to act as emergency coordinators (see Sections 264.55 and 265.55–Emergency Coordinators). (b) List of emergency coordinators must be kept up to date. (c) Where More than One Emergency Coordinator Is Listed One must be named as primary emergency coordinator and others listed in order which they will assume responsibility as alternates.
254.52(d)	(d) For New Facilities Above information must be supplied to Regional Administrator at time of certification, rather than at time of permit application.

264.52(e)
and 265.52(e)
	(5) List of Emergency Equipment at Facility
	(a) Equipment and Where Required
	   (i) Fire extinguishing systems.
	   (ii) Spill control equipment.
	   (iii) Communications and alarm systems (internal and external).
	   (iv) Decontamination equipment.
	(b) List must be kept up to date.
	(c) Location, physical description, and brief outline of capabilities of each piece of equipment.

264.52(e)
and 265.52(e)
	(6) Evacuation Plan for Facility Personnel
	(a) Where there is a possibility that evacuation could be necessary.
	(b) Content of Plan
	   (i) Description of signal(s) to be used to begin evacuation.
	   (ii) Evacuation routes.
	   (iii) Alternate Evacuation Routes
	      In cases where primary routes could be blocked by release of hazardous waste or fires.

264.53 and 265.53	C. Copies of Contingency Plan A copy of contingency plan and all revisions must be:
264.53(a) and 265.53(a)	(1) Maintained at facility.
264.53(b) and 265.53(b)	(2) Submitted to All State and Local Authorities That May Be Called Upon to Provide Emergency Services May include local police and fire departments, hospitals and State and local emergency response teams.
264.53(b) Comment	(3) For Final Permitting Copy must be submitted to Regional Administrator with Part B of permit application under Part 122 (Subpart A and B), and after modification or approval, will become a condition of any permit issued.
264.54 and 265.54	D. Amendment of Contingency Plan Plan must be reviewed and immediately amended, if necessary, whenever:
264.54(a) 265.54(a)	(1) Facility permit is revised.
	(2) Applicable regulations are revised.
264.54(b) . and 265.54(b)	(3) Plan fails in an emergency.
264.54(c) and 265.54(c)	(4) Facility Changes In its design, construction, operation, maintenance, or other circumstances in a way that materially increases potential for fires, explosions, or releases of hazardous waste or hazardous waste constituents, or changes the response necessary in an emergency.
264.54(d) and 265.54(d)	(5) List of emergency coordinators changes.
264.54(e) and 265.54(e)	(6) List of emergency equipment changes.
264.54(e) Comment	(7) Regarding Final Permitting A change in the lists of facility emergency coordinators or equipment in the contingency plan constitutes a minor modification to the facility permit.
	X. Emergency Coordinator and Emergency Procedures
264.55 and 265.55	A. Emergency Coordinator (1) General Requirement At all times, there must be at least one employee either on the facility premises or on call (e.g., available to respond to an emergency by reaching the facility within a short period of time), with the responsibility for coordinating all emergency response measures.

Table 6-1. Final (i.e., Part 264) and Interim (i.e., Part 265) Status Standards for Owners and Operators of Hazardous Waste Treatment, Storage, and Disposal Facilities. (Continued)

Section No.	Description
	(2) Thorough Familiarity with All Aspects of Facility's Hazardous Waste Operation Specifically: (a) All aspects of facility's contingency plan. (b) All operations and activities at facility. (c) Location and characteristics of waste handled. (d) Location of all records within facility. (3) Authority The emergency coordinator must have the authority to commit the resources needed to carry out the contingency plan.
264.56 and 265.56 264.55 Comment and 265.55 Comment	B. Emergency Procedures (1) Emergency Coordinator Responsibilities (a) General Applicable responsibilities for emergency coordinator vary, depending on factors such as type and variety of waste(s) handled by facility and type and complexity of facility.
264.56(a) and 265.56(a) 264.56(a)(1) and 265.56(a)(1) 264.56(a)(2) and 265.56(a)(2) 264.56(b) and 265.56(b)	(b) Immediate Required Action for Imminent or Actual Emergency Situation Emergency coordinator or designee must: (i) Activate internal facility alarms or communication systems, where applicable, to notify all facility personnel. (ii) Notify appropriate State or local agencies with designated response roles if their help is needed. (c) For Hazardous Waste Release, Fire, or Explosion (i) Immediately identify character, exact source, amount, and real extent of any released material. (ii) Possible Approaches to Identifying Character, Source, Amount, and Extent of Released Material • By observation • Review of facility records or manifests • Chemical analysis.
264.56(c) and 265.56(c)	(iii) Concurrent Assessment of Possible Hazards to Human Health or Environment Assessment must consider following: • In general, both direct and indirect effects of waste release, fire, and explosion. • Effects of any toxic, irritating, or asphyxiating gases generated. • Effects of any hazardous surface water runoff from water or chemical agents used to control fire and heat-induced explosions.
264.56(d) and 265.56(d) 264.56(d)(1) and 265.56(d)(1) 264.56(d)(2) and 265.56(d)(2)	(d) If Emergency Coordinator Determines that Hazardous Waste Release, Fire or Explosion Could Threaten Human Health or Environment Outside of Facility Report findings as follows: (i) If Assessment Indicates that Evacuation of Local Areas May be Advisable • He must immediately notify appropriate local authorities. • He must be available to help appropriate local officials decide whether local areas should be evacuated. (ii) Notification of Government Official(s) • Designated on-scene coordinator for that geographical area in applicable regional contingency plan under 40 CFR 1510; or • National Response Center, using their 24-hour toll free number (800-424-8802). • Content of Report
264.56(d)(2)(i) and 265.56(d)(2)(i) 264.56(d)(2)(ii) and 265.56(d)(2)(ii) 264.56(d)(2)(iii) and 265.56(d)(2)(iii) 264.56(d)(2)(iv) and 265.56(d)(2)(iv) 264.56(d)(2)(v) and 265.56(d)(2)(v) 264.56(d)(2)(vi) and 265.56(d)(2)(vi) 264.56(e) and 265.56	– Name and telephone number of reporter. – Name and address of facility. – Time of incident. – Type of incident (e.g., release, fire). – Name and quantity of material(s) involved, to extent known. – Extent of injuries, if any. – Possible hazards to human health, or environment outside facility. (e) Additional Measures to Be Taken by Emergency Coordinator during Emergency (i) Ensure that fires, explosions, and releases do not occur, recur, or spread to other hazardous waste at facility.

Table 6-1. Final (i.e., Part 264) and Interim (i.e., Part 265) Status Standards for Owners and Operators of Hazardous Waste Treatment, Storage, and Disposal Facilities. (Continued)

Section No.	Description
	(ii) Where applicable, measures include:
	• Stopping processes and operations.
	• Collecting and containing released waste.
	• Removing or isolating containers.
264.56(f) and 265.56(f)	(f) If Facility Stops Operations in Response to Fire, Explosion, or Waste Release Emergency coordinator must monitor for leaks, pressure buildup, gas generation, or ruptures in valves, pipes, or other equipment, wherever this is appropriate.
264.56(g) and 265.56(g)	(g) Immediately after Emergency Emergency coordinator must provide for treating, storing, or disposing of recovered waste, contaminated soil, or surface water, or any other material that results from a release, fire, or explosion at facility.
264.56(g) Comment and 265.56(g) Comment	(h) Owner/Operator May Become Generator Unless he can demonstrate per Section 261.3(c) or (d) (e.g., Definition of Hazardous Waste) that recovered material is not hazardous, the owner/operator becomes a generator of hazardous waste and must manage it per all applicable requirements of 40 CFR 262, 263, 264, or 265.
264.56(h) and 265.56(h)	(i) Care in Affected Area(s) of Facility Emergency coordinator must ensure that:
264.56(h)(1) and 265.56(h)(1)	(i) No waste that may be incompatible with released material is treated, stored, or disposed of until cleanup procedures are complete.
264.56(h)(2) and 265.56(h)(2)	(ii) All emergency equipment listed in contingency plan is cleaned and fit for its intended use before operations are resumed.
	(2) Facility Owner/Operator Responsibilities
264.56(i) and 265.56(i)	(a) Before Operations Are Resumed in Affected Areas of Facility Owner/operator must notify Regional Administrator and appropriate State and local authorities that facility is in compliance with Section 264.56(h) or 265.56(h).
264.56(j) and 265.56(j)	(b) Operating Record Owner/operator must note time, date, and details of any incident requiring implementation of contingency plan in the operating record.
	(c) Written Report on Incident to Regional Administrator
	(i) Due Date Within 15 days after incident.
	(ii) Contents
264.56(j)(1) and 265.56(j)(1)	• Name, address, and telephone number of owner or operator.
264.56(j)(2) and 265.56(j)(2)	• Name, address, and telephone number of facility.
264.56(j)(3) and 265.56(j)(3)	• Date, time, and type of incident (e.g., fire, explosion).
264.56(j)(4) and 265.56(j)(4)	• Name and quantity of material(s) involved.
264.56(j)(5) and 265.56(j)(5)	• Extent of injuries, if any.
264.56(j)(6) and 265.56(j)(6)	• Assessment of actual or potential hazards to human health or environment, where applicable.
264.56(j)(7) and 265.56(j)(7)	• Estimated quantity and disposition of recovered material that resulted from incident.
	XI. Manifest System
264.70 and 265.70	A. Applicability Apart from exceptions of Sections 264.1 and 265.1, these manifest regulations apply to owners/operators of both on-site and off-site facilities, except for following Sections, which do not apply to on-site facilities that do not receive any hazardous waste from off-site sources: (1) Sections 264.71 and 265.71—Use Of Manifest System. (2) Sections 264.72 and 265.72—Manifest Discrepancies. (3) Sections 264.76 and 265.76—Unmanifested Waste Report.
264.71 and 265.71	B. Use of Manifest System
264.71(a) and 265.71(a)	(1) Where Facility Receives Hazardous Waste Accompanied by Manifest Owner/operator, or his agent, must:

Table 6-1. Final (i.e., Part 264) and Interim (i.e., Part 265) Status Standards for Owners and Operators of Hazardous Waste Treatment, Storage, and Disposal Facilities. (Continued)

Section No.	Description
264.71(a)(1) and 265.71(a)(1)	(a) Sign and date each copy of manifest to certify that hazardous waste covered by manifest was received.
264.71(a)(2) and 265.71(a)(2)	(b) Note any significant discrepancies in manifest, as defined in Sections 264.72(a) or 265.72(a) (Manifest Discrepancies), on each copy of manifest.
264.71(a)(2) Comment and 265.71(a)(2) Comment	(c) Waste Analysis EPA does not intend facility owner/operator to perform waste analysis procedures of Sections 264.13(c) or 265.13(c) (General Waste Analysis) prior to signing manifest and giving it to transporter. (d) Unreconciled Discrepancy Sections 264.72(b) and 265.72(b) (Manifest Discrepancies) requires reporting unreconciled discrepancies found during later waste analysis.
264.71(a)(3) and 265.71(a)(3)	(e) Immediately give transporter at least one copy of signed manifest.
264.71(a)(4) and 265.71(a)(4)	(f) Within 30 days after delivery, send copy of manifest to generator.
264.71(a)(5) and 265.71(a)(5)	(g) Retain copy of each manifest at facility for at least three years from date of delivery.
264.71(b) and 265.71(b)	(2) Where Facility Receives Hazardous Waste from Rail or Water (Bulk Shipment) Transporter If accompanying shipping paper contains all information required on manifest (excluding EPA identification numbers, generator's certification, and signatures), the owner/operator, or his agent, must:
264.71(b)(1) and 265.71(b)(1)	(a) Sign and date each copy of shipping paper to certify that hazardous waste covered by shipping paper was received.
264.71(b)(2) and 265.71(b)(2)	(b) Note any significant discrepancies in shipping paper as defined in Sections 264.72(a) or 265.72(a) (Manifest or Shipping Paper Discrepancies) on each copy of shipping paper.
264.71(b)(2) Comment and 265.71(b)(2) Comment	(c) Waste Analysis EPA does not intend facility owner/operator to perform waste analysis procedures of Sections 264.13(c) or 265.13(c) (General Waste Analysis) prior to signing shipping paper and giving it to transporter. (d) Unreconciled Discrepancy Sections 264.72(b) and 265.72(b) (Manifest or Shipping Paper Discrepancies) requires reporting unreconciled discrepancies found during later waste analysis.
264.71(b)(3) and 265.71(b)(3)	(e) Immediately give rail or water (bulk shipment) transporter at least one copy of shipping paper.
264.71(b)(4) and 265.71(b)(4)	(f) Within 30 days after delivery, send copy of shipping paper to generator. (g) If Manifest Is Received within 30 Days after Delivery Owner/operator, or his agent, must sign and date manifest and return it to generator in place of shipping paper.
264.71(b)(4) Comment and 265.71(b)(4) Comment	(h) Manifest Copy Requirements When Hazardous Waste Is Sent by Rail or Water (Bulk Shipment) Section 262.23(c) (Manifest Handling Requirements) requires generator to send three copies of manifest to facility.
264.71(b)(5) and 265.71(b)(5)	(i) Retain copy of each shipping paper and manifest at facility for at least three years from date of delivery.
264.72 and 265.72	C. Manifest Discrepancies
264.72(a) and 265.72(a)	(1) Definition Manifest discrepancies are differences between quantity or type of hazardous waste designated on manifest or shipping paper, and quantity or type of hazardous waste a facility actually receives. (2) Significant Discrepancies in Quantity (a) For Bulk Waste Variations greater than 10 percent in weight. (b) For Batch Waste Any variation in piece count, such as a discrepancy of one drum in a truckload. (3) Significant Discrepancies in Type Obvious differences which can be discovered by inspection or waste analysis. Examples follow: (a) Waste solvent substituted for waste acid. (b) Toxic constituents not reported on manifest or shipping paper.

Table 6-1. Final (i.e., Part 264) and Interim (i.e., Part 265) Status Standards for Owners and Operators of Hazardous Waste Treatment, Storage, and Disposal Facilities. (Continued)

Section No.	Description

264.72(b)
and 265.72(b)
	(4) Owner/Operator Response to Discovering Significant Discrepancy
	    (a) He must attempt to reconcile discrepancy with waste generator or transporter (e.g., by telephone conversations).
	    (b) If Discrepancy Not Resolved within 15 Days after Receiving Waste
	        Letter must be submitted to Regional Administrator describing discrepancy and attempts to reconcile it, with copy of manifest or shipping paper at issue.

	XII. Recordkeeping Requirements
264.73 and 265.73	A. Operating Record

264.73(a)
and 265.73(a)
 	
264.73(b)
and 265.73(b)
	(1) General Requirements
	    (a) Owner/operator must keep a written operating record at his facility.
	    (b) Information must be recorded as it becomes available.
	    (c) Recorded information must be maintained in operating record until closure of facility.
	(2) Content of Operating Record

264.73(b)(1) and 265.73(b)(1)	(a) Description, Quantity, Method(s) and Date(s) of Treatment, Storage, or Disposal for Each Hazardous Waste Received at Facility As required by Appendix I (Recordkeeping Instructions) and associated Tables 1 and 2, for units of measure and handling codes for treatment, storage, and disposal methods, respectively.

264.73(b)(2)
and 265.73(b)(2)
	(b) Locational Information
	      (i) Location of each hazardous waste within facility and quantity at each location.
	     (ii) For Disposal Facilities
	         Record location and quantity of each hazardous waste on map or diagram of each cell or disposal area.
	    (iii) For All Facilities
	         Include cross reference to specific manifest document numbers, if waste was accompanied by a manifest.

265.73(b)(2) Comment	(iv) Related Requirements Via Interim Status Regulations
	       ● Section 265.119–Notice To Local Land Authority (for closure and post-closure).
	       ● Section 265.279–Recordkeeping (for land treatment).
	       ● Section 265.309–Surveying and Recordkeeping (for landfills).

264.73(b)(3)
and 265.73(b)(3)
	(c) Records and Results of Waste Analyses and/or Trial Tests Performed
	      (i) For Both Interim Status and Final Permitting
	         Sections 264.13 and 265.13–General Waste Analysis.
	     (ii) For Interim Status Only
	       ● Section 265.193–Waste Analysis and Trial Tests (for tanks).
	       ● Section 265.225–Waste Analysis and Trial Tests (for surface impoundments).
	       ● Section 265.252–Waste Analysis (for waste piles).
	       ● Section 265.273–Waste Analysis (for land treatment).
	       ● Section 265.345–Waste Analysis (for incinerators).
	       ● Section 265.375–Waste Analysis (for thermal treatment).
	       ● Section 265.402–Waste Analysis and Trial Tests (for chemical, physical, and biological treatment).

264.73(b)(4) and 265.73(b)(4)	(d) Summary Reports and Details of Incidents Requiring Implementation of Contingency Plan As specified in Sections 264.56(j) or 265.56(j) (Emergency Procedures).
264.73(b)(5) and 265.73(b)(5)	(e) Records and Results Of Inspections As required by Section 264.15(d) or 265.15(d) (General Inspection Requirements), except that data must be kept for only three years rather than until closure of facility.

265.73(b)(6)	(f) For Interim Status Only
	      (i) Monitoring, Testing or Analytical Data
	         Where required by:
	       ● Sections 265.90 and 265.94–Applicability and Recordkeeping and Reporting (for ground-water monitoring).
	       ● Sections 265.276 and 265 278–Food Chain Crops and Unsaturated Zone (Zone of Aeration) Monitoring (for land treatment).
	       ● Section 265.280(d)(1)–Closure and Post-Closure (for land treatment).
	       ● Section 265.347–Monitoring and Inspections (for incinerators).
	       ● Section 265.377–Monitoring and Inspections (for thermal treatment).

Table 6-1. Final (i.e., Part 264) and Interim (i.e., Part 265) Status Standards for Owners and Operators of Hazardous Waste Treatment, Storage, and Disposal Facilities. (Continued)

Section No.	Description
265.73(b)(6) Comment	(ii) Monitoring Data at Disposal Facilities Must be kept throughout post-closure period as required by Section 265.94 (Record-keeping and Reporting for ground-water monitoring).
265.73(b)(7)	(iii) Facility Closure Cost Estimates Under Section 265.142 (Cost Estimate for Facility Closure for financial requirements). (iv) Post-Closure Cost Estimates for Disposal Facilities Under Section 265.144 (Cost Estimate for Post-Closure Monitoring and Maintenance for financial requirements).
264.73(b)(6)	(g) For Final Permitting Only (i) Notices To Generators By Off-Site Facilities As specified in Section 264.12(b) (Required Notices).
264.74 and 265.74	B. Availability, Retention, and Disposition of Records
264.74(a) and 265.74(a)	(1) Availability of Records and Plans to EPA Must be furnished upon request, and made available at reasonable times for inspection by officer, employee, or representative of EPA duly designated by Administrator.
264.74(b) and 265.74(b)	(2) Extension of Record Retention Period (a) Occurs automatically during unresolved enforcement action. (b) If requested by Administrator.
264.74(c) and 265.74(c)	(3) Upon Closure of Facility Copy of records of waste disposal locations and quantities under Section 264.73(b)(2) or 265.73(b)(2) (Operating Record) must be submitted to Regional Administrator and local land authority.
265.74(c)	(a) For Interim Status Requirements Only See Section 265.119 (Notice to Local Land Authority for closure and post-closure).
	XIII. Reporting Requirements
264.70 and 265.70	A. Applicability (1) Apart from exceptions of Sections 264.1 and 265.1, these reporting regulations apply to owners/operators of both on-site and off-site facilities except for Section 265.76 (Unmanifested Waste Report), which does not apply to on-site facilities that do not receive any hazardous waste from off-site sources.
264.76(g) Comment and 265.76(g) Comment	(2) Small Quantities of Hazardous Waste Excluded from regulation under Parts 264 and 265.
264.75 and 265.75	B. Annual Report (1) General Requirements (a) Owner/operator must prepare and submit single copy of annual report to Regional Administrator by March 1 of each year. (b) Report form and instructions of Fig. 6-1 must be used for report. (c) Annual report must cover facility activities during previous calendar year. (2) Content
264.75(a) and 265.75(a) 264.75(b) and 265.75(b) 264.75(c) and 265.75(c)	(a) EPA identification number. (b) Name and address of facility. (c) Calendar year covered by report. (d) For Off-Site Facilities EPA identification number of each hazardous waste generator from which the facility received a hazardous waste during the year. (e) For Imported Shipments Name and address of foreign generator.
264.75(d) and 265.75(d) 264.75(e) and 265.75(e) 265.75(f)	(f) Description and Quantity of Each Hazardous Waste Facility Received during Year For off-site facilities, list information by EPA identification number of each generator. (g) Method of treatment, storage, or disposal for each hazardous waste. (h) For Interim Status Only (i) Monitoring Data Under Sections 265.94(a)(2)(ii) and (iii), and (b)(2) (Recordkeeping and Reporting of ground-water monitoring), where required.

Table 6-1. Final (i.e., Part 264) and Interim (i.e., Part 265) Status Standards for Owners and Operators of Hazardous Waste Treatment, Storage, and Disposal Facilities. (Continued)

Section No.	Description
265.75(g)	(ii) Cost Estimates • Most recent closure cost estimate under Section 265.142 (Cost Estimate For Facility Closure for financial requirements). • Most recent disposal facility post-closure estimate under Section 265.144 (Cost Estimate For Post-Closure Monitoring and Maintenance for financial requirements).
264.75(f) and 265.75(f)	(i) The certification signed by owner/operator of facility or his authorized representative.
264.76 and 265.76	C. Unmanifested Waste Report (1) General Requirements (a) Conditions for Submitting Unmanifested Waste Report to Regional Administrator (i) If facility accepts hazardous waste for treatment, storage, or disposal from off-site source without accompanying manifest or shipping papers per Section 263.20(e) (Bulk Shipment by Rail or Water); and (ii) If waste is not excluded from manifest requirements under Section 261.5 (Small Quantity Generators). (b) Preparation and submittal of single copy of unmanifested waste report to Regional Administrator within 15 days after receiving waste. (c) Report form and instructions of Fig. 6-1 must be used. (2) Contents
264.76(a) and 265.76(a)	(a) EPA identification number.
264.76(b) and 265.76(b)	(b) Name and address of facility. (c) Date facility received waste.
264.76(c) and 265.76(c)	(d) EPA identification number, name, and address of generator and transporter, if available.
264.76(d) and 265.76(d)	(e) Description and quantity of each unmanifested hazardous waste facility received.
264.76(e) and 265.76(e)	(f) Method of treatment, storage, or disposal for each hazardous waste.
264.76(f) and 265.76(f)	(g) Certification signed by owner/operator of facility or his authorized representative.
264.76(g) and 265.76(g)	(h) Brief explanation why waste was unmanifested, if known.
264.76(g) Comment and 265.76(g) Comment	(3) Agency Suggestions Where Facility Receives Unmanifested Hazardous Waste (a) Owner/operator should obtain certification from each generator that waste qualifies for exclusion. (b) Otherwise, file an unmanifested waste report for the hazardous waste movement.
264.77 and 265.77	D. Additional Reports Owner/operator must also report to Regional Administrator:
264.77(a) and 265.77(a)	(1) Releases, Fires, and Explosions As specified in Sections 264.56(j) and 265.56(j) (Emergency Procedures). (2) For Interim Status Only
265.77(b)	(a) Ground-Water Contamination and Monitoring Data As specified in: (i) Section 265.93—Preparation, Evaluation and Response. (ii) Section 265.94—Recordkeeping and Reporting.
265.77(c)	(b) Facility Closure As specified in Section 265.115 (Certification of Closure).
Subpart F	XIV. Ground-Water Monitoring
265.90 265.90(a)	A. Applicability (1) Types of Hazardous Waste Facilities Requiring Ground-Water Monitoring (a) Surface impoundment. (b) Landfill. (c) Land treatment. (2) Effectivity Within one year after effective date of regulations (November 19, 1981).

Table 6-1. Final (i.e., Part 264) and Interim (i.e., Part 265) Status Standards for Owners and Operators of Hazardous Waste Treatment, Storage, and Disposal Facilities. (Continued)

Section No.	Description
	B. General Requirements
265.90(a)	(1) Purpose of Ground-Water Monitoring Program
	To determine facility's impact on quality of ground water in uppermost aquifer underlying facility.
265.90(b)	(2) Responsibility For Implementation
	Owner/operator must install, operate, and maintain ground-water monitoring system in compliance with:
	(a) Section 265.91–Ground-Water Monitoring System.
	(b) Section 265.92–Sampling and Analysis.
	(c) Section 265.93–Preparation, Evaluation and Response.
	(d) Section 265.94–Recordkeeping and Reporting.
	(e) Exemptions
	(i) If waiver obtained per Section 265.90(c).
	(ii) If alternate ground-water monitoring system per Section 265.90(d) is used.
	(3) Duration of Ground-Water Monitoring Program
	(a) During active life of facility.
	(b) For Disposal Facilities
	During post-closure care period as well.
265.90(c)	C. Waiver for All or Part of Ground-Water Monitoring Requirements
	(1) Owner/operator must be able to demonstrate that low potential exists for migration of hazardous waste/constituents from facility:
	(a) Via uppermost aquifer to water supply wells (domestic, industrial or agricultural); or
	(b) To surface water.
	(2) Demonstration must be in writing and kept at facility.
	(3) Demonstration must be certified by qualified geologist or geotechnical engineer.
	(4) Detailed Demonstration Requirements
265.90(c)(1)	(a) Must establish potential for migration of hazardous waste/constituents from facility to uppermost aquifer by evaluation of:
265.90(c)(1)(i)	(i) A water balance of precipitation, evapotranspiration, runoff, and infiltration; and
265.90(c)(1)(ii)	(ii) Unsaturated zone characteristics (e.g., geologic materials, physical properties, and depth of ground water).
265.90(c)(2)	(b) Must establish potential for hazardous waste/constituents which enter uppermost aquifer to migrate to a water supply well or surface water by evaluation of:
265.90(c)(2)(i)	(i) Saturated zone characteristics (e.g., geologic materials, physical properties, and rate of ground-water flow); and
265.90(c)(2)(ii)	(ii) Proximity of facility to water supply wells or surface water.
265.90(d)	D. Alternate Ground-Water Monitoring System
	(1) General
	Such an alternate system may be used by owner/operator if he assumes or knows that ground-water monitoring of indicator–parameters of Sections 265.91 and 265.92 would show his facility to be contributing hazardous waste constituents to the ground water.
	(2) Requirements When Using Alternate Plan
	Owner/operator must:
265.90(d)(1)	(a) Submit plan to Regional Administrator within one year after effective date of regulations (11/19/81).
	(b) Obtain Certification of Plan
	Must be done by qualified geologist or geotechnical engineer.
	(c) Assure that plan satisfies all requirements of Section 265.93(d)(3).
265.90(d)(2) and 265.90(d)(4)	(d) Initiate determination of ground-water conditions specified in Section 265.93(d)(4) not later than 11/19/81, and continue on quarterly basis until final closure of facility.
265.90(d)(3)	(e) Prepare and submit written report in accordance with Section 265.93(d)(5).
265.90(d)(5)	(f) Comply with recordkeeping and reporting requirements in Section 265.94(b).
265.91	E. Ground-Water Monitoring System
	(1) General
265.91(a)	(a) Overall Requirement
	A ground-water monitoring system must be capable of yielding ground-water samples for analysis.

Table 6-1. Final (i.e., Part 264) and Interim (i.e., Part 265) Status Standards for Owners and Operators of Hazardous Waste Treatment, Storage, and Disposal Facilities. (Continued)

Section No.	Description
265.91(b)	(b) Waste Management Area
265.91(b)(1)	(i) For Facility Consisting of Only One Surface Impoundment, Landfill, or Land Treatment Area
	The waste management area is described by the waste boundary (perimeter).
265.91(b)(2)	(ii) For Facility Consisting of More than One Surface Impoundment, Landfill, or Land Treatment Area
	The waste management area is described by an imaginary boundary line which circumscribes the several waste management components.
265.91(a)	(2) Detailed Requirements for Ground-Water Monitoring System
265.91(a)(1)	(a) Hydraulically Upgradient Monitoring Wells
	(i) Install in direction of increasing static head (e.g., hydraulically upgradient).
	(ii) At least one monitoring well required.
	(iii) Install at limit of waste management area.
	(iv) Number, location(s), and depth(s) must be sufficient to yield ground-water samples that are:
265.91(a)(1)(i)	• Representative of background ground-water quality in uppermost aquifer near facility.
265.91(a)(1)(ii)	• Not affected by facility.
265.91(a)(2)	(b) Hydraulically Downgradient Monitoring Wells
	(i) Install in direction of decreasing static head (e.g., hydraulically downgradient).
	(ii) At least three monitoring wells required.
	(iii) Install at limit of waste management area.
	(iv) Number, location(s), and depth(s) must ensure immediate detection of any statistically significant amounts of hazardous waste that migrate from waste management area to uppermost aquifer.
265.91(b)	(c) Separate Monitoring System for Each Waste Mangement Component of Facility
	Not required provided that provisions for sampling upgradient and downgradient water quality will detect any discharge from the waste management area.
265.91(c)	(d) Monitoring Well Design/Performance Requirements
	(i) Casing of Monitoring Wells
	• Must maintain integrity of monitoring well bore hole.
	• Casing must be screened or perforated and packed with gravel or sand, where necessary, to enable collection at depths where appropriate aquifer flow zones exist.
	(ii) Annular Space
	• That space between bore hole and well casting.
	• The annular space above the sampling depth must be sealed with suitable material such as cement grout or bentonite slurry to prevent contamination of samples and ground water.
265.92	F. Sampling and Analysis Requirements
265.92(a)	(1) General Requirements
	Owner/operator must:
	(a) Develop and follow ground-water sampling and analysis plan.
	(b) Keep plan at facility.
	(c) Obtain and analyze samples from installed ground-water monitoring system.
	(2) Procedures and Techniques to Be Included in Plan
265.92(a)(1)	(a) Sample collection.
265.92(a)(2)	(b) Sample preservation and shipment.
265.92(a)(3)	(c) Analytical procedures.
265.92(a)(4)	(d) Chain of custody control.
265.92(a) Comment	*Note:* For discussions of sampling and analysis procedures see:
	• "Procedures Manual For Ground-Water Monitoring at Solid Waste Disposal Facilities," EPA-530/SW-611, August 1977.
	• "Methods for Chemical Analysis of Water and Wastes," EPA-600/4-79-020, March 1979.
265.92(b)	(3) Determination of Concentration or Values of Specified Parameters in Ground-Water Samples
	Owner/operator must make such determinations in accordance with timing and protocols of Section 265.92(c) (during first year) and with Section 265.92(d) (after first year).

Table 6-1. Final (i.e., Part 264) and Interim (i.e., Part 265) Status Standards for Owners and Operators of Hazardous Waste Treatment, Storage, and Disposal Facilities. (Continued)

Section No.	Description
265.92(b)(1)	(a) Parameters Characterizing Suitability of Ground-Water as Drinking Water Supply As specified in the 21 parameters of Appendix III (EPA Interim Primary Drinking Water Standards).
265.92(b)(2)	(b) Parameters Establishing Ground-Water Quality
265.92(b)(2)(i)	(i) Chloride.
265.92(b)(2)(ii)	(ii) Iron.
265.92(b)(2)(iii)	(iii) Manganese.
265.92(b)(2)(iv)	(iv) Phenols.
265.92(b)(2)(v)	(v) Sodium.
265.92(b)(2)(vi)	(vi) Sulfate.
265.92(b)(3)	(c) Parameters Used as Indicators of Ground-Water Contamination
265.92(b)(3)(i)	(i) pH.
265.92(b)(3)(ii)	(ii) Specific Conductance.
265.92(b)(3)(iii)	(iii) Total Organic Carbon.
265.92(b)(3)(iv)	(iv) Total Organic Halogen.
265.92(c)(1)	(4) Establish Background Concentrations of Section 265.92(b) Parameters for Monitoring Wells during First Year (a) Must be done quarterly for one year.
265.92(c)(2)	(b) For Each Contamination Indicator Parameter of Section 265.92(b)(3) (i) At least four replicate measurements must be obtained for each sample. (ii) Initial background arithmetic mean and variance must be determined by pooling replicate measurements for respective parameter concentrations or values in samples obtained from upgradient wells during first year.
265.92(d)	(5) Frequency of Sampling Monitoring Wells after First Year
265.92(d)(1)	(a) Samples Collected to Establish Ground-Water Quality Must be obtained and analyzed for parameters of Section 265.92(b)(2) at least annually.
265.92(d)(2)	(b) Samples Collected to Indicate Ground-Water Contamination Must be obtained and analyzed for parameters of Section 265.92(b)(3) at least semi-annually.
265.92(e)	(6) Elevation of Ground-Water Surface at Each Monitoring Well Must be determined each time a sample is obtained.
265.93	G. Preparation, Evaluation, and Response
265.93(a)	(1) Outline of Ground-Water Quality Assessment Program Such an outline must be prepared by owner/operator within one year after effective date of regulations (11/19/81). (a) General Outline must describe a more comprehensive ground-water monitoring program than described in Sections 265.91 (Ground-Water Monitoring System) and 265.92 (Sampling and Analysis). (b) Outline must describe a program capable of determining:
265.93(a)(1)	(i) Whether hazardous waste/constituents have entered the ground water.
265.93(a)(2)	(ii) Rate and extent of migration of hazardous waste/constituents in the ground water.
265.93(a)(3)	(iii) Concentrations of hazardous waste/constituents in ground water.
265.93(b)	(2) Evaluation of Each Contamination Indicator Parameter of Section 265.92(b)(3) (a) Calculate Arithmetic Mean and Variance Based on at least four replicate measurements on each sample for each well monitored. (b) Compare Results with Initial Background Arithmetic Mean Values (i) Comparison must consider each well in monitoring system. (ii) Must use Student's t-test at the 0.01 level of significance (see Appendix IV) to determine statistically significant increases (and decreases in case of pH) over initial background values.
265.93(c)(1)	(c) If Comparisons for Upgradient Wells Show Significant Contamination Increase(s) (Or pH Decrease) Owner/operator must submit information in accordance with report requirements of Section 265.94(a)(2)(ii).
265.93(c)(2)	(d) If Comparisons For Downgradient Wells Show Significant Contamination Increase(s) (Or pH Decrease) Owner/operator must: (i) Immediately obtain additional ground water samples for those downgradient wells where significant difference was detected. (ii) Split samples in two.

Table 6-1. Final (i.e., Part 264) and Interim (i.e., Part 265) Status Standards for Owners and Operators of Hazardous Waste Treatment, Storage, and Disposal Facilities. (Continued)

Section No.	Description
	(iii) Obtain analyses of all additional samples to determine whether the significant difference was a result of laboratory error.
265.93(d)(1)	• If Analysis Confirms Significant Contamination Increase (or pH Decrease) – Provide written notice to Regional Administrator within 7 days that facility may be affecting ground-water quality.

265.93(d)(2)	– Develop and Submit Ground-Water Quality Assessment Program Plan to Regional Administrator
	  * Plan based on outline of Section 265.93(a).
	  * Submit within 15 days of notification of Section 265.93(d)(1).
	  * Have plan certified by qualified geologist or geotechnical engineer.

265.93(d)(3)	– Items Plan Must Specify
265.93(d)(3)(i)	* Number, location, and depth of wells.
265.93(d)(3)(ii)	* Sampling and analytical methods for those hazardous wastes/constituents in facility.
265.93(d)(3)(iii)	* Evaluation procedures, including use of previously gathered ground-water quality information.
265.93(d)(3)(iv)	* Schedule of implementation.
265.93(d)(4)	(3) Implementation of Ground-Water Quality Assessment Plan (a) Minimum Required Determinations
265.93(d)(4)(i)	(i) Rate and extent of migration of hazardous waste/constituents in ground water.
265.93(d)(4)(ii)	(ii) Concentrations of hazardous waste/constituents in ground-water.

265.93(d)(5)	(b) First Determination of Extent of Contamination
	  (i) Do as soon as technically feasible.
	  (ii) Within 15 days after determination submit written report to Regional Administrator containing assessment of ground-water quality.

265.93(d)(6)	(c) If Results Indicate that No Hazardous Waste/Constituents from Facility Entered Ground Water
	  (i) Owner/operator may reinstate contamination indicator program of Sections 265.92 and 265.93(b).
	  (ii) If indicator program is reinstated, notify Regional Administrator in report submitted under Section 265.93(d)(5).

265.93(d)(7)	(d) If Results Indicate that Hazardous Waste/Constituents from Facility Entered Ground Water
265.93(d)(7)(i)	(i) Owner/operator must continue to make determinations under Section 265.93(d)(4) on a quarterly basis until final closure of facility, if the ground-water quality assessment plan was implemented prior to final closure of facility; or
265.93(d)(7)(ii)	(ii) Owner/operator may cease to make determinations under Section 265.93(d)(4) if ground-water quality assessment plan was implemented during post-closure care period.

265.93(f)	(4) If Ground-Water Monitoring to Satisfy Requirements of Section 265.93(d)(4) Not Done Annually
	  (a) Data on ground-water surface elevations obtained under Section 265.92(e) must be evaluated to determine whether requirements under Section 265.91(a) for locating monitoring wells continues to be satisfied.
	  (b) If Section 265.91(a) requirements are no longer satisfied then immediate modification in number, location, or depth of monitoring wells must be made to bring ground-water monitoring system into compliance with this requirement.

265.94	H. Recordkeeping and Reporting
265.94(a)	(1) If Ground-Water Is Monitored to Satisfy Requirements Other than for Section 265.93(d)(4) Owner/operator must:

265.94(a)(1)	(a) Keep Records
	  (i) Of analyses required in Section 265.92(c) and (d).
	  (ii) Associated ground-water elevations required in Section 265.92(e).
	  (iii) Evaluations required in Section 265.93(b).
	  (iv) Duration
	   • Through active life of facility.
	   • For Disposal Facilities
	    Throughout post-closure period as well.

265.94(a)(2)	(b) Report Ground-Water Monitoring Information to Regional Administrator

265.94(a)(2)(i)	(i) Background Concentrations During First Year
	   • Concentrations or values of parameters listed in Section 265.92(b)(1) (Drinking water suitability parameters of Appendix III) for each ground-water monitoring well.
	   • Report within 15 days after completing each quarterly analysis.
	   • Separately identify for each monitoring well any parameters whose concentrations or values exceed contaminant levels listed in Appendix III.

Table 6-1. Final (i.e., Part 264) and Interim (i.e., Part 265) Status Standards for Owners and Operators of Hazardous Waste Treatment, Storage, and Disposal Facilities. (Continued)

Section No.	Description
265.94(a)(2)(ii)	(ii) Annual Reporting • Concentrations or values of contamination parameters listed in Section 265.92 (b)(3) for each ground-water monitoring well, along with required evaluations under Section 265.93(b). • Separately identify significant differences from initial background found in up-gradient wells, per Section 265.93(c)(1). • During active life of facility, this information must be submitted as part of annual report under Section 265.75 (Annual Report).
265.94(a)(2)(iii)	• Results of evaluation of ground-water surface evaluations under Section 265.93(f) and a description of response to that evaluation.
265.94(b)	(2) If Ground Water Is Monitored to Satisfy Requirements of Section 265.93(d)(4) (Ground-Water Quality Assessment Program)
265.94(b)(1)	(a) Keep Records of Analysis and Evaluation Specified In Plan And Satisfying Section 265.93(d)(3) (i) Throughout active life of facility. (ii) For Disposal Facility Throughout post-closure care period as well.
265.94(b)(2)	(b) Annual Report of Results of Ground-Water Quality Assessment Program (i) Submit to Regional Administrator. (ii) Minimal Required Content Calculated or measured rate of migration of hazardous waste/constituents in ground water during reporting period. (iii) Submit as part of annual report under Section 265.75 (Annual Report).

XV. General Closure Requirements

265.111	A. Closure Performance Standard Owner/operator must close facility in manner that: (1) Minimizes need for future maintenance. (2) Controls, minimizes, or eliminates, to extent necessary to protect human health and environment, post-closure escape of hazardous waste/constituents, leachate, contaminated rainfall, or waste decomposition products to ground water, or surface waters, or to atmosphere.
265.112 265.112(a)	B. Closure Plan (1) General Requirement Plan must identify steps necessary to completely close facility at any point during its intended life and at end of its intended life. (2) Timing Owner/operator must have a written closure plan on effective date of regulations (11/19/80). (3) Plan must be kept at facility. (4) Contents
265.112(a)(1)	(a) Description of How and When Facility Will Be Partially Closed (If Applicable) and Ultimately Closed (i) Identify maximum extent of operation which will be open during life of facility. (ii) Describe how performance requirements of Section 265.11 will be met. (iii) Describe How Site-Specific Technical Closure Requirements Will Be Met • Section 265.197—Closure for tanks. • Section 265.228—Closure and Post-closure for surface impoundments. • Section 265.280—Closure and Post-closure for land treatment. • Section 265.310—Closure and Post-closure for landfills. • Section 265.351—Closure for incinerators. • Section 265.381—Closure for thermal treatment. • Section 265.404—Closure for chemical, physical, and biological treatment.
265.112(a)(2)	(b) Estimate of maximum inventory of wastes in storage or in treatment at any given time during life of facility.
265.112(a)(3)	(c) Description of steps needed to decontaminate facility equipment during closure.
265.112(a)(4)	(d) Schedule for Final Closure Must include following, at a minimum: (i) Anticipated date when wastes will no longer be received. (ii) Date when completion of final closure is anticipated. (iii) Intervening milestone dates which will allow tracking of progress of closure.

Table 6-1. Final (i.e., Part 264) and Interim (i.e., Part 265) Status Standards for Owners and Operators of Hazardous Waste Treatment, Storage, and Disposal Facilities. (Continued)

Section No.	Description
265.112(c)	(e) Plan Submittal and Approval Requirements Relative to Actual Facility Closure (i) Submittal Owner/operator must submit closure plan to Regional Administrator at least 180 days before date expected to begin closure. (ii) Regional Administrator Actions • Modify, approve, or disapprove plan within 90 days of receipt. • Provide owner/operator and affected public (through a newspaper notice) opportunity to submit written comments. (iii) If Owner/Operator Plans to Begin Closure Within 180 Days After Effective Date of Regulations (11/19/80–5/18/81) He must submit necessary plans on effective date of regulations (11/19/80).
265.112(b)	C. Amendment of Closure Plan (1) Owner/operator may amend closure plan any time during active life of facility. (2) Mandatory Changes to Closure Plan (a) If changes in operating plans affect closure plan. (b) If facility's design affect closure plan.
265.113 265.113(a)	D. Time Allowed for Closure (1) Within 90 Days after Receipt of Final Volume of Hazardous Waste Owner/operator must do following, in accordance with approved closure plan: (a) Treat all hazardous wastes in storage or in treatment; or (b) Remove them from the site; or (c) Dispose of them on-site.
265.113(b)	(2) Completion of Closure Activities (a) Must be done in accordance with approved closure plan. (b) Must be completed within 6 months after receipt of final volume of hazardous wastes. (i) Exception Regional Administrator may approve a longer closure period if owner/operator can demonstrate that:
265.113(b)(1)	• Required or planned closure activities will, of necessity, take longer than 6 months to complete; and
265.113(b)(2)	• He has taken all steps to eliminate any significant threat to human health and environment from unclosed but inactive facility.
265.114	E. Disposal or Decontamination of Equipment When closure is completed, all facility equipment and structures must have been properly disposed of, or decontaminated by removing all hazardous waste and residues.
265.115	F. Certification of Closure When closure is completed, owner/operator must submit to Regional Administrator certification that facility was closed in accordance with specifications in approved closure plan. (1) Certification Parties (a) Owner/operator; and (b) Independent registered professional engineer.
	XVI. General Post-Closure Requirements
265.117(a) 265.117(a)(1)	A. Minimal Efforts Comprising Post-Closure Care (1) Ground-Water Monitoring and Reporting In accordance with requirements of Subpart F of Part 265 (Ground-Water Monitoring).
265.117(a)(2)	(2) Maintenance of Monitoring and Waste Containment Systems As specified in, and where applicable: (a) Section 265.91–Ground-Water Monitoring System. (b) Section 265.223–Containment System for surface impoundments. (c) Section 265.228–Closure and Post-Closure for surface impoundments. (d) Section 265.280–Closure and Post-Closure for land treatment. (e) Section 265.310–Closure and Post-Closure for landfills.
265.117(b)	B. Security Requirements during Post-Closure Regional Administrator may require some or all security requirements of Section 265.14 (Security) during post-closure period, when:
265.117(b)(1)	(1) Wastes may remain exposed after completion of closure; or

Table 6-1. Final (i.e., Part 264) and Interim (i.e., Part 265) Status Standards for Owners and Operators of Hazardous Waste Treatment, Storage, and Disposal Facilities. (Continued)

Section No.	Description
265.117(b)(2)	(2) Short term, incidental access by public or domestic livestock may pose hazard to human health.
265.117(c)	C. Post-Closure Use of Property on or in Which Hazardous Waste Remains after Closure Such use must never be allowed to disturb: (1) Integrity of final cover, liner(s), or any other components of any containment system; or (2) Function of facility's monitoring systems. (3) Exceptions If owner/operator can demonstrate to Regional Administrator, either in the post-closure plan or by petition, that the disturbance:
265.117(c)(1)	(a) Is necessary to proposed use of property, and will not increase potential hazard to human health or environment; or
265.117(c)(2)	(b) Is necessary to reduce a threat to human health or environment.
265.117(d)	D. 30-Year Post-Closure Care Period (1) Post-Closure Care in Accordance with Approved Plan Owner/operator must provide such care for at least 30 years after completing closure. (2) Exemption Owner/operator may petition Regional Administrator to allow some or all post-closure requirements to be discontinued or altered prior to end of 30-year period. (a) Content of Petition It must include evidence demonstrating secure nature of facility that makes continuing specified post-closure requirements unnecessary, e.g.: (i) No detected leaks and none likely to occur. (ii) Characteristics of waste. (iii) Application of advanced technology, or alternative disposal, treatment, or reuse techniques. (b) Regional Administrator Response He may require owner or operator to continue one or more of the post-closure care and maintenance requirements contained in facility's post-closure plan for specified period of time, if: (i) There has been noncompliance with any applicable standards or requirements; or (ii) Such continuation is necessary to protect human health or environment. (3) At End of 30-Year Post-Closure Care Period (a) Regional Administrator will determine whether to continue or terminate post-closure care and maintenance of facility. (b) Petitioning (i) Anyone (public or owner/operator) may petition Regional Administrator for extension or reduction of post-closure care period based on cause. (ii) When Regional Administrator Will Consider Petition • At time post-closure plan is submitted, and • At 5-year intervals after completion of closure.
265.118 265.118(a)	E. Post-Closure Plan (1) General Requirements Plan must identify activities which will be carried on after final closure and frequency of those activities. (2) Timing Owner/operator of disposal facility must have a written post-closure plan on effective date of regulations (11/19/80). (3) Plan must be kept at facility. (4) Contents
265.118(a)(1)	(a) Ground-Water Monitoring Activities and Frequencies As specified in Subpart F of Part 265 (Ground Water Monitoring) for post-closure period.
265.118(a)(2)	(b) Maintenance Activities and Frequencies To ensure: (i) Integrity of cap and final cover or other containment structures as specified in: • Section 265.223–Containment System for surface impoundment. • Section 265.228–Closure and Post-Closure for surface impoundment. • Section 265.280–Closure and Post-Closure for land treatment. • Section 265.310–Closure and Post-Closure for landfills. (ii) Function of facility's monitoring equipment as specified in Section 265.91 (Ground-Water Monitoring System).

Table 6-1. Final (i.e., Part 264) and Interim (i.e., Part 265) Status Standards for Owners and Operators of Hazardous Waste Treatment, Storage, and Disposal Facilities. (Continued)

Section No.	Description

265.118(c)	(c) Plan Submittal and Approval Requirements Relative to Actual Facility Closure
	  (i) Submittal
	    Owner/operator of disposal facility must submit post-closure plan to Regional Administrator at least 180 days before date expected to begin closure.
	  (ii) Regional Administrator Actions
	    • Modify, approve, or disapprove plan within 90 days of receipt.
	    • Provide owner/operator and affected public (through a newspaper notice) opportunity to submit written comments.
	  (iii) Security Equipment Maintenance
	    Plan may be modified to include such maintenance under Section 265.117(b).
	  (iv) If Owner/Operator of Disposal Facility Plans to Begin Closure within 180 Days after Effective Date of Regulations (11/19/80–5/18/81)
	    He must submit necessary plans on effective date of regulations (11/19/80).

265.118(b)
 	
265.118(c)
	F. Amendment of Post-Closure Plan
	  (1) Owner/operator may amend post-closure plan any time during active life of disposal facility or during post-closure care period.
	  (2) Mandatory Changes in Post-Closure Plan
	    (a) If changes in operating plans affect post-closure plan.
	    (b) If facility's design affect post-closure plan.
	  (3) Any amendments to plan which occur after approval of plan must also be approved by Regional Administrator before they may be implemented.

265.119	G. Notices to Local Land Authority and Regional Administrator
	  (1) Within 90 Days after Closure Is Completed
	    Owner/operator of disposal facility must submit a survey plat indicating location and dimensions of landfill cells or other disposal areas with respect to permanently surveyed benchmarks.
	    (a) To local land authority.
	    (b) To Regional Administrator.
	  (2) Plat Requirements
	    (a) Plat must be prepared and certified by a professional land surveyor.
	    (b) Plat Filed with Local Land Authority
	      Must contain note prominently displayed, stating owner's/operator's obligation to restrict disturbance of site as specified in Section 265.117(c).
	  (3) Record of Type, Location, and Quantity of Hazardous Wastes Disposed of Within Each Cell or Area of Facility
	    Owner/operator must submit such a record to Regional Administrator and local land authority.
	  (4) For Wastes Disposed of before These Regulations Were Promulgated
	    Owner/operator must identify type, location, and quantity of wastes to best of his knowledge and in accordance with any records he kept.

265.120	H. Notice in Deed to Property
	  Owner of property on which disposal facility is located must record, per State law, a notation on deed to facility property (or other instrument normally examined during title search) that will in perpetuity notify any potential purchaser of property that:
	  (1) Land has been used to manage hazardous waste, and
	  (2) Its use is restricted under Section 265.117(c).

Subpart H	XVII. Financial Requirements

265.140
265.140(a)
 	
265.140(b)
 	
265.140(c)
	A. Applicability
	  (1) Owners/Operators of All Hazardous Waste Facilities
	    Section 265.142 (Cost Estimate for facility closure) applies, except as this Section or Section 265.1 provides otherwise.
	  (2) Owners/Operators of Disposal Facilities Only
	    Section 265.144 (Cost Estimate for Post-Closure Monitoring and Maintenance) applies.
	  (3) States and Federal Government
	    Exempt from requirements of this Subpart.

265.142
265.142(a)
	B. Cost Estimate for Facility Closure
	  (1) Facility Closure Cost Estimate Requirements
	    (a) Timing
	      Owner/operator of each facility must have written estimate of cost of closing facility on effective date of regulations (11/19/80).

Table 6-1. Final (i.e., Part 264) and Interim (i.e., Part 265) Status Standards for Owners and Operators of Hazardous Waste Treatment, Storage, and Disposal Facilities. (Continued)

Section No.	Description

	(b) Make estimate in accord with Subpart G (Closure and Post-Closure) requirements that follow:
	   (i) Section 265.111–Closure Performance Standard.
	   (ii) Section 265.112–Closure Plan; Amendments of Plan.
	   (iii) Section 265.113–Time Allowed for Closure.
	   (iv) Section 265.114–Disposal or Decontamination of Equipment.
	   (v) Section 265.115–Certification of Closure.
	(c) Also estimate in accord with applicable closure requirements that follow:
	   (i) Section 265.197–Closure for tanks.
	   (ii) Section 265.228–Closure and Post-Closure for surface impoundments.
	   (iii) Section 265.280–Closure and Post-Closure for land treatment.
	   (iv) Section 265.310–Closure and Post-Closure for landfills.
	   (v) Section 265.351–Closure for incinerators.
	   (vi) Section 265.381–Closure for thermal treatment.
	   (vii) Section 265.404–Closure for chemical, physical, and biological treatment.
	(d) Initial estimate, and all subsequent estimates must be kept at facility.
	(e) Maximum Closure Cost
	   Estimate must equal cost of closure at point in facility's operating life when extent and manner of its operation would make closure the most expensive, as indicated by its closure plan of Section 265.112(a).

265.142(b)	(2) New Closure Cost Estimate for Changes in Facility Closure Plan Required of owner/operator whenever change in closure plan effects cost of closure.

265.142(c)	(3) Annual Inflation Factor Adjustment of Closure Cost Estimate
	Required of owner/operator on each anniversary of effective date of regulations (19 November).
	(a) Inflation Factor
	   (i) Derived from annual Implicit Price Deflator for Gross National Product published by U.S. Department of Commerce
	   (ii) Inflation factor must be calculated by dividing latest published annual Deflator by Deflator for previous year.
	(b) Adjusted Closure Cost Estimate
	   Must equal latest closure cost estimate times inflation factor.

265.142 (Comment)	(i) Sample Calculation
	   • Latest closure cost estimate for a facility is $50,000.
	   • Latest published annual Deflator is 152.05.
	   • Annual Deflator for previous year is 141.70.
	   • Deflators may be rounded to nearest whole number.
	   • Dividing 152 by 142 gives inflation factor of 1.07.
	   • Multiplying $50,000 by 1.07 for a product of $53,500–the adjusted closure cost estimate.

265.144
265.144(a)
	C. Cost Estimate for Post-Closure Monitoring and Maintenance for Disposal Facility
	(1) Disposal Facility Estimate of Annual Cost of Post-Closure Monitoring and Maintenance.
	(a) Timing
	   Owner/operator must have written estimate on effective date of regulations (11/19/80).
	(b) Make estimate in accord with applicable post-closure regulations that follow:
	   (i) Section 265.117–Post-Closure Care and Use of Property; Period of Care.
	   (ii) Section 265.118–Post-Closure Plan; Amendment of Plan.
	   (iii) Section 265.119–Notice to Local Land Authority.
	   (iv) Section 265.120–Notice in Deed to Property.
	   (v) Section 265.228–Closure and Post-Closure for surface impoundments.
	   (vi) Section 265.280–Closure and Post-Closure for land treatment.
	   (vii) Section 265.310–Closure and Post-Closure for landfills.

265.144(b)	(2) New Annual Post-Closure Cost Estimate for Change in Post-Closure Plan (a) Required of owner/operator whenever change in post-closure plan affects cost of post-closure care provided in Section 265.118(b). (b) Latest post-closure cost estimate calculated by multiplying latest annual post-closure cost estimate by 30.
265.144(c)	(3) Annual Inflation Factor Adjustment of Post-Closure Cost Estimate Required of owner/operator on each anniversary of effective date of regulations (19 November). (a) Inflation factor calculated in accordance with Section 265.142(c).

Table 6-1. Final (i.e., Part 264) and Interim (i.e., Part 265) Status Standards for Owners and Operators of Hazardous Waste Treatment, Storage, and Disposal Facilities. (Continued)

Section No.	Description
265.144(a)	(b) Adjusted post-closure cost estimate must equal latest post-closure cost estimate of Section 265.144(b) times inflation factor.
	(4) Location of Cost Estimates Owner/operator must keep initial and subsequent estimates at the facility.
FR pp. 33265–33278, 5/19/80	D. Proposed Rules to Modify Subpart H, Financial Requirements See Appendix D of this book.
Subpart I.	XVIII. Use and Management of Containers
265.170	A. Applicability Regulations in this Subpart apply to owners/operators of all hazardous waste facilities storing containers of hazardous waste, except as provided otherwise in Section 265.1.
265.171	B. Container Holding Hazardous Waste That Leaks or Is Not in Good Condition Owner/operator must: (1) Transfer hazardous waste from this container to a container that is in good condition; or (2) Manage waste in some other way that complies with requirements of this Part.
265.172	C. Compatibility of Waste With Container Owner/operator must use container made of or lined with materials which will not react with, and are otherwise compatible with hazardous waste to be stored, so that ability of container to contain waste is not impaired.
265.173 265.173(a)	D. Management of Containers (1) Container holding hazardous waste must always be closed during storage. (a) Exception When it is necessary to add or remove wastes.
265.173(b)	(2) Opening, Handling, or Storing Container Holding Hazardous Waste Must be done in manner that will not rupture container or cause it to leak.
265.173(b) Comment	(3) Container Listed as Hazardous Waste in Section 261.33 (Discarded Containers That Held Acutely Toxic Waste) Such a container must be managed in compliance with regulations of this Subpart. (4) Re-use of Containers in Transportation Governed by U.S. Department of Transportation regulations including 49 CFR 173.28 (see Appendix E of this book for D.O.T. regulation contents only).
265.174	E. Inspections Owner/operator must inspect areas where containers are stored, at least weekly, looking for leaks and for deterioration caused by corrosion or other factors.
265.176	F. Special Requirements for Ignitable or Reactive Waste (1) Separation from Facility's Property Line At least 15 meters (50 feet) required. (2) See general requirements of Section 265.17(a) for ignitable, reactive, or incompatible wastes.
265.177 265.177(a)	G. Special Requirements for Incompatible Wastes (1) Placement of Incompatible Wastes, or Incompatible Wastes and Materials in Same Container Must not be done unless Section 265.17(b) (General Requirements for Incompatible Wastes) complied with.
265.177(a)	(2) Examples of Incompatible Wastes See Appendix V.
265.177(b)	(3) Placement of Hazardous Waste into Unwashed Container Previously Holding Incompatible Waste or Material Must not be done unless Section 265.17(b) (General Requirements for Incompatible Wastes) complied with.
265.177(c)	(4) Storage Containers Holding Hazardous Waste Incompatible with Waste or Other Materials Stored Nearby in Containers, Piles, Open Tanks, or Surface Impoundments Must be separated from the other materials or protected from them via a dike, berm, wall or other device.
265.177(c) Comment	(5) Purpose of This Section (265.177) To prevent fires, explosions, gaseous emissions, leaching, or other discharge of hazardous waste/constituents which could result from mixing of imcompatible wastes or materials if containers break or leak.

Table 6-1. Final (i.e., Part 264) and Interim (i.e., Part 265) Status Standards for Owners and Operators of Hazardous Waste Treatment, Storage, and Disposal Facilities. (Continued)

Section No.	Description
Subpart J	XIX. Tanks
265.190	A. Applicability Regulations in this Subpart apply to owners/operators of facilities that use tanks to treat or store hazardous waste, except as provided otherwise in Section 265.1.
265.192 265.192(a)	B. General Operating Requirements (1) Treatment or Storage of Hazardous Waste in Tanks Must be in compliance with Section 265.17(b) (General Requirements for Ignitable, Reactive, or Incompatible Wastes).
265.192(b)	(2) Placement of Hazardous Waste or Treatment Reagents in Tank Must not be done if they could cause the tank or its liner to rupture, leak, corrode, or otherwise fail before end of its intended life.

265.192(c)	(3) Uncovered Tanks
	(a) Must be operated to ensure at least 60 centimeters (2 feet) of freeboard.
	(b) Not required if tank is equipped with:
	    (i) Containment structure such as a dike or trench; or
	    (ii) Drainage control system; or
	    (iii) Diversion structure such as a standby tank, with capacity equal to or greater than volume of top 60 centimeters (2 feet) of the tank.

265.192(d)	(4) Where Hazardous Waste Is Continuously Fed into a Tank Tank must be equipped with means to stop inflow—e.g., waste feed cutoff system or bypass system to standby tank.
265.192(d) Comment	(a) Purpose of Systems To be used in event of leak or overflow from tank due to system failure—e.g., malfunction in treatment process, crack in the tank, etc.
265.193	C. Waste Analysis and Trial Tests (1) Definition
265.193(a)(2)(i)	(a) Trial Treatment or Storage Tests These are bench scale or pilot plant scale tests. (2) Conditions and Requirements for Owner/Operator to Conduct Waste Analyses and Trial Tests (a) Conditions
265.193(a)	Whenever a tank is to be used to:
265.193(a)(1)	(i) Chemically treat or store a hazardous waste which is substantially different from waste previously treated or stored in that tank; or
265.193(a)(2)	(ii) Chemically treat hazardous waste with a substantially different process than any previously used in that tank. (b) Required Analyses and Trial Tests before Treating or Storing Different Waste or Using Different Process
265.193(a)(2)(i) 265.193(a)(2)(ii)	(i) Conduct waste analyses and trial treatment or storage tests; or (ii) Obtain written, documented information on similar storage or treatment of similar waste under similar operating conditions.
265.193(a)(2)	(iii) Purpose To show that proposed treatment or storage will meet all applicable general operating requirements of Sections 265.192(a) and (b).

265.193(a)
265.193(a) Comment
	(c) Comply with General Waste Analyses Requirements of Section 265.13
	In particular:
	(i) Waste analysis plan must include analyses needed to comply with:
	    • Section 265.198—Special Requirements for Ignitable or Reactive Waste.
	    • Section 265.199—Special Requirements for Incompatible Wastes.
	(ii) As required by Section 265.73 (Operating Record) owner/operator must place results from each waste analysis and trial test, or documented information, in operating record of facility.

265.194 265.194(a) 265.194(a)(1)	D. Inspections Where present, owner/operator of a tank must inspect following: (1) Discharge Control Equipment (a) Includes waste feed cutoff systems, bypass systems, and drainage systems. (b) Frequency of Inspection and Purpose At least once each operating day to ensure good working order.

Table 6-1. Final (i.e., Part 264) and Interim (i.e., Part 265) Status Standards for Owners and Operators of Hazardous Waste Treatment, Storage, and Disposal Facilities. (Continued)

Section No.	Description
265.194(a)(2)	(2) Data Gathered from Monitoring Equipment (a) Includes pressure and temperature gauges. (b) Frequency of Inspection and Purpose At least once each operating day to ensure that tank is being operated according to design.
265.194(a)(3)	(3) Level of Waste In Tank (a) Frequency of Inspection and Purpose At least once each operating day to ensure compliance with Section 265.192(c) of general operating requirements.
265.194(a)(4)	(4) Construction Materials of Tank (a) Frequency of Inspection and Purpose At least weekly to detect corrosion or leaking of fixtures or seams.
265.194(a)(5)	(5) Construction Materials of, and Area Immediately Surrounding Discharge Confinement Structures (a) Includes dikes or trenches. (b) Frequency of Inspection and Purpose At least weekly to detect erosion or obvious signs of leakage (e.g., wet spots or dead vegetation).
265.194(a)(5) Comment	*Note:* As required by general inspection requirements of Section 265.15(c), owner/operator must remedy any deterioration or malfunction he finds.
265.197	E. Closure (1) Hazardous Waste/Residues from Tanks, Discharge Control Equipment, and Discharge Confinement Structures Must be removed at closure.
265.197 Comment	(2) Removal of Hazardous Waste from Tank At closure, as throughout operating period, owner/operator becomes generator of hazardous waste and must manage it in accordance with applicable requirements of Parts 262, 263, and 265. (a) Exemption Owner/operator can demonstrate that his waste is not hazardous, in accordance with Sections 261.3(c) or (d) (Definition Of Hazardous Waste).
265.198 265.198(a) 265.198(a)(1)	F. Special Requirements for Ignitable or Reactive Waste (1) Conditions by Which Ignitable or Reactive Waste May Be Placed in a Tank (a) Waste is treated, rendered, or mixed before or immediately after placement in tank so that:
265.198(a)(1)(i)	(i) Resulting waste, mixture, or dissolution of material no longer meets definition of ignitable or reactive waste under: • Section 261.21–Characteristics of Ignitability. • Section 261.23–Characteristics of Reactivity.
265.198(a)(1)(ii) and 265.198(b) Comment	(ii) Compliance with Section 265.17–General Requirements for Ignitable, Reactive, or Incompatible Wastes; or
265.198(a)(2)	(b) Waste is stored or treated in manner that protects it from any material or conditions which may cause waste to ignite or react; or
265.198(a)(3)	(c) Tank is used solely for emergencies.
265.198(b)	(2) Compliance with National Fire Protection Association's (NFPA's) Buffer Zone Requirements for Tanks Owner/operator of facility treating or storing ignitable or reactive waste in covered tanks must comply with NFPA's requirements contained in Tables 2-1 through 2-6 of "Flammable and Combustible Code–1977."
265.199 265.199(a)	G. Special Requirements for Incompatible Wastes (1) Placement of Incompatible Wastes or Incompatible Wastes and Materials in Same Tank Must not be done unless Section 265.17(b) (General Requirements For Ignitable, Reactive, or Incompatible Wastes) is complied with.
265.199(b)	(2) Placement of Hazardous Waste in Unwashed Tank Which Held an Incompatible Waste or Material Must not be done unless Section 265.17(b) is complied with.
265.199(a)	(3) See Appendix V for examples of potentially incompatible waste.

Table 6-1. Final (i.e., Part 264) and Interim (i.e., Part 265) Status Standards for Owners and Operators of Hazardous Waste Treatment, Storage, and Disposal Facilities. (Continued)

Section No.	Description
Subpart K	XX. Surface Impoundments
265.220	A. Applicability Regulations in this Subpart apply to owners/operators of facilities that use surface impoundments to treat, store, or dispose of hazardous waste, except as provided otherwise in Section 265.1.

265.222
 	
265.222 Comment
	B. General Operating Requirements
	(1) Freeboard
	   (a) General
	      Surface impoundment must maintain enough freeboard to prevent any overtopping of dike by overfilling, wave action, or a storm.
	   (b) Amount of Freeboard
	      At least 60 centimeters (2 feet).
	(2) Point Source Discharge from Surface Impoundment to Waters of United States
	   (a) Such discharges subject to requirements of Section 402 of Clean Water Act, as amended.
	   (b) Spills may be subject to Section 311 of that Act.

265.223	C. Containment System
	(1) Protective Cover
	    All earthen dikes must have a protective cover such as grass, shale, or rock.
	(2) Purpose
	    To minimize wind and water erosion and preserve structural integrity.

265.225
 	
265.225(a)
265.225(a)(1)
 	
265.225(a)(2)
 	
265.225(a)(2)(i)
265.225(a)(2)(ii)
 	
265.225(a)(2)
 	
265.225(a)
265.225(a) Comment
	D. Waste Analysis and Trial Tests
	(1) Conditions and Requirements for Owner/Operator to Conduct Waste Analyses and Trial Tests
	   (a) Conditions
	      Whenever a surface impoundment is to be used to:
	      (i) Chemically treat a hazardous waste which is substantially different from waste previously treated in that impoundment; or
	      (ii) Chemically treat hazardous waste with a substantially different process than any previously used in that impoundment.
	   (b) Required Analyses and Trial Tests before Testing Different Waste or Using Different Process
	      (i) Conduct waste analyses and trial treatment tests; or
	      (ii) Obtain written, documented information on similar treatment of similar waste under similar operating conditions.
	      (iii) Purpose
	        To show that this treatment will comply with Section 265.17(b) (General Requirements for Ignitable, Reactive, or Incompatible Wastes).
	   (c) Comply with General Waste Analyses Requirements of Section 265.13
	      In particular:
	      (i) Waste analysis plan must include analyses needed to comply with:
	        • Section 265.229–Special Requirements for Ignitable or Reactive Waste.
	        • Section 265.230–Special Requirements for Incompatible Wastes.
	      (ii) As required by Section 265.73 (Operating Record) owner/operator must place results from each waste analysis and trial test, or documented information, in operating record of facility.

265.226
265.226(a)
265.226(a)(1)
 	
265.226(a)(2)
 	
265.226(a)(2) Comment
	E. Inspections
	Owner/operator must inspect:
	(1) Freeboard Level
	   (a) Frequency
	      At least once each operating day.
	   (b) Purpose
	      To ensure compliance with Section 265.222 (General Operating Requirements).
	(2) Surface Impoundment Including Dikes and Vegetation Surrounding Dike
	   (a) Frequency
	      At least once a week.
	   (b) Purpose
	      To detect leaks, deterioration, or failures in impoundment.
	Note: As required by general inspection requirements of Section 265.15(c), owner/operator must remedy any deterioration or malfunction he finds.

Table 6-1. Final (i.e., Part 264) and Interim (i.e., Part 265) Status Standards for Owners and Operators of Hazardous Waste Treatment, Storage, and Disposal Facilities. (Continued)

Section No.	Description
265.228	F. Closure
265.228(a)	(1) Impoundment Materials That Owner/Operator May Elect to Remove at Closure
265.228(a)(1)	(a) Standing liquids.
265.228(a)(2)	(b) Waste and waste residues.
265.228(a)(3)	(c) Liner, if any.
265.228(a)(4)	(d) Underlying and surrounding contaminated soil.
265.228(b)	(2) Conditions for Impoundment Not Being Subject to Further Requirements of This Part (265)
	(a) All impoundment materials have been removed; or
	(b) Owner/operator can demonstrate under Section 261.3(c) and (d) (Definition of Hazardous Waste) that none of the impoundment materials at any stage of removal are hazardous wastes.
265.228(b) Comment	(3) Removal of Hazardous Waste from Surface Impoundment
	At closure, as throughout operating period, owner/operator becomes generator of hazardous waste and must manage it in accordance with all applicable requirements of Parts 262, 263, and 265.
	(a) Exemption
	Owner/operator can demonstrate that his waste is not hazardous, in accordance with Section 261.3(c) or (d) (Definition of Hazardous Waste).
	(4) Surface impoundment may be subject to Part 257 (Solid Waste Disposal Facilities) regulations even if not subject to regulations of this Part.
265.228(c)	(5) Support of Final Cover
	Owner/operator must treat remaining liquids, residues, and soils by removal of liquids, drying or other means, if support of the final cover specified in the closure plan makes it necessary.
265.228(c) Comment	(6) Variation in Closure Requirements
	These will vary with the amount and nature of the residue remaining, if any, and the degree of contamination of the underlying and surrounding soil.
265.228	G. Post-Closure
265.228(c)	(1) When Closed Impoundment Must Undergo Post-Closure Care as a Landfill (Under Subpart G and Section 265.310)
	(a) If owner/operator does not remove all impoundment materials in Section 265.228(a); or
	(b) If owner/operator does not make demonstration in Section 265.228(b) that waste is not hazardous.
265.228(c) Comment	(2) Regional Administrator May Vary Post-Closure Care Requirements
	Allowed under Section 265.117(d) (Post-Closure Care and Use of Property).
265.229	H. Special Requirements for Ignitable or Reactive Wastes
265.229(a)	(1) Conditions by Which Ignitable or Reactive Waste May Be Placed in a Surface Impoundment
265.229(a)(1)	(a) Waste is treated, rendered, or mixed before or immediately after placement in impoundment so that:
265.229(a)(1)(i)	(i) Resulting waste, mixture, or dissolution of material no longer meets definition of ignitable or reactive waste under:
	• Section 261.21—Characteristics of Ignitability.
	• Section 261.23—Characteristics of Reactivity.
265.229(a)(1)(ii)	(ii) Compliance with Section 265.17(b)—General Requirements for Ignitable, Reactive, or Incompatible Wastes; or
265.229(a)(2)	(b) Surface impoundment is used solely for emergencies.
265.230	I. Special Requirements for Incompatible Wastes
	(1) Placement of Incompatible Waste or Incompatible Wastes and Materials in Same Surface Impoundment
	Must not be done unless Section 265.17(b) (General Requirements for Ignitable, Reactive, or Incompatible Wastes) is complied with.
	(2) See Appendix V for examples of potentially incompatible waste.
Subpart L	XXI. Waste Piles
265.250	A. Applicability
	(1) Regulations in this Subpart apply to owners/operators of facilities that treat or store hazardous wastes in piles, except as provided otherwise in Section 265.1.
	(2) Pile of hazardous waste may be managed as a landfill under Subpart N.

Table 6-1. Final (i.e., Part 264) and Interim (i.e., Part 265) Status Standards for Owners and Operators of Hazardous Waste Treatment, Storage, and Disposal Facilities. (Continued)

Section No.	Description
265.251	B. Protection from Wind (1) Hazardous Waste Pile Subject to Dispersal by Wind Owner/operator must control wind dispersal of hazardous waste. (a) Cover pile; or (b) Use other means.

265.252
 	
265.252 Comment
 	
265.252
	C. Waste Analysis
	(1) Comply with General Waste Analyses Requirements of Section 265.13
	In particular:
	(a) Waste analysis plan must include analyses needed to comply with:
	   • Section 265.256–Special Requirements for Ignitable or Reactive Wastes.
	   • Section 265.257–Special Requirements for Incompatible Wastes.
	(b) As required by Section 265.73 (Operating Record), owner/operator must place results of this analysis in operating record of facility.
	(2) Analysis of Representative Sample from Each Incoming Movement
	Owner/operator must do such an analysis before adding waste to any existing pile.
	(a) Exceptions
	   (i) The only wastes the facility receives which are amenable to piling are compatible with each other; or
	   (ii) Waste received is compatible with waste in pile to which it is to be added.
	(3) Waste Analysis Requirement
	Analysis conducted must be capable of differentiating between types of hazardous waste owner/operator places in piles, so that mixing of incompatible waste does not inadvertently occur.
	(a) Analysis must include visual comparison of color and texture.

265.253
 	
265.253(a)
 	
265.253(b)
265.253(b)(1)
265.253(b)(2)
265.253(b)(2) Comment
 	
265.253(c)
	D. Containment
	(1) If Leachate or Run-Off from a Pile Is a Hazardous Waste
	(a) Approach No. 1
	   (i) Pile must be placed on impermeable base that is compatible with waste under conditions of treatment or storage.
	   (ii) Run-on must be diverted away from pile.
	   (iii) Any leachate and run-off from pile must be collected and managed as a hazardous waste; or
	(b) Approach No. 2
	   (i) Pile must be protected from precipitation and run-on by some other means; and
	   (ii) No liquids or wastes containing free liquids may be placed in the pile.
	(2) Where Collected Leachate or Run-Off Is Discharged through a Point Source to Waters of the United States
	It is subject to requirements of Section 402 of Clean Water Act, as amended.
	(3) Date of Compliance with Sections 265.253(a) and (b)(1)
	Twelve (12) months after effective date of this Part (11/19/81).

265.256
265.256(a)
265.256(a)(1)
265.256(a)(1)(i)
 	
265.256(a)(1)(ii)
 	
265.256(a)(2)
	E. Special Requirements for Ignitable or Reactive Waste
	(1) Conditions by Which Ignitable or Reactive Waste May Be Placed in a Pile
	(a) Addition of waste to existing pile:
	   (i) Results in waste or mixture no longer meeting definition of ignitable or reactive waste under:
	     • Section 261.21–Characteristic of Ignitability.
	     • Section 261.23–Characteristics of Reactivity.
	   (ii) Compliance with Section 265.17(b)–General Requirements for Ignitable, Reactive, or Incompatible Wastes; or
	(b) Waste is managed in a way that it is protected from any material or conditions which may cause it to ignite or react.

265.257 265.257(a)	F. Special Requirements For Incompatible Wastes (1) Placement of Incompatible Wastes or Incompatible Wastes and Materials in Same Pile Must not be done unless Section 265.17(b) (General Requirements for Ignitable, Reactive, or Incompatible Wastes) is complied with. (2) See Appendix V for examples of potentially incompatible waste.

Table 6-1. Final (i.e., Part 264) and Interim (i.e., Part 265) Status Standards for Owners and Operators of Hazardous Waste Treatment, Storage, and Disposal Facilities. (Continued)

Section No.	Description
265.257(b)	(3) Separation of Hazardous Waste Pile from Incompatible Wastes or other Materials Stored Nearby These other material, be they stored in containers, piles, open tanks, or surface impoundments, must be: (a) Separated from the other materials; or (b) Protected from them by means of a dike, berm, wall, or other device.
265.257(b) Comment	*Purpose:* To prevent fires, explosions, gaseous emissions, leaching, or other discharge of hazardous waste/constituents which could result from contact or mixing of incompatible wastes or materials.
265.257(c)(4)	(4) Piling Hazardous Waste on Same Area Where Incompatible Wastes or Materials Were Previously Piled Not to be done unless area has been decontaminated sufficiently to ensure compliance with Section 265.17(b) (General Requirements for Ignitable, Reactive, Or Incompatible Waste).
Subpart M	XXII. Land Treatment
265.272	A. General Operating Requirements
265.272(a)	(1) Placement of Hazardous Waste on Land Treatment Facility Prohibited unless waste can be made less hazardous or non-hazardous by biological degradation or chemical reactions occuring in or on soil.
265.272(b)	(2) Run-On Must be diverted away from active portions of land treatment facility.
265.272(c)	(3) Run-Off from Active Portions of Land Treatment Facility (a) Must be collected.
265.272(c) Comment	(b) If Collected Run-Off Is Hazardous Waste under Part 261 It must be managed as hazardous waste in accordance with applicable requirements of Parts 262, 263, and 265. (c) If Collected Run-Off Is Discharged through a Point Source to Waters of United States It is subject to requirements of Section 402 of Clean Water Act, as amended.
265.272(d)	(4) Timing Date of compliance for Sections 265.272(b) and (c) is 12 months after effective date of regulations (11/19/81).
265.273	B. Waste Analysis (1) Comply with General Waste Analysis Requirements under Section 265.13
265.273 Comment	In particular: (a) Waste analysis plan must include analyses needed to comply with: (i) Section 265.281—Special Requirements for Ignitable or Reactive Waste. (ii) Section 265.282—Special Requirements for Incompatible Wastes. (b) As required by Section 265.73 (Operating Record), owner/operator must place results from each waste analysis, or documented information, in operating record at facility.
265.273	(2) Waste Analysis Requirements Prior to Placement of Hazardous Waste in or on Land Treatment Facility Owner/operator must:
265.273(a)	(a) Determine Concentrations of Wastes Exhibiting EP Toxicity Characteristic Waste that exceeds maximum concentrations shown in Table I (Maximum Concentration of Contaminants for Characteristics of EP Toxicity) of Section 261.24.
265.273(b)	(b) Determine concentrations for any waste listed in Part 261, Subpart D (Lists of Hazardous Wastes).
265.273(c)	(c) Determine Concentrations of Arsenic, Cadmium, Lead and Mercury if Food Chain Crops are Grown (i) Exception If owner/operator has written, documented data that shows that constituents are not present.
265.276	C. Food Chain Crops
265.276(a)	(1) Notification of Regional Administrator (a) Required of owner/operator of hazardous waste land treatment facility on which food chain crops: (i) Have been grown. (ii) Are being grown. (iii) Will be grown in future.

Table 6-1. Final (i.e., Part 264) and Interim (i.e., Part 265) Status Standards for Owners and Operators of Hazardous Waste Treatment, Storage, and Disposal Facilities. (Continued)

Section No.	Description
	(b) Timing Within 60 days after effective date of these regulations (1/16/81).
265.276(a) Comment	(2) Growth of Food Chain Crops at Facility Which Had Never Before Been Used for That Purpose (a) Considered a significant change in process under Section 122.23(c)(3) (Changes During Interim Status for permitting regulations). (b) If After Effective Date Of Regulations (11/19/80) Owner/operator must comply with Section 122.23(c)(3).
265.276(b)(1)	(3) Food Chain Crops Grown on Treated Area of Hazardous Waste Land Treatment Facility (a) Arsenic, Lead, Mercury and Listed Hazardous Waste (Subpart D of Part 261) Considerations (i) Food chain crops prohibited unless field testing demonstrates that toxic metals and listed wastes:
265.276(b)(1)(i)	• Will not be transferred to food portion of crop by plant uptake or direct contact. • Will not otherwise be ingested by food chain animals (e.g., by grazing).
265.276(b)(1)(ii)	• Will not occur in greater concentrations in crops grown in land treatment facility than in same crops grown on untreated soils under similar conditions in same region.
265.276(b)(2)	(ii) Information Necessary For Demonstration Required by Section 265.276(b)(1) Must be kept at facility and must, at a minimum:
265.276(b)(2)(i)	• Be based on tests for specific waste and application rates being used at facility; and
265.276(b)(2)(ii)	• Include descriptions of: – Crop characteristics. – Soil characteristics. – Sample selection criteria. – Sample size determination. – Analytical methods. – Statistical procedures.
265.276(c)	(b) Cadmium Considerations (i) Food chain crops receiving cadmium wastes prohibited unless requirements of following Sections met: • 265.276(c)(1)(i)–(iii) • 265.276(c)(2)(i)–(iv).
265.276(c)(1)(i)	(ii) pH The pH of waste and soil mixture must be 6.5 or greater at time of each waste application, except for waste containing cadmium at concentrations of 2 milligrams per kilogram (dry weight) or less.
265.276(c)(1)(ii)	(iii) Annual Application of Cadmium from Waste • For Production Of Tobacco, Leafy Vegetables, Or Root Crops Grown For Human Consumption Not to exceed 0.5 kilograms per hectare (kg/ha). • For Other Food Chain Crops Annual cadmium application rate not to exceed:

Time period	Annual Cd application rate (kg/ha)
Present to June 30, 1984	2.0
July 1, 1984 to Dec 31, 1986	1.25
Beginning Jan 1, 1987	0.5

Section No.	Description
265.276(c)(1)(iii)	(iv) Cumulative Application of Cadmium from Waste Not to exceed levels in Section 265.276(c)(1)(iii)(A) or 265.276(c)(1)(iii)(B) that follow:
265.276(c)(1)(iii)(A)	•

Soil cation exchange capacity (meq/100 g)	Maximum cumulative application (kg/ha)	
	Background soil pH less than 6.5	Background soil pH greater than 6.5
Less than 5	5	5
5–15	5	10
Greater than 15	5	20

Table 6-1. Final (i.e., Part 264) and Interim (i.e., Part 265) Status Standards for Owners and Operators of Hazardous Waste Treatment, Storage, and Disposal Facilities. (Continued)

Section No.	Description
265.276(c)(1)(iii)(B)	• For Soils with Background *p*H Less Than 6.5 Cumulative cadmium application rate must not exceed levels below, provided that *p*H of waste and soil mixture is adjusted to and maintained at 6.5 or greater whenever food chain crops are grown.

Soil cation exchange capacity (meq/100 g)	Maximum cumulative application (kg/ha)
Less than 5	5
5–15	10
Greater than 15	20

Section No.	Description
265.276(c)(2)(i)	(v) Only food chain crop that can be produced is animal feed.
265.276(c)(2)(ii)	(vi) Other *p*H Factors • *p*H of waste and soil mixture must be 6.5 or greater at time of waste application or when crop is planted, whichever occurs later. • This *p*H level must be maintained whenever food chain crops are grown.
265.276(c)(2)(iii)	(vii) Facility Operating Plan Content • Demonstrates how animal feed will be distributed to preclude ingestion by humans. • Describes measures to be taken to safeguard against possible health hazards from cadmium entering food chain.
265.276(c)(2)(iv)	(viii) Future Property Owner Must be notified by stipulation in land record or property deed stating that: • Property has received waste at high cadmium application rates; and • Food chain crops should not be grown due to possible health hazard.
265.276(c)(2)(iv) Comment	(4) Operating Record As required by Section 265.73 (Operating Record), if owner/operator grows food chain crops on land treatment facility, information developed in this Section must be placed in operating record of facility.
265.278	D. Unsaturated Zone (Zone of Aeration) Monitoring
265.278(a)	(1) Unsaturated Zone Monitoring Plan (a) Owner/operator must have such a plan in writing.
265.278(a)(1)	(b) Purposes (i) To detect vertical migration of hazardous waste/constituents under active portion of land treatment facility.
265.278(a)(2)	(ii) To provide information on background concentrations of hazardous waste/constituents in similar but untreated soils nearby. • This background monitoring must be conducted before or in conjunction with monitoring required under Section 265.278(a)(1) above.
265.278(b)	(c) Minimum Content of Unsaturated Zone Monitoring Plan
265.278(b)(1)	(i) Soil monitoring using soil pores.
265.278(b)(2)	(ii) Soil-pore water monitoring using devices such as lysimeters.
265.278(c)	(iii) Demonstration of Compliance with Section 265.278(a)(1)
265.278(c)(1)	• That depth at which soil and soil-pore water samples are to be taken is below depth to which waste is incorporated into soil.
265.278(c)(2)	• That number of soil and soil-pore water samples to be taken is based on variability of:
265.278(c)(2)(i)	– Hazardous waste constituents (as identified in Sections 265.273(a) and (b)) in waste and in soil; and
265.278(c)(2)(ii)	– Soil type(s).
265.278(c)(3)	• That frequency and timing of soil and soil-pore sampling is based on frequency, time, and rate of waste application, proximity to ground and soil permeability.
265.278(d)	(iv) Rationale used in developing plan. (d) Owner/operator must keep plan at facility.
265.278(e)	(2) Analysis of Soil and Soil-Pore Water Samples Owner/operator must do this analysis for hazardous waste constituents found in waste during waste analysis under Section 265.273(a) and (b).
265.278(e) Comment	(3) Operating Record As required by Section 265.73 (Operating Record) all data and information developed by owner/operator under this Section must be placed in operating record of facility.

Table 6-1. Final (i.e., Part 264) and Interim (i.e., Part 265) Status Standards for Owners and Operators of Hazardous Waste Treatment, Storage, and Disposal Facilities. (Continued)

Section No.	Description
265.279	E. Recordkeeping Owner/operator must keep following records in operating record required in Section 265.73 (Operating Record): (1) Application dates. (2) Application rates. (3) Quantities. (4) Location of each hazardous waste placed in facility.
265.280	F. Closure and Post-Closure
265.280(a)	(1) Closure Plan (Section 265.112) and Post-Closure Plan (Section 265.118) Objectives. Owner/operator must address following objectives and indicate how they will be achieved:
265.280(a)(1)	(a) Control of migration of hazardous waste/constituents from treated area into ground water.
265.280(a)(2)	(b) Control of release of contaminated run-off from facility into surface water.
265.280(a)(3)	(c) Control of release of airborne particulate contaminants caused by wind erosion.
265.280(a)(4)	(d) Compliance with Section 265.276 (Food Chain Crops) concerning growth of food-chain crops.
265.280(b)	(2) Factors to Be Considered in Addressing Closure and Post-Closure Care Objectives of Section 265.280(a)
265.280(b)(1)	(a) Type and amount of hazardous waste/constituents applied to land treatment facility.
265.280(b)(2)	(b) Mobility and expected rate of migration of hazardous waste/constituents.
265.280(b)(3)	(c) Site location, topography, and surrounding land use, with respect to potential effects of pollution migration (e.g., proximity to ground water, surface water, and drinking water sources).
265.280(b)(4)	(d) Climate, including amount, frequency, and pH of precipitation.
265.280(b)(5)	(e) Geological and soil profiles and surface and subsurface hydrology of site, soil characteristics, including cation exchange capacity, total organic carbon, and pH.
265.280(b)(6)	(f) Unsaturated zone monitoring information obtained under Section 265.278.
265.280(b)(7)	(g) Type, concentration, and depth of migration of hazardous waste constituents in soil as compared to their background concentrations.
265.280(c)	(3) Methods to Be Considered in Addressing Closure and Post-Closure Objectives of Section 265.280(a)
265.280(c)(1)	(a) Removal of contaminated soils.
265.280(c)(2)	(b) Placement of Final Cover Consider:
265.280(c)(2)(i)	(i) Functions of Cover • Infiltration control. • Erosion control. • Run-off control. • Wind Erosion control.
265.280(c)(2)(ii)	(ii) Characteristics of Cover • Material. • Final surface contours. • Thickness. • Porosity. • Permeability. • Slope. • Length of run of slope. • Type of vegetation on cover.
265.280(c)(3)	(c) Collection and treatment of run-off.
265.280(c)(4)	(d) Diversion structures to prevent surface water run-on from entering treated area.
265.280(c)(5)	(e) Monitoring of soil, soil-pore water, and ground water.
265.280(d)	(4) Post-Closure Care Requirements in Addition to That Specified in Section 265.117
265.280(d)(1)	(a) Maintain any unsaturated zone monitoring system. (b) Collect and analyze samples from this system.
265.280(d)(2)	(c) Restrict access to facility as appropriate for its post-closure use.
265.280(d)(3)	(d) Ensure that growth of food chain crops complies with Section 265.276 (Food Chain Crops).
265.281	G. Special Requirements for Ignitable or Reactive Waste Ignitable or reactive wastes must not be land treated, unless waste is immediately incorporated into soil so that:

Table 6-1. Final (i.e., Part 264) and Interim (i.e., Part 265) Status Standards for Owners and Operators of Hazardous Waste Treatment, Storage, and Disposal Facilities. (Continued)

Section No.	Description
	(1) Resulting waste, mixture, or dissolution of material no longer meets definition of ignitable or reactive waste under:
	(a) Section 261.21–Characteristic of Ignitability.
	(b) Section 261.23–Characteristic of Reactivity.
	(2) Section 265.17(b) (General Requirements for Ignitable, Reactive, or Incompatible Wastes) is complied with.
265.282	H. Special Requirements for Incompatible Wastes
	(1) Incompatible wastes, or incompatible wastes and materials must not be placed in same land treatment area, unless Section 265.17(b) is complied with.
	(2) See Appendix V for examples of incompatible wastes.
Subpart N	XXIII. Landfills
265.300	A. Applicability
	(1) Regulations in this Subpart apply to owners/operators of facilities that dispose of hazardous waste in landfills, except as provided otherwise in Section 265.1.
	(2) A waste pile used as a disposal facility is a landfill and is goverened by this Subpart.
265.302	B. General Operating Requirements
265.302(a)	(1) Run-On
	Must be diverted away from active portions of landfill.
265.302(b)	(2) Run-Off from Active Portions of Landfill
	(a) Must be collected.
265.302(b) Comment	(b) If Collected Run-Off Is Hazardous Waste under Part 261
	It must be managed as hazardous waste in accordance with applicable requirements of Parts 262, 263, and 265.
	(c) If Collected Run-Off Is Discharged through a Point Source to Waters of United States
	It is subject to requirements of Section 402 of Clean Water Act, as amended.
265.302(c)	(3) Timing
	Date of compliance for Sections 265.302(a) and (b) is 12 months after effective date of regulations (11/19/81).
265.302(d)	(4) Protection from Wind
	(a) Landfill Subject to Dispersal by Wind
	Owner/operator must control wind dispersal of hazardous waste.
	(i) Cover landfill; or
	(ii) Use other means.
265.302 Comment	(5) Waste Analysis
	(a) Comply with General Waste Analysis Requirements of Section 265.13
	In particular:
	(i) Waste analysis plan must include analyses needed to comply with:
	• Section 265.312–Special Requirements for Ignitable or Reactive Waste.
	• Section 265.313–Special Requirements for Incompatible Wastes.
	(ii) As required by Section 265.73 (Operating Record), owner/operator must place results of these analyses in operating record of facility.
265.309	C. Surveying and Recordkeeping
	Owner/operator of landfill must maintain following items in operating record required in Section 265.73:
265.309(a)	(1) Survey
	On a map, exact location and dimensions, including depth, of each cell with respect to permanently surveyed benchmarks.
265.309(b)	(2) Records
	Content of each cell and approximate location of each hazardous waste type within each cell.
265.310	D. Closure and Post-Closure
265.310(a)	(1) Final Cover over Landfill
	(a) Such a cover is required of owner/operator upon closure of landfill.
	(b) Closure Plan of Section 265.112
	Must specify function and design of cover.
265.310(b)	(2) Closure and Post-Closure Plan Objectives
	Owner/operator must address following objectives and indicate how they will be achieved:
265.310(b)(1)	(a) Control of pollutant migration from facility via ground water, surface water, and air.

Table 6-1. Final (i.e., Part 264) and Interim (i.e., Part 265) Status Standards for Owners and Operators of Hazardous Waste Treatment, Storage, and Disposal Facilities. (Continued)

Section No.	Description
265.310(b)(2)	(b) Control of surface water infiltration, including prevention of pooling.
265.310(b)(3)	(c) Prevention of erosion.
265.310(c)	(3) Factors to Be Considered in Addressing Closure and Post-Closure Objectives of Section 265.310(b)
265.310(c)(1)	(a) Type and amount of hazardous waste/constituents in landfill.
265.310(c)(2)	(b) Mobility and expected rate of migration of hazardous waste/constituents.
265.310(c)(3)	(c) Site location, topography, and surrounding land use, with respect to potential effects of pollutant migration (e.g., proximity to ground water, surface water, and drinking water sources).
265.310(c)(4)	(d) Climate, including amount, frequency and pH of precipitation.
265.310(c)(5)	(e) Characteristics of Cover (i) Material. (ii) Final surface contours. (iii) Thickness. (iv) Porosity. (v) Permeability. (vi) Slope. (vii) Length of run of slope. (viii) Type of vegetation on cover.
265.310(c)(6)	(f) Geological and soil profiles and surface and subsurface hydrology of site.
265.310(d)	(4) Post-Closure Care Requirements (in Addition to Requirements Specified in Section 265.117)
265.310(d)(1)	(a) Maintain function and integrity of final cover as specified in approved closure plan.
265.310(d)(2)	(b) Maintain and monitor leachate collection, removal, and treatment system (if there is one present in the landfill) to prevent excess accumulation of leachate in system.
265.310(d)(2) Comment	(i) If Collected Leachate Is Hazardous Waste under Part 261 It must be managed as hazardous waste in accordance with applicable requirements of Parts 262, 263, and 265. (ii) If Collected Leachate Is Discharged through a Point Source to Waters of United States It is subject to requirements of Section 402 of Clean Water Act, as amended.
265.310(d)(3)	(c) Maintain and monitor gas collection and control system (if one is present in landfill) to control vertical and horizontal escape of gases.
265.310(d)(4)	(d) Protect and maintain surveyed benchmarks.
265.310(d)(5)	(e) Restrict access to landfill as appropriate for its post-closure use.
265.310(a)	*Note:* All of above post-closure care requirements must be included in post-closure plan under Section 265.118.
265.312	E. Special Requirements for Ignitable or Reactive Waste Ignitable or reactive wastes must not be placed in a landfill, unless waste is treated, rendered, or mixed before or immediately after placement in landfill so that: (1) Resulting waste, mixture, or dissolution of material no longer meets definition of ignitable or reactive waste under: (a) Section 261.21–Characteristic of Ignitability. (b) Section 261.23–Characteristic of Reactivity. (2) Section 265.17(b) (General Requirements for Ignitable, Reactive, or Incompatible Wastes) is complied with.
265.313	F. Special Requirements for Incompatible Wastes (1) Incompatible wastes, or incompatible wastes and materials must not be placed in same landfill cell, unless Section 265.17(b) is complied with. (2) See Appendix V for examples of incompatible wastes.
265.314	G. Special Requirements for Liquid Waste
265.314(a)	(1) Bulk or Non-Containerized Liquid Waste or Waste Containing Free Liquids Must not be placed in a landfill unless:
265.314(a)(1)	(a) Landfill has a liner which is chemically and physically resistant to added liquid. (b) Landfill has a functioning leachate collection and removal system with sufficient capacity to remove all leachate produced.
265.314(a)(2)	(c) Before disposal, liquid waste or waste containing free liquids is treated or stabalized, chemically or physically (e.g., by mixing with an absorbent solid), so that free liquids are no longer present.

Table 6-1. Final (i.e., Part 264) and Interim (i.e., Part 265) Status Standards for Owners and Operators of Hazardous Waste Treatment, Storage, and Disposal Facilities. (Continued)

Section No.	Description
265.314(b)	(2) Container Holding Liquid Waste or Waste Containing Free Liquids Must not be placed in a landfill unless:
265.314(b)(1)	(a) Container is designed to hold liquids or free liquids for a use other than storage, such as a battery or capacitor; or
265.314(b)(2)	(b) Container is very small, such as an ampule.
265.314(c)	(3) Timing Date of compliance with this Section is 12 months after effective date of regulations (11/19/81).
265.315	H. Special Requirements for Containers
265.316(a)	(1) Empty Containers Such containers must be crushed flat, shredded, or similarly reduced in volume before being buried beneath surface of a landfill.
265.316(b)	(2) Date of compliance with this Section is 11/19/81.
Subpart O	XXIV. Incinerators
265.340	A. Applicability Regulations in this Subpart apply to owners/operators of facilities that treat hazardous waste in incinerators, except as provided otherwise in Section 265.1.
265.343	B. General Operating Requirements (1) Steady State (Normal) Conditions of Operation Before adding hazardous wastes to incinerator, owner/operator must bring it to steady state conditions—including steady state operating temperature and air flow—by using auxiliary fuel or other means.
265.345	C. Waste Analysis (1) Comply with general waste analysis requirements under Section 265.13. (2) Analyze Wastes Not Previously Burned in Incinerator Required of owner/operator to enable: (a) Establishment of steady state (normal) operating conditions (including waste and auxiliary fuel feed and air flow). (b) Determination of type pollutants which might be emitted. (3) Minimum Determinations Required of Analysis
265.345(a)	(a) Heating value of waste.
265.345(b)	(b) Halogen content in waste. (c) Sulfur content in waste.
265.345(c)	(d) Lead and mercury concentrations in waste. (i) Exception Owner/operator has written documented data showing that metals are not present.
265.345(c) Comment	(4) Operating Record As required by Section 265.73 (Operating Record), owner/operator must place results from each waste analysis, or documented information, in operating record of facility.
265.347 265.347(a) 265.347(a)(1)	D. Monitoring and Inspections When Incinerating Hazardous Wastes Owner/operator must conduct, as a minimum, following monitoring and inspections: (1) Existing Instruments Relating to Combustion and Emission Control Must be monitored at least every 15 minutes. (a) Instruments Relating to Combustion and Emission Control Normally includes instruments measuring: (i) Waste feed. (ii) Auxiliary fuel feed. (iii) Air flow. (iv) Incinerator temperatures. (v) Scrubber flow. (vi) Scrubber pH. (vii) Relevant level controls. (2) Appropriate Corrections to Maintain Steady State Combustion Conditions Must be made immediately, either automatically or by operator.
265.347(a)(2)	(3) Stack-Plume (Emissions) (a) Must be observed visually at least hourly for normal appearance (color and opacity). (b) Operator must immediately make any indicated operating corrections necessary to return visible emissions to their normal appearance.

Table 6-1. Final (i.e., Part 264) and Interim (i.e., Part 265) Status Standards for Owners and Operators of Hazardous Waste Treatment, Storage, and Disposal Facilities. (Continued)

Section No.	Description
265.347(a)(3)	(4) Inspection of Complete Incinerator and Associated Equipment (Pumps, Valves, Conveyors, Pipes, etc.) Must be inspected at least daily for leaks, spills and fugative emissions. (5) Emergency Shutdown Controls and System Alarms Must be checked daily to ensure proper operation.
265.351	E. Closure (1) At closure, owner/operator must remove all hazardous waste and hazardous waste residues (including but not limited to ash, scrubber waters, and scrubber sludges) from incinerator.
265.351 Comment	(2) Removal of Solid Waste from Incinerator At closure, as throughout operating period, owner/operator becomes a generator of hazardous waste and must manage it in accordance with applicable requirements of Parts 262, 263, and 265. (a) Exception Owner/operator can demonstrate that solid waste is not hazardous, in accordance with Section 261.3(c) or (d) (Definition of Hazardous Waste).
Subpart P	XXV. Thermal Treatment
265.370	A. Applicability (1) Regulations in this Subpart apply to owners/operators of facilities that thermally treat hazardous waste in devices other than incinerators, except as otherwise provided in Section 265.1. (2) Thermal treatment in incinerators is subject to requirements of Subpart O.
265.373	B. General Operating Requirements (1) Steady State (Normal) Conditions of Operation Before adding hazardous waste, the owner/operator must bring the thermal treatment process to steady state conditions—including steady state operating temperature—by using auxiliary fuel or other means. (a) Exception If process is a non-continuous (batch) thermal treatment process requiring a complete thermal cycle to treat a discrete quantity of hazardous waste.

265.375
 	
265.375(a)
265.375(b)
 	
265.375(c)
	C. Waste Analysis
	(1) Comply with general waste analysis requirements under Section 265.13.
	(2) Analyze Wastes Not Previously Treated in Thermal Process
	Required of owner/operator to enable:
	(a) Establishment of a steady state (normal) or other appropriate (for a non-continuous process) operating conditions (including waste and auxiliary fuel feed).
	(b) Determination of type pollutant which might be emitted.
	(3) Minimum Determinations Required of Analysis
	(a) Heating value of waste.
	(b) Halogen content in waste.
	(c) Sulfur content in waste.
	(d) Lead and mercury concentrations in waste.
	  (i) Exception
	    Owner/operator has written documented data showing that metals are not present.

265.375 Comment	(4) Operating Record As required by Section 265.73 (Operating Record), owner/operator must place results from each waste analysis, or documented information, in operating record of facility.

265.377
265.377(a)
265.377(a)(1)
	D. Monitoring and Inspections When Thermally Treating Hazardous Wastes
	Owner/operator must conduct, as a minimum, following monitoring and inspections:
	(1) Existing Instruments Relating to Temperature and Emission Control (If Present)
	Must be monitored at least every 15 minutes.
	(a) Instruments Relating to Combustion and Emission Control
	Normally includes instruments measuring:
	  (i) Waste feed.
	  (ii) Auxiliary fuel feed.
	  (iii) Treatment process temperature.
	  (iv) Relevant process flow.
	  (v) Level controls.

Table 6-1. Final (i.e., Part 264) and Interim (i.e., Part 265) Status Standards for Owners and Operators of Hazardous Waste Treatment, Storage, and Disposal Facilities. (Continued)

Section No.	Description
	(2) Appropriate Corrections to Maintain Steady State or Other Appropriate Thermal Treatment Conditions Must be made immediately, either automatically or by operator.
265.377(a)(2)	(3) Stack Plume (Emissions) Where present: (a) Must be observed visually at least hourly for normal appearance (color and opacity). (b) Operator must immediately make any indicated operating corrections necessary to return visible emissions to normal appearance.
265.377(a)(3)	(4) Inspection of Complete Thermal Treatment Process and Associated Equipment (Pumps, Valves, Conveyers, Pipes, Etc.) Must be inspected at least daily for leaks, spills and fugitive emissions. (5) Emergency Shutdown Controls and System Alarms Must be checked daily to assure proper operation.
265.381	E. Closure (1) At closure, owner/operator must remove all hazardous waste and hazardous waste residues (including but not limited to ash) from thermal treatment process or equipment.
265.381 Comment	(2) Removal of Solid Waste from Thermal Treatment Process or Equipment At closure, as throughout operating period, owner/operator becomes a generator of hazardous waste and must manage it in accordance with applicable requirements of Parts 262, 263, and 265. (a) Exception Owner/operator can demonstrate that solid waste is not hazardous, in accordance with Section 261.3(c) or (d) (Definition of Hazardous Waste).
265.382	F. Open Burning; Waste Explosives (1) Open Burning of Hazardous Waste Prohibited. (a) Exception Open burning and detonation of waste explosives. (2) Definitions (a) Detonation An explosion in which chemical transformation passes through the material faster than the speed of sound (0.33 kilometers per second at sea level). (b) Waste Explosives (i) Waste which has potential to detonate. (ii) Bulk military propellants which cannot safely be disposed of through other modes of treatment. (3) Open Burning or Detonation of Waste Explosives Owners/operators choosing to dispose of explosives by open burning or detonation must do so: (a) In a manner not threatening human health or environment, and (b) In accordance with following table:

Pounds of waste explosives or propellants	Minimum distance from open burning or detonation to the property of others
0–100	204 meters (670 feet)
101–1,000	380 meters (1,250 feet)
1,001–10,000	530 meters (1,730 feet)
10,001–30,000	690 meters (2,260 feet)

Section No.	Description
Subpart Q	XXVI. Chemical, Physical, and Biological Treatment
265.400	A. Applicability (1) Regulations in this Subpart apply to owners/operators of facilities which treat hazardous wastes by chemical, physical, or biological methods in other than tanks, surface impoundments and land treatment facilities, except as provided otherwise in Section 265.1. (2) Chemical, physical, and biological treatment of hazardous waste in tanks, surface impoundments, and land treatment facilities must be conducted in accordance with Subparts J, K, and M, respectively.

Table 6-1. Final (i.e., Part 264) and Interim (i.e., Part 265) Status Standards for Owners and Operators of Hazardous Waste Treatment, Storage, and Disposal Facilities. (Continued)

Section No.	Description
265.401	B. General Operating Requirements
265.401(a)	(1) Treatment of Hazardous Wastes by Chemical, Physical, or Biological Means Must be in compliance with Section 265.17(b) (General Requirements for Ignitable, Reactive, or Incompatible Wastes).
265.401(b)	(2) Placement of Hazardous Waste or Treatment Reagents in Treatment Process or Equipment Must not be done if they could cause treatment process or equipment to rupture, leak, corrode or otherwise fail before the end of its intended life.
265.401(c)	(3) Where Hazardous Waste Is Continuously Fed into a Treatment Process or Equipment Process or equipment must be equipped with means to stop inflow—e.g., waste feed cut-off system or by-pass system to standby containment system.
265.401(c) Comment	(a) Purpose of Systems To be used in event of malfunction in treatment process or equipment.
265.402	C. Waste Analysis and Trial Tests (1) Definition
265.402(a)(?)(i)	(a) Trial Treatment Tests These are bench scale or pilot plant scale tests. (2) Conditions and Requirements for Owner/Operator to Conduct Waste Analyses and Trial Tests (a) Conditions
265.402(a)	Whenever a treatment process or equipment is used to:
265.402(a)(1)	(i) Treat a hazardous waste which is substantially different from waste previously treated in that process or equipment; or
265.402(a)(2)	(ii) Chemically treat hazardous waste with a substantially different process than any previously used at the facility. (b) Required Analyses and Trial Tests before Testing Different Waste or Using Different Process
265.402(a)(2)(i)	(i) Conduct waste analyses and trial treatment tests; or
265.402(a)(2)(ii)	(ii) Obtain written, documented information on similar treatment of similar waste under similar operating conditions.
265.402(a)(2)	(iii) Purpose To show that this treatment will meet all applicable general operating requirements of Sections 265.401(a) and (b).
265.402(a)	(c) Comply with General Waste Analysis Requirements of Section 265.13
265.402(a) Comment	In particular: (i) Waste analysis plan must include analyses needed to comply with: • Section 265.405—Special Requirements for Ignitable or Reactive Waste. • Section 265.406—Special Requirements for Incompatible Wastes. (ii) As required by Section 265.73 (Operating Record) owner/operator must place results from each waste analysis and trial test, or documented information, in operating record of facility.
265.403	D. Inspections
265.403(a)	Where present, owner/operator of treatment facility must inspect following:
265.403(a)(1)	(1) Discharge Control and Safety Equipment (a) Includes waste feed cut-off systems, by-pass systems, drainage systems, and pressure relief systems. (b) Frequency of Inspection and Purpose At least once each operating day to ensure that it is in good working order.
265.403(a)(2)	(2) Data Gathered from Monitoring Equipment (a) Includes pressure and temperature gauges. (b) Frequency of Inspection and Purpose At least once each operating day to ensure that treatment process or equipment is being operated according to design.
265.403(a)(3)	(3) Construction Materials of Treatment Process or Equipment (a) Frequency of Inspection and Purpose At least weekly to detect corrosion or leaking of fixtures or seams.
265.403(a)(4)	(4) Construction Materials of, and Area Immediately Surrounding, Discharge Confinement Structures (a) Includes dikes. (b) Frequency of Inspection and Purpose At least weekly, to detect erosion or obvious signs of leakage (e.g., wet spots or dead vegetation).

Table 6-1. Final (i.e., Part 264) and Interim (i.e., Part 265) Status Standards for Owners and Operators of Hazardous Waste Treatment, Storage, and Disposal Facilities. (Continued)

Section No.	Description
265.403(a)(4) Comment	*Note:* As required by general inspection requirements of Section 265.15(c), owner/operator must remedy any deterioration or malfunction he finds.
265.404	E. Closure (1) Hazardous Waste/Residues from Treatment Processes or Equipment, Discharge Control Equipment, and Discharge Confinement Structures Must be removed at closure.
265.404 Comment	(2) Removal of Solid Waste from Treatment Process or Equipment At closure, as throughout operating period, owner/operator becomes generator of hazardous waste and must manage it in accordance with applicable requirements of Parts 262, 263, and 265. (a) Exception Owner/operator can demonstrate that solid waste is not hazardous, in accordance with Sections 261.3(c) or (d) (Definition of Hazardous Waste).
265.405 265.405(a)	F. Special Requirements for Ignitable or Reactive Waste (1) Conditions by Which Ignitable or Reactive Waste May Be Placed in a Treatment Process or Equipment
265.405(a)(1)	(a) Waste is treated, rendered, or mixed before or immediately after placement in treatment process or equipment so that:
265.405(a)(1)(i)	(i) Resulting waste, mixture, or dissolution of material no longer meets definition of ignitable or reactive waste under: ● Section 261.21–Characteristic of Ignitability. ● Section 261.23–Characteristic of Reactivity.
265.405(a)(1)(ii)	(ii) Compliance with Section 265.17(b)–General Requirements for Ignitable, Reactive, or Incompatible Wastes; or
265.405(a)(2)	(b) Waste is treated in manner that protects it from any material or conditions which may cause waste to ignite or react.
265.406 265.406(a)	G. Special Requirements for Incompatible Wastes (1) Placement of Incompatible Wastes or Incompatible Wastes and Materials in Same Treatment Process or Equipment Must not be done unless Section 265.17(b) (General Requirements for Ignitable, Reactive, Or Incompatible Wastes) is complied with.
265.406(b)	(2) Placement of Hazardous Waste in Unwashed Treatment Equipment Which Held an Incompatible Waste or Material Must not be done unless Section 265.17(b) is complied with.
265.406(a)	(3) See Appendix V for examples of potentially incompatible wastes.
Subpart R	XXVII. Underground Injection
265.430	A. Applicability Except as Section 265.1 provides otherwise, owners/operators of wells used to dispose of hazardous wastes:
265.430(b)	(1) Must meet requirements of this Subpart for following well classifications: (a) Class I wells under Section 122.32(a). (b) Class IV wells under Section 122.32(d).
265.430(b) Comment	(2) Must meet general requirements of: (a) Subpart A–General. (b) Subpart B–General Facility Standards. (c) Subpart C–Preparedness and Prevention. (d) Subpart D–Contingency Plan and Emergency Procedures. (e) Subpart E–Manifest System, Recordkeeping, and Reporting.
265.430(a)	(3) Are exempt from requirements of: (a) Subpart G–Closure and Post-Closure. (b) Subpart H–Financial Requirements.

Please print or type with ELITE type (12 characters per inch).

GSA No. 12345-XX
Form Approved OMB No. 158-R00XX

&EPA   **U.S. ENVIRONMENTAL PROTECTION AGENCY**
HAZARDOUS WASTE REPORT

I. TYPE OF HAZARDOUS WASTE REPORT

PART A: GENERATOR ANNUAL REPORT

THIS REPORT IS FOR THE YEAR ENDING DEC. 31.   1 9

PART B: FACILITY ANNUAL REPORT

THIS REPORT FOR YEAR ENDING DEC. 31,   1 9

PART C: UNMANIFESTED WASTE REPORT

THIS REPORT IS FOR A WASTE RECEIVED (day, mo., & yr.)   — — 1 9

PLEASE PLACE LABEL IN THIS SPACE

INSTRUCTIONS: You may have received a preprinted label attached to the front of this pamphlet; affix it in the designated space above—left. If any of the information on the label is incorrect, draw a line through it and supply the correct information in the appropriate section below. If the label is complete and correct, leave Sections II, III, and IV below blank. If you did not receive a preprinted label, complete all sections. "Installation" means a single site where hazardous waste is generated, treated, stored, or disposed of. Please refer to the specific instructions for generators or facilities before completing this form. The information requested herein is required by law *(Section 3002/3004 of the Resource Conservation and Recovery Act)*.

II. INSTALLATION'S EPA I.D. NUMBER

F

III. NAME OF INSTALLATION

IV. INSTALLATION MAILING ADDRESS

STREET OR P.O. BOX

3

CITY OR TOWN   ST.   ZIP CODE

4

V. LOCATION OF INSTALLATION

STREET OR ROUTE NUMBER

5

CITY OR TOWN   ST.   ZIP CODE

6

VI. INSTALLATION CONTACT

NAME (last and first)   PHONE NO. (area code & no.)

2

VII. TRANSPORTATION SERVICES USED (for Part A reports only)

List the EPA Identification Numbers for those transporters whose services were used during the reporting year represented by this report.

VIII. COST ESTIMATES FOR FACILITIES (for Part B reports only)

A. COST ESTIMATE FOR FACILITY CLOSURE

$

B. COST ESTIMATE FOR POST CLOSURE MONITORING AND MAINTENANCE (disposal facilities only)

$

IX. CERTIFICATION

I certify under penalty of law that I have personally examined and am familiar with the information submitted in this and all attached documents, and that based on my inquiry of those individuals immediately responsible for obtaining the information, I believe that the submitted information is true, accurate, and complete. I am aware that there are significant penalties for submitting false information, including the possibility of fine and imprisonment.

A. PRINT OR TYPE NAME   B. SIGNATURE   C. DATE SIGNED

EPA Form 8700-13 (4-80)

PAGE ___1___ OF ___

Fig. 6-1. EPA Form 8700-13: Facility Annual Report for On-Site or Off-Site Treatment, Storage, or Disposal of Hazardous Waste; and Unmanifested Waste Report.

Please print or type with ELITE type *(12 characters/inch).*

GSA No. 12345-XX
Form Approved OMB No. 158-R00XX

EPA

U.S. ENVIRONMENTAL PROTECTION AGENCY
FACILITY REPORT — PARTS B & C
(Collected under the authority of Section 3004 of RCRA.)

FOR OFFICIAL USE ONLY *(Items 1 & 2)*	1. DATE RECEIVED · · 19	XVI. TYPE OF REPORT *(enter an "X")*	XVII. FACILITY'S E.I.D. NO.
	2. RECEIVED BY	☐ PART B ☐ PART C	G

XVIII. GENERATOR'S EPA I.D. NO.

XX. GENERATOR ADDRESS *(street or P.O. box, city, state, & zip code)*

XIX. GENERATOR NAME *(specify)*

XXI. WASTE IDENTIFICATION

LINE NUMBER	A. DESCRIPTION OF WASTE	B. EPA HAZARDOUS WASTE NUMBER *(see instructions)*	C. HANDLING METHOD *(enter code)*	D. AMOUNT OF WASTE	UNIT OF MEASURE *(enter code)*
1					
2					
3					
4					
5					
6					
7					
8					
9					
10					
11					
12					

XXII. COMMENTS *(enter information by line number — see instructions)*

EPA Form 8700-13B (5-80)
BILLING CODE 6560-01-C

PAGE ____ OF ____

Fig. 6-1. EPA Form 8700-13B: Facility Annual Report for On-Site or Off-Site Treatment, Storage, or Disposal of Hazardous Waste; and Unmanifested Waste Report. (Continued)

General Instructions Hazardous Waste Report (EPA Form 8700-13)

Important: Read all instructions before completing this form.

Section I.—Type of Hazardous Waste Report

Part A—Generator Annual Report

For generators who ship their waste off-site to facilities which they do not own or operate; fill in the reporting year for this report (e.g., 1982).

Note.—Generators who ship hazardous waste off-site to a facility which they own or operate must complete the facility (Part B) report instead of the Part A report.

Part B—Facility Annual Report

For owners or operators of on-site or off-site facilities that treat, store, or dispose of hazardous waste; fill in the reporting year for this report (e.g., 1982).

Part C—Unmanifested Waste Report

For facility owners or operators who accept for treatment, storage, or disposal any hazardous waste from an off-site source without an accompanying manifest; fill in the date the waste was received at the facility (e.g. 04–12–1982).

Section II. thru Section IV.—Installation I.D. Number, Name of Installation, and Installation Mailing Address

If you received a preprinted lable from EPA, attach it in the space provided and leave Sections II through IV blank. If there is an error or omission on the label, cross out the incorrect information and fill in the appropriate item(s). If you did not receive a preprinted label, complete Section II through Section IV.

Section V.—Location of Installation

If your installation location address is different than the mailing address, enter the location address of your installation.

Section VI.—Installation Contact

Enter the name (last and first) and telephone number of the person whom may be contacted regarding information contained in this report.

Section VII.—Transportation Services Used (For Part A reports ONLY)

List the EPA Identification Number for each transporter whose services you used during the reporting year.

Section VIII.—Cost Estimates for Facilities (For Part B reports ONLY)

A. Enter the most recent cost estimate for facility closure in dollars. See subpart H of 40 CFR part 264 or 265 for more detail.

B. For disposal facilities only, enter the most recent cost estimate for post closure monitoring and maintenance. See Subpart H of 40 CFR Part 264 or 265 for more detail.

Section IX.—Certification

The generator or his authorized representative (Part A reports) or the owner or operator of the facility or his authorized representative (Parts B and C reports) must sign and date the certification where indicated. The printed or typed name of the person signing the report must also be included where indicated.

Note.—Since more than one page is required for each report, enter the page number of each sheet in the lower right corner as well as the total number of pages.

Facility Annual Report Part B Instructions (EPA Form 8700–13B)

Facility Annual Report for owners or operators of on-site or off-site facilities that treat, store, or dispose of hazardous waste.

Note.—Generators who ship hazardous waste off-site to a facility they own or operate must complete this Part B report instead of the Generator (Part A) Annual Report.

Important: Read All Instructions Before Completing This Form

Section XVI.—Type of Report

Put an "X" in the box marked Part B.

Section XVII.—Facility's EPA Identification Number

Enter the EPA identification number for your facility.

Example:

Section XVIII.—Generator's EPA Identification Number

Enter the EPA identification number of the generator of the waste described under Section XXI which was received by your facility during the reporting year. A separate sheet must be used for each generator. If the waste came from a foreign generator, enter the EPA identification number of the importer in this section and enter the name and address of the foreign generator in Section XXII, Comments. If the waste was generated and treated, stored, or disposed of at the same installation, leave this section blank.

Section XIX.—Generator's Name

Enter the name of the generator corresponding to the generator's EPA identification number in Section XVIII.

If the waste was generated and treated, stored, or disposed of at the same installation, enter "ON–SITE".

If the waste came from a foreign generator, enter the name of the importer corresponding to the EPA identification number in Section XVIII.

Section XX.—Generator's Address

Enter the address of the generator corresponding to the generator's EPA identification number in Section XVIII. If the waste was generated and treated, stored, or disposed of at the same installation, leave this section blank. If the waste came from a foreign generator, enter the address of the importer corresponding to the EPA identification number in Section XVIII.

Section XXI.—Waste Identification

All information in this section must be entered by line number. A separate line entry is required for each different waste or mixture of wastes that your facility received during the reporting year. The handling code applicable to that waste at the end of the reporting year should be reported. If a different handling code applies to portions of the same waste, (e.g., part of the waste is stored while the remainder was "chemically fixed" during the year), use a separate line entry for each portion.

Example:

XXI. WASTE IDENTIFICATION					
LINE NUMBER	A. DESCRIPTION OF WASTE	B. EPA HAZARDOUS WASTE NUMBER (see instructions)	C. HAND-LING METHOD (enter code)	D. AMOUNT OF WASTE	UNIT OF MEASURE (enter code)
1	Steel Finishing Sludge	K060 K061	S02	25000	T
2	Steel Finishing Sludge	K060 K061	T21	157245	T

Section XXI-A.—Description of Waste

For hazardous wastes that are listed under 40 CFR Part 261, Subpart D, enter the EPA listed name, abbreviated if necessary. Where mixtures of listed wastes were received, enter the description which you believe best describes the waste.

For unlisted hazardous waste identified under 40 CFR Part 261, Subpart C, enter the description which you believe best describes the waste. Include the specific manufacturing or other process generating the waste (e.g., green sludge from widget manufacturing) and if known, the chemical or generic chemical name of the waste.

Section XXI-B.—EPA Hazardous Waste Number

For listed waste, enter the four digit EPA Hazardous Waste Number from 40 CFR Part 261, Subpart D, which identifies the waste.

For a mixture of more than one listed waste, enter each of the applicable EPA Hazardous Waste Numbers.

Four spaces are provided. If more space is needed, continue on the next line(s) and leave all other information on that line blank.

Fig. 6-1. EPA Form 8700-13: Facility Annual Report for On-Site or Off-Site Treatment, Storage, or Disposal of Hazardous Waste; and Unmanifested Waste Report. (Continued)

Example:

XXI WASTE IDENTIFICATION					
LINE NUMBER	A DESCRIPTION OF WASTE	B. EPA HAZARDOUS WASTE NUMBER (see instructions)	C. HANDLING METHOD (enter code)	D. AMOUNT OF WASTE	UNIT OF MEASURE (enter code)
1	Steel Finishing Sludge	K060 K061 / K062 K063	T21	2917455	T
2		K064			

For unlisted hazardous wastes, enter the EPA Hazardous Waste Numbers from 40 CFR Part 261, Subpart C, applicable to the waste. If more than four spaces are required, follow the procedure described above.

Section XXI–C—Handling Code

Enter one EPA handling code for each waste line entry. Where several handling steps have occurred during the year, report only the handling code representing the waste's status at the end of the reporting year or its final disposition. EPA handling codes are given in Appendix I of this Part.

Section XXI–D—Amount of Waste

Enter the total amount of waste described on this line which you received during this reporting year.

Section XXI–E—Unit of Measure

Enter the unit of measure code for the quantity of waste described on this line. Units of measure which must be used in this report and the appropriate codes are:

Units of measure	Code
Pounds	P
Short tons (2,000 lbs)	T
Kilograms	K
Tonnes (1,000 kg)	M

Units of volume may not be used for reporting but must be converted into one of the above units of weight, taking into account the appropriate density or specify gravity of the waste.

Section XXII—Comments

This space may be used to explain or clarify any entry. If used, enter a cross-reference to the appropriate Section number.

Note.—Since more than one page is required for each report, enter the page number of each sheet in the lower right hand corner as well as the total number of pages.

Where required by 40 CFR 264 or 265, Subparts F or R, attach ground-water monitoring data to this report.

Unmanifested Waste Report—Part C Instructions (EPA Form 8700–13B)

Unmanifested Waste Report for facility owners or operators who accept for treatment, storage, or disposal any hazardous waste from an off site source without an accompanying manifest.

Important: Read all instructions before completing this form.

For the Unmanifested Waste Report, EPA Forms 8700–13 and 8700–13B must be filled out according to the directions for the Part B Facility Annual Report except that: (1) blocks for which information is not available to the owner or operator of the reporting facility

may be marked "UNKNOWN," and (2) the following special instructions apply:

Section VIII—Cost Estimates for Facilities

Do not enter closure or post-closure cost estimates.

Section XVI—Type of Report

Put an "X" in the box marked Part C.

Section XXI–A—Description of Waste

Use as many line numbers as are needed to describe the waste.

Section XXI–C—Handling Code

Enter the handling code which describes the status of the waste on the date the report is filed.

Section XXI–D—Amount of Waste

Enter the amount of waste received, rather than a total annual aggregate.

Section XXII—Comments

a. Enter the EPA Identification number, name, and address of the transporter, if known. If the transporter is not known to you, enter the name and chauffeur license number of the driver and the State and license number of the transporting vehicle which presented the waste to your facility, if known.

b. Enter an explanation of how the waste movement was presented to your facility; why you believe the waste is hazardous; and how your facility plans to manage the waste. Continue on a separate blank sheet of paper if additional space is needed.

Monitoring data

Do not attach monitoring data.

Appendix I.— Recordkeeping Instructions

The recordkeeping provisions of § 264.73 specify that an owner or operator must keep a written operating record at his facility. This appendix provides additional instructions for keeping *portions* of the operating record. See § 264.73(b) for additional recordkeeping requirements.

The following information must be recorded, as it becomes available, and maintained in the operating record until closure of the facility in the following manner:

Records of each hazardous waste received, treated, stored, or disposed of at the facility which include the following:

(1) A description by its common name and the EPA Hazardous Waste Number(s) from Part 261 of this Chapter which apply to the waste. The waste description also must include the waste's physical form, i.e., liquid,

sludge, solid, or contained gas. If the waste is not listed in Part 261, Subpart D, of this Chapter, the description also must include the process that produced it (for example, solid filter cake from production of —, EPA Hazardous Waste Number W051).

Each hazardous waste listed in Part 261, Subpart D, of this Chapter, and each hazardous waste characteristic defined in Part 261, Subpart C, of this Chapter, has a four-digit EPA Hazardous Waste Number assigned to it. This number must be used for recordkeeping and reporting purposes. Where a hazardous waste contains more than one listed hazardous waste, or where more than one hazardous waste characteristic applies to the waste, the waste description must include all applicable EPA Hazardous Waste Numbers.

(2) The estimated or manifest-reported weight, or volume and density, where applicable, in one of the units of measure specified in Table 1;

(3) The method(s) (by handling code(s) as specified in Table 2) and date(s) of treatment, storage, or disposal.

Table 1

Unit of measure	Symbol	[1]Density
Pounds	P	
Short tons (2000 lbs)	T	
Gallons (U.S.)	G	P/G
Cubic yards	Y	T/Y
Kilograms	K	
Tonnes (1000 kg)	M	
Liters	L	K/L
Cubic meters	C	M/C

[1] Single digit symbols are used here for data processing purposes.

Table 2.—Handling Codes for Treatment, Storage, and Disposal Methods

Enter the handling code(s) listed below that most closely represents the technique(s) used at the facility to treat, store, or dispose of each quantity of hazardous waste received.

1. Storage
   - S01 Container (barrel, drum, etc.)
   - S02 Tank
   - S03 Waste pile
   - S04 Surface impoundment
   - S05 Other (specify)
2. Treatment
   - (a) Thermal Treatment
   - T06 Liquid injection incinerator
   - T07 Rotary kiln incinerator
   - T08 Fluidized bed incinerator
   - T09 Multiple hearth incinerator
   - T10 Infrared furnace incinerator
   - T11 Molten salt destructor
   - T12 Pyrolysis
   - T13 Wet Air oxidation
   - T14 Calcination
   - T15 Microwave discharge
   - T16 Cement kiln
   - T17 Lime kiln
   - T18 Other (specify)
   - (b) Chemical Treatment
   - T19 Absorption mound
   - T20 Absorption field
   - T21 Chemical fixation
   - T22 Chemical oxidation
   - T23 Chemical precipitation
   - T24 Chemical reduction

Fig. 6-1. EPA Form 8700-13: Facility Annual Report for On-Site or Off-Site Treatment, Storage, or Disposal of Hazardous Waste; and Unmanifested Waste Report. (Continued)

T25 Chlorination
T26 Chlorinolysis
T27 Cyanide destruction
T28 Degradation
T29 Detoxification
T30 Ion exchange
T31 Neutralization
T32 Ozonation
T33 Photolysis
T34 Other (specify)
  (c) Physical Treatment
    (1) Separation of components
T35 Centrifugation
T36 Clarification
T37 Coagulation
T38 Decanting
T39 Encapsulation
T40 Filtration
T41 Flocculation
T42 Flotation
T43 Foaming
T44 Sedimentation
T45 Thickening
T46 Ultrafiltration
T47 Other (specify)
    (2) Removal of Specific Components
T48 Absorption-molecular sieve
T49 Activated carbon
T50 Blending
T51 Catalysis
T52 Crystallization
T53 Dialysis
T54 Distillation
T55 Electrodialysis
T56 Electrolysis
T57 Evaporation
T58 High gradient magnetic separation
T59 Leaching
T60 Liquid ion exchange
T61 Liquid-liquid extraction
T62 Reverse osmosis
T63 Solvent recovery
T64 Stripping
T65 Sand filter
T66 Other (specify)
  (d) Biological Treatment
T67 Activated sludge
T68 Aerobic lagoon
T69 Aerobic tank
T70 Anaerobic lagoon
T71 Composting
T72 Septic tank
T73 Spray irrigation
T74 Thickening filter
T75 Tricking filter
T76 Waste stabilization pond
T77 Other (specify)
T78–79 [Reserved]
3. Disposal
D80 Underground injection
D81 Landfill
D82 Land treatment
D83 Ocean disposal
D84 Surface impoundment (to be closed
    as a landfill)
D85 Other (specify)

Appendix II.—EPA Report Form and Instructions

BILLING CODE 6560-01-M

Appendix III.—EPA interim primary drinking water standards

Parameter	Maximum level (mg/1)
Arsenic	0.05
Barium	1.0
Cadmium	0.01
Chromium	0.05
Fluoride	1.4–2.4
Lead	0.05
Mercury	0.002
Nitrate (as N)	10
Selenium	0.01
Silver	0.05
Endrin	0.0002
Lindane	0.004
Methoxychlor	0.1
Toxaphene	0.005
2,4-D	0.1
2,4,5-TP Silver	0.01
Radium	5 pCi/1
Gross Alpha	15 pCi/1
Gross Beta	4 millirem/yr
Turbidity	1/TU
Coliform Bacteria	1/100 ml

[*Comment:* Turbidity is applicable only to surface water supplies.]

Appendix IV—Tests for Significance

As required in § 265.93(b) the owner or operator must use the Student's t-test to determine statistically significant changes in the concentration or value of an indicator parameter in periodic ground-water samples when compared to the initial background concentration or value of that indicator parameter. The comparison must consider individually each of the wells in the monitoring system. For three of the indicator parameters (specific conductance, total organic carbon, and total organic halogen) a single-tailed Student's t-test must be used to test at the 0.01 level of significance for significant increases over background. The difference test for pH must be a two-tailed Student's t-test at the overall 0.01 level of significance.

The student's t-test involves calculation of the value of a t-statistic for each comparison of the mean (average) concentration or value (based on a minimum of four replicate measurements) of an indicator parameter with its initial background concentration or value. The calculated value of the t-statistic must then be compared to the value of the t-statistic found in a table for t-test of significance at the specified level of significance. A calculated value of t which exceeds the value of t found in the table indicates a statistically significant change in the concentration or value of the indicator parameter.

Formulae for calculation of the t-statistic and tables for t-test of significance can be found in most introductory statistics texts.

Appendix V—Examples of Potentially Incompatible Waste

Many hazardous wastes, when mixed with other waste or materials at a hazardous waste facility, can produce effects which are harmful to human health and the environment, such as (1) heat or pressure, (2) fire or explosion, (3) violent reaction, (4) toxic dusts, mists, fumes, or gases, or (5) flammable fumes or gases.

Below are examples of potentially incompatible wastes, waste components, and materials, along with the harmful consequences which result from mixing materials in one group with materials in another group. The list is intended as a guide to owners or operators of treatment, storage, and disposal facilities, and to enforcement and permit granting officials, to indicate the need for special precautions when managing these potentially incompatible waste materials or components.

This list is not intended to be exhaustive. An owner or operator must, as the regulations require, adequately analyze his wastes so that he can avoid creating uncontrolled substances or reactions of the type listed below, whether they are listed below or not.

It is possible for potentially incompatible wastes to be mixed in a way that precludes a reaction (e.g., adding acid to water rather than water to acid) or that neutralizes them (e.g., a strong acid mixed with a strong base), or that controls substances produced (e.g., by generating flammable gases in a closed tank equipped so that ignition cannot occur, and burning the gases in an incinerator).

In the lists below, the mixing of a Group A material with a Group B material may have the potential consequence as noted.

Group 1-A	Group 1-B
Acetylene sludge	Acid sludge
Akaline caustic liquids	Acid and water
Alkaline cleaner	Battery acid
Alkaline corrosive liquids	Chemical cleaners
Alkaline corrosive battery fluid	Electrolyte, acid
Caustic wastewater	Etching acid liquid or solvent
Lime sludge and other corrosive alkalies	Pickling liquor and other corrosive acids
Lime wastewater	Spent acid
Lime and water	Spent mixed acid
Spent caustic	Spent sulfuric acid

Potential consequences: Heat generation; violent reaction.

Group 2-A	Group 2-B
Aluminum	Any waste in Group 1-A or 1-B
Beryllium	
Calcium	
Lithium	
Magnesium	
Potassium	
Sodium	
Zinc powder	
Other reactive metals and metal hydrides	

Potential consequences: Fire or explosion; generation of flammable hydrogen gas.

Fig. 6-1. EPA Form 8700-13: Facility Annual Report for On-Site or Off-Site Treatment, Storage, or Disposal of Hazardous Waste; and Unmanifested Waste Report. (Continued)

Group 3-A	Group 3-B	Group 6-A	Group 6-B
Alcohols	Any concentrated waste	Chlorates	Acetic acid and other
Water	in Groups 1-A or 1-B	Chlorine	organic acids
	Calcium	Chlorites	Concentrated mineral
	Lithium	Chromic acid	acides
	Metal hydrides	Hyphochlorites	Group 2-A wastes
	Potassium	Nitrates	Group 4-A wastes
	SO_2Cl_2, $SOCl_2$, PCl_3,	Nitric acid, fuming	Other flammable and
	CH_3SiCl_3	Perchlorates	combustible wastes
	Other water-reactive	Permanganates	
	waste	Peroxides	
		Other strong oxidizers	

Potential consequences: Fire, explosion, or heat generation; generation of flammable or toxic gases.

Potential consequences: Fire, explosion, or violent reaction.

Source: "Law, Regulations, and Guidelines for Handling of Hazardous Waste." California Department of Health, February 1975.

[FR Doc. 80–14309 Filed 5–16–80; 8:45 am]

BILLING CODE 6560–01–M

Group 4-A	Group 4-B
Alcohols	Concentrated Group 1-A
Aldehydes	or 1-B wastes
Halogenated hydrocarbons	Group 2-A wastes
Nitrated hydrocarbons	
Unsaturated hydrocarbons	
Other reactive organic compounds	
and solvents	

Potential consequences: Fire, explosion, or violent reaction.

Group 5-A	Group 5-B
Spent cyanide and sulfide	Group 1-B wastes
solutions	

Potential consequences: Generation of toxic hydrogen cyanide or hydrogen sulfide gas.

Fig. 6-1. EPA Form 8700-13: Facility Annual Report for On-Site or Off-Site Treatment, Storage, or Disposal of Hazardous Waste; and Unmanifested Waste Report. (Continued)

7
PERMIT REQUIREMENTS AND PROCEDURES FOR DECISIONMAKING: PARTS 122 AND 124

EPA and State governments share responsibility for implementing and monitoring the progress to regulate hazardous waste under Sections 3005 and 3006 of RCRA. EPA regulates hazardous waste only in States that choose not to implement their own programs or fail to qualify for EPA authorization.

Part 122 includes definitions and basic requirements for EPA to issue permits under the RCRA (Section 3005), NPDES, and UIC programs. It also provides certain requirements applicable to State programs. Part 122 covers in detail who must apply for a permit, content of the applications, conditions that must be incorporated into permits, when permits may be revised, reissued or terminated, and other requirements.

Part 124 establishes the procedures EPA will use in issuing RCRA (Section 3005), NPDES, UIC, and PSD permits. Included are procedures for public participation, for consolidated review and issuance of two or more permits to the same facility or activity, for appealing permit decisions, and other requirements. Most requirements for Part 124 are only applicable where EPA is the permit-issuing authority. However, Part 123 requires States to comply with some of the Part 124 provisions, such as basic public participation requirements of permit issuance.

The following tables address the detailed requirements of Parts 122 and 124 that are applicable to the hazardous waste management program:

- Part 122, EPA-Administered Permit Programs
  - Table 7-1A: General Program Requirements—Subpart A
  - Table 7-1B: Additional Requirements for Hazardous Waste Programs under RCRA—Subpart B
  - Table 7-1C: Additional Requirements for Underground Injection Control Programs under Safe Drinking Water Act—Subpart C
- Part 124, Procedures for Decisionmaking
  - Table 7-2A: General Program Requirements—Subpart A
  - Table 7-2B: Evidentiary Hearings for EPA-Issued NPDES Permits and EPA-Terminated RCRA Permits—Subpart E
  - Table 7-2C: Non-Adversary Panel Procedures—Subpart F

EPA has developed a single set of permit application forms for programs covered by these regulations. They consist of a single general form (see Fig. 7-1) to collect basic information from all applicants, followed by separate program-specific forms (see Fig. 7-2 for hazardous waste form) which collect additional information needed to issue permits under each program.

Table 7-1A. Part 122, EPA-Administered Permit Programs: General Program Requirements—Subpart A.

Section No.	Description
	I. Applicability
122.1	A. Consolidated Permit Regulations
122.1(a)	(1) Five Programs Covered
122.1(a)(1)(i)	(a) Hazardous Waste Management Program under Subtitle C of RCRA.
122.1(a)(1)(ii)	(b) Underground Injection Control (UIC) Program under Safe Drinking Water Act.
122.1(a)(1)(iii)	(c) National Pollutant Discharge Elimination System (NPDES) Program under Clean Water Act.
122.1(a)(1)(iv)	(d) Dredge or Fill (404) Program under Clean Water Act.
122.1(a)(1)(v)	(e) Prevention of Significant Deterioration (PSD) Program under Clean Air Act.
122.1(b) and 122.1(b)(1)	(2) Overall Structure and Coverage of Consolidated Permit Regulations (Parts 122, 123, and 124)
122.2(a) and (b)	(a) Scope
	Permitting requirements apply to:
	(i) EPA-administered RCRA (UIC and NPDES) programs.
	(ii) State-administered programs to extent specified by cross-reference in Section 123.7 (Requirements for Permitting).
122.1(b)(1)(i)	(b) Part 122
	(i) Contains definitions for all programs except PSD.
	(ii) Contains basic permitting requirements for EPA-administered RCRA, UIC, and NPDES programs:
	• Application requirements.
	• Standard permit conditions.
	• Monitoring and reporting requirements.
122.1(b)(1)(ii)	(c) Part 123
	(i) Describes what States must do to obtain EPA approval of their RCRA, UIC, NPDES, or 404 programs.
	(ii) Sets forth minimum requirements for administering these permit programs after approval.
122.1(b)(1)(iii)	(d) Part 124
	(i) Establishes procedures for EPA issuance of RCRA, UIC, NPDES, and PSD permits.
	(ii) Establishes procedures for administrative appeals of EPA permit decisions.
122.1(b)(2)	(3) Subpart Structure of Parts 122, 123, and 124
	(a) Subpart A
	Each Part has a general Subpart A containing requirements that apply to all programs covered by that Part.
	(b) Other Subparts
	Additional Subparts supplement the general provisions with requirements applicable to one or more specified programs.
	(c) Inconsistency between Subpart A and Program-Specific Subpart
	Program-specific Subpart is controlling.
122.1(b)(3)	(4) Parts 122 and 124 Requirements Applicable to Approved State Programs (also see Section 123.7)
122.4	(a) Application for a Permit
122.6	(b) Signatories to Permit Applications and Reports
122.7	(c) Conditions Applicable to All Permits
122.8	(d) Establishing Permit Conditions
122.9	(e) Duration of Permits
122.10(a)	(f) Schedules of Compliance
122.11	(g) Requirements for Recording and Reporting of Monitoring Results
122.13(a and &(b)	(h) Effect of a Permit
122.14	(i) Transfer of Permits
122.15	(j) Modification or Revocation and Reissuance of Permits
122.16	(k) Termination of Permits
122.18	(l) Noncompliance and Program Reporting by the Director
122.19(b)–(d)	(m) Confidentiality of Information
122.21(d)(1)	(n) Specific Inclusions
122.22	(o) Application for a Permit
122.24	(p) Contents of Part A
122.25	(q) Contents of Part B
122.26	(r) Permits by Rule
122.27	(s) Emergency Permits
122.28	(t) Additional Conditions Applicable to All RCRA Permits
122.29	(u) Establishing RCRA Permit Conditions

Table 7-1A. Part 122, EPA-Administered Permit Programs: General Program Requirements—Subpart A. (Continued)

Section No.	Description
122.30	(v) Interim Permits for UIC Wells
124.3(a)	(w) Application for a Permit
124.5(a),(c), and (d)	(x) Modification, Revocation and Reissuance, or Termination of Permits
124.6(a),(d), and (e)	(y) Draft Permits
124.8	(z) Fact Sheet
124.10(a)(1)(ii), (iii), (b)–(e)	(aa) Public Notice of Permit Actions and Public Comment Period
124.11	(ab) Public Comments and Requests for Public Hearings
124.12(a)	(ac) Public Hearing
124.17(a) and (c)	(ad) Response to Comments
122.1(b)(4)	(5) If State is Permitting Authority
	Applicant or permittee should read State laws and program regulations.
122.1(c)	(6) Relation to Other Requirements
122.1(c)(1)	(a) Consolidated Permit Application Forms
	(i) Applicants for EPA-Issued RCRA Permit or Seeking Interim Status
	Must use EPA forms for Part A of permit application.
	• EPA Form 1 General (No. 3510-1) (See Fig. 7-1.) Essential information in this form is listed in Section 122.4 (Application for a Permit).
	• EPA Form 3 RCRA (No. 3510-3) (See Fig. 7-2.) Additional RCRA information required for Part A applications is listed in Section 122.24 (Contents of Part A)
	(ii) RCRA Part B Applications
	There will be no EPA Form. See Section 122.25 for requirements (Contents of Part B).
	(iii) There Is No "Consolidated Permit"
	Each permit and application under a program is a separate document.
	(iv) Applicants for State-Issued Permits
	Must use State forms which require at a minimum the information listed in these Sections.
122.1(c)(2)	(b) Technical Regulations
	Parts 260–266 are the technical regulations for RCRA and are used by permit-issuing authorities to determine what requirements must be placed in permits if they are issued.
122.1(e)	(7) Public Participation
	(a) General
	Rules of Parts 122, 123, and 124 establish requirements for public participation in EPA and State permit issuance, enforcement, and related variance proceedings, and in approval of State RCRA (UIC, NPDES, and 404) programs.
	(b) Relationship of Parts 122, 123, and 124 Rules to 40 CFR 25 (Public Participation)
	(i) Public participation objectives of 40 CFR 25 are carried out.
	(ii) 40 CFR 25 requirements are superseded by Parts 122, 123, and 124.
122.1(f)	(8) State Authority
	(a) General
	Nothing in Parts 122, 123, and 124 precludes more stringent State regulation of any activity covered by these regulations.
	(b) RCRA Exception
	Section 123.32 (Consistency) requires that State RCRA programs under final authorization be consistent with the Federal program and other State programs.
122 Authority	B. Authority
	Resource Conservation and Recovery Act, 42 U.S.C. Section 6901 et seq.; Safe Drinking Water Act, 42 U.S.C. Section 300f et seq.; and Clean Water Act, 33 U.S.C. Section 1251 et seq.
122.4	II. Application for a Permit
	A. General
122.4(a)	(1) For Person Required to Have a Permit (New applicant and Permittees with Expiring Permits)
	Must complete, sign, and submit application to Director, as described in this Section and Section 122.23 (Interim Status).
	(2) For Person with RCRA Interim Status
	Apply for permits when required by Director.
	(3) For Persons Covered by RCRA Permits By Rule (Section 122.26)
	Need not apply for a permit.
	(4) Emergency Permits
	Procedures for applications, issuance, and administration are covered in Section 122.27.

Table 7-1A. Part 122, EPA-Administered Permit Programs: General Program Requirements—Subpart A. (Continued)

Section No.	Description
122.4(b)	(5) For Facility Owned and Operated by Different Persons Owner must also sign RCRA permit application.

122.4(c)	(6) Completeness
	  (a) Director shall not issue a permit before receiving complete application for permit.
	    (i) Exceptions
	      • RCRA permit by rule.
	      • Emergency permits.
	  (b) When complete?
	    Director is satisfied with received application form and any supplemental information.
	  (c) Completeness for any one application shall be judged independently of status of other permit applications for same facility or activity.

122.4(d)	B. Information Requirements Provide following information to Director using application form provided by him:
122.4(d)(1)	(1) Activities conducted by applicant which require RCRA permit.
122.4(d)(2)	(2) Name, mailing address, and location of facility for which application is submitted.
122.4(d)(3)	(3) Up to four SIC codes which best reflect principal products or services provided by facility.
122.4(d)(4)	(4) Operator's name, address, telephone number, and ownership status. (5) Status as Federal, State, private, public, or other entity.
122.4(d)(5)	(6) Whether facility is located on Indian lands.
122.4(d)(6)	(7) Listing of all permits or construction approvals received or applied for under any of following programs:
122.4(d)(6)(i)	(a) Hazardous Waste Management.
122.4(d)(6)(ii)	(b) UIC.
122.4(d)(6)(iii)	(c) NPDES.
122.4(d)(6)(iv)	(d) PSD.
122.4(d)(6)(v)	(e) Non-attainment program under Clean Air Act.
122.4(d)(6)(vi)	(f) National Emission Standards for Hazardous Pollutants (NESHAPS) preconstruction approval under Clean Air Act.
122.4(d)(6)(vii)	(g) Ocean dumping permits under Marine Protection Research and Sanctuaries Act.
122.4(d)(6)(viii)	(h) Dredge or Fill (404) program.
122.4(d)(6)(ix)	(i) Other relevant environmental permits, including State permits.

122.4(d)(7)	(8) Topographic Map
	  (a) May be other map if topographic map unavailable.
	  (b) Map must extend one mile beyond property boundaries of source.
	  (c) Map must depict facility and each intake and discharge structure.
	  (d) Include each hazardous waste treatment, storage, or disposal facility.
	  (e) Include each well where fluids from facility are injected underground.
	  (f) Include those wells, springs, other surface water bodies, and drinking water wells listed in public records or otherwise known to applicant in map area.

122.4(d)(8)	(9) Brief description of nature of business.
122.4(d)	(10) Additional information set forth in Section 122.24 (Contents of Part A of RCRA Permit Application). (11) Additional information set forth in Section 122.25 (Contents of Part B of RCRA Permit Application).

122.4(e)	C. Recordkeeping
	(1) Applicant shall keep records of all data used to complete permit application.
	(2) Keep any supplemental information submitted under:
	  (a) Section 122.4(d)–Application for a Permit.
	  (b) Section 122.24–Contents of Part A of RCRA Permit Application.
	  (c) Section 122.25–Contents of Part B of RCRA Permit Application.
	(3) Duration
	  Keep records for period of at least 3 years from date application is signed.

122.5	III. Continuation of Expiring Permits
122.5(a)	A. For EPA as Permit-Issuing Authority (1) Expired Permit Under 5 U.S.C. Section 558(c), the permit remains valid if:
122.5(a)(1)	(a) Permittee has submitted timely application per Section 122.25 (Contents of Part B of RCRA permit application). (b) Application for new permit is complete per Section 122.4(c) (Application for a Permit).

Table 7-1A. Part 122, EPA-Administered Permit Programs: General Program Requirements—Subpart A. (Continued)

Section No.	Description
122.5(a)(2)	(c) Regional Administrator, through no fault of permittee, does not issue new permit on or before expiration date.
122.5(d)(1)	B. For State as Permit-Issuing Authority (1) Expired Permit Initially Issued By EPA with State Presently Permit-Issuing Authority (a) Expired permit does not remain valid under Federal law. (b) Authorized State May Continue EPA-Issued (or State-Issued) Permit after Expiration Date (i) Provided State law so allows. (ii) Otherwise, facility or activity is operating without a permit (from time of expiration of old permit to effective date of State-issued new permit).
122.5(b)	C. Permits Issued Under This Section (122.5) They remain fully effective and enforceable.
122.5(c)	D. Enforcement (1) When Permittee Not in Compliance with Conditions of Expiring or Expired Permit Regional Administrator, may choose to do any or all of following:
122.5(c)(1)	(a) Initiate enforcement action.
122.5(c)(2)	(b) Issue Notice of Intent to Deny New Permit under Section 124.6 (Draft Permit) (i) If Permit is Denied • Owner/operator would be required to cease activities; or • Be subject to enforcement action for operating without a permit.
122.5(c)(3)	(c) Issue a new permit under Part 124 with appropriate conditions; or
122.5(c)(4)	(d) Take other actions authorized by these regulations.
122.6	IV. Signatories to Permit Applications and Reports
122.6(a)	A. Applications RCRA permit applications shall be signed as follows:
122.6(a)(1)	(1) For a Corporation By a principal executive officer of at least level of vice-president.
122.6(a)(2)	(2) For a Partnership or Sole Proprietorship By a general partner or proprietor, respectively.
122.6(a)(3)	(3) For a Municipality, State, Federal, or Other Public Agency By either a principal executive officer or ranking elected official.
122.6(b)	B. Reports Required for Permits and Other Information Requested by Director (1) Must be signed by person described in Section 122.6(a) above; or (2) Signed by Duly Authorized Representative of Person Described in Section 122.6(a) (a) Person is duly authorized representative only if:
122.6(b)(1)	(i) Authorization is made in writing by person described in Section 122.6(a).
122.6(b)(2)	(ii) Authorization specifies either individual or position having responsibility for overall operation of regulated facility such as: • Plant manager. • Superintendent. • Position of equivalent responsibility. *Note:* Thus, a duly authorized representative may be either: • A named individual; or • Any individual occupying a named position.
122.6(b)(3)	(iii) Written authorization is submitted to Director.
122.6(c)	(3) Changes to Authorization (a) If Different Individual or Position Has Responsibility for Overall Operation of Facility New authorization satisfying requirements of Section 122.6(b) must be submitted to Director prior to or together with any reports, information, or applications to be signed by an authorized representative.

Table 7-1A. Part 122, EPA-Administered Permit Programs: General Program Requirements—Subpart A. (Continued)

Section No.	Description
122.6(d)	C. Certification Any person signing a document under Sections 122.6(a) or (b) shall make following certification: "I certify under penalty of law that I have personally examined and am familiar with the information submitted in this document and all attachments and that, based on my inquiry of those individuals immediately responsible for obtaining the information, I believe that the information is true, accurate, and complete. I am aware that there are significant penalties for submitting false information, including the possibility of fine and imprisonment."
122.7	V. Conditions Applicable to All Permits A. General (1) Conditions that follow apply to all RCRA (UIC, NPDES, and 404) permits. (2) Additional Conditions Applicable to RCRA Permits See Section 122.28 (Additional Conditions Applicable to all RCRA Permits). (3) Incorporate All Conditions into Permit (a) Conditions applicable to all permits. (b) Additional conditions applicable to RCRA permits. (c) Conditions may be incorporated into permits expressly or by reference. (i) If incorporated by reference, a specific citation to these regulations (or corresponding approved State regulations) must be given in permit. B. Specific Conditions
122.7(a)	(1) Duty to Comply (a) Permittee must comply with all conditions of permit. (b) Any permit noncompliance constitutes a violation of RCRA and is grounds for: (i) Enforcement action. (ii) Permit termination, revocation and reissuance, or modification; or (iii) Denial of permit renewal application.
122.7(b)	(2) Duty to Reapply If permittee wishes to continue activity regulated by permit after expiration date he must apply for and obtain a new permit.
122.7(c)	(3) Duty to Halt or Reduce Activity It shall not be a defense for a permittee in an enforcement action that it would have been necessary to halt or reduce permitted activity in order to maintain compliance with conditions of permit.
122.7(d)	(4) Duty to Mitigate Permittee shall take all reasonable steps to minimize or correct any adverse impact on environment resulting from noncompliance with permit.
122.7(e)	(5) Proper Operation and Maintenance (a) What Proper Operation and Maintenance Includes (i) Effective performance. (ii) Adequate funding. (iii) Adequate operator staffing. (iv) Adequate operator training. (v) Adequate laboratory and process controls, including appropriate quality assurance procedures. (b) Permittee Compliance Requirements He must always properly operate and maintain all facilities and systems of treatment and control (and related appurtenances) which are installed or used by permittee to achieve compliance with conditions of permit. (c) Operation of Back-Up or Auxiliary Facilities or Similar Systems Required only when necessary to achieve compliance with conditions of permit.
122.7(f)	(6) Permit Actions (a) Permit may be modified, revoked and reissued, or terminated for cause. (b) Stay of Permit Condition(s) Filing of a request for a permit modification, revocation and reissuance, or termination, or a notification of planned changes or anticipated noncompliance, does not stay any permit condition.
122.7(g)	(7) Property Rights Permit does not convey any property rights of any sort, or any exclusive privilege.

Table 7-1A. Part 122, EPA-Administered Permit Programs: General Program Requirements—Subpart A. (Continued)

Section No.	Description
122.7(h)	(8) Duty to Provide Information
	(a) Information Director May Request to Determine Whether Cause Exists for Modifying, Revoking and Reissuing, or Terminating Permit
	Permittee shall furnish Director such information within a reasonable time.
	(b) Information Director May Request to Determine Compliance with Permit
	Permittee shall furnish Director such information within a reasonable time.
	(c) Copies of Records Required to Be Kept for Permit
	Must be furnished to Director upon request.
122.7(i)	(9) Inspection and Entry
	Permittee shall allow Director or authorized representative to perform following activities upon presentation of credentials and other documents as may be required by law:
122.7(i)(1)	(a) Enter permittee's premises where a regulated facility or activity is located or conducted.
	(b) Enter permittee's premises where records must be kept under conditions of permit.
122.7(i)(2)	(c) Have access to and copy, at reasonable times, any records that must be kept under conditions of permit.
122.7(i)(3)	(d) Inspect at reasonable times any facilities, equipment (including monitoring and control equipment), practices, or operations regulated or required under permit.
122.7(i)(4)	(e) Sample or monitor at reasonable times, for purpose of assuring permit compliance or as otherwise authorized by RCRA, any substances or parameters at any location.
122.7(j)	(10) Monitoring and Records
122.7(j)(1)	(a) Samples and Measurements Taken for Purpose of Monitoring
	Shall be representative of monitoring activity.
122.7(j)(2)	(b) Permittee Must Retain Records of All Monitoring Information
	Includes:
	(i) Calibration records.
	(ii) Maintenance records.
	(iii) All original strip chart recordings for continuous monitoring instrumentation.
	(iv) Copies of all reports required by permit.
	(v) Records of all data used to complete application for permit.
	(c) Record Retention
	(i) Records must be kept for at least 3 years from date of sample, measurement, report or application.
	(ii) Period may be extended by request of Director at any time.
122.7(j)(3)	(d) Content of Monitoring Information Records
122.7(j)(3)(i)	(i) Date, exact place, and time of sampling or measurements.
122.7(j)(3)(ii)	(ii) Individual(s) who performed sampling or measurements.
122.7(j)(3)(iii)	(iii) Date(s) analyses were performed.
122.7(j)(3)(iv)	(iv) Individual(s) who performed analyses.
122.7(j)(3)(v)	(v) Analytical techniques or methods used.
122.7(j)(3)(vi)	(vi) Results of analyses.
122.7(k)	(11) Signatory Requirement
	All applications, reports, or information submitted to Director shall be signed and certified in accordance with requirements of Section 122.6 (Signatories to Permit Applications and Reports).
122.7(1)	(12) Reporting Requirements
122.7(1)(1)	(a) Planned Changes
	Permittee shall give notice to Director as soon as possible of any planned physical alterations or additions to permitted facility.
122.7(1)(2)	(b) Anticipated Noncompliance
	Permittee shall give advance notice to Director of any planned changes in permitted facility or activity which may result in noncompliance with permit requirements.
122.7(1)(3)	(c) Transfers of Permit
	(i) Not transferable to any person except after notice to Director.
	(ii) Director may require modification or revocation and reissuance of permit to change name of permittee and incorporate other requirements as may be necessary under RCRA.
	• Also see Section 122.14—Transfer of Permits.
	• In some cases, modification or revocation and reissuance is mandatory.
122.7(1)(4)	(d) Monitoring Reports
	Monitoring results shall be reported at intervals specified elsewhere in permit.
122.7(1)(5)	(e) Compliance Schedules
	(i) Reports must be submitted within 14 days of any compliance schedule date contained in permit.

Table 7-1A. Part 122, EPA-Administered Permit Programs: General Program Requirements—Subpart A. (Continued)

Section No.	Description
	(ii) Compliance schedule reports include: • Reports of compliance. • Reports of noncompliance. • Progress reports on interim and final requirements.

122.7(1)(6)	(f) Reporting Noncompliance Which May Endanger Health or Environment
	(i) Permittee must report any such noncompliance.
	(ii) 24-Hour Oral Report
	Required from time permittee becomes aware of circumstances.
	(iii) 5-Day Written Report
	• Required within 5 days from time permittee becomes aware of circumstances.
	• Content follows:
	  – Description of noncompliance and its cause.
	  – Period of noncompliance, including exact dates and times.
	  – If noncompliance not corrected, anticipated time it is expected to continue.
	  – Steps taken or planned to reduce, eliminate, and prevent reoccurrence of noncompliance.

122.7(1)(7)	(g) Other Noncompliance (i) Permittee shall report all instances of noncompliance not reported in following Sections, at time monitoring reports are submitted: • Section 122.7(1)(1)–Planned changes. • Section 122.7(1)(4)–Monitoring Reports. • Section 122.7(1)(5)–Compliance Schedules. • Section 122.7(1)(6)–Twenty-Four Hour Reporting. (ii) Reports shall include information listed in 122.7(1)(6).
122.7(1)(8)	(h) Other Information Where permittee becomes aware of failure to submit relevant facts in permit application, or submitted incorrect information in permit application or in any report to Director, such facts or information shall be promptly submitted.
122.8	VI. Establishing Permit Conditions
122.8(a)	A. For All Programs In addition to conditions required under Section 122.7 (Conditions Applicable To All Permits), the Director shall establish conditions on a case-by-case basis for permits under following Sections: (1) 122.9–Duration of Permits. (2) 122.10(a)–Schedules of Compliance. (3) 122.11–Requirements for Recording and Reporting of Monitoring Results. (4) 122.10(b)–Alternate Schedules of Compliance. (5) 122.12–Considerations under Federal Law.
122.8(b) 122.8(b)(1)	B. For RCRA Programs (1) See Section 122.28–Additional Conditions Applicable to All RCRA Permits. (2) Director Shall Establish Conditions In Permit on Case-by-Case Basis To provide for and assure compliance with applicable requirements of RCRA law and regulations.
122.8(b)(2)	(3) For State-Issued Permit An applicable requirement is a State statutory or regulatory requirement which takes effect prior to final administrative disposition of a permit. (4) For EPA-Issued Permit (a) An applicable requirement is a statutory or regulatory requirement (including any interim regulations) which takes effect prior to issuance of permit. (i) Exception As provided in Section 124.86(c) (Motions) for RCRA permits being processed under following Subparts of Part 124: • Subpart E–Evidentiary Hearings for EPA-Issued NPDES Permits and EPA-Terminated RCRA Permits. • Subpart F–Non-Adversary Panel Procedures. (b) Reopening EPA Permit Proceedings May be done at discretion of Director under Section 124.14 (Reopening of Public Comment Period) where new requirements become effective during permitting process and are of sufficient magnitude to make additional proceedings desirable.

Table 7-1A. Part 122, EPA-Administered Permit Programs: General Program Requirements—Subpart A. (Continued)

Section No.	Description
	(5) For State and EPA-Administered Programs An applicable requirement is also any requirement which takes effect prior to modification or revocation and reissuance of a permit, to extent allowed in Section 122.15 (Modification or revocation and Reissuance of Permits).
122.8(b)(3)	(6) New or Reissued Permits and Modified or Revoked and Reissued Permits (As Allowed Under Section 122.15) Applicable requirements shall be incorporated as referenced in Section 122.29 (Establishing RCRA Permit Conditions).
122.8(c)	C. Incorporating All Conditions into Permit (1) Conditions shall be incorporated into permits expressly or by reference. (2) If Incorporated by Reference A specific citation to applicable regulations or requirements must be given in permit.
122.9	VII. Duration of Permits
122.9(b) 122.9(e) 122.9(d) 122.9(b)	A. RCRA Permits Effective for Fixed Term Not Greater than 10 Years (1) Director may issue permit for duration less than full allowable term. (2) Permit shall not be extended beyond maximum specified duration, except as provided in Section 122.5 (Continuation of Expiring Permits). (3) Also see Section 122.30—Interim Permits for UIC Wells.
122.10	VIII. Schedules of Compliance
122.10(a) 122.10(a)(1) 122.10(a)(3)	A. General (1) Permit may specify schedule of compliance leading to compliance with RCRA laws and regulations. (2) Time for Compliance Any schedules of compliance requires compliance as soon as possible. (3) Interim Dates Required if permit establishes schedule of compliance that exceeds date of permit issuance by more than one year. (a) Schedule Shall Set Forth Interim Requirements and Dates for Achievement
122.10(a)(3)(i) 122.10(a)(3)(ii) Note 122.10(a)(3)(ii)	(i) Time between interim dates not to exceed 1 year. (ii) Examples of Interim Requirements • Let a contract for construction of required facilities. • Commence construction. • Complete construction. (iii) If interim dates exceed one year (e.g., construction of control facility) and work is not readily divisible into stages for completion then: • Permit shall specify interim dates for submission of progress reports toward completion of interim requirements. • Permit shall indicate projected completion date.
122.10(a)(4)	(4) Reporting on Interim and Final Completion Dates (a) Permit shall be prepared requiring notification of Director in writing of compliance or noncompliance with interim and final requirements, no later than 14 days following each schedule date; or (b) To submit progress reports if Section 122.10(a)(1)(ii) is applicable.
122.10(b) 122.10(b)(1) 122.10(b)(1)(i) 122.10(b)(1)(ii)	B. Alternative Schedules of Compliance An RCRA permit applicant or permittee may cease conducting regulated activities (rather than continue to operate and meet permit requirements) as follows: (1) Within Term of Permit Which Has Already Been Issued (a) Permit may be modified to contain new or additional schedule leading to timely cessation of activities; or (b) Permittee shall cease conducting permitted activities before noncompliance with any interim or final compliance schedule requirement already specified in permit.
122.10(b)(2)	(2) Prior to Issuance of Permit Permit shall contain schedule leading to termination which will ensure timely compliance with applicable requirements.
122.10(b)(3) 122.10(b)(3)(ii)	(3) If Permittee Undecided Whether to Cease Conducting Regulated Activities Director may issue or modify permit to contain two schedules as follows: (a) Schedule No. 1 Shall lead to timely compliance with applicable requirements.

Table 7-1A. Part 122, EPA-Administered Permit Programs: General Program Requirements—Subpart A. (Continued)

Section No.	Description
122.10(b)(3)(iii)	(b) Schedule No. 2 Shall lead to cessation of regulated activities by date which will ensure timely compliance with applicable requirements. (c) Conditions Applicable to Both Schedules No. 1 and 2
122.10(b)(3)(i)	(i) Both schedules shall contain identical interim deadline requiring final decision on whether to cease conducting regulated activities no later than a date which ensures sufficient time to comply with applicable requirements in a timely manner if decision is to continue conducting regulated activities.
122.10(b)(3)(iv)	(ii) Each permit containing two schedules shall include requirement that after permittee has made final decision on continuing or ceasing regulated activity that compliance schedule will be followed.
122.10(b)(4)	(4) Applicant's or Permittee's Decision to Cease Conducting Regulated Activities Shall be Evidenced by a Firm Public Commitment The commitment must be satisfactory to the Director, such as a resolution of the board of directors of a corporation.
122.11	IX. Requirements for Recording and Reporting of Monitoring Results
	A. Information Permits Must Specify
122.11(a)	(1) Requirements Concerning Proper Use, Maintenance, and Installation of Monitoring Equipment or Methods Includes biological monitoring methods when appropriate.
122.11(b)	(2) Required Monitoring (a) Includes type, intervals, and frequency sufficient to yield data which are representative of monitored activity. (b) Includes continuous monitoring when appropriate.
122.11(c)	(3) Applicable Reporting Requirements (a) Based on regulated activity specified in Parts 264 and 266. (b) Reporting shall be no less frequent than specified in regulations of Parts 264 and 266.
122.12	X. Considerations under Federal Law
	A. Issuance of Permits Shall be done in a manner and shall contain conditions consistent with requirements of applicable Federal laws, which may include:
122.12(a)	(1) The Wild and Scenic Rivers Act.
122.12(b)	(2) The National Historic Preservation Act of 1966.
122.12(c)	(3) The Endangered Species Act.
122.12(d)	(4) The Coastal Zone Management Act.
122.12(e)	(5) The Fish and Wildlife Coordination Act.
122.13	XI. Effect of a Permit
122.13(a)	A. Compliance with a Permit during Its Term Constitutes Compliance, for Purposes of Enforcement, with Subtitle C of RCRA However, a permit may be modified, revoked and reissued, or terminated during its term for cause as set forth in: (1) Section 122.15—Modification or Revocation and Reissuance of Permits. (2) Section 122.16—Termination of Permits.
122.13(b)	B. Issuance of a permit does not convey any property rights of any sort, or any exclusive privilege.
122.13(c)	C. Issuance of a permit does not authorize any injury to persons or property or invasion of other private rights, or any infringement of State or local law or regulations.
122.14	XII. Transfer of Permits
122.14(a)	A. Transfers by Modification (1) Permit may be transferred by permittee to new owner or operator only if permit has been modified or revoked and reissued under: (a) Section 122.15(b)(2)—Modification or Revocation and Reissuance of Permits; or (b) Section 122.17(d)—Minor Modifications of Permits. (2) Purpose of Modification or Revocation and Reissue of Permit To identify new permittee and incorporate such other requirements as may be necessary under RCRA.

Table 7-1A. Part 122, EPA-Administered Permit Programs: General Program Requirements—Subpart A. (Continued)

Section No.	Description
122.15	XIII. Modification or Revocation and Reissuance of Permits

A. General
   (1) Actions That May Make Director Consider Cause to Exist (under Section 122.15(a) and (b)) to Modify or Revoke and Reissue a Permit
      (a) Receives Certain Information
         (i) Facility inspection data.
         (ii) Permit-required information under Section 122.7—Conditions Applicable to All Permits.
      (b) Receives Request for Modification or Revocation and Reissue of Permit
         Under Section 124.5—Modification, Revocation and Reissuance, or Termination of Permits.
      (c) Conducts review of permit file.
   (2) If Cause Exists to Modify or Revoke and Reissue a Permit
      (a) Director may modify or revoke and reissue permit accordingly.
      (b) Director's action subject to limitations of Section 122.15(c).
   (3) If Permit Is Modified
      Only conditions subject to modification are reopened.
   (4) If Permit Is Revoked and Reissued
      Entire permit is reopened and subject to revision and permit is issued for a new term (see Section 124.5(c)(2)).
   (5) If Cause Does Not Exist to Modify or Revoke and Reissue a Permit
      (a) Director shall not modify or revoke and reissue the permit
      (b) Cause is based on:
         (i) This Section; and
         (ii) Section 122.17—Minor Modification of Permits.
   (6) If Permit Modification Satisfies Criteria of Section 122.17 for "Minor Modifications"
      Permit may be modified without draft permit or public review.
      (a) Otherwise draft permit must be prepared and other procedures in Part 124 (or procedures of an approved State program) followed.

Section No.	Description
122.15(a)	B. Causes for Modification

The following are causes for modification but not revocation and reissuance of permits:

122.15(a)(1)	(1) Alterations Substantial alterations or additions to permitted facility or activity occurred after permit issuance.

122.15(a)(2)	(2) Information
	(a) Director has received pertinent information.
	(b) Permits may be modified during their terms for this cause:
	   (i) Only if the information was not available at time of permit issuance (other than revised regulations, guidance, or test methods); and
	   (ii) If it would have justified different permit conditions at time of issuance.

122.15(a)(3)	(3) New Regulations
	(a) After permit was issued, regulations on which permit was based have been changed by:
	   (i) Promulgation of amended regulations; or
	   (ii) Judicial decision.
	(b) Permits may be modified during their terms for this cause only as follows:

122.15(a)(3)(i)	(i) For promulgation of amended regulations when:
122.15(a)(3)(i)(A)	• Permit condition requested to be modified was based on promulgated Parts 260–266.
122.15(a)(3)(i)(B)	• EPA has revised, withdrawn or modified that portion of the regulation on which the permit condition was based.
122.15(a)(3)(i)(C)	• Permittee requests modification in accordance with Section 124.5 (Modification, Revocation and Reissuance, or Termination of Permits) within 90 days after *Federal Register* notice of action on which request is based.
122.15(a)(3)(ii)	(ii) For Judicial Decisions • Court of competent jurisdiction has remanded and stayed EPA promulgated regulations. • Remand and stay must concern that portion of regulations on which permit condition was based. • Request is filed by permittee in accordance with Section 124.5 within 90 days of judicial remand.

Table 7-1A. Part 122, EPA-Administered Permit Programs: General Program Requirements—Subpart A. (Continued)

Section No.	Description

122.15(a)(4)	(4) Compliance Schedules
	(a) Director determines if good cause exists for modification of compliance schedule.
	(b) Examples of Good Cause
	  (i) An act of God; or
	  (ii) Strike; or
	  (iii) Flood; or
	  (iv) Materials shortage; or
	  (v) Other events over which permittee has little or no control and for which there is no reasonable available remedy.

122.15(b)	C. Causes for Modification or Revocation and Reissuance The following are causes to modify or, alternatively, revoke and reissue a permit:
122.15(b)(1)	(1) Cause Exists for Termination under Section 122.16 (Termination of Permits) Director determines that modification or revocation and reissuance is appropriate.
122.15(b)(2)	(2) Director has received notification of proposed transfer of permit.
122.15(c)	D. Facility Siting Suitability of facility location will not be considered at time of permit modification or revocation and reissuance. (1) Exception New information or standards indicate that threat to human health or environment exists which was unknown at time of permit issuance.
122.16	XIV. Termination of Permits
122.16(a) 122.16(a)(1) 122.16(a)(2)	A. Causes for Terminating a Permit during Its Term or for Denying a Permit Renewal Application (1) Noncompliance by permittee with any condition of permit. (2) Permittee's failure in application or during permit issuance process to disclose fully all relevant facts. (3) Permittee's misrepresentation of any relevant facts at any time.
122.16(a)(3)	(4) Determination that permitted activity endangers human health or environment and can only be regulated to acceptable levels by permit modification or termination.
122.16(b)	B. Procedures Director Must Use in Terminating RCRA Permit (1) Applicable procedures in Part 124 (Procedures for Decisionmaking); or (2) State procedures.
122.17	XV. Minor Modifications of Permits A. General (1) Permit Modification Processed as Minor Modification under This Section (122.17) (a) Upon consent of permittee, Director may modify permit to make corrections or allowances for changes in permitted activity listed in this Section. (b) May be done without following procedures of Part 124 (Procedures for Decisionmaking). (2) Permit Modification Not Processed as Minor Modification under This Section Must be made for cause and with Part 124 draft permit and public notice as required in Section 122.15 (Modification or Revocation and Reissuance of Permits). B. Scope of Modifications Allowable as "Minor Modifications"
122.17(a)	(1) Correct typographical errors.
122.17(b)	(2) Require more frequent monitoring or reporting by permittee.
122.17(c)	(3) Change Interim Compliance Date in Schedule of Compliance (a) Provided new date is not more than 120 days after date specified in existing permit; and (b) Does not interfere with attainment of final compliance date requirement.
122.17(d)	(4) Allow for Change in Ownership or Operational Control of Facility (a) Where Director determines that no other change in permit is necessary. (b) Provided that written agreement containing specific date for transfer of permit responsibility, coverage, and liability between current and new permittees has been submitted to Director.
122.17(e)	(5) Change list of facility emergency coordinators in permit's contingency plan. (6) Change list of equipment in permit's contingency plan.

Table 7-1A. Part 122, EPA-Administered Permit Programs: General Program Requirements—Subpart A. (Continued)

Section No.	Description
122.18	XVI. Noncompliance and Program Reporting by the Director
	A. General
	(1) Quarterly and Annual Reports
	Shall be prepared by Director as detailed in Sections 122.18(a) and (c).
	(2) When State Is Permit-Issuing Authority
	State Director shall submit any reports required under this Section to Regional Administrator.
	(3) When EPA Is Permit-Issuing Authority
	Regional Administrator shall submit any report required under this Section to EPA Headquarters.
	(4) RCRA Interim Status Facilities
	For purposes of this Section only, RCRA permittees shall include RCRA interim status facilities when appropriate.
122.18(a)	B. Quarterly Reports
	The Director shall submit quarterly narrative reports for major facilities as follows:
122.18(a)(1)	(1) Format for Report
122.18(a)(1)(i)	(a) Separate lists for RCRA (UIC and NPDES) permittees.
122.18(a)(1)(ii)	(b) For Facilities or Activities with Permits under More than One Program
	Provide an additional list combining information on noncompliance for each such facility.
122.18(a)(1)(iii)	(c) Alphabetize Each List by Permittee Name
	When two or more permittees have same name, lowest permit number shall be entered first.
122.18(a)(1)(iv)	(d) For Each Entry on a List
	Include following information in following order:
122.18(a)(1)(iv)(A)	(i) Name, location, and permit number of noncomplying permittee.
122.18(a)(1)(iv)(B)	(ii) Brief Description and Date of Each Instance of Noncompliance for Permittee
	• Instances of noncompliance may include types set forth in Section 122.18(a)(2).
	• When permittee has noncompliance of more than one kind under single program, combine information into single entry for each such permittee.
122.18(a)(1)(iv)(C)	(iii) Date(s) and brief description of action(s) taken by Director to ensure compliance.
122.18(a)(1)(iv)(D)	(iv) Status of Instance(s) of Noncompliance
	• Date of review of status; or
	• Date of resolution.
122.18(a)(1)(iv)(E)	(v) Any details which tend to explain or mitigate the instance(s) of noncompliance.
122.18(a)(2)	(2) Instances of Noncompliance to Be Reported
	(a) General
	(i) Report instances of noncompliance in successive reports until noncompliance is reported as resolved.
	(ii) Once noncompliance is reported as resolved it need not appear in subsequent reports.
122.18(a)(2)(i)	(b) Failure to Complete Construction Elements of Compliance Schedule by Date Specified in Permit.
	(i) Also, permittee has not returned to compliance within 30 days from date a compliance schedule report is due under the permit.
	(ii) "Construction element" may include either:
	• Planning for construction (e.g., award of a contract, preliminary plans); or
	• Construction step (e.g., begin construction, attain operation level).
122.18(a)(2)(ii)	(c) Modifications to Schedules of Compliance
	When schedule of compliance in permit has been modified under:
	(i) Section 122.15—Modification or Revocation and Reissuance of Permit; or
	(ii) Section 122.17—Minor Modifications of Permit
	because of permittee's noncompliance.
122.18(a)(2)(iii)	(d) Failure to Complete or Provide Report Required in Permit Compliance Schedule or Monitoring Report
	(i) Also, permittee has not submitted complete report:
	• Within 30 days from due date under permit for compliance schedules; or
	• From date specified in permit for monitoring reports.
	(ii) Examples of Reports Required in Permit Compliance Schedule
	• Progress report; or
	• Notice of noncompliance or compliance.

Table 7-1A. Part 122, EPA-Administered Permit Programs: General Program Requirements—Subpart A. (Continued)

Section No.	Description
122.18(a)(2)(iv)	(e) Deficient Reports When required reports provided by permittee are so deficient as to cause misunderstanding by Director and thus impede review of compliance status.
122.18(a)(2)(v)	(f) Noncompliance with Other Permit Requirements Noncompliance shall be reported in following circumstances:
122.18(a)(2)(v)(A)	(i) Whenever Permittee Has Violated a Permit Requirement • Other than reported under Sections 122.18(a)(2)(i) and (ii); and • Has not returned to compliance within 45 days from date reporting of noncompliance was due under the permit.

122.18(a)(2)(v)(B)	(ii) When Director determines that a pattern of noncompliance exists for a major facility permittee over the most recent four consecutive reporting periods.
	• Pattern includes:
	  – Any violation of same requirement in two consecutive reporting periods; and
	  – Any violation of one or more requirements in each of four consecutive reporting periods.

122.18(a)(2)(v)(C)	(iii) When Director determines significant event has occurred, such as a fire or explosion at an RCRA facility.
122.18(a)(2)(vi)	(g) All Other Statistical information shall be reported quarterly on all other instances of noncompliance by major facilities with permit requirements not otherwise reported under Section 122.18(a).
122.18(a)(3)	(3) RCRA Reports from Director to Administrator (a) In manner and form prescribed by Administrator. (b) Noncompliance by transporters (e.g., recordkeeping requirements). (c) Noncompliance by generators that send their wastes to off-site treatment, storage, or disposal facilities.
122.18(c)	C. Annual Reports Required of Director (1) Noncompliance report. (2) Program Report (a) Contains information on generators and transporters. (b) Manner and form of report prescribed by Administrator. (3) Permit status of regulated facilities. (4) Summary information on quantities and types of hazardous wastes generated, treated, stored, transported and disposed of during preceding year. (a) Summary information shall be reported according to EPA characteristics and lists of hazardous wastes under 40 CFR 261.
122.18(e) 122.18(e)(1)	D. Schedule (1) For All Quarterly Reports (a) State Director report to Regional Administrator concerning noncompliance with RCRA permits. (b) Regional Administrator report to EPA Headquarters on EPA-issued permits. (c) Timing Submit reports on last working day of May, August, November, and February, covering quarterly periods shown:

Quarter	Completion Date
January, February, and March	May 31
April, May, and June	Aug. 31
July, August, and September	Nov. 30
October, November, and December	Feb. 28

Section No.	Description
122.18(e)(2)	(2) For All Annual Reports Period for annual reports shall be for calendar year ending December 31, with reports completed and available to public no more than 60 days later.
122.19	XVII. Confidentiality of Information
122.19(a)	A. General (1) Information Submitted to EPA Pursuant to These Regulations May Be Claimed as Confidential by Submitter In accordance with 40 CFR 2 (Public Information).

Table 7-1A. Part 122, EPA-Administered Permit Programs: General Program Requirements—Subpart A. (Continued)

Section No.	Description
	(2) Assertion of Confidentiality (a) Must be done at time of submission in manner prescribed on application form or instructions; or (b) In case of other submissions, do by stamping words "confidential business information" on each page containing such information. (3) If No Claim of Confidentiality Is Made at Time of Submission EPA may make information available to public without further notice. (4) If Claim of Confidentiality Is Asserted Information will be treated in accord with procedures of 40 CFR 2.
122.19(b) 122.19(b)(1)	B. Information Not Allowed a Claim of Confidentiality by EPA Name and address of any permit applicant or permittee.
122.19(d) 122.19(d)(1) 122.19(d)(2)	C. For RCRA Only (1) Claims of Confidentiality for Permit Application Information must be substantiated at time application is submitted and in manner prescribed in application instructions. (2) If Submitter Does Not Provide Substantiation (a) Director will notify submitter by certified mail of requirement to do so. (b) If Director Does Not Receive Substantiation Within 10 Days after Submitter Receives Notice Director shall place the unsubstantiated information in the public file.

Table 7-1B. Part 122, EPA-Administered Permit Programs: Additional Requirements for Hazardous Waste Programs under RCRA—Subpart B.

Section No.	Description
122.21	I. Purpose, Scope, and Applicability of Subpart B
122.21(a)	A. Content of Subpart B (1) Overall (a) Regulations in this Subpart set forth specific requirements for the RCRA permit program. (b) These regulations supplement requirements of Subpart A of Part 122, which contains requirements for all programs. (2) Applicability Regulations apply to: (a) EPA (b) Approved States to extent set forth in Part 123 (State Program Requirements).
122.21(c)	B. Overview of the RCRA Permit Program (1) Notification of Hazardous Waste Activity Within 90 days of promulgation or revision of 40 CFR 261 regulations, generators and transporters of hazardous waste, and owners or operators of hazardous waste treatment, storage, or disposal facilities must file notification of that activity. (2) RCRA Permit Application Consists of Two Parts (a) Part A (i) See Fig. 7-1 for EPA Consolidated Permit Application Forms 1 and 3. (ii) See Section 122.24—Contents of Part A of the RCRA Permit Application. (b) Part B See Section 122.25—Contents of Part B of RCRA Permit Application (no forms). (3) Applying for a Permit Owners and operators of treatment, storage or disposal facilities must apply for permit within 6 months of promulgation of Part 261 regulations (11/18/80). (a) For Existing Treatment, Storage, or Disposal Facilities Requirement to submit an application is satisfied by submitting only Part A of permit application, until date Director sets for submitting Part B of application.

**Table 7-1B. Part 122, EPA-Administered Permit Programs: Additional Requirements for
Hazardous Waste Programs under RCRA—Subpart B. (Continued)**

Section No.	Description

	(4) Interim Status Requirements for Owners/Operators of Existing Hazardous Waste Treatment, Storage, or Disposal Facilities
	  (a) Timely submission of notification under Section 3010 of RCRA (by 8/18/80).
	  (b) Timely submission of Part A of permit application (by 11/18/80).
	(5) Interim Status Highlights
	  (a) Effect of Having Interim Status
	    Facility owners/operators with interim status are treated as having been issued a permit until EPA or State makes a final determination on the permit application.
	  (b) Compliance with Interim Status Standards of Part 265
	    Required of facility owner/operator, or with equivalent provisions of a State program which has received interim or final authorization under Part 123.
	  (c) Facility owners and operators with interim status are not relieved from complying with State requirements.
	(6) Submission of Part B of Application
	  (a) For Existing Treatment, Storage, or Disposal Facilities
	    Director shall set a date for submission of Part B, giving at least 6 months notice.
	  (b) For New Treatment, Storage, or Disposal Facilities
	    Owners/operators must submit Part A and Part B of permit application at least 180 days before physical construction is expected to commence.
	  (c) Submission of Part B is in narrative form and contains information set forth in Section 122.25 (Contents of Part B of RCRA permit application).

122.21(d)	C. Scope of the RCRA Permit Requirement (1) General RCRA requires a permit for treatment, storage, or disposal of any hazardous waste as identified or listed in 40 CFR 261 (Identification and Listing of Hazardous Waste).

122.21(d)(1)	(2) Specific Inclusions
	  (a) General
	    Owners/operators of certain facilities require RCRA permits as well as permits under other programs for certain aspects of the facility operation.

122.21(d)(1)(i)	(b) Underground Injection Wells (UIJ)
	    (i) RCRA permits are required for injection wells that dispose of hazardous waste, and associated surface facilities that treat, store, or dispose of hazardous waste (see Section 122.30–Interim Permits for UIC wells)
	    (ii) Permit by Rule for Injection Wells
	      Owner/operator with UIC permit in state with approved or promulgated UIC program will be deemed to have a RCRA permit for the injection well itself if they comply with requirements of Section 122.26(b) (Permit by Rule for injection wells).

122.21(d)(1)(ii)	(c) Publicly Owned Treatment Works (POTWs)
	    (i) RCRA permits are required for treatment, storage, or disposal of hazardous waste at facilities requiring an NPDES permit.
	    (ii) Permit by Rule for POTWs
	      Owners/operators of POTWs receiving hazardous waste will be deemed to have a RCRA permit for that waste if they comply with requirements of Section 122.26(c) (Permit by Rule for POTWs).

122.21(d)(1)(iii)	(d) Ocean Disposal Barges and Vessels
	    (i) RCRA permits are required for barges or vessels that dispose of hazardous waste by ocean disposal and onshore hazardous waste treatment or storage facilities associated with an ocean disposal operation.
	    (ii) Permits by Rule for Ocean Disposal Barges and Vessels
	      Owners/operators will be deemed to have a RCRA permit for Ocean disposal from barge or vessel if they comply with requirements of Section 122.26(a) (Permit by Rule for ocean disposal barges and vessels).

122.21(d)(2)	(2) Specific Exclusions Following persons are not required to obtain a RCRA permit:
122.21(d)(2)(i)	(a) Generators Who Accumulate Hazardous Waste On-Site for Less than 90 Days As provided in 40 CFR 262.34 (Accumulation Time of pre-transport requirements).
122.21(d)(2)(ii)	(b) Farmers Who Dispose of Hazardous Waste Pesticides from Their Own Use As provided in 40 CFR 262.51 (Farmers).
122.21(d)(2)(iii)	(c) Small Facility Persons who own or operate facilities solely for treatment, storage, or disposal of hazardous waste are excluded from regulations under this Part by 40 CFR (261.4 or 261.5)

Table 7-1B. Part 122, EPA-Administered Permit Programs: Additional Requirements for Hazardous Waste Programs under RCRA—Subpart B. (Continued)

Section No.	Description
	(Exclusions and Special Requirements for Hazardous Waste Generated by Small Quantity Generators, respectively).
122.21(d)(2)(iv)	(d) Owners or Operators of Totally Enclosed Treatment Facilities As defined in 40 CFR 260.10 (Totally Enclosed Treatment Facility).
122.22	II. Application for a Permit
122.22(a) 122.22(a)(1)	A. Existing Hazardous Waste Management (HWM) Facilities (1) Within 6 Months After Promulgation of 40 CFR 261 (11/18/80) Part A of permit application must be submitted to Regional Administrator by owner/operator of existing hazardous waste treatment, storage, or disposal facility.
122.22(a)(2)	(2) After Promulgation of Phase II Standards for 40 CFR 264 (End of 1980 or Beginning of 1981) (a) Owner/operator of HWM facility will be required to submit Part B of permit application at some time. (b) Submittal of Part B to State Director If State in which facility is located has received interim authorization for Phase II or final authorization. (c) Submittal of Part B to Regional Administrator If no State authorization. (d) Timing Owner/operator will be allowed at least 6 months from date of request to submit Part B. (e) Part B of application may be voluntarily submitted at any time.
122.22(a)(3)	(3) Failure to Furnish a Requested Part B Application on Time or to Furnish Full Information Required by Part B Constitutes grounds for termination of interim status under Part 124.
122.22(b) 122.22(b)(1)	B. New Hazardous Waste Management (HWM) Facilities (1) Start of Construction Not permitted without having submitted Part A and Part B of permit application and received a finally effective RCRA permit.
122.22(b)(2)	(2) Timing (a) Permit application for new HWM facility (including both Part A and Part B) may be filed any time after promulgation of Part 264 Phase II standards. (b) All applications must be submitted at least 180 days before physical construction is expected to commence. (3) Submittal to State Director If State in which facility is located has received interim authorization for Phase II or final authorization. (4) Submittal to Regional Administrator If no State authorization.
122.22(c) 122.22(c)(1)	C. Updating Permit Applications (1) Reasons for Owner/Operator to File Amended Part A Application When Part B Has Not Yet Been Filed
122.22(c)(1)(i) and (ii)	(a) Promulgation of Revised Regulations under Part 261 Listing or Identifying New Hazardous Wastes Handled by Facility (i) File amended permit with Regional Administrator if in State without interim authorization for Phase II or final authorization. (ii) File amended permit with State Director if in State with Phase II interim authorization or final authorization. (iii) File within 6 months of promulgation.
122.22(c)(1)(iii)	(b) To Comply with Provisions of Section 123.22, (Interim Status) for Changes during Interim Status (or Analogous Provisions of State Approved Program) (i) File with Regional Administrator or State Director, as may be required.
122.22(c)(2)	(2) Failure to Comply with Above Updated Requirements of Section 122.22(c)(1) Owner/operator does not receive interim status for wastes not covered by duly filed Part A application(s).
122.22(d)	D. Reapplications (1) Timing (a) Any HWM facility with an effective permit shall submit a new application at least 180 days before expiration date of effective permit. (i) Exception Permission for a later date has been granted by Director. (b) Director shall not grant permission for applications to be submitted later than expiration date of existing permit.

Table 7-1B. Part 122, EPA-Administered Permit Programs: Additional Requirements for Hazardous Waste Programs under RCRA—Subpart B. (Continued)

Section No.	Description
122.23	III. Interim Status
122.23(a)	A. Qualifying for Interim Status Any person who owns or operates an existing hazardous waste management (HWM) facility shall have interim status and shall be treated as having been issued a permit by complying with the following requirements.
122.23(a)(1)	(1) Notification Administrator is notified within 90 days from promulgation or revision of Part 261, as required in Section 3010 of RCRA. (a) This may be done by completing EPA form 8700-12 (see Fig. 9-1).
122.23(a)(2)	(2) Part A of Permit Application Comply with requirements of Section 122.22(a) and (c) (Application For A Permit) governing submission of Part A applications.
122.23(a)(3)	B. Disqualifying for Interim Status and Effect (1) EPA Examination or Re-examination of Part A of Application (a) Should this reveal that application fails to meet regulations, EPA may notify owner/operator of deficiency and lack of entitlement to interim status. (b) Owner/operator will then be subject to EPA enforcement for operating without permit.
122.23(b)	C. Coverage During interim status period the facility shall not:
122.23(b)(1)	(1) Treat, store, or dispose of hazardous waste not specified in Part A of permit application.
122.23(b)(2)	(2) Employ processes not specified in Part A of permit application.
122.23(b)(3)	(3) Exceed design capacities specified in Part A of permit application.
122.23(c)	D. Changes during Interim Status
122.23(c)(1)	(1) New Hazardous Wastes Not Previously Identified in Part A of Permit Application May be treated, stored, or disposed of at facility if owner/operator submits revised Part A permit application prior to such a change.
122.23(c)(2)	(2) Increases in Design Capacity of Processes Used at Facility May be made if: (a) Owner/operator submits revised Part A permit application prior to such a change. (b) Justification explaining need for change is also submitted. (c) Director approves change because of lack of available treatment, storage, or disposal capacity at other hazardous waste management facilities.
122.23(c)(3)	(3) Changes in Processes for Treatment, Storage, or Disposal of Hazardous Waste May be made at facility or additional processes may be added if: (a) Owner/operator submits revised Part A permit application prior to such a change. (b) Justification explaining need for change is also submitted. (c) Director approves change because:
122.23(c)(3)(i)	(i) It is necessary to prevent a threat to human health or environment because of emergency situation.
122.23(c)(3)(ii)	(ii) It is necessary to comply with: • Federal regulations, including interim status standards under 40 CFR 265; or • State or local laws.
122.23(c)(4)	(4) Changes in Ownership or Operational Control of a Facility (a) New owner/operator must submit revised Part A permit application no later than 90 days prior to scheduled change. (b) Financial Requirements (i) Upon transfer of ownership or operational control of facility, old owner/operator shall comply with Subpart H (Financial Requirements) of Part 265, until new owner/operator has demonstrated to Director that he is complying with that Subpart. (ii) Demonstration by New Owner/Operator of Compliance with Subpart H of Part 265 Upon such demonstration, Director shall notify old owner/operator in writing that he no longer needs to comply with that Part as of date of demonstration. (c) All other interim status duties are transferred effective immediately upon date of change of ownership or operational control of facility.
122.23(c)(5)	(5) Reconstruction of HWM Facility (a) Prohibited during interim status.

**Table 7-1B. Part 122, EPA-Administered Permit Programs: Additional Requirements for
Hazardous Waste Programs under RCRA—Subpart B. (Continued)**

Section No.	Description
	(b) Definition of Reconstruction Reconstruction occurs when capital investment in changes to facility exceeds fifty percent of capital cost of comparable entirely new HWM facility.
122.23(d)	E. Interim Status Standards During interim status, owners/operators shall comply with interim status facility standards of 40 CFR 265.

122.23(e)	F. Grounds for Termination of Interim Status
	   Interim status terminates when:
	   (1) Final administrative disposition of a permit application is made; or
	   (2) Interim status is terminated as provided in Section 122.22(a)(3) (Application For A Permit).

122.24	IV. Contents of Part A
	A. Information Required in Part A of Permit Application In addition to information in Section 122.4(d) (Application For A Permit), Part A of RCRA application shall include following information.
122.24(a)	(1) Latitude and longitude of facility.
122.24(b)	(2) Name, address, and telephone number of owner of facility.
122.24(c)	(3) Indication of whether facility is new or existing. (4) Indication of whether application is first one submitted or revised.
122.24(d)	(5) For Existing Facilities (a) Scale drawing of facility showing location of all past, present, and future treatment, storage, and disposal areas.

122.24(e)	(b) Photographs of facility clearly delineating all:
	         (i) Existing structures.
	         (ii) Existing treatment, storage, and disposal areas.
	         (iii) Sites of future treatment, storage, and disposal areas.

122.24(f)	(6) Description of processes to be used for treating, storing, and disposing of hazardous waste, and design capacity of these items.

122.24(g)	(7) Hazardous Wastes Listed or Designated under 40 CFR 261 to Be Treated, Stored, or Disposed of at the Facility
	      (a) Specification of such wastes.
	      (b) Estimate of quantity of such wastes to be treated, stored, or disposed of annually.
	      (c) General description of processes to be used for such wastes.

122.25	V. Contents of Part B
122.25(a)	A. General Information Required in Part B of Permit Application
122.25(a)(1)	(1) General description of facility.
122.25(a)(2)	(2) Chemical and Physical Analysis of Hazardous Wastes to Be Handled at Facility At a minimum, these analyses shall contain all information which must be known to treat, store, or dispose of wastes in accordance with Part 264.
122.25(a)(3)	(3) Copy of waste analysis plan required in Section 264.13(b) (General Waste Analysis) and if applicable Section 264.13(c).
122.25(a)(4)	(4) Description of security procedures and equipment required by Section 264.14 (Security), or a justification demonstrating reasons for requesting waiver of this requirement.
122.25(a)(5)	(5) Copy of general inspection schedule required by 264.15(b) (General Inspection Requirements).
122.25(a)(6)	(6) Justification of any request for waiver(s) of preparedness and prevention requirements of Subpart C of Part 264.
122.25(a)(7)	(7) Copy of contingency plan required in Subpart D of Part 264.
122.25(a)(8)	(8) Description of procedures, structures, or equipment used at facility to:
122.25(a)(8)(i)	(a) Prevent uncontrolled reaction of incompatible waste (e.g., procedures to avoid fires, explosions, or toxic gases).
122.25(a)(8)(ii)	(b) Prevent hazards in unloading operations (e.g., ramps, special forklifts).
122.25(a)(8)(iii)	(c) Prevent runoff from hazardous waste handling areas to other areas of facility or environment, or to prevent flooding (e.g., berms, dikes, trenches).
122.25(a)(8)(iv)	(d) Prevent contamination of water supplies.
122.25(a)(8)(v)	(e) Mitigate effects of equipment failure and power outages.
122.25(a)(8)(vi)	(f) Prevent undue exposure of personnel to hazardous waste (e.g., protective clothing).

Table 7-1B. Part 122, EPA-Administered Permit Programs: Additional Requirements for Hazardous Waste Programs under RCRA—Subpart B. (Continued)

Section No.	Description
122.25(a)(9)	(9) Traffic Pattern, Volume, and Control If Appropriate
	(a) For example, show turns across traffic lanes and stacking lanes.
	(b) Provide access road surfacing and load bearing capacity.
	(c) Show traffic control signals.
	(d) Provide estimates of traffic volume (number and type vehicles).
	NOTES: (1) Requirements set forth in Section 122.25(a) above reflect those permit application requirements related to the initial promulgation of Part 264. (2) Additional permit application requirements including specific design and operating data, financial plans, and site engineering information will be promulgated when remaining portions of Part 264 are promulgated.
122.26	VI. Permits by Rule
	A. General
	Notwithstanding any other provision of this Part or Part 124, the facilities of Sections 122.26(a), (b) and (c) that follow shall be deemed to have a RCRA permit if the conditions listed are met.
122.26(a)	B. Ocean Disposal Barges or Vessels
	Owner/operator accepts hazardous waste for ocean disposal and complies with following conditions:
122.26(a)(1)	(1) Has permit for ocean dumping issued under 40 CFR 220 (Ocean Dumping, authorized by Marine Protection, Research, and Sanctuaries Act, as amended, 33 U.S.C. Section 1420 et seq.).
122.26(a)(2)	(2) Complies with conditions of that permit.
122.26(a)(3)	(3) Complies with following hazardous waste regulations:
122.26(a)(3)(i)	(a) 40 CFR 264.11–Identification Number.
122.26(a)(3)(ii)	(b) 40 CFR 264.71–Use of Manifest System.
122.26(a)(3)(iii)	(c) 40 CFR 264.72–Manifest Discrepancies.
122.26(a)(3)(iv)	(d) 40 CFR 264.73(a) and (b)(1)–Operating Record.
122.26(a)(3)(v)	(e) 40 CFR 264.75–Annual Report.
122.26(a)(3)(vi)	(f) 40 CFR 264.76–Unmanifested Waste Report.
122.26(b)	C. Injection Wells
	Owner/operator uses injection well(s) to dispose of hazardous waste and complies with following conditions:
122.26(b)(1)	(1) Has permit for undergound injection issued under Subpart C of Part 122 or Subpart C of Part 123.
122.26(b)(2)	(2) Complies with conditions of that permit.
	(3) Complies with requirements of Section 122.45 (Requirements for Wells Injecting Hazardous Waste).
122.26(c)	D. Publicly Owned Treatment Works (POTWs)
	Owner/operator accepts hazardous waste for treatment and complies with following conditions:
122.26(c)(1)	(1) Has an NPDES permit.
122.26(c)(2)	(2) Complies with conditions of that permit.
122.26(c)(3)	(3) Complies with following hazardous waste regulations:
122.26(c)(3)(i)	(a) 40 CFR 264.11–Identification Number.
122.26(c)(3)(ii)	(b) 40 CFR 264.71–Use of Manifest System.
122.26(c)(3)(iii)	(c) 40 CFR 264.72–Manifest Discrepancies.
122.26(c)(3)(iv)	(d) 40 CFR 264.73(a) and (b)(1)–Operating Record.
122.26(c)(3)(v)	(e) 40 CFR 264.75–Annual Report.
122.26(c)(3)(vi)	(f) 40 CFR 264.76–Unmanifested Waste Report.
122.26(c)(4)	(4) The waste meets all Federal, State and local pretreatment requirements which would be applicable to the waste if it were being discharged into the POTW through a sewer, pipe, or similar conveyance.
122.27	VII. Emergency Permits
	A. General
	Notwithstanding any other provision of this Part or Part 124, if Director finds an imminent and substantial endangerment to human health or environment he may issue a temporary emergency permit to a facility to allow treatment, storage, or disposal of hazardous waste for:
	(1) A nonpermitted facility; or
	(2) An interim permitted or permitted facility to perform operations not covered by permit (e.g., treatment-permitted facility to temporarily store hazardous wastes).

Table 7-1B. Part 122, EPA-Administered Permit Programs: Additional Requirements for Hazardous Waste Programs under RCRA—Subpart B. (Continued)

Section No.	Description
	B. Emergency Permit Particulars
122.27(a)	(1) May Be Oral or Written
	If oral, it shall be followed within 5 days by a written emergency permit.
122.27(b)	(2) Shall not exceed 90 days in duration.
122.27(c)	(3) Shall clearly specify hazardous wastes to be received, and manner and location of their treatment, storage, or disposal.
122.27(d)	(4) May be terminated by Director at any time without process if he determines that termination is appropriate to protect human health and environment.
122.27(e)	(5) Shall be accomplished by a public notice published under Section 124.11(b) (Public Comments and Requests for Public Hearings) including:
122.27(e)(1)	(a) Name and address of office granting emergency authorization.
122.27(e)(2)	(b) Name and location of permitted HWM facility.
122.27(e)(3)	(c) Brief description of wastes involved.
122.27(e)(4)	(d) Brief description of action authorized and reasons for authorizing it.
122.27(e)(5)	(e) Duration of emergency permit.
122.27(f)	(6) Shall incorporate, to extent possible and not inconsistent with the emergency situation, all applicable requirements of this Part and Parts 264 and 266.
122.28	VIII. Additional Conditions Applicable to All RCRA Permits
	A. Conditions That Apply to RCRA Permits in Addition to Those of Section 122.7 (Conditions Applicable to All Permits)
122.28(a)	(1) In Addition to Section 122.7(a) (Duty to Comply)
	Permittee need not comply with conditions of permit to extent and for duration such non-compliance is authorized in an emergency permit (see Section 122.27—Emergency Permits).
122.28(b)	(2) In Addition to Section 122.7(j) (Monitoring)
	Permittee shall maintain records from all ground monitoring wells and associated ground-water surface elevations, for active life of facility, and for disposal facilities for post-closure care period as well.
122.28(c)	(3) In Addition to Section 122.7(1)(1) (Notice of Planned Changes)
	For new or modified portion of HWM facility, permittee may not begin treatment, storage or disposal of hazardous waste until:
122.28(c)(1)	(a) Permittee has submitted to Director by certified mail or hand delivery a letter signed by permittee and registered professional engineer stating that facility has been constructed or modified in accordance with permit; and
	(b) Inspection Requirement Met
122.18(c)(2)(i)	(i) Director has inspected modified or newly constructed facility and finds it is in compliance with conditions of permit; or
122.28(c)(2)(ii)	(ii) Inspection is waived and permittee may commence treatment, storage, or disposal of hazardous waste, if within 15 days of date of submission of letter of Section 122.28(c)(1) permittee has not received notice from Director of intent to inspect.
122.28(d)	(4) Information Which Must Be Reported Orally within 24 Hours under Section 122.7(1)(6) (Twenty-Four Hour Reporting)
122.28(d)(1)	(a) Information concerning release of any hazardous waste that may cause an endangerment to public drinking water supplies.
122.28(d)(2)	(b) Information of release or discharge of hazardous waste or fire or explosion from HWM facility, which could threaten environment or human health outside facility.
122.28(d)(2)(i)	(i) Name, address, and telephone number of owner or operator.
122.28(d)(2((ii)	(ii) Name, address, and telephone number of facility.
122.28(d)(2)(iii)	(iii) Date, time, and type of incident.
122.28(d)(2)(iv)	(iv) Name and quantity of material(s) involved.
122.28(d)(2)(v)	(v) Extent of injuries, if any.
122.28(d)(2)(vi)	(vi) Assessment of actual or potential hazards to environment and human health outside the facility.
122.28(d)(2)(vii)	(vii) Estimated quantity and disposition of recovered material that resulted from incident.
	(viii) Director may waiver the five-day written notice requirement in favor of a written report within 15 days.
122.28(e)	(5) Reports Required by Part 264 in Addition to Those Required by Section 122.7(1) (Reporting Requirements)
122.28(e)(1)	(a) Manifest Discrepancy Report
	(i) If significant discrepancy in manifest is discovered, permittee must attempt to reconcile discrepancy.

Table 7-1B. Part 122, EPA-Administered Permit Programs: Additional Requirements for Hazardous Waste Programs under RCRA—Subpart B. (Continued)

Section No.	Description
	(ii) If Not Resolved in 15 Days Permittee must submit letter report to Director including copy of manifest (see 40 CFR 264.72—Manifest Discrepancies).
122.28(e)(2)	(b) Unmanifested Waste Report Must be submitted to director within 15 days of receipt of unmanifested waste (see Section 264.76—Unmanifested Waste Report).
122.28(e)(3)	(c) Annual Report This report must be submitted covering facility activities during previous calendar year (see 40 CFR 264.75—annual Report).
122.28(e)(3) Note	*Notes:* (1) Above reports are required in Part 264 as initially promulgated. (2) Additional reports will be required and added to this section when remaining portions of Part 264 are promulgated.
122.29	IX. Establishing RCRA Permit Conditions
	A. RCRA Permit Shall Include Each Applicable Requirement Specified in 40 CFR 264 and 266 In addition to conditions established under Section 122.8(a) (Establishing Permit Condition).
122.30	X. Interim Permits for UIC Wells
	A. For State with No Approved or Promulgated UIC Program (1) Director may issue a permit under this Part to any Class I UIC well (see Section 122.32—Classification of Injection Wells). (2) Any such permit shall apply and ensure compliance with all applicable requirements of 40 CFR 264, Subpart R (RCRA Standards for Wells). (3) Term of Permit Not to exceed two years.
	B. For State with Approved or Promulgated UIC Program No such permit shall be issued.
	C. Condition of Permit under This Section It terminates upon final action by the Director under a UIC program to issue or deny a UIC permit for the facility.

Table 7-1C. Part 122, EPA-Administered Permit Programs: Additional Requirements for Underground Injection Control Programs under Safe Drinking Water Act—Subpart C.

Section No.	Description
122.45	I. Requirements for Wells Injecting Hazardous Waste
122.45(a)	A. Applicability Regulations in this Section apply to all generators of hazardous waste, and to owners/operators of hazardous waste management facilities, using any class of well to inject hazardous wastes accompanied by a manifest (See also Section 122.36—Elimination of Certain Class IV Wells).
122.45(b)	B. Authorization Owner/operator of any well used to inject hazardous wastes accompanied by a manifest or delivery document shall apply for authorization to inject as specified in Section 122.38 (Application for a Permit; Authorization by Permit) within 6 months after approval of an applicable State program.
122.45(c)	C. Requirements In addition to requiring compliance with applicable requirements of this Part and 40 CFR 146, Subparts B–F, the Director shall, for each facility meeting requirements of Section 122.45(b) require owner/operator to comply with following:
122.45(c)(1)	(1) Notification Owner/operator shall comply with notification requirements of Section 3010 of Public law 94-580 (RCRA).

Table 7-1C. Part 122, EPA-Administered Permit Programs: Additional Requirements for Underground Injection Control Programs under Safe Drinking Water Act—Subpart C. (Continued)

Section No.	Description
122.45(c)(2)	(2) Identification Number Owner/operator shall comply with requirements of 40 CFR 264.11 (Identification Number).
122.45(c)(3)	(3) Manifest System Owner/operator shall comply with applicable recordkeeping and reporting requirements for manifested wastes in 40 CFR 264.71 (Use of Manifest System).
122.45(c)(4)	(4) Manifest Discrepancies Owner/operator shall comply with 40 CFR 264.72 (Manifest Discrepancies).
122.45(c)(5)	(5) Operating Record Owner/operator shall comply with 40 CFR 264.73(a), (b)(1), and (b)(2) (Operating Record).
122.45(c)(6)	(6) Annual Report Owner/operator shall comply with 40 CFR 264.75 (Annual Report).
122.45(c)(7)	(7) Unmanifested Waste Report Owner/operator shall comply with 40 CFR 264.76 (Unmanifested Waste Report).
122.45(c)(8)	(8) Personnel Training Owner/operator shall comply with applicable personnel training requirements of 40 CFR 264.16 (Personnel Training).
122.45(c)(9)	(9) Certification of Closure When abandonment is completed, owner/operator must submit to Director certification by himself and an independent registered professional engineer that facility had been closed in accordance with specifications in Section 122.42(f) (Plugging and Abandonment).

Table 7-2A. Part 124, Procedures for Decisionmaking: General Program Requirements—Subpart A.

Section No.	Description
124.1	I. Purpose, Scope and Applicability
124.1(a)	A. Content of Part 124 (1) Overall (a) This Part contains EPA procedures for: (i) Issuing (ii) Modifying (iii) Revoking and reissuing (iv) Terminating RCRA permits. (b) "Permits" Not Included (i) Emergency permits under Section 122.27 (Emergency Permits). (ii) Permits by rule under Section 122.26 (Permits by Rule).
124.1(b)	B. Organization of Part 124 It is organized into six Subparts, four of which may relate to hazardous waste. (1) Subpart A Contains general procedural requirements applicable to all permit programs covered by these regulations, and describes steps EPA will follow in: (a) Receiving permit applications. (b) Preparing draft permits. (c) Issuing public notice. (d) Inviting public comment. (e) Holding public hearings on draft permits. (f) Assembling an administrative record. (g) Responding to comments. (h) Issuing a final permit decision. (i) Allowing for administrative appeal of the final permit decision. (2) Subpart B Reserved for specific procedural requirements for RCRA permits, none of which presently exist (may be added in future). (3) Subpart E These procedures take over for EPA-issued NPDES permits and EPA-terminated RCRA permits. (4) Subpart F Based on provisions of Administrative Procedures Act (APA), and can be used instead of Subparts A–E in appropriate cases.
124.1(c)	C. Hearings Available under This Part Part 124 offers an opportunity for three kinds of hearings: (1) Public Hearing under Subpart A On draft permit at Director's discretion or on request (Section 124.12–Public Hearing). (2) Evidentiary Hearing under Subpart E Permit termination under Section 3008 of RCRA. (3) Panel Hearing under Subpart F At Regional Administrator's discretion in lieu of public hearing (Section 124.12–Public Hearing and Section 124.111(a)(3)–Applicability).
124.1(d)	D. Multiple Permits at Single Facility (1) This Part allows multiple permits to be processed separately or together at choice of Regional Administrator. (2) Regional Administrator Consolidating Permit Processing May be done when: (a) Permit applications are submitted; (b) Draft permits are prepared; or (c) Final permit decisions are issued. (3) Permit applicants may recommend whether or not their applications should be consolidated in any given case.
	E. Hearings This Part allows consolidated permits to be subject to a single hearing. (1) Public hearing under 124.12 (Public Hearing); or (2) Evidentiary hearing under 124.75 (Decision on Request for a Hearing); or (3) Non-adversary panel hearing under Section 124.120 (Panel Hearing).

Table 7-2A. Part 124, Procedures for Decisionmaking: General Program Requirements—Subpart A. (Continued)

Section No.	Description
124.1(e)	F. Certain procedural requirements set forth in Part 124 must be adopted by States in order to gain EPA approval to operate RCRA permit programs (listed in Section 123.7).
124.1(f)	G. Coordination of Decisionmaking When Different Permits Will Be Issued by EPA and Approved State Programs This part allows: (1) Applications to be jointly processed. (2) Joint comment periods and hearings to be held. (3) Final permits to be issued on a cooperative basis whenever EPA and a State agree to take such steps in general or in individual cases. (a) These joint processing agreements may be provided in the Memorandum of Agreement developed under Section 123.6 (Memorandum of Agreement with Regional Administrator).
124.21 124.21(b)	H. Effective Date of Part 124 All provisions of Part 124 pertaining to the RCRA program will become effective November 19, 1980.
124.3	II. Application for a Permit
124.3(a)(1)	A. General (1) Person Requiring a RCRA Permit He shall complete, sign, and submit to Director an application under Section 122.21 (Purpose and Scope of Subpart B). (2) Applications not required for RCRA permits by rule (Section 122.26—Permits by Rule).
122.3(a)(2)	(3) Start of Processing of Permit by Director Shall not begin until applicant has fully complied with application requirements for the permit under Section 122.22 (Application for a Permit).
122.3(a)(3)	(4) Compliance with Signature and Certification Requirements Required of permit applicants under Section 122.6 (Signatories to Permit Applications and Reports).

122.3(c)	B. Review of Application Completeness by Regional Administrator for EPA-Issued Permits
	(1) For New Hazardous Waste Management (HWM) Facility
	Completeness review by Regional Administrator required within 30 days of receipt of application.
	(2) For Existing HWM Facility
	Review required within 60 days of receipt of application.
	(3) Written notification to applicant required by Regional Administrator as to whether application is complete.
	(4) If Application Considered Incomplete by Regional Administrator
	    (a) He must list information necessary to make application complete.
	    (b) For Existing HWM Facility
	    Regional Administrator shall specify in notice of deficiency, a date for submitting necessary information.
	        (i) Regional Administrator shall notify applicant that application is complete upon receiving this information.
	
	C. Request for Additional Information by Regional Administrator after Application is Complete
	(1) To be done only when necessary to clarify, modify, or supplement previously submitted material.
	(2) Requests for such material will not render an application incomplete.

122.3(d)	D. If Applicant Fails or Refuses to Correct Deficiencies In Application (1) Permit may be denied; and (2) Appropriate enforcement actions may be taken under applicable statutory provision including RCRA Section 3008.
122.3(e)	E. If Regional Administrator Decides that Site Visit Is Necessary for Processing Application He shall notify applicant and a date shall be scheduled.
122.3(f)	F. Effective Date of an Application The date on which Regional Administrator notifies applicant that application is complete as provided in Section 122.3(c).

Table 7-2A. Part 124, Procedures for Decisionmaking: General Program Requirements—Subpart A. (Continued)

Section No.	Description
122.3(g)	G. Project Decision Schedule For each application from a major new HWM facility, Regional Administrator shall, no later than effective date of application, prepare and mail to applicant a project decision schedule which shall specify dates by which he intends to:
122.3(g)(1)	(1) Prepare a draft permit.
122.3(g)(2)	(2) Give public notice.
122.3(g)(3)	(3) Complete the public comment period, including any public hearing.
122.3(g)(4)	(4) Issue a final permit.
124.4	III. Consolidation of Permit Processing
124.4(a)(1)	A. Whenever Facility or Activity Requires a Permit under More than One Statute Covered by These Regulations (1) Processing of two or more applications for those permits may be consolidated. (2) First Step in Consolidation Prepare each draft permit at same time.
124.4(a)(2)	(3) Next Steps in Consolidation Whenever draft permits are prepared at same time, the following should also be consolidated: (a) Statements of Basis Required under Section 124.7 (Statement of Basis for EPA-issued permits only); or (b) Fact Sheets Under Section 124.8 (Fact Sheet). (c) Administrative Records Required under Section 124.9 (Administrative Record for Draft Permits When EPA is the Permitting Authority) for EPA-issued permits only. (d) Public Comment Periods Under Section 124.10 (Public Notice of Permit Actions and Public Comment Period). (e) Public Hearings on Those Permits Under Section 124.12 (Public Hearings). B. Final Permits May Be Issued Together However, they need not be issued together if in judgment of Regional Administrator or State Director(s) joint processing would result in unreasonable delay in issuance of one or more permits.
124.4(b)	C. Coordination of Expiration Date(s) of New Permits and Existing Permits (1) For statutes covered by these regulations, the permitting authority may coordinate the various permits so that they all expire simultaneously. (2) Processing of subsequent applications for renewal permits may then be consolidated.
124.4(c) 124.4(c)(1)	D. Administrative Alternates in Consolidation of Processing of Permit Applications (1) Director may consolidate permit processing at his discretion whenever facility or activity requires all permits either from: (a) EPA; or (b) An approved state.
124.4(c)(2)	(2) Regional Administrator and State Director(s) may agree to consolidate draft permits whenever facility of activity requires permits from both EPA and an approved State.
124.4(c)(3)	(3) Permit applicants may recommend whether or not processing of their applications should be consolidated.
124.7(d)	E. When Permit Processing Is Consolidated and Regional Administrator Invokes Provisions of Subpart F for Single Type Permit (RCRA, UIC, NPDES) Other type permit(s) shall likewise be processed under Subpart F.
124.5	IV. Modification, Revocation and Reissuance, or Termination of Permits
124.5(a)	A. General (1) Who May Request Permits To Be Modified, Revoked and Reissued, or Terminated (a) Any interested person, including permittee; or (b) Director. (2) Reasons By Which Permits May Be Modified, Revoked and Reissued, or Terminated Only as specified in: (a) Section 122.15—Modification or Revocation and Reissuance of Permits; or (b) Section 122.16—Termination of Permits. (3) All requests shall be in writing and shall contain facts or reasons supporting request.

Table 7-2A. Part 124, Procedures for Decisionmaking: General Program Requirements—Subpart A. (Continued)

Section No.	Description

124.5(b)	B. If Director Decides Request Is Not Justified
	   (1) He shall send requester brief written response giving reason for decision.
	   (2) Denials of requests not subject to public notice, comment, or hearings.
	
	C. Denials of Request by Regional Administrator
	   (1) May be informally appealed to Administrator.
	   (2) Provide letter briefly setting forth the relevant facts.
	   (3) Administrator may direct Regional Administrator to begin modification, revocation and reissuance, or termination proceedings under Section 124.5(c).
	   (4) If Administrator Takes No Action on Letter within 60 Days of Receiving It
	     (a) The appeal shall be considered denied.
	     (b) This informal appeal is, under 5 U.S.C. Section 704, a prerequisite to seeking judicial review of EPA action.

124.5(c)(1)	D. If Director Tentatively Decides to Modify or Revoke and Reissue a Permit Under Section 122.15
	   (1) Director shall prepare draft permit under Section 124.6 (Draft Permit) incorporating proposed changes.
	   (2) He may request additional information.
	   (3) For Modified Permit
	     Director may require submission of updated permit application.
	   (4) For Revoked and Reissued Permit(s)
	     Director shall require submission of new application.

124.5(c)(2)	E. For Actual Permit Modification
	   (1) Only those conditions to be modified shall be reopened when a new draft permit is prepared.
	   (2) All other aspects of existing permit shall remain in effect for duration of the unmodified permit.
	
	F. For Actual Permit Revocation and Reissuance
	   (1) Entire permit is reopened just as if permit had expired and was being reissued.
	   (2) During Revocation and Reissuance Proceeding
	     Permittee shall comply with all conditions of existing permit until new final permit is reissued.

124.5(c)(3)	G. Minor Modifications As Defined In Section 122.17 (Minor Modification of Permits) Not subject to requirements of this Section.

124.5(d)	H. If Director Tentatively Decides to Terminate Permit under Section 122.16 (Termination of Permits)
	   (1) Director shall issue notice of intent to terminate permit.
	     (a) This document is a type of draft permit which follows same procedures as any draft permit prepared under Section 124.6 (Draft Permit).
	   (2) For EPA-Issued Permits
	     A notice of intent to terminate shall not be issued if Regional Administrator and permittee agree to termination in the course of transferring permit responsibility to an approved State under Section 123.6(b) (1) (Memorandum of Agreement with Regional Administrator).

124.5(e)	I. When EPA Is Permitting Authority All draft permits (including notices of intent to terminate) prepared under this Section shall be based on administrative record defined in Section 124.9 (Administrative Record For Draft Permits When EPA Is Permitting Authority).
124.6	V. Draft Permits

124.6(a)	A. Once Permit Application Is Complete
	   Director shall tentatively decide whether to:
	   (1) Prepare a draft permit; or
	   (2) Deny the application.

124.6(b)	(a) Director must issue notice of intent to deny. (b) Notice of intent to deny permit application is a type of draft permit which follows same procedures as any draft permit prepared under Section 124.6(c).

Table 7-2A. Part 124, Procedures for Decisionmaking: General Program Requirements—Subpart A. (Continued)

Section No.	Description
	(c) If Director's final decision under Section 124.15 (Issuance and Effective Date of Permit) is that the tentative decision to deny the permit application was incorrect, the notice of intent to deny shall be withdrawn and draft permit under Section 124.6(d) shall be prepared.
124.6(d)	B. Content of Draft Permit
124.6(d)(1)	(1) All conditions under Section 122.7 (Conditions Applicable to All Permits).
	(2) All conditions under Section 122.8 (Establishing Permit Conditions).
124.6(d)(2)	(3) All compliance schedules under Section 122.10 (Schedules of Compliance).
124.6(d)(3)	(4) All monitoring requirements under Section 122.11 (Requirements for Recording and Reporting of Monitoring Results).
124.6(d)(4)(i)	(5) Standards for treatment, storage, and/or disposal and other permit conditions under Section 122.28 (Additional Conditions Applicable to All RCRA Permits).
124.6(e)	C. Draft Permits Prepared by EPA
	(1) Shall be accompanied by a:
	(a) Statement of basis under Section 124.7 (Statement of Basis); or
	(b) Fact sheet under Section 124.8 (Fact Sheet).
	(2) Shall be based on administrative record under Section 124.9 (Administrative Record for Draft Permits When EPA Is the Permitting Authority).
	(3) Shall be publicly noticed under Section 124.10 (Public Notice of Permit Actions and Public Comment Period).
	(4) Shall be made available for public comment under Section 124.11 (Public Comments and Requests for Public Hearings).
	(5) Regional Administrator shall:
	(a) Give notice of opportunity for a public hearing under Section 124.12 (Public Hearing).
	(b) Issue a final decision under Section 124.15 (Issuance and Effective Date of Permit).
	(c) Respond to comments under Section 124.17 (Response to Comments).
	(6) An appeal may be taken under Section 124.19 (Appeal of RCRA, UIC and PSD Permits).
	D. Draft Permits Prepared by a State Shall be accompanied by a fact sheet if required under Section 124.8 (Fact Sheet).
124.7	VI. Statement of Basis
	A. When EPA Must Prepare Statement of Basis Required for every draft permit for which a fact sheet under Section 124.8 (Fact Sheet) is not prepared.
	B. Content
	(1) Briefly describes derivation of conditions of draft permit and reasons for them; or
	(2) In case of notice of intent to deny or terminate, reasons supporting the tentative decision.
	C. Who Statement of Basis Is Sent To:
	(1) To applicant
	(2) On request, to any other person.
124.8	VII. Fact Sheet
124.8(a)	A. When Fact Sheet Must Be Prepared
	(1) For every draft permit for a major hazardous waste management (HWM) facility.
	(2) When Director finds:
	(a) Draft permit is subject to widespread public interest; or
	(b) Raises major issues.
	B. Purpose Fact sheet shall briefly set forth:
	(1) Principal facts.
	(2) Significant factual, legal, methodological and policy questions considered in preparing draft permit.
	C. Who Fact Sheet Is Sent To
	(1) To Applicant
	(2) On request, to any other person.

Table 7-2A. Part 124, Procedures for Decisionmaking: General Program Requirements—Subpart A. (Continued)

Section No.	Description
124.8(b)	D. Content (Where Applicable)
124.8(b)(1)	(1) Brief description of type facility or activity which is subject of draft permit.
124.8(b)(2)	(2) Type and quantity of wastes which are proposed to be or are being treated, stored, or disposed of.
124.8(b)(4)	(3) Summary of the Basis for Draft Permit Conditions
	(a) Include references to applicable statutory or regulatory provisions; and
	(b) Appropriate supporting references to administrative record required by Section 124.9 (Administrative Record for Draft Permits When EPA is the Permitting Authority) for EPA-issued permits.
124.8(b)(5)	(4) Reasons why any requested variances or alternatives to required standards do or do not appear justified.
124.8(b)(6)	(5) Description of Procedures for Reaching Final Decision on Draft Permit Includes:
124.8(b)(6)(i)	(a) Beginning and ending dates of comment period under Section 124.10 (Public Notice of Permit Actions and Public Comment Period).
	(b) Address where comments will be received.
124.8(b)(6)(ii)	(c) Procedures for requesting a hearing and nature of that hearing.
124.8(b)(6)(iii)	(d) Any other procedures by which public may participate in final decision.
124.8(b)(7)	(6) Name and telephone number of person to contact for additional information.
124.9	VIII. Administrative Record for Draft Permits When EPA Is Permitting Authority
124.9(a)	A. Provisions of Draft Permit Prepared by EPA under Section 124.6 (Draft Permit) Shall be based on administrative record defined in this Section.
124.9(b)	B. Content of Administrative Record in Preparing Draft Permit under Section 124.6 (Draft Permit)
124.9(b)(1)	(1) Application, if required, and any supporting data furnished by applicant.
124.9(b)(2)	(2) Draft permit or notice of intent to deny the application or to terminate permit.
124.9(b)(3)	(3) Statement of basis under Section 124.7 (Statement of Basis); or
	(4) Fact sheet under Section 124.8 (Fact Sheet)
124.9(b)(4)	(5) All documents cited in statement of basis or fact sheet.
124.9(b)(5)	(6) Other documents contained in supporting file for the draft permit.
124.9(b)(6)	C. Environmental Impact Statement Provisions of Section 102(2)(C) of National Environmental Policy Act, 42 U.S.G. 4321 RCRA permits not subject to these provisions.
124.9(c)	D. Material Included in Administrative Record under Sections 124.9(b) and (c) Material need not be physically included with rest of record if:
	(1) It is readily available at issuing Regional Office; or
	(2) Is published material that is generally available; and
	(3) It is specifically referred to in the statement of basis or fact sheet.
124.9(d)	E. Applicability This Section applies to all draft permits when public notice was given after effective date of these regulations.
124.10	IX. Public Notice of Permit Actions and Public Comment Period
124.10(a)	A. Scope
124.10(a)(1)	(1) Actions for Which Director Will Give Public Notice
124.10(a)(1)(i)	(a) A Permit Application Has Been Tentatively Denied Under Section 124.6(b) (Draft Permit).
124.10(a)(1)(ii)	(b) A Draft Permit Has Been Prepared Under Section 124.6(d) (Draft Permit).
124.10(a)(1)(iii)	(c) A Hearing Has Been Scheduled Under:
	(i) Section 124.12–Public Hearing
	(ii) Subpart E–Evidentiary Hearing for EPA-Terminated RCRA Permits; or
	(iii) Subpart F–Non-Adversary Panel Procedures.
124.10(a)(1)(iv)	(d) An Appeal Has Been Granted Under Section 124.19(c) (Appeal of RCRA, UIC, and PSD Permits).

Table 7-2A. Part 124, Procedures for Decisionmaking: General Program Requirements—Subpart A. (Continued)

Section No.	Description
124.10(a)(2)	(2) Denial of Request for Permit Modification, Revocation and Reissuance, or Termination Under Section 124.5(b)
	(a) No public notice is required.
	(b) Written notice of denial shall be given to requester and permittee.
124:10(a)(3)	(3) Public notices may describe more than one permit or permit action.
124.10(b)	B. Timing
124.10(b)(1)	(1) Public Notice of Preparation of Draft Permit
	(a) Including notice of intent to deny a permit application.
	(b) At least 30 days shall be allowed for public comment.
124.10(b)(2)	(2) Public Notice of a Public Hearing
	(a) Shall be given at least 30 days before hearing.
	(b) Public notice of hearing may be given at same time as public notice of draft permit.
	(i) Two notices may be combined.
124.10(c)	C. Methods for Giving Public Notice of Activities Described in Section 124.10(a)(1)
124.10(c)(1)	(1) Mailing Copy of Notice to Interested Persons
124.10(c)(1)(i)	(a) The applicant.
124.10(c)(1)(ii)	(b) Any Other Agency Which Has Issued or Is Required to Issue RCRA, PSD, NPDES, or 404 Permit for Same Facility or Activity
	Includes EPA when draft permit is prepared by State.
124.10(c)(1)(iii)	(c) Federal And State Agencies
	(i) With jurisdiction over fish, shellfish, and wildlife resources.
	(ii) With jurisdiction over coastal zone management plans.
	(iii) Advisory Council on Historic Preservation.
	(iv) State Historic Preservation Officers.
	(v) Other appropriate government authorities, including any affected States.
124.10(c)(1)(viii)	(d) Other Persons on Mailing List
124.10(c)(1)(viii)(A)	(i) Those who request in writing to be on list.
124.10(c)(1)(viii)(B)	(ii) Participants from past permit proceeding in that area.
124.10(c)(1)(viii)(C)	(iii) Those responding to opportunity to be put on mailing list through periodic publication in public press, Regional and State newsletters, environmental bulletins, and others.
124.10(c)(2)	(2) For Major Permits
	Publication of a notice in a daily or weekly newspaper within area affected by facility or activity.
124.10(c)(3)	(3) For Program Being Administered by an Approved State
	Notice shall be given in a manner constituting legal notice to the public under State law.
124.10(c)(4)	(4) Any Other Method Reasonably Calculated to Give Actual Notice of Action in Question to Persons Potentially Affected by It
	Includes:
	(a) Press releases; or
	(b) Any other forum or medium to elicit public participation.
124.10(d)	D. Contents
124.10(d)(1)	(1) All Public Notices: Minimum Information Requirements
124.10(d)(1)(i)	(a) Name and address of office processing the permit for which notice is being given.
124.10(d)(1)(ii)	(b) Name and address of permittee or permit applicant and, if different, of facility regulated by permit.
124.10(d)(1)(iv)	(c) Name, Address, and Telephone Number of Person from whom Interested Persons May Obtain Further Information
	Including copies of:
	(i) Draft permit or draft general permit.
	(ii) Statement of basis or fact sheet.
	(iii) Application.
124.10(d)(1)(v)	(d) Procedures
	(i) Brief description of comment procedures required by:
	• Section 124.11—Public Comments and Requests for Public Hearings.
	• Section 124.12—Public Hearings.
	(ii) Time and place of any hearing that will be held, including a statement of procedures to request a hearing.
	(iii) Other procedures by which the public may participate in the final permit decision.

Table 7-2A. Part 124, Procedures for Decisionmaking: General Program Requirements—Subpart A. (Continued)

Section No.	Description
124.10(d)(1)(vi)	(e) For EPA-Issued Permits 　　(i) Location of administrative record required by Section 124.9 (Administrative Record for Draft Permits When EPA Is Permitting Authority). 　　(ii) The times at which the record will be open for public inspection. 　　(iii) Statement that all data submitted by applicant is available as part of the administrative record.
124.10(d)(1)(vi)	(f) Any additional information considered necessary or proper.
124.10(d)(2)	(2) Public Notice for Hearings The public notice of a hearing under Section 124.12 (Public Hearings), Subpart E, or Subpart F shall contain following information, in addition to general public notice described in Section 124.10(d)(1) above.
124.10(d)(2)(i)	(a) References to date of previous public notices relating to permit.
124.10(d)(2)(ii)	(b) Date, time, and place of hearing.
124.10(d)(2)(iii)	(c) Brief description of nature and purpose of hearing, including applicable rules and procedures.
124.10(e)	E. Persons to Be Mailed Documents and the Documents 　(1) Documents 　　(a) General public notice described in Section 124.10(d)(1). 　　(b) Fact sheet or statement of basis for EPA-issued permits. 　　(c) Permit application, if any. 　　(d) Draft permit, if any. 　(2) Persons 　　Those persons identified in Section 124.10(c)(1)(i), (ii), and (iii).
124.11	X. Public Comments and Requests for Public Hearings A. During Public Comment Period Provided under Section 124.10 (Public Notice of Permit Actions and Public Comment Period) 　(1) Public Comments 　　(a) Any interested person may submit written comments on the draft permit or the permit. 　　(b) All comments shall be considered in making the final decision and shall be answered as provided in Section 124.17 (Response to Comments). 　(2) Public Hearing 　　(a) Any interested person may request a public hearing, if no hearing has already been scheduled. 　　(b) Request for public hearing shall be in writing and shall state nature of issues proposed to be raised in hearing.
124.12	XI. Public Hearing
124.12(a)	A. When Director Shall Hold Public Hearing 　(1) Whenever he finds, on basis of requests, a significant degree of public interest in draft permit(s). 　(2) At discretion of Director, whenever, for instance, such a hearing might clarify one or more issues involved in the permit decision. B. Public Notice of the Hearing 　Shall be given as specified in Section 124.10 (Public Notice of Permit Actions and Public Comment Period).
124.12(b)	C. Public Hearing with EPA as Permitting Authority 　Regional Administrator shall designate a Presiding Officer for the hearing who shall be responsible for its scheduling and orderly conduct.
124.12(c)	D. Submittal of Oral or Written Statements and Data Concerning Draft Permit 　(1) Any person can submit such statement or data. 　(2) Reasonable limits may be set upon the time allowed for oral statements, and submission of statements in writing may be required. 　(3) Extension of Public Comment Period. 　　(a) Under Section 124.10 (Public Notice of Permit Actions and Public Comment Period), public comment period is automatically extended to the close of any public hearing under this Section. 　　(b) Hearing officer may also extend comment period by so stating at hearing.

Table 7-2A. Part 124, Procedures for Decisionmaking: General Program Requirements—Subpart A. (Continued)

Section No.	Description
124.12(d)	E. Tape Recording or Written Transcript of Hearing Shall be made available to public.
124.12(e)	F. Processing of Permits under Procedure in Subpart F (Non-Adversary Panel Procedures) May be done at discretion of Regional Administrator.
124.13	XII. Obligation to Raise Issues and Provide Information during Public Comment Period

	A. If Draft Permit Condition(s) or Director's Tentative Decision to Deny an Application, Terminate a Permit, or Prepare a Draft Permit Is Believed Inappropriate
	   All persons harboring such beliefs, including applicant, must:
	   (1) Raise all reasonable issues; and
	   (2) Submit all reasonably available arguments and factual grounds supporting their position including all supporting material.
	    (a) All supporting materials shall be included in full and not by reference.
	      Exceptions:
	       (i) Unless materials are already part of administrative record in same proceeding; or
	       (ii) Material consists of State or Federal statutes and regulations, EPA documents of general applicability; or
	       (iii) Other generally available reference material.
	    (b) Commenters shall make supporting material not already included in administrative record available to EPA as directed by Regional Administrator.

	B. Comment Period
	   (1) General
	     All arguments and supporting material shall be inputted by close of public comment period, under Section 124.10 (Public Notice of Permit Actions and Public Comment Period).
	   (2) A comment period longer than 30 days will often be necessary in complicated proceedings to give commenters a reasonable opportunity to comply with requirements of this Section.
	   (3) Commenters may request longer comment periods and they should be freely established under Section 124.10 to extent they appear necessary.

124.14	XIII. Reopening of Public Comment Period
124.14(a)	A. If Arguments, Information, or Data Submitted during Public Comment Period Raise Substantial New Questions Concerning Permit Regional Administrator may take one or more actions that follow:

124.14(a)(1)
124.14(a)(2)
	(1) Prepare new draft permit appropriately modified under Section 124.6 (Draft Permit)
	   (2) Prepare revised statement of basis under Section 124.7 (Statement of Basis)
	   (3) Prepare fact sheet or revised fact sheet under Section 124.8 (Fact Sheet)
	   (4) Reopen or Extend Comment Period
	    (a) Under Section 124.14 (Reopening of the Public Comment Period).

124.14(a)(3)	(b) Under Section 124.10 (Public Notice of Permit Actions and Public Comment Period) to give interested persons an opportunity to comment on information or arguments submitted.

124.14(b)	B. Comments Filed During Reopened Comment Period
	   (1) Shall be limited to substantial new questions that caused its reopening.
	   (2) Public notice under Section 124.10 shall define scope of reopening.

124.14(c)	C. Proceedings under Subpart F
	   (1) Regional Administrator may elect to hold further proceedings under Subpart F.
	   (2) This decision may be combined with any actions enumerated in Section 124.14(a).

124.14(d)	D. Public Notice of Any of Above Actions Shall be issued under Section 124.10.
124.15	XIV. Issuance and Effective Date of Permit

124.15(a)	A. Issuance of Final Permit Decision by Regional Administrator
	   (1) Definition of "Final Permit Decision"
	     Final decision to issue, deny, modify, revoke and reissue, or terminate a permit.
	   (2) Issuance of permit shall be done after close of public comment period under Section 124.10 (Public Notice of Permit Actions and Public Comment Period).

Table 7-2A. Part 124, Procedures for Decisionmaking: General Program Requirements—Subpart A. (Continued)

Section No.	Description
	(3) Who Regional Administrator Shall Notify (a) Applicant. (b) Each person who submitted written comments. (c) Each person who requested notice of final permit decision. (4) Content of Notice (a) Shall include reference to procedures for appealing a decision on a RCRA permit; or (b) Procedures for contesting a decision to terminate a RCRA permit.
124.15(b)	B. Effective Date of Permit Final permit decision becomes effective 30 days after service of notice of decision under Section 124.15(a). Exceptions:
124.15(b)(1) 124.15(b)(2) 124.15(b)(3)	(1) Later effective date is specified in decision; or (2) Review is requested under Section 124.19 (Appeal of RCRA, UIC and PSD Permits); or (3) Evidentiary hearing is requested under Section 124.74 (Requests for Evidentiary Hearing); or (4) No Comments Requesting Change in Draft Permit Thus, permit becomes effective immediately upon issuance.
124.16	XV. Stays of Contested Permit Conditions
124.16(a) 124.16(a)(1) 124.16(a)(2)	A. Stays (1) Contested permit conditions shall be stayed and not subject to judicial review pending final Agency action if: (a) Request for review of RCRA permit under Section 124.19 (Appeal of RCRA, UIC and PSD Permits) is granted; or (b) Conditions of RCRA permit are consolidated for reconsideration in an evidentiary hearing. (2) If Permit Involves New Facility Applicant shall be without a permit for proposed new facility pending final Agency action. (3) Uncontested Conditions Which Are Not Separable From Contested Conditions They shall be stayed together with the contested conditions. (4) Stayed Provisions of Permits for Existing Facilities (a) These shall be identified by Regional Administrator. (b) All other provisions of permit shall remain fully effective and enforceable.
124.16(b) 124.16(b)(1) 124.16(b)(2)	B. Stays Based on Cross Effects (1) Granting a Stay on Grounds that an Appeal to Administrator (under Section 124.19) for One Permit May Result in Change to Another EPA-Issued Permit Only possible when each of the permits involved has been appealed to Administrator who accepts each appeal. (2) No stay of an EPA-issued RCRA permit shall be granted based on the staying of any State-issued permit. Exception: (a) At discretion of Regional Administrator; and (b) Only upon written request from State Director.
124.16(c) 124.16(c)(1) 124.16(c)(2)	C. Any Facility or Activity Holding an Existing Permit (1) Must comply with conditions of that permit during any modification or revocation and re-issuance proceeding under Section 124.5 (Modification, Revocation and Reissuance, or Termination of Permits). (2) Must to extent conditions of any new permit are stayed under this Section, comply with conditions of existing permit which correspond to stayed conditions. Exception: (a) Unless compliance with existing conditions would be technologically incompatible with compliance with other conditions of the new permit which have not been stayed.
124.17	XVI. Response to Comments
124.17(a)	A. Issuance of Response to Comments (1) When to Issue and by Whom (a) At time any final permit decision is issued under Section 124.15 (Issuance and Effective Date of Permit). (b) Director shall issue response to comments. (c) States are only required to issue a response to comments when a final permit is issued.

Table 7-2A. Part 124, Procedures for Decisionmaking: General Program Requirements—Subpart A. (Continued)

Section No.	Description
	(2) Content
124.17(a)(1)	(a) Specify which provisions if any, of draft permit have been changed in final permit decision, and reasons for change.
124.17(a)(2)	(b) Briefly describe and respond to all significant comments on the draft permit raised during the public comment period or during any hearing.
124.17(b)	B. For EPA-Issued Permits
	(1) Any documents cited in response to comments shall be included in administrative record for final permit decision as defined in Section 124.18 (Administrative Record for Final Permit When EPA Is the Permitting Authority).
	(2) If New Points Are Raised or New Material Supplied during Public Comment Period EPA may document its response to those matters by adding new materials to administrative record.
124.17(c)	C. Response to comment shall be available to public.
124.18	XVII. Administrative Record for Final Permit When EPA Is the Permitting Authority
124.18(d)	A. Applicability Section 124.18 applies to all final RCRA permits when the draft permit was subject to administrative record requirements of Section 124.9 (Administrate Record for Draft Permits when EPA is the Permitting Authority).
124.18(a)	B. Final Permit Decision by Regional Administrator Shall be based on administrative record defined in this Section (124.18).
124.18(b)	C. Content of Administrative Record for Any Final Permit
	(1) Administrative record for draft permit.
124.18(b)(1)	(2) All comments received during public comment period provided under Section 124.10 (Public Notice of Permit Actions and Public Comment Period).
	(a) Includes comments received via any extension or reopening under Section 124.14 (Reopening of the Public Comment Period).
124.18(b)(2)	(3) Tape or transcript of any hearing(s) held under Section 124.12 (Public Hearings).
124.18(b)(3)	(4) Any written materials submitted at such a hearing.
124.18(b)(4)	(5) Response to comment required by Section 124.17 (Response to Comments) and any new material placed in the record.
124.18(b)(6)	(6) Other documents contained in supporting file for permit.
124.18(b)(7)	(7) The final permit.
124.18(c)	D. Additional Documents Required under Section 124.18(b) Above
	(1) Should be added to administrative record as soon as possible after their receipt or publication by the Agency.
	(2) The record shall be complete on date final permit is issued.
124.18(e)	E. Materials that Need Not Be Physically Included in Same File As Rest of Administrative Record
	(1) Materials readily available at issuing Regional Office; or
	(2) Published materials which are generally available and which are included in administrative record under standards of this Section or Section 124.17 (Response to Comments).
	(3) However, material must be specifically referred to in statement of basis or fact sheet or in response to comments.
124.19	XVIII. Appeal of RCRA, UIC, and PSD Permits
124.19(a)	A. Petitioning Administrator to Review Any Condition of Permit Decision
	(1) When to Petition
	(a) Within 30 days after RCRA final permit decision has been issued under Section 124.15 (Issuance and Effective Date of Permit).
	(b) Thirty (30) day period begins with service of notice of Regional Administrator's action, unless a later date is specified in that notice.
	(2) Who May Petition
	(a) Any person who filed comments on that draft permit; or
	(b) Participated in the public hearing.

Table 7-2A. Part 124, Procedures for Decisionmaking: General Program Requirements—Subpart A. (Continued)

Section No.	Description
	(c) Person(s) Who Failed to File Comments or Failed to Participate in Public Hearing on Draft Permit May petition for administrative review only to extent of changes from draft to final permit decision.
	(3) Content of Petition
	(a) Statement of reasons supporting that review.
	(b) Demonstration that any issues being raised were raised during comment period (including any public hearing) to extent required by these regulations.
	(c) When appropriate, a showing that the condition in question is based on:
124.19(a)(1)	(i) A finding of fact or conclusions of law which is clearly erroneous; or
124.19(a)(2)	(ii) An exercise of discretion or an important policy consideration which Administrator should review.
124.19(b)	B. Administrator's Review of Any Conditions of RCRA Permit (1) May be done by Administrator on his initiative. (2) Administrator must act within 30 days of service date of notice of Regional Administrator's action.
124.19(c)	C. Administrator's Order either Granting or Denying Petition (1) Must be done in a reasonable time following filing of petition for review. (2) To Extent Review Is Denied The conditions of the final permit decision become final Agency action. (3) Public Notice (a) Public notice of any grant or review by Administrator under Sections 124.19(a) or (b) shall be given as provided in Section 124.10 (Public Notice of Permit Actions and Public Comment Period). (b) Public notice shall set forth a briefing schedule for the appeal and shall state that any interested person may file an amicus brief. (4) Notice of Denial of Review Shall be sent only to person(s) requesting review.
124.19(d)	D. Administrator's Deferral of Consideration of an Appeal of RCRA Permit May be done until completion of formal proceedings under Subpart E or F relating to an NPDES permit issued to the same facility or activity upon concluding that: (1) The NPDES permit is likely to raise issues relevant to a decision of the RCRA appeal.
124.19(e)	E. Judicial Review of Final Agency Action (1) A petition to Administrator under Section 124.19(a) is, under 5 U.S.C. Section 704, a prerequisite to seeking of judicial review of final Agency action.
124.19(f)(1)	(2) When Final Agency Action Occurs This occurs, for purposes of judicial review, when a final RCRA permit is issued or denied by EPA and Agency review procedures are exhausted. (3) When Final Permit Decision Shall Be Issued by Regional Administrator
124.19(f)(1)(i)	(a) When Administrator issues notice to parties that review has been denied.
124.19(f)(1)(ii)	(b) When Administrator issues decision on merits of appeal and decision does not include a remand of the proceedings.
124.19(f)(1)(iii)	(c) Upon completion of remand proceedings if proceedings are remanded, unless Administrator's remand order specifically provides that appeal of the remand decision will be required to exhaust administrative remedies.
124.20	XIX. Computation of Time
124.20(a)	A. Any Time Period Scheduled to Begin on the Occurrence of an Act or Event Shall begin on the day after the act or event.
124.20(b)	B. Any Time Period Scheduled to Begin before the Occurrence of an Act or Event Shall be computed so that the period ends on the day before the act or event.
124.20(c)	C. If Final Day of Any Time Period Falls on a Weekend or Legal Holiday The time period shall be extended to the next working day.
124.20(d)	D. Three (3) days shall be added to prescribed response time after being served notice by mail.

Table 7-2B. Part 124, Procedures for Decisionmaking: Evidentiary Hearings for EPA-Issued NPDES Permits and EPA-Terminated RCRA Permits—Subpart E.

Section No.	Description
124.71	I. Applicability
124.71(a)	A. Regulations in This Subpart (E) (1) Govern evidentiary hearings conducted under RCRA Section 3008 in connection with termination of RCRA permit. (2) This includes termination of interim status for failure to furnish information needed to make a final decision.
124.71(b)	B. Evidentiary Hearing for RCRA Permits Which Are Being Issued, Modified, or Revoked and Reissued (Not Terminated or Suspended) Will occur when conditions of RCRA permit in question are closely linked with conditions of NPDES permit.
124.73	II. Filing and Submission of Documents
124.73(a)	A. General (1) File All Submissions to Agency with Regional Hearing Clerk Unless otherwise provided by regulation. (2) Date Submissions Considered to Be Filed (a) Date on which they are mailed; or (b) Date delivered in person to Regional Hearing Clerk.
124.73(b)	(3) Who Can Sign Submissions (a) Person making submission; or (b) Attorney; or (c) Other authorized agent or representative.
124.73(c)(1)	B. Data and Information Referred To or Relied Upon in Any Submission (1) Shall be included in full and may not be incorporated by reference. Exceptions: (a) Data/information previously submitted as part of administrative record in same proceedings. (b) State or Federal statutes and regulations, judicial decisions published in national reporter system, and officially issued EPA documents of general applicability. (2) Any party incorporating materials by reference shall provide copies upon request by Regional Administrator or Presiding Officer.
124.73(c)(2)	(3) Material Submitted in Foreign Language (a) Shall be accompanied by an English translation verified under oath to be complete and accurate. (b) Name and address and brief statement of qualifications of person making translation. (c) Translations of literature or other materials in foreign language shall be accompanied by copies of original publication.
124.73(c)(3)	(4) Where Relevant Data or Information Is Contained in Document Containing Irrelevant Matter Either: (a) Irrelevant matter shall be deleted; or (b) Relevant portions shall be indicated.
124.73(c)(4)	(5) Failure to Comply with Requirements of This Section or Any Other Requirements in This Subpart (a) May result in noncomplying portions of submissions being excluded from consideration. (b) If Regional Administrator or Presiding Officer Determines that a Submission Fails to Meet Requirements of This Subpart (i) Regional Hearing Clerk shall be directed to return submission. (ii) Also, reference to applicable regulations will be provided. (c) Party Whose Materials Have Been Rejected (i) Has 14 days to correct errors and resubmit. (ii) Unless Regional Administrator or Presiding Officer finds good cause to allow a longer time.
124.73(d)	C. What Filing a Submission Shall Not Mean or Imply (1) That it in fact meets all applicable requirements; or (2) That it contains reasonable grounds for the action requested; or (3) That the action requested is in accordance with law.

Table 7-2B. Part 124, Procedures for Decisionmaking: Evidentiary Hearings for EPA-Issued NPDES Permits and EPA-Terminated RCRA Permits—Subpart E. (Continued)

Section No.	Description
124.73(e)	D. The Original Copy of Statements and Documents Containing Factual Material Data or Other Information (1) Shall be signed in ink. (2) Shall state name, address, and representative capacity of person making submission.
124.74	III. Requests for Evidentiary Hearing
124.74(a)	A. General (1) Timing of Request For Evidentiary Hearing May be requested of Regional Administrator under Section 124.74(b) within 30 days of serving of notice of his final permit decision. (2) Purpose of Request for Evidentiary Hearing To contest or reconsider Regional Administrator's decision. (3) If Request Is Submitted by Person Other than Permittee Person shall simultaneously serve a copy of request to permittee.
124.74(b)(1)	B. Content of Request (1) Statement of Each Legal or Factual Question Alleged to Be an Issue (a) Under Section 124.76 (Obligation to Submit Evidence and Raise Issues Before a Final Permit Is Issued). (b) Their relevance to permit decision. (c) Designation of specific factual areas to be adjudicated. (d) Hearing time estimated to be necessary for adjudication. (2) Information Supporting Request or Other Written Documents Relied Upon to Support Request (a) Shall be submitted as required by Section 124.73 (Filing and Submission of Documents). (b) Unless they are already part of administrative record required by Section 124.18 (Administrative Record for Final Permit When EPA is the Permitting Authority).
124.74(c)(1)	(3) Name, mailing address, and telephone number of person making request.
124.74(c)(2)	(4) Clear and concise factual statement of nature and scope of interest of requester.
124.74(c)(3)	(5) Names and addresses of all persons whom the requestor represents.
124.74(c)(4)	(6) Statement by requester to make available for appearance and testimony at motion/order of Presiding Officer, following parties without cost or expense to any other party: (a) The requester. (b) All persons represented by requester. (c) All officers, directors, employees, consultants, and agents of requester. (d) All persons represented by requester.
124.74(c)(7)	(7) Identification of Permit Obligations That Are Contested or Are Inseverable from Contested Conditions Should be stayed if request is granted by reference to particular contested conditions warranting the stay.
124.74(c)(8)	(8) Hearing Requests May Ask that Formal Hearing Be Held Under Procedures of Subpart F (Non-Adversary Panel Procedures) Applicant may make such a request even if proceeding does not constitute "initial licensing" as defined in Section 124.111 (Applicability for Subpart F).
124.74(b)(2)	C. Person Requesting an Evidentiary Hearing on an NPDES Permit (1) May also request an evidentiary hearing on an RCRA permit. (2) This request subject to all requirements of Section 124.74(b)(1) and will be granted only if:
124.74(b)(2)(i)	(a) Processing of RCRA permit at issue was consolidated with processing of NPDES permit as provided in Section 124.4 (Consolidation of Permit Processing).
124.74(b)(2)(ii)	(b) Standards for granting a hearing on NPDES permit are met.
124.74(b)(2)(iii)	(c) Resolution of NPDES permit issues is likely to make necessary modification of RCRA permit.
124.74(d)	D. If Regional Administrator Grants an Evidentiary Hearing Request in Whole or in Part (1) He shall identify permit conditions which have been contested by requester and for which evidentiary hearing has been granted. (2) Permit Conditions Which Are Not Contested or for Which Regional Administrator Has Denied the Hearing Request These shall not be affected by, or considered at, the evidentiary hearing and will be specified in writing by Regional Administrator.

**Table 7-2B. Part 124, Procedures for Decisionmaking: Evidentiary Hearings for EPA-Issued
NPDES Permits and EPA-Terminated RCRA Permits—Subpart E. (Continued)**

Section No.	Description
124.74(e)	E. Regional Administrator Must Grant or Deny All Requests for Evidentiary Hearing on Particular Permit All requests that are granted for a particular permit shall be combined in a single evidentiary hearing.
124.74(f)	F. Extension of Time Allowed for Submitting Hearing Requests May be done by Regional Administrator for good cause.
124.75	IV. Decision on Request for a Hearing
124.75(a)(1)	A. Timing of Decision by Regional Administrator on Evidentiary Hearing Request Within 30 days following expiration of time allowed by Section 124.74 (Requests for Evidentiary Hearing) for submitting an evidentiary hearing request.
	B. Provisions Required of Request for Regional Administrator to Make Decision (1) Request conforms to requirements of Section 124.74. (2) Request sets forth the material issues of fact relevant to issuance of permit.
124.75(b)	C. If Request for Hearing is Denied (1) Regional Administrator shall briefly state reasons. (2) Denial is subject to review by Administrator under Section 124.91 (Appeal to Administrator).
124.76	V. Obligation to Submit Evidence and Raise Issues before a Final Permit Is Issued
	A. Submitting New Evidence or Raising New Issues Not permitted if evidence or issues were not submitted to Administrative Record required by Section 124.18 (Administrative Record for Final Permit When EPA Is Permitting Authority) as part of preparation of comment on draft permit. Exception: (1) Good cause can be shown for failure to submit it.
	B. Good Cause (1) Includes case where party seeking to raise new issues or introduce new information shows that it could not reasonably have ascertained the issues or made the information available within the time required by Section 124.15 (Issuance and Effective Date of Permit); or (2) That it could not have reasonably anticipated the relevance or materiality of the information sought to be introduced.
124.77	VI. Notice of Hearing
	A. Public Notice of the Grant of an Evidentiary Hearing Regarding a Permit Shall be given as provided in Section 124.57(b) (Public Notice).
	B. Mailing Copies of Public Notice (1) To all persons who commented on draft permit. (2) Testified at public hearing. (3) Submitted a request for a hearing.
	C. Before Issuance of the Notice Regional Administrator shall designate Agency trial staff and members of the decisional body (as defined in Section 124.78—Ex Parte Communications).
124.78	VII. Ex Parte Communications
124.78(b)(1)	A. Interested Person outside Agency or Member of Agency Trial Staff Shall not make or knowingly cause to be made to any members of decisional body, an ex parte communication on the merits of the proceedings.
124.78(b)(2)	B. Member of Decisional Body (1) Shall not make or knowingly cause to be made to any interested person outside the Agency or member of the Agency trial staff, an ex parte communication on the merits of the proceedings.

Table 7-2B. Part 124, Procedures for Decisionmaking: Evidentiary Hearings for EPA-Issued NPDES Permits and EPA-Terminated RCRA Permits—Subpart E. (Continued)

Section No.	Description
124.78(b)(3)	(2) Member Who Receives, Makes, or Knowingly Causes to Be Made a Communication Prohibited by This Section Shall file with Regional Hearing Clerk: (a) All written communications or memoranda stating substance of all oral communications. (b) All written responses and memoranda stating substance of all oral responses.
124.78(d)	C. Prohibitions of This Section (124.78) (1) Begin to apply upon issuance of notice of grant of a hearing under: (a) Section 124.77—Notice of Hearing for evidentiary hearing. (b) Section 124.116—Notice of Hearing for non-adversary panel procedures. (2) This prohibition terminates at the date of final agency action.
124.79	VIII. Additional Parties and Issues
124.79(a)	A. Request to Be Admitted As a Party to Evidentiary Hearing (1) Any person may submit such a request within 15 days after mailing, publication, or posting of notice of grant of an evidentiary hearing, whichever occurs last. (2) Presiding Officer shall grant requests that meet requirements of: (a) Section 124.74—Requests for Evidentiary Hearing. (b) Section 124.76—Obligation to Submit Evidence and Raise Issues before a Final Permit Is Issued.
124.79(b)	B. Filing a Motion for Leave to Intervene As a Party (1) Such a motion may be filed after the 15 day expiration period of Section 124.79(a). (2) Motion must: (a) Meet requirements of Sections 124.74 and 124.76. (b) Set forth grounds for proposed intervention. (c) Contain verified statement showing good cause for failure to file a timely request to be admitted as a party. (3) Findings Required For Presiding Officer to Grant the Motion.
124.79(b)(1) 124.79(b)(2) 124.79(b)(2)(i) 124.79(b)(2)(ii) 124.79(b)(3)	(a) Extraordinary circumstances justifying granting the motion. (b) The intervener has consented to be bound by: (i) Prior written agreements and stipulations by and between the existing parties; and (ii) All orders previously entered in the proceedings. (c) Intervention will not cause undue delay or prejudice the rights of the existing parties.
124.80	IX. Filing and Service
124.80(a)	A. Written Submissions Relating to an Evidentiary Hearing Filed after Notice is Published An original and one (1) copy shall be filed with Regional Hearing Clerk.
124.80(b)	B. Additional Copies by Party Filing Submission(s) Via mail or personal delivery: (1) One copy for Presiding Officer. (2) One copy for each party of record.
124.80(c)	C. Proof of Submission Each submission shall be: (1) Accompanied by an acknowledgement of service by the person served; or (2) Certificate of service citing the date, place, time, and manner of service and names of persons serviced.
124.80(d)	D. Availability of List of All Parties upon Request The Regional Hearing Clerk shall maintain and furnish a list containing name, service address, and telephone number of all parties and their attorneys or duly authorized representatives to any person upon request.
124.81	X. Assignment of Administrative Law Judge A. No later than the date of mailing, publication, or posting of the notice of a grant of an evidentiary hearing, whichever occurs last: (1) Regional Administrator shall refer the proceeding to the Chief Administrative Law Judge. (2) This judge shall assign an Administrative Law Judge to serve as Presiding Officer for the hearing.

Table 7-2B. Part 124, Procedures for Decisionmaking: Evidentiary Hearings for EPA-Issued NPDES Permits and EPA-Terminated RCRA Permits—Subpart E. (Continued)

Section No.	Description
124.82	XI. Consolidation and Severance

124.82(a)	A. Consolidation of Proceedings
	    (1) Who Can Consolidate Proceedings
	        (a) Administrator.
	        (b) Regional Administrator.
	        (c) Presiding Officer.
	    (2) Extent of Consolidation
	    Two or more proceedings to be held under this Subpart (F) may be consolidated in whole or in part.
	    (3) When Consolidation Considered
	        (a) Whenever it appears that a joint hearing on any or all of the matters in issue would expedite or simplify consideration of the issues.
	        (b) No party would be prejudiced thereby.
	    (4) Consolidation shall not affect the right of any party to raise issues that might have been raised had there been no consolidation.

124.82(b)	B. Severence of Party or Issues If Presiding Officer determines consolidation is not conducive to an expeditious, full, and fair hearing, any party or issues may be severed and heard in a separate proceeding.
124.83	XII. Prehearing Conferences

124.83(a)	A. Prehearing Conference(s) or Submittal of Suggestions
	Presiding Officer may direct the parties or their attorneys or duly authorized representatives:
	    (1) To appear at a specified time and place for one or more conferences before or during a hearing; or
	    (2) To submit written proposals or correspond, for purpose of considering any of the matters set forth in Section 124.83(c).

124.83(b)	B. Allowance of Reasonable Period before Hearing Begins for Orderly Completion of Prehearing Procedures and for Submission and Disposition of Prehearing Motions
	    (1) Presiding Officer's responsibility.
	    (2) Call for Prehearing Conference
	    May be done by Presiding Officer in order to:
	        (a) Inquire into use of available procedures contemplated by parties and time required for their completion.
	        (b) Establish schedule for their completion.
	        (c) Set tentative date for beginning the hearing.

124.83(c)	C. Matters That May Be Considered in Conferences Held or Suggestions Submitted under Section 124.83(a)
124.83(c)(1)	(1) Simplification, clarification, amplification, or limitation of the issues.
124.83(c)(2)	(2) Admission of facts and of the genuineness of documents, and stipulations of facts.

124.83(c)(3)	(3) Objections to Introduction into Evidence at Hearing of Any Written Testimony, Documents, Papers, Exhibits, or Other Submissions Proposed by a Party
	Exceptions:
	    (a) The Administrative record required by Section 124.19 (Appeal of RCRA, UIC, and PSD Permits) shall be received in evidence subject to provisions of Section 124.85(d)(2) (Hearing Procedures).
	    (b) Motions to Strike Testimony or Other Evidence Other than Administrative Record
	    At any time before end of hearing any party may make such motions, and Presiding Officer shall consider and rule upon motion(s) based on grounds of relevancy, competency, or materiality.

124.83(c)(4)	(4) Matters subject to official notice may be taken.
124.83(c)(5)	(5) Scheduling as many of following as deemed necessary and proper by Presiding Officer:
124.83(c)(5)(i)	(a) Submission of narrative statements of position on each factual issue in controversy.
124.83(c)(5)(ii)	(b) Submission of written testimony and documentary evidence (e.g., affidavits, data, studies, reports, and any other written material) in support of those statements.
124.83(c)(5)(iii)	(c) Requests by any party for production of additional documentation, data, or other information relevant and material to facts in issue.
124.83(c)(6)	(6) Grouping Participants with Substantially Similar Interests In order to eliminate redundant evidence, motions and objections.
124.83(c)(7)	(7) Such other matters that may expedite hearing or aid in disposition of matter.

Table 7-2B. Part 124, Procedures for Decisionmaking: Evidentiary Hearings for EPA-Issued NPDES Permits and EPA-Terminated RCRA Permits—Subpart E. (Continued)

Section No.	Description
124.83(d)	D. Witnesses, Testimony, and Documentation (1) Each Party to Make Available to Other Parties the Names of Expert and Other Witnesses It Expects to Call To be done at prehearing conference or at some other reasonable time set by Presiding Officer. (2) Inclusion of Brief Narrative Summary of Any Witness's Anticipated Testimony May be done by party at its discretion, or at request of Presiding Officer. (3) Marking Evidence for Identification Copies of any written testimony, documents, papers, exhibits, or materials which a party expects to introduce into evidence, and administrative record required by Section 124.18 shall be marked for identification as ordered by Presiding Officer. (4) Addition or Amendment of Witnesses, Proposed Written Testimony, and Other Evidence Only upon order of Presiding Officer for good cause. (5) Agency Employees and Consultants as Witnesses Shall be made available as witnesses by the Agency to same extent as other parties under: (a) Section 124.74(c)(4)–Requests for Evidentiary Hearing (b) Section 124.85(b)(16)–Hearing Procedure.
124.83(e)	E. Presiding Officer's Written Prehearing Order (1) Shall be prepared giving actions taken at each prehearing conference and setting forth schedule for hearing. Exception: (a) Unless transcript has been taken and accurately reflects these matters. (2) Content of Order (a) Written statement of areas of factual agreement and disagreement. (b) Methods and procedures to be used in developing evidence. (c) Respective duties of parties in connection therewith. (3) Order Shall Control Subsequent Course of Hearing Unless modified by Presiding Officer For Good Cause.
124.84	XIII. Summary Determination
124.84(a)	A. Move for a Summary Determination in Its Favor on Any Issued Being Adjudicated (1) Any party to an evidentiary hearing may make such a move with or without supporting affidavits and briefs. (2) Basis of Move That there is no genuine issue of material fact for determination. (3) Timing (a) Motion shall be filed at least 45 days before date set for hearing. (b) Upon showing good cause, motion may be filed at any time before close of hearing.
124.84(b)	B. Response to Move for a Summary Determination (1) Timing Within 30 days after service of the motion, any other party may file and serve a response to it or a countermotion for summary determination. (2) Party Opposing Motion for Summary Determination (a) May not rest upon mere allegations or denials. (b) Must show by affidavit or other materials subject to consideration by Presiding Officer, that there is a genuine issue of material fact for determination at the hearing.
124.84(c)	C. Affidavits (1) Shall be made on personal knowledge. (2) Shall set forth facts that would be admissible in evidence. (3) Shall show affirmatively that the affiant is competent to testify to the matters stated therein.
124.84(d)	D. Ruling on Motion for Summary Determination (1) Presiding Officer may set the matter for oral argument and call for submission of proposed findings, conclusions, briefs, or memoranda of law. (2) Timing Presiding Officer shall rule on the motion not more than 30 days after date responses to motion are filed under Section 124.84(b).

Table 7-2B. Part 124, Procedures for Decisionmaking: Evidentiary Hearings for EPA-Issued NPDES Permits and EPA-Terminated RCRA Permits—Subpart E. (Continued)

Section No.	Description
124.84(e)	E. If All Factual Issues Are Decided by Summary Determination No hearing will be held and Presiding Officer shall prepare an initial decision under Section 124.89 (Decisions).
	F. If Summary Determination Is Denied or If Partial Summary Determination Is Granted (1) Presiding Officer shall issue a memorandum opinion and order, interlocutory in character. (2) Hearing will proceed on remaining issues.
	G. Appeals from Interlocutory Rulings Governed by Section 124.90 (Interlocutory Appeal).
124.84(f)	H. Where Affidavits of Party Opposing Motion for Summary Determination Cannot Present Facts Essential to Justify His Opposition Presiding Officer may: (1) Deny the motion; or (2) Order a continuance to allow additional affidavits or other information to be obtained; or (3) May make such other order as is just and proper.
124.85	XIV. Hearing Procedure
124.85(a)(1)	A. Burden of Proof (1) Permit Applicant (a) He always bears burden of persuading Agency that a permit authorizing pollutants to be discharged should be issued and not denied. (b) This burden does not shift.
124.85(a)(1) Note	(c) In many cases the documents contained in the administrative record, in particular the fact sheet or statement of basis and the response to comments, should adequately discharge this burden.
124.85(a)(2)	(2) The Agency They have the burden of going forward to present an affirmative case in support of any challenged condition of a final permit.
124.85(a)(3) 124.85(a)(3)(i)	(3) What Hearing Participant Contends by Raising Material Issues of Fact (a) That particular conditions or requirements in the permit are improper or invalid and who desires either:
124.85(a)(3)(i)(A) 124.85(a)(3)(i)(B) 124.85(a)(3)(ii)	(i) The inclusion of new or different conditions or requirements; or (ii) The deletion of those conditions or requirements. (b) That the denial or issuance of a permit is otherwise improper or invalid. (i) He shall have burden of going forward to present an affirmative case at the conclusion of the Agency case on the challenged requirement.
124.85(b)	B. Presiding Officer (1) Responsibilities (a) Shall conduct a fair and impartial hearing. (b) Take action to avoid unnecessary delay in the disposition of the proceedings. (c) Maintain order. (2) Actions Presiding Officer May Take in Carrying Out Responsibilities
124.85(b)(1)	(a) Arrange and issue notice of date, time, and place of hearings and conferences.
124.85(b)(2)	(b) Establish methods and procedures to be used in development of evidence.
124.85(b)(3)	(c) Prepare, after considering views of participants, written statements of areas of factual disagreement among participants.
124.85(b)(4)	(d) Hold conferences to settle, simplify, determine, or strike any issues in a hearing, or to consider other matters that may facilitate the expeditious disposition of the hearing.
124.85(b)(5)	(e) Administer oaths and affirmations.
124.85(b)(6)	(f) Regulate course of hearing and govern conduct of participants.
124.85(b)(7)	(g) Examine witnesses.
124.85(b)(8)	(h) Identify and refer issues for interlocutory decision under Section 124.90 (Interlocutory Appeal).
124.85(b)(9)	(i) Rule on, admit, exclude, or limit evidence.
124.85(b)(10)	(j) Establish the time for filing motions, testimony, and other written evidence, briefs, findings, and other submissions.
124.85(b)(11)	(k) Rule on motions and other procedural matters, including but not limited to motions for summary determination in accordance with Section 124.84 (Summary Determination).

Table 7-2B. Part 124, Procedures for Decisionmaking: Evidentiary Hearings for EPA-Issued NPDES Permits and EPA-Terminated RCRA Permits—Subpart E. (Continued)

Section No.	Description
124.85(b)(12)	(l) Order that the hearing be conducted in stages whenever the number of parties is large or the issues are numerous and complex.
124.85(b)(13)	(m) Take any action not inconsistent with provisions of this Subpart (E) for maintenance of order at the hearing and for expeditious, fair, and impartial conduct of the proceeding.
124.85(b)(14)	(n) Provide for testimony of opposing witnesses to be heard simultaneously or for such witnesses to meet outside the hearing to resolve or isolate issues or conflicts.
124.85(b)(15)	(o) Order that trade secrets be treated as confidential business information in accordance with Section 122.19 (Confidentiality of Information) and 40CFR2 (Public Information).

124.85(b)(16)	(p) Cross-Examination
	  (i) Allow such cross-examination as may be required for a full and true disclosure of the facts.
	  (ii) No cross-examination shall be allowed on questions of policy except:
	    • To extent required to disclose the factual basis for permit requirements; or
	    • On questions of law; or
	    • Regarding matters such as the validity of effluent limitations guidelines that are not subject to challenge in an evidentiary hearing.
	  (iii) No Agency witnesses shall be required to testify or be made available for cross-examination on such matters.
	  (iv) In deciding whether or not to allow cross-examination, the Presiding Officer shall consider the likelihood of clarifying or resolving a disputed issue of material fact compared to other available methods.
	  (v) The party seeking cross-examination has the burden of demonstrating that this standard has been met.

124.85(c)	C. All Direct and Rebuttal Evidence at Evidentiary Hearing
	  (1) Shall be submitted in written form.
	    Exception:
	    (a) Unless, upon motion and good cause shown, the Presiding Officer determines that oral presentation of the evidence on any particular fact will materially assist in the efficient identification and clarification of the issues.
	  (2) Written testimony shall be prepared in narrative form.

124.85(d)(1)	D. Evidence
	  (1) Admissibility
	    (a) Presiding Officer shall admit all relevant, competent, and material evidence.
	    Exception:
	      (i) Evidence that is unduly repetitious.
	    (b) Evidence may be received at any hearing even though inadmissible under the rules of evidence applicable to judicial proceedings.
	  (2) Weight
	    The weight to be given evidence shall be determined by its reliability and probative value.

124.85(d)(2)	(3) Administrative Record
	    (a) The administrative record required by Section 124.18 (Administrative Record for Final Permit When EPA Is the Permitting Authority) shall be admitted and received in evidence.
	    (b) Upon motion by any party the Presiding Officer may direct that a witness be provided to sponsor a portion or portions of the administrative record.

124.85(d)(2)	(i) The Presiding Officer, upon finding that standards in Section 124.85(b)(3) have been met, shall direct the appropriate party to produce the witness for cross-examination. (ii) If a sponsoring witness cannot be provided, the Presiding Officer may reduce the weight accorded the appropriate portion of the record.

124.85(d)(2) Note	(c) Receiving Administrative Record into Evidence Automatically Serves Several Purposes
	      (i) It documents the prior course of the proceedings.
	      (ii) It provides a record of the views of affected persons for consideration by the agency decisionmaker.
	      (iii) It provides factual material for use by the decisionmaker.

124.85(d)(3)	(4) When Any Evidence Or Testimony Is Excluded By Presiding Officer As Inadmissible
	    (a) All such evidence or testimony existing in written form shall remain a part of the record as an offer of proof.
	    (b) Party Seeking Admission of Oral Testimony
	      May make an offer of proof, by means of a brief statement on the record describing the testimony excluded.

Table 7-2B. Part 124, Procedures for Decisionmaking: Evidentiary Hearings for EPA-Issued NPDES Permits and EPA-Terminated RCRA Permits—Subpart E. (Continued)

Section No.	Description

124.85(d)(4)	(5) When Two or More Parties Have Substantially Similar Interests and Positions
	    (a) Presiding Officer may limit number of attorneys or other party representatives who will be permitted to cross-examine and to make and argue motions and objections on behalf of those parties.
	    (b) Attorneys may, however, engage in cross-examination relevant to matters not adequately covered by previous cross-examination.

124.95(d)(5)	(6) Rulings of Presiding Officer on Admissibility of Evidence or Testimony, Propriety of Cross-Examination, and Other Procedural Matters Shall appear in the record and shall control further proceedings. Exception: (a) Unless reversed as a result of an interlocutory appeal taken under Section 124.90 (Interlocutory Appeal).

124.85(d)(6)	(7) Objections
	    (a) All objections shall be made promptly or be deemed waived.
	    (b) Parties shall be presumed to have taken exception to an adverse ruling.
	    (c) No objection shall be deemed waived by further participation in the hearing.

124.86	XV. Motions

124.86(a)	A. General
	  (1) Filing a Motion
	    Any party may file a motion (including a motion to dismiss a particular claim on a contested issue), with the Presiding Officer on any matter relating to the proceedings.
	  (2) All motions shall be in writing.
	  (3) All motions shall be served as provided in Section 124.80 (Filing and Service).
	    Exception:
	    (a) Those motions made on the record during an oral hearing before the Presiding Officer.

124.86(b)	B. Response to the Motion
	  (1) Within 10 days after service of any written motion, any party to the proceeding may file a response to the motions.
	  (2) The time for response may be shortened to 3 days or extended for an additional 10 days by the Presiding Officer for good cause shown.

124.86(c)	C. Motion Seeking to Apply to Permit Any Regulatory or Statutory Provision Issued or Made Available after Issue of Permit under Section 124.15
	  (1) Any party may file such a motion with Presiding Officer.
	  (2) Presiding Officer shall grant any motion to apply a new statutory or regulatory provision unless he finds it contrary to legislative intent.

124.87	XVI. Record of Hearing

124.87(a)	A. Content of Hearing Record
	  (1) All orders issued by Presiding Officer.
	  (2) Transcripts of oral hearing or arguments.
	  (3) Written statements of position.
	  (4) Written direct and rebuttal testimony.
	  (5) Any other data, studies, reports, documentation, information and other written material of any kind submitted in the proceeding.
	
	B. Availability of Hearing Record
	  (1) It shall be available to public except as provided in Section 122.19 (Confidentiality of Information).
	  (2) It shall be located in Office of Regional Hearing Clerk.

124.87(b)
 	
124.87(b)(1)
124.87(b)(2)
	C. Transcription of Hearing Tape Recording or Stenographical Reporting
	  (1) Required for evidentiary hearing.
	  (2) After the hearing, the reporter shall certify and file with Regional Hearing Clerk:
	    (a) Original of transcript.
	    (b) Exhibits received or offered into evidence at hearing.

124.87(c)	D. Prompt Notification of Each Party of Filing of Certified Proceeding Transcript Required of Regional Hearing Clerk.

Table 7-2B. Part 124, Procedures for Decisionmaking: Evidentiary Hearings for EPA-Issued NPDES Permits and EPA-Terminated RCRA Permits—Subpart E. (Continued)

Section No.	Description
	E. Availability of Transcript Any party desiring copy of transcript of hearing may obtain copy from Regional Hearing Clerk upon payment of costs.
124.87(d)	F. Correction of Transcript (1) Presiding Officer shall allow witnesses, parties, and their counsel an opportunity to submit written proposed corrections of transcript of oral testimony taken at hearing. (2) Except in unusual cases, no more than 30 days should be allowed for submitting corrections, from day a complete transcript of hearing becomes available.
124.88	XVII. Proposed Findings of Fact and Conclusions: Brief A. Filing Proposed Findings of Fact and Conclusion of Law and Brief in Support Thereof (1) Any party may so file with Regional Hearing Clerk, within 45 days after certified transcript is filed. (2) Briefs shall contain appropriate references to the record. (3) Copy of these findings, conclusions, and brief shall be served upon all other parties and Presiding Officer. (4) Presiding Officer, for good cause shown, may extend time for filing proposal findings and conclusions and/or brief. (5) Presiding Officer may allow reply briefs.
124.89	XVIII. Decisions
124.89(a)	A. Review and Evaluation of Record by Presiding Officer Includes: (1) Proposed findings and conclusions. (2) Any briefs filed by parties. (3) Any interlocutory decisions under Section 124.90 (Interlocutory Appeal).
	B. Initial Decision (1) Presiding Officer shall issue and file his initial decision with Regional Hearing Clerk. (2) Regional Hearing Clerk shall immediately serve copies of initial decision upon all parties (or their counsel of record) and the Administrator.
124.89(b)	(3) Initial Decision of Presiding Officer Automatically Becomes Final Decision 30 Days after Its Service Exceptions:
124.89(b)(1)	(a) A party files a petition for review by Administrator pursuant to Section 124.91 (Appeal to the Administrator); or
124.89(b)(2)	(b) Administrator files notice that he will review decision pursuant to Section 124.91.
124.90	XIX. Interlocutory Appeal
	A. General
124.90(a)	(1) Except as provided in this Section, appeals to Administrator may be taken only under Section 124.91 (Appeal to the Administrator).
124.90(c)	(2) Ordinarily, the interlocutory appeal will be decided on the basis of submissions made to Presiding Officer. (a) The Administrator may however allow briefs and oral arguments.
124.90(a)	B. How Appeals from Orders or Rulings May Be Taken under This Section (124.90) Only if Presiding Officer, upon motion of a party, certifies those orders or rulings to the Administrator for appeal on the record.
	C. Certification (1) Requests to the Presiding Officer for certification must be filed in writing within 10 days of service of notice of the order, ruling, or decision. (2) It shall state briefly the grounds relied on.
124.90(b)	(3) Presiding Officer may certify an order or ruling for appeal to Administrator if:
124.90(b)(1)	(a) Order or ruling involves important question on which there is substantial ground for difference of opinion; and

Table 7-2B. Part 124, Procedures for Decisionmaking: Evidentiary Hearings for EPA-Issued NPDES Permits and EPA-Terminated RCRA Permits—Subpart E. (Continued)

Section No.	Description
124.90(b)(2)	(b) Either:
124.90(b)(2)(i)	(i) An immediate appeal of the order or ruling will materially advance the ultimate completion of the proceeding; or
124.90(b)(2)(ii)	(ii) A review after final order is issued will be inadequate or ineffective.
124.90(c)	(4) If Administrator Decides Certification Was Improperly Granted He shall decline to hear the appeal.
	(5) When Presiding Officer Declines to Certify an Order or Ruling to Administrator for an Interlocutory Appeal It may be reviewed by Administrator only upon appeal from the initial decision of Presiding Officer. Exception: (a) When Administrator determines, upon motion of a party and in exceptional circumstances, that to delay review would not be in public interest. (i) Such motion may be made within 5 days after receipt of notification that Presiding officer has refused to certify an order or ruling for interlocutory appeal to Administrator.
	C. Acceptance or Rejection of Interlocutory Appeal by Administrator (1) Within 30 days of submission. (2) If Administrator takes no action within that time, the appeal shall be automatically dismissed.
124.90(d)	D. Stay of Proceeding In exceptional circumstances, the Presiding Officer may stay the proceeding pending a decision by the Administrator upon an order or ruling certified by the Presiding Officer for an interlocutory appeal, or upon the denial of such certification by the Presiding Officer.
124.90(e)	E. Failure to Request an Interlocutory Appeal Shall not prevent taking exception to an order or ruling in an appeal under Section 124.91 (Appeal to the Administrator).
124.91	XX. Appeal to the Administrator
124.91(a)(1)	A. Filing Notice of Appeal and Petition for Review with Administrator (1) Timing Should be filed within 30 days after service of an initial decision, or denial in whole or in part of a request for an evidentiary hearing. (2) What May Be Appealed (a) Any matter set forth in initial decision or denial; or (b) Any adverse order or ruling to which the party objected during the hearing. (3) What Petition Shall Include (a) Statement of supporting reasons. (b) When appropriate, a showing that initial decision contains:
124.91(a)(1)(i)	(i) A finding of fact or conclusion of law which is clearly erroneous; or
124.91(a)(1)(ii)	(ii) An exercise of discretion or policy which is important and which Administrator should review.
124.91(a)(2)	B. Responsive Petition Within 15 days after service of a petition for review under Section 124.91(a)(1), any other party to proceeding may file a responsive petition.
124.91(a)(3)	C. Policy Decisions Made or Legal Conclusions Drawn in the Course of Denying a Request for an Evidentiary Hearing May be reviewed and changed by Administrator in an appeal under this Section (124.91).
124.91(b)	D. Initial Decision or Denial of Request for Evidentiary Hearing (1) Administrator may review such decision within 30 days of request. (2) Within 7 days of Administrator's decision, notice shall be served by mail upon all affected parties and Regional Administrator.
124.91(c)(1)	E. Issuance of Order either Granting or Denying Petition for Review (1) Such an order shall be issued within a reasonable time following filing of petition for review.

Table 7-2B. Part 124, Procedures for Decisionmaking: Evidentiary Hearings for EPA-Issued NPDES Permits and EPA-Terminated RCRA Permits—Subpart E. (Continued)

Section No.	Description
124.91(c)(2)	(2) When Administrator Grants Petition for Review or Determines under Section 124.91(b) to Review a Decision (a) Administrator may notify the parties that only certain issues shall be briefed. (b) Regional Hearing Clerk shall promptly forward copy of record to Judicial Officer and shall retain a complete duplicate copy of record in Regional Office.
124.91(d)	F. Summarily Affirming without Opinion an Initial Decision or Denial of Request for Evidentiary Hearing May be done by Administrator notwithstanding the grant of a petition for review or determination under Section 124.91(b) to review a decision.
124.91(e)	G. Seeking Judicial Review of Final Decision of Agency A prerequisite to such action under 5 U.S.C. Section 704, is a petition to Administrator under Section 124.91(a) for review of any initial decision or denial of an evidentiary hearing.
124.91(f) 124.91(f)(1) 124.91(f)(2) 124.91(f)(3)	H. Final Agency Action on an Issue For purposes of judicial review, such action follows: (1) If Administrator Denies Review or Summarily Affirms without Opinion (Section 124.91(d)) (a) The initial decision or denial becomes final Agency action. (b) It occurs upon service of notice of the Administrator's action. (2) If Administrator Issues a Decision without Remanding the Proceeding Then final permit, redrafted as required by Administrator's original decision, shall be re-issued and served upon all parties to the appeal. (3) If Administrator Issues a Decision Remanding the Proceeding Then final Agency action occurs upon completion of remanding proceeding, including any appeals to Administrator from results of remanded proceeding.
124.91(g)	I. Briefs (1) Filing a Brief in Support of Petition Petitioner may do so within 21 days after Administrator has granted a petition for review. (2) Responsive Brief Any other party may file a responsive brief within 21 days of service of petitioner's brief. (3) Petitioner's Reply Brief Petitioner may then file a reply brief within 14 days of service of responsive brief. (4) Amicus Brief Any person may file an amicus brief for consideration of Administrator within same time periods that govern reply briefs. (5) If Administrator Reviews an Initial Regional Administractor's Decision or Denial of Request for Evidentiary Hearing Administrator shall notify the parties of the schedule for filing briefs.
124.91(h)	J. Review by Administrator of an Initial Decision or Denial of an Evidentiary Hearing Shall be limited to issues specified under Section 124.91(a). Exception: (1) After notice to all parties, Administrator may raise and decide other matters which he considers material on the basis of the record.

Table 7-2C. Part 124, Procedures for Decisionmaking: Non-Adversary Panel Procedures—Subpart F.

Section No.	Description
124.111	I. Applicability
124.111(a)	A. Except as set forth in this Subpart (in particular see Section 124.112), this Subpart applies in place of, and to the complete exclusion of Subparts A–E in the following cases:
124.111(a)(2)	(1) In any proceeding for which a hearing under this Subpart: (a) Was granted under Section 124.75(a)(2) (Decision on Request for a Hearing). (b) Following a request for a formal hearing under Section 124.74(c)(8) (Requests for Evidentiary Hearing).

Table 7-2C. Part 124, Procedures for Decisionmaking: Non-Adversary Panel Procedures—Subpart F. (Continued)

Section No.	Description
124.111(a)(3)	(2) Whenever Regional Administrator determines as a matter of discretion that the more formalized mechanisms of this Subpart should be used to process a draft RCRA permit.
124.112	II. Relation of Subpart F to Other Subparts
	A. Following Provisions of Subparts A through E Apply to Proceedings under This Subpart
124.112(a)(1)	(1) Section 124.1—Purpose and Scope.
	(2) Section 124.2—Definitions.
	(3) Section 124.3—Application for a Permit.
	(4) Section 124.4—Consolidation of Permit Processing.
	(5) Section 124.5—Modification, Revocation and Reissuance, or Termination of Permits.
	(6) Section 124.6—Draft Permit.
	(7) Section 124.7—Statement of Basis.
	(8) Section 124.8—Fact Sheet.
	(9) Section 124.9—Administrative Record for Draft Permits When EPA Is Permitting Authority.
	(10) Section 124.10—Public Notice of Permit Actions and Public Comment Period.
124.112(a)(2)	(11) Section 124.14—Reopening of the Public Comment Period.
124.112(a)(3)	(12) Section 124.16—Stays of Contested Permit Conditions.
124.112(a)(4)	(13) Section 124.20—Computation of Time.
124.112(d)(1)	(14) Section 124.72—Definitions, except for definition of "Presiding Officer" (see Section 124.119—Presiding Officer).
124.112(d)(2)	(15) Section 124.73—Filing and Submission of Documents.
124.112(d)(3)	(16) Section 124.78—Ex Parte Communications.
124.112(d)(4)	(17) Section 124.80—Filing and Service.
124.112(d)(5)	(18) Section 124.85(a)—Burden of Proof.
124.113(d)(6)	(19) Section 124.86—Motions.
124.112(d)(7)	(20) Section 124.87—Record of Hearings.
124.112(d)(8)	(21) Section 124.90—Interlocutory Appeal.
124.112(e)	(22) In Case of Permits to Which This Subpart Is Made Applicable after Final Permit Has Been Issued Under Section 124.15 Either by:
	(a) The grant under Section 124.75 (Decision on Request for a Hearing) of a hearing request under Section 124.74 (Decision on Request for a Hearing); or
	(b) By notice of supplemental proceedings under following Sections shall also apply:
	(i) 124.14—Reopening of Public Comment Period.
	(ii) 124.13—Obligation to Raise Issues and Provide Information during the Public Comment Period.
	(iii) 124.76—Obligation to Submit Evidence and Raise Issues Before a Final Permit is Issued.
124.113	III. Public Notice of Draft Permits and Public Comment Period
	A. Public Notice of a Draft Permit under This Subpart (F) Shall be given as provided in Section 124.10 (Public Notice of Permit Actions and Public Comment Period).
	B. Public Comment Period At discretion of Regional Administrator, the public comment period specified in this notice may include an opportunity for a public hearing under Section 124.12 (Public Hearings).
124.114	IV. Request for Hearing
124.114(a)	A. Request for Panel Hearing on Draft Permit By close of comment period under Section 124.113 any person may request Regional Administrator to hold a panel hearing by submitting written request containing following:
124.114(a)(1)	(1) Brief statement of interest of person requesting hearing.
124.114(a)(2)	(2) Statement of any objections to draft permit.
124.114(a)(3)	(3) Statement of issues person proposes to raise for consideration at hearing.
124.114(a)(4)	(4) Statements meeting requirements of Section 124.74(c)(1)–(5) (Requests for Evidentiary Hearing).
124.114(b)	B. Conditions for Regional Administrator to Conduct Hearing
	(1) Written request satisfying requirements of Section 124.114(a) has been received and presents genuine issues of material fact; or

Table 7-2C. Part 124, Procedures for Decisionmaking: Non-Adversary Panel Procedures—Subpart F. (Continued)

Section No.	Description
	(2) Regional Administrator determines that a hearing under this Subpart is necessary or appropriate.
	C. If Regional Administrator Decides upon a Hearing (1) He shall notify any person(s) requesting hearing. (2) He shall provide public notice under Section 124.57(c) (Public Notice).
	D. If Regional Administrator Denies Request for Hearing He shall serve written notice of that determination on all persons requesting the hearing.

124.114(c)	E. Regional Administrator May Also Decide that Hearing Should Be Held under This Section before Draft Permit Is Prepared under Section 124.6 (Draft Permit)
	(1) In such cases public notice of draft permit:
	    (a) Shall explicitly so state; and
	    (b) Shall contain information required by Section 124.57(c) (Public Notice).
	(2) This notice may also provide for a hearing under Section 124.12 (Public Hearing) before a hearing is conducted under this Section (124.114).

124.115	V. Effect of Denial of or Absence of Request for Hearing
	A. Preparation of Recommended Decision under Section 124.124 (Recommended Decision) by Regional Administrator Shall be done: (1) If no request for a hearing is made under Section 124.114 (Request for a Hearing); or (2) If all such requests are denied under this Section.
	B. Person Whose Hearing Request Has Been Denied He may appeal that recommended decision to Administrator as provided in Section 124.91 (Appeal to Administrator).
124.116	VI. Notice of Hearing
124.116(a)	A. Publication of Notice of Hearing under Section 124.57(c) (Public Notice) (1) Regional Administrator shall promptly publish this notice, upon granting request for hearing under Section 124.114 (Request For a Hearing). (2) Mailed notice shall include statement which indicates whether Presiding Officer or Regional Administrator will issue recommended decision. (3) Mailed notice shall also allow participants at least 30 days to submit written comments as provided under Section 124.118 (Submission of Written Comments on Draft Permit).
124.116(b)	B. Regional Administrator May Also Give Notice of Hearing under This Section at Same Time As Notice of Draft Permit Under Section 124.113 In that case the comment periods under: (1) Section 124.113—Public Notice of Draft Permits and Public Comment Period; and (2) Section 124.118—Submission of Written Comments on Draft Permit shall be merged and held as a single public comment period.
124.116(c)	C. Notice of Hearing under This Section in Response to Hearing Request under Section 124.74 As Provided in Section 124.75 Regional Administrator may give such notice.
124.117	VII. Request to Participate in Hearing
124.117(a)	A. Persons Desiring to Participate in Hearing Where Notice Was Given under This Section (124.117) Shall file a request to participate with Regional Hearing Clerk before deadline set forth in notice of hearing.
	B. Any Person Filing Such a Request He becomes a party to proceedings within meaning of Administrative Procedures Act.
124.117(a)(1) 124.117(a)(2)	C. Content of Request to Participate in Hearing (1) Brief statement of interest of person in proceeding. (2) Brief outline of points to be addressed.

Table 7-2C. Part 124, Procedures for Decisionmaking: Non-Adversary Panel Procedures—Subpart F. (Continued)

Section No.	Description
124.117(a)(3)	(3) Estimate of time required.
124.117(a)(4)	(4) Requirements of Section 124.74(c)(1)–(5) (Requests for Evidentiary Hearing).
124.117(a)(5)	(5) If request is submitted by an organization, a nonbinding list of persons to take part in the presentation.
124.117(b)	D. Hearing Schedule (1) As soon as practicable, but no later than 2 weeks before scheduled date of hearing, the Presiding Officer shall make a hearing schedule available to public. (2) He shall mail it to each person who requests to participate in the hearing.
124.118	VIII. Submission of Written Comments on Draft Permit
124.118(a)	A. Filing Comments on Draft Permit (1) No later than 30 days before scheduled start of hearing (or such other date set forth in notice of hearing), each party shall file all of its comments. (2) What Comments Should Be Based On (a) Information in administrative record. (b) Any other information which is or reasonably could have been available to that party. (3) What Shall Be Included in Comments Any affidavits, studies, data, tests, or other materials relied upon for making any factual statements in the comments.
124.118(b)(1) 124.118(c)	B. Written Comments (1) Such comments filed under Section 124.118(a) shall constitute the bulk of evidence submitted at the hearing. (2) Written Material in Response to Comments Filed by Other Parties under Section 124.118(a) May be submitted by any parties to hearing at time they appear at the panel stage of the hearing under Section 124.120 (Panel Hearing).
124.118(b)(1) 124.118(b)(2)	C. Oral Statements at Hearing (1) Should be brief and in nature of argument. (2) They shall be restricted either: (a) To points that could not have been made in written comments; or (b) To emphasize points which are made in the comments, but which the party believes can more effectively be argued in the hearing context. (3) Move by Party to Submit All or Part of Its Comments Orally at Hearing Instead of Submitting Written Comments (a) Any party may move to take this approach within 2 weeks prior to deadline specified in Section 124.118(a) for filing comments. (b) Presiding Officer shall, within one week grant such motion if he finds that party will be prejudiced if required to submit comments in written form.
124.119	IX. Presiding Officer
124.119(b)	A. Duty of Presiding Officer To conduct a fair and impartial hearing. B. Authority of Presiding Officer
124.119(b)(1)	(1) That conferred by: (a) Sections 124.85(b)(1)–(15)–Hearing Procedure. (b) Sections 124.83(b) and (c)–Prehearing Conferences.
124.119(b)(2)	(2) To Receive Relevant Evidence (a) Under Section 124.113–Public Notice to Draft Permits and Public Comment Period; or (b) Under Section 124.118–Submission of Written Comments on Draft Permit; or (c) Under Section 124.120–Panel Hearing; or (d) Under Section 124.9–Administrative Record for Draft Permits When EPA Is the Permitting Authority; or (e) Under Section 124.18–Administrative Record for Final Permit When EPA is the Permitting Authority.
124.119(b)(3)	(3) To Change Date of Hearing under Section 124.120 (Panel Hearing) or to Recess Hearing Until Future Date (a) May be done upon motion or on Presiding Officer's initiative. (b) In any such case the notice required in Section 124.10 (Public Notice of Permit Actions and Public Comment Period) shall be given.

Table 7-2C. Part 124, Procedures for Decisionmaking: Non-Adversary Panel Procedures—Subpart F. (Continued)

Section No.	Description
124.120	X. Panel Hearing
124.120(a)	A. Presiding Officer (1) Shall preside at each hearing held under this Subpart (F).
124.120(c)	(2) May Request Appearance and Testimony of Knowledgable Persons At any time before close of hearing, after consultation with EPA panel (see below).
124.120(a)	B. EPA Panel (1) Such a panel shall also take part in the hearing. (2) Make-up of Panel (a) Three or more EPA temporary or permanent employees having special expertise or responsibility in areas related to the hearing issue. (b) At least two of the EPA panel members shall not have taken part in writing the draft permit. (3) Change in Panel Membership If appropriate for evaluation of new or different issues presented at the hearing, the panel membership, at discretion of Regional Administrator, may change or may include persons not employed by EPA.
124.120(b)	(4) When Regional Administrator Shall Designate Persons Who Shall Serve on Panel At time of hearing notice under Section 124.116 (Notice of Hearing). (5) Regional Administrator shall file with Regional Hearing Clerk the name and address of each panel member. (6) Staff Support for Panel (a) Regional Administrator may designate EPA employees who will provide staff support. (b) These EPA employees may or may not serve as panel members. (c) Designated persons shall be subject to ex parte rules in Section 124.78 (Ex Parte Communications).
	C. Agency Trial Staff Regional Administrator may also designate Agency trial staff (as defined in Section 124.78) for the hearing.
124.120(d)	D. Questioning Hearing Participants (1) Panel members may question any person participating in the panel hearing. (2) Cross-Examination (a) By Persons Other than Panel Members Not permitted except when Presiding Officer determines, after consultation with panel, that cross-examination would expedite consideration of the issues. (b) Written Questions (i) Parties may submit written questions to Presiding Officer for him to ask the participants. (ii) Presiding Officer may, after consultation with panel, and at his sole discretion, ask these questions.
124.120(e)	E. Written Questions by Any Party (1) May be submitted at any time before close of hearing. (2) Presiding Officer, after consultation with panel, may at his sole discretion, ask the written questions so submitted.
124.120(f)	F. Submittal of Additional Written Testimony, Affidavits, Information after Close of Hearing (1) May be done within 10 days after close of hearing by any party if they consider information relevant. (2) These additional submissions shall be filed with Regional Hearing Clerk and shall be a part of hearing record.
124.121	XI. Opportunity for Cross-Examination
124.121(a)	A. Request to Cross-Examine Any Issue of Material Fact (1) Any party to a panel hearing may submit such a written request. (2) Timing The motion shall be submitted to Presiding Officer within 15 days after full transcript of panel hearing is filed with Regional Hearing Clerk.

Table 7-2C. Part 124, Procedures for Decisionmaking: Non-Adversary Panel Procedures—Subpart F. (Continued)

Section No.	Description
	(3) What Request Shall Specify
124.121(a)(1)	(a) Disputed Issue(s) of Material Fact
	This shall include:
	(i) Explanation of why questions at issue are factual rather than of an analytical or policy nature.
	(ii) Extent to which they are in dispute in light of the then-existing record.
	(iii) Extent to which they are material to the decision on the application.
124.121(a)(2)	(b) Person(s) to be cross-examined.
	(c) Estimate of time necessary to conduct cross-examination.
	(d) Statement explaining how cross-examination will resolve disputed issues of material fact.
124.121(b)	B. Issuance of Order by Presiding Officer Granting or Denying Each Request for Cross-Examination
	(1) Shall be promptly issued after receipt of all motions for cross-examination under Section 124.121(a), and after consultation with hearing panel.
	(2) Order Granting Requests for Cross-Examination
	Shall be served on all parties and shall specify:
124.121(b)(1)	(a) Issues on which cross-examination is granted.
124.121(b)(2)	(b) Persons to be cross-examined on each issue.
124.121(b)(3)	(c) Persons allowed to conduct cross-examination.
124.121(b)(4)	(d) Time limits for examination of witnesses by each cross-examiner.
124.121(b)(5)	(e) Date, time, and place of supplementary hearing at which cross-examination shall take place.
124.121(c)	C. If Presiding Officer Determines that Two or More Parties Have the Same or Similar Interests
	(1) He may require them to choose a single representative for pusposes of cross-examination, in order to prevent unduly repetitious cross-examination.
	(2) In that case, the order shall simply assign time for cross-examination without further identifying the representative.
	(3) If Designated Parties Fail to Choose a Single Representative
	Presiding Officer may divide the assigned time among the representatives or issue any other order which justice may require.
124.121(d)	D. Supplementary Hearing
	(1) Presence at Supplementary Hearing
	Presiding Officer, and to extent possible, members of hearing panel.
	(2) Authority to Modify Any Order Issued under Section 124.121(b)
	Presiding Officer shall have such authority.
	(3) A record shall be made under Section 124.87 (Record of Hearing).
124.121(e)(1)	E. Alternative Method(s) of Clarifying the Record
	(1) A party may request such an approach be used in lieu of or in addition to cross-examination.
	(2) Example of Alternative Method
	Submittal of additional written information.
	(3) Timing
	Request no later than time set for requesting cross-examination.
	(4) Issuance of Order by Presiding Officer Granting or Denying Request
	Shall be issued at same time he issues (or would have issued) an order granting or denying a request for cross-examination under Section 124.121(b).
	(5) If Request for Alternative Method Granted
	Order shall specify:
	(a) The alternative.
	(b) Any other relevant information, such as due date for submitting written information.
	(6) Requirement for Alternative Method(s) as Precondition to Ruling on Merits of Request for Cross-Examination (Section 124.121(a))
	(a) Presiding Officer may so require whether or not a request to do so has been made.
	(b) Timing
	Party requesting cross-examination shall have one week to comment on results of using the alternative method.
	(c) After considering these comments, Presiding Officer shall issue an order granting or denying request for cross-examination.
124.121(f)	F. Provisions of Section 124.85(d)(2) (Hearing Procedure) apply to proceedings under this Subpart.

Table 7-2C. Part 124, Procedures for Decisionmaking: Non-Adversary Panel Procedures—Subpart F. (Continued)

Section No.	Description
124.122	XII. Record for Final Permit
	A. Content of the Record on Which Final Permit Shall Be Based in Any Proceeding Under This Subpart
124.122(a)	(1) Administrative Record Compiled under: (a) Section 124.9—Administrative Record for Draft Permits When EPA is the Permitting Authority; or (b) Section 124.18—Administrative Record for Final Permit When EPA Is the Permitting Authority.
124.122(b)	(2) Ex Parte Contacts Any material submitted under Section 124.78 (Ex parte Communications) relating to ex parte contacts.
124.122(c)	(3) All Notices Issued Under Section 124.113—Public Notice of Draft Permits and Public Comment Period.
124.122(d)	(4) All Requests for Hearings and Rulings on Those Requests Received or Issued Under Section 124.114—Request for Hearing.
124.122(e)	(5) Any Notice of Hearing Issued under Section 124.116—Notice of Hearing.
124.122(f)	(6) Any Request to Participate in the Hearing Received under Section 124.117—Request to Participate in Hearing.
124.122(g)	(7) All Comments Submitted, Motions and Rulings Made Under Section 124.118—Submission of Written Comments on Draft Permit. (8) Any Comments Filed Under Section 124.113—Public Notice of Draft Permits and Public Comment Period.
124.122(h)	(9) Full Transcript and Other Material Received into Record of Panel Hearing Under Section 124.120—Panel Hearing.
124.122(i)	(10) Any Motion for or Ruling on Cross-Examination Filed or issued under Section 124.121—Opportunity for cross-examination.
124.122(j)	(11) Any Motions for, Orders for, and Results of, Any Alternatives to Cross-Examination Under Section 124.121—Opportunity for Cross-Examination.
124.122(k)	(12) Full transcript of any cross-examination held.
124.123	XIII. Filing of Brief, Proposed Finding of Fact and Conclusions of Law and Proposed Modified Permit
	A. Within 20 Days after All Requests for Cross-Examination Are Denied or After Transcript (Including Cross-Examination) of Full Hearing Becomes Available Each party may submit following items, unless otherwise ordered by Presiding Officer: (1) Proposed finding of fact. (2) Conclusions regarding material issues of law, fact or discretion. (3) A proposed modified permit, if such person is urging that the draft or final permit be modified. (4) A brief in support thereof. (5) References to relevant pages of transcript and to relevant exhibits.
	B. Within 10 Days Thereafter Each party may file a reply brief concerning matters contained in opposing briefs and containing alternative: (1) Findings of fact. (2) Conclusions regarding material issue of law, fact, or discretion. (3) Proposed modified permit where appropriate.
	C. Oral Arguments May be held at discretion of Presiding Officer on motion of any party or on his own initiative.
124.124	XIV. Recommended Decision
	A. Person Named to Prepare Decision (1) Shall as soon as practicable after conclusion of hearing evaluate the record of the hearing and prepare a recommended decision. (2) He shall file recommended decision with Regional Hearing Clerk. (3) He may consult with and receive assistance from any member of hearing panel in drafting the recommended decision.

Table 7-2C. Part 124, Procedures for Decisionmaking: Non-Adversary Panel Procedures—Subpart F. (Continued)

Section No.	Description
	(4) He may delegate preparation of recommended decision to panel or to any member or members of it.
	B. Content of Decision (1) Findings of fact. (2) Conclusions regarding all material issues of law. (3) Recommendations as to whether and in what respect the draft or final permit should be modified.
	C. After Recommended Decision Has Been Filed Regional Hearing Clerk shall serve a copy of that decision on each party and Administrator.
124.125	XV. Appeal from or Review of Recommended Decision A. Exception to and Appeal of Recommended Decision (1) Any party may take exception to and appeal: (a) Any matter set forth in recommended decision; or (b) Any adverse order or ruling of the Presiding Officer to which that party objected. (2) Appeal to Administrator May be made as provided under Section 124.91 (Appeal to Administrator). (3) References to "initial decision" will mean recommended decision under Section 124.124 (Recommended Decision). (4) Timing of Appeal Within 30 days after service of recommended decision.
124.126	XVI. Final Decision A. Timing As soon as practicable after all appeal proceedings have been completed, Administrator shall issue a final decision. B. Content of Final Decision (1) Finding of fact. (2) Conclusions regarding material issues of law, fact, or discretion, as well as reasons thereof. (3) Modified permit to extent appropriate. C. Relation to Recommended Decision Administrator may accept or reject all or part of it. D. Delegation of Work of Preparing Final Decision by Administrator He may delegate some or all of the work to a person or persons without substantial prior connection with the matter. E. Consultation by Administrator or His Designee in Preparation of Final Decision He may consult with Presiding Officer, members of hearing panel, or any other EPA employee other than members of Agency Trial Staff under Section 124.78 (Ex parte Communications). F. Hearing Clerk shall file a copy of decision to all parties.
124.127	XVII. Final Decision If There Is No Review A. Conditions for Recommended Decision Becoming Final Decision of Agency (1) If no party appeals a recommended decision to Administrator. (2) If Administrator does not elect to review it. (3) Expiration of time for filing any appeals.
124.128	XVIII. Delegation of Authority: Time Limitations
124.128(a)	A. Delegation of Authority Administrator may delegate to a Judicial Officer any or all of his or her authority under this Subpart.

Table 7-2C. Part 124, Procedures for Decisionmaking: Non-Adversary Panel Procedures—Subpart F. (Continued)

Section No.	Description
124.128(b)	B. Failure of Administrator, Regional Administrator, or Presiding Officer to Do Any Act Within the Time Periods Specified under This Part (1) This failure shall not waive or diminish any right, power or authority of the United States Environmental Protection Agency.
124.128(c)	(2) Upon Showing by Any Party that It Has Been Prejudiced by Such Failure The Administrator, Regional Administrator, or Presiding Officer may grant that party such relief of a procedural nature as may be appropriate, including extension of any time for compliance or other action.

Please print or type in the unshaded areas only
(fill-in areas are spaced for elite type, i.e., 12 characters/inch).

Form Approved OMB No. 158-R0175

FORM 1 GENERAL	⊕EPA	U.S. ENVIRONMENTAL PROTECTION AGENCY **GENERAL INFORMATION** *Consolidated Permits Program* *(Read the "General Instructions" before starting.)*	I. EPA I.D. NUMBER

LABEL ITEMS

I. EPA I.D. NUMBER

III. FACILITY NAME

V. FACILITY MAILING ADDRESS

VI. FACILITY LOCATION

PLEASE PLACE LABEL IN THIS SPACE

GENERAL INSTRUCTIONS

If a preprinted label has been provided, affix it in the designated space. Review the information carefully; if any of it is incorrect, cross through it and enter the correct data in the appropriate fill-in area below. Also, if any of the preprinted data is absent *(the area to the left of the label space lists the information that should appear)*, please provide it in the proper fill-in area(s) below. If the label is complete and correct, you need not complete Items I, III, V, and VI *(except VI-E which must be completed regardless)*. Complete all items if no label has been provided. Refer to the instructions for detailed item descriptions and for the legal authorizations under which this data is collected.

II. POLLUTANT CHARACTERISTICS

INSTRUCTIONS: Complete A through J to determine whether you need to submit any permit application forms to the EPA. If you answer "yes" to any questions, you must submit this form and the supplemental form listed in the parenthesis following the question. Mark "X" in the box in the third column if the supplemental form is attached. If you answer "no" to each question, you need not submit any of these forms. You may answer "no" if your activity is excluded from permit requirements; see Section C of the instructions. See also, Section D of the instructions for definitions of bold-faced terms.

SPECIFIC QUESTIONS	YES	NO	FORM ATTACHED	SPECIFIC QUESTIONS	YES	NO	FORM ATTACHED
A. Is this facility a **publicly owned treatment works** which results in a **discharge to waters of the U.S.**? (FORM 2A)				B. Does or will this facility *(either existing or proposed)* include a **concentrated animal feeding operation** or **aquatic animal production facility** which results in a **discharge to waters of the U.S.**? (FORM 2B)	19	20	21
C. Is this a facility which currently results in **discharges** to **waters of the U.S.** other than those described in A or B above? (FORM 2C)				D. Is this a **proposed facility** *(other than those described in A or B above)* which will result in a **discharge to waters of the U.S.**? (FORM 2D)	25	26	27
E. Does or will this facility treat, store, or dispose of **hazardous wastes**? (FORM 3)	28	29	30	F. Do you or will you inject at this facility industrial or municipal effluent below the lowermost stratum containing, within one quarter mile of the well bore, underground sources of drinking water? (FORM 4)	31	32	3
G. Do you or will you inject at this facility any produced water or other fluids which are brought to the surface in connection with conventional oil or natural gas production, inject fluids used for enhanced recovery of oil or natural gas, or inject fluids for storage of liquid hydrocarbons? (FORM 4)	34	35	36	H. Do you or will you inject at this facility fluids for special processes such as mining of sulfur by the Frasch process, solution mining of minerals, in situ combustion of fossil fuel, or recovery of geothermal energy? (FORM 4)	37	38	39
I. Is this facility a proposed **stationary source** which is one of the 28 industrial categories listed in the instructions and which will potentially emit 100 tons per year of any air pollutant regulated under the Clean Air Act and may affect or be located in an **attainment area**? (FORM 5)	40	41	42	J. Is this facility a proposed **stationary source** which is NOT one of the 28 industrial categories listed in the instructions and which will potentially emit 250 tons per year of any air pollutant regulated under the Clean Air Act and may affect or be located in an **attainment area**? (FORM 5)	43	44	45

III. NAME OF FACILITY

C 1 SKIP

IV. FACILITY CONTACT

A. NAME & TITLE *(last, first, & title)*   B. PHONE *(area code & no.)*

C 2

V. FACILITY MAILING ADDRESS

A. STREET OR P.O. BOX

C 3

B. CITY OR TOWN   C. STATE   D. ZIP CODE

C 4

VI. FACILITY LOCATION

A. STREET, ROUTE NO. OR OTHER SPECIFIC IDENTIFIER

C 5

B. COUNTY NAME

C. CITY OR TOWN   D. STATE   E. ZIP CODE   F. COUNTY CODE *(if known)*

C 6

EPA Form 3510-1 (5-80)   **CONTINUE ON REVERSE**

Fig. 7-1 EPA Form 1 General (No. 3510-1), General Information for Consolidated Permits Program.

CONTINUED FROM THE FRONT

VII. SIC CODES (4-digit, in order of priority)

A. FIRST

C 7 | (specify)

B. SECOND

C 7 | (specify)

C. THIRD

C 7 | (specify)

D. FOURTH

C 7 | (specify)

VIII. OPERATOR INFORMATION

A. NAME

C 8

B. Is the name listed in Item VIII-A also the owner?

☐ YES   ☐ NO

C. STATUS OF OPERATOR (Enter the appropriate letter into the answer box, if "Other", specify.)

F = FEDERAL   M = PUBLIC (other than federal or state)
S = STATE   O = OTHER (specify)
P = PRIVATE

(specify)

D. PHONE (area code & no.)

C A

E. STREET OR P.O. BOX

F. CITY OR TOWN

C B

G. STATE   **H. ZIP CODE**

IX. INDIAN LAND

Is the facility located on Indian lands?

☐ YES   ☐ NO

X. EXISTING ENVIRONMENTAL PERMITS

A. NPDES (Discharges to Surface Water)

C 9 N

D. PSD (Air Emissions from Proposed Sources)

C 9 P

B. UIC (Underground Injection of Fluids)

C 9 U

E. OTHER (specify)

C 9

(specify)

C. RCRA (Hazardous Wastes)

C 9 R

E. OTHER (specify)

C 9

(specify)

XI. MAP

Attach to this application a topographic map of the area extending to at least one mile beyond property bounderies. The map must show the outline of the facility, the location of each of its existing and proposed intake and discharge structures, each of its hazardous waste treatment, storage, or disposal facilities, and each well where it injects fluids underground. Include all springs, rivers and other surface water bodies in the map area. See instructions for precise requirements.

XII. NATURE OF BUSINESS (provide a brief description)

XIII. CERTIFICATION (see instructions)

I certify under penalty of law that I have personally examined and am familiar with the information submitted in this application and all attachments and that, based on my inquiry of those persons immediately responsible for obtaining the information contained in the application, I believe that the information is true, accurate and complete. I am aware that there are significant penalties for submitting false information, including the possibility of fine and imprisonment.

A. NAME & OFFICIAL TITLE (type or print)

B. SIGNATURE

C. DATE SIGNED

COMMENTS FOR OFFICIAL USE ONLY

C C

EPA Form 3510-1 (5-80)   REVERSE

BILLING CODE 6560-01-C

Fig. 7-1 EPA Form 1 General (No. 3510-1), General Information for Consolidated Permits Program. (Continued)

Instructions for Consolidated Permit Application Forms

Section A. General Instructions

Who Must Apply?

With the exceptions described in section C of these instructions, Federal laws prohibit you from conducting any of the following activities without a permit.

NPDES (National Pollutant Discharge Elimination System under the Clean Water Act, 33 U.S.C. 1251). Discharge of pollutants into the waters of the United States.

RCRA (Resource Conservation and Recovery Act, 42 U.S.C. 6901). Treatment, storage, or disposal of hazardous wastes.

UIC (Underground Injection Control under the Safe Drinking Water Act, 42 U.S.C. 300f). Injection of fluids underground by gravity flow or pumping.

PSD (Prevention of Significant Deterioration under the Clean Air Act, 72 U.S.C. 7401). Emission of an air pollutant by a new or modified facility in or near an area which has attained the National Ambient Air Quality Standards for that pollutant.

Each of the above permit programs is operated in any particular State by either the United States Environmental Protection Agency (EPA) or by an approved State agency. You must use this application form to apply for a permit for those programs administered by EPA. For those programs administered by approved States, contact the State environmental agency for the proper forms.

If you have any questions about whether you need a permit under any of the above programs, or if you need information as to whether a particular program is administered by EPA or a State agency or if you need to obtain application forms, contact your EPA Regional office (listed in Table 1).

Upon your request, and based upon information supplied by you, EPA will determine whether you are required to obtain a permit for a particular facility. Contact your EPA Regional office (listed in Table 1). Be sure to contact EPA if you have a question, because Federal laws provide that *you may be heavily penalized if you do not apply for a permit when a permit is required.*

Form 1 of the EPA consolidated application forms (attached to these instructions) collects general information applying to all programs. You must fill out Form 1 regardless of which permit you are applying for. In addition, you must fill out one of the supplementary forms (Forms 2–5) for each permit needed under each of the above programs. Item II of Form 1 will guide you to the appropriate supplementary forms.

You should note that there are certain exclusions to the permit requirements listed above. The exclusions are described in detail in section C of these instructions. If your activities are excluded from permit requirements then you do not need to complete and return any forms.

Note: Certain activities not listed above also are subject to EPA-administered environmental permit requirements. These include permits for ocean dumping, dredged or fill material discharging, and certain types of air emissions. Contact your EPA Regional office for further information.

Table 1.—*Addresses of EPA Regional Offices and States Within Their Jurisdiction*

Region I

Permit Contact, Environmental and Economic Impact Office, U.S. Environmental Protection Agency, John F. Kennedy Building, Boston, Massachusetts 02203, (617) 223–4635, FTS 223–4635. Connecticut, Maine, Massachusetts, New Hampshire, Rhode Island, Vermont.

Region II

Permit Contact, Permits Administration Branch, Room 432, U.S. Environmental Protection Agency, 26 Federal Plaza, New York, New York 10007, (212) 264–9880, FTS 264–9880. New Jersey, New York, Virgin Islands, Puerto Rico.

Region III

Permit Contact (3 EN 23), U.S. Environmental Protection Agency, 6th & Walnut Streets, Philadelphia, Pennsylvania 19106, (215) 597–8816, FTS 597–8816. Delaware, District of Columbia, Maryland, Pennsylvania, Virginia, West Virginia.

Region IV

Permit Contact, Permits Section, U.S. Environmental Protection Agency, 345 Courtland Street, N.E., Atlanta, Georgia 30365, (404) 881–2017, FTS 257–2017. Alabama, Florida, Georgia, Kentucky, Mississippi, North Carolina, South Carolina, Tennessee.

Region V

Permit Contact (5EP), U.S. Environmental Protection Agency, 230 South Dearborn Street, Chicago, Illinois 60604, (312) 353–2105, FTS 353–2105. Illinois, Indiana, Michigan, Minnesota, Ohio, Wisconsin.

Region VI

Permit Contact (6AEP), U.S. Environmental Protection Agency, First International Building, 1201 Elm Street, Dallas, Texas 75270, (214) 767–2765, FTS 729–2765. Arkansas, Louisiana, New Mexico, Oklahoma, Texas.

Region VII

Permit Contact, Permits Branch, U.S. Environmental Protection Agency, 324 East 11th Street, Kansas City, Missouri 64106, (816) 758–5955, FTS 758–5955. Iowa, Kansas, Missouri, Nebraska.

Region VIII

Permit Contact (8E–WE), Suite 103, U.S. Environmental Protection Agency, 1816 Lincoln Street, Denver, Colorado 80203, (303) 837–4901, FTS 837–4901. Colorado, Montana, North Dakota, South Dakota, Utah, Wyoming.

Region IX

Permit Contact Permits Branch (E–4), U.S. Environmental Protection Agency, 215 Freemont Street, San Francisco, California 94105, (415) 556–3450, FTS 556–3450. Arizona, California, Hawaii, Nevada, Guam, American Samoa, Trust Territories.

Region X

Permit Contact. (M/S 521), U.S. Environmental Protection Agency, 1200 6th Avenue, Seattle, Washington 98101, (206) 442–7176, FTS 399–7176. Alaska, Idaho, Oregon, Washington.

Where To File

The application forms should be mailed to the EPA Regional office whose Region includes the State in which the facility is located (see Table 1).

If the State in which the facility is located administers a Federal permit program under which you need a permit, you should contact the appropriate State agency for the correct forms. Your EPA Regional Office (Table 1) can tell you to whom to apply and can provide the appropriate address and phone number.

When To File

Because of statutory requirements, the deadlines for filing applications vary according to the type of facility you operate and the type of permit you need. These deadlines are as follows: [1]

Table 2.—*Filing Dates for Permits*

Form (permit)	When to file
2a (NPDES)	180 days before your present NPDES permit expires
2b (NPDES)	180 days before your present NPDES permit expires[2], or 180 days prior to startup if you are a new facility

[1] Please note that some of these forms are not yet available for use and are listed as "Reserved" at the beginning of these instructions. Contact your EPA Regional office for information on current application requirements and forms.

Fig. 7-1 EPA Form 1 General (No. 3510-1), General Information for Consolidated Permits Program. (Continued)

Table 2.—*Filing Dates for Permits*—Continued

Form (permit)	When to file
2c (NPDES)	180 days before your present NPDES permit expires[2]
2d (NPDES)	180 days prior to startup
3 (Hazardous Waste)	Existing facility: 180 days following publication of regulations listing hazardous wastes. New facility: 180 days before commencing physical construction
4 (UIC)	A reasonable time prior to construction for new wells; as directed by the Director for existing wells
5 (PSD)	Prior to commencement of construction

[2] If your present permit expires on or before November 30, 1980, the filing date is the date on which your permit expires. If your permit expires during the period December 1, 1980–May 31, 1981, the filing date is 90 days before your permit expires.

Federal regulations provide that you may not begin to construct a new source in the NPDES program, a new hazardous waste management facility, a new injection well or a facility covered by the PSD program before the issuance of a permit under the applicable program. Please note that if you are required to obtain a permit before beginning construction, as described above, you may need to submit your permit application well in advance of an applicable deadline listed in Table 2.

Fees

The U.S. EPA does not require a fee for applying for any permit under the consolidated permit programs. (However, some States which administer one or more of these programs require fees for the permits which they issue.)

Availability of Information to Public

Information contained in these application forms will, upon request, be made available to the public for inspection and copying. However, you may request confidential treatment for certain information which you submit on certain supplementary forms. The specific instructions for each supplementary form state what information on the form, if any, may be claimed as confidential and what procedures govern the claim. No information on Forms 1 and 2 may be claimed as confidential.

Completion of Forms

Unless otherwise specified in instructions to the forms, each item in each form must be answered. To indicate that each item has been considered, enter "NA," for not applicable, if a particular item does not fit the circumstances or characteristics of your facility or activity.

If you have previously submitted information to EPA or to an approved State agency which answers a question, you may either repeat the information in the space provided or attach a copy of the previous submission. Some items in the form require narrative explanation. If more space is necessary to answer a question, attach a separate sheet entitled "Additional Information."

Financial Assistance for Pollution Control

There are a number of direct loans, loan guarantees, and grants available to firms and communities for pollution control expenditures. These are provided by the Small Business Administration, the Economic Development Administration, the Farmers Home Administration, and the Department of Housing and Urban Development. Each EPA Regional office (Table 1) has an economic assistance coordinator who can provide you with additional information.

EPA's construction grants program under Title II of the Clean Water Act is an additional source of assistance to publicly owned treatment works. Contact your EPA Regional office for details.

Section B. Instructions for Form 1— General Information

This form must be completed by all applicants.

Completing this form. Please type or print in the unshaded areas only. Some items have small graduation marks in the fill-in spaces. These marks indicate the number of characters that may be entered into our data system. The marks are spaced at ⅛" intervals which accommodate elite type (12 characters per inch). If you use another type you may ignore the marks. If you print, place each character between the marks. Abbreviate if necessary to stay within the number of characters allowed for each item. Use one space for breaks between words, but not for punctuation marks unless they are needed to clarify your response.

Item I. Space is provided at the upper right hand corner of Form 1 for insertion of your EPA Identification Number. If you have an existing facility, enter your Identification Number. If you don't know your EPA Identification Number, please contact your EPA Regional office (table 1), which will provide you with your number. If your facility is new (not yet constructed), leave this item blank.

Item II. Answer each question to determine which supplementary forms you need to fill out. Be sure to check the glossary in section D of these instructions for the legal definitions of the bold faced words. Check section C of these instructions to determine whether your activity is excluded from permit requirements.

If you answer "no" to every question, then you do not need a permit, and you do not need to complete and return any of these forms.

If you answer "yes" to any question, then you must complete and file the supplementary form by the deadline listed in Table 2 along with this form. (The applicable form number follows each question and is enclosed in parentheses.) You need not submit a supplementary form if you already have a permit under the appropriate Federal program, unless your permit is due to expire and you wish to renew your permit.

Questions (I) and (J) of Item II refer to major new or modified sources subject to Prevention of Significant Deterioration (PSD) requirements under the Clean Air Act. For the purpose of the PSD program, major sources are defined as (1) sources listed in Table 3 which have the potential to emit 100 tons or more per year emissions, and (2) all other sources with the potential to emit 250 tons or more per year. See section C of these instructions for discussion of exclusions of certain modified sources.

Table 3.—*28 Industrial Categories Listed in Section 169(1) of the Clean Air Act of 1977*

Fossil fuel-fired steam generators of more than 250 million BTU per hour heat input
Coal cleaning plants (with thermal dryers)
Kraft pulp mills
Portland cement plants
Primary zinc smelters
Iron and steel mill plants
Primary aluminum ore reduction plants
Primary copper smelters.

Municipal incinerators capable of charging more than 250 tons of refuse per day

Hydrofluoric acid plants
Nitric acid plants
Sulfuric acid plants
Petroleum refineries
Lime plants
Phosphate rock processing plants
Coke oven batteries
Sulfur recovery plants
Carbon black plants (furnace process)
Primary lead smelters
Fuel conversion plants
Sintering plants
Secondary metal production plants
Chemical process plants.

Fig. 7-1 EPA Form 1 General (No. 3510-1), General Information for Consolidated Permits Program. (Continued)

Fossil fuel boilers (or combination thereof) totaling more than 250 million BTU per hour heat input

Petroleum storage and transfer units with a total storage capacity exceeding 300,000 barrels

Taconite ore processing plants

Glass fiber processing plants

Charcoal production plants.

Item III. Enter the facility's official or legal name. Do not use a colloquial name.

Item IV. Give the name, title, and work telephone number of a person who is thoroughly familiar with the operation of the facility and with the facts reported in this application and who can be contacted by reviewing offices if necessary.

Item V. Give the complete mailing address of the office where correspondence should be sent. This often is not the address used to designate the location of the facility or activity.

Item VI. Give the address or location of the facility identified in Item III of this form. If the facility lacks a street name or route number, give the most accurate alternative geographic information (e.g., section number, quarter section number, or description).

Item VII. List, in descending order of significance, the four 4-digit standard industrial classification (SIC) codes which best describe your facility in terms of the principal products or services you produce or provide. Also, specify each classification in words. These classifications may differ from the SIC codes describing the operation generating the discharge, air emissions, or hazardous wastes.

SIC code numbers are descriptions which may be found in the "Standard Industrial Classification Manual" prepared by the Executive Office of the President, Office of Management and Budget, which is available from the Government Printing Office, Washington, D.C. Use the current edition of the manual. If you have any questions concerning the appropriate SIC code for your facility, contact your EPA Regional office (see Table 1).

Item VIII–A. Give the name, as it is legally referred to, of the person, firm, public organization, or any other entity which operates the facility described in this application. This may or may not be the same name as the facility. The operator of the facility is the legal entity which controls the facility's operation rather than the plant or site manager. Do not use a colloquial name.

Item VIII–B. Indicate whether the entity which operates the facility also owns it by marking the appropriate box.

Item VIII–C. Enter the appropriate letter to indicate the legal status of the operator of the facility. Indicate "public" for a facility solely owned by local government(s) such as a city, town, county, parish, etc.

Items VIII–D–H. Enter the telephone number and address of the operator identified in item VIII–A.

Item IX. Indicate whether the facility is located on Indian lands.

Item X. Give the number of each presently effective permit issued to the facility for each program or, if you have previously filed an application but have not yet received a permit, give the number of the application, if any. Fill in the unshaded area only. If you have more than one currently effective permit for your facility under a particular permit program, you may list additional permit numbers on a separate sheet of paper. List any relevant environmental Federal (e.g., permits under the Ocean Dumping Act, section 404 of the Clean Water Act or the Surface Mining Control and Reclamation Act), State (e.g., State permits for new air emission sources in nonattainment areas under Part D of the Clean Air Act or State permits under section 404 of the Clean Water Act) or local permits or applications under "other."

Item XI. Provide a topographic map or maps of the area extending at least to one mile beyond the property boundaries of the facility which clearly show the following:
- The legal boundaries of the facility;
- The location and serial number of each of your existing and proposed intake and discharge structures;
- All hazardous waste management facilities;
- Each well where you inject fluids underground; and
- All springs and surface water bodies in the area, plus all drinking water wells within ¼ mile of the facility which are identified in the public record or otherwise known to you.

If an intake or discharge structure, hazardous waste disposal site, or injection well associated with the facility is located more than one mile from the plant, include it on the map, if possible. If not, attach additional sheets describing the location of the structure, disposal site, or well, and identify the U.S. Geological Survey (or other) map corresponding to the location.

On each map, include the map scale, a meridian arrow showing north, and latitude and longitude at the nearest whole second. On all maps of rivers, show the direction of the current, and in tidal waters, show the directions of the ebb and flow tides. Use a 7½ minute series map published by the U.S.

Geological Survey, which may be obtained through the U.S. Geological Survey Offices in Washington, D.C., Denver, Colorado, or Anchorage, Alaska. If a 7½ minute series map has not been published for your facility site, then you may use a 15 minute series map from the U.S. Geological Survey. If neither a 7½ nor 15 minute series map has been published for your facility site, use a plat map or other appropriate map, including all the requested information; in this case, briefly describe land uses in the map area (e.g., residential, commercial).

You may trace your map from a geological survey chart, or other map meeting the above specifications. If you do, your map should bear a note showing the number or title of the map or chart it was traced from. Include the names of nearby towns, water bodies, and other prominent points. An example of an acceptable location map is shown in Figure A of these instructions.

(Note—Figure A is provided for purposes of illustration only, and does not represent any actual facility.)

BILLING CODE 6560-01-M

Fig. 7-1 EPA Form 1 General (No. 3510-1), General Information for Consolidated Permits Program. (Continued)

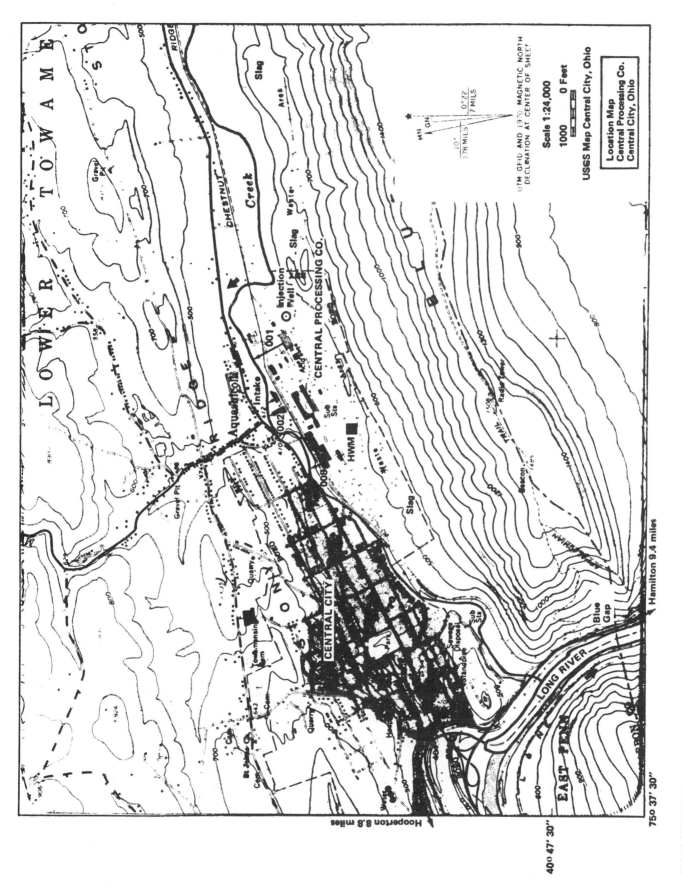

Fig. 7-1 EPA Form 1 General (No. 3510-1), General Information for Consolidated Permits Program.   (Continued)

Item XII. Briefly describe the nature of your business (e.g., products produced or services provided).

Item XIII. Federal statues provide for severe penalties for submitting false information on this application form.

18 U.S.C. section 1001 provides that "Whoever, in any matter within the jurisdiction of any department or agency of the United States knowingly and willfully falsifies, conceals or covers up by any trick, scheme, or device a material fact, or makes or uses any false writing or document knowing same to contain any false, fictitious or fraudulent statement or entry, shall be fined not more than $10,000 or imprisoned not more than five years, or both."

Section 309(c)(2) of the Clean Water Act and section 113(c)(2) of the Clean Air Act each provide that "Any person who knowingly makes any false statement, representation, or certification in any application, . . . shall upon conviction, be punished by a fine of no more than $10,000 or by imprisonment for not more than six months, or both."

In addition, section 3008(d)(3) of the Resource Conservation and Recovery Act provides for a fine up to $25,000 or imprisonment up to one year for a first conviction for making a false statement in any application under the Act, and for double these penalties upon subsequent convictions.

Federal regulations require this application to be signed as follows:

(1) For a corporation, by a principal executive officer of a least the level of vice president. However, if the only activity in item II which is marked "yes" is Question G, the officer may authorize a person having responsibility for the overall operations of the well or well field to sign the certification. In that case, the authorization must be written and submitted to the permitting authority.

(2) For partnership or sole proprietorship, by a general partner or the proprietor, respectively; or

(3) For a municipality, State, Federal, or other public facility, by either a principal executive officer or ranking elected official.

Section C. Activities Which Do Not Require Permits

I. *National Pollutant Discharge Elimination System Permits under the Clean Water Act.* You are not required to obtain an NPDES permit if your discharge is in one of the following categories, as provided by the Clean Water Act (CWA) and by the NPDES regulations (40 CFR Parts 122–125). However, under section 510 of CWA a discharge exempted from the federal

NPDES requirements may still be regulated by a State authority; contact your State environmental agency to determine whether you need a State permit.

A. *Discharges from Vessels.* Discharges of sewage from vessels, effluent from properly functioning marine engines, laundry, shower, and galley sink wastes, and any other discharge incidental to the normal operation of a vessel do not require NPDES permits. However, discharges of rubbish, trash, garbage, or other such materials discharged overboard require permits, and so do other discharges when the vessel is operating in a capacity other than as a means of transportation, such as when the vessel is being used as an energy or mining facility, a storage facility, or a seafood processing facility, or is secured to the bed of the ocean, contiguous zone, or waters of the United States for the purpose of mineral or oil exploration or development.

B. *Dredged or Fill Material.* Discharges of dredged or fill material into waters of the United States do not need NPDES permits if the dredging or filling is authorized by a permit issued by the U.S. Army Corps of Engineers or an EPA-approved State under section 404 of CWA.

C. *Discharges into Publicly Owned Treatment Works (POTW).* The introduction of sewage, industrial wastes, or other pollutants into a POTW does not need an NPDES permit. You must comply with all applicable pretreatment standards promulgated under section 307(b) of CWA, which may be included in the permit issued to the POTW. If you have a plan or an agreement to switch to a POTW in the future, this does not relieve you of the obligation to apply for and receive an NPDES permit until you have stopped discharging pollutants into waters of the United States.

[Note: Dischargers into privately owned treatment works do not have to apply for or obtain NPDES permits except as otherwise required by the EPA Regional Administrator. The owner or operator of the treatment works itself, however, must apply for a permit and identify all users in its application. Users so identified will receive public notice of actions taken on the permit for the treatment works.]

D. *Discharges from Agricultural and Silvicultural Activities.* Most discharges from agricultural and silvicultural activities to waters of the United States do not require NPDES permits. These include runoff from orchards, cultivated crops, pastures, range lands, and forest lands. However, the discharges listed

below do require NPDES permits. Definitions of the terms listed below are contained in the Glossary section of these instructions.

(1) Discharges from Concentrated Animal Feeding Operations. (See Glossary for definitions of "animal feeding operations" and "concentrated animal feeding operations." Only the latter require permits.)

(2) Discharges from Concentrated Aquatic Animal Production Facilities. (See Glossary for size cutoffs.)

(3) Discharges associated with approved Aquaculture Projects.

(4) Discharges from Silvicultural Point Sources. (See Glossary for the definition of "silvicultural point source.") Non-point source silvicultural activities are excluded from NPDES permit requirements. However, some of these activities, such as stream crossings for roads, may involve point source discharges of dredged or fill material which may require a section 404 permit. See 33 CFR 209.120.

E. *Discharges in Compliance with an On-Scene Coordinator's Instructions.*

II. *Hazardous Waste Permits under the Resource Conservation and Recovery Act.* You may be excluded from the requirement to obtain a permit under this program if you fall into one of the following categories.

Generators who accumulate their own hazardous waste on-site for less than 90 days;

Certain small generators;

Owners or operators of totally enclosed treatment facilities; or

Farmers who dispose of waste pesticide from their own use.

Check with your Regional office for details. Please note that even if you are excluded from permit requirements, you may be required by Federal regulations to handle your waste in a particular manner.

III. *Underground Injection Control Permits under the Safe Drinking Water Act.* You are not required to obtain a permit under this program if you:

Inject into existing wells used to enhance recovery of oil and gas or to store hydrocarbons (note, however, that these underground injections are regulated by Federal rules); or

Inject into or above a stratum which contains, within ¼ mile of the well bore, an underground source of drinking water (unless your injection is the type identified in item II-H, for which you do need a permit). However, you must notify EPA of your injection and submit certain required information on forms supplied by the Agency, and your operation may be phased out if you are a generator of hazardous wastes or a hazardous waste management facility

Fig. 7-1 EPA Form 1 General (No. 3510-1), General Information for Consolidated Permits Program. (Continued)

which uses wells or septic tanks to dispose of hazardous waste.

IV. *Prevention of Significant Deterioration Permits under the Clean Air Act.* The PSD program applies to newly constructed or modified facilities (both of which are referred to as "new sources") which increase air emissions. The Clean Air Act Amendments of 1977 exclude small new sources of air emissions from the PSD review program. Any new source in an industrial category listed in Table 3 of these instructions whose potential to emit is less than 100 tons per year is not required to get a PSD permit. In addition, any new source in an industrial category not listed in Table 3 whose potential to emit is less than 250 tons per year is exempted from the PSD requirements.

Modified sources which increase their net emissions (the difference between the total emission increases and total emission decreases at the source) less than the significant amount set forth in EPA regulations are also exempt from PSD requirements. Contact your EPA Regional office (Table 1) for further information.

Section D. Glossary

Note: This Glossary includes terms used in the instructions and in Forms 1, 2b, 2c, and 3. Additional terms will be included in the future when other forms are developed to reflect the requirements of other parts of the consolidated permit program. If you have any questions concerning the meaning of any of these terms, please contact your EPA Regional office (Table 1).

"Aliquot" means a sample of specified volume used to make up a total composite sample.

"Animal feeding operation" means a lot or facility (other than an aquatic animal production facility) where the following conditions are met:

1. Animals (other than aquatic animals) have been, are, or will be stabled or confined and fed or maintained for a total of 45 days or more in any 12-month period, and

2. Crops, vegetation, forage growth, or post-harvest residues are not sustained in the normal growing season over any portion of the lot or facility.

Two or more animal feeding operations under common ownership are a single animal feeding operation if they adjoin each other or if they use a common area or system for the disposal of wastes.

"Animal unit" means a unit of measurement for any animal feeding operation calculated by adding the following numbers: the number of slaughter and feeder cattle multiplied by 1.0, plus the number of mature dairy cattle multiplied by 1.4, plus the number of swine weighing over 25 kilograms (approximately 55 pounds) multiplied by 0.4, plus the number of sheep multiplied by 0.1, plus the number of horses multiplied by 2.0.

"Application" means the EPA standard national forms for applying for a permit, including any additions, revisions, or modifications to the forms; or forms approved by EPA for use in approved States, including any approved modifications or revisions. For RCRA, "application" also means "Application, Part B."

"Application, Part A" means that part of the consolidated permit application forms which a RCRA permit applicant must complete to qualify for interim status under section 3005(e) of RCRA and for consideration for a permit. Part A consists of Form 1 (General Information) and Form 3 (Hazardous Waste Application Form).

"Application, Part B", means that part of the application which a RCRA permit applicant must complete to be issued a permit. (*Note:* EPA is not developing a specific form for Part B of the permit application, but an instruction booklet explaining what information must be supplied is available from the EPA Regional office.)

"Approved program" or "approved State" means a State program which has been approved or authorized by EPA under 40 CFR Part 123.

"Aquaculture project" means a defined managed water area which uses discharges of pollutants into that designated area for the maintenance or production of harvestable freshwater, estuarine, or marine plants or animals. "Designated area" means the portions of the waters of the United States within which the applicant plans to confine the cultivated species, using a method of plan or operation (including, but not limited to, physical confinement) which, on the basis of reliable scientific evidence, is expected to ensure the specific individual organisms comprising an aquaculture crop will enjoy increased growth attributable to the discharge of pollutants and be harvested within a defined geographic area.

"Aquifer" means a geological formation, group of formations, or part of a formation that is capable of yielding a significant amount of water to a well or spring.

"Area of review" means the area surrounding an injection well which is described according to the criteria set forth in 40 CFR § 146.06.

"Area permit" means a UIC permit applicable to all or certain wells within a geographic area, rather than to a specified well, under 40 CFR § 122.37.

"Attainment area" means, for any air pollutant, an area which has been designated under section 107 of the Clean Air Act as having ambient air quality levels better than any national primary or secondary ambient air quality standard for that pollutant. Standards have been set for sulfur oxides, particulate matter, nitrogen dioxide, carbon monoxide, ozone, lead and hydrocarbons. For purposes of the Glossary, "attainment area" also refers to "unclassifiable area," which means, for any pollutants, an area designated under section 107 as unclassifiable with respect to that pollutant due to insufficient information.

"Best Management Practices" ("BMP") means schedules of activities, prohibitions of practices, maintenance procedures, and other management practices to prevent or reduce the pollution of waters of the United States. BMPs include treatment requirements, operating procedures, and practices to control plant site runoff, spillage or leaks, sludge or waste disposal, or drainage from raw material storage.

"Biological monitoring test" means any test which includes the use of aquatic algal, invertebrate, or vertebrate species to measure acute or chronic toxicity, and any biological or chemical measure of bioaccumulation.

"Bypass" means the intentional diversion of wastes from any portion of a treatment facility.

"Concentrated animal feeding operation" means an animal feeding operation which meets the criteria set forth in either (1) or (2) or which the Director designates as such on a case-by-case basis:

1. More than the numbers of animals specified in any of the following categories are confined:

(A) 1,000 slaughter or feeder cattle,

(B) 700 mature dairy cattle (whether milked or dry cows),

(C) 2,500 swine each weighing over 25 kilograms (approximately 55 pounds),

(D) 500 horses,

(E) 10,000 sheep or lambs,

(F) 55,000 turkeys,

(G) 100,000 laying hens or broilers (if the facility has a continuous overflow watering)

(H) 30,000 laying hens or broilers (if the facility has a liquid manure handling system),

(I) 5,000 ducks, or

(J) 1,000 animal units; or

2. More than the following numbers and types of animals are confined:

(A) 300 slaughter or feeder cattle,

(B) 200 mature dairy cattle (whether milked or dry cows),

Fig. 7-1 EPA Form 1 General (No. 3510-1), General Information for Consolidated Permits Program. (Continued)

(C) 750 swine each weighing over 25 kilograms (approximately 55 pounds).

(D) 150 horses,

(E) 3,000 sheep or lambs,

(F) 16,500 turkeys,

(G) 30,000 laying hens or broilers (if the facility has continuous overflow watering),

(H) 9,000 laying hens or broilers (if the facility has a liquid manure handling system),

(I) 1,500 ducks, or

(J) 300 animal units;

and either one of the following conditions are met: pollutants are discharged into waters of the United States through a manmade ditch, flushing system or other similar manmade device ("manmade" means constructed by man and used for the purpose of transporting wastes); or pollutants are discharged directly into waters of the United States which originate outside of and pass over, across, or through the facility or otherwise come into direct contact with the animals confined in the operation.

Provided, however, that no animal feeding operation is a concentrated animal feeding operation as defined above if such animal feeding operation discharges only in the event of a 25 year, 24 hour storm event.

"Concentrated aquatic animal production facility" means a hatchery, fish farm, or other facility which contains, grows or holds aquatic animals in either of the following categories, or which the Director designates as such on a case-by-case basis:

1. Cold water fish species or other cold water aquatic animals including, but not limited to, the *Salmonidae* family of fish (e.g., trout and salmon) in ponds, raceways or other similar structures which discharge at least 30 days per year but does not include:

(a) Facilities which produce less than 9,090 harvest weight kilograms (approximately 20,000 pounds) of aquatic animals per year; and

(b) Facilities which feed less than 2,272 kilograms (approximately 5,000 pounds) of food during the calendar month of maximum feeding.

2. Warm water fish species or other warm water aquatic animals including, but not limited to, the *Ameiuridae*, *Cetrarchidae*, and *Cyprinidae* families of fish (e.g., respectively, catfish, sunfish and minnows) in ponds, raceways, or other similar structures which discharge at least 30 days per year, but does not include:

(a) Closed ponds which discharge only during periods of excess runoff; or

(b) Facilities which produce less than 45,454 harvest weight kilograms (approximately 100,000 pounds) of aquatic animals per year.

"Contact cooling water" means water used to reduce temperature which comes into contact with a raw material, intermediate product, waste product other than heat, or finished product.

"Contiguous zone" means the entire zone established by the United States under article 24 of the convention of the Territorial Sea and the Contiguous Zone.

"CWA" means the Clean Water Act (formerly referred to the Federal Water Pollution Control Act) Pub. L. 92–500, as amended by Pub. L. 95–217 and Pub. L. 95–576, 33 U.S.C. 1251 *et seq*.

"Direct discharge" means the discharge of a pollutant as defined below.

"Director" means the EPA Regional Administrator or the State Director as the context requires.

"Discharge (of a pollutant)" means:

(1) Any addition of any pollutant or combination of pollutants to waters of the United States from any point source, or

(2) Any addition of any pollutant or combination of pollutants to the waters of the contiguous zone or the ocean from any point source other than a vessel or other floating craft which is being used as a means of transportation.

This definition includes discharges into waters of the United States from: surface runoff which is collected or channelled by man; discharges through pipes, sewers, or other conveyances owned by a State, municipality, or other person which do not lead to POTW's; and discharges through pipes, sewers, or other conveyances, leading into privately owned treatment works. This term does not include an addition of pollutants by any indirect discharger.

"Disposal" (in the RCRA program) means the discharge, deposit, injection, dumping, spilling, leaking, or placing of any hazardous waste into or on any land or water so that the hazardous waste or any constituent of it may enter the environment or be emitted into the air or discharged into any waters, including ground water.

"Disposal facility" means a facility or part of a facility at which hazardous waste is intentionally placed into or on land or water, and at which hazardous waste will remain after closure.

"Effluent limitation" means any restriction imposed by the Director on quantities, discharge rates, and concentrations of pollutants which are discharged from point sources into waters of the United States, the waters of the contiguous zone, or the ocean.

"Effluent limitation guideline" means a regulation published by the Administrator under section 304(b) of the Clean Water Act to adopt or revise effluent limitations.

"Environmental Protection Agency" ("EPA") means the United States Environmental Protection Agency.

"Exempted aquifer" means an aquifer or its portion that meets the criteria in the definition of USDW, but which has been exempted according to the procedures in 40 CFR § 122.35(b).

"Existing HWM facility" means a Hazardous Waste Management facility which was in operation, or for which construction had commenced, on or before October 21, 1976. Construction had commenced if (1) the owner or operator had obtained all necessary Federal, State and local preconstruction approvals or permits, and either (2a) a continuous on-site, physical construction program had begun, or (2b) the owner or operator had entered into contractual obligations, which could not be cancelled or modifed without substantial loss, for construction of the facility to be completed within a reasonable time.

[Note: This definition reflects the literal language of the statute. However, EPA believes that amendments to RCRA now in conference will shortly be enacted and will change the date for determining when a facility is an "existing facility" to one no earlier than May of 1980; indications are the conferees are considering October 30, 1980. Accordingly, EPA encourages every owner or operator of a facility which was built or under construction as of the promulgation date of the RCRA program regulations to file Part A of its permit application so that it can be quickly processed for interim status when the change in the law takes effect. When those amendments are enacted, EPA will amend this definition.]

"Existing source" or "existing discharger" (in the NPDES program) means any source which is not a new source or a new discharger.

"Existing injection well" means an injection well other than a new injection well.

"Facility" means any HWM facility, UIC underground injection well, NPDES point source, PSD stationary source, or any other facility or activity (including land or appurtenances thereto) that is subject to regulation under the RCRA, UIC, NPDES or PSD programs.

"Fluid" means material or substance which flows or moves whether in a semisolid, liquid, sludge, gas, or any other form or state.

"Generator" means any person by site location, whose act or process produces hazardous waste identified or listed in 40 CFR Part 261.

Fig. 7-1 EPA Form 1 General (No. 3510-1), General Information for Consolidated Permits Program. (Continued)

"Groundwater" means water below the land surface in a zone of saturation.

"Hazardous substance" means any of the substances designated under 40 CFR Part 116 pursuant to section 311 of CWA. [*Note:* These substances are listed in Table 2c–4 of the instructions to Form 2c.]

"Hazardous waste" means a hazardous waste as defined in 40 CFR § 261.3.

"Hazardous waste management facility" ("HWM facility") means all contiguous land, structures, appurtenances, and improvements on the land, used for treating, storing, or disposing of hazardous wastes. A facility may consist of several treatment, storage or disposal operational units (for example, one or more landfills, surface impoundments, or combinations of them).

"In operation" means a facility which is treating, storing, or disposing of hazardous waste.

"Indirect discharger" means a non-domestic discharger introducing pollutants to a publicly owned treatment works.

"Injection well" means a well into which fluids are being injected.

"Interim authorization" means approval by EPA of a State hazardous waste program which has met the requirements of section 3006(c) of RCRA and applicable requirements of 40 CFR Part 123, Subparts A, B, and F.

"Listed State" means a State listed by the Administrator under section 1422 of SDWA as needing a State UIC program.

"MGD" means millions of gallons per day.

"Municipality" means a city, village, town, borough, county, parish, district; association, or other public body created by or under State law and having jurisdiction over disposal of sewage, industrial wastes, or other wastes, or an Indian tribe or an authorized Indian tribal organization, or a designated and approved managment agency under section 208 of CWA.

"National Pollutant Discharge Elimination System" ("NPDES") means the national program for issuing, modifying, revoking and reissuing, terminating, monitoring, and enforcing permits and imposing and enforcing pretreatment requirements, under sections 307, 318, 402 and 405 of CWA. The term includes an approved program.

"New discharger" means any building, structure, facility, or installation: (1) from which there is or may be a new or additional discharge of pollutants at a site at which on October 18, 1972, it had never discharged pollutants; (2) which has never received a finally effective NPDES permit for discharges at that site;

and (3) which is not "new source." This definition includes an indirect discharger which commences discharging into waters of the United States. It also includes any existing mobile point source, such as an offshore oil drilling rig, seafood processing vessel, or aggregate plant that begins discharging at a location for which it does not have an existing permit.

"New HWM facility" means a Hazardous Waste Management facility which began operation or for which construction commenced after October 21, 1976.

"New injection well" means a well which begins injection after a UIC program for the State in which the well is located is approved.

"New source" (in the NPDES program) means any building, structure, facility, or installation from which there is or may be a discharge of pollutants, the construction of which commenced:

(i) After promulgation of standards of performance under section 306 of CWA which are applicable to such source, or

(ii) After proposal of standards of performance in accordance with section 306 of CWA which are applicable to such source, but only if the standards are promulgated in accordance with section 306 within 120 days of their proposal.

"Non-contact cooling water" means water used to reduce temperature which does not come into direct contact with any raw material, intermediate product, waste product (other than heat), or finished product.

"Off-site" means any site which is not "on-site."

"On-site" means on the same or geographically contiguous property which may be divided by public or private right(s)-of-way, provided the entrance and exit between the properties is at a cross-roads intersection, and access is by crossing as opposed to going along, the right(s)-of-way. Non-Contiguous properties owned by the same person, but connected by a right-of-way which the person controls and to which the public does not have access, is also considered on-site property.

"Outfall" means a point source.

"Permit" means an authorization, license, or equivalent control document issued by EPA or an approved State to implement the requirements of 40 CFR Parts 122, 123, and 124.

"Physical construction" (in the RCRA program) means excavation, movement of earth, erection of forms or structures, or similar activity to prepare a HWM facility to accept hazardous waste.

"Point source" means any discernible, confined, and discrete conveyance,

including but not limited to any pipe, ditch, channel, tunnel, conduit, well, discrete fissure, container, rolling stock, concentrated animal feeding operation, vessel or other floating craft from which pollutants are or may be discharged. This term does not include return flows from irrigated agriculture.

"Pollutant" means dredged spoil, solid waste, incinerator residue, filter backwash, sewage, garbage, sewage sludge, munitions, chemical waste, biological materials, radioactive materials (except those regulated under the Atomic Energy Act of 1954, as amended (42 U.S.C. § 2011 et seq.)), heat, wrecked or discarded equipment, rocks, sand, cellar dirt and industrial, municipal, and agriculture waste discharged into water. It does not mean:

(1) Sewage from vessels; or

(2) Water, gas, or other material which is injected into a well to facilitate production of oil or gas, or water derived in association with oil and gas production and disposed of in a well, if the well used either to facilitate production or for disposal purposes is approved by authority of the State in which the well is located, and if the State determines that the injection or disposal will not result in the degradation of ground or surface water resources.

[*Note:* Radioactive materials covered by the Atomic Energy Act are those encompassed in its definition of source, byproduct, or special nuclear materials. Examples of materials not covered include radium and accelerator produced isotopes. See *Train* v. *Colorado Public Interest Research Group, Inc.,* 426 U.S. 1 (1976).]

"Prevention of significant deteriorioration" (PSD) means the national permitting program under 40 CFR 52.21 to prevent emissions of certain pollutants regulated under the Clean Air Act from significantly deteriorating air quality in attainment areas.

"Primary industry category" means any industry category listed in the NRDC Settlement Agreement (*Natural Resources Defense Council* v. *Train*, 8 ERC 2120 (D.D.C. 1976), *modified* 12 ERC 1833 (D.D.C. 1979)).

"Privately owned treatment works" means any device or system which is (1) used to treat wastes from any facility whose operator is not the operator of the treatment works and (2) not a POTW.

"Process wastewater" means any water which, during manufacturing or processing, comes into direct contact with or results from the production or use of any raw material, intermediate product, finished product, by-product, or waste product.

Fig. 7-1 EPA Form 1 General (No. 3510-1), General Information for Consolidated Permits Program. (Continued)

"Publicly owned treatment works" or "POTW" means any device or system used in the storage, treatment, recycling, and reclamation of municipal sewage or industrial waste of a liquid nature which is owned by a State or municipality. This definition includes any sewers that convey wastewater to a POTW, but does not include pipes, sewers, or other conveyances not connected to a POTW.

"Rent" means use of another's property in return for regular payment.

"RCRA" means the Solid Waste Disposal Act as amended by the Resource Conservation and Recovery Act of 1976 (Pub. L. 94–580, as amended by Pub. L. 95–609, 42 U.S.C. § 6901 *et seq.*).

"Rock crushing and gravel washing facilities" are facilities which process crushed and broken stone, gravel, and riprap (see 40 CFR Part 436, Subpart B, and the effluent limitations guidelines for these facilities).

"SDWA" means the Safe Drinking Water Act (Pub. L. 95–523, as amended by Pub. L. 95–1900, 42 U.S.C. § 300(f) *et seq.*).

"Secondary industry category" means any industry category which is not a primary industry category.

"Sewage from vessels" means human body wastes and the wastes from toilets and other receptacles intended to receive or retain body wastes that are discharged from vessels and regulated under section 312 of CWA, except that with respect to commercial vessels on the Great Lakes this term includes graywater. For the purposes of this definition, "graywater" means galley, bath, and shower water.

"Sewage sludge" means the solids, residues, and precipitate separated from or created in sewage by the unit processes of a POTW. "Sewage" as used in this definition means any wastes, including wastes from humans, households, commercial establishments, industries, and storm water runoff, that are discharged to or otherwise enter a publicly owned treatment works.

"Silvicultural point source" means any discernable, confined, and discrete conveyance related to rock crushing, gravel washing, log sorting, or log storage facilities which are operated in connection with silvicultural activities and from which pollutants are discharged into waters of the United States. This term does not include non-point source silvicultural activities such as nursery operations, site preparation, reforestation and subsequent cultural treatment, thinning, prescribed burning, pest and fire control, harvesting operations, surface drainage, or road construction and maintenance from which there is natural runoff. However,

some of these activities (such as stream crossing for roads) may involve point source discharges of dredged or fill material which may require a CWA section 404 permit. "Log sorting and log storage facilities" are facilities whose discharges result from the holding of unprocessed wood, e.g., logs or roundwood with bark or after removal of bark in self-contained bodies of water (mill ponds or log ponds) or stored on land where water is applied intentionally on the logs (wet decking). (See 40 CFR Part 429, Subpart J, and the effluent limitations guidelines for these facilities.)

"State" means any of the 50 States, the District of Columbia, Guam, the Commonwealth of Puerto Rico, the Virgin Islands, American Samoa, the Trust Territory of the Pacific Islands (except in the case of RCRA), and the Commonwealth of the Northern Mariana Islands (except in the case of CWA).

"Stationary source" (in the PSD program) means any building, structure, facility, or installation which emits or may emit any air pollutant regulated under the Clean Air Act. "Building, structure, facility, or installation" means any grouping of pollutant-emitting activities which are located on one or more contiguous or adjacent properties and which are owned or operated by the same person (or by persons under common control).

"Storage" (in the RCRA program) means the holding of hazardous waste for a temporary period at the end of which the hazardous waste is treated, disposed, or stored elsewhere.

"Storm water runoff" means water discharged as a result of rain, snow, or other precipitation.

"Toxic pollutant" means any pollutant listed as toxic under section 307(a)(1) of CWA.

"Transporter" (in the RCRA program) means a person engaged in the off-site transportation of hazardous waste by air, rail, highway, or water.

"Treatment" (in the RCRA program) means any method, technique, or process, including neutralization, designed to change the physical, chemical, or biological character or composition of any hazardous waste so as to neutralize such waste, or so as to recover energy or material resources from the waste, or so as to render such waste non-hazardous, or less hazardous; safer to transport, store, or dispose of; or amenable for recovery, amenable for storage, or reduced in volume.

"Underground injection" means well injection.

"Underground source of drinking water" or "USDW" means an aquifer or

its portion which is not an exempted aquifer and:

(1) Which supplies drinking water for human consumption, or

(2) In which the ground water contains fewer than 10,000 mg/l total dissolved solids.

"Upset" means an exceptional incident in which there is unintentional and temporary noncompliance with technology-based permit effluent limitations because of factors beyond the reasonable control of the permittee. An upset does not include noncompliance to the extent caused by operational error, improperly designed treatment facilities, inadequate treatment facilities, lack of preventive maintenance, or careless or improper operation.

"Waters of the United States" means:

1. All waters which are currently used, were used in the past, or may be susceptible to use in interstate or foreign commerce, including all waters which are subject to the ebb and flow of the tide;

2. All interstate waters, including interstate wetlands;

3. All other waters such as intrastate lakes, rivers, streams (including intermittent streams), mudflats, sandflats, wetlands, sloughs, prairie potholes, wet meadows, playa lakes, and natural ponds, the use, degradation, or destruction of which would or could affect interstate or foreign commerce including any such waters:

(a) Which are or could be used by interstate or foreign travelers for recreational or other purposes;

(b) From which fish or shellfish are or could be taken and sold in interstate or foreign commerce;

(c) Which are used or could be used for industrial purposes by industries in interstate commerce;

4. All impoundments of waters otherwise defined as waters of the United States under this definition;

5. Tributaries of waters identified in paragraphs (1)–(4) above;

6. The territorial sea; and

7. Wetlands adjacent to waters (other than waters that are themselves wetlands) identified in paragraphs (1)–(6) of this definition.

Waste treatment systems, including treatment ponds or lagoons designed to meet requirement of CWA (other than cooling ponds as defined in 40 CFR § 423.11(m) which also meet the criteria of this definition) are not waters of the United States. This exclusion applies only to manmade bodies of water which neither were originally created in waters of the United States (such as a disposal area in wetlands) nor resulted from the

Fig. 7-1 EPA Form 1 General (No. 3510-1), General Information for Consolidated Permits Program. (Continued)

impoundments of waters of the United States.

"Well injection" or "underground injection" means the subsurface emplacement of fluids through a bored, drilled, or driven well; or through a dug well, where the depth of the dug well is greater than the largest surface dimension.

"Wetlands" means those areas that are inundated or saturated by surface or ground water at a frequency and duration sufficient to support, and that under normal circumstances do support, a prevalence of vegetation typically adapted for life in saturated soil conditions. Wetlands generally include swamps, marshes, bogs, and similar areas.

BILLING CODE 6560-01-M

Fig. 7-1 EPA Form 1 General (No. 3510-1), General Information for Consolidated Permits Program. (Continued)

Please print or type in the unshaded areas only
(fill—in areas are spaced for elite type, i.e., 12 characters/inch).

Form Approved OMB No. 158-S80004

FORM **3** RCRA	⊕EPA	U.S. ENVIRONMENTAL PROTECTION AGENCY **HAZARDOUS WASTE PERMIT APPLICATION** *Consolidated Permits Program* (This information is required under Section 3005 of RCRA.)	I. EPA I.D. NUMBER

F | | | | | — | | | | | | | T/A | C 1

FOR OFFICIAL USE ONLY

APPLICATION APPROVED	DATE RECEIVED (yr., mo., & day)	COMMENTS

II. FIRST OR REVISED APPLICATION

Place an "X" in the appropriate box in A or B below (mark one box only) to indicate whether this is the first application you are submitting for your facility or a revised application. If this is your first application and you already know your facility's EPA I.D. Number, or if this is a revised application, enter your facility's EPA I.D. Number in item I above.

A. FIRST APPLICATION (place an "X" below and provide the appropriate date)

☐ 1. EXISTING FACILITY (See instructions for definition of "existing" facility. Complete item below.) ☐ 2. NEW FACILITY (Complete item below.)

C 8 | YR. | MO. | DAY | FOR EXISTING FACILITIES, PROVIDE THE DATE (yr., mo., & day) OPERATION BEGAN OR THE DATE CONSTRUCTION COMMENCED (use the boxes to the left)

FOR NEW FACILITIES, PROVIDE THE DATE (yr., mo., & day) OPERA-TION BEGAN OR IS EXPECTED TO BEGIN | YR. | MO. | DAY

B. REVISED APPLICATION (place an "X" below and complete Item I above)

☐ 1. FACILITY HAS INTERIM STATUS ☐ 2. FACILITY HAS A RCRA PERMIT

III. PROCESSES — CODES AND DESIGN CAPACITIES

A. PROCESS CODE — Enter the code from the list of process codes below that best describes each process to be used at the facility. Ten lines are provided for entering codes. If more lines are needed, enter the code(s) in the space provided. If a process will be used that is not included in the list of codes below, then describe the process (including its design capacity) in the space provided on the form (Item III-C).

B. PROCESS DESIGN CAPACITY — For each code entered in column A enter the capacity of the process.
1. AMOUNT — Enter the amount.
2. UNIT OF MEASURE — For each amount entered in column B(1), enter the code from the list of unit measure codes below that describes the unit of measure used. Only the units of measure that are listed below should be used.

PROCESS	PRO-CESS CODE	APPROPRIATE UNITS OF MEASURE FOR PROCESS DESIGN CAPACITY	PROCESS	PRO-CESS CODE	APPROPRIATE UNITS OF MEASURE FOR PROCESS DESIGN CAPACITY
Storage:			**Treatment:**		
CONTAINER (barrel, drum, etc.)	S01	GALLONS OR LITERS	TANK	T01	GALLONS PER DAY OR LITERS PER DAY
TANK	S02	GALLONS OR LITERS	SURFACE IMPOUNDMENT	T02	GALLONS PER DAY OR LITERS PER DAY
WASTE PILE	S03	CUBIC YARDS OR CUBIC METERS	INCINERATOR	T03	TONS PER HOUR, METRIC TONS PER HOUR, GALLONS PER HOUR OR LITERS PER HOUR
SURFACE IMPOUNDMENT	S04	GALLONS OR LITERS			
Disposal:			OTHER (Use for physical, chemical, thermal or biological treatment processes not occurring in tanks, surface impoundments or incinerators. Describe the processes in the space provided; Item III-C.)	T04	GALLONS PER DAY OR LITERS PER DAY
INJECTION WELL	D79	GALLONS OR LITERS			
LANDFILL	D80	ACRE-FEET (the volume that would cover one acre to a depth of one foot) OR HECTARE-METER			
LAND APPLICATION	D81	ACRES OR HECTARES			
OCEAN DISPOSAL	D82	GALLONS PER DAY OR LITERS PER DAY			
SURFACE IMPOUNDMENT	D83	GALLONS OR LITERS			

UNIT OF MEASURE	UNIT OF MEASURE CODE	UNIT OF MEASURE	UNIT OF MEASURE CODE	UNIT OF MEASURE	UNIT OF MEASURE CODE
GALLONS	G	LITERS PER DAY	V	ACRE-FEET	A
LITERS	L	TONS PER HOUR	D	HECTARE-METER	F
CUBIC YARDS	Y	METRIC TONS PER HOUR	W	ACRES	B
CUBIC METERS	C	GALLONS PER HOUR	E	HECTARES	Q
GALLONS PER DAY	U	LITERS PER HOUR	H		

EXAMPLE FOR COMPLETING ITEM III (shown in line numbers X-1 and X-2 below): A facility has two storage tanks, one tank can hold 200 gallons and the other can hold 400 gallons. The facility also has an incinerator that can burn up to 20 gallons per hour.

C				T/A	C 1

LINE NUMBER	A. PRO-CESS CODE (from list above)	B. PROCESS DESIGN CAPACITY		2. UNIT OF MEA-SURE (enter code)	FOR OFFICIAL USE ONLY	LINE NUMBER	A. PRO-CESS CODE (from list above)	B. PROCESS DESIGN CAPACITY	2. UNIT OF MEA-SURE (enter code)	FOR OFFICIAL USE ONLY
		1. AMOUNT (specify)						1. AMOUNT		
X-1	S 0 2	600		G		5				
X-2	T 0 3	20		E		6				
1						7				
2						8				
3						9				
4						10				

EPA Form 3510-3 (5-80) PAGE 1 OF 5 CONTINUE ON REVERSE

Fig. 7-2 EPA Form 3 RCRA (No. 3510-3), Hazardous Waste Permit Application for Consolidated Permits Program.

Continued from the front.

III. PROCESSES (continued)

C. SPACE FOR ADDITIONAL PROCESS CODES OR FOR DESCRIBING OTHER PROCESSES (code "T04"). FOR EACH PROCESS ENTERED HERE INCLUDE DESIGN CAPACITY.

IV. DESCRIPTION OF HAZARDOUS WASTES

A. **EPA HAZARDOUS WASTE NUMBER** — Enter the four—digit number from 40 CFR, Subpart D for each listed hazardous waste you will handle. If you handle hazardous wastes which are not listed in 40 CFR, Subpart D, enter the four—digit number (s) from 40 CFR, Subpart C that describes the characteristics and/or the toxic contaminants of those hazardous wastes.

B. **ESTIMATED ANNUAL QUANTITY** — For each listed waste entered in column A estimate the quantity of that waste that will be handled on an annual basis. For each characteristic or toxic contaminant entered in column A estimate the total annual quantity of all the non—listed waste(s) that will be handled which possess that characteristic or contaminant.

C. **UNIT OF MEASURE** — For each quantity entered in column B enter the unit of measure code. Units of measure which must be used and the appropriate codes are:

ENGLISH UNIT OF MEASURE	CODE	METRIC UNIT OF MEASURE	CODE
POUNDS	P	KILOGRAMS	K
TONS	T	METRIC TONS	M

If facility records use any other unit of measure for quantity, the units of measure must be converted into one of the required units of measure taking into account the appropriate density or specific gravity of the waste.

D. **PROCESSES**

1. PROCESS CODES:

   For listed hazardous waste: For each listed hazardous waste entered in column A select the code(s) from the list of process codes contained in Item III to indicate how the waste will be stored, treated, and/or disposed of at the facility.

   For non—listed hazardous wastes: For each characteristic or toxic contaminant entered in column A, select the code(s) from the list of process codes contained in Item III to indicate all the processes that will be used to store, treat, and/or dispose of all the non—listed hazardous wastes that possess that characteristic or toxic contaminant.

   Note: Four spaces are provided for entering process codes. If more are needed: (1) Enter the first three as described above; (2) Enter "000" in the extreme right box of Item IV-D(1); and (3) Enter in the space provided on page 4, the line number and the additional code(s).

2. PROCESS DESCRIPTION: If a code is not listed for a process that will be used, describe the process in the space provided on the form.

NOTE: **HAZARDOUS WASTES DESCRIBED BY MORE THAN ONE EPA HAZARDOUS WASTE NUMBER** — Hazardous wastes that can be described by more than one EPA Hazardous Waste Number shall be described on the form as follows:

1. Select one of the EPA Hazardous Waste Numbers and enter it in column A. On the same line complete columns B,C, and D by estimating the total annual quantity of the waste and describing all the processes to be used to treat, store, or dispose of the waste.

2. In column A of the next line enter the other EPA Hazardous Waste Number that can be used to describe the waste. In column D(2) on that line enter "included with above" and make no other entries on that line.

3. Repeat step 2 for each other EPA Hazardous Waste Number that can be used to describe the hazardous waste.

EXAMPLE FOR COMPLETING ITEM IV (shown in line numbers X-1, X-2, X-3, and X-4 below) — A facility will treat and dispose of an estimated 900 pounds per year of chrome shavings from leather tanning and finishing operation. In addition, the facility will treat and dispose of three non—listed wastes. Two wastes are corrosive only and there will be an estimated 200 pounds per year of each waste. The other waste is corrosive and ignitable and there will be an estimated 100 pounds per year of that waste. Treatment will be in an incinerator and disposal will be in a landfill.

LINE NO.	A. EPA HAZARD. WASTE NO. (enter code)	B. ESTIMATED ANNUAL QUANTITY OF WASTE	C. UNIT OF MEASURE (enter code)	D. PROCESSES 1. PROCESS CODES (enter)	D. PROCESSES 2. PROCESS DESCRIPTION (if a code is not entered in D(1))
X-1	K 0 5 4	900	P	T 0 3 D 8 0	
X-2	0 1 0 0	400	P	T 0 3 D 8 0	
X-3	0 1 0 0	100	P	T 0 3 D 8 0	
X-4	1 0 0 0				included with above

EPA Form 3510-3 (5-80)

CONTINUE ON PAGE 3

Fig. 7-2 EPA Form 3 RCRA (No. 3510-3), Hazardous Waste Permit Application for Consolidated Permits Program. (Continued)

Continued from page 2.
NOTE: Photocopy this page before completing if you have more than 26 wastes to list.   *Form Approved OMB No. 158-S80004*

EPA I.D. NUMBER (enter from page 1)		FOR OFFICIAL USE ONLY

IV. DESCRIPTION OF HAZARDOUS WASTES *(continued)*

LINE NO.	A. EPA HAZARD. WASTE NO. (enter code)	B. ESTIMATED ANNUAL QUANTITY OF WASTE	C. UNIT OF MEASURE (enter code)	D. PROCESSES	
				1. PROCESS CODES (enter)	2. PROCESS DESCRIPTION (if a code is not entered in D(1))
1					
2					
3					
4					
5					
6					
7					
8					
9					
10					
11					
12					
13					
14					
15					
16					
17					
18					
19					
20					
21					
22					
23					
24					
25					
26					

EPA Form 3510-3 (5-80)   PAGE 3 _____ OF 5   CONTINUE ON REVERSE
(enter "A", "B", "C", etc. behind the "3" to identify photocopied pages)

Fig. 7-2 EPA Form 3 RCRA (No. 3510-3), Hazardous Waste Permit Application for Consolidated Permits Program. (Continued)

Continued from the front.

IV. DESCRIPTION OF HAZARDOUS WASTES (continued)

E. USE THIS SPACE TO LIST ADDITIONAL PROCESS CODES FROM ITEM B(1) ON PAGE 3.

EPA I.D. NO. (enter from page 1)

F												T/A	C
9											13	14	6

V. FACILITY DRAWING

All existing facilities must include in the space provided on page 5 a scale drawing of the facility (see instructions for more detail).

VI. PHOTOGRAPHS

All existing facilities must include photographs (aerial or ground—level) that clearly delineate all existing structures; existing storage, treatment and disposal areas; and sites of future storage, treatment or disposal areas (see instructions for more detail).

VII. FACILITY GEOGRAPHIC LOCATION

LATITUDE (degrees, minutes, & seconds)	LONGITUDE (degrees, minutes, & seconds)
65 66 67 68 69 70	71 - 73 74 75 76 77

VIII. FACILITY OWNER

☐ A. If the facility owner is also the facility operator as listed in Section VIII on Form 1, "General Information", place an "X" in the box to the left and skip to Section IX below.

B. If the facility owner is not the facility operator as listed in Section VIII on Form 1, complete the following items:

1. NAME OF FACILITY'S LEGAL OWNER		2. PHONE NO. (area code & no.)
D		55 56 - 58 59 - 61 62 - 65

3. STREET OR P.O. BOX	4. CITY OR TOWN	5. ST.	6. ZIP CODE
E	F	40 41 42	47 - 51

IX. OWNER CERTIFICATION

I certify under penalty of law that I have personally examined and am familiar with the information submitted in this and all attached documents, and that based on my inquiry of those individuals immediately responsible for obtaining the information, I believe that the submitted information is true, accurate, and complete. I am aware that there are significant penalties for submitting false information, including the possibility of fine and imprisonment.

A. NAME (print or type)	B. SIGNATURE	C. DATE SIGNED

X. OPERATOR CERTIFICATION

I certify under penalty of law that I have personally examined and am familiar with the information submitted in this and all attached documents, and that based on my inquiry of those individuals immediately responsible for obtaining the information, I believe that the submitted information is true, accurate, and complete. I am aware that there are significant penalties for submitting false information, including the possibility of fine and imprisonment.

A. NAME (print or type)	B. SIGNATURE	C. DATE SIGNED

EPA Form 3510-3 (5-80)   **PAGE 4 OF 5**   CONTINUE ON PAGE 5

Fig. 7-2 EPA Form 3 RCRA (No. 3510-3), Hazardous Waste Permit Application for Consolidated Permits Program. (Continued)

Continued from page 4.   *Form Approved OMB No. 158-S80004*

V. FACILITY DRAWING *(see page 4)*

EPA Form 3510-3 (5-80)   **PAGE 5 OF 5**

[FR Doc. 80–14313 Filed 5–16–80; 8:45 am]

BILLING CODE 6560-01-C

Fig. 7-2 EPA Form 3 RCRA (No. 3510-3), Hazardous Waste Permit Application for Consolidated Permits Program. (Continued)

Instructions.—Form 3—RCRA Hazardous Waste Permit Application

This form must be completed by all applicants who check "yes" to Item II–E in Form 1.

General Instructions

Permit Application Process.—There are two parts to a RCRA permit application—Part A and Part B. Part A consists of this form and Form 1 of the Consolidated Permit Application. Part B requires detailed site-specific information such as geology, hydrology, and engineering data. 40 CFR 122.25 specifies the information that will be required from hazardous waste management facilities in Part B.

RCRA established a procedure for obtaining "interim status" which allows existing hazardous waste management facilities to continue their operations until a final hazardous waste permit is issued. In order to qualify for interim status, existing hazardous waste management facilities must submit Part A of the permit application to EPA within six months after the promulgation of regulations under section 3001 of RCRA (40 CFR Part 261). In order to receive a hazardous waste permit, existing facilities must submit a complete Part B within six months after it is requested by EPA. New facilities must submit both Part A and Part B to EPA at least 180 days before physical construction is expected to commence.

Operation During Interim Status.—As provided in 40 CFR 122.23(b), Part A of the permit application defines the processes to be used for treatment, storage, and disposal of hazardous wastes; the design capacity of such processes; and the specific hazardous wastes to be handled at a facility during the interim status period. Once Part A is submitted to EPA, changes in the hazardous wastes handled, changes in design capacities, changes in processes, and changes in ownership or operational control at a facility during the interim status period may only be made in accordance with the procedures in 40 CFR 123.23(c). Changes in design capacity and changes in processes require prior EPA approval. Changes in the quantity of waste handled at a facility during interim status can be made without submitting a revised Part A provided the quantity does not exceed the design capacities of the processes specified in Part A of the permit application. Failure to furnish all information required to process a permit application is grounds for termination of interim status.

Confidential Information.—All information submitted in this form will be subject to public disclosure, to the extent provided by RCRA and the Freedom of Information Act, 5 U.S.C. Section 552, and EPA's Business Confidentiality Regulations, 40 CFR Part 2 (see especially 40 CFR 2.305). Persons filing this form may make claims of confidentiality. Such claims must be clearly indicated by marking "confidential" on the specific information on the form for which confidential treatment is requested or on any attachments, and must be accompanied, *at the time of filing,* by a written substantiation of the claim, by answering the following questions:

1. Which portions of the information do you claim are entitled to confidential treatment?

2. For how long is confidential treatment desired for this information?

3. What measures have you taken to guard against undesired disclosure of the information to others?

4. To what extent has the information been disclosed to others, and what precautions have been taken in connection with that disclosure?

5. Has EPA or any other Federal agency made a pertinent confidentiality determination? If so, include a copy of such determination or reference to it, if available.

6. Will disclosure of the information be likely to result in substantial harmful effects on your competitive position? If so, what would those harmful effects be and why should they be viewed as substantial? Explain the causal relationship between disclosure and the harmful effects.

Information covered by a confidentiality claim and the above substantiation will be disclosed by EPA only to the extent and by means of the procedures set forth in 40 CFR Part 2.

If no claim of confidentiality or no substantiation accompanies the information when it is submitted, EPA may make the information available to the public without further notice to the submitter.

Definitions.—Terms used in these instructions and in this form are defined in the Glossary section of these instructions. For additional definitions and procedures to use in applying for a permit for a hazardous waste management facility, refer to the regulations promulgated under Section 3005 of RCRA and published in 40 CFR Parts 122 and 124.

Line by Line Instructions

Completing this form. Please type or print in the unshaded areas only. Some items have small graduation marks or boxes in the fill in spaces. These marks indicate the number of characters that may be inputted into our data system. The marks are spaced at 1/8" intervals which accommodate elite type (12 characters per inch—one space between letters). If you do not have a typewriter with elite type then please print, placing each character between the marks. Abbreviate if necessary to stay within the number of characters allowed for each item. Use one space for breaks between words, but not for punctuation marks unless the space is needed to clarify your information.

Item I. Existing hazardous waste management facilities should enter their EPA Identification Number (if known). New facilities should leave this item blank.

Item II. A. First Application.—If this is the first application that is being filed for the facility place an "X" in either the Existing Facility box or the New Facility box.

1. *Existing Facility.*—Existing facilities are:

(1) Those facilities which received hazardous waste for treatment, storage, and/or disposal on or before October 21, 1976, or

(2) Those facilities for which construction had commenced on or before October 21, 1976. Construction had "commenced" only if:

(a) The owner or operator had obtained all necessary Federal, State, and local pre-construction approvals or permits; *and*

(b1) A continuous physical, on-site construction program had begun (facility design or other preliminary non-physical and non-site specific preparatory activities do not constitute an on-site construction program), or

(b2) The owner or operator had entered into contractual obligations (options to purchase or contracts for feasibility, engineering, and design studies do not constitute contractual obligations) which could not be cancelled or modified without substantial loss. Generally, a loss is deemed substantial if the amount an owner or operator must pay to cancel construction agreements or stop construction exceeds 10% of the total project cost.

(Note—This definition of "existing facility" reflects the literal language of the statute. However, EPA believes that amendments to RCRA now in conference will shortly be enacted and will change the date for determining when a facility is an "existing facility" to one no earlier than May of 1980; indications are the conferees are considering October 30, 1980. When those amendments are enacted, EPA

Fig. 7-2 EPA Form 3 RCRA (No. 3510-3), Hazardous Waste Permit Application for Consolidated Permits Program. (Continued)

will amend the definition of "existing facility."

Accordingly, EPA encourages every facility built or under construction on the promulgation date of the RCRA program regulations to notify EPA and file Part A of the permit application so that it can be quickly processed for interim status when the change in the law takes effect.)

Existing Facility Date.—If the Existing Facility box is marked, enter the date hazardous waste operations began (i.e., the date the facility began treating, storing, or disposing of hazardous waste) or the date construction commenced.

2. *New Facility.*—New facilities are all facilities for which construction commenced, or will commence, after October 21, 1976.

New Facility Date.—If the New Facility box is marked, enter the date that operation began or is expected to begin.

B. *Revised Application.*—If this is a subsequent application that is being filed to amend data filed in a previous application, place an "X" in the appropriate box to indicate whether the facility has interim status or a permit.

1. *Facility Has Interim Status.*—Place an "X" in this box if this is a revised application to make changes at a facility during the interim status period.

2. *Facility Has a Permit.*—Place an "X" in this box if this is a revised application to make changes at a facility for which a permit has been issued.

(Note—When submitting a revised application, applicants must resubmit in their entirety each item on the application for which changes are requested. In addition, items I and IX (and item X if applicable) must be completed. It is not necessary to resubmit information for other items that will not change).

Item III. The information in item III describes all the processes that will be used to treat, store, or dispose of hazardous waste at existing facilities during the interim status period, and at new facilities after a permit is issued. The design capacity of each process must be provided as part of the description. The design capacity of injection wells and landfills at existing facilities should be measured as the remaining, unused capacity. See the form for the detailed instructions to item III.

Item IV. The information in item IV describes all the hazardous wastes that will be treated, stored, or disposed at existing facilities during the interim status period, and at new facilities after a permit is issued. In addition, the processes that will be used to treat,

store, or dispose of each waste and the estimated annual quantity of each waste must be provided. See the form for the detailed instructions to item IV.

Item V. All existing facilities must include a drawing showing the general layout of the facility during interim status. This drawing should be approximately to scale and fit in the space provided on the form. This drawing should show the following:

• The property boundaries of the facility;

• The areas occupied by all storage, treatment, or disposal operations that will be used during interim status;

• The name of each operation. (Example-multiple hearth incinerator, drum storage area, etc.);

• Areas of past storage, treatment, or disposal operations;

• Areas of future storage, treatment, or disposal operations; and

• The approximate dimensions of the property boundaries and all areas

See Figure 3–1 for an example of a facility drawing. New facilities do not have to complete this item.

BILLING CODE 6560-01-M

Fig. 7-2 EPA Form 3 RCRA (No. 3510-3), Hazardous Waste Permit Application for Consolidated Permits Program. (Continued)

Continued from page 4.

Form Approved OMB No. 158-S8004

V. FACILITY DRAWING *(see page 4)*

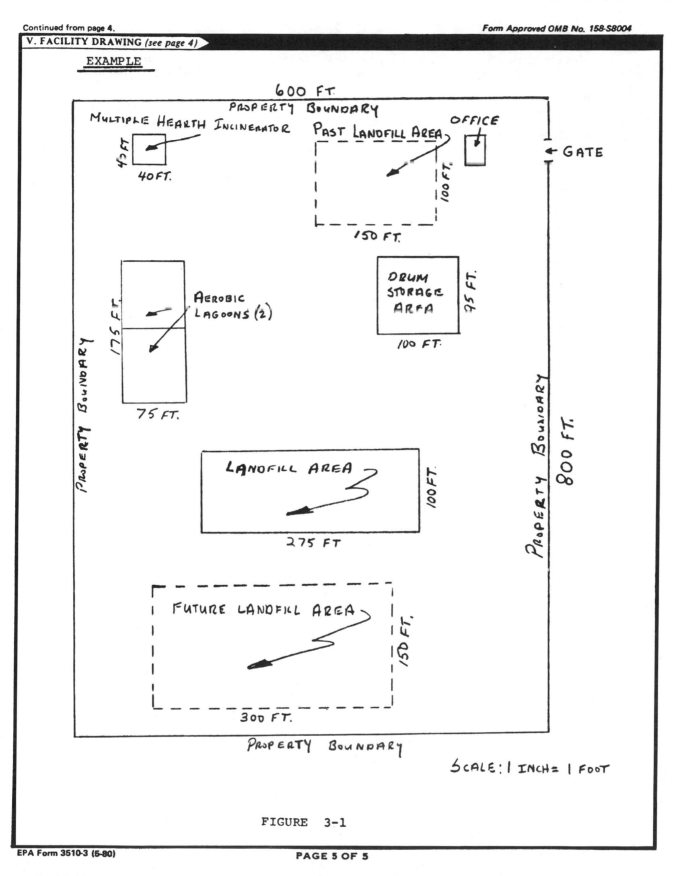

EXAMPLE

FIGURE   3-1

EPA Form 3510-3 (5-80)   **PAGE 5 OF 5**

Fig. 7-2 EPA Form 3 RCRA (No. 3510-3), Hazardous Waste Permit Application for Consolidated Permits Program. (Continued)

Item VI. All existing facilities must include photographs that clearly delineate *all* existing structures; *all* existing areas for storing, treating, or disposing of hazardous waste; and *all* known sites of future storage, treatment, or disposal operations. Photographs may be color or black and white, ground-level or aerial. Indicate the date the photograph was taken on the back of each photograph.

Item VII. Enter the latitude and longitude of the facility in degrees, minutes, and seconds. For larger facilities, enter the latitude and longitude at the approximate mid-point of the facility. You may use the map you provided for Item XI of Form 1 to determine latitude and longitude. Latitude and longitude information is also available from Regional Offices of the U.S. Department of Interior,

Geological Survey; from State Agencies, such as the Department of Natural Resources; and from the National Cartographic Information Center, U.S. Geological Survey, 12202 Sunrise Valley Dr., Reston, VA. 22092.

Item VIII. See the form for the instructions to item VIII.

Item IX and Item X. All facility owners must sign Item IX. If the facility will be operated by someone other than the owner, then the operator must sign Item X. Federal regulations require the certification to be signed as follows.

(1) For a corporation, by a principal executive officer at least the level of vice president;

(2) For a partnership or sole proprietorship, by a general partner or the proprietor, respectively; or

(3) For a municipality, State, Federal, or other public facility, by either a

principal executive officer or ranking elected official.

The Resource Conservation and Recovery Act provides for severe penalties for submitting false information on this application form.

Section 3008(d) of the Resource Conservation and Recovery Act provides that "Any person who knowingly makes any false statement or representation in any application, . . . shall, upon conviction be subject to a fine of not more than $25,000 for each day of violation, or to imprisonment not to exceed one year, or both."

BILLING CODE 6560-01-M

Fig. 7-2 EPA Form 3 RCRA (No. 3510-3), Hazardous Waste Permit Application for Consolidated Permits Program. (Continued)

8
STATE PROGRAM REQUIREMENTS: PART 123

In order to receive authorization to manage a hazardous waste program under Section 3006 of RCRA, each State must develop its own regulatory program, which must then be authorized by EPA. To receive authorization, the State program must be able to meet certain criteria EPA has developed. EPA will regulate hazardous waste only in States that choose not to implement their own programs or fail to qualify for EPA authorization.

Part 123 establishes the requirements that must be met by a State seeking approval from EPA to administer a hazardous waste program under Section 3006 of RCRA. Part 123 also references applicable requirements of Parts 122 and 124 that State programs must meet, and outlines the process by which EPA approves, revises, or withdraws approval of State programs.

RCRA's approach in approving State hazardous waste programs is similar to that of the NPDES and UIC programs in that it sets minimum requirements and allows States, in general, to be more stringent. States are prohibited from imposing any requirements that would prevent hazardous waste from moving between States.

The following tables address the detailed requirements of Part 123 that are applicable to the hazardous waste management program:

- Table 8-1A: General Program Requirements—Subpart A
- Table 8-1B: Additional Requirements for State Hazardous Waste Programs—Subpart B
- Table 8-1C: Requirements for Interim Authorization of State Hazardous Waste Programs—Subpart F.

Table 8-1A. General Program Requirements—Subpart A.

Section No.	Description
123.1	I. Purpose, Scope, and Applicability
123.1(a)	A. What Part 123 Specifies Relative to RCRA, UIC, NPDES, and Section 404 Programs (1) Requirements State programs must meet to be approved by Administrator. (2) Procedures EPA will follow in approving, revising and withdrawing State programs.
123.1(b)	B. Subpart Structure Relative to Hazardous Waste Program (1) Subpart A (a) Contains requirements applicable to all progams under Section 123.1(a). Exception: (i) Hazardous waste programs operating under interim authorization (Subpart F). (b) Includes elements which must be part of submissions to EPA for program approval. (c) Substantive provisions which must be present in State programs for them to be approved. (d) Procedures EPA will follow in approving, revising, and withdrawing State programs. (2) Subpart B Contains additional requirements for States seeking final authorization under RCRA.
123.1(c)	C. State Submission for Program Approval (1) Must be made in accordance with procedures of Subpart A. (2) State submission for interim authorization for hazardous waste shall be in accordance with Subpart F. (3) Includes development and submittal of following to EPA: (a) Section 123.4—Program Description. (b) Section 123.5—Attorney General's Statement. (c) Section 123.6—Memorandum of Agreement with Regional Administrator.
123.1(d)	D. Substantive Provisions Which Must Be Included in State Program for Them to Be Approved (1) Section 123.7—Requirements for Permitting.

Table 8-1A. General Program Requirements—Subpart A. (Continued)

Section No.	Description
	(2) Section 123.8—Requirements for Compliance Evaluation Programs.
	(3) Section 123.9—Requirements for Enforcement Authority.
	(4) Section 123.10—Sharing of Information.
	(5) Section 123.11—Coordination with Other Programs.
	(6) Program Specific Subparts
	(a) Subpart B—Additional Requirements for State Hazardous Waste Programs.
	(b) Subpart F—Requirements for Interim Authorization of State Hazardous Waste Programs.
	(7) Cross-Referencing Regulations
	(a) Many of the requirements for State programs are made applicable to States by cross-referencing other EPA regulations.
	(b) In particular, many of the provisions of Parts 122 and 124 are made applicable to States by references contained in Section 123.7 (Requirements for Permitting).
123.1(e)	E. Approval or Disapproval of State Program
	(1) Upon submission of a complete program, EPA will conduct a public hearing, if interest is shown, and determine whether to approve or disapprove the program considering:
	(a) Requirements of this Part.
	(b) RCRA.
	(c) Any comments received.
123.1(f)	(2) State Programs Which Conform to Applicable Requirements of This Part
	Shall be approved by Administrator.
123.1(g)	(3) Upon Approval of State Program
	Administrator shall suspend issuance of Federal permits for those activities subject to the approved RCRA program.
123.1(h)	(4) Any State Program Approved by Administrator
	Shall at all times be conducted in accordance with requirements of this Part.
123.12	(5) Approval Process
	The process for EPA approval of State RCRA programs is set out in Section 123.39 (Approval Process).
123.1(i)	F. These Regulations Do Not Require States to Consolidate Their Permitting Activities
	(1) However, States are encouraged by EPA to do so.
	(2) Each of four programs under this Part may be applied for separately.
123.1(j)	G. Partial State Programs Are Not Allowed under RCRA
	However, in many cases States will lack authority to regulate activities on Indian lands.
123.1(k)(1) and (2)	H. Adoption and Enforcement by State of More Stringent, Broader, or More Extensive Requirements than Required under This Part
	(1) Nothing precludes a State from so doing except as provided in Section 123.32 (Consistency).
	(2) Where an approved State program has a greater scope of coverage than required by Federal law the additional coverage is not part of the Federally approved program.
123 Authority	I. Authority
	Resource Conservation and Recovery Act, 42 U.S.C. 6901 et seq.; Safe Drinking Water Act, 42 U.S.C. 300(f) et seq.; Clean Water Act, 33 U.S.C. 1251 et seq.
123.3	II. Element of a Program Submission
123.3(a)	A. Content of Program Submission
	Any State seeking to administer a hazardous waste program under this Part (123) shall submit at least three copies of a program submission, the content of which follows:
123.3(a)(1)	(1) Letter from Governor of State requesting program approval.
123.3(a)(2)	(2) Complete Program Description
	As required by Section 123.4 (Program Description), describing how State intends to carry out its responsibilities under this Part.
123.3(a)(3)	(3) Attorney General's Statement
	As required by Section 123.5 (Attorney General's Statement).
123.3(a)(4)	(4) Memorandum of Agreement with Regional Administrator
	As required by Section 123.6 (Memorandum of Agreement with Regional Administrator).
123.3(a)(5)	(5) Copies of all applicable State statutes and regulations, including those governing State administrative procedures.
123.3(a)(6)	(6) State's Public Participation Activities Prior to Program Submission
	Required by Section 123.39(c) (Approval Process).

Table 8-1A. General Program Requirements—Subpart A. (Continued)

Section No.	Description
123.3(b)	B. Upon Receipt of State Program Submission by EPA (1) Within 30 days EPA will notify State whether submission is complete. (2) If EPA Finds that State's Submission Is Complete The statutory review period shall be deemed to have begun on date of receipt of State's submission. (3) If EPA Finds that State's Submission Is Incomplete The statutory review period shall not begin until all the necessary information is received by EPA.
123.3(c)	C. If State's Submission Is Materially Changed during Statutory Period The statutory review period shall begin again upon receipt of revised submission.
123.3(d)	D. The State and EPA may extend the statutory review period by agreement.
123.4	III. Program Description A. General (1) Any State Seeking to Administer a Hazardous Waste Program under This Part (123) Shall submit a description of program it proposes to administer in lieu of the Federal program under: (a) State law; or (b) An interstate compact.
123.4(a)	B. Content of Program Description (1) Description of Scope, Structure, Coverage, and Processes of State Program In narrative form.

123.4(b)	(2) Description of Organization and Structure of State Agency Or Agencies Which Will Have Responsibility for Administering The Program
	(a) Include organization charts.
	(b) If More than One Agency is Responsible for Administration of Program
	   (i) Each agency must have statewide jurisdiction over a class of activities.
	   (ii) Responsibilities of each agency must be delineated.
	   (iii) Their procedures for coordination set forth.

123.4(f)(4)	(iv) For RCRA programs only, an identification of a "lead agency" is required, and a description of how the State agencies will coordinate their activities.
123.4(b)	(c) When State Proposes to Administer a Program of Greater Scope of Coverage than Required by Federal Law State must indicate resources dedicated to administering the Federally required portion of the program.

123.4(b)(1)	(d) Description of State Agency Staff Who Will Carry Out State Program
	   (i) Include number, occupation, and general duties of employees.
	   (ii) State need not submit complete job descriptions for every employee carrying out State program.

123.4(b)(2)	(e) Itemization of Estimated Cost of Establishing and Administering the Program for First Two Years after Approval
	Include cost of:
	   (i) Personnel listed in Section 123.4(b)(1).
	   (ii) Administrative support.
	   (iii) Technical support.

123.4(b)(3)	(f) Itemization of Sources and Amounts of Funding
	   (i) Include estimate of Federal grant money available to State Director for first two years after approval to meet costs listed in Section 123.4(b)(2).
	   (ii) Identify any restrictions or limitations upon this funding.

123.4(c)	(3) Description of Applicable State Procedures Includes: (a) Permitting procedures. (b) Administrative procedures. (c) Judicial review procedures.

123.4(d)	(4) State Forms
	(a) Copies of Forms State Intends to Employ in Its Hazardous Waste Program
	   (i) Permit form(s).
	   (ii) Application form(s).
	   (iii) Reporting form(s).
	   (iv) Manifest format.

Table 8-1A. General Program Requirements—Subpart A. (Continued)

Section No.	Description
	(b) Forms Used by States Need Not Be Identical to Forms Used by EPA But same basic information is required.
	(c) Uniform National Forms
	(i) State need not provide copies of such forms it intends to use but should note its intention to use such forms.
123.4(d) Note	(ii) States are encouraged to use uniform national forms established by Adminstrator.
	(iii) If uniform national forms are used, they may be modified to include the State Agency's name, address, logo, and other similar information, in place of EPA's.
123.4(e)	(5) Complete description of State's compliance tracking and enforcement program.
123.4(f)	(6) Special Requirements for RCRA Programs Only
	(a) Description of State manifest tracking system.
	(b) Interstate and International Shipments Description of procedures State will use to coordinate information with other approved State and Federal programs regarding interstate and international shipments.
123.4(f)(2)	(c) Estimate of Number of Hazardous Waste Program Participants
123.4(f)(2)(i)	(i) Generators.
123.4(f)(2)(ii)	(ii) Transporters.
123.4(f)(2)(iii)	(iii) On-Site and Off-Site Storage, Treatment and Disposal Facilities Brief description of types of facilities and indication of permit status of these facilities.
123.4(f)(3)	(d) Estimate (If Available) of Annual Quantities of Hazardous Wastes Handled by State
123.4(f)(3)(i)	(i) Generated within State.
123.4(f)(3)(ii)	(ii) Transported into and out of State.
123.4(f)(3)(iii)	(iii) Stored, Treated, or Disposed Of within the State
123.4(f)(3)(iii)(A)	• On-site; and
123.4(f)(3)(iii)(B)	• Off-site.
123.5	IV. Attorney General's Statement
123.5(a)	A. For State That Seeks to Administer a Program under This Part
	(1) State shall submit a statement from:
	(a) State Attorney General; or
	(b) Attorney for those States or interstate agencies which have independent legal counsel.
	(2) Overall Substance of Statement
	(a) That the laws of the State, or an interstate compact, provide adequate authority to carry out the program described under Section 123.4 (Program Description); and
	(b) To meet requirements of this Part (123).
	(3) Specific Content of Statement
	(a) Citations to specific statutes.
	(b) Administrative regulations.
	(c) Where appropriate, judicial decisions which demonstrate adequate authority.
	B. State Statutes and Regulations Cited by State Attorney General or Independent Legal Counsel
	(1) Shall be in the form of lawfully adopted state statutes and regulations at time statement is signed.
	(2) Shall be fully effective by the time program is approved.
	C. To Qualify As "Independent Legal Counsel" The attorney signing the statement required by this Section must have full authority to independently represent the State agency in court on all matters pertaining to the State program.
123.5(a) Note	D. Attorney General's Statement Format EPA will supply States with the format on request.
123.5(b)	E. When a State Seeks Authority over Activities on Indian Lands The statement shall contain an appropriate analysis of the State's authority.
123.6	V. Memorandum of Agreement with Regional Administrator
123.6(a)	A. General
	(1) For State That Seeks to Administer a Program under This Part Shall submit a Memorandum of Agreement.
	(2) Who Shall Execute Memorandum of Agreement? State Director and Regional Administrator.

Table 8-1A. General Program Requirements—Subpart A. (Continued)

Section No.	Description
	(3) When Agreement Shall Become Effective When approved by Administrator.
	(4) Content of Memorandum of Agreement Apart from Requirements of Section 123.6(b) It may include other terms, conditions, or agreements consistent with this Part and relevant to administration and enforcement of State's regulatory program.
	(5) Memorandum of Agreement Containing Provisions Which Restrict EPA's Statutory Oversight Responsibility Administrator shall not approve such an agreement.
123.6(b) 123.6(b)(1)	B. Content of Memorandum of Agreement (1) Provisions for Prompt Transfer from EPA to State of Information Relevant to Program Operation and Not in Possession of State Director Includes: (a) Pending permit applications. (b) Support files for permit issuance. (c) Compliance reports, etc.
	(2) When Existing Permits Are Transferred from EPA to State for Administration (a) Memorandum of Agreement shall contain provisions specifying a procedure for transferring the administration of these permits. (b) If State Lacks Authority to Directly Administer Permits Issued by Federal Government (i) A procedure may be established to transfer responsibility for these permits.
123.6(b)(2) Note	(ii) For example, EPA and State and permittee could agree that State would issue a permit(s) identical to the outstanding Federal permit, which would simultaneously be terminated.
123.6(b)(2)	(3) Permit Applications, Draft Permits, and Proposed Permits That State Will Send to Regional Administrator for Review, Comment, and Objection Provisions shall be made for specifying classes and categories of each permit grouping.
123.6(b)(3)	(4) Provisions Specifying Frequency and Content of Reports, Documents, and Other Information Which State Is Required to Submit to EPA (a) State shall allow EPA to routinely review State records, reports, and files relevant to administration and enforcement of the approved program. (b) State reports may be combined with grant reports where appropriate.
123.6(b)(4) 123.6(b)(4)(i)	(5) Provision on State's Compliance Monitoring and Enforcement Program Includes: (a) Provisions For Coordination of Compliance Monitoring Activities by State and by EPA (i) These may specify the basis on which Regional Administrator will select facilities or activities within State for EPA inspection. (ii) Regional Administrator will normally notify State at least 7 days before any such inspection.
123.6(b)(4)(ii) 123.6(b)(5)	(b) Procedures to ensure coordination of enforcement activities. (6) Provisions for Joint Processing of Permits by State and EPA (a) When appropriate, such provisions shall be made for facilities or activities which require permits from both EPA and the State under different programs (see Section 124.4— Consolidation of Permit Processing).
123.6(b)(5) Note	(b) To promote efficiency and to avoid duplication and inconsistency, States are encouraged to enter into joint processing agreements with EPA for permit issuance. (c) States are encouraged (but not required) to consider steps to coordinate or consolidate their own permit programs and activities.
123.6(b)(6)	(7) Provision for Modification of Memorandum of Agreement Shall be in accordance with this Part.
123.6(c)	C. Consistency (1) Memorandum of Agreement, the annual program grant and State/EPA Agreement should be consistent. (2) If State/EPA Agreement Indicates Change Is Needed in Memorandum of Agreement Memorandum of Agreement may be amended through procedures set forth in this Part.
123.6(c) Note	(3) Detailed program priorities and specific arrangements for EPA support of the State program will change and are therefore more appropriately negotiated in the context of annual agreements rather than in the Memorandum of Agreement (MOA). (4) However, it may still be appropriate to specify in the MOA the basis for such detailed agreements.

Table 8-1A. General Program Requirements—Subpart A. (Continued)

Section No.	Description
123.6(d)	D. State RCRA Programs Only
	In the case of State RCRA programs the Memorandum of Agreement shall also provide:
123.6(d)(1)	(1) Compliance Inspections
	(a) EPA may conduct compliance inspections of all generators, transporters, and HWM facilities in each year for which the State is operating under final authorization.
	(b) Limitations on Compliance Inspections of Generators, Transporters, and Non-Major HWM Facilities
	(i) Regional Administrator and State may agree to such limitations.
123.6(d)(2)	(ii) Such limitations shall not restrict EPA's right to inspect any generator, transporter or HWM facility which it has cause to believe is not in compliance with RCRA.
	• However, before conducting such an inspection, EPA will normally allow the State a reasonable opportunity to conduct a compliance evaluation inspection.
123.6(d)(3)	(2) Draft Permits and Permit Applications for Major HWM Facilities
	(a) State Director shall promptly forward to EPA copies of such permits/permit applications for review and comment.
	(b) State Director shall supply EPA copies of final permits for all major HWM facilities.
	(3) Draft Permits and Permit Applications for Non-Major HWM Facilities
	Regional Administrator and State Director may agree to limitations regarding review of and comment on draft permits and/or permit applications for non-major HWM facilities.
123.6(d)(4)	(4) Information Obtained in Notifications Provided under Section 3010(a) of RCRA
	(a) Regional Administrator shall promptly forward to State Director information obtained prior to program approval via notifications.
	(b) Assignment of EPA Identification Numbers for New Generators, Transporters, Treatment, Storage, and Disposal Facilities
	Regional Administrator and State Director shall agree on procedures for assignment of EPA Identification Numbers.
123.6(d)(5)	(5) State Director Review of All Permits Issued Under State Law
	(a) Shall be done prior to date of program approval.
	(b) Director shall modify or revoke and reissue them to require compliance with requirements of this Part.
	(c) Regional Administrator and State Director shall establish a time within which this review must take place.
123.7	VI. Requirements for Permitting
123.7(a)	A. General
	(1) All State programs under this Part must have legal authority to implement each of the provisions from Parts 122 and 124 that follow.
	(2) The State programs must be administered in accordance with each of the Part 122 and 124 provisions.
	(3) Imposing More Stringent Requirements
	States are not precluded from omitting or modifying any provisions to impose more stringent requirements than the Federal requirements.
	B. Part 122 and Part 124 Requirements for All State Programs
123.7(a)(1)	(1) Section 122.4—Application for a Permit.
123.7(a)(2)	(2) Section 122.6—Signatories to Permit Applications and Reports.
123.7(a)(3)	(3) Section 122.7—Conditions Applicable to All Permits.
123.7(a)(4)	(4) Section 122.8—Establishing Permit Conditions.
123.7(a)(5)	(5) Section 122.9—Duration of Permits.
123.7(a)(6)	(6) Section 122.10(a)—Schedules of Compliance.
123.7(a)(7)	(7) Section 122.11—Requirements for Recording and Reporting of Monitoring Results.
123.7(a)(8)	(8) Section 122.13(a) and (b)—Effect of a Permit.
123.7(a)(9)	(9) Section 122.14—Transfer of Permits.
123.7(a)(10)	(10) Section 122.15—Modification or Revocation and Reissuance of Permits.
123.7(a)(11)	(11) Section 122.16—Termination of Permits.
123.7(a)(12)	(12) Section 122.18—Noncompliance and Program Reporting by the Director.
123.7(a)(13)	(13) Section 122.19(b)–(d)—Confidentiality of Information.
123.7(a)(14)	(14) Section 124.3(a)—Application for a Permit.
123.7(a)(15)	(15) Section 124.5(a), (c), and (d)—Modification, Revocation and Reissuance, or Termination of Permits.
123.7(a)(16)	(16) Section 124.6(a), (d), and (e)—Draft Permits.
123.7(a)(17)	(17) Section 124.8—Fact Sheet.

Table 8-1A. General Program Requirements—Subpart A. (Continued)

Section No.	Description
123.7(a)(18)	(18) Section 124.10(a)(i)(ii), (a)(i)(iii), (b), (c), (d), and (e)—Public Notice of Permit Actions and Public Comment Period.
123.7(a)(19)	(19) Section 124.11—Public Comments and Requests for Public Hearings.
123.7(a)(20)	(20) Section 124.12(a)—Public Hearing.
123.7(a)(21)	(21) Section 124.17(a) and (c)—Response to comments.
123.7(a) Note	*Notes.* (1) States need not implement provisions identical to above listed provisions or provisions listed in Section 123.7(b) (State RCRA Programs Only). (2) Implemented provisions must, however, establish requirements at least as stringent as the corresponding listed provisions. (3) While States may impose more stringent requirements, they may not make one requirement more lenient as a tradeoff for making another one more stringent.
123.7(b)	C. Part 122 Requirements for State RCRA Programs Only
123.7(b)(1)	(1) Section 122.21(d)(1)—Specific Inclusions.
123.7(b)(2)	(2) Section 122.22—Application for a Permit.
123.7(b)(3)	(3) Section 122.24—Contents of Part A.
123.7(b)(4)	(4) Section 122.25—Contents of Part B.
123.7(b)(4) Note	*Notes:* (1) States need not use a two-part permit application process. (2) The State application process must, however, require information in sufficient detail to satisfy requirements of Sections 122.24 and 122.25.
123.7(b)(5)	(5) Section 122.26—Permits by Rule.
123.7(b)(6)	(6) Section 122.27—Emergency Permits.
123.7(b)(7)	(7) Section 122.28—Additional Conditions Applicable to All RCRA Permits.
123.7(b)(8)	(8) Section 122.29—Establishing RCRA Permit Conditions.
123.6(b)(9)	(9) Section 122.30—Interim Permits for UIC Wells.
123.8	VII. Requirements for Compliance Evaluation Programs
123.8(a)	A. Procedures for Receipt, Evaluation, Retention, and Investigation of Notices and Reports Required of Permittees (and Other Regulated Persons) (1) State programs shall have such procedures for possible enforcement action. (2) They shall also have procedures for investigation for possible enforcement of those failing to submit required notices and reports.
123.8(b)	B. Inspection and Surveillance Procedures (1) State programs shall have such procedures to determine, independent of information supplied by regulated persons, compliance or noncompliance with applicable program requirements. (2) Content of State Programs
123.8(b)(1)	(a) Comprehensive Surveys of All Facilities and Activities Subject to State Director's Authority (i) Program shall be capable of making such surveys. (ii) Purpose To identify persons subject to regulation who have failed to comply with permit application or other program requirements. (iii) Any compilation, index, or inventory of such facilities and activities shall be made available to the Regional Administrator upon request.
123.8(b)(2)	(b) Program for Periodic Inspections of Facilities and Activities Subject to Regulation These inspections shall be conducted in a manner designed to:
123.8(b)(2)(i)	(i) Determine Compliance or Noncompliance with: • Issued permit conditions. • Other program requirements.
123.8(b)(2)(ii)	(ii) Verify Accuracy of Information Submitted by Permittees and Other Regulated Persons • In reporting forms; and • Other forms supplying monitoring data.
123.8(b)(2)(iii)	(iii) Verify adequacy of sampling, monitoring, and other methods used by permittees and other regulated persons to develop that information.
123.8(b)(3)	(c) Program for investigating information obtained regarding violations of applicable program and permit requirements.

Table 8-1A. General Program Requirements—Subpart A. (Continued)

Section No.	Description

123.8(b)(4)	(d) Procedures for Receiving and Ensuring Proper Consideration of Information Submitted by Public about Violations
	    (i) Public effort in reporting violations shall be encouraged.
	    (ii) State Director shall make available information on reporting procedures.

123.8(c)	C. State Director and State Officers Engaged in Compliance Evaluation
	    (1) Shall have authority to:
	        (a) Enter any site or premises subject to regulation.
	        (b) Enter site or premises in which records relevant to program operation are kept.
	    (2) Purpose of entry authority is to copy any records, inspect, monitor, or otherwise investigate compliance with State program including compliance with permit conditions and other program requirements.
	    (3) States Whose Law Requires a Search Warrant before Entry
	    Those entering shall conform with this requirement.

123.8(d)	D. Investigatory Inspections Such inspections shall be conducted, samples taken, and other information gathered in a manner (e.g., using proper "chain of custody" procedures) that will produce evidence admissible in an enforcement proceeding or in court.
123.9	VIII. Requirements for Enforcement Authority
123.9(a)	A. Legal Remedies for Violations of State Program Requirements The following such remedies must be available to State agency administering a program:
123.9(a)(1)	(1) Restraining Order/Suit (a) To restrain immediately and effectively any person by order or by suit in State court from engaging in any unauthorized activity which is endangering or causing damage to public health or the environment.
123.9(a)(1) Note	(b) The above requires that States have a mechanism (e.g., an administrative cease and desist order or the ability to seek a temporary restraining order) to stop any unauthorized activity endangering public health or the environment.

123.9(a)(2)	(2) To Sue in Courts to Enjoin Any Threatened or Continuing Violation of Any Program Requirement
	        (a) Includes permit conditions.
	        (b) Authority available without necessity of prior revocation of permit.

123.9(a)(3)	(3) To Assess or Sue to Recover in Court Civil Penalties and to Seek Criminal Remedies, Including Fines As follows for State RCRA programs:
123.9(a)(3)(i)(A)	(a) Civil Penalties Shall be recoverable for any program violation in at least the amount of $10,000 per day.

123.9(a)(3)(i)(B)	(b) Criminal Remedies
	        (i) Shall be obtainable against any person:
	            • Who knowingly transports hazardous waste to an unpermitted facility; or
	            • Who treats, stores, or disposes of hazardous waste without a permit; or
	            • Who makes any false statement or representation in any application, label, manifest, record, report, permit, or other document filed, maintained, or used for purposes of program compliance.
	        (ii) Criminal Fines
	            • Shall be recoverable in at least the amount of $10,000 per day for each violation; and
	            • Imprisonment for at least six months shall be available.

123.9(b)(1)	B. Maximum Civil Penalty or Criminal Fine
	    The penalty/fine provided in Section 123.9(a)(3) shall be:
	    (1) Assessable for each instance of violation; and
	    (2) If the violation is continuous, it shall be assessable up to the maximum amount for each day of violation.

123.9(b)(2)	C. Burden of Proof and Degree of Knowledge or Intent Required under State Law for Establishing Violations under Section 123.9(a)(3) (1) Shall be no greater than the burden of proof or degree of knowledge or intent EPA must provide when it brings an action under RCRA.

Table 8-1A. General Program Requirements—Subpart A. (Continued)

Section No.	Description
123.9(b)(2) Note	(2) For example, this requirement is not met if State law includes mental state as an element of proof for civil violations.
123.9(c)	D. Civil Penalty (1) Penalty Appropriate to Violation Such shall be the case for any civil penalty assessed, sought, or agreed upon by State Director under Section 123.9(a)(3). (2) Civil Penalty Agreed upon by State Director in Settlement of Administrative or Judicial Litigation May be adjusted by a percentage which represents the likelihood of success in establishing the underlying violation(s) in such litigation. (3) If Civil Penalty Plus Compliance Costs Jeopardize Continuance of Business (a) Payment of penalty may be deferred; or (b) Penalty may be forgiven in whole or part. (4) "Appropriate to the Violation" Penalty for Failure to Meet a Statutory or Final Permit Compliance Deadline Means a penalty which is equal to:
123.9(c)(1)	(a) An amount appropriate to redress the harm or risk to public health or the environment; plus
123.9(c)(2)	(b) An amount appropriate to remove the economic benefit gained or to be gained from delayed compliance; plus
123.9(c)(3)	(c) An amount appropriate as a penalty for the violator's degree of recalcitrance, defiance, or indifference to requirements of the law; plus
123.9(c)(4)	(d) An amount appropriate to recover unusual or extraordinary enforcement costs thrust upon the public; minus
123.9(c)(5)	(e) An amount, if any, appropriate to reflect any part of the noncompliance attributable to the government itself; and minus
123.9(c)(6)	(f) An amount appropriate to reflect any part of the noncompliance caused by factors completely beyond the violator's control (e.g., floods, fires).
123.9(c) Note	*Note:* In addition to requirements of this Section, the State may have other enforcement remedies.
123.9(d)	E. Public Participation in State Enforcement Process Any State administering a program shall provide for such public participation by providing either:
123.9(d)(1)	(1) Authority which allows intervention in any civil or administrative action to obtain remedies specified in Sections 123.9(a)(1), (2), or (3) by any citizen having an interest which is or may be adversely affected; or
123.9(d)(2)	(2) Assurance that State agency or enforcement authority will:
123.9(d)(2)(i)	(a) Investigate and provide written responses to all citizen complaints submitted pursuant to procedures specified in Section 123.8(b)(4).
123.9(d)(2)(ii)	(b) Not oppose intervention by any citizen when permissive intervention may be authorized by statute, rule, or regulation.
123.9(d)(2)(iii)	(c) Publish notice of and provide at least 30 days for public comment on any proposed settlement of a State enforcement action.
123.10	IX. Sharing of Information
123.10(a)	A. Information Obtained or Used in Administration of State Program (1) Shall be available to EPA upon request without restriction. (2) If Information Has Been Submitted to State under Claim of Confidentiality State must submit that claim to EPA when providing information under this Section. (3) Information Obtained from State and Subject to Claim of Confidentiality Will be treated in accordance with regulations in 40 CFR 2 (Public Information). (4) If EPA Obtains Non-Confidential Information from a State EPA may make that information available to the public without further notice.
123.10(b)	B. Information EPA Will Furnish to States with Approved Programs (1) Non-Confidential That information which State needs to implement its approved program. (2) Confidential That information which State needs to implement its approved program, subject to conditions of 40 CFR 2 (Public Information).

Table 8-1A. General Program Requirements—Subpart A. (Continued)

Section No.	Description
123.11	X. Coordination with Other Programs
123.11(a)	A. Coordination of Permit Issuance (1) Issuance of State permits under this Part may be coordinated with issuance of RCRA, UIC, NPDES, and 404 permits whether they are controlled by State, EPA, or the Corps of Engineers. (2) See Section 124.4—Consolidation of Permit Processing.
123.11(b)	B. State Director of Any Approved Program Which May Affect Planning and Development of HWM Facilities and Practices He shall consult and coordinate with agencies designated under Section 4006 (b) of RCRA (40 CFR 255) as responsible for development and implementation of State solid waste management plans.
123.13	XI. Procedures for Revision of State Programs
123.13(a)	A. Revision of State Programs: General (1) Who May Initiate Program Revision (a) EPA; or (b) Approved State. (2) When Program Revision May Be Necessary When controlling Federal or State statutory or regulatory authority is modified or supplemented. (3) State Shall Keep EPA Fully Informed in Areas That Can Affect State's Approval Status Report proposed modifications to: (a) Its basic statutory or regulatory authority. (b) Its forms, procedures or priorities.
123.13(b) 123.13(b)(1)	B. Approach to Revision of State Program (1) Documents State Shall Submit (a) Modified program description. (b) Attorney General's statement. (c) Memorandum of Agreement. (d) Such other documents as EPA determines to be necessary.
123.13(b)(2)	(2) Whenever EPA Determines that Proposed Program Revision Is Substantial (a) EPA shall issue public notice and provide an opportunity to comment for a period of at least 30 days. (b) Who Public Notice Shall Be Mailed To Interested parties. (c) Publication (i) In *Federal Register;* and (ii) In enough of largest newspapers in State to provide statewide coverage. (d) Content of Public Notice (i) Shall summarize proposed revisions. (ii) Shall Provide for Opportunity to Request a Public Hearing Such a hearing will be held if there is significant public interest based on requests received.
123.13(b)(3)	(3) Administrator shall approve or disapprove program revisions based on requirements of this Part (123) and of RCRA.
123.13(b)(4)	(4) Effectivity A program revision shall become effective upon approval of Administrator. (5) Notice of Approval (a) For Substantial Program Revisions Notice published in *Federal Register.* (b) For Non-Substantial Program Revisions Notice may be given by a letter from Administrator to State Governor or his designee.
123.13(c)	C. When State Proposes To Transfer All or Part of Any Program from Approved State Agency to Any Other State Agency (1) State shall notify EPA. (2) State shall identify any new division of responsibilities among the agencies involved. (3) New agency is not authorized to administer the program until approved by Administrator under Section 123.13(b).

Table 8-1A. General Program Requirements—Subpart A. (Continued)

Section No.	Description

	(4) Organization Charts Shall be revised and resubmitted (see Section 123.4(b)) (Program Description).
123.13(d)	D. Whenever Administrator Has Reason to Believe that Circumstances Have Changed with Respect to State Program He may request, and State shall provide: (1) Supplemental Attorney General's statement; or (2) Program description; or (3) Such other documents or information as are necessary.
123.13(e)	E. State RCRA Programs Only (1) All New Programs Must comply with these regulations immediately upon approval. (2) Approved Program Which Requires Revision Because of Modification to This Part or to 40 CRF 122, 124, 260–265. Program shall be so revised within one year of date of promulgation of such regulations. Exception: (a) Unless State must amend or enact a statute in order to make the required revision, in which case such revision shall take place within two years.
123.14	XII. Criteria for Withdrawal of State Programs
123.14(a)	A. Withdrawal of Approved Program When State Program No Longer Complies with Requirements of This Part and State Fails to Take Corrective Action Such circumstances include following:
123.14(a)(1)	(1) When State's Legal Authority No Longer Meets Requirements of This Part Includes:
123.14(a)(1)(i)	(a) Failure of State to promulgate or enact new authorities when necessary; or
123.14(a)(1)(ii)	(b) Action by State legislature or court striking down or limiting State authorities.
123.14(a)(2)	(2) When Operation of State Program Fails to Comply With Requirements of This Part Includes:
123.14(a)(2)(i)	(a) Failure to exercise control over activities required to be regulated under this Part— including failure to issue permits; or
123.14(a)(2)(ii)	(b) Repeated issuance of permits which do not conform to requirements of this Part; or
123.14(a)(2)(iii)	(c) Failure to comply with public participation requirements of this Part.
123.14(a)(3)	(3) When State's Enforcement Program Fails to Comply with Requirements of This Part Includes:
123.14(a)(3)(i)	(a) Failure to act on violations of permits or other program requirements; or
123.14(a)(3)(ii)	(b) Failure to seek adequate enforcement penalties; or (c) Failure to collect administrative fines when imposed; or
123.14(a)(3)(iii)	(d) Failure to inspect and monitor activities subject to regulation.
123.14(a)(4)	(4) When State Program Fails to Comply with Terms of Memorandum of Agreement Under Section 123.6 (Memorandum of Agreement with Regional Administrator).
123.15	XIII. Procedures for Withdrawal of State Programs
123.15(a)	A. Voluntary Transfer of Program Responsibilities to EPA for State with Approved Program State shall take following actions, or others, as may be agreed upon with Administrator.
123.15(a)(1)	(1) State shall give Administrator 180 days notice of proposed transfer. (2) State Shall Submit Plan for Orderly Transfer of Relevant Program Information Not in Possession of EPA and Needed to Administer Program Includes permits, permit files, compliance files, reports, and permit applications.
123.15(a)(2)	(3) Within 60 Days of Receiving Notice and Transfer Plan Administrator shall: (a) Evaluate State's transfer plan. (b) Identify any additional information needed by Federal government for program administration and/or identify any other deficiencies in the plan.
123.15(a)(3)	(4) At Least 30 Days before Transfer Is to Occur Administrator shall: (a) Publish notice of transfer: (i) In the *Federal Register;* and (ii) In enough of the largest newspapers in the State to provide Statewide coverage.

Table 8-1A. General Program Requirements—Subpart A. (Continued)

Section No.	Description
	(b) Mail notice to:
	(i) All permit holders and applicants.
	(ii) Other regulated persons.
	(iii) Other interested persons on appropriate EPA and State mailing lists.
123.15(b)	B. Procedures That Apply When Administrator Orders Proceedings to Determine Whether to Withdraw Approval of a State Program
123.15(b)(1)	(1) Order
	(a) How Administrator May be Prompted to Order Commencement of Withdrawal Proceedings
	(i) On his own initiative; or
	(ii) In response to a petition from an interested person alleging failure of State to comply with requirements of this Part as set forth in Section 123.14 (Criteria For Withdrawal of State Programs).
	(b) Administrator shall respond in writing to any petition to commence withdrawal proceedings.
	(c) He May Conduct an Informal Investigation of Allegations in the Petition
	To determine whether cause exists to commence proceedings under this Section (123.15).
	(d) Administrator's Order Commencing Proceedings
	(i) Shall fix a time and place for commencement of hearing.
	(ii) Shall specify allegations against the State which are to be considered at the hearing.
	(e) Within 30 Days after Order
	State shall admit or deny these allegations in a written answer.
	(f) Burden of Proof
	Party seeking withdrawal of State's program shall have the burden of coming forward with the evidence in a hearing.
123.15(b)(3)	(2) Procedures
	The following provisions of 40 CFR 22 (Consolidated Rules of Practice) are applicable to proceedings under this Section:
123.15(b)(3)(i)	(a) Section 22.02—Use of Number/Gender.
123.15(b)(3)(ii)	(b) Section 22.04—Authorities of Presiding Officer.
123.15(b)(3)(iii)	(c) Section 22.06—Filing/Service of Rulings and Orders.
123.15(b)(3)(iv)	(d) Section 22.07(a) and (b)—Computation/Extension of Time (except that the time for commencement of the hearing shall not be extended beyond the date set in the Administrator's order without approval of the Administrator).
123.15(b)(3)(v)	(e) Section 22.08—Ex Parte Contacts (however, substitute "order commencing proceedings" for "complaint").
123.15(b)(3)(vi)	(f) Section 22.09—Examination of Filed Documents.
123.15(b)(3)(vii)	(g) Section 22.11(a), (c), and (d)—Intervention (however, motions to intervene must be filed within 15 days from date notice of Administrator's order is first published).
123.15(b)(3)(viii)	(h) Section 22.16—Motions (except that service shall be in accordance with Sections 123.15 (b)(4), the first sentence in Section 22.16(c) shall be deleted, and word "recommended" substituted for word "initial" in Section 22.16(c)).
123.15(b)(3)(ix)	(i) Section 22.19(a), (b), and (c)—Prehearing Conference.
123.15(b)(3)(x)	(j) Section 22.22—Evidence.
123.15(b)(3)(xi)	(k) Section 22.23—Objections/Offers of Proof.
123.15(b)(3)(xii)	(l) Section 22.25—Filing the Manuscript.
123.15(b)(3)(xiii)	(m) Section 22.26—Findings/Conclusions.
123.15(b)(4)	(3) Record of Proceedings
123.15(b)(4)(i)	(a) Hearing Shall Be either Stenographically Reported Verbatim or Tape Recorded
	Thereupon, it shall be transcribed by an official reporter designated by Presiding Officer.
123.15(b)(4)(ii)	(b) Written Material of Any Kind Submitted in the Hearing Shall Be a Part of the Record
	(i) Types of Written Material
	• All orders issued by Presiding Officer.
	• Transcripts of testimony.
	• Written statements of position, stipulations, exhibits, motions, and briefs.
	• Any other written material of any kind.
	(ii) Availability
	Record shall be available for inspection or copying in Office of Hearing Clerk, upon payment of costs.
	(iii) Inquiries
	May be made at Office of Administrative Law Judges, Hearing Clerk, 401 M Street, S.W., Washington, D.C. 20460

Table 8-1A. General Program Requirements—Subpart A. (Continued)

Section No.	Description
123.15(b)(4)(iii)	(c) Corrections to Transcript Upon notice to all parties, Presiding Officer may authorize corrections which involve matters of substance. (d) Handling of Submissions
123.15(b)(4)(iv)	(i) An original and two copies of all written submissions to hearing shall be filed with Hearing Clerk.
123.15(b)(4)(v)	(ii) A copy of each such submission shall be served by person making submission, to Presiding Officer and each party of record. (iii) Service shall take place by mail or personal delivery.
123.15(b)(4)(vi)	(iv) Every submission shall be: • Accompanied by an acknowledgement of service by person served; or • Proof of service in form of statement of date, time, and manner of service and names of persons served, certified by person who made service.
123.15(b)(4)(vii)	(e) List of Name, Service Address, Telephone Number of All Parties and Their Attorneys or Duly Authorized Representatives Hearing Clerk shall maintain such a list and furnish it to any person upon request.
123.15(b)(5)	(4) Participation by person Not a Party (a) Such a person may, at discretion of Presiding Officer, be permitted to make a limited appearance and oral or written statement of his position on the issues. (b) He may not otherwise participate in the proceeding.
123.15(b)(6)	(5) Rights of Parties All parties to the proceeding may: (a) Appear by counsel or other representative in all hearing and prehearing proceedings. (b) Agree to stipulations of facts which shall be made a part of the record.
123.15(b)(7) 123.15(b)(7)(i)	(6) Recommended Decision (a) Within 30 Days after Filing of Proposed Findings and Conclusion, and Reply Briefs Presiding Officer shall: (i) Evaluate record before him, the proposed findings and conclusions and any briefs filed by the parties. (ii) Prepare a recommended decision. (iii) Certify the entire record, including the recommended decision to the Administrator. (b) Copies of Recommended Decision Shall be served upon all parties. (c) Within 20 Days after Certification and Filing of Record and Recommended Decision All parties may file with Administrator exceptions to recommended decision and a supporting brief.
123.15(b)(8) 123.15(b)(8)(i)	(7) Decision by Administrator (a) Within 60 Days after Certification of Record and Filing of Presiding Officer's Recommended Decision Administrator shall review the record before him and issue his own decision.
123.15(b)(8)(ii)	(b) If Administrator Concludes That State Has Administered the Program in Conformity with RCRA and Regulations His decision shall constitute "final agency action" within meaning of 5 U.S.C. Section 704.
123.15(b)(8)(iii)	(c) If Administrator Concludes that State Has Not Administered Program in Conformity with RCRA and Regulations He shall: (i) List the deficiencies in the program. (ii) Provide the State a reasonable time, not to exceed 90 days, to take appropriate corrective action Administrator deems necessary.
123.15(b)(8)(ii)	(d) Corrective Action (i) Within time prescribed by Administrator, State shall take corrective action required by him. (ii) State shall file with Administrator and all parties a statement certified by State Director that appropriate corrective action has been taken.
123.15(b)(8)(v)	(iii) Administrator may require a further showing, in addition to certified statement, that corrective action has been taken.
123.15(b)(8)(vi)	(iv) If State Fails to Take Appropriate Corrective Action and File Certified Statement within Time Prescribed by Administrator Administrator shall issue a supplementary order withdrawing approval of State program. (v) If State Takes Appropriate Corrective Action Administrator shall issue a supplementary order stating that approval of authority is not withdrawn.

Table 8-1A. General Program Requirements—Subpart A. (Continued)

Section No.	Description
123.15(b)(8)(vii)	(e) Administrator's Supplementary Order Shall constitute final Agency action within meaning of 5 U.S.C. Section 704.
123.15(c)	C. Withdrawal of State Authorization under This Section (123.15) and RCRA (1) Does not relieve any person from complying with requirements of State law. (2) Nor does it affect the validity of actions by the State prior to withdrawal.

Table 8-1B. Additional Requirements for State Hazardous Waste Programs—Subpart B.

Section No.	Description
123.31	I. Purpose, Scope, and Applicability
	A. Subpart B—Final, Authoritization
123.31(a)	(1) This Subpart specifies additional requirements a State program must meet in order to obtain final authoritization under Section 3006(b) of RCRA.
123.31(b)	(2) States approved under this Subpart are authorized to administer and enforce their hazardous waste program in lieu of the Federal program.
123.31(a)	B. Subpart F—Interim Authorization (1) All requirements a State program must meet in order to obtain interim authorization under Section 3006(c) of RCRA are specified in Subpart F.
123.31(e)	(2) States need not have been approved under Subpart F in order to qualify for final authorization.
123.31(c)	C. When States May Apply for Final Authorization At any time after initial promulgation of Phase II.
	D. When State Programs under Final Authorization May Take Effect Not until effective date of Phase II.
123.31(d)	E. States Operating under Interim Authorization (1) May apply for and receive final authorization as specified in Section 123.31(c). (2) Notwithstanding approval under Subpart F, such States must meet all requirements of Subpart A and this Subpart in order to qualify for final authorization.
123.32	II. Consistency

	A. General
	(1) To obtain approval, a State program:
	    (a) Must be consistent with the Federal program and State programs applicable in other States.
	    (b) In particular, State programs must comply with provisions of Sections 123.32(a), (b), and (c) below.
	(2) "State Programs Applicable in Other States"
	Refers only to those State hazardous waste programs which have received final authorization under this Part (123).

	B. Examples of Inconsistency
123.32(a)	(1) Any aspect of the State program which unreasonably restricts, impedes, or operates as a ban on the free movement across the State border of hazardous wastes from or to other States for treatment, storage, or disposal at facilities authorized to operate under the Federal or an approved State program shall be deemed inconsistent.
123.32(b)	(2) Any aspect of State law or of the State program which has no basis in human health or environmental protection and which acts as a prohibition on the treatment, storage, or disposal of hazardous waste in the State may be deemed inconsistent.
123.32(c)	(3) If the State manifest system does not meet the requirements of this Part, the State program shall be deemed inconsistent.

Table 8-1B. Additional Requirements for State Hazardous Waste Programs—Subpart B. (Continued)

Section No.	Description
123.33	III. Requirements for Identification and Listing of Hazardous Wastes

A. Control of Hazardous Wastes
State program must control all hazardous wastes controlled under 40 CFR 261 (Identification and Listing of Hazardous Waste).

B. Adoption of List of Hazardous Wastes and Set of Characteristics for Identifying Hazardous Wastes
State must adopt such a list and characteristics equivalent to those under 40 CFR 261.

123.34   IV. Requirements for Generators of Hazardous Wastes

123.34(a)   A. Generator Coverage
The State program must cover all generators covered by 40 CFR 262 (Standards Applicable to Generators of Hazardous Waste).

B. New Generators
States must require new generators to contact the State and obtain an EPA identification number before they perform any activity subject to regulation under the approved State hazardous waste program.

123.34(b)   C. Reporting and Recordkeeping Requirements
(1) The State shall have authority to require and shall require all generators to comply with reporting and recordkeeping requirements equivalent to those under:
(a) 40 CFR 262.40—Recordkeeping.
(b) 40 CFR 262.41—Annual Reporting.
(2) States must require that generators keep these records at least 3 years.

123.34(c)   D. Hazardous Waste Containers and Tanks
The State program must require that generators who accumulate hazardous wastes for short periods of time prior to shipment off-site do so:
(1) In containers meeting DOT shipping requirements under:
(a) 49 CFR 173—Shippers—General Requirements for Shipments and Packaging. (See Appendix E of this book for DOT regulation contents only).
(b) 49 CFR 178—Shipping Container Specifications. (See Appendix E of this book for DOT regulation contents only).
(c) 49 CFR 179—Specifications for Tank Cars. (See Appendix E of this book for DOT regulation contents only).
(2) In tanks in accordance with State storage standards authorized by EPA under the approved State program.

123.34(d)   E. Packaging, Labeling, Marking, and Placarding Hazardous Waste
The State program must require that generators comply with requirements that are:
(1) Equivalent to requirements for packaging, labeling, marking, and placarding of hazardous waste under:
(a) 40 CFR 262.30—Packaging.
(b) 40 CFR 263.31—Labeling.
(c) 40 CFR 263.32—Marking.
(d) 40 CFR 263.33—Placarding.
(2) Consistent with relevant DOT regulations under:
(a) 49 CFR 172—Hazardous Materials Table and Hazardous Materials Communicating Regulations. (See Appendix E of this book for DOT regulation contents only.)
(b) 49 CFR 173. (See Appendix E of this book for DOT regulation contents only.)
(c) 49 CFR 178. (See Appendix E of this book for DOT regulation contents only.)
(d) 49 CFR 179. (See Appendix E of this book for DOT regulation contents only.)

123.34(e)   F. International Shipments
(1) The State program shall provide requirements respecting international shipments which are equivalent to those under 40 CFR 262.50 (International Shipments).
(2) Exception
Advance notification of international shipments, as required by 40 CFR 262.50(b)(1), shall be filed with Administrator.
(3) State may require that a copy of such advance notice be filed with the State Director, or may require equivalent reporting procedures.

Table 8-1B. Additional Requirements for State Hazardous Waste Programs—Subpart B. (Continued)

Section No.	Description
123.34(e) Note	(4) Such notices shall be mailed to:
	Hazardous Waste Export Division for Oceans and Regulatory Affairs (A-107) U.S. Environmental Protection Agency Washington, D.C. 20460
123.34(f)	G. Manifest System (1) The State must require that all generators of hazardous waste who transport (or offer for transport) such hazardous waste off-site:

123.34(f)(1)	(a) Use a Manifest System
	   (i) That ensures that interstate and intrastate shipments of hazardous waste are designated for delivery; and
	   (ii) In case of intrastate shipments, are delivered to facilities that are authorized to operate under an approved State program or Federal program.

123.34(f)(2)	(b) Initiate the manifest. (c) Designate on the manifest the storage, treatment, or disposal facility to which the waste is to be shipped.

123.34(f)(3)	(d) Ensure that all wastes offered for transport are accompanied by the manifest.
	Exception:
	   (i) Case of shipments by rail or water specifed in:
	    • 40 CFR 262.23(c)—Use of the Manifest.
	    • 40 CFR 263.20(e)—The Manifest System.
	(2) Shipments by Rail or Water
	The State program shall provide requirements for shipments by rail or water equivalent to those under:
	(a) 40 CFR 262.23(c).
	(b) 40 CFR 263.20(e).

123.34(f)(4)	(3) Instances Where Manifests Have Not Been Returned by Owner/Operator of Designated Facility (a) Investigate such instances. (b) Report such instances to State in which shipment originated.
123.34(g)	(4) Case of Interstate Shipments for Which Manifest Has Not Been Returned (a) The State program must provide for notification to State in which facility designated on manifest is located; and (b) Must provide notification to the State in which the shipment may have been delivered (or to EPA in case of unauthorized States).
123.34(h)	(5) State Must Follow the Federal Manifest Format (40 CFR 262.21—Required Information) State may supplement format to a limited extent subject to consistency requirements of Hazardous Materials Transportation Act (49 U.S.C. 1801 et seq.).
123.35	V. Requirements for Transporters of Hazardous Wastes
123.35(a)	A. Transporter Coverage The State program must cover all transporters covered by 40 CFR 263 (Standard Applicable to Transporters of Hazardous Waste). B. New Transporters New transporters must be required to contact the State and obtain an EPA identification number from the State before they accept hazardous waste for transport.
123.35(b)	C. Recordkeeping Requirements (1) The State shall have authority to require and shall require all transporters to comply with recordkeeping requirements equivalent to those found at 40 CFR 263.22 (Recordkeeping). (2) States must require that records be kept at least 3 years.
123.35(c)	D. Manifest System (1) The State must require the transporter to carry the manifest during transport. Exception: (a) Case of shipments by rail or water specified in 40 CFR 263.20(e) (The Manifest System). (2) Transporter must deliver wates only to facility designated on manifest. (3) Shipments by Rail or Water The State program shall provide requirements for shipments by rail or water equivalent to those under 40 CFR 263.20(e).

Table 8-1B. Additional Requirements for State Hazardous Waste Programs—Subpart B. (Continued)

Section No.	Description

123.35(d)	E. For Hazardous Wastes That are Discharged in Transit
	    (1) The State program must require that transporters:
	        (a) Notify appropriate State, local, and Federal agencies of such discharges; and
	        (b) Clean up such wastes; or
	        (c) Take action so that such wastes do not present a hazard to human health or the environment.
	    (2) These requirements shall be equivalent to those found at 40 CFR 263.30 (Immediate Action).

123.36	VI. Requirements for Hazardous Waste Management Facilities
	
	A. The State Shall Have Standards for Hazardous Waste Management Facilities
	    (1) They shall be equivalent to:
	        (a) 40 CFR 264—Standards for Owners and Operators of Hazardous Waste Treatment, Storage, and Disposal Facilities.
	        (b) 40 CFR 266—Waste Specific Standards (for Phase II release).
	    (2) What Standards Shall Include

123.36(a)	(a) Technical Standards
	            For:
	            (i) Tanks.
	            (ii) Containers.
	            (iii) Waste piles.
	            (iv) Incineration.
	            (v) Chemical, physical, and biological treatment facilities.
	            (vi) Surface impoundments.
	            (vii) Landfills.
	            (viii) Land treatment facilities.

123.36(b)
123.36(c)
	(b) Financial responsibility during facility operation.
	        (c) Preparedness for the prevention of discharges or releases of hazardous waste.
	        (d) Contingency plans and emergency procedures to be followed in event of discharge or release of hazardous waste.

123.36(d)	(e) Closure and Post-Closure Requirements Includes financial requirements to ensure that money will be available for closure and post-closure monitoring and maintenance.

123.36(e)
123.36(f)
	(f) Ground water monitoring.
	        (g) Security
	            To prevent unauthorized access to the facility.

123.36(g)
123.36(h)
123.36(i)
	(h) Facility personnel training.
	        (i) Inspections, monitoring, recordkeeping, and reporting.
	        (j) Compliance with Manifest System
	            Including requirement that facility owners or operators return a signed copy of manifest to generator to certify delivery of that hazardous waste shipment.

123.36(j)	(k) Other requirements to extent that they are included in 40 CFR (264 and 266).
123.37	VII. Requirements with Respect to Permits and Permit Applications
123.37(a)	A. Owners/Operators of Hazardous Waste Management Facilities Required to Obtain a Permit Under 40 CFR 122 State law must require permits for such facility owners/operators. B. State Law Must Prohibit Operation of Any Hazardous Waste Management Facility without Such a Permit Exception: (1) States may, if adequate legal authority exists, authorize owners and operators of any facility which would qualify for interim status under the Federal program to remain in operation until a final decision is made on the permit application. C. Where State Law Authorizes Interim Status It shall require compliance by owners/operators of such facilities with standards at least as stringent as EPA's interim status standards under 40 CFR 265 (Interim Status Standards for Owners and Operators of Hazardous Waste Treatment, Storage, and Disposal Facilities).
123.37(b)	D. New HWM Facilities The State must require all such facilities to contact the State and obtain an EPA identification number before commencing treatment, storage, or disposal of hazardous waste.

Table 8-1B. Additional Requirements for State Hazardous Waste Programs—Subpart B. (Continued)

Section No.	Description
123.37(c)	E. All Permits Issued by State Shall require compliance with the standards adopted by the State under Section 123.36 (Requirements for Hazardous Waste Management Facilities).
123.37(d)	F. All Permits Issued under State Law Prior to Date of Approval of Final Authorization Shall be reviewed by State Director and Modified or revoked and reissued to require compliance with requirements of this Part.
123.38	VIII. EPA Review of State Permits
123.38(a)	A. Regional Administrator May Comment on Permit Applications and Draft Permits As provided in Memorandum of Agreement under Section 123.6 (Memorandum of Agreement with Regional Administrator).
123.38(b)	B. Where EPA Indicates in a Comment that Issuance of the Permit Would Be Inconsistent with the Approved State Program EPA shall include in the comment:
123.38(b)(1)	(1) Statement of Reasons for Comment Includes Section of RCRA or regulations promulgated that support comment.
123.38(b)(2)	(2) Actions That Should Be Taken by State Director in Order to Address Comments Include conditions which permit would include if it were issued by Regional Administrator.
123.38(c)	C. Copy of Any Comment Shall be sent to permit applicant by Regional Administrator.
123.38(d)	D. When Regional Administrator Shall Withdraw Such a Comment When satisfied that State has met or refuted his concerns.
123.38(e)	E. Section 3008(a)(3) of RCRA (1) Under this statute: (a) EPA may terminate a State-issued permit in accordance with procedures of Part 124, Subpart E (Evidentiary Hearing); or (b) EPA may bring enforcement action in accordance with procedures of 40 CFR 22 (Consolidated Rules of Practice) in the case of a violation of a State program requirement. (2) Grounds by Which Regional Administrator May Take Action Against Holder of State-Issued Permit under Section 3008(a)(3) of RCRA
123.38(e)(1) 123.38(e)(2)	(a) Permittee is not complying with condition of that permit. (b) Permittee is not complying with condition that Regional Administrator stated was necessary to implement approved State program requirements.
123.38(e)(3)	(c) Permittee is not complying with a condition necessary to implement approved State program requirements.
123.38(e)(4)	F. Action under Section 7003 of RCRA Regional Administrator may take action under this statute against a permit holder at any time whether or not the permit holder is complying with permit conditions.
123.39	IX. Approval Process
123.39(a)	A. Prior to Submitting an Application to EPA for Approval of a State Program (1) State shall issue public notice of its intent to seek program approval from EPA. (2) What Public Notice Shall Do
123.39(a)(1) 123.39(a)(1)(i) 123.39(a)(1)(ii)	(a) Be circulated in manner calculated to attract attention of interested persons by: (i) Publication in enough of largest newspapers in State to attract statewide attention. (ii) Mailing to persons on State agency mailing list. (iii) Mailing to any other persons whom Agency has reason to believe are interested.
123.39(a)(2) 123.39(a)(3) 123.39(a)(4)	(b) Indicate when and where State's proposed submission may be reviewed by public. (c) Indicate cost of obtaining copy of submission. (d) Provide for comment period of not less than 30 days during which time interested members of public may express their views on proposed program.
123.39(a)(5)	(3) Public Hearing (a) Provide that a public hearing be held by State or EPA if sufficient public interest is shown; or (b) Schedule such a public hearing.

Table 8-1B. Additional Requirements for State Hazardous Waste Programs—Subpart B. (Continued)

Section No.	Description

123.39(a)(6)
123.39(a)(7)
	(c) Any public hearing to be held by State on its application for authorization shall be scheduled no earlier than 30 days after notice of hearing is published.
	   (i) Briefly outline fundamental aspects of State program.
	   (ii) Identify a person that an interested member of the public may contact with any questions.

123.39(b)	D. **If Proposed State Program Is Substantially Modified after Public Comment Period Provided in Section 123.39(a)(4)**
	  (1) State shall prior to submitting program to Administrator provide opportunity for public comment in accordance with procedure of Section 123.39(a).
	  (2) Opportunity for further public comment may be limited to those portions of State's application which has been changed since the prior public notice.

123.39(c)	C. State Submittal of Proposed Program to EPA for Approval
	  (1) After complying with requirements of Sections 123.39(a) and (b), State may make submittal.
	  (2) Such formal submission may only be made after date of promulgation of Phase II.
	  (3) Content of Program Submission
	    (a) Copies of all written comments received by the State.
	    (b) Transcript, recording, or summary of any public hearing which was held by the State.
	    (c) Responsiveness Summary
	      (i) Identifies public participation activities conducted.
	      (ii) Describes matters presented to public.
	      (iii) Summarizes significant comments received and response to these comments.

123.39(d)
 	
123.39(d)(1)
 	
123.39(d)(2)
 	
123.39(d)(3)
	D. Within 90 Days from Date of Receipt of Complete Program Submission for Final Authorization
	  (1) Administrator shall make a tentative determination as to whether or not he expects to grant authorization to State program.
	  (2) If Administrator Indicates that He May Not Approve State Program
	    (a) He shall include a general statement of his areas of concern.
	    (b) He shall give notice of his tentative determination in the Federal Register in accordance with Section 123.39(a)(1).
	    (c) Notice of Tentative Determination of Authorization
	      (i) Shall indicate that a public hearing will be held by EPA no earlier than 30 days after notice of tentative determination of authorization.
	      (ii) Notice may require persons wishing to present testimony to file request with Regional Administrator who may cancel public hearing if sufficient public interest in a hearing is not expressed.
	    (d) Administrator may afford public 30 days after notice to comment on State's submission and tentative determination.
	    (e) He will note the availability of the State submission for inspection and copying by public.

123.39(e)	E. Within 90 Days of Notice Given Pursuant to Section 123.39(d)
	  (1) Administrator shall make a final determination whether or not to approve the State's program, taking into account any comments submitted.
	  (2) Administrator will grant final authorization only after effective date of Phase II.
	  (3) Administrator shall give notice of this final determination in the Federal Register in accordance with Section 123.39(a)(1).
	  (4) The notification shall include a concise statement of the reasons for this determination, and a response to significant comments received.

Table 8-1C. Requirements for Interim Authorization of State Hazardous Waste Programs—Subpart F.

Section No.	Description
123.121	I. Purpose, Scope, and Applicability

123.121(a)	A. General
	  (1) Subpart F—Interim Authorization
	    This Subpart specifies all the requirements a State must meet in order to obtain interim authorization under Section 3006(c) of RCRA.
	  (2) Subparts A and B—Final Authorization
	    The requirements a State program must meet in order to obtain final Authorization under Section 3006(b) of RCRA are specified in Subparts A and B.

Table 8-1C. Requirements for Interim Authorization of State Hazardous Waste Programs—Subpart F. (Continued)

Section No.	Description

123.121(b)	B. Interim Authorization Phasing
	Interim authorization of State programs under this Subpart may occur in two phases.
	  (1) Phase I
	    This first phase allows States to administer a hazardous waste program in lieu of and corresponding to that portion of the Federal program which covers:
	    (a) Identification and Listing of Hazardous Waste
	      Under 40 CFR 261.
	    (b) Generators of Hazardous Waste
	      Under 40 CFR 262.
	    (c) Transporters of Hazardous Waste
	      Under 40 CFR 263.
	    (d) Preliminary (Interim Status) Standards for Hazardous Waste Treatment, Storage, and Disposal Facilities
	      Under 40 CFR 265.
	  (2) Phase II
	    This second phase allows States with interim authority for Phase I to establish a permit program for hazardous waste treatment, storage and disposal facilities in lieu of and corresponding to Federal hazardous waste permit program under 40 CFR (264 and 265).
	  (3) Applying for Interim Authorization
	    States may apply for interim authorization in the following ways, as long as they adhere to schedule in Section 123.122 (Schedule):
	    (a) Sequentially
	      Application submitted for interim authorization for Phase I followed by an amendment of that application for Phase II; or
	    (b) All at Once
	      Application submitted for interim authorization for both Phase I and Phase II at same time.

123.121(c)	C. Approval of State Program Administrator shall so do if State program meets applicable requirements of this Subpart.
123.121(d)	D. Upon Approval of a State Program for Phase II Administrator shall suspend issuance of Federal permits for these activities subject to approved State program.
123.121(e)	E. Any State Program Approved by Administrator under This Subpart Shall at all times be conducted in accordance with this Subpart.

123.121(f)
 	
123.121(f) Note
	F. Regulating Activities on Indian Lands
	  (1) Lack of State's Authority to Regulate Activities on Indian Lands
	    Does not impair a State's ability to obtain interim authorization under this Subpart.
	  (2) EPA will administer the program on Indian lands if State does not seek this authority.
	  (3) States are advised to contact United States Department of Interior, Bureau of Indian Affairs, concerning authority over Indian lands.

123.121(g)(1)
and (2)
	G. Adoption And Enforcement by State of More Stringent, Broader, or More Extensive Requirements Than Required under This Subpart
	  (1) Nothing precludes a State from so doing.
	  (2) Where an approved program has a greater scope of coverage than required by Federal law, the additional coverage is not part of the Federally approved program.

123.122	II. Schedule
123.122(a)	A. Interim Authorization for Phase I Shall not take effect until Phase I commences (11/19/80).

123.122(b)	B. Interim Authorization for Phase II
	  (1) Shall not take effect until Phase II commences.
	  (2) Interim Authorization May Extend for 24 Month Period from Start of Phase II
	    At end of this period:
	    (a) All interim authorizations automatically expire.
	    (b) EPA shall administer the Federal program in any State which has not received final authorization.

Table 8-1C. Requirements for Interim Authorization of State Hazardous Waste Programs—Subpart F. (Continued)

Section No.	Description
123.122(c)	C. States Applying for Interim Authorization (1) May do so at any time prior to expiration of 6th month of 24-month period beginning with commencement of Phase II.
123.122(c)(1)	(2) States Applying for Interim Authorization Prior to Promulgation of Phase II Shall apply only for interim authorization for Phase I.

123.122(c)(2)	(3) States Applying for Interim Authorization after Promulgation of Phase II but Before Start of Phase II
	        May apply for:
	        (a) Interim authorization for both Phase I and Phase II; or
	        (b) Only for interim authorization for Phase I.

123.122(c)(3)	(4) States Applying for Interim Authoritization after Start of Phase II Shall apply for interim authorization for both Phase I and Phase II, unless they have already applied for interim authorization for Phase I.
123.122(c)(4)	(5) States Which have Received Interim Authorization for Phase I Shall amend their original submission to meet the requirements for interim authorization for Phase II not later than 6 months after effective date of Phase II.

123.122(d)	D. Applying for Interim Authorization for Phase II
	    No State may so apply unless:
	    (1) It has received interim authoritization for Phase I; or
	    (2) Is simultaneously applying for interim authorization for both Phase I and Phase II.

123.123	III. Elements of a Program Submission
123.123(a)	A. Content of Program Submission States applying for interim authorization shall submit at least 3 copies of a program submission, the content of which follows:
123.123(a)(1)	(1) Letter from Governor of State requesting program approval.
123.123(a)(2)	(2) Complete Program Discription As required by Section 123.124 (Program Description), describing how State intends to carry out its responsibilities under this Subpart (F).
123.123(a)(3)	(3) Attorney General's Statement As required by Section 123.125 (Attorney General's Statement).
123.123(a)(4)	(4) Memorandum of Agreement with Regional Administrator As required by Section 123.126 (Memorandum of Agreement).
123.123(a)(5)	(5) Authorization Plan As required by Section 123.127 (Authorization Plan).
123.123(a)(6)	(6) Copies of all applicable State statutes and regulations, including those governing State administrative procedures.

123.123(b)	B. Upon Receipt of State Program Submission by EPA
	    (1) Within 30 days EPA will notify State whether submission is complete.
	    (2) If EPA Finds that State's Submission Is Complete
	        EPA's formal review of proposed State program shall be deemed to have begun on date of receipt of State's submission (see Section 123.135—Approval Process).
	    (3) If EPA Finds that State's Submission Is Incomplete
	        Formal review shall not begin until all necessary information is received by EPA.

123.123(c)	C. If State's Submission Is Materially Changed During Formal Review Period The formal review period shall recommence upon receipt of revised submission.
123.123(d)	D. States Simultaneously Applying for Interim Authorization for Both Phase I and Phase II Shall prepare a single submission.

123.123(e)	E. States Applying for Interim Authorization for Phase II
	    Shall amend their submission for interim authorization for Phase I as specified in:
	    (1) Section 123.124—Program Description.
	    (2) Section 123.125—Attorney General's Statement.
	    (3) Section 123.126—Memorandum of Agreement.
	    (4) Section 123.127—Authorization Plan.

Table 8-1C. Requirements for Interim Authorization of State Hazardous Waste Programs—Subpart F. (Continued)

Section No.	Description
123.124	IV. Program Description
	A. General
	(1) Any State Wishing to Administer a Program under This Subpart (F) Shall submit to Regional Administrator a complete description of program it proposes to administer under State law, in lieu of the Federal program.
	(2) State Applying Only for Interim Authorization for Phase II Shall amend its program description for interim authorization for Phase I to reflect program it proposes to administer to meet requirements for interim authorization for Phase II.
	B. Content of Program Description
123.124(a)	(1) Description of Scope, Structure, Coverage, and Processes of State Program In narrative form.

123.124(b)	(2) Description of Organization and Structure of State Agency or Agencies Which Will Have Responsibility for Administering the Program
	    (a) Include organization charts.
	    (b) If More Than One Agency Is Responsible for Administration of Program
	      (i) Each agency must have statewide jurisdiction over a class of activities.
	      (ii) Responsibilities of each agency must be delineated.
	      (iii) Their procedures for coordination must be set forth.
	      (iv) One of the agencies must be designated a "lead agency" to facilitate communications between EPA and State agencies having program responsibility.
	    (c) Where State Proposes to Administer a Program of Greater Scope of Coverage than Required by Federal Law
	    Must indicate resources dedicated to administering the Federally required portion of program.

123.124(b)(1)	(d) Description of State Agency Staff Who Will Carry Out State Program
	      (i) Include number, occupations, and general duties of employees.
	      (ii) State need not submit complete job descriptions for every employee carrying out State program.

123.124(b)(2)	(e) Itemization of Proposed or Actual Costs of Establishing and Administering Program
	    Include cost of:
	      (i) Personnel listed in Section 123.124(b)(1).
	      (ii) Administrative support.
	      (iii) Technical support.

123.124(b)(3)	(f) Itemization of Sources and Amounts of Funding
	      (i) Include estimate of Federal grant money available to State Director to meet costs listed in Section 123.124(b)(2).
	      (ii) Identify any restrictions or limitations upon this funding.

123.124(c)	(3) Description of Applicable State Procedures
	    Includes:
	    (a) Permitting Procedures

123.124(c) Note	States applying only for interim authorization for Phase I need describe permitting procedures only to extent they will be used to ensure compliance with standards substantially equivalent to 40 CFR 265. (b) State appellate review procedures.

123.124(d)	(4) State Forms and Manifest Format
	    (a) Copies of forms and manifest format State intends to use in its program.
	    (b) Forms Used by States Need Not Be Identical to Forms Used by EPA
	    But same basic information is required.
	    (c) Uniform National Forms
	    If State chooses to use uniform national forms it should so note.

123.124(e)	(5) Complete description of State's compliance monitoring and enforcement program.

123.124(f)	(6) Manifest System
	    (a) Description of State manifest system if State has such a system.
	    (b) Procedures State will use to coordinate information with other approved State programs and the Federal program regarding interstate and international shipments.

123.124(g)	(7) Estimate of Number of Hazardous Waste Program Participants
	    (a) Generators.
	    (b) Transporters.
	    (c) On-Site and Off-Site Storage, Treatment, and Disposal Facilities
	      (i) Brief description of types of facilities.
	      (ii) Indication, if applicable, of permit status of these facilities.

Table 8-1C. Requirements for Interim Authorization of State Hazardous Waste Programs—Subpart F. (Continued)

Section No.	Description
123.125	V. Attorney General's Statement
123.125(a)	A. For State That Seeks to Administer a Program under This Subpart
	(1) Shall submit a statement from:
	(a) State Attorney General; or
	(b) Attorney for those States or interstate agencies which have independent legal counsel.
	(2) Overall Substance of Statement
	(a) That laws of the State, or interstate compact, provide adequate authority to carry out the program described under Section 123.124 (Program Description).
	(3) Specific Content of Statement
	(a) Citations to specific statutes.
	(b) Administrative regulations.
	(c) Where appropriate, judicial decisions which demonstrate adequate authority.
	(4) Certification that Enabling Legislation for Program for Phase I Was in Existence Within 90 Days of Promulgation of Phase I
	State Attorney General or independent legal counsel must so certify.
	Exception:
	(a) As provided in Section 123.128(d) (Elements of a Program Submission).
	(5) For State Applying for Interim Authorization for Phase II
	State Attorney General or independent legal counsel must:
	(a) Certify that enabling legislation for the program for Phase II was in existence within 90 days of promulgation of Phase II.
	(b) Amend and recertify statement submitted for interim authorization for Phase I to demonstrate adequate authority to carry out all the requirements of this Subpart.
123.125(b)(2)	(c) Certify that authorization plan under Section 123.127(b) (Authorization Plan), if carried out, would provide State with enabling authority and regulations adequate to meet all requirements for final authorization.
123.125(b)(1)	(6) For State Applying for Interim Authorization for Phase I
	Attorney General's statement shall certify that the authorization plan under Section 123.127 (a), is carried out, would provide the State with enabling authority and regulations adequate to meet the requirements for final authorization contained in Phase I.
123.125(a)	B. State Statutes and Regulations Cited by State Attorney General or Independent Legal Counsel
	(1) Shall be lawfully adopted at time statement is signed.
	(2) Shall be fully effective by time program is approved.
	C. To Qualify As "Independent Legal Counsel"
	The attorney signing the statement required by this Section must have full authority to independently represent the State agency in court on all matters pertaining to the State program.
123.125(c)	D. When a State Seeks Authority over Activities on Indian Lands
	The statement shall contain an appropriate analysis of State's authority.
123.126	VI. Memorandum of Agreement
	A. General
123.126(a)	(1) Who Shall Execute Memorandum of Agreement (MOA)
	State Director and Regional Administrator.
	(2) In Addition to Meeting Requirements of Sections 123.126(b) and (c)
	The MOA may include other terms, conditions, or agreements relevant to administration and enforcement of State's regulatory program which are not inconsistent with this Subpart.
	(3) Memorandum of Agreement Containing Provisions Which Restrict EPA's Statutory Oversight Responsibility
	Administrator shall not approve such an agreement.
	(4) For State Applying for Interim Authorization for Phase II
	The MOA shall be amended and re-executed to include requirements of Section 123.126(c) and any revisions to requirements of Section 123.126(b).
123.126(b)	B. Contents of Memorandum of Agreement
123.126(b)(1)	(1) Provisions for Prompt Transfer from EPA to State of Section 3010 Notification Information Prior to Approval of State Program
	Includes:
	(a) EPA identification numbers for new generators, transporters, and treatment, storage, and disposal facilities.

Table 8-1C. Requirements for Interim Authorization of State Hazardous Waste Programs—Subpart F. (Continued)

Section No.	Description
	(b) Pending permit applications.
	(c) Compliance reports, etc.
123.126(b)(2)	(2) Provisions Specifying Frequency and Content of Reports, Documents, and Other Information Which State Is Required to Submit to EPA
	(a) State shall allow EPA to routinely review State records, reports, and files relevant to administration and enforcement of the approved program.
	(b) State reports may be combined with grant reports when appropriate.
123.126(b)(3)	(3) Provisions on State's Compliance Monitoring and Enforcement Program
	Includes:
123.126(b)(3)(i)	(a) Provisions for Coordination of Compliance Monitoring Activities by State and by EPA
	(i) These may specify the basis on which Regional Administrator will select facilities or activities within State for EPA inspection.
	(ii) Regional Administrator will normally notify State at least 7 days before any such inspection.
123.126(b)(3)(ii)	(b) Procedures to ensure coordination of enforcement activities.
123.126(b)(4)	(4) Provisions for Modification of Memorandum of Agreement
	In accordance with this Part (123).
123.126(b)(5)	(5) Compliance Inspections
	(a) Provision allowing EPA to conduct compliance inspections of all generators, transporters, and HWM facilities during interim authorization.
	(b) Limitations on Compliance Inspections of Generators, Transporters, and Non-Major HWM Facilities
	(i) Regional Administrator and State Director may agree to such limitations.
123.126(b)(6)	(ii) Such limitations shall not restrict EPA's right to inspect any generator, transporter, or HWM facility which it has cause to believe is not in compliance with RCRA.
	• However, before conducting such an inspection, EPA will normally allow the State a reasonable opportunity to conduct a compliance evaluation inspection.
123.126(b)(7)	(6) Provisions delineating respective State and EPA responsibilities during interim authorization period.
123.126(c)	C. Additional Content of Memorandum of Understanding for State Applying for Interim Authorization for Phase II
123.126(c)(1)	(1) Provisions for Prompt Transfer from EPA to State of Pending Permit Applications and Support Files for Permit Issuance
	(a) Where Existing Permits Are Transferred to State for Administration
	MOA shall contain provisions specifying a procedure for transferring responsibility for these permits.
	(b) If State Lacks Authority to Directly Administer Permits Issued by Federal Government
	A procedure may be established to transfer responsibility for these permits.
123.126(c)(2)	D. Provisions Specifying Classes and Categories of Permit Applications and Draft Permits
	(1) State Director will send to Regional Administrator for review and comment.
	(2) Major HWM Facilities.
	State Director shall promptly forward to EPA copies of permit applications and draft permits for all major HWM facilities.
	(3) Limitations Regarding Review of and Comment on Permit Applications and Draft Permits for Non-Major HWM Facilities
	Regional Administrator and State may agree to such limitations.
	(4) State Director shall supply EPA copies of final permits for all major HWM facilities.
123.126(c)(3)	E. Joint Processing of Permits by State and EPA
	Where appropriate, such provisions shall be made for facilities or activities which require permits under different programs, from both EPA and the State.
123.127	VII. Authorization Plan
	A. General
	(1) State must submit authorization plan.
	(2) What Authorization Plan Shall Describe
	The additions and modifications necessary for the State program to qualify for final authorization.
	(3) Timing
	(a) Submit as soon as practicable.
	(b) But no later than end of interim authorization period.

Table 8-1C. Requirements for Interim Authorization of State Hazardous Waste Programs—Subpart F. (Continued)

Section No.	Description
	B. Content of Authorization Plan
	(1) The nature of and schedules for any changes in State legislation and regulations.
	(2) Resource levels.
	(3) Actions State must take to control the complete universe of hazardous waste listed or designated under Section 3001 of RCRA, as soon as possible.
	(4) Manifest system.
	(5) Permit system.
	(6) Surveillance and enforcement program necessary for State to become eligible for final authorization.
123.127(a)	C. For State Applying Only for Interim Authorization for Phase I Authorization plan shall describe additions and modifications necessary for State program to meet requirements for final authorization contained in Phase I.
123.127(b)	D. For State Applying for Interim Authorization for Phase II The authorization plan under Section 123.127(a) shall be amended to describe the further additions and modifications necessary for the State program to meet all the requirements for final authorization.
123.128	VIII. Program Requirements for Interim Authorization for Phase I
	A. General
	(1) The requirements in Sections 123.128(a)–(g) are applicable to States applying for interim authorization for Phase I.
	(2) If State Does Not Have Legislative Authority or Regulatory Control over Certain Activities That Occur in the State The State may be granted interim authorization for Phase I provided the state authorization plan under Section 123.127 (Authorization Plan) provides for development of a complete program as soon as practicable after receiving interim authorization.
123.128(a)	B. Requirements for Identification and Listing of Hazardous Waste The State program must control a universe of hazardous wastes generated, transported, treated, stored, and disposed of in the State which is nearly identical to that which would be controlled by the Federal program under 40 CFR 261.
123.128(b) 123.128(b)(1)	C. Requirements for Generators of Hazardous Waste (1) Section 123.128(b) applies unless the State comes within the exceptions described under Section 123.128(d).
123.128(b)(2)	(2) State program must cover all generators of hazardous wastes controlled by the State.

123.128(b)(3)	(3) Reporting and Recordkeeping Requirements
	State shall have authority to require and shall require all generators covered by State program to comply with reporting and recordkeeping requirements substantially equivalent to:
	     (a) 40 CFR 262.40–Recordkeeping.
	     (b) 40 CFR 262.41–Annual Reporting.

123.128(b)(4)	(4) Generators Who Accumulate Hazardous Wastes for Short Periods of Time Prior to Shipment State program must require that such generators do so in a manner that does not present a hazard to human health or the environment.

123.128(b)(5)	(5) International Shipments
	     (a) State program shall provide requirements respecting international shipments which are substantially equivalent to those of 40 CFR 262.50 (International Shipments).
	     (b) Advance Notification of International Shipments
	       (i) As required by 40 CFR 261.50(b)(1), notification shall be filed with the Administrator.
	       (ii) State may require that a copy of such advance notice be filed with State Director; or
	       (iii) May require equivalent reporting procedures.

123.128(b)(5) Note	(iv) Such notices shall be mailed to:
	
	     Hazardous Waste Export
	     Division for Oceans and Regulatory Affairs (A-107)
	     U.S. Environmental Protection Agency
	     Washington, D.C. 20460.

Table 8-1C. Requirements for Interim Authorization of State Hazardous Waste Programs—Subpart F. (Continued)

Section No.	Description
123.128(b)(6)	(6) Transportation of Hazardous Waste Off-Site State program must require: (a) That such generators who transport or offer for transport hazardous waste off-site use a manifest system that ensures that interstate and intrastate shipments are designated for delivery. (b) That intrastate shipments are delivered to facilities that are authorized to operate under an approved State program or the Federal Program.
123.128(b)(7)	(7) State Manifest System It must require that:
123.128(b)(7)(i)	(a) Manifest itself identify the generator, transporter, designated facility to which the hazardous waste will be transported, and the hazardous waste being transported.
123.128(b)(7)(ii)	(b) Manifest accompany all wastes offered for transport. Exception: (i) Shipments by rail or water specified in: • Section 262.23(c)—Use of Manifest. • Section 263.20(e)—The Manifest System.
123.128(b)(7)(iii)	(c) Shipments of Hazardous Waste Not Delivered To Designated Facility (i) Must be identified and reported by generator to State in which shipment originated; or (ii) Are independently identified by the State in which the shipment originated.
123.128(b)(8)	(d) Interstate Shipments for Which Manifest Has Not Been Returned State program must provide: (i) For notification to State in which facility designated on manifest is located; and (ii) To State in which the shipment may have been delivered; or (iii) To EPA in case of unauthorized States.
123.128(c)	D. Requirements for Transporters of Hazardous Wastes
123.128(c)(1)	(1) Section 123.128(c) applies unless the State comes within the exceptions described under Section 123.128(d).
123.128(c)(2)	(2) The State program must cover all transporters of hazardous waste controlled by the State.
123.128(c)(3)	(3) Recordkeeping Requirements State shall have authority to require and shall require all transporters covered by State program to comply with recordkeeping requirements substantially equivalent to those found at 40 CFR 263.22 (Recordkeeping).
123.128(c)(4)	(4) Manifest System (a) State program must require transporters of hazardous waste to use a manifest system that ensures that interstate and intrastate shipments are delivered only to facilities that are authorized under an approved State program or the Federal program.
123.128(c)(5)	(b) State program must require that transporters carry the manifest with all shipments. Exception: (i) Shipments by rail or water specified in 40 CFR 263.20(e) (Manifest System).
123.128(c)(6)	(5) For Hazardous Wastes Discharged in Transit (a) State program must require: (i) That transporters notify appropriate State, local, and Federal agencies of discharges; and (ii) Clean up the wastes; or (iii) Take action so that wastes do not present a hazard to human health or the environment. (b) These requirements shall be substantially equivalent to those found at: (i) 40 CFR 263.30—Immediate Action. (ii) 40 CFR 263.31—Discharge Clean Up.
123.128(d)	E. Limited Exceptions from Generator, Transporter, and Related Manifest Requirements (1) State Applying for Interim Authorization for Phase I (a) Must meet all requirements for such interim authorization except: (i) It does not have statutory or regulatory authority for the manifest system; or (ii) Other generator or transporter requirements of Sections 123.128(b) and (c). (b) State May Still Be Granted Interim Authorization If State authorization plan under Section 123.127 delineates the necessary steps in obtaining this authority no later than end of interim authorization period under Section 123.122(b) (Schedule). (c) State May Apply for Interim Authorization to Implement the Manifest System and Other Generator and Transporter Requirements (i) If enabling legislation for that part of the program was in existence within 90 days of promulgation of Phase I.

Table 8-1C. Requirements for Interim Authorization of State Hazardous Waste Programs—Subpart F. (Continued)

Section No.	Description
	(ii) If such application is made, it shall be made as part of the State's submission for interim authorization for Phase II.
	(2) Until State Manifest System and Other Generator and Transporter Requirements Are Approved by EPA
	(a) All Federal requirements for generators and transporters (including use of manifest system) shall apply in such States, and enforcement responsibility for that part of the program shall remain with the Federal Government.
	(b) The universe of wastes for which these Federal requirements apply shall be the universe of wastes controlled by the State under Section 123.128(a).
123.128(e)	F. Requirements for Hazardous Waste Treatment, Storage, and Disposal Facilities
	(1) States must have standards applicable to HWM facilities substantially equivalent to 40 CFR 265.
	(2) State law shall prohibit the operation of facilities not in compliance with such standards.
	(3) Content of Part 265 HWM Facility Standards
123.128(e)(1)	(a) Preparedness for and Prevention of Hazardous Waste Controlled by State under Section 123.128(a).
	(b) Contingency Plans and Emergency Procedures To be followed in event of release of hazardous waste.
123.128(e)(2)	(c) Closure and post-closure requirements.
123.128(e)(3)	(d) Ground water monitoring.
123.128(e)(4)	(e) Security To prevent unknowing and unauthorized access to facility.
123.128(e)(5)	(f) Facility personnel training.
123.128(e)(6)	(g) Inspection and monitoring.
	(h) Recordkeeping.
	(i) Reporting.
123.128(e)(7)	(j) Compliance with Manifest System Including requirement that facility owner/operator or State in which facility is located must return a copy of manifest to generator or to State in which the generator is located indicating delivery of waste shipment.
123.128(e)(8)	(k) Other Facility Standards
	(i) To extent they are included in 40 CFR 265.
	(ii) Subpart R of Part 265 (standard for injection wells) may be included in the State standards at the State's option.
123.128(f)	G. Requirements for Enforcement Authority
123.128(f)(1)	(1) Legal Remedies for Violations of State Program Requirements The following such remedies must be available to State agency, under this Subpart, for administering a program:
123.128(f)(1)(i)	(a) Restraining Order/Suit Authority to restrain immediately and effectively any person by order or by suit in State court from engaging in any unauthorized activity which is endangering or causing damage to public health or the environment.
123.128(f)(1)(ii)	(b) To Sue in Courts to Enjoin Any Threatened or Continuing Violation of Any Program Requirement
	(i) Includes permit conditions.
	(ii) Authority without necessity of prior revocation of permit.
123.128(f)(1)(iii)	(c) To Assess or Sue to Recover in Civil Court Penalties and to Seek Criminal Fines
	(i) Civil Penalties $1000 per day.
	(ii) Criminal Fines $1000 per day.
123.128(f)(2)	(d) Public Participation in State Enforcement Process Any State administering a program shall provide for such public participation by providing either:
123.128(f)(2)(i)	(i) Authority which allows intervention in any civil or administrative action to obtain remedies specified in Section 123.128(f)(1) by any citizen having an interest which is or may be adversely affected; or
123.128(f)(2)(ii)	(ii) Assurance that State agency or enforcement authority will:
123.128(f)(2)(ii)(A)	• Investigate and provide written responses to all citizen complaints submitted pursuant to procedures specified in Section 123.128(g)(2)(iv).

Table 8-1C. Requirements for Interim Authorization of State Hazardous Waste Programs—Subpart F. (Continued)

Section No.	Description
123.128(f)(2)(ii)(B)	• Not oppose intervention by any citizen where permissive intervention may be authorized by statute, rule, or regulation.
123.128(f)(2)(ii)(C)	• Publish and provide at least 30 days for public comment on any proposed settlement of a State enforcement action.
123.128(g)	H. Requirements for Compliance Evaluation Programs
123.128(g)(1)	(1) Procedures for Receipt, Evaluation, and Investigation of All Notices and Reports State programs shall have such procedures for possible enforcement action.
123.128(g)(2)	(2) Inspection and Surveillance Authority and Procedures (a) State programs shall have such authority and procedures to determine compliance or noncompliance with applicable program requirements. (b) Content of State Program
123.128(g)(2)(i)	(i) Comprehensive Survey of Any Activities Subject to State Director's Authority • Program shall be capable of making such surveys. • Purpose To identify persons subject to regulation who have failed to comply with program requirements.
123.128(g)(2)(ii)	(ii) Program for periodic inspections of activities subject to regulations.
123.128(g)(2)(iii)	(iii) Capability to investigate evidence of violations of applicable program and permit requirements.
123.128(g)(2)(iv)	(iv) Procedures for Receiving and Ensuring Proper Consideration of Information Submitted by Public about Violations • Public effort in reporting violations shall be encouraged. • State Director shall make available information on reporting procedures.
123.128(g)(3)	(3) State Officers Engaged in Compliance Evaluation Activities (a) Shall have authority to: (i) Enter any conveyance, vehicle, facility or premises subject to regulation; and (ii) Enter where records relevant to program are kept. (b) Purpose of entry authority is to inspect, monitor, or otherwise investigate compliance with State program. (c) States Whose Law Requires a Search Warrant before Entry Shall conform with this requirement.
123.128(g)(4)	(4) Investigatory Inspections Such inspections shall be conducted, samples taken and other information gathered in a manner (e.g., using proper "chain of custody" procedures) that will produce evidence admissible in an enforcement proceeding or in court.
123.129	IX. Additional Program Requirements for Interim Authorization for Phase II A. Requirements Applicable to States Applying for Interim Authorization for Phase II (1) All requirements of Section 123.128 (Program Requirements for Interim Authorization for Phase I) apply.
123.129(a)	(2) HWM Facility Standards under 40 CFR (264 and 265) State programs must have standards applicable to HWM facilities that provide substantially the same degree of human health and environmental protection as standards promulgated under: (a) 40 CFR 264—Standards for Owners and Operators of Hazardous Waste Treatment, Storage, and Disposal Facilities. (b) 40 CFR 266—Waste Specific Standards.
123.129(b)	(3) Permits (a) State programs shall require a permit for owners and operators of those hazardous waste treatment, storage, and disposal facilities which handle wastes: (i) Controlled by the State under Section 123.128(a) (Program Requirements for Interim Authorization for Phase I); and (ii) For which a permit is required under 40 CFR 122. (b) Facility Owners with Interim Status (i) States may authorize owners/operators of facilities which would qualify for interim status under the Federal program (if State law authorizes) to remain in operation pending permit action. (ii) Where State law authorizes such continued operation it shall require compliance by owners/operators of such facilities with standards substantially equivalent to EPA's interim status standards under 40 CFR 265.
123.129(c)	(c) All Permits Issued by State under This Section Shall require compliance with the standards adopted by the State in accordance with Section 123.129(a).

Table 8-1C. Requirements for Interim Authorization of State Hazardous Waste Programs—Subpart F. (Continued)

Section No.	Description
123.129(d)	(d) State programs shall have requirements for permitting which are substantially equivalent to the provisions listed in Section 123.7(a) and (b) (Requirements for Permitting).
123.129(e)	(e) Duration of Permit No permit may be issued by a State with interim authorization for Phase II with a term greater than 10 years.
123.130	X. Interstate Movement of Hazardous Waste
123.130(a)	A. Waste Transported from State Where It Is Listed or Designated as Hazardous into State with Interim Authorization Where It Is Not Listed or Designated (1) The waste must be manifested in accordance with the laws of the State where the waste was generated; and (2) Wastes must be treated, stored, or disposed of as required by laws of State into which it has been transported.
123.130(b)	B. Waste Transported from State with Interim Authorization Where It Is Not Listed or Designated as Hazardous into State Where It Is Listed or Designated as Hazardous Waste must be treated, stored, or disposed of in accordance with law applicable in the State into which it has been transported.
123.130(c)	C. DOT Requirements In all cases of interstate movement of hazardous waste, as defined by 40 CFR 261, generators and transporters must meet DOT requirements in 49 CFR (172, 173, 178 and 179) (e.g., for shipping paper, packaging, labeling, marking, and placarding). (1) See Appendix E of this book for DOT regulation contents only.
123.131	XI. Progress Reports A. Semi-Annual Progress Reports from State Director to EPA Regional Administrator (1) Timing (a) Report shall be submitted within 6 months after Phase I commences, plus a 4 week grace period. (b) Continue at 6-month intervals until expiration of interim authorization. (2) Content The report, in manner and form prescribed by Regional Administrator, shall: (a) Briefly summarize the State's compliance in meeting requirements of authorization plan. (b) Provide reasons and proposed remedies for any delay in meeting milestones. (c) Provide anticipated problems and solutions for next reporting period.
123.132	XII. Sharing of Information
123.132(a)	A. Information Obtained or Used in Administration of State Program (1) Shall be available to EPA upon request without restriction. (2) If Information Has Been Submitted to State Under Claim of Confidentiality State must submit that claim to EPA when providing information under this Subpart. (3) Information Obtained from State and Subject to Claim of Confidentiality Will be treated in accordance with regulations in 40 CFR 2 (Public Information). (4) If EPA Obtains Non-Confidential Information from a State EPA may make that information available to the public without further notice.
123.1(b)	B. Information EPA Will Furnish to States with Approved Programs (1) Non-Confidential That information which State needs to implement its approved program. (2) Confidential That information which State needs to implement its approved program, subject to conditions of 40 CFR 2.
123.133	XIII. Coordination with Other Programs
123.133(a)	A. Coordination of Permit Issuance Issuance of State permits under this Part may be coordinated, as provided in Part 124, with issuance of NPDES, 404, and UIC permits whether they are controlled by the State, EPA, or the Corps of Engineers.

Table 8-1C. Requirements for Interim Authorization of State Hazardous Waste Programs—Subpart F. (Continued)

Section No.	Description
123.133(b)	B. State Director of Any Approved Program Which May Affect Planning and Development of HWM Facilities and Practices He shall consult and coordinate with agencies designated under Section 4006(b) of RCRA (40 CFR 255) as being responsible for development and implementation of State solid waste management plans.
123.134	XIV. EPA Review of State Permits
123.134(a)	A. Regional Administrator May Comment on Permit Applications and Draft Permits As provided in Memorandum of Agreement under Section 123.126 (Memorandum of Agreement with Regional Administrator).
123.134(b)	B. Where EPA Indicates in a Comment that Issuance of the Permit Would Be Inconsistent with the Approved State Program EPA shall include in the comment:
123.134(b)(1)	(1) Statement of Reasons for Comment Includes Section of RCRA or regulations that support comment.
123.134(b)(2)	(2) Actions That Should Be Taken by State Director in Order to Address Comments Include conditions which permit would include if it were issued by Regional Administrator.
123.134(c)	C. Copy of Any Comment Shall be sent to permit applicant by Regional Administrator.
123.134(d)	D. When Regional Administrator Shall Withdraw Such a Comment When satisfied that State has met or refuted his concerns.

123.134(e)	E. Section 3008(a)(3) of RCRA
	(1) Under this statute:
	    (a) EPA may terminate a State-issued permit in accordance with procedures of Part 124, Subpart E (Evidentiary Hearing); or
	    (b) EPA may bring enforcement action in accordance with procedures of 40 CFR 22 (Consolidated Rules of Practice) in the case of a violation of a State program requirement.
	(2) Grounds by Which Regional Administrator May Take Action against Holder of State-Issued Permit under Section 3008(a)(3) of RCRA

123.134(e)(1)	(a) Permittee is not complying with condition of that permit.
123.134(e)(2)	(b) Permittee is not complying with condition that Regional Administrator stated was necessary to implement approved State program requirements.
123.134(e)(3)	(c) Permittee is not complying with a condition necessary to implement approved State program requirements.
123.134(e)(4)	F. Action under Section 7003 of RCRA Regional Administrator may take action under this statute against a permit holder at any time whether or not the permit holder is complying with permit conditions.
123.135	XV. Approval Process
123.135(a)	A. Action Regional Administrator Shall Take upon Receipt of a Complete Program Submission for Interim Authorization

123.135(a)(1)	(1) Within 30 days he shall issue public notice on State's application for interim authorization.
	    (a) In *Federal Register*.
	    (b) Notice of public hearing in accordance with Section 123.39(a)(1) (Approval Process).
	(2) Public Hearing by EPA
	    (a) Such hearing will be held no earlier than 30 days after notice of hearing.
	    (b) There must be significant public interest in a hearing, or it may be cancelled, if a statement to that effect is included in the public notice.
	    (c) The State shall participate in any public hearing held by EPA.

123.135(a)(2)	(3) Comment Period Public shall be afforded 30 days after the notice to comment on State's submission.

123.135(a)(3)	(4) Availability of State's Submission for Inspection and Copying by Public
	    (a) Regional Administrator shall so note.
	    (b) The State submission shall, at a minimum, be available in:
	        (i) Main office of lead State agency; and
	        (ii) EPA Regional Office.

Table 8-1C. Requirements for Interim Authorization of State Hazardous Waste Programs—Subpart F. (Continued)

Section No.	Description
123.135(b)	B. Final Determination by Administrator Whether or Not to Approve State's Program (1) Shall be done within 90 days of notice in *Federal Register* (see Section 123.135(a)(1)). (2) Administrator shall take into account any comments received. (3) Public Notice of Final Determination (a) He will give such notice in the *Federal Register* and in accordance with Section 123.39 (a)(1) (Approval Process). (b) The notification shall include a concise statement of reasons for this determination (c) It shall also include a response to significant comments received.
123.135(c)	C. Where a State Has Received Interim Authorization for Phase I The same procedures required in Sections 123.135(a) and (b) shall be used in determining whether this amended program submission meets the requirements of the Federal program.
123.136	XVI. Withdrawal of State Programs
123.136(a)	A. Criteria and Procedures for Withdrawal of State Program (1) Same criteria and procedures that follow apply to this Section (123.136): (a) Section 123.14—Criteria for Withdrawal of State Programs. (b) Section 123.15—Procedures for Withdrawal of State Programs.
123.136(6)	(2) Additional Criteria If a State which has obtained interim authorization fails to meet the schedule for or accomplish the additions or revisions of its program set forth in its authorization plan.
123.137	XVII. Reversion of State Programs
123.137(a)	A. Termination of State Program Approved for Interim Authorization for Phase I It shall terminate: (1) On last day of 6th month after effective date of Phase II. (2) If State has failed to submit by that date an amended submission pursuant to Section 123.122(c)(4) (Schedule).
123.137(b)	(3) If Regional Administrator determines pursuant to Section 123.135(c) (Approval Process) that a program submission amended pursuant to Section 123.122(c)(4) does not meet the requirements of the Federal program.
123.137(a) and (b)	B. Upon Termination of State Program EPA shall administer and enforce the Federal program in the State.

9
NOTIFICATION OF HAZARDOUS WASTE ACTIVITY: RCRA SECTION 3010

The *Federal Register* public notice of February 26, 1980 establishes the requirements for notification of EPA of hazardous waste activity. This notice provides procedures and a form which should be used when filing a notification of hazardous waste activity.

Any person generating or transporting hazardous waste, or owners/operators of facilities for treatment, storage, or disposal of hazardous waste must notify EPA of such activity within 90 days of promulgation or revision of RCRA Section 3001 (40 CFR 261) regulations. Federal agencies must also comply with notification requirements.

Unless notification has been given, no identified or listed hazardous waste under 40 CFR 261 may be transported, treated, stored or disposed of. Further, owners/ operators of existing treatment, storage, or disposal facilities who fail to notify EPA of their activities become ineligible for "interim status"—that is, the temporary authority to continue their operation until a final permit is issued.

Table 9-1 provides detailed notification requirements and contains the following six areas:

    I. Applicability
   II. Information Required for Filing
  III. Number of Forms to be Filed
  IV. When to File Notification
   V. Where to File
  VI. Claims of Confidentiality.

Appropriate *Federal Register* page numbers ("*FR* Page No.") of February 26, 1980 are referenced in Table 9-1.

Figure 9-1 is a copy of EPA's Form 8700-12, including instructions, which is to be used by generators, transporters and owners/operators of treatment, storage, or disposal facilities to inform EPA of their activities.

Table 9-1. Notification of Hazardous Waste Activity

FR Page No.	Description
	I. Applicability
12747 and 12752	A. General This notice establishes detailed requirements for notifying EPA of hazardous waste activities for following: (1) Generators. (2) Transporters. (3) Owners/Operators of treatment, storage or disposal facilities.
12752	B. Specific Entities Covered (1) Individuals. (2) Trusts. (3) Firms. (4) Joint stock companies. (5) Corporations (including government corporations). (6) Partnerships. (7) Associations. (8) States. (9) Municipalities. (10) Commissions. (11) Interstate bodies. (12) Federal agencies.

Table 9-1. Notification of Hazardous Waste Activity (Continued)

FR Page No.	Description
12747	C. Exemptions Under 40 CFR 261 (Identification and Listing of Hazardous Waste), persons generating small quantities of hazardous waste are exempt from notification.
12752	D. Penalties for Engaging in Hazardous Waste Activities without Filing a Notification Person may be subject to civil and criminal penalties.
12746	E. Federal Statutes/Authority This notice is issued under authority of Sections 2002(a) and 3010 of the Solid Waste Disposal Act, as amended by the Resource Conservation Act of 1976 and as amended by the Quiet Communities Act of 1978 ("RCRA" or "The Act"), 42 U.S.C. 6912(a) and 6930.
	II. Information Required for Filing
12747, 12748, and 12752	A. Overall Requirements Section 3010 of RCRA requires all persons handling hazardous wastes to notify EPA and provide following information: (1) General description of activity. (2) Location of activity. (3) Identity of listed and non-listed hazardous wastes handled.
12747	B. Description of Activity Very simply, designate which of following types of hazardous waste activity is being conducted: (1) Generation. (2) Transportation. (3) Treatment, storage or disposal.
12753	C. Location of Activity Provide following information regarding installation: (1) Name. (2) Mailing address. (3) Location.
12750	*Note:* "Installation" refers to: ● Single site where hazardous waste is generated, treated, stored and/or disposed of; or ● Transporter's principle place of business.

12754
12751
	D. Identity of Listed and Non-Listed Hazardous Waste Handled
	(1) Read Part 261 to Identify Solid Waste EPA Defines as Hazardous Waste
	(a) Section 261.31—Listed hazardous wastes from non-specific sources.
	(b) Section 261.32—Listed hazardous wastes from specific sources.
	(c) Section 261.33—Listed commercial chemical product hazardous wastes.
	(d) Section 261.34—Listed infectious wastes from hospitals, veterinary hospitals, medical and research laboratories (regulations not yet promulgated).
	(e) Non-listed hazardous wastes:
	   (i) Section 261.20—Ignitable.
	   (ii) Section 261.21—Corrosive.
	   (iii) Section 261.22—Reactive.
	   (iv) Section 261.23—Toxic.

12754	(2) Identify Each Listed Hazardous Waste Handled Use appropriate four-digit number assigned by EPA.
12748 and 12754	(3) Indicate If Any Non-Listed Wastes Meet Standard for Ignitability, Corrosivity, Reactivity, or Toxicity (a) Notifiers are not required to identify every non-listed waste that is ignitable, corrosive, reactive, or toxic—only if single waste handled exhibits such characteristics. (b) However, six months after promulgation of Part 261 regulations, generators will be required to make such a determination for each waste (per Part 262, Standards Applicable to Generators of Hazardous Waste).

12748 and 12754	(4) Time Period as Consideration for Selecting Which Wastes to Report
	(a) For Existing Waste Handlers
	   (i) Report all wastes handled during 3-month period prior to date of filing notification.
	   (ii) Notifiers may also include other wastes which they anticipate will be handled.

Table 9-1. Notification of Hazardous Waste Activity (Continued)

FR Page No.	Description
	(iii) Notifiers may include waste occasionally handled, but not handled during previous three month period.
	(b) For New Generators
	Describe wastes expected to be generated.
	(c) For New Transporters
	Description of wastes to be handled not required.
12747 and 12752	E. Underground Injection Wells
	Per RCRA, generators, owners/operators of facilities for treating, storing, or disposing of hazardous waste who operate Class IV underground injection wells must notify EPA of such wells.
12748	F. Use of EPA Form 8700-12 for Notification
	(1) Not Mandatory
	While EPA prefers use of their form for standardized notification information and orderly data management, use of Form 8700-12 (see Fig. 9-1) is not mandatory.
	(2) Computerized Printouts Acceptable
	If printouts are used notifiers must provide:
	(a) Appropriate EPA hazardous waste number for each listed waste.
	(b) The general characteristic of the waste for non-listed hazardous waste.
	(c) All required information including certification signature.
	III. Number of Forms to be Filed
12752	A. For Single Site or Location
	Only one notification form per site or location required, but all activities must be described.
	B. For More than One Site or Location
	Separate notification form required for each site or location where hazardous waste is conducted.
	C. For Hazardous Waste Transporters
	Single form covering all transportation activities conducted by company required, provided transporter does not generate, treat, store, or dispose of hazardous waste.
	IV. When to File Notification
	A. File Notification Within 90- Days of Publication of Regulations Under Section 3001 of RCRA (40 CFR 261)
	(1) For anyone conducting hazardous waste activity.
	(2) For Existing Owners/Operators of Facilities That Treat, Store, or Dispose of Hazardous Waste
	(a) Notification is necessary to qualify for "interim status"—that is, the temporary authority to continue their operations until a final permit is issued.
	(b) Failure to notify within the 90-day period will cause prohibition of activities until a permit is received.
	B. File Notification within 90 Days of Any Amendments to Section 3001 Regulations (40 CFR 261)
	If EPA amends procedures for identifying hazardous waste, or revises list of hazardous wastes, a notification must be filed if these wastes are handled.
12747	*Note:* If after proper notification of hazardous waste activity a person later begins to handle additional hazardous wastes not included in the original notification, it is not necessary to file a new certification unless these wastes have been added by EPA amendment.
12752	C. New Generators
	If generation of hazardous waste has begun without having filed a notification, then it is necessary to obtain an EPA Identification Number (per 40 CFR 262) before transporting or offering for transport any wastes.
	D. New Transporters
	If it is desired to transport hazardous waste without having filed a notification, then it is necessary to obtain an EPA Identification Number (per 40 CFR 263) before moving any wastes.

Table 9-1. Notification of Hazardous Waste Activity (Continued)

FR Page No.	Description
	E. Exceptions
	(1) New Hazardous Waste Treatment, Storage, or Disposal Facility
	A hazardous waste permit must be obtained before commencing operations, and the permit application fulfills notification requirements.
	V. Where to File
	A. Notification should be sent to EPA Regional Office for area where hazardous waste activity is located.
	VI. Claims of Confidentiality
	A. General
12753	(1) Freedom of Information Act (FIA)
	All information submitted in a notification can be disclosed via FIA.
12748 and 12753	(2) Confidentiality Claim Option
	Although EPA believes notification information to be very general, a confidentiality claim per 40 CFR 2 (Subpart B) can be made.
12753	(3) "Confidential" Markings on Notification Form
	This marking is required on both sides of form and on any attachments.
12748 and 12753	(4) Failure to Include Substantiation for Claim of Confidentiality
	If such a failure exists at time of submittal of notification form, the claim is waived and the information is available to public.
	B. Questions to Be Answered in Seeking Confidentiality
	Written answers to each of the following questions is required at time of notification:
12753	(1) Which portions of the information do you claim are entitled to confidential treatment?
	(2) How long do you want this information treated confidential?
	(3) What measures have you taken to guard against undesired disclosure of the information to others?
	(4) To what extent has the information been disclosed to others, and what precautions have you taken in connection with that disclosure?
	(5) Has EPA or any other Federal agency made a pertinent confidentiality determination? If so, include a copy of this determination or reference to it, if available.
	(6) Will disclosure of the information be likely to substantially harm your competitive position?
	(a) If so, what would the harm be?
	(b) Why should it be viewed as substantial?
	(c) What is the relationship between disclosure and the harm?

GSA No. 12345-XX
Form Approved OMB No. 158-R00XX

Please print or type with ELITE type (12 characters/inch) in the shaded area only.

⊕EPA
U.S. ENVIRONMENTAL PROTECTION AGENCY
NOTIFICATION OF HAZARDOUS WASTE ACTIVITY

INSTALLATION'S EPA I.D. NO.

I. NAME OF INSTALLATION

II. INSTALLATION MAILING ADDRESS

III. LOCATION OF INSTALLATION

PLEASE PLACE LABEL IN THIS SPACE

INSTRUCTIONS: If you received a preprinted label, affix it in the space at left. If any of the information on the label is incorrect, draw a line through it and supply the correct information in the appropriate section below. If the label is complete and correct, leave Items I, II, and III below blank. If you did not receive a preprinted label, complete all items. "Installation" means a single site where hazardous waste is generated, treated, stored and/or disposed of, or a transporter's principal place of business. Please refer to the INSTRUCTIONS FOR FILING NOTIFICATION before completing this form. The information requested herein is required by law (Section 3010 of the Resource Conservation and Recovery Act).

FOR OFFICIAL USE ONLY

COMMENTS

C

INSTALLATION I.D. NUMBER | APPROVED | DATE RECEIVED (mo., da., & yr.)

F

I. NAME OF INSTALLATION

II. INSTALLATION MAILING ADDRESS
STREET OR P.O. BOX

3

CITY OR TOWN | ST. | ZIP CODE

4

III. LOCATION OF INSTALLATION
STREET OR ROUTE NUMBER

5

CITY OR TOWN | ST. | ZIP CODE

6

IV. INSTALLATION CONTACT
NAME AND TITLE (last, first, & job title) | PHONE NO. (area code & no.)

2

V. OWNERSHIP
A. NAME OF INSTALLATION'S LEGAL OWNER

7

B. TYPE OF OWNERSHIP (enter the appropriate letter into box)

F = FEDERAL   M = NON-FEDERAL

VI. TYPE OF HAZARDOUS WASTE ACTIVITY
57 ☐ A. GENERATION   58 ☐ B. TRANSPORTATION (complete item VII)
59 ☐ C. TREAT/STORE/DISPOSE 60 ☐ D. UNDERGROUND INJECTION

VII. MODE OF TRANSPORTATION (transporters only)

☐ A. AIR   ☐ B. RAIL   ☐ C. HIGHWAY   ☐ D. WATER   ☐ E. OTHER (specify)
61   62   63   64   65

VIII. FIRST OR SUBSEQUENT NOTIFICATION

Mark 'X' in the appropriate box to indicate whether this is your installation's first notification of hazardous waste activity or a subsequent notification. If this is not your first notification, enter your installation's EPA I.D. number in the space provided below.

☐ A. FIRST NOTIFICATION   ☐ B. SUBSEQUENT NOTIFICATION (complete item C)

C. INSTALLATION'S EPA I.D. NO.

IX. DESCRIPTION OF HAZARDOUS WASTES

Please go to the reverse of this form and provide the requested information.

EPA Form 8700-12 (2-80)   CONTINUE ON REVERSE

Fig. 9-1. EPA Form 8700-12, Notification of Hazardous Waste Activity.

I.D. NO. — FOR OFFICIAL USE ONLY

IX. DESCRIPTION OF HAZARDOUS WASTES (continued from front)

A. HAZARDOUS WASTES FROM NON-SPECIFIC SOURCES. Enter the four-digit number from 40 CFR Part 261.31 for each listed hazardous waste from non-specific sources your installation handles. Use additional sheets if necessary.

1	2	3	4	5	6
7	8	9	10	11	12

B. HAZARDOUS WASTES FROM SPECIFIC SOURCES. Enter the four-digit number from 40 CFR Part 261.32 for each listed hazardous waste from specific industrial sources your installation handles. Use additional sheets if necessary.

13	14	15	16	17	18
19	20	21	22	23	24
25	26	27	28	29	30

C. COMMERCIAL CHEMICAL PRODUCT HAZARDOUS WASTES. Enter the four-digit number from 40 CFR Part 261.33 for each chemical substance your installation handles which may be a hazardous waste. Use additional sheets if necessary.

31	32	33	34	35	36
37	38	39	40	41	42
43	44	45	46	47	48

D. LISTED INFECTIOUS WASTES. Enter the four-digit number from 40 CFR Part 261.34 for each listed hazardous waste from hospitals, veterinary hospitals, medical and research laboratories your installation handles. Use additional sheets if necessary.

49	50	51	52	53	54

E. CHARACTERISTICS OF NON-LISTED HAZARDOUS WASTES. Mark 'X' in the boxes corresponding to the characteristics of non-listed hazardous wastes your installation handles. (See 40 CFR Parts 261.20 – 261.23.)

☐ 1. IGNITABLE   ☐ 2. CORROSIVE   ☐ 3. REACTIVE   ☐ 4. TOXIC

X. CERTIFICATION

I certify under penalty of law that I have personally examined and am familiar with the information submitted in this and all attached documents, and that based on my inquiry of those individuals immediately responsible for obtaining the information, I believe that the submitted information is true, accurate, and complete. I am aware that there are significant penalties for submitting false information, including the possibility of fine and imprisonment.

SIGNATURE	OFFICIAL TITLE	DATE SIGNED

EPA Form 8700-12 (2-80) REVERSE

BILLING CODE 6560-01-C

Fig. 9-1. EPA Form 8700-12, Notification of Hazardous Waste Activity. (Continued)

Line-by-Line Instructions—EPA Form 8700-12

Type or print in ink all items except X(A), SIGNATURE, leaving a blank box between words. If you must use additional sheets, indicate clearly the number of the item on the form to which the information on the separate sheet applies.

Items I through III

Name, Mailing Address, and Location of Installation: If you received a preprinted label from EPA, attach it in the space provided and leave items I, II, and III blank. If there is an error or omission on the label, cross out the incorrect information and fill in the appropriate item(s). If you did not receive a preprinted label, complete items I, II, and III.

Example:

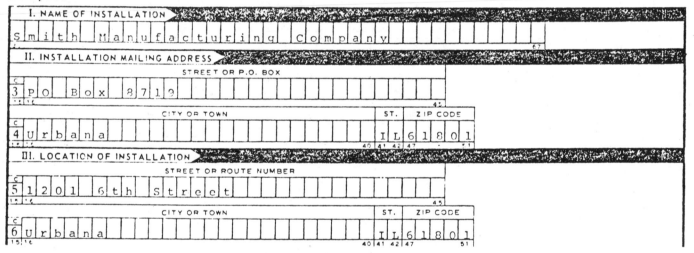

Item IV.

Installation Contact: Enter the name, title, and business telephone number of the person who should be contacted regarding information submitted on this form.

Item V.

Ownership: (A) Enter the name of the legal owner of the installation. Use additional sheets if necessary to list more than one owner.

(B) Enter an F in the box if the installation is owned by a Federal Agency. Enter an M if the installation is not owned by a Federal Agency. An

installation is Federally owned if the owner is the Federal Government, even if it is operated by a private contractor.

Item VI.

Type of Hazardous Waste Activity: Mark "X" in the appropriate box(es) to indicate the hazardous waste activity or activities at the installation. If you mark item C, you are reminded that you should mail the enclosed post card to request a RCRA Permit Application. Generators, owners and operators of facilities for treating, storing, or disposing of hazardous waste must mark item D if an injection well is located at their installation. An injection well is

defined as any hole in the ground that is deeper than it is wide and that is used for the subsurface placement of fluid, *including septic tanks.*

Item VII.

Mode of Transportation: Complete this item only if you are a transporter of hazardous waste to indicate the mode(s) of transportation you use.

Item VIII

First or Subsequent Notification: If you handle any hazardous waste that is identified in an amendment to Part 261 you will have to file a notification on that waste within 90 days after the

Fig. 9-1. EPA Form 8700-12, Notification of Hazardous Waste Activity. (Continued)

amendment is published. Place an "X" in the appropriate box to indicate whether this is your first or a subsequent notification. If you have filed a previous notification, enter your EPA Identification Number in the boxes provided.

Note.—If you have filed a notification before, you only need enter the four-digit numbers of those wastes that were identified in the amendment to Part 261.

Item IX

Description of Hazardous Waste: You need to read Title 40, Code of Federal Regulations Part 261 in order to complete this item. Part 261 identifies those solid wastes that EPA defines to be hazardous wastes. Part 261 identifies hazardous wastes in two ways:

(1) A number of hazardous wastes are listed by name in various tables and appendices. EPA has assigned a four-digit number to each waste that is listed to make it easier to identify the wastes.

(2) Part 261 also lists the general characteristics of hazardous wastes. EPA has also assigned a four-digit number to these characteristics.

As you will note, Item IX on the form is divided into five sections. You should use Sections A through D to identify any listed hazardous wastes which you handle; use Section E to identify those characteristics of the non-listed hazardous wastes which you handle.

You should include in Sections A through E all hazardous wastes you handled during the three-month period preceding the date of notification. If you occasionally handle a hazardous waste but did not handle that waste during the three-month period preceding the date of notification, you may also include that waste (or wastes) in Section A through E.

If you are a new generator applying for an EPA Identification Number under the provisions of 40 CFR Part 262, you should describe the wastes which you believe you will be generating.

If you are a new transporter applying for an EPA Identification Number under the provisions of 40 CFR Part 263, you are not required to complete Item IX.

The specific instructions for Sections A through E are:

Section A

If you handle hazardous wastes from the non-specific sources listed in Part 261.31, enter the appropriate four-digit numbers in the boxes provided.

Section B

If you handle hazardous wastes from the specific industrial sources listed in Part 261.32, enter the appropriate four-digit numbers in the boxes provided.

Section C

If you handle any of the commercial chemical products or manufacturing intermediate or material listed in Part 261.33 as wastes, enter the appropriate four-digit numbers in the boxes provided. Manufacturers may include the products or raw materials that can be reasonably anticipated to require treatment, storage, or disposal as wastes from time to time even though you may not have handled them in the past three months.

Section D

If you handle any of the hazardous wastes from hospitals, veterinary hospitals, or medical and research laboratories listed in Part 261.34, enter the appropriate four-digit numbers in the boxes provided.

Section E

If you handle hazardous wastes which are not listed in Subpart D of Part 261, you should describe these wastes by the characteristics in Subpart C of Part 261. For purposes of notification, it is not necessary to use the four-digit numbers for each characteristic. Rather, you should place an "X" in the box next to the characteristic of those non-listed wastes which you handle.

Item X

Certification: This certification must be signed by the owner or operator or an authorized representative of your installation. An "authorized representative" is a person responsible for the overall operation of the facility— for example—a plant manager or superintendent, or a person of equivalent responsibilty.

[FR Doc. 80-5950 Filed 2-25-80; 8:45 am]
BILLING CODE 6560-01-M

Fig. 9-1. EPA Form 8700-12, Notification of Hazardous Waste Activity. (Continued)

10
SCENARIOS

Five scenarios have been postulated that affect the regulated community to varying degrees. These scenarios follow:

SCENARIO 1: SMALL GENERATOR

- Generates about 700 kilograms per month of hazardous waste, and accumulates less than 1000 kilograms of such waste at any time.

Notes: (1) Although essentially excluded from these regulations, the generator must still dispose of his hazardous wastes at any approved facility.

(2) Since it is only a matter of time before this small generator comes under the regulatory umbrella, it would be useful for him to get to understand the requirements and impact of these regulations.

SCENARIO 2: MODERATE-SIZED PARTS MANUFACTURER

- In single plant.
- Generates over 1000 kilograms per month of hazardous wastes.
- Some wastes are stored in drums for less than 90 days, for shipment by outside trucker to disposal or treatment facilities.

SCENARIO 3: LARGE EQUIPMENT MANUFACTURER

- Multiple plants and sites in several States.
- Generates well over 1000 kilograms per month of hazardous wastes at many but not all sites.
- Some hazardous wastes are moved between sites in company-owned tank trucks and flat-bed trucks (for waste drums), for treatment and/or storage at company facilities.
- No wastes are stored for more than 90 days. Some wastes are temporarily stored in tanks.
- Other wastes are given to outside truckers for delivery to disposal and/or treatment facilities.
- Several unlined surface impoundments comprised of clay are used for on-site disposal of hazardous waste.

SCENARIO 4: FARMER/RANCHER

- Several abandoned ICBM underground steel and concrete silos (approximately 30 feet in diameter and 300 feet deep) are on property.
- These underground injection wells are used strictly for immediate hazardous waste disposal (e.g., no storage, just dumping).

SCENARIO 5: LARGE CHEMICAL MANUFACTURER

- Multiple plants and sites in several states.
- Generates well over 1000 kilograms per month of hazardous wastes at most but not all sites.
- Hazardous wastes are moved between sites in company-owned tank trucks and flat-bed trucks (for waste drums) for treatment, storage and disposal.
- There are some hazardous wastes stored for periods greater than 90 days.
- Other wastes are given to outside truckers for delivery to disposal and/or treatment facilities.
- Treatment facilities include several incinerators. Hazardous wastes from other companies are handled, but no international shipments.
- Microwave treatment and destruction is used for some organic hazardous wastes.
- Some wastes are treated in totally enclosed facilities.
- Storage tanks as well as drums are used for temporary storage of hazardous wastes.
- Numerous surface impoundments (lined and unlined) are used for disposal of hazardous wastes for company use only.
- Some wastes are recycled, some with and some without storage and transportation.

The detailed requirements of Parts 260–265 and Part 122, as applicable to each scenario, are listed in Table 10-1, giving respective table number and description number. An R indicates that the listed item for a particular scenario is, or may be, a regulatory requirement, while an I indicates that the item is of interest but is not necessarily a compliance requirement.

Table 10-1. Scenarios.

Table Description No.	Scenario ①	②	③	④	⑤
Identification and List (261)					
Table 3-1					
I. A–C		I	I	I	I
II. A–C		I	I	I	I
II. D		I	I		I
II. E	I	R	R	R	R
III. A, B		I	I		I
IV. A–D	I	I	I		I
IV. E	R	I	I/R		I/R
IV. F		I	I		I
V. A, B					R
VI. A		I	I		I
VII. A–C		I	I	I	I
VIII. A–E	R	R	R	R	R
IX. A–D	R	R	R	R	R

Table Description No.	Scenario ①	②	③	④	⑤
Generators (262)					
Table 4-1					
I. A		I	I		I
I. B		R	I/R		I/R
I. C			I		I
I. F, G		I	I		I
II. A–C		R	R		R
III. A–C		R	R		R
IV. A–D		R	R		R
IV. C (3), D (5)				R	
V. A–D		R	R		R
VI. A		R	R		R
VI. B					R
VII. A–E		R	R		R
VIII. A, B		R	R		R
VIII. C		I	I		I
IX. A			I		I

Table Description No.	Scenario ①	②	③	④	⑤
Notification (3010)					
Table 9-1					
I. A, B		I	I	I	I
I. C	I	I	I		I
I. D, E	I	I	I	I	I
II. A–D		R	R	R	R
II. E			R		
II. F		R	R	R	R
III. A		R		R	
III. B			R		R
IV. A		R	R	R	R
IV. B	I	R	R	R	R
IV. E			I		I
V. A		R	R	R	R
V. B		I	I	I	I
VI. A, B		I	I		I

Table Description No.	Scenario ①	②	③	④	⑤
Transporters (263)					
Table 5-1					
I. A			I		I
I. B			R		R
I. C		I	I		I
I. E			I		I
I. F			R		R
I. G			I		I
II. A			R		R
II. B			I		I
III. A–D			R		R
III. E			I		I
III. G			R		R
IV. A			R		R
IV. D			I		I
V. A, B			R		R

Table Description No.	Scenario ①	②	③	④	⑤
General Provisions (261)					
Table 2-1					
I. A, B		I	I	I	I
II. A, B		I	I	I	I
III. A–D		I	I	I	I
IV. A		I	I	I	I
V. A–E		I	I	I	I
VI. A–E		I	I	I	I
VII. A–N		I	I		I

Table Description No.	Scenario ①	②	③	④	⑤
Facilities (264 and 265)					
Table 6-1					
I. A			I	I	I
I. B			R	R	R
I. C (2)				R	
I. C (4)(a)			I		I
I. C (4)(b)				I	
I. C (5)			I		I
I. C (6)					I

Table 10-1. Scenarios. (Continued)

Table Description No.	①	②	③	④	⑤
I. C (7)		I	I		I
I. C (9)					I
I. D			R	R	R
I. E			I	I	I
II. A			R	R	R
II. B (2)				R	R
II. B (3)			I	I	I
III. A, B			R		R
IV. A–F			R		R
V. A–D			R		R
VI. A–E		R	R		R
VII. A–D			R		R
VIII. A, B		R	R	R	R
VIII. C		R	R		R
VIII. D, E		R	R	R	R
IX. A–D		R	R	R	R
X. A, B		R	R	R	R
XI. A, B (1)			R	R	R
XI. B (2)					I
XI. C			R	R	R
XII. A, B			R	R	R
XIII. A–D			R	R	R
XIV. A–H			R		R
XV. A–F			R		R
XVI. A–H			R		R
XVII. A–C			R		R
XVII. D			I		I
XVIII. A–D			R		R
XVIII. E, F		R	R		R
XVIII. G			R		R
XIX. A–G		I	R		R
XX. A–I			R		R
XXIV. A–E					R
XXV. A–E					R
XXVI. A–G					R
XXVII. A				R	

Permits (122 and 124)					
Table 7–1A					
I. A, B			I	I	I
II. A–C			R	R	R
III. A–D			I	I	I
IV. A–C			R	R	R
V. A, B			R	R	R

Table Description No.	①	②	③	④	⑤
VI. A–C			R	R	R
VII. A (1), (2)			I	I	I
VII. A (3)				I	
VIII. A, B			R	R	R
IX. A			R	R	R
X. A			R	R	R
XI. A			R	R	R
XI. B, C			I	I	I
XII. A			I	I	I
XIII. A–D			I	I	I
XIV. A, B			I	I	I
XV. A, B			I	I	I
XVI. A–D			I	I	I
XVII. A–C			I	I	I

Table 7–1B					
I. A			I	I	I
I. B–C (2)(a)			R	R	R
I. C (2)(b)			R		
I. C (3)(a)		I	I		I
I. C (3)(d)					I
II. A			R	R	R
II. B			I		I
II. C, D			I	I	I
III. A–F			R	R	R
IV. A			R	R	R
V. A			R	R	R
VI. A			I	I	I
VI. C			I/R		
VII. A, B			I	I	I
VIII. A			R	R	R
IX. A			R	R	R
X. A–C			I/R		

Table 7–1C					
I. A–C			R		

Table 7–2A					
I. A–G			I	I	I
II. A–G			I	I	I
III. A–E			I	I	I
IV. A–I			I	I	I
V. A–D			I	I	I
VI. A–C			I	I	I

Table 10-1. Scenarios. (Continued)

Table Description No.	Scenario ①	②	③	④	⑤
VII. A–D			I	I	I
VIII. A–E			I	I	I
IX. A–E			I	I	I
X. A			I	I	I
XI. A–F			I	I	I
XII. A, B			I	I	I
XIII. A–D			I	I	I
XIV. A, B			I	I	I
XV. A–C			I	I	I
XVI. A–C			I	I	I
XVII. A–E			I	I	I
XVIII. A–E			I	I	I
XIX. A–D			I	I	I
XX. A			I	I	I

Table 7-2B					
I. A, B			I	I	I
II. A–D			I	I	I
III. A–F			I	I	I
IV. A–C			I	I	I
V. A, B			I	I	I
VI. A–C			I	I	I
VII. A–C			I	I	I
VIII. A, B			I	I	I
IX. A–D			I	I	I
X. A			I	I	I
XI. A, B			I	I	I
XII. A–E			I	I	I

Table Description No.	Scenario ①	②	③	④	⑤
XIII. A–H			I	I	I
XIV. A–D			I	I	I
XV. A–C			I	I	I
XVI. A–F			I	I	I
XVII. A			I	I	I
XVIII. A, B			I	I	I
XIX. A–F			I	I	I
XX. A–J			I	I	I

Table 7-2C					
I. A			I	I	I
II. A			I	I	I
III. A, B			I	I	I
IV. A–E			I	I	I
V. A, B			I	I	I
VI. A–C			I	I	I
VII. A–D			I	I	I
VIII. A–C			I	I	I
IX. A, B			I	I	I
X. A–F			I	I	I
XI. A–F			I	I	I
XII. A, B			I	I	I
XIII. A–C			I	I	I
XIV. A–C			I	I	I
XV. A			I	I	I
XVI. A–F			I	I	I
XVII. A			I	I	I
XVIII. A, B			I	I	I

APPENDIX A
DEFINITIONS

Part 260 provides definitions of words and phrases which appear in Parts 260-265. These definitions are included in this appendix. Additional definitions covered in Appendix A are those defined in Part 122 and used in that Part as well as Parts 123 and 124.

The Resource Conservation and Recovery Act defines a number of terms, many of which apply to the management of hazardous wastes, and are also included in Appendix A. These definitions have not been modified for purposes of the regulations and are not repeated in Part 260.

Act or RCRA The Solid Waste Disposal Act, as amended by the Resource Conservation and Recovery Act of 1976, as amended, 42 U.S.C. Section 6901 et seq.

Active life of facility That period during which wastes are periodically received.

Active portion That portion of a facility where treatment, storage, or disposal operations are being or have been conducted after the effective date of Part 261 (i.e., November 19, 1980) and which is not a closed portion. (See also *Closed portion* and *Inactive portion.*)

Administrator The Administrator of the Untied States Environmental Protection Agency (EPA), or his designee.

Agency trial staff Those Agency employees, whether temporary or permanent, who have been designated by the Agency under Sections 124.77 or 124.116 as available to investigate, litigate, and present the evidence, arguments, and position of the Agency in the evidentiary hearing or nonadversary panel hearing. Appearance as a witness does not necessarily require a person to be designated as a member of the Agency trial staff.

Application The EPA standard national forms for applying for a permit, including any additions, revisions or modifications to the forms; or forms approved by EPA for use in "approved States," including any approved modifications or revisions. For RCRA, application also includes the information required by the Director under Section 122.25 (contents of Part B of the RCRA application).

Appropriate act and regulations The Clean Water Act (CWA); the Solid Waste Disposal Act, as amended by the Resource Conservation and Recovery Act (RCRA); or the Safe Drinking Water Act (SDWA), whichever is applicable; and applicable regulations promulgated under those statutes. In the case of an approved State program, appropriate act and regulations includes State program requirements.

Approved program or Approved State program A State or interstate program which has been approved or authorized by EPA under Part 123.

Aquifer A geologic formation, group of formations, or part of a formation capable of yielding a significant amount of ground water to wells or springs.

Authorized representative The person responsible for the overall operation of a facility or an operational unit (i.e., part of a facility), e.g., the plant manager, superintendent, or person of equivalent responsibility.

Bulk shipment by water Any movement where the waste is not placed in individual packages or containers but is instead loaded into the vessel or barge in bulk.

Class IV wells Wells used by generators of hazardous wastes or of radioactive wastes, by owners or operators of hazardous waste management facilities, or by owners or operators of radioactive waste disposal sites to dispose of hazardous wastes or radioactive wastes into or above a formation which within one-quarter mile of the well contains an underground source of drinking water.

Closed portion That portion of a facility which an owner or operator has closed in accordance with the approved facility closure plan and all applicable closure requirements. (See also *Active portion* and *Inactive portion.*)

Closure The act of securing a hazardous waste management facility pursuant to the requirements of 40 CFR Part 264.

Confined aquifer An aquifer bounded above and below by impermeable beds or by beds of distinctly lower permeability than that of the aquifer itself; an aquifer containing confined ground water.

Constituent or Hazardous waste constituent A constituent which caused the Administrator to list the hazardous waste in Part 261, Subpart D, or a constituent listed in Table 1 of Section 261.24.

Contingency plan A document setting out an organized, planned, and coordinated course of action to be followed in case of a fire, explosion, or release of hazardous waste or hazardous waste constituents which could threaten human health or the environment.

Container Any portable device in which a material is stored, transported, treated, disposed of, or otherwise handled.

CWA The Clean Water Act (formerly referred to as the Federal Water Pollution Control Act or Federal Water Pollution Control Act Amendments of 1972) Public Law 92-500, as amended by Public Law 95-217 and Public Law 95-576; 33 U.S.C. Section 1251 et seq.

Decisional body Any Agency employee who is or may reasonably be expected to be involved in the decisional process of the proceeding, including the Administrator, Judicial Officer, Presiding Officer, the Regional Administrator (if he or she does not designate himself or herself as a member of the Agency trial staff), and any of their staff participating in the decisional process. In the case of a nonadversary panel hearing, the decisional body shall also include the panel members, whether or not permanently employed by the Agency.

Demonstration The initial exhibition of a new technology process or practice or a significantly new combination or use of technologies, processes, or practices, subsequent to the development stage, for the purpose of proving technological feasibility and cost effectiveness.

Designated facility A hazardous waste treatment, storage or disposal facility which has received an EPA permit or a facility with interim status in accordance with the requirements of 40 CFR 122 and 124 or a permit from a State authorized in accordance with Part 123 that has been designated on the manifest by the generator pursuant to Section 262.20.

Dike An embankment or ridge of either natural or man-made materials used to prevent the movement of liquids, sludges, solids, or other materials.

Director The Regional Administrator or the State Director, as the context requires, or an authorized representative.

- When there is no approved State program, and there is an EPA-administered program, *Director* means the Regional Administrator.
- When there is an approved State program, *Director* normally means the State Director.
- In some circumstances, however, EPA retains the authority to take certain actions even when there is an approved State program. (For example, when EPA has issued an NPDES permit prior to the approval of a State program, EPA may retain jurisdiction over that permit after program approval; see Section 123.71.) In such cases, the term *Director* means the Regional Administrator and not the State Director.

Discharge or Hazardous waste discharge The accidental or intentional spilling, leaking, pumping, pouring, emitting, emptying, or dumping of hazardous waste into or on any land or water.

Disposal The discharge, deposit, injection, dumping, spilling, leaking, or placing of any solid waste or hazardous waste into or on any land or water so that such solid waste or hazardous waste or any constituent thereof may enter the environment or be emitted into the air or discharge into any waters, including ground waters.

Disposal facility A facility or part of a facility at which hazardous waste is intentionally placed into or on any land or water, and at which waste will remain after closure.

Domestic sewage Untreated sanitary wastes that pass through a sewer system.

Draft permit A document prepared under Section 124.6 indicating the Director's tentative decision to issue or deny, modify, revoke and reissue, terminate, or reissue a permit. A notice of intent to terminate a permit and a notice of intent to deny a permit, as discussed in Section 124.5, are types of draft permits. A denial of a request for modification, revocation and reissuance, or termination, as discussed in Section 124.5, is not a draft permit. A proposed permit is not a draft permit.

EPA or The Agency United States Environmental Protection Agency.

EPA hazardous waste number The number assigned by EPA to each hazardous waste listed in Part 261, Subpart D and to each characteristics identified in Part 261, Subpart C.

EPA identification number The unique number assigned by EPA to each generator, transporter and treatment, storage, or disposal facility.

EPA region The states and territories found in any one of the following ten regions:

- Region I. Maine, Vermont, New Hampshire, Massachusetts, Connecticut, and Rhode Island.
- Region II. New York, New Jersey, Commonwealth of Puerto Rico, and the U. S. Virgin Islands.
- Region III. Pennsylvania, Delaware, Maryland, West Virginia, Virginia, and the District of Columbia.
- Region IV. Kentucky, Tennessee, North Carolina, Mississippi, Alabama, Georgia, South Carolina, and Florida.
- Region V. Minnesota, Wisconsin, Illinois, Michigan, Indiana, and Ohio.
- Region VI. New Mexico, Oklahoma, Arkansas, Louisiana, and Texas.
- Region VII. Nebraska, Kansas, Missouri, and Iowa.
- Region VIII. Montana, Wyoming, North Dakota, South Dakota, Utah, and Colorado.
- Region IX. California, Nevada, Arizona, Hawaii, Guam, American Samoa, Commonwealth of the Northern Mariana Islands.
- Region X. Washington, Oregon, Idaho, and Alaska.

Equivalent method Any testing or analytical method approved by the Administrator under Sections 260.20 and 260.21.

Existing hazardous waste management facility (HWM) or Existing facility A facility which was in operation, or for which construction had commenced, on or before October 21, 1976. Construction had commenced if:

(i) The owner or operator has obtained all necessary Federal, State, and local preconstruction approvals or permits; and

(ii) either
    (a) A continuous physical, on-site construction pro-
        gram has begun; or
    (b) The owner or operator has entered into contrac-
        tual obligations—which cannot be cancelled or
        modified without substantial loss—for construc-
        tion of the facility to be completed within a
        reasonable time.

Ex parte communication Any communication, written or oral, relating to the merits of the proceeding between the decisional body and an interested person outside the Agency or the Agency trial staff which was not originally filed or stated in the administrative record or in the hearing. Ex parte communications do not include:

- Communications between Agency employees other than between the Agency trial staff and the members of the decisional body.
- Discussions between the decisional body and either:
  - Interested persons outside the Agency; or
  - The Agency trial staff, if all parties have received prior written notice of the proposed communications and have been given the opportunity to be present and participate therein.

Facility All contiguous land, and structures, other appurtenances, and improvements on the land, used for treating, storing, or disposing of hazardous waste. A facility may consist of several treatment, storage, or disposal operational units (e.g., one or more landfills, surface impoundments, or combinations of them). Also, a facility or activity means any UIC injection well, NPDES point source, or State 404 dredge or fill activity, or any other facility or activity (including land or appurtenances thereto) that is subject to regulation under the RCRA, UIC, NPDES, or 404 programs.

Federal agency Any department, agency, or other instrumentality of the Federal Government, any independent agency or establishment of the Federal Government including any Government corporation, and the Government Printing Office.

Final authorization Approval by EPA of a State program which has met the requirements of Section 3006(b) of RCRA and the applicable requirements of Part 123, Subparts A and B.

Final permit decision A final decision to issue, deny, modify, revoke and reissue, or terminate a permit.

Food-chain crops Tobacco, crops grown for human consumption, and crops grown for feed for animals whose products are consumed by humans.

Formal hearing Any evidentiary hearing under Subpart E or any panel hearing under Subpart F but does not mean a public hearing conducted under Section 124.12.

Freeboard The vertical distance between the top of a tank or surface impoundment dike, and the surface of the waste contained therein.

Free liquids Liquids which readily separate from the solid portion of a waste under ambient temperature and pressure.

Generator Any person, by site location, whose act or process produces hazardous waste identified or listed in 40 CFR 261.

Ground water Water below the land surface in a zone of saturation.

Hazardous waste A waste as defined in 40 CFR 261.3.

Hazardous waste discharge The accidental or intentional spilling, leaking, pumping, pouring, emitting, emptying, or dumping of hazardous waste or a material listed or identified in 40 CFR 261 which, because it is discharged, becomes a hazardous waste, into or onto the land or water. Hazardous waste discharges do not include discharges permitted under the Clean Water Act, Safe Drinking Water Act, or the Marine Protection, Research, and Sanctuaries Act.

Hazardous waste generation The act or process of producing a hazardous waste.

Hazardous waste management or Management The systematic control of the collection, source separation, storage, transportation, processing, treatment, recovery, and disposal of hazardous wastes.

Hazardous waste management facility or HWM facility All contiguous land, and structures, other appurtenances, and improvements on the land, used for treating, storing, or disposing of hazardous waste. A facility may consist of several treatment, storage, or disposal operational units (for example, one or more landfills, surface impoundments, or combinations of them).

Hearing Clerk The Hearing Clerk, U. S. Environmental Protection Agency, 401 M. Street, S. W., Washington, D. C. 20460.

Household waste Any waste material including garbage, trash and sanitary wastes in septic tanks derived from households (including single and multiple residences, hotels, and motels).

Injection well Any hole in the ground that is deeper than it is wide that is used for the subsurface placement of fluid, including septic tanks. (See also *Underground injection.*)

Inactive portion That portion of a facility which is not operated after the effective date of Part 261 of this Chapter. (See also *Active portion* and *Closed portion.*)

Incinerator An enclosed device using controlled flame combustion, the primary purpose of which is to thermally break down hazardous waste. Examples of incinerators are rotary kiln, fluidized bed, and liquid injection incinerators.

Incompatible waste A hazardous waste which is unsuitable for:

    (i) Placement in a particular device or facility because it may cause corrosion or decay of containment materials (e.g., container inner liners or tank walls); or
    (ii) Commingling with another waste or material under uncontrolled conditions because the commingling might produce heat or pressure, fire or explosion, violent reaction, toxic dusts, mists, fumes, or gases, or flammable fumes or gases.

(See Part 265, Appendix V, for examples.)

Individual generation site The contiguous site at or on which one or more hazardous wastes are generated. An individual generation site, such as a large manufacturing plant, may have one or more sources of hazardous waste but is considered a single or individual generation site if the site or property is contiguous.

In operation Refers to a facility which is treating, storing, or disposing of hazardous waste.

Inner liner A continuous layer of material placed inside a tank or container which protects the construction materials of the tank or container from the contained waste or reagents used to treat the waste.

Installation A single site where hazardous waste is generated, treated, stored, and/or disposed of, or a transporter's principal place of business.

Interim authorization Approval by EPA of a State hazardous waste program which has met the requirements of Section 3006(c) of RCRA and applicable requirements of Part 123, Subpart F.

Interim status The temporary authority of owners or operators of existing facilites that treat, store, or dispose of hazardous waste to continue their operations until a final permit is issued.

Intermunicipal agency An agency established by two or more municipalities with responsibility for planning or administration of solid waste.

International shipment The transportation of hazardous waste into or out of the jurisdiction of the United States.

Interested person outside the Agency Includes the permit applicant, any person who filed written comments in the proceeding, any person who requested the hearing, any person who requested to participate or intervene in the hearing, any participant in the hearing and any other interested person not employed by the Agency at the time of the communications, and any attorney of record for those persons.

Interstate agency An agency of two or more States established by or under an agreement or compact approved by the Congress, or any other agency of two or more States having substantial powers or duties pertaining to the control of pollution as determined and approved by the Administrator under the appropriate Act and regulations.

Judicial Officer A permanent or temporary employee of the Agency appointed as a Judicial Officer by the Administrator under these regulations.

Landfill A disposal facility or part of a facility where hazardous waste is placed in or on land and which is not a land treatment facility, a surface impoundment, or an injection well.

Landfill cell A discrete volume of a hazardous waste landfill which uses a liner to provide isolation of wastes from adjacent cells or wastes. Examples of landfill cells are trenches and pits.

Land treatment facility A facility or part of a facility at which hazardous waste is applied onto or incorporated into the soil surface; such facilities are disposal facilities if the waste will remain after closure.

Leachate Any liquid, including any suspended components in the liquid, that has percolated through or drained from hazardous waste.

Liner A continuous layer of natural or man-made materials, beneath or on the sides of a surface impoundment, landfill, or landfill cell, which restricts the downward or lateral escape of hazardous waste, hazardous waste constituents, or leachate.

Major facility Any RCRA, UIC, NPDES, or 404 facility or activity classified as such by the Regional Administrator, or, in the case of approved State programs, the Regional Administrator in conjunction with the State Director.

Manifest The shipping document originated and signed by the generator which contains the information required by Subpart B of 40 CFR 262.

Manifest document number The serially increasing number assigned to the manifest by the generator for recordkeeping and reporting purposes.

Mining overburden returned to the mine site Any material overlying an economic mineral deposit which is removed to gain access to that deposit and is then used for reclamation of a surface mine.

Movement That hazardous waste transported to a facility in an individual vehicle.

Municipality A city, town, borough, county, parish, district, or other public body created by or pursuant to State law, with responsibility for the planning or administration of solid waste management, or an Indian tribe or authorized tribal organization or Alaska Native village or organization. It includes any rural community or unincorporated town or village or any other public entity for which an application for assistance is made by a State or political subdivision thereof.

National Pollutant Discharge Elimination System The national program for issuing, modifying, revoking and reissuing, terminating, monitoring, and enforcing permits, and imposing and enforcing pretreatment requirements, under sections 307, 402, 318, and 405 of CWA. The term includes an approved program.

New hazardous waste management (HWM) facility or New facility A facility which began operation, or for which construction commenced after October 21, 1976. (See also *Existing hazardous waste management facility*.)

NPDES National Pollutant Discharge Elimination System.

Off-site Any site which is not "on-site."

On-site The same or geographically contiguous property which may be divided by public or private right-of-way, provided the entrance and exit between the properties is at a cross-roads intersection, and access is by crossing as opposed to going along, the right-of-way. Non-contiguous properties owned by the same person but connected by a right-of-way which he controls and to which the public does not have access, is also considered on-site property.

Open burning The combustion of any material without the following characteristics:

(i) Control of combustion air to maintain adequate temperature for efficient combustion;
(ii) Containment of the combustion reaction in an enclosed device to provide sufficient residence time and mixing for complete combustion; and
(iii) Control of emission of the gaseous combustion products.

(See also *Incineration* and *Thermal treatment.*)

Open dump A site for the disposal of solid waste which is not a sanitary landfill within the meaning of Section 4004 of RCRA.

Operator The person responsible for the overall operation of a facility and subject to regulations under RCRA.

Owner The person who owns a facility or part of a facility and is subject to regulation under RCRA.

Partial closure The closure of a discrete part of a facility in accordance with the applicable closure requirements of 40 CFR 264 or 265. For example, partial closure may include the closure of a trench, a unit operation, a landfill cell, or a pit, while other parts of the same facility continue in operation or will be placed in operation in the future.

Party The petitioner, State, Agency, and any other person whose request to participate as a party to a withdrawal proceeding is granted.

Permit An authorization, license, or equivalent control document issued by EPA or an approved State to implement the requirements of this Part and Parts 123 and 124. *Permit* includes RCRA permit by rule (Section 122.26), UIC area permit (Section 122.39), NPDES or 404 general permit (Sections 122.59 and 123.95), and RCRA, UIC, or 404 emergency permit (Sections 122.27, 122.40, and 123.96). Permit does not include RCRA interim status (Section 122.23), UIC authorization by rule (Section 122.37), or any permit which has not yet been the subject of final agency action, such as a draft permit or a proposed permit.

Permit by rule A provision of these regulations stating that a facility or activity is deemed to have a RCRA permit if it meets the requirements of the provision.

Person An individual, trust, firm, joint stock company, Federal Agency, corporation (including a government corporation), partnership, association, State, municipality, commission, political subdivision of a State, or any interstate body.

Personnel or Facility personnel All persons who work at, or oversee the operations of, a hazardous waste facility, and whose actions or failure to act may result in noncompliance with the requirements of 40 CFR (264 or 265).

Phase I That phase of the Federal hazardous waste management program commencing on the effective date of the last of the following to be initially promulgated: 40 CFR Parts 122, 123, 260, 261, 262, 263, and 265. Promulgation of Phase I refers to promulgation of the regulations necessary for Phase I to begin.

Phase II That phase of Federal hazardous waste management program commencing on the effective date of the first Subpart of 40 CFR Part 264, Subparts F-R to be initially promulgated. Promulgation of Phase II refers to promulgation of the regulations necessary for Phase II to begin.

Physical construction Excavation, movement of earth, erection of forms or structures, or similar activity to prepare an HWM facility to accept hazardous waste.

Pile Any noncontainerized accumulation of solid, nonflowing hazardous waste that is used for treatment or storage.

Point source Any discernible, confined, and discrete conveyance, including, but not limited to any pipe, ditch, channel, tunnel, conduit, well, discrete fissure, container, rolling stock, concentrated animal feeding operation, or vessel or other floating craft, from which pollutants are or may be discharged. This term does not include return flows from irrigated agriculture.

Post-closure The period after closure during which owners or operators of disposal facilities must conduct monitoring and maintenance activities.

POTW Publicly owned treatment works.

Petitioner Any person whose petition for commencement of withdrawal proceedings has been granted by the Administrator.

Presiding officer An Administrative Law Judge appointed under 5 U.S.C. 3105 and designated to preside at the hearing. Under Subpart F of Part 124 other persons may also serve as hearing officers. See Section 124.119.

Publicly owned treatment works or POTW Any device or system used in the treatment (including recycling and reclamation) of municipal sewage or industrial wastes of a liquid nature which is owned by a State or municipality (as defined by Section 502(4) of the CWA). This definition includes sewers, pipes, or other conveyances only if they convey wastewater to a POTW providing treatment.

RCRA The Solid Waste Disposal Act as amended by the Resource Conservation and Recovery Act of 1976 (Public Law 94-580, as amended by Public Law 95-609, 42 U.S.C. Section 6901 et seq).

Regional Administrator The Regional Administrator of the appropriate Regional Office of the Environmental Protection Agency or the authorized representative of the Regional Administrator.

Regional authority The authority established or designated under Section 4006 of RCRA.

Regional hearing clerk An employee of the Agency designated by a Regional Administrator to establish a repository for all books, records, documents, and other materials relating to hearings under Subpart E of Part 124.

Representative sample A sample of a universe or whole (e.g., waste pile, lagoon, ground water) which can be expected to exhibit the average properties of the universe or whole.

Run-off Any rainwater, leachate, or other liquid that drains over land from any part of a facility.

Run-on Any rainwater, leachate, or other liquid that drains over land onto any part of a facility.

Sanitary landfill A facility for the disposal of solid waste which meets the criteria published under Section 4004 of RCRA.

Saturated zone or Zone of saturation That part of the earth's crust in which all voids are filled with water.

Schedule of compliance A schedule of remedial measures included in a permit, including an enforceable sequence of interim requirements (for example, actions, operations, or milestone events) leading to compliance with the appropriate Act and regulations.

SDWA The Safe Drinking Water Act (Public Law 95-253, as amended by Public Law 95-1900; 42 U.S.C. Section 300f et seq.).

Site The land or water area where any facility or activity is physically located or conducted, including adjacent land used in connection with the facility or activity.

Sludge Any solid, semi-solid, or liquid waste generated from a municipal, commercial, or industrial wastewater treatment plant, water supply treatment plant, or air pollution control facility exclusive of the treated effluent from a wastewater treatment plant.

Solid waste A solid waste as defined in 40 CFR 261.2.

Solid waste management The systematic administration of activities which provide for the collection, source separation, storage, transportation, transfer, processing, treatment, and disposal of solid waste.

Statutory preview period The period of time allotted for formal EPA review of a proposed State program under the appropriate act (e.g., RCRA, Safe Drinking Water Act, etc.)

Solid waste management facility Includes:

1. Any resource recovery system or component thereof.
2. Any system, program, or facility for resource conservation.
3. Any facility for the treatment of solid waste, including hazardous wastes, whether such facility is associated with facilities generating such wastes or otherwise.

State Any of the several States, the District of Columbia, Guam, the Commonwealth of Puerto Rico, the Virgin Islands, American Samoa, and the Commonwealth of the Northern Mariana Islands.

State authority The agency established or designated under Section 4007 of RCRA.

State Director The chief administrative officer of any State or interstate agency operating an approved program, or the delegated representative of the State Director. If responsibility is divided among two or more State or interstate agencies,

State Director means the chief administrative officer of the State or interstate agency authorized to perform the particular procedure or function to which reference is made.

State/EPA agreement An agreement between the Regional Administrator and the State which coordinates EPA and State activities, responsibilities, and programs including those under RCRA, SDWA, and CWA programs.

Storage The holding of hazardous waste for a temporary period at the end of which the hazardous waste is treated, disposed of, or stored elsewhere.

Surface impoundment or Impoundment A facility or part of a facility which is a natural topographic depression, man-made excavation, or diked area formed primarily of earthen materials (although it may be lined with man-made materials), which is designed to hold an accumulation of liquid wastes or wastes containing free liquids, and which is not an injection well. Examples of surface impoundments are holding, storage, settling, and aeration pits, ponds, and lagoons.

Tank A stationary device, designed to contain an accumulation of hazardous waste which is constructed primarily of non-earthen materials (e.g., wood, concrete, steel, plastic) which provide structural support.

Thermal treatment The treatment of hazardous waste in a device which uses elevated temperatures as the primary means to change the chemical, physical or biological character or composition of the hazardous waste. Examples of thermal treatment processes are incineration, molten salt, pyrolysis, calcination, wet air oxidation, and microwave discharge. (See also *Incinerator* and *Open burning*.)

Totally enclosed treatment facility A facility for the treatment of hazardous waste which is directly connected to an industrial production process and which is constructed and operated in a manner which prevents the release of any hazardous waste or any constituent thereof into the environment during treatment. An example is a pipe in which waste acid is neutralized.

Transportation The movement of hazardous waste by air, rail, highway, or water.

Transporter A person engaged in the off-site transportation of hazardous waste by air, rail, highway, or water.

Treatment Any method, technique, or process, including neutralization, designed to change the physical, chemical, or biological character or composition of any hazardous waste so as to neutralize such waste, or so as to recover energy or material resources from the waste, or so as to render such waste non-hazardous, or less hazardous; safer to transport, store, or dispose of; or amenable for recovery, amenable for storage, or reduced in volume.

Triple rinsed Refers to containers which have been flushed three times, each time using a volume of diluent at least equal to 10% of the container's capacity.

Underground injection The subsurface emplacement of fluids through a bored, drilled, or driven well; or through a dug well, where the depth of the dug well is greater than the largest surface dimension. (See also *Injection well*.)

UIC The Underground Injection Control Program under Part C of the Safe Drinking Water Act, including an approved program.

Underground source of drinking water (USDW) An aquifer or its portion:

(a) (1) Which supplies drinking water for human consumption; or
    (2) In which the ground water contains fewer than 10,000 milligrams per liter total dissolved solids; and
(b) Which is not an exempted aquifer.

United States The 50 States, District of Columbia, the Commonwealth of Puerto Rico, the U. S. Virgin Islands, Guam, American Samoa, and the Commonwealth of the North Mariana Islands.

Unsaturated zone or Zone of aeration The zone between the land surface and the water table.

USDW Underground source of drinking water.

Water (bulk shipment) The bulk transportation of hazardous waste which is loaded or carried on board a vessel without containers or labels.

Waters of the United States or Waters of the U. S.

(a) All waters which are currently used, were used in the past, or may be susceptible to use in interstate or foreign commerce, including all waters which are subject to the ebb and flow of the tide;
(b) All interstate waters, including interstate wetlands;
(c) All other waters such as interstate lakes, rivers, streams (including intermittent streams), mudflats, sandflats, wetlands, sloughs, prairie potholes, wet meadows, playa lakes, or natural ponds the use, degradation, or destruction of which would affect or could affect interstate or foreign commerce including any such waters:
    (1) Which are or could be used by interstate or foreign travelers for recreational or other purposes;
    (2) From which fish or shellfish are or could be taken and sold in interstate or foreign commerce; or
    (3) Which are used or could be used for industrial purposes by industries in interstate commerce;
(d) All impoundments of waters otherwise defined as waters of the United States under this definition;
(e) Tributaries of waters identified in paragraphs (a)–(d) of this definition;
(f) The territorial sea; and
(g) Wetlands adjacent to waters (other than waters that are themselves wetlands) identified in paragraphs (a)–(f) of this definition.

Well Any shaft or pit dug or bored into the earth, generally of a cylindrical form, and often walled with bricks or tubing to prevent the earth from caving in.

Well injection See *Underground injection.*

Wetlands Those areas that are inundated or saturated by surface or ground water at a frequency and duration sufficient to support, and that under normal circumstances do support, a prevalence of vegetation typically adapted for life in saturated soil conditions. Wetlands generally include swamps, marshes, bogs, and similar areas.

Zone of incorporation Three times the depth to which the waste was tilled into the soil.

APPENDIX B
RESOURCE CONSERVATION AND RECOVERY ACT (RCRA)

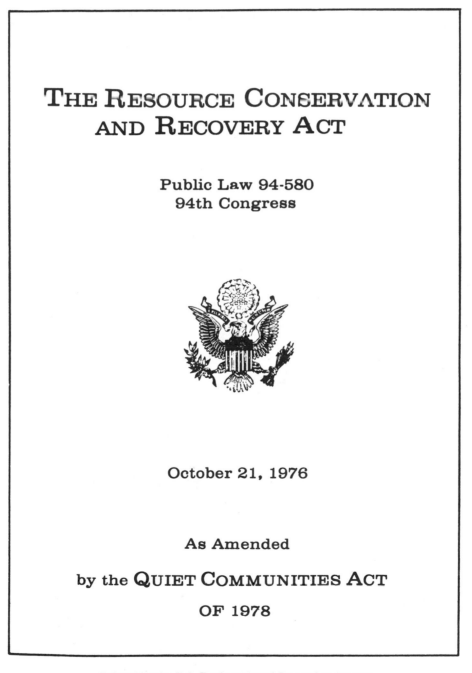

THE RESOURCE CONSERVATION
AND RECOVERY ACT

Public Law 94-580
94th Congress

October 21, 1976

As Amended

by the QUIET COMMUNITIES ACT

OF 1978

Printed by the U.S. Environmental Protection Agency
1978

PUBLIC LAW 94–580—OCT. 21, 1976   90 STAT. 2795

Public Law 94–580
94th Congress

An Act

To provide technical and financial assistance for the development of management plans and facilities for the recovery of energy and other resources from discarded materials and for the safe disposal of discarded materials, and to regulate the management of hazardous waste.

Oct. 21, 1976
[S. 2150]

Be it enacted by the Senate and House of Representatives of the United States of America in Congress assembled,

Resource Conservation and Recovery Act of 1976.

SHORT TITLE

SECTION 1. This Act may be cited as the "Resource Conservation and Recovery Act of 1976".

42 USC 6901 note.

AMENDMENT OF SOLID WASTE DISPOSAL ACT

SEC. 2. The Solid Waste Disposal Act (42 U.S.C. 3251 and following) is amended to read as follows:

"TITLE II—SOLID WASTE DISPOSAL

"Subtitle A—General Provisions

"SHORT TITLE AND TABLE OF CONTENTS

"SEC. 1001. This title (hereinafter in this title referred to as 'this Act'), together with the following table of contents, may be cited as the 'Solid Waste Disposal Act':

42 USC 6901 note.

90 STAT. 2796   PUBLIC LAW 94–580—OCT. 21, 1976

"CONGRESSIONAL FINDINGS

42 USC 6901.

"SEC. 1002. (a) SOLID WASTE.—The Congress finds with respect to solid waste—

"(1) that the continuing technological progress and improvement in methods of manufacture, packaging, and marketing of consumer products has resulted in an ever-mounting increase, and in a change in the characteristics, of the mass material discarded by the purchaser of such products;

"(2) that the economic and population growth of our Nation, and the improvements in the standard of living enjoyed by our population, have required increased industrial production to meet

PUBLIC LAW 94–580—OCT. 21, 1976   90 STAT. 2797

our needs, and have made necessary the demolition of old buildings, the construction of new buildings, and the provision of highways and other avenues of transportation, which, together with related industrial, commercial, and agricultural operations, have resulted in a rising tide of scrap, discarded, and waste materials;

"(3) that the continuing concentration of our population in expanding metropolitan and other urban areas has presented these communities with serious financial, management, intergovernmental, and technical problems in the disposal of solid wastes resulting from the industrial, commercial, domestic, and other activities carried on in such areas;

"(4) that while the collection and disposal of solid wastes should continue to be primarily the function of State, regional, and local agencies, the problems of waste disposal as set forth above have become a matter national in scope and in concern and necessitate Federal action through financial and technical assistance and leadership in the development, demonstration, and application of new and improved methods and processes to reduce the amount of waste and unsalvageable materials and to provide for proper and economical solid waste disposal practices.

"(b) ENVIRONMENT AND HEALTH.—The Congress finds with respect to the environment and health, that—

"(1) although land is too valuable a national resource to be needlessly polluted by discarded materials, most solid waste is disposed of on land in open dumps and sanitary landfills;

"(2) disposal of solid waste and hazardous waste in or on the land without careful planning and management can present a danger to human health and the environment;

"(3) as a result of the Clean Air Act, the Water Pollution Control Act, and other Federal and State laws respecting public health and the environment, greater amounts of solid waste (in the form of sludge and other pollution treatment residues) have been created. Similarly, inadequate and environmentally unsound practices for the disposal or use of solid waste have created greater amounts of air and water pollution and other problems for the environment and for health;   42 USC 1857 note. 33 USC 1251 note.

"(4) open dumping is particularly harmful to health, contaminates drinking water from underground and surface supplies, and pollutes the air and the land;

"(5) hazardous waste presents, in addition to the problems associated with non-hazardous solid waste, special dangers to health and requires a greater degree of regulation than does non-hazardous solid waste; and

"(6) alternatives to existing methods of land disposal must be developed since many of the cities in the United States will be running out of suitable solid waste disposal sites within five years unless immediate action is taken;

"(c) MATERIALS.—The Congress finds with respect to materials, that—

"(1) millions of tons of recoverable material which could be used are needlessly buried each year;

"(2) methods are available to separate usable materials from solid waste; and

"(3) the recovery and conservation of such materials can reduce the dependence of the United States on foreign resources and reduce the deficit in its balance of payments.

90 STAT. 2798   PUBLIC LAW 94–580—OCT. 21, 1976

"(d) ENERGY.—The Congress finds with respect to energy, that—
"(1) solid waste represents a potential source of solid fuel, oil, or gas that can be converted into energy;
"(2) the need exists to develop alternative energy sources for public and private consumption in order to reduce our dependence on such sources as petroleum products, natural gas, nuclear and hydroelectric generation; and
"(3) technology exists to produce usable energy from solid waste.

"OBJECTIVES

42 USC 6902.

"SEC. 1003. The objectives of this Act are to promote the protection of health and the environment and to conserve valuable material and energy resources by—
"(1) providing technical and financial assistance to State and local governments and interstate agencies for the development of solid waste management plans (including resource recovery and resource conservation systems) which will promote improved solid waste management techniques (including more effective organizational arrangements), new and improved methods of collection, separation, and recovery of solid waste, and the environmentally safe disposal of nonrecoverable residues;
"(2) providing training grants in occupations involving the design, operation, and maintenance of solid waste disposal systems;
"(3) prohibiting future open dumping on the land and requiring the conversion of existing open dumps to facilities which do not pose a danger to the environment or to health;
"(4) regulating the treatment, storage, transportation, and disposal of hazardous wastes which have adverse effects on health and the environment;
"(5) providing for the promulgation of guidelines for solid waste collection, transport, separation, recovery, and disposal practices and systems;
"(6) promoting a national research and development program for improved solid waste management and resource conservation techniques, more effective organizational arrangements, and new and improved methods of collection, separation, and recovery, and recycling of solid wastes and environmentally safe disposal of nonrecoverable residues;
"(7) promoting the demonstration, construction, and application of solid waste management, resource recovery, and resource conservation systems which preserve and enhance the quality of air, water, and land resources; and
"(8) establishing a cooperative effort among the Federal, State, and local governments and private enterprise in order to recover valuable materials and energy from solid waste.

"DEFINITIONS

42 USC 6903.

"SEC. 1004. As used in this Act:
"(1) The term 'Administrator' means the Administrator of the Environmental Protection Agency.
"(2) The term 'construction,' with respect to any project of construction under this Act, means (A) the erection or building of new structures and acquisition of lands or interests therein, or the acquisition, replacement, expansion, remodeling, alteration, modernization.

or extension of existing structures, and (B) the acquisition and installation of initial equipment of, or required in connection with, new or newly acquired structures or the expanded, remodeled, altered, modernized or extended part of existing structures (including trucks and other motor vehicles, and tractors, cranes, and other machinery) necessary for the proper utilization and operation of the facility after completion of the project; and includes preliminary planning to determine the economic and engineering feasibility and the public health and safety aspects of the project, the engineering, architectural, legal, fiscal, and economic investigations and studies, and any surveys, designs, plans, working drawings, specifications, and other action necessary for the carrying out of the project, and (C) the inspection and supervision of the process of carrying out the project to completion.

"(2A) The term 'demonstration' means the initial exhibition of a new technology process or practice or a significantly new combination or use of technologies, processes or practices, subsequent to the development stage, for the purpose of proving technological feasibility and cost effectiveness.

"(3) The term 'disposal' means the discharge, deposit, injection, dumping, spilling, leaking, or placing of any solid waste or hazardous waste into or on any land or water so that such solid waste or hazardous waste or any constituent thereof may enter the environment or be emitted into the air or discharged into any waters, including ground waters.

"(4) The term 'Federal agency' means any department, agency, or other instrumentality of the Federal Government, any independent agency or establishment of the Federal Government including any Government corporation, and the Government Printing Office.

"(5) The term 'hazardous waste' means a solid waste, or combination of solid wastes, which because of its quantity, concentration, or physical, chemical, or infectious characteristics may—

"(A) cause, or significantly contribute to an increase in mortality or an increase in serious irreversible, or incapacitating reversible, illness; or

"(B) pose a substantial present or potential hazard to human health or the environment when improperly treated, stored, transported, or disposed of, or otherwise managed.

"(6) The term 'hazardous waste generation' means the act or process of producing hazardous waste.

"(7) The term 'hazardous waste management' means the systematic control of the collection, source separation, storage, transportation, processing, treatment, recovery, and disposal of hazardous wastes.

"(8) For purposes of Federal financial assistance (other than rural communities assistance), the term 'implementation' does not include the acquisition, leasing, construction, or modification of facilities or equipment or the acquisition, leasing, or improvement of land.   *Post*, p. 2813.

"(9) The term 'intermunicipal agency' means an agency established by two or more municipalities with responsibility for planning or administration of solid waste.

"(10) The term 'interstate agency' means an agency of two or more municipalities in different States, or an agency established by two or more States, with authority to provide for the *management* of solid wastes and serving two or more municipalities located in different States.

"(11) The term 'long-term contract' means, when used in relation to solid waste supply, a contract of sufficient duration to assure the viability of a resource recovery facility (to the extent that such viability depends upon solid waste supply).

"(12) The term 'manifest' means the form used for identifying the quantity, composition, and the origin, routing, and destination of hazardous waste during its transportation from the point of generation to the point of disposal, treatment, or storage.

"(13) The term 'municipality' (A) means a city, town, borough, county, parish, district, or other public body created by or pursuant to State law, with responsibility for the planning or administration of solid waste management, or an Indian tribe or authorized tribal organization or Alaska Native village or organization, and (B) includes any rural community or unincorporated town or village or any other public entity for which an application for assistance is made by a State or political subdivision thereof.

"(14) The term 'open dump' means a site for the disposal of solid waste which is not a sanitary landfill within the meaning of section 4004.

"(15) The term 'person' means an individual, trust, firm, joint stock company, corporation (including a government corporation), partnership, association, State, municipality, commission, political subdivision of a State, or any interstate body.

"(16) The term 'procurement item' means any device, good, substance, material, product, or other item whether real or personal property which is the subject of any purchase, barter, or other exchange made to procure such item.

"(17) The term 'procuring agency' means any Federal agency, or any State agency or agency of a political subdivision of a State which is using appropriated Federal funds for such procurement, or any person contracting with any such agency with respect to work performed under such contract.

"(18) The term 'recoverable' refers to the capability and likelihood of being recovered from solid waste for a commercial or industrial use.

"(19) The term 'recovered material' means material which has been collected or recovered from solid waste.

"(20) The term 'recovered resources' means material or energy recovered from solid waste.

"(21) The term 'resource conservation' means reduction of the amounts of solid waste that are generated, reduction of overall resource consumption, and utilization of recovered resources.

"(22) The term 'resource recovery' means the recovery of material or energy from solid waste.

"(23) The term 'resource recovery system' means a solid waste management system which provides for collection, separation, recycling, and recovery of solid wastes, including disposal of nonrecoverable waste residues.

"(24) The term 'resource recovery facility' means any facility at which solid waste is processed for the purpose of extracting, converting to energy, or otherwise separating and preparing solid waste for reuse.

Post, p. 2816.

"(25) The term 'regional authority' means the authority established or designated under section 4006.

"(26) The term 'sanitary landfill' means a facility for the disposal of solid waste which meets the criteria published under section 4004.

"(26A) The term 'sludge' means any solid, semisolid or liquid waste generated from a municipal, commercial, or industrial wastewater treatment plant, water supply treatment plant, or air pollution control facility or any other such waste having similar characteristics and effects.

"(27) The term 'solid waste' means any garbage, refuse, sludge from a waste treatment plant, water supply treatment plant, or air pollution control facility and other discarded material, including solid, liquid, semisolid, or contained gaseous material resulting from industrial, commercial, mining, and agricultural operations, and from community activities, but does not include solid or dissolved material in domestic sewage, or solid or dissolved materials in irrigation return flows or industrial discharges which are point sources subject to permits under section 402 of the Federal Water Pollution Control Act, as amended (86 Stat. 880), or source, special nuclear, or byproduct material as defined by the Atomic Energy Act of 1954, as amended (68 Stat. 923). 33 USC 1342. 42 USC 2011 note.

"(28) The term 'solid waste management' means the systematic administration of activities which provide for the collection, source separation, storage, transportation, transfer, processing, treatment, and disposal of solid waste.

"(29) The term 'solid waste management facility' includes—

"(A) any resource recovery system or component thereof,

"(B) any system, program, or facility for resource conservation, and

"(C) any facility for the collection, source separation, storage, transportation, transfer, processing, treatment or disposal of solid wastes, including hazardous wastes, whether such facility is associated with facilities generating such wastes or otherwise.

"(30) The terms 'solid waste planning', 'solid waste management', and 'comprehensive planning' include planning or management respecting resource recovery and resource conservation.

"(31) The term 'State' means any of the several States, the District of Columbia, the Commonwealth of Puerto Rico, the Virgin Islands, Guam, American Samoa, and the Commonwealth of the Northern Mariana Islands.

"(32) The term 'State authority' means the agency established or designated under section 4007. *Post*, p. 2817.

"(33) The term 'storage', when used in connection with hazardous waste, means the containment of hazardous waste, either on a temporary basis or for a period of years, in such a manner as not to constitute disposal of such hazardous waste.

"(34) The term 'treatment', when used in connection with hazardous waste, means any method, technique, or process, including neutralization, designed to change the physical, chemical, or biological character or composition of any hazardous waste so as to neutralize such waste or so as to render such waste nonhazardous, safer for transport, amenable for recovery, amenable for storage, or reduced in volume. Such term includes any activity or processing designed to change the physical form or chemical composition of hazardous waste so as to render it nonhazardous.

"(35) The term 'virgin material' means a raw material, including previously unused copper, aluminum, lead, zinc, iron, or other metal or metal ore, any undeveloped resource that is, or with new technology will become, a source of raw materials.

"GOVERNMENTAL COOPERATION

"SEC. 1005. (a) INTERSTATE COOPERATION.—The provisions of this Act to be carried out by States may be carried out by interstate agencies and provisions applicable to States may apply to interstate regions where such agencies and regions have been established by the respective 42 USC 6904.

States and approved by the Administrator. In any such case, action required to be taken by the Governor of a State, respecting regional designation shall be required to be taken by the Governor of each of the respective States with respect to so much of the interstate region as is within the jurisdiction of that State.

"(b) CONSENT OF CONGRESS TO COMPACTS.—The consent of the Congress is hereby given to two or more States to negotiate and enter into agreements or compacts, not in conflict with any law or treaty of the United States, for—

"(1) cooperative effort and mutual assistance for the management of solid waste or hazardous waste (or both) and the enforcement of their respective laws relating thereto, and

"(2) the establishment of such agencies, joint or otherwise, as they may deem desirable for making effective such agreements or compacts.

No such agreement or compact shall be binding or obligatory upon any State a party thereto unless it is agreed upon by all parties to the agreement and until it has been approved by the Administrator and the Congress.

"APPLICATION OF ACT AND INTEGRATION WITH OTHER ACTS

42 USC 6905.

"SEC. 1006. (a) APPLICATION OF ACT.—Nothing in this Act shall be construed to apply to (or to authorize any State, interstate, or local authority to regulate) any activity or substance which is subject to the

33 USC 1251 note.

Federal Water Pollution Control Act (33 U.S.C. 1151 and following), the Safe Drinking Water Act (42 U.S.C. 300f and following), the Marine Protection, Research and Sanctuaries Act of 1972 (33 U.S.C. 1401 and following), or the Atomic Energy Act of 1954 (42 U.S.C. 2011 and following) except to the extent that such application (or regulation) is not inconsistent with the requirements of such Acts.

"(b) INTEGRATION WITH OTHER ACTS.—The Administrator shall integrate all provisions of this Act for purposes of administration and enforcement and shall avoid duplication, to the maximum extent practicable, with the appropriate provisions of the Clean Air Act (42 U.S.C. 1857 and following), the Federal Water Pollution Control Act

33 USC 1251 note.

(33 U.S.C. 1151 and following), the Federal Insecticide, Fungicide, and Rodenticide Act (7 U.S.C. 135 and following), the Safe Drinking Water Act (42 U.S.C. 300f and following), the Marine Protection, Research and Sanctuaries Act of 1972 (33 U.S.C. 1401 and following) and such other Acts of Congress as grant regulatory authority to the Administrator. Such integration shall be effected only to the extent that it can be done in a manner consistent with the goals and policies expressed in this Act and in the other acts referred to in this subsection.

"FINANCIAL DISCLOSURE

42 USC 6906.

"SEC. 1007. (a) STATEMENT.—Each officer or employee of the Administrator who—

"(1) performs any function or duty under this Act; and

"(2) has any known financial interest in any person who applies for or receives financial assistance under this Act

shall, beginning on February 1, 1977, annually file with the Administrator a written statement concerning all such interests held by such officer or employee during the preceding calendar year. Such statement shall be available to the public.

PUBLIC LAW 94–580—OCT. 21, 1976   90 STAT. 2803

"(b) ACTION BY ADMINISTRATOR.—The Administrator shall—
"(1) act within ninety days after the date of enactment of this Act—
"(A) to define the term 'known financial interest' for purposes of subsection (a) of this section; and
"(B) to establish the methods by which the requirement to file written statements specified in subsection (a) of this section will be monitored and enforced, including appropriate provision for the filing by such officers and employees of such statements and the review by the Administrator of such statements; and

"(2) report to the Congress on June 1, 1978, and of each succeeding calendar year with respect to such disclosures and the actions taken in regard thereto during the preceding calendar year.

Report to Congress.

"(c) EXEMPTION.—In the rules prescribed under subsection (b) of this section, the Administrator may identify specific positions within the Environmental Protection Agency which are of a nonpolicy-making nature and provide that officers or employees occupying such positions shall be exempt from the requirements of this section.

"(d) PENALTY.—Any officer or employee who is subject to, and knowingly violates, this section shall be fined not more than $2,500 or imprisoned not more than one year, or both.

"SOLID WASTE MANAGEMENT INFORMATION AND GUIDELINES

"SEC. 1008. (a) GUIDELINES.—Within one year of enactment of this section, and from time to time thereafter, the Administrator shall, in cooperation with appropriate Federal, State, municipal, and inter-municipal agencies, and in consultation with other interested persons, and after public hearings, develop and publish suggested guidelines for solid waste management. Such suggested guidelines shall—

42 USC 6907.

"(1) provide a technical and economic description of the level of performance that can be attained by various available solid waste management practices (including operating practices) which provide for the protection of public health and the environment;

"(2) not later than two years after the enactment of this section, describe levels of performance, including appropriate methods and degrees of control, that provide at a minimum for (A) protection of public health and welfare; (B) protection of the quality of ground waters and surface waters from leachates; (C) protection of the quality of surface waters from runoff through compliance with effluent limitations under the Federal Water Pollution Control Act, as amended; (D) protection of ambient air quality through compliance with new source performance standards or requirements of air quality implementation plans under the Clean Air Act, as amended; (E) disease and vector control; (F) safety; and (G) esthetics; and

33 USC 1251 note.

42 USC 1857 note.

"(3) provide minimum criteria to be used by the States to define those solid waste management practices which constitute the open dumping of solid waste or hazardous waste and are to be prohibited under *subtitle D* of this Act.

Minimum criteria of management practices.

Where appropriate, such suggested guidelines also shall include minimum information for use in deciding the adequate location, design, and construction of facilities associated with solid waste management

90 STAT. 2804   PUBLIC LAW 94–580—OCT. 21, 1976

practices, including the consideration of regional, geographic, demographic, and climatic factors.

Notification to
congressional
committees.

"(b) NOTICE.—The Administrator shall notify the Committee on Public Works of the Senate and the Committee on Interstate and Foreign Commerce of the House of Representatives a reasonable time before publishing any suggested guidelines *or proposed regulations under this Act*, of the content of such proposed suggested guidelines *or proposed regulations under this Act.*

"Subtitle B—Office of Solid Waste; Authorities of the Administrator

"OFFICE OF SOLID WASTE

Establishment.
42 USC 6911

"SEC. 2001. The Administrator shall establish within the Environmental Protection Agency an Office of Solid Waste (hereinafter referred to as the 'Office') to be headed by a Deputy Assistant Administrator of the Environmental Protection Agency. The duties and responsibilities (other than duties and responsibilities relating to research and development) of the Administrator under this Act (as modified by applicable reorganization plans) shall be carried out through the Office.

"AUTHORITIES OF ADMINISTRATOR

42 USC 6912.

"SEC. 2002. (a) AUTHORITIES.—In carrying out this Act, the Administrator is authorized to—

"(1) prescribe, in consultation with Federal, State, and regional authorities, such regulations as are necessary to carry out his functions under this Act;

"(2) consult with or exchange information with other Federal agencies undertaking research, development, demonstration projects, studies, or investigations relating to solid waste;

"(3) provide technical and financial assistance to States or regional agencies in the development and implementation of solid waste plans and hazardous waste management programs;

"(4) consult with representatives of science, industry, agriculture, labor, environmental protection and consumer organizations, and other groups, as he deems advisable; and

"(5) utilize the information, facilities, personnel and other resources of Federal agencies, including the National Bureau of Standards and the National Bureau of the Census, on a reimbursable basis, to perform research and analyses and conduct studies and investigations related to resource recovery and conservation and to otherwise carry out the Administrator's functions under this Act.

"(b) REVISION OF REGULATIONS.—Each regulation promulgated under this Act shall be reviewed and, where necessary, revised not less frequently than every three years.

"RESOURCE RECOVERY AND CONSERVATION PANELS

Technical
assistance by
personnel teams.
42 USC 6913.

"SEC. 2003. The Administrator shall provide teams of personnel, including Federal, State, and local employees or contractors (hereinafter referred to as 'Resource Conservation and Recovery Panels') to provide *Federal agencies,* States and local governments upon request with technical assistance on solid waste management, resource recovery, and resource conservation. Such teams shall include technical, marketing, financial, and institutional specialists, and the services of such teams shall be provided without charge to States or local governments.

"GRANTS FOR DISCARDED TIRE DISPOSAL

"SEC. 2004. (a) GRANTS.—The Administrator shall make available grants equal to 5 percent of the purchase price of tire shredders (including portable shredders attached to tire collection trucks) to those eligible applicants best meeting criteria promulgated under this section. An eligible applicant may be any private purchaser, public body, or public-private joint venture. Criteria for receiving grants shall be promulgated under this section and shall include the policy to offer any private purchaser the first option to receive a grant, the policy to develop widespread geographic distribution of tire shredding facilities, the need for such facilities within a geographic area, and the projected risk and viability of any such venture. In the case of an application under this section from a public body, the Administrator shall first make a determination that there are no private purchasers interested in making an application before approving a grant to a public body.

Eligible applicants.
42 USC 6914.

"(b) AUTHORIZATION.—There is authorized to be appropriated $750,000 for each of the fiscal years 1978 and 1979 to carry out this section.

"ANNUAL REPORT

"SEC. 2005. The Administrator shall transmit to the Congress and the President, not later than ninety days after the end of each fiscal year, a comprehensive and detailed report on all activities of the Office during the preceding fiscal year. Each such report shall include—

42 USC 6915.

"(1) a statement of specific and detail objectives for the activities and programs conducted and assisted under this Act;

"(2) statements of the Administrator's conclusions as to the effectiveness of such activities and programs in meeting the stated objectives and the purposes of this Act, measured through the end of such fiscal year;

"(3) a summary of outstanding solid waste problems confronting the Administrator, in order of priority;

"(4) recommendations with respect to such legislation which the Administrator deems necessary or desirable to assist in solving problems respecting solid waste;

"(5) all other information required to be submitted to the Congress pursuant to any other provision of this Act; and

"(6) the Administrator's plans for activities and programs respecting solid waste during the next fiscal year.

"GENERAL AUTHORIZATION

"SEC. 2006. (a) GENERAL ADMINISTRATION.—There are authorized to be appropriated to the Administrator for the purpose of carrying out the provisions of this Act, $35,000,000 for the fiscal year ending September 30, 1977, $38,000,000 for the fiscal year ending September 30, 1978, and $42,000,000 for the fiscal year ending September 30, 1979.

42 USC 6916.

"(b) RESOURCE RECOVERY AND CONSERVATION PANELS.—Not less than 20 percent of the amount appropriated under subsection (a) shall be used only for purposes of Resource Recovery and Conservation Panels established under section 2003 (including travel expenses incurred by such panels in carrying out their functions under this Act).

90 STAT. 2806 PUBLIC LAW 94–580—OCT. 21, 1976

Infra.

"(c) HAZARDOUS WASTE.—Not less than 30 percent of the amount appropriated under subsection (a) shall be used only for purposes of carrying out subtitle C of this Act (relating to hazardous waste) other than section 3011.

"Subtitle C—Hazardous Waste Management

"IDENTIFICATION AND LISTING OF HAZARDOUS WASTE

42 USC 6921.

"SEC. 3001. (a) CRITERIA FOR IDENTIFICATION OR LISTING.—Not later than eighteen months after the date of the enactment of this Act, the Administrator shall, after notice and opportunity for public hearing, and after consultation with appropriate Federal and State agencies, develop and promulgate criteria for identifying the characteristics of hazardous waste, and for listing hazardous waste, which should be subject to the provisions of this subtitle, taking into account toxicity, persistence, and degradability in nature, potential for accumulation in tissue, and other related factors such as flammability, corrosiveness, and other hazardous characteristics. Such criteria shall be revised from time to time as may be appropriate.

Regulations.

Ante, p. 2799.

"(b) IDENTIFICATION AND LISTING.—Not later than eighteen months after the date of enactment of this section, and after notice and opportunity for public hearing, the Administrator shall promulgate regulations identifying the characteristics of hazardous waste, and listing particular hazardous wastes (within the meaning of section 1004(5)), which shall be subject to the provisions of this subtitle. Such regulations shall be based on the criteria promulgated under subsection (a) and shall be revised from time to time thereafter as may be appropriate.

"(c) PETITION BY STATE GOVERNOR.—At any time after the date eighteen months after the enactment of this title, the Governor of any State may petition the Administrator to identify or list a material as a hazardous waste. The Administrator shall act upon such petition within ninety days following his receipt thereof and shall notify the Governor of such action. If the Administrator denies such petition because of financial considerations, in providing such notice to the Governor he shall include a statement concerning such considerations.

"STANDARDS APPLICABLE TO GENERATORS OF HAZARDOUS WASTE

Regulations.
42 USC 6922.

"SEC. 3002. Not later than eighteen months after the date of the enactment of this section, and after notice and opportunity for public hearings and after consultation with appropriate Federal and State agencies, the Administrator shall promulgate regulations establishing such standards, applicable to generators of hazardous waste identified or listed under this subtitle, as may be necessary to protect human health and the environment. Such standards shall establish requirements respecting—

"(1) recordkeeping practices that accurately identify the quantities of such hazardous waste generated, the constituents thereof which are significant in quantity or in potential harm to human health or the environment, and the disposition of such wastes;

"(2) labeling practices for any containers used for the storage, transport, or disposal of such hazardous waste such as will identify accurately such waste;

"(3) use of appropriate containers for such hazardous waste;

"(4) furnishing of information on the general chemical composition of such hazardous waste to persons transporting, treating, storing, or disposing of such wastes;

"(5) use of a manifest system to assure that all such hazardous waste generated is designated for treatment, storage, or disposal in treatment, storage, or disposal facilities (other than facilities on the premises where the waste is generated) for which a permit has been issued as provided in this subtitle, *or pursuant to title I of the Marine Protection, Research, and Sanctuaries Act (86 Stat. 1052); and*

"(6) submission of reports to the Administrator (or the State agency in any case in which such agency carries out an authorized permit program pursuant to this subtitle) at such times as the Administrator (or the State agency if appropriate) deems necessary, setting out— *Reports.*

"(A) the quantities of hazardous waste identified or listed under this subtitle that he has generated during a particular time period; and

"(B) the disposition of all hazardous waste reported under subparagraph (A).

"STANDARDS APPLICABLE TO TRANSPORTERS OF HAZARDOUS WASTE

"Sec. 3003. (a) STANDARDS.—Not later than eighteen months after the date of enactment of this section, and after opportunity for public hearings, the Administrator, after consultation with the Secretary of Transportation and the States, shall promulgate regulations establishing such standards, applicable to transporters of hazardous waste identified or listed under this subtitle, as may be necessary to protect human health and the environment. Such standards shall include but need not be limited to requirements respecting— *Regulations. 42 USC 6923.*

"(1) recordkeeping concerning such hazardous waste transported, and their source and delivery points;

"(2) transportation of such waste only if properly labeled;

"(3) compliance with the manifest system referred to in section 3002(5); and

"(4) transportation of all such hazardous waste only to the hazardous waste treatment, storage, or disposal facilities which the shipper designates on the manifest form to be a facility holding a permit issued under this subtitle, *or pursuant to title I of the Marine Protection, Research, and Sanctuaries Act (86 Stat. 1052).*

"(b) COORDINATION WITH REGULATIONS OF SECRETARY OF TRANSPORTATION.—In case of any hazardous waste identified or listed under this subtitle which is subject to the Hazardous Materials Transportation Act (88 Stat. 2156; 49 U.S.C. 1801 and following), the regulations promulgated by the Administrator under this *section* shall be consistent with the requirements of such Act and the regulations thereunder. The Administrator is authorized to make recommendations to the Secretary of Transportation respecting the regulations of such hazardous waste under the Hazardous Materials Transportation Act and for addition of materials to be covered by such Act. *Recommendations.*

"STANDARDS APPLICABLE TO OWNERS AND OPERATORS OF HAZARDOUS WASTE TREATMENT, STORAGE, AND DISPOSAL FACILITIES

"Sec. 3004. Not later than eighteen months after the date of enactment of this section, and after opportunity for public hearings and after consultation with appropriate Federal and State agencies, the Administrator shall promulgate regulations establishing such performance standards, applicable to owners and operators of facilities for the treatment, storage, or disposal of hazardous waste identified or *Regulations. 42 USC 6924.*

listed under this subtitle, as may be necessary to protect human health and the environment. Such standards shall include, but need not be limited to, requirements respecting—

"(1) maintaining records of all hazardous wastes identified or listed under this title which is treated, stored, or disposed of, as the case may be, and the manner in which such wastes were treated, stored, or disposed of;

"(2) satisfactory reporting, monitoring, and inspection and compliance with the manifest system referred to in section 3002(5);

"(3) treatment, storage, or disposal of all such waste received by the facility pursuant to such operating methods, techniques, and practices as may be satisfactory to the Administrator;

"(4) the location, design, and construction of such hazardous waste treatment, disposal, or storage facilities;

"(5) contingency plans for effective action to minimize unanticipated damage from any treatment, storage, or disposal of any such hazardous waste;

"(6) the maintenance of operation of such facilities and requiring such additional qualifications as to ownership, continuity of operation, training for personnel, and financial responsibility as may be necessary or desirable; and

"(7) compliance with the requirements of section 3005 respecting permits for treatment, storage, or disposal.

No private entity shall be precluded by reason of criteria established under paragraph (6) from the ownership or operation of facilities providing hazardous waste treatment, storage, or disposal services where such entity can provide assurances of financial responsibility and continuity of operation consistent with the degree and duration of risks associated with the treatment, storage, or disposal of specified hazardous waste.

"PERMITS FOR TREATMENT, STORAGE, OR DISPOSAL OF HAZARDOUS WASTE

Regulations.
42 USC 6925.

"SEC. 3005. (a) PERMIT REQUIREMENTS.—Not later than eighteen months after the date of the enactment of this section, the Administrator shall promulgate regulations requiring each person owning or operating a facility for the treatment, storage, or disposal of hazardous waste identified or listed under this subtitle to have a permit issued pursuant to this section. Such regulations shall take effect on the date provided in section 3010 and upon and after such date the *treatment, storage, or* disposal of any such hazardous waste is prohibited except in accordance with such a permit.

"(b) REQUIREMENTS OF PERMIT APPLICATION.—Each application for a permit under this section shall contain such information as may be required under regulations promulgated by the Administrator, including information respecting—

"(1) estimates with respect to the composition, quantities, and concentrations of any hazardous waste identified or listed under this subtitle, or combinations of any such hazardous waste and any other solid waste, proposed to be disposed of, treated, transported, or stored, and the time, frequency, or rate of which such waste is proposed to be disposed of, treated, transported, or stored; and

"(2) the site at which such hazardous waste or the products of treatment of such hazardous waste will be disposed of, treated, transported to, or stored.

"(c) PERMIT ISSUANCE.—Upon a determination by the Administrator (or a State, if applicable), of compliance by a facility for which a permit is applied for under this section with the requirements of this section and section 3004, the Administrator (or the State) shall issue a permit for such facilities. In the event permit applicants propose modification of their facilities, or in the event the Administrator (or the State) determines that modifications are necessary to conform to the requirements under this section and section 3004, the permit shall specify the time allowed to complete the modifications.

"(d) PERMIT REVOCATION.—Upon a determination by the Administrator (or by a State, in the case of a State having an authorized hazardous waste program under section 3006) of noncompliance by a facility having a permit under this title with the requirements of this section or section 3004, the Administrator (or State, in the case of a State having an authorized hazardous waste program under section 3006) shall revoke such permit.

"(e) INTERIM STATUS.—Any person who—

"(1) owns or operates a facility required to have a permit under this section which facility is in existence on the date of enactment of this Act,

"(2) has complied with the requirements of section 3010(a), and

"(3) has made an application for a permit under this section

shall be treated as having been issued such permit until such time as final administrative disposition of such application is made, unless the Administrator or other plaintiff proves that final administrative disposition of such application has not been made because of the failure of the applicant to furnish information reasonably required or requested in order to process the application.

"AUTHORIZED STATE HAZARDOUS WASTE PROGRAMS

"SEC. 3006. (a) FEDERAL GUIDELINES.—Not later than eighteen months after the date of enactment of this Act, the Administrator, after consultation with State authorities, shall promulgate guidelines to assist States in the development of State hazardous waste programs.

42 USC 6926.

"(b) AUTHORIZATION OF STATE PROGRAM.—Any State which seeks to administer and enforce a hazardous waste program pursuant to this subtitle may develop and, after notice and opportunity for public hearing, submit to the Administrator an application, in such form as he shall require, for authorization of such program. Within ninety days following submission of an application under this subsection, the Administrator shall issue a notice as to whether or not he expects such program to be authorized, and within ninety days following such notice (and after opportunity for public hearing) he shall publish his findings as to whether or not the conditions listed in items (1), (2), and (3) below have been met. Such State is authorized to carry out such program in lieu of the Federal program under this subtitle in such State and to issue and enforce permits for the storage, treatment, or disposal of hazardous waste unless, within ninety days following submission of the application the Administrator notifies such State that such program may not be authorized and, within ninety days following such notice and after opportunity for public hearing, he finds that (1) such State program is not equivalent to the Federal program under this subtitle, (2) such program is not consistent with the Federal or State programs applicable in other States, or (3) such

Notice and hearing.

90 STAT. 2810   PUBLIC LAW 94–580—OCT. 21, 1976

program does not provide adequate enforcement of compliance with the requirements of this subtitle.

"(c) INTERIM AUTHORIZATION.—Any State which has in existence a hazardous waste program pursuant to State law before the date ninety days after the date of promulgation of regulations under sections 3002, 3003, 3004, and 3005, may submit to the Administrator evidence of such existing program and may request a temporary authorization to carry out such program under this subtitle. The Administrator shall, if the evidence submitted shows the existing State program to be substantially equivalent to the Federal program under this subtitle, grant an interim authorization to the State to carry out such program in lieu of the Federal program pursuant to this subtitle for a twenty-four month period beginning on the date six months after the date of promulgation of regulations under sections 3002 through 3005.

"(d) EFFECT OF STATE PERMIT.—Any action taken by a State under a hazardous waste program authorized under this section shall have the same force and effect as action taken by the Administrator under this subtitle.

"(e) WITHDRAWAL OF AUTHORIZATION —Whenever the Administrator determines after public hearing that a State is not administering and enforcing a program authorized under this section in accordance with requirements of this section, he shall so notify the State and, if appropriate corrective action is not taken within a reasonable time, not to exceed ninety days, the Administrator shall withdraw authorization of such program and establish a Federal program pursuant to this subtitle. The Administrator shall not withdraw authorization of any such program unless he shall first have notified the State, and made public, in writing, the reasons for such withdrawal.

"INSPECTIONS

42 USC 6927.

"SEC. 3007. (a) ACCESS ENTRY.—For purposes of developing or assisting in the development of any regulation or enforcing the provisions of this subtitle, any person who generates, stores, treats, transports, disposes of, or otherwise handles hazardous wastes shall, upon request of any officer or employee of the Environmental Protection Agency, duly designated by the Administrator, or upon request of any duly designated officer employee of a State having an authorized hazardous waste program, furnish or permit such person at all reasonable times to have access to, and to copy all records relating to such wastes. For the purposes of developing or assisting in the development of any regulation or enforcing the provisions of this title, such officers or employees are authorized—

"(1) to enter at reasonable times any establishment or other place maintained by any person where hazardous wastes are generated, stored, treated, disposed of, or transported from;

"(2) to inspect and obtain samples from any person of any such wastes and samples of any containers or labeling for such wastes. Each such inspection shall be commenced and completed with reasonable promptness. If the officer or employee obtains any samples, prior to leaving the premises, he shall give to the owner, operator, or agent in charge a receipt describing the sample obtained and if requested a portion of each such sample equal in volume or weight to the portion retained. If any analysis is made of such samples, a copy of the results of such analysis shall be furnished promptly to the owner, operator, or agent in charge.

"(b) AVAILABILITY TO PUBLIC.—Any records, reports, or information obtained from any person under this section shall be available to the public, except that upon a showing satisfactory to the Administrator (or the State, as the case may be) by any person that records, reports, or information, or particular part thereof, to which the Administrator (or the State, as the case may be) has access under this section if made public, would divulge information entitled to protection under section 1905 of title 18 of the United States Code, the Administrator (or the State, as the case may be) shall consider such information or particular portion thereof confidential in accordance with the purposes of that section, except that such record, report, document, or information may be disclosed to other officers, employees, or authorized representatives of the United States concerned with carrying out this Act, or when relevant in any proceeding under this Act.

"FEDERAL ENFORCEMENT

"SEC. 3008. (a) COMPLIANCE ORDERS.—(1) Except as provided in paragraph (2), whenever on the basis of any information the Administrator determines that any person is in violation of any requirement of this subtitle, the Administrator shall give notice to the violator of his failure to comply with such requirement. If such violation extends beyond the thirtieth day after the Administrator's notification, the Administrator may issue an order requiring compliance within a specified time period or the Administrator may commence a civil action in the United States district court in the district in which the violation occurred for appropriate relief, including a temporary or permanent injunction. **42 USC 6928.**

"(2) In the case of a violation of any requirement of this subtitle where such violation occurs in a State which is authorized to carry out a hazardous waste program under section 3006, the Administrator shall give notice to the State in which such violation has occurred thirty days prior to issuing an order or commencing a civil action under this section.

"(3) If such violator fails to take corrective action within the time specified in the order, he shall be liable for a civil penalty of not more than $25,000 for each day of continued noncompliance and the Administrator may suspend or revoke any permit issued to the violator (whether issued by the Administrator or the State). **Penalty.**

"(b) PUBLIC HEARING.—Any order or any suspension or revocation of a permit shall become final unless, no later than thirty days after the order or notice of the suspension or revocation is served, the person or persons named therein request a public hearing. Upon such request the Administrator shall promptly conduct a public hearing. In connection with any proceeding under this section the Administrator may issue subpenas for the attendance and testimony of witnesses and the production of relevant papers, books, and documents, and may promulgate rules for discovery procedures. **Subpenas.**

"(c) REQUIREMENTS OF COMPLIANCE ORDERS.—Any order issued under this section shall state with reasonable specificity the nature of the violation and specify a time for compliance and assess a penalty, if any, which the Administrator determines is reasonable taking into account the seriousness of the violation and any good faith efforts to comply with the applicable requirements. **Penalty.**

"(d) CRIMINAL PENALTY.—Any person who knowingly—

"(1) transports any hazardous waste identified or listed under this subtitle to a facility which does not have a permit under Section 3005 (or 3006 in the case of a State program), or pursuant to title I of the Marine Protection, Research, and Sanctuaries Act (86 Stat. 1052),

"(2) treats, stores, or disposes of any hazardous waste identified or listed under this subtitle without having obtained a permit under section 3005 (or 3006 in the case of a State program) or pursuant to title I of the Marine Protection, Research, and Sanctuaries Act (86 Stat. 1052).

"(3) makes any false statement or representation in any application, label, manifest, record, report, permit or other document filed, maintained, or used for purposes of compliance with this subtitle.

shall, upon conviction, be subject to a fine of not more than $25,000 for each day of violation, or to imprisonment not to exceed one year, or both. If the conviction is for a violation committed after a first conviction of such person under this paragraph, punishment shall be by a fine of not more than $50,000 per day of violation, or by imprisonment for not more than two years, or by both.

"RETENTION OF STATE AUTHORITY

42 USC 6929.

"SEC. 3009. Upon the effective date of regulations under this subtitle no State or political subdivision may impose any requirements less stringent than those authorized under this subtitle respecting the same matter as governed by such regulations, except that if application of a regulation with respect to any matter under this subtitle is postponed or enjoined by the action of any court, no State or political subdivision shall be prohibited from acting with respect to the same aspect of such matter until such time as such regulation takes effect.

"EFFECTIVE DATE

42 USC 6930.

"SEC. 3010. (a) PRELIMINARY NOTIFICATION.—Not later than ninety days after promulgation or revision of regulations under section 3001 identifying by its characteristics or listing any substance as hazardous waste subject to this subtitle, any person generating or transporting such substance or owning or operating a facility for treatment, storage, or disposal of such substance shall file with the Administrator (or with States having authorized hazardous waste permit programs under section 3006) a notification stating the location and general description of such activity and the identified or listed hazardous wastes handled by such person. Not more than one such notification shall be required to be filed with respect to the same substance. No identified or listed hazardous waste subject to this subtitle may be transported, treated, stored, or disposed of unless notification has been given as required under this subsection.

"(b) EFFECTIVE DATE OF REGULATION.—The regulations under this subtitle respecting requirements applicable to the generation, transportation, treatment, storage, or disposal of hazardous waste (including requirements respecting permits for such treatment, storage, or disposal) shall take effect on the date six months after the date of promulgation thereof (or six months after the date of revision in the case of any regulation which is revised after the date required for promulgation thereof).

"AUTHORIZATION OF ASSISTANCE TO STATES   42 USC 6931.

SEC. 3011. (a) AUTHORIZATION.—There is authorized to be appropriated $25,000,000 for each of the fiscal years 1978 and 1979 to be used to make grants to the States for purposes of assisting the States in the development and implementation of authorized State hazardous waste programs.

"(b) ALLOCATION.—Amounts authorized to be appropriated under subsection (a) shall be allocated among the States on the basis of regulations promulgated by the Administrator, after consultation with the States, which take into account, the extent to which hazardous waste is generated, transported, treated, stored, and disposed of within such State, the extent of exposure of human beings and the environment within such State to such waste, and such other factors as the Administrator deems appropriate.

"Subtitle D—State or Regional Solid Waste Plans

"OBJECTIVES OF SUBTITLE

"SEC. 4001. The objectives of this subtitle are to assist in developing   42 USC 6941.
and encouraging methods for the disposal of solid waste which are environmentally sound and which maximize the utilization of valuable resources and to encourage resource conservation. Such objectives are to be accomplished through Federal technical and financial assistance to States or regional authorities for comprehensive planning pursuant to Federal guidelines designed to foster cooperation among Federal, State, and local governments and private industry.

"FEDERAL GUIDELINES FOR PLANS

"SEC. 4002. (a) GUIDELINES FOR IDENTIFICATION OF REGIONS.—For   Publication.
purposes of encouraging and facilitating the development of regional   42 USC 6942.
planning for solid waste management, the Administrator, within one hundred and eighty days after the date of enactment of this section and after consultation with appropriate Federal, State, and local authorities, shall by regulation publish guidelines for the identification of those areas which have common solid waste management problems and are appropriate units for planning regional solid waste management services. Such guidelines shall consider—
          "(1) the size and location of areas which should be included,
          "(2) the volume of solid waste which should be included, and
          "(3) the available means of coordinating regional planning with other related regional planning and for coordination of such regional planning into the State plan.
    "(b) GUIDELINES FOR STATE PLANS.—Not later than eighteen months   Regulations.
after the date of enactment of this section and after notice and hearing, the Administrator shall, after consultation with appropriate Federal, State, and local authorities, promulgate regulations containing guidelines to assist in the development and implementation of State solid waste management plans (hereinafter in this title referred to as 'State plans'). The guidelines shall contain methods for achieving the objectives specified in section 4001. Such guidelines shall be reviewed from   Review.
time to time, but not less frequently than every three years, and revised as may be appropriate.
    "(c) CONSIDERATIONS FOR STATE PLAN GUIDELINES.—The guidelines promulgated under subsection (b) shall consider—
          "(1) the varying regional, geologic, hydrologic, climatic, and other circumstances under which different solid waste practices are required in order to insure the reasonable protection of the quality of the ground and surface waters from leachate contamination,

90 STAT. 2814 PUBLIC LAW 94-580—OCT. 21, 1976

the reasonable protection of the quality of the surface waters from surface runoff contamination, and the reasonable protection of ambient air quality;

"(2) characteristics and conditions of collection, storage, processing, and disposal operating methods, techniques and practices, and location of facilities where such operating methods, techniques, and practices are conducted, taking into account the nature of the material to be disposed;

"(3) methods for closing or upgrading open dumps for purposes of eliminating potential health hazards;

"(4) population density, distribution, and projected growth;

"(5) geographic, geologic, climatic, and hydrologic characteristics;

"(6) the type and location of transportation;

"(7) the profile of industries;

"(8) the constituents and generation rates of waste;

"(9) the political, economic, organizational, financial, and management problems affecting comprehensive solid waste management;

"(10) types of resource recovery facilities and resource conservation systems which are appropriate; and

"(11) available new and additional markets for recovered material.

"MINIMUM REQUIREMENTS FOR APPROVAL OF PLANS

42 USC 6943.

"SEC. 4003. In order to be approved under section 4007, each State plan must comply with the following minimum requirements—

"(1) The plan shall identify (in accordance with section 4006(b)) (A) the responsibilities of State, local, and regional authorities in the implementation of the State plan, (B) the distribution of Federal funds to the authorities responsible for development and implementation of the State plan, and (C) the means for coordinating regional planning and implementation under the State plan.

"(2) The plan shall, in accordance with section 4005(c), prohibit the establishment of new open dumps within the State, and contain requirements that all solid waste (including solid waste originating in other States, but not including hazardous waste) shall be (A) utilized for resource recovery or (B) disposed of in sanitary landfills (within the meaning of section 4004(a)) or otherwise disposed of in an environmentally sound manner.

"(3) The plan shall provide for the closing or upgrading of all existing open dumps within the State pursuant to the requirements of section 4005.

"(4) The plan shall provide for the establishment of such State regulatory powers as may be necessary to implement the plan.

Long-term contracts.

"(5) The plan shall provide that no local government within the State shall be prohibited under State or local law from entering into long-term contracts for the supply of solid waste to resource recovery facilities.

Resource conservation and disposal of solid waste.

"(6) The plan shall provide for such resource conservation or recovery and for the disposal of solid waste in sanitary landfills or any combination of practices so as may be necessary to use or dispose of such waste in a manner that is environmentally sound.

"CRITERA FOR SANITARY LANDFILLS; SANITARY LANDFILLS REQUIRED FOR ALL DISPOSAL

"SEC. 4004. (a) CRITERIA FOR SANITARY LANDFILLS.—Not later than one year after the date of enactment of this section, after consultation with the States, and after notice and public hearings, the Administrator shall promulgate regulations containing criteria for determining which facilities shall be classified as sanitary landfills and which shall be classified as open dumps within the meaning of this Act. At a minimum, such criteria shall provide that a facility may be classified as a sanitary landfill and not an open dump only if there is no reasonable probability of adverse effects on health or the environment from disposal of solid waste at such facility. Such regulations may provide for the classification of the types of sanitary landfills.

Regulations. 42 USC 6944.

"(b) DISPOSAL REQUIRED TO BE IN SANITARY LANDFILLS, ETC.—For purposes of complying with section 4003(2) each State plan shall prohibit the establishment of open dumps and contain a requirement that disposal of all solid waste within the State shall be in compliance with such section 4003(2).

"(c) EFFECTIVE DATE.—The prohibition contained in subsection (b) shall take effect on the date six months after the date of promulgation of regulations under subsection (a) or on the date of approval of the State plan, whichever is later.

"UPGRADING OF OPEN DUMPS

"SEC. 4005. (a) OPEN DUMPS.—For purposes of this Act, the term 'open dump' means any facility or site where solid waste is disposed of which is not a sanitary landfill which meets the criteria promulgated under section 4004 and which is not a facility for disposal of hazardous waste.

"Open dump." 42 USC 6945.

"(b) INVENTORY.—Not later than one year after promulgation of regulations under section 4004, the Administrator, with the cooperation of the Bureau of the Census shall publish an inventory of all disposal facilities or sites in the United States which are open dumps within the meaning of this Act.

Publication.

"(c) CLOSING OR UPGRADING OF EXISTING OPEN DUMPS.—Any solid waste management practice or disposal of solid waste or hazardous waste which constitutes the open dumping of solid waste or hazardous waste is prohibited, except in the case of any practice or disposal of solid waste under a timetable or schedule for compliance established under this section. For purposes of complying with section 4003(2), each State plan shall contain a requirement that all existing disposal facilities or sites for solid waste in such State which are open dumps listed in the inventory under subsection (b) shall comply with such measures as may be promulgated by the Administrator to eliminate health hazards and minimize potential health hazards. Each such plan shall establish, for any entity which demonstrates that it has considered other public or private alternatives for solid waste management to comply with the prohibition on open dumping and is unable to utilize such alternatives to so comply, a timetable or schedule for compliance for such practice or disposal of solid waste which specifies a schedule of remedial measures, including an enforceable sequence of actions or operations, leading to compliance with the prohibition on open dumping of solid waste within a reasonable time (not to exceed 5 years from the date of publication of the inventory under subsection (b)).

Schedule of remedial measures.

PUBLIC LAW 94–580—OCT. 21, 1976

"PROCEDURE FOR DEVELOPMENT AND IMPLEMENTATION OF STATE PLAN

Regulations.
42 USC 6946.

"SEC. 4006. (a) IDENTIFICATION OF REGIONS.—Within one hundred and eighty days after publication of guidelines under section 4002(a) (relating to identification of regions), the Governor of each State, after consultation with local elected officials, shall promulgate regulations based on such guidelines identifying the boundaries of each area within the State which, as a result of urban concentrations, geographic conditions, markets, and other factors, is appropriate for carrying out regional solid waste management. Such regulations may be modified from time to time (identifying additional or different regions) pursuant to such guidelines.

State plan.

"(b) IDENTIFICATION OF STATE AND LOCAL AGENCIES AND RESPONSIBILITIES.—(1) Within one hundred and eighty days after the Governor promulgates regulations under subsection (a), for purposes of facilitating the development and implementation of a State plan which will meet the minimum requirements of section 4003, the State, together with appropriate elected officials of general purpose units of local government, shall jointly (A) identify an agency to develop the State plan and identify one or more agencies to implement such plan, and (B) identify which solid waste functions will, under such State plan, be planned for and carried out by the State and which such functions will, under such State plan, be planned for and carried out by a regional or local authority or a combination of regional or local and State authorities. If a multi-functional regional agency authorized by

Multi-functional regional agency.

State law to conduct solid waste planning and management (the members of which are appointed by the Governor) is in existence on the date of enactment of this Act, the Governor shall identify such authority for purposes of carrying out within such region clause (A) of this paragraph. Where feasible, designation of the agency for the affected area designated under section 208 of the Federal Water Pollution

33 USC 1288.

Control Act (86 Stat. 839) shall be considered. A State agency identified under this paragraph shall be established or designated by the Governor of such State. Local or regional agencies identified under this paragraph shall be composed of individuals at least a majority of whom are elected local officials.

"(2) If planning and implementation agencies are not identified and designated or established as required under paragraph (1) for any affected area, the governor shall, before the date two hundred and seventy days after promulgation of regulations under subsection (a), establish or designate a State agency to develop and implement the State plan for such area.

"(c) INTERSTATE REGIONS.—(1) In the case of any region which, pursuant to the guidelines published by the Administrator under section 4002(a) (relating to identification of regions), would be located in two or more States, the Governors of the respective States, after consultation with local elected officials, shall consult, cooperate, and enter into agreements identifying the boundaries of such region pursuant to subsection (a).

"(2) Within one hundred and eighty days after an interstate region is identified by agreement under paragraph (1), appropriate elected officials of general purpose units of local government within such region shall jointly establish or designate an agency to develop a plan for such region. If no such agency is established or designated within such period by such officials, the Governors of the respective States may, by agreement, establish or designate for such purpose a single representative organization including elected officials of general purpose units of local government within such region.

"(3) Implementation of interstate regional solid waste management plans shall be conducted by units of local government for any portion of a region within their jurisdiction, or by multijurisdictional agencies or authorities designated in accordance with State law, including those designated by agreement by such units of local government for such purpose. If no such unit, agency, or authority is so designated, the respective Governors shall designate or establish a single interstate agency to implement such plan.

Regional solid waste management plans.

"(4) For purposes of this subtitle, so much of an interstate regional plan as is carried out within a particular State shall be deemed part of the State plan for such State.

"APPROVAL OF STATE PLAN; FEDERAL ASSISTANCE

"SEC. 4007. (a) PLAN APPROVAL.—The Administrator shall, within six months after a State plan has been submitted for approval, approve or disapprove the plan. The Administrator shall approve a plan if he determines that—

42 USC 6947.

"(1) it meets the requirements of paragraphs (1), (2), (3), and (5) of section 4003; and

"(2) it contains provision for revision of such plan, after notice and public hearing, whenever the Administrator, by regulation, determines—

"(A) that revised regulations respecting minimum requirements have been promulgated under paragraphs (1), (2), (3), and (5) of section 4003 with which the State plan is not in compliance;

"(B) that information has become available which demonstrates the inadequacy of the plan to effectuate the purposes of this subtitle; or

"(C) that such revision is otherwise necessary.

The Administrator shall review approved plans from time to time and if he determines that revision or corrections are necessary to bring such plan into compliance with the minimum requirements promulgated under section 4003 (including new or revised requirements), he shall, after notice and opportunity for public hearing, withdraw his approval of such plan. Such withdrawal of approval shall cease to be effective upon the Administrator's determination that such complies with such minimum requirements.

Review; withdrawal of approval.

"(b) ELIGIBILITY OF STATES FOR FEDERAL FINANCIAL ASSISTANCE.— (1) The Administrator shall approve a State application for financial assistance under this subtitle, and make grants to such State, if such State and local and regional authorities within such State have complied with the requirements of section 4006 within the period required under such section and if such State has a State plan which has been approved by the Administrator under this subtitle.

"(2) The Administrator shall approve a State application for financial assistance under this subtitle, and make grants to such State, for fiscal years 1978 and 1979 if the Administrator determines that the State plan continues to be eligible for approval under subsection (a) and is being implemented by the State.

"(3) Upon withdrawal of approval of a State plan under subsection (a), the Administrator shall withhold Federal financial and technical assistance under this subtitle (other than such technical assistance as

Withholding of financial and technical assistance.

90 STAT. 2818   PUBLIC LAW 94–580—OCT. 21, 1976

may be necessary to assist in obtaining the reinstatement of approval) until such time as such approval is reinstated.

"(c) EXISTING ACTIVITIES.—Nothing in this subtitle shall be construed to prevent or affect any activities respecting solid waste planning or management which are carried out by State, regional, or local authorities unless such activities are inconsistent with a State plan approved by the Administrator under this subtitle.

"FEDERAL ASSISTANCE

42 USC 6948.   "SEC. 4008. (a) AUTHORIZATION OF FEDERAL FINANCIAL ASSISTANCE.—(1) There are authorized to be appropriated $30,000,000 for fiscal year 1978 and $40,000,000 for fiscal year 1979 for purposes of making grants to the States for the development and implementation of State plans under this subtitle.

"(2)(A) The Administrator is authorized to provide financial assistance to States, counties, municipalities, and intermunicipal agencies and State and local public solid waste management authorities for implementation of programs to provide solid waste management, resource recovery, and resource conservation services and hazardous waste management. Such assistance shall include assistance for facility planning and feasibility studies; expert consultation; surveys and analyses of market needs; marketing of recovered resources; technology assessments; legal expenses; construction feasibility studies; source separation projects; and fiscal or economic investigations or studies; but such assistance shall not include any other element of construction, or any acquisition of land or interest in land, or any subsidy for the price of recovered resources. Agencies assisted under this subsection shall consider existing solid waste management and hazardous waste management services and facilities as well as facilities proposed for construction.

Compliance with project or program.   "(B) An applicant for financial assistance under this paragraph must agree to comply with respect to the project or program assisted with the applicable requirements of section 4005 and Subtitle C of this Act and apply applicable solid waste management practices, methods, and levels of control consistent with any guidelines published pursuant to section 1008 of this Act. Assistance under this paragraph shall be available only for programs certified by the State to be consistent with any applicable State or areawide solid waste management plan or program.

Ante, p. 2803.

Appropriation authorization.   "(C) There are authorized to be appropriated $15,000,000 for each of the fiscal years 1978 and 1979 for purposes of this section.

"(b) STATE ALLOTMENT.—The sums appropriated in any fiscal year under subsection (a)(1) shall be allotted by the Administrator among all States, in the ratio that the population in each State bears to the population in all of the States, except that no State shall receive less than one-half of 1 per centum of the sums so allotted in any fiscal year. No State shall receive any grant under this section during any fiscal year when its expenditures of non-Federal funds for other than non-recurrent expenditures for solid waste management control programs will be less than its expenditures were for such programs during fiscal year 1975, except that such funds may be reduced by an amount equal to their proportionate share of any general reduction of State spending ordered by the Governor or legislature of such State. No State shall receive any grant for solid waste management programs unless the Administrator is satisfied that such grant will be so used as to supplement and, to the extent practicable, increase the level of State, local, regional, or other non-Federal funds that would in the absence of such grant be made available for the maintenance of such programs.

"(c) DISTRIBUTION OF FEDERAL FINANCIAL ASSISTANCE WITHIN THE STATE.—The Federal assistance allotted to the States under subsection (b) shall be allocated by the State receiving such funds to State, local, regional, and interstate authorities carrying out planning and implementation of the State plan. Such allocation shall be based upon the responsibilities of the respective parties as determined pursuant to section 4006(b).

"(d) TECHNICAL ASSISTANCE.—The Administrator may provide technical assistance to State and local governments for purposes of developing and implementing State plans. Technical assistance respecting resource recovery and conservation may be provided through resource recovery and conservation panels, established in the Environmental Protection Agency under subtitle B, to assist the State and local governments with respect to particular resource recovery and conservation projects under consideration and to evaluate their effect on the State plan.

Ante, p. 2804.

"(e) SPECIAL COMMUNITIES.—(1) The Administrator, in cooperation with State and local officials, shall identify communities within the United States (A) having a population of less than twenty-five thousand persons, (B) having solid waste disposal facilities in which more than 75 per centum of the solid waste disposal of is from areas outside the jurisdiction of the communities, and (C) which have serious environmental problems resulting from the disposal of such solid waste.

"(2) There is authorized to be appropriated to the Administrator $2,500,000 for each of the fiscal years 1978 and 1979 to make grants to be used for the conversion, improvement, or consolidation of existing solid waste disposal facilities, or for the construction of new solid waste disposal facilities, or for both, within communities identified under paragraph (1). Not more than one community in any State shall be eligible for grants under this paragraph and not more than one project in any State shall be eligible for such grants.

Appropriation authorization.

"(3) Grants under this subsection shall be made only to projects which the Administrator determines will be consistent with an applicable State plan approved under this subtitle and which will assist in carrying out such plan.

"RURAL COMMUNITIES ASSISTANCE

"SEC. 4009. (a) IN GENERAL.—The Administrator shall make grants to States to provide assistance to municipalities with a population of five thousand or less, or counties with a population of ten thousand or less or less than twenty persons per square mile and not within a metropolitan area, for solid waste management facilities (including equipment) necessary to meet the requirements of section 4005 of this Act or restrictions on open burning or other requirements arising under the Clean Air Act or the Federal Water Pollution Control Act. Such assistance shall only be available—

42 USC 6949.

42 USC 1857 note.
33 USC 1251 note.

"(1) to any municipality or county which could not feasibly be included in a solid waste management system or facility serving an urbanized, multijurisdictional area because of its distance from such systems;

"(2) where existing or planned solid waste management services or facilities are unavailable or insufficient to comply with the requirements of section 4005 of this Act; and

"(3) for systems which are certified by the State to be consistent with any plans or programs established under any State or areawide planning process.

"(b) ALLOTMENT.—The Administrator shall allot the sums appropriated to carry out this section in any fiscal year among the States in accordance with regulations promulgated by him on the basis of the average of the ratio which the population of rural areas of each State bears to the total population of rural areas of all the States, the ratio which the population of counties in each State having less than twenty persons per square mile bears to the total population of such counties in all the States, and the ratio which the population of such low-density counties in each State having 33 per centum or more of all families with incomes not in excess of 125 per centum of the poverty level bears to the total population of such counties in all the States.

Land acquisition, prohibition.

"(c) LIMIT.—The amount of any grant under this section shall not exceed 75 per centum of the costs of the project. No assistance under this section shall be available for the acquisition of land or interests in land.

"(d) APPROPRIATIONS.—There are authorized to be appropriated $25,000,000 for each of the fiscal years 1978 and 1979 to carry out this section.

"Subtitle E—Duties of the Secretary of Commerce in Resource and Recovery

"FUNCTIONS

42 USC 6951.

"SEC. 5001. The Secretary of Commerce shall encourage greater commercialization of proven resource recovery technology by providing—

"(1) accurate specifications for recovered materials;

"(2) stimulation of development of markets for recovered materials;

"(3) promotion of proven technology; and

"(4) a forum for the exchange of technical and economic data relating to resource recovery facilities.

"DEVELOPMENT OF SPECIFICATIONS FOR SECONDARY MATERIALS

Publication of guidelines.
42 USC 6952.

"SEC. 5002. The Secretary of Commerce, acting through the National Bureau of Standards, and in conjunction with national standards-setting organizations in resource recovery, shall, after public hearings, and not later than two years after the date of the enactment of this Act, publish guidelines for the development of specifications for the classification of materials recovered from waste which were destined for disposal. The specifications shall pertain to the physical and chemical properties and characteristics of such materials with regard to their use in replacing virgin materials in various industrial, commercial, and governmental uses. In establishing such guidelines the Secretary shall also, to the extent feasible, provide such information as may be necessary to assist Federal agencies with procurement of items containing recovered materials. The Secretary shall

Cooperation with national standards-setting organizations.

continue to cooperate with national standards-setting organizations, as may be necessary, to encourage the publication, promulgation and updating of standards for recovered materials and for the use of recovered materials in various industrial, commercial, and governmental uses.

"DEVELOPMENT OF MARKETS FOR RECOVERED MATERIALS

"SEC. 5003. The Secretary of Commerce shall within two years after the enactment of this Act take such actions as may be necessary to—

> "(1) identify the geographical location of existing or potential markets for recovered materials;
> "(2) identify the economic and technical barriers to the use of recovered materials; and
> "(3) encourage the development of new uses for recovered materials.

42 USC 6953.

"TECHNOLOGY PROMOTION

"SEC. 5004. The Secretary of Commerce is authorized to evaluate the commercial feasibility of resource recovery facilities and to publish the results of such evaluation, and to develop a data base for purposes of assisting persons in choosing such a system.

42 USC 6954.

"Subtitle F—Federal Responsibilities

"APPLICATION OF FEDERAL, STATE, AND LOCAL LAW TO FEDERAL FACILITIES

"SEC. 6001. Each department, agency, and instrumentality of the executive, legislative, and judicial branches of the Federal Government (1) having jurisdiction over any solid waste management facility or disposal site, or (2) engaged in any activity resulting, or which may result, in the disposal *or management* of solid waste or hazardous waste shall be subject to, and comply with, all Federal, State, interstate, and local requirements, both substantive and procedural (including any requirement for permits or reporting or any provisions for injunctive relief and such sanctions as may be imposed by a court to enforce such relief), respecting control and abatement of solid waste or hazardous waste disposal in the same manner, and to the same extent, as any person is subject to such requirements, including the payment of reasonable service charges. Neither the United States, nor any agent, employee, or officer thereof, shall be immune or exempt from any process or sanction of any State or Federal Court with respect to the enforcement of any such injunctive relief. The President may exempt any solid waste management facility of any department, agency, or instrumentality in the executive branch from compliance with such a requirement if he determines it to be in the paramount interest of the United States to do so. No such exemption shall be granted due to lack of appropriation unless the President shall have specifically requested such appropriation as a part of the budgetary process and the Congress shall have failed to make available such requested appropriation. Any exemption shall be for a period not in excess of one year, but additional exemptions may be granted for periods not to exceed one year upon the President's making a new determination. The President shall report each January to the Congress all exemptions from the requirements of this section granted during the preceding calendar year, together with his reason for granting each such exemption.

42 USC 6961.

Exemptions.

Presidential report to Congress.

90 STAT. 2822 PUBLIC LAW 94–580—OCT. 21, 1976

"FEDERAL PROCUREMENT

42 USC 6962.
"SEC. 6002. (a) APPLICATION OF SECTION.—Except as provided in subsection (b), a procuring agency shall comply with the requirements set forth in this section and any regulations issued under this section, with respect to any purchase or acquisition of a procurement item where the purchase price of the item exceeds $10,000 or where the quantity of such items or of functionally equivalent items purchased or acquired in the course of the preceding fiscal year was $10,000 or more.

"(b) PROCUREMENT SUBJECT TO OTHER LAW.—Any procurement, by any procuring agency, which is subject to regulations of the Administrator under section 6004 (as promulgated before the date of enactment of this section under comparable provisions of prior law) shall not be subject to the requirements of this section to the extent that such requirements are inconsistent with such regulations.

"(c) REQUIREMENTS.—(1) After two years after the date of enactment of this section, each procuring agency shall procure items composed of the highest percentage of recovered materials practicable consistent with maintaining a satisfactory level of competition. The decision not to procure such items shall be based on a determination that such procurement items—

"(A) are not reasonably available within a reasonable period of time;

"(B) fail to meet the performance standards set forth in the applicable specifications or fail to meet the reasonable performance standards of the procuring agencies; or

"(C) are only available at an unreasonable price. Any determination under clause (ii) shall be made on the basis of the guidelines of the Bureau of Standards in any case in which such material is covered by such guidelines.

"(2) Agencies that generate heat, mechanical, or electrical energy from fossil fuel in systems that have the technical capability of using recovered material and recovered-material-derived fuel as a primary or supplementary fuel shall use such capability to the maximum extent practicable.

"(3)After the date specified in any applicable guidelines prepared pursuant to subsection (e) of this section, contracting officers shall require that vendors certify the percentage of the total material utilized for the performance of the contract which is recovered materials.

"(d) SPECIFICATIONS.—(1) All Federal agencies that have the responsibility for drafting or reviewing specifications for procurement item procured by Federal agencies shall, in reviewing those specifications, ascertain whether such specifications violate the prohibitions contained in subparagraphs (A) through (C) of paragraph (2). Such review shall be undertaken not later than eighteen months after the date of enactment of this section.

"(2) In drafting or revising such specifications, after the date of enactment of this section—

"(A) any exclusion of recovered materials shall be eliminated;

"(B) such specification shall not require the item to be manufactured from virgin materials; and

"(C) such specifications shall require reclaimed materials to the maximum extent possible without jeopardizing the intended end use of the item.

"(e) GUIDELINES.—The Administrator, after consultation with the Administrator of General Services, the Secretary of Commerce (acting through the Bureau of Standards), and the Public Printer, shall prepare, and from time to time revise, guidelines for the use of procuring agencies in complying with the requirements of this section. Such guidelines shall set forth recommended practices with respect to the procurement of recovered materials and items containing such materials *and with respect to certification by vendors of the percentage of recovered materials used*, and shall provide information as to the availability, sources of supply, and potential uses of such materials and items.

"(f) PROCUREMENT OF SERVICES.—A procuring agency shall, to the maximum extent practicable, manage or arrange for the procurement of solid waste management services in a manner which maximizes energy and resource recovery.

"(g) EXECUTIVE OFFICE.—The Office of Procurement Policy in the Executive Office of the President, in cooperation with the Administrator, shall implement the policy expressed in this section. It shall be the responsibility of the Office of Procurement Policy to coordinate this policy with other policies for Federal procurement, in such a way as to maximize the use of recovered resources, and to annually report to the Congress on actions taken by Federal agencies and the progress made in the implementation of such policy.

"COOPERATION WITH ENVIRONMENTAL PROTECTION AGENCY

"SEC. 6003. All Federal agencies having functions relating to solid waste or hazardous waste shall cooperate to the maximum extent permitted by law with the Administrator in carrying out his functions under this Act and shall make all appropriate information, facilities, personnel, and other resources available, on a reimbursable basis, to the Administrator upon his request.   42 USC 6963.

"APPLICABILITY OF SOLID WASTE DISPOSAL GUIDELINES TO EXECUTIVE AGENCIES

"SEC. 6004. (a) COMPLIANCE.—(1) If—   42 USC 6964.

"(A) an Executive agency (as defined in section 105 of title 5, United States Code) has jurisdiction over any real property or facility the operation or administration of which involves such agency in solid waste *management* activities, or

"(B) such an agency enters into a contract with any person for the operation by such person of any Federal property or facility, and the performance of such contract involves such person in solid waste *management* activities,

then such agency shall insure compliance with the guidelines recommended under section 1008 and the purposes of this Act in the operation or administration of such property or facility, or the performance of such contract, as the case may be.   *Ante,* p. 2803.

"(2) Each Executive agency which conducts any activity—

"(A) which generates solid waste, and

"(B) which, if conducted by a person other than such agency, would require a permit or license from such agency in order to dispose of such solid waste,

shall insure compliance with such guidelines and the purposes of this Act in conducting such activity.

"(3) Each Executive agency which permits the use of Federal property for purposes of disposal of solid waste shall insure compliance with such guidelines and the purposes of this Act in the disposal of such waste.

"(4) The President shall prescribe regulations to carry out this subsection.   Regulations.

90 STAT. 2824 PUBLIC LAW 94–580—OCT. 21, 1976

Ante, p. 2803.

"(b) LICENSES AND PERMITS.—Each Executive agency which issues any license or permit for disposal of solid waste shall, prior to the issuance of such license or permit, consult with the *Administrator* to insure compliance with guidelines recommended under section 1008 and the purposes of this Act.

"Subtitle G—Miscellaneous Provisions

"EMPLOYEE PROTECTION

42 USC 6971.

"SEC. 7001. (a) GENERAL.—No person shall fire, or in any other way discriminate against, or cause to be fired or discriminated against, any employee or any authorized representative of employees by reason of the fact that such employee or representative has filed, instituted, or caused to be filed or instituted any proceeding under this Act or under any applicable implementation plan, or has testified or is about to testify in any proceeding resulting from the administration or enforcement of the provisions of this Act or of any applicable implementation plan.

Application to Secretary for review.

"(b) REMEDY.—Any employee or a representative of employees who believes that he has been fired or otherwise discriminated against by any person in violation of subsection (a) of this section may, within thirty days after such alleged violation occurs, apply to the Secretary of Labor for a review of such firing or alleged discrimination. A copy of the application shall be sent to such person who shall be the respondent. Upon receipt of such application, the Secretary of Labor shall cause such investigation to be made as he deems appropriate. Such

Hearing.

investigation shall provide an opportunity for a public hearing at the request of any party to such review to enable the parties to present information relating to such alleged violation. The parties shall be

Notice.

given written notice of the time and place of the hearing at least five days prior to the hearing. Any such hearing shall be of record and shall be subject to section 554 of title 5 of the United States Code. Upon receiving the report of such investigation, the Secretary of Labor shall make findings of fact. If he finds that such violation did occur, he shall issue a decision, incorporating an order therein and his findings, requir-

Rehiring or reinstatement of employee.

ing the party committing such violation to take such affirmative action to abate the violation as the Secretary of Labor deems appropriate, including, but not limited to, the rehiring or reinstatement of the employee or representative of employees to his former position with compensation. If he finds that there was no such violation, he shall issue

Judicial review.

an order denying the application. Such order issued by the Secretary of Labor under this subparagraph shall be subject to judicial review in the same manner as orders and decisions of the Administrator or subject to judicial review under this Act.

"(c) COSTS.—Whenever an order is issued under this section to abate such violation, at the request of the applicant, a sum equal to the aggregate amount of all costs and expenses (including the attorney's fees) as determined by the Secretary of Labor, to have been reasonably incurred by the applicant for, or in connection with, the institution and prosecution of such proceedings, shall be assessed against the person committing such violation.

"(d) EXCEPTION.—This section shall have no application to any employee who, acting without direction from his employer (or his agent) deliberately violates any requirement of this Act.

"(e) EMPLOYMENT SHIFTS AND LOSS.—The Administrator shall conduct continuing evaluations of potential loss or shifts of employment which may result from the administration or enforcement of the provisions of this Act and applicable implementation plans, including, where appropriate, investigating threatened plant closures or reductions in employment allegedly resulting from such administration or enforcement. Any employee who is discharged, or laid off, threatened with discharge or layoff, or otherwise discriminated against by any person because of the alleged results of such administration or enforcement, or any representative of such employee, may request the Administrator to conduct a full investigation of the matter. The Administrator shall thereupon investigate the matter and, at the request of any party, shall hold public hearings on not less than five days' notice, and shall at such hearings require the parties, including the employer involved, to present information relating to the actual or potential effect of such administration or enforcement on employment and on any alleged discharge, layoff, or other discrimination and the detailed reasons or justification therefor. Any such hearing shall be of record and shall be subject to section 554 of title 5 of the United States Code. Upon receiving the report of such investigation, the Administrator shall make findings of fact as to the effect of such administration or enforcement on employment and on the alleged discharge, layoff, or discrimination and shall make such recommendations as he deems appropriate. Such report, findings, and recommendations shall be available to the public. Nothing in this subsection shall be construed to require or authorize the Administrator or any State to modify or withdraw any standard, limitation, or any other requirement of this Act or any applicable implementation plan.

Request for investigation.

Hearing.

Information, availability to public.

"CITIZEN SUITS

"SEC. 7002. (a) IN GENERAL.—Except as provided in subsection (b) or (c) of this section, any person may commence a civil action on his own behalf—

42 USC 6972.

"(1) against any person (including (a) the United States, and (b) any other governmental instrumentality or agency, to the extent permitted by the eleventh amendment to the Constitution) who is alleged to be in violation of any permit, standard, regulation, condition, requirement, or order which has become effective pursuant to this Act; or

"(2) against the Administrator where there is alleged a failure of the Administrator to perform any act or duty under this Act which is not discretionary with the Administrator.

Any action under paragraph (a)(1) of this subsection shall be brought in the district court for the district in which the alleged violation occurred. Any action brought under paragraph (a)(2) of this subsection may be brought in the district court for the district in which the alleged violation occurred or in the District Court of the District of Columbia. The district court shall have jurisdiction, without regard to the amount in controversy or the citizenship of the parties, to enforce such regulation or order, or to order the Administrator to perform such act or duty as the case may be.

"(b) ACTIONS PROHIBITED.—No action may be commenced under paragraph (a)(1) of this section—

"(1) prior to sixty days after the plaintiff has given notice of the violation (A) to the Administrator; (B) to the State in which the alleged violation occurs; and (C) to any alleged violator of such permit, standard, regulation, condition, requirement, or order; or

90 STAT. 2826   PUBLIC LAW 94-580—OCT. 21, 1976

"(2) if the Administrator or State has commenced and is diligently prosecuting a civil or criminal action in a court of the United States or a State to require compliance with such permit, standard, regulation, condition, requirement, or order: *Provided, however,* That in any such action in a court of the United States, any person may intervene as a matter of right.

"(c) NOTICE.—No action may be commenced under paragraph (a)(2) of this section prior to sixty days after the plaintiff has given notice to the Administrator that he will commence such action, except that such action may be brought immediately after such notification in the case of an action under this section respecting a violation of *subtitle C* of this Act. Notice under this subsection shall be given in

42 USC 3254f.

such manner as the Administrator shall prescribe by regulation. Any action respecting a violation under this Act may be brought under this section only in the judicial district in which such alleged violation occurs.

"(d) INTERVENTION.—In any action under this section the Administrator, if not a party, may intervene as a matter of right.

"(e) COSTS.—The court, in issuing any final order in any action brought pursuant to this section, may award costs of litigation (including reasonable attorney and expert witness fees) to any party, whenever the court determines such an award is appropriate. The court may, if a temporary restraining order or preliminary injunction is sought, *require* the filing of a bond or equivalent security in accord-

28 USC app.

ance with the Federal Rules of Civil Procedure.

"(f) OTHER RIGHTS PRESERVED.—Nothing in this section shall restrict any right which any person (or class of persons) may have under any statute or common law to seek enforcement of any standard or requirement relating to the management of solid waste or hazardous waste, or to seek any other relief (including relief against the Administrator or a State agency).

"IMMINENT HAZARD

42 USC 6973.

"SEC. 7003. Notwithstanding any other provision of this Act, upon receipt of evidence that the handling, storage, treatment, transportation or disposal of any solid waste or hazardous waste is presenting an imminent and substantial endangerment to health or the environment, the Administrator may bring suit on behalf of the United States in the appropriate district court to immediately restrain any person

contributing to the alleged disposal to stop such handling, storage, treatment, transportation, or disposal or to take such other action as may be necessary. The Administrator shall provide notice to the affected State of any such suit.

"PETITION FOR REGULATIONS; PUBLIC PARTICIPATION

Publication in Federal Register. 42 USC 6974.

"SEC. 7004. (a) PETITION.—Any person may petition the Administrator for the promulgation, amendment, or repeal of any regulation under this Act. Within a reasonable time following receipt of such petition, the Administrator shall take action with respect to such petition and shall publish notice of such action in the Federal Register, together with the reasons therefor.

"(b) PUBLIC PARTICIPATION.—Public participation in the development, revision, implementation, and enforcement of any regulation, guideline, information, or program under this Act shall be provided for, encouraged, and assisted by the Administrator and the States.

The Administrator, in cooperation with the States, shall develop and publish minimum guidelines for public participation in such processes.

"SEPARABILITY

"SEC. 7005. If any provision of this Act, or the application of any provision of this Act to any person or circumstance, is held invalid, the application of such provision to other persons or circumstances, and the remainder of this Act, shall not be affected thereby.

42 USC 6975.

"JUDICIAL REVIEW

"SEC. 7006. Any judicial review of final regulations promulgated pursuant to this Act shall be in accordance with sections 701 through 706 of title 5 of the United States Code, except that—

42 USC 6976.

"(1) a petition for review of action of the Administrator in promulgating any regulation, or requirement under this Act may be filed only in the United States Court of Appeals for the District of Columbia. Any such petition shall be filed within ninety days from the date of such promulgation, or after such date if such petition is based solely on grounds arising after such ninetieth day. Action of the Administrator with respect to which review could have been obtained under this subsection shall not be subject to judicial review in civil or criminal proceedings for enforcement; and

"(2) in any judicial proceeding brought under this section in which review is sought of a determination under this Act required to be made on the record after notice and opportunity for hearing, if a party seeking review under this Act applies to the court for leave to adduce additional evidence, and shows to the satisfaction of the court that the information is material and that there were reasonable grounds for the failure to adduce such evidence in the proceeding before the Administrator, the court may order such additional evidence (and evidence in rebuttal thereof) to be taken before the Administrator, and to be adduced upon the hearing in such manner and upon such terms and conditions as the court may deem proper. The Administrator may modify his findings as to the facts, or make new findings, by reason of the additional evidence so taken, and he shall file with the court such modified or new findings and his recommendation, if any, for the modification or setting aside of his original order, with the return of such additional evidence.

"GRANTS OR CONTRACTS FOR TRAINING PROJECTS

"SEC. 7007. (a) GENERAL AUTHORITY.—The Administrator is authorized to make grants to, and contracts with any eligible organization. For purposes of this section the term "eligible organization" means a State or interstate agency, a municipality, educational institution, and any other organization which is capable of effectively carrying out a project which may be funded by grant under subsection (b) of this section.

42 USC 6977.

"Eligible organization."

"(b) PURPOSES.—(1) Subject to the provisions of paragraph (2), grants or contracts may be made to pay all or a part of the costs, as may be determined by the Administrator, of any project operated or to be operated by an eligible organization, which is designed—

"(A) to develop, expand, or carry out a program (which may

combine training, education, and employment) for training persons for occupations involving the management, supervision, design, operation, or maintenance of solid waste *management* and resource recovery equipment and facilities; or

"(B) to train instructors and supervisory personnel to train or supervise persons in occupations involving the design, operation, and maintenance of solid waste *management* and resource recovery equipment and facilities.

"(2) A grant or contract authorized by paragraph (1) of this subsection may be made only upon application to the Administrator at such time or times and containing such information as he may prescribe, except that no such application shall be approved unless it provides for the same procedures and reports (and access to such reports and to other records) as required by section 207(b) (4) and (5) (as in effect before the date of the enactment of Resource Conservation and Recovery Act of 1976) with respect to applications made under such section (as in effect before the date of the enactment of Resource Conservation and Recovery Act of 1976).

"(c) STUDY.—The Administrator shall make a complete investigation and study to determine—

"(1) the need for additional trained State and local personnel to carry out plans assisted under this Act and other solid waste and resource recovery programs;

"(2) means of using existing training programs to train such personnel; and

"(3) the extent and nature of obstacles to employment and occupational advancement in the solid waste *management* and resource recovery field which may limit either available manpower or the advancement of personnel in such field.

He shall report the results of such investigation and study, including his recommendations to the President and the Congress.

42 USC 3254a.
Ante, p. 2795.

Report to President and Congress.

"PAYMENTS

42 USC 6978.

"SEC. 7008. (a) GENERAL RULE.—Payments of grants under this Act may be made (after necessary adjustment on account of previously made underpayments or overpayments) in advance or by way of reimbursement, and in such installments and on such conditions as the Administrator may determine.

"(b) PROHIBITION.—No grant may be made under this Act to any private profitmaking organization.

"LABOR STANDARDS

42 USC 6979.

"SEC. 7009. No grant for a project of construction under this Act shall be made unless the Secretary finds that the application contains or is supported by reasonable assurance that all laborers and mechanics employed by contractors or subcontractors on projects of the type covered by the Davis-Bacon Act, as amended (40 U.S.C. 276a—276a-5), will be paid wages at rates not less than those prevailing on similar work in the locality as determined by the Secretary of Labor in accordance with that Act; and the Secretary of Labor shall have with respect to the labor standards specified in this section the authority and functions set forth in Reorganization Plan Numbered 14 of 1950 (15 F.R. 3176; 5 U.S.C. 133z-5) and section 2 of the Act of June 13, 1934, as amended (40 U.S.C. 276c).

5 USC app. II.

PUBLIC LAW 94–580—OCT. 21, 1976   90 STAT. 2829

"Subtitle H—Research, Development, Demonstration, and
Information

"RESEARCH, DEMONSTRATIONS, TRAINING, AND OTHER ACTIVITIES

"SEC. 8001. (a) GENERAL AUTHORITY.—The Administrator, alone or   42 USC 6981.
after consultation with the Administrator of the Federal Energy
Administration, the Administrator of the Energy Research and
Development Administration, or the Chairman of the Federal Power
Commission, shall conduct, and encourage, cooperate with, and render
financial and other assistance to appropriate public (whether Federal,
State, interstate, or local) authorities, agencies, and institutions,
private agencies and institutions, and individuals in the conduct of,
and promote the coordination of, research, investigations, experiments, training, demonstrations, surveys, public education programs,
and studies relating to—

"(1) any adverse health and welfare effects of the release into
the environment of material present in solid waste, and methods
to eliminate such effects;

"(2) the operation and financing of solid waste *management*
programs;

"(3) the planning, implementation, and operation of resource
recovery and resource conservation systems and hazardous waste
management systems, including the marketing of recovered
resources;

"(4) the production of usable forms of recovered resources,
including fuel, from solid waste;

"(5) the reduction of the amount of such waste and unsalvageable waste materials;

"(6) the development and application of new and improved
methods of collecting and disposing of solid waste and processing
and recovering materials and energy from solid wastes;

"(7) the identification of solid waste components and potential
materials and energy recoverable from such waste components;

"(8) small scale and low technology solid waste management
systems, including but not limited to, resource recovery source
separation systems;

"(9) methods to improve the performance characteristics of
resources recovered from solid waste and the relationship of such
performance characteristics to available and potentially available
markets for such resources;

"(10) improvements in land disposal practices for solid waste
(including sludge) which may reduce the adverse environmental
effects of such disposal and other aspects of solid waste disposal
on land, including means for reducing the harmful environmental
effects of earlier and existing landfills, means for restoring areas
damaged by such earlier or existing landfills, means for rendering
landfills safe for purposes of construction and other uses, and
techniques of recovering materials and energy from landfills;

"(11) methods for the sound disposal of, or recovery of
resources, including energy, from, sludge (including sludge from
pollution control and treatment facilities, coal slurry pipelines,
and other sources);

"(12) methods of hazardous waste management, including
methods of rendering such waste environmentally safe; and

"(13) any adverse effects on air quality (particularly with

regard to the emission of heavy metals) which result from solid waste which is burned (either alone or in conjunction with other substances)for purposes of *treatment,* disposal or energy recovery.

"(b) MANAGEMENT PROGRAM.—(1)(A) In carrying out his functions pursuant to this Act, and any other Federal legislation respecting solid waste or discarded material research, development, and demonstrations, the Administrator shall establish a management program or system to insure the coordination of all such activities and to facilitate and accelerate the process of development of sound new technology (or other discoveries) from the research phase, through development, and into the demonstration phase.

"(B) The Administrator shall (i) assist, on the basis of any research projects which are developed with assistance under this Act or without Federal assistance, the construction of pilot plant facilities for the purpose of investigating or testing the technological feasibility of any promising new fuel, energy, or resource recovery or resource conservation method or technology; and (ii) demonstrate each such method and technology that appears justified by an evaluation at such pilot plant stage or at a pilot plant stage developed without Federal assistance. Each such demonstration shall incorporate new or innovative technical advances or shall apply such advances to different circumstances and conditions, for the purpose of evaluating design concepts or to test the performance, efficiency, and economic feasibility of a particular method or technology under actual operating conditions. Each such demonstration shall be so planned and designed that, if successful, it can be expanded or utilized directly as a full-scale operational fuel, energy, or resource recovery or resource conservation facility.

"(2) Any energy-related research, development, or demonstration project for the conversion including bioconversion, of solid waste carried out by the Environmental Protection Agency or by the Energy Research and Development Administration pursuant to this or any other Act shall be administered in accordance with the May 7, 1976, Interagency Agreement between the Environmental Protection Agency and the Energy Research and Development Administration on the Development of Energy from Solid Wastes and specifically, that in accordance with this agreement, (A) for those energy-related projects of mutual interest, planning will be conducted jointly by the Environmental Protection Agency and the Energy Research and Development Administration, following which project responsibility will be assigned to one agency; (B) energy-related portions of projects for recovery of synthetic fuels or other forms of energy from solid waste shall be the responsibility of the Energy Research and Development Administration; (C) the Environmental Protection Agency shall retain responsibility for the environmental, economic, and institutional aspects of solid waste projects and for assurance that such projects are consistent with any applicable suggested guidelines published pursuant to section 1008, and any applicable State or regional solid waste management plan; and (D) any activities undertaken under provisions of sections 8002 and 8003 as related to energy; as related to energy or synthetic fuels recovery from waste; or as related to energy conservation shall be accomplished through coordination and consultation with the Energy Research and Development Administration.

"(c) AUTHORITIES.—(1) In carrying out subsection (a) of this section respecting solid waste research, studies, development, and demonstration, except as otherwise specifically provided in section 8004(d), the Administrator may make grants to or enter into contracts (including contracts for construction) with, public agencies and authorities or private persons.

"(2) Contracts for research, development, or demonstrations or for both (including contracts for construction) shall be made in accordance with and subject to the limitations provided with respect to research contracts of the military departments in title 10, United States Code, section 2353, except that the determination, approval, and certification required thereby shall be made by the Administrator.

"(3) Any invention made or conceived in the course of, or under, any contract under this Act shall be subject to section 9 of the Federal Nonnuclear Energy Research and Development Act of 1974 to the same extent and in the same manner as inventions made or conceived in the course of contracts under such Act, except that in applying such section, the Environmental Protection Agency shall be substituted for the Energy Research and Development Administration and the words 'solid waste' shall be substituted for the word 'energy' where appropriate.

42 USC 5908.

"(4) For carrying out the purpose of this Act the Administrator may detail personnel of the Environmental Protection Agency to agencies eligible for assistance under this section.

Detail of EPA personnel to other agencies.

"SPECIAL STUDIES; PLANS FOR RESEARCH, DEVELOPMENT, AND DEMONSTRATIONS

"SEC. 8002. (a) GLASS AND PLASTIC.—The Administrator shall undertake a study and publish a report on resource recovery from glass and plastic waste, including a scientific, technological, and economic investigation of potential solutions to implement such recovery.

42 USC 6982.

"(b) COMPOSITION OF WASTE STREAM.—The Administrator shall undertake a systematic study of the composition of the solid waste stream and of anticipated future changes in the composition of such stream and shall publish a report containing the results of such study and quantitatively evaluating the potential utility of such components.

"(c) PRIORITIES STUDY.—For purposes of determining priorities for research on recovery of materials and energy from solid waste and developing materials and energy recovery research, development, and demonstration strategies, the Administrator shall review, and make a study of, the various existing and promising techniques of energy recovery from solid waste (including, but not limited to, waterwall furnace incinerators, dry shredded fuel systems, pyrolysis, densified refuse-derived fuel systems, anerobic digestion, and fuel and feedstock preparation systems). In carrying out such study the Administrator shall investigate with respect to each such technique—

"(1) the degree of public need for the potential results of such research, development, or demonstration,

"(2) the potential for research, development, and demonstration without Federal action, including the degree of restraint on such potential posed by the risks involved, and

"(3) the magnitude of effort and period of time necessary to develop the technology to the point where Federal assistance can be ended.

"(d) SMALL-SCALE AND LOW TECHNOLOGY STUDY.—The Administrator shall undertake a comprehensive study and analysis of, and publish a report on, systems of small-scale and low technology solid waste management, including household resource recovery and resource recovery systems which have special application to multiple dwelling units and high density housing and office complexes. Such study and analysis shall include an investigation of the degree to which such systems could contribute to energy conservation.

"(e) FRONT-END SOURCE SEPARATION.—The Administrator shall undertake research and studies concerning the compatibility of front-end source separation systems with high technology resource recovery systems and shall publish a report containing the results of such research and studies.

"(f) MINING WASTE.—The Administrator, in consultation with the Secretary of the Interior, shall conduct a detailed and comprehensive study on the adverse effects of solid wastes from active and abandoned surface and underground mines on the environment. including, but not limited to, the effects of such wastes on humans, water, air, health, welfare, and natural resources, and on the adequacy of means and measures currently employed by the mining industry, Government agencies, and others to dispose of and utilize such solid wastes and to prevent or substantially mitigate such adverse effects. Such study shall include an analysis of—

"(1) the sources and volume of discarded material generated per year from mining;

"(2) present disposal practices;

"(3) potential dangers to human health and the environment from surface runoff of leachate and air pollution by dust;

"(4) alternatives to current disposal methods;

"(5) the cost of those alternatives in terms of the impact on mine product costs; and

"(6) potential for use of discarded material as a secondary source of the mine product.

In furtherance of this study, the Administrator shall, as he deems appropriate, review studies and other actions of other Federal agencies concerning such wastes with a view toward avoiding duplication of effort and the need to expedite such study. The Administrator shall publish a report of such study and shall include appropriate findings and recommendations for Federal and non-Federal actions concerning such effects.

"(g) SLUDGE.—The Administrator shall undertake a comprehensive study and publish a report on sludge. Such study shall include an analysis of—

"(1) what types of solid waste (including but not limited to sewage and pollution treatment residues and other residues from industrial operations such as extraction of oil from shale, liquefaction and gasification of coal and coal slurry pipeline operations) shall be classified as sludge;

"(2) the effects of air and water pollution legislation on the creation of large volumes of sludge;

"(3) the amounts of sludge originating in each State and in each industry producing sludge;

"(4) methods of disposal of such sludge, including the cost, efficiency, and effectiveness of such methods;

"(5) alternative methods for the use of sludge, including agricultural applications of sludge and energy recovery from sludge; and

"(6) methods to reclaim areas which have been used for the disposal of sludge or which have been damaged by sludge.

"(h) TIRES.—The Administrator shall undertake a study and publish a report respecting discarded motor vehicle tires which shall include an analysis of the problems involved in the collection, recovery of resources including energy, and use of such tires.

"(i) RESOURCE RECOVERY FACILITIES.—The Administrator shall conduct research and report on the economics of, and impediments, to the effective functioning of resource recovery facilities.

"(j) RESOURCE CONSERVATION COMMITTEE.—(1) The Administrator shall serve as Chairman of a Committee composed of himself, the Secretary of Commerce, the Secretary of Labor, the Chairman of the Council on Environmental Quality, the Secretary of Treasury, the Secretary of the Interior, *the Secretary of Energy, the Chairman of the Council of Economic Advisors,* and a representative of the Office of Management and Budget, which shall conduct a full and complete investigation and study of all aspects of the economic, social, and environmental consequences of resource conservation with respect to—

"(A) the appropriateness of recommended incentives and disincentives to foster resource conservation;

"(B) the effect of existing public policies (including subsidies and economic incentives and disincentives, percentage depletion allowances, capital gains treatment and other tax incentives and disincentives) upon resource conservation, and the likely effect of the modification or elimination of such incentives and disincentives upon resource conservation;

"(C) the appropriateness and feasibility of restricting the manufacture or use of categories of consumer products as a resource conservation strategy;

"(D) the appropriateness and feasibility of employing as a resource conservation strategy the imposition of solid waste management charges on consumer products, which charges would reflect the costs of solid waste management services, litter pickup, the value of recoverable components of such product, final disposal, and any social value associated with the nonrecycling or uncontrolled disposal of such product; and

"(E) the need for further research, development, and demonstration in the area of resource conservation.

"(2) The study required in paragraph *(1)(D)* may include pilot scale projects, and shall consider and evaluate alternative strategies with respect to— **Pilot scale projects.**

"(A) the product categories on which such charges would be imposed;

"(B) the appropriate state in the production of such consumer product at which to levy such charge;

"(C) appropriate criteria for establishing such charges for each consumer product category;

"(D) methods for the adjustment of such charges to reflect actions such as recycling which would reduce the overall quantities of solid waste requiring disposal; and

"(E) procedures for amending, modifying, or revising such charges to reflect changing conditions.

"(3) The design for the study required in paragraph *(1)* of this subsection shall include timetables for the completion of the study. A preliminary report putting forth the study design shall be sent to the President and the Congress within six months following enactment of this section and followup reports shall be sent six months thereafter. Each recommendation resulting from the study shall include at least two alternatives to the proposed recommendation. **Study design.** **Report to President and Congress.**

Report to
President and
Congress.

Appropriation
authorization.

"(4) The results of such investigation and study, including recommendations, shall be reported to the President and the Congress not later than two years after enactment of this subsection.

"(5) There are authorized to be appropriated not to exceed $2,000,000 to carry out this subsection.

"(k) AIRPORT LANDFILLS.—The Administrator shall undertake a comprehensive study and analysis of and publish a report on systems to alleviate the hazards to aviation from birds congregating and feeding on landfills in the vicinity of airports.

"(l) COMPLETION OF RESEARCH AND STUDIES.—The Administrator shall complete the research and studies, and submit the reports, required under subsections (b), (c), (d), (e), (f), (g), and (k) not later than October 1, 1978. The Administrator shall complete the research and studies, and submit the reports, required under sub-sections *(a), (h), and (i)*   not later than October 1, 1979. Upon completion, each study specified in subsections (a) through (k) of this section, the Administrator shall prepare a plan for research, development, and demonstration respecting the findings of the study and shall submit any legislative recommendations resulting from such study to appropriate committees of Congress.

"(m) AUTHORIZATION OF APPROPRIATIONS.—There are authorized to be appropriated not to exceed $8,000,000 for the fiscal years 1978 and 1979 to carry out this section other than subsection (j).

"COORDINATION, COLLECTION, AND DISSEMINATION OF INFORMATION

42 USC 6983.

"SEC. 8003. (a) INFORMATION.—The Administrator shall develop, collect, evaluate, and coordinate information on—

"(1) methods and costs of the collection of solid waste;

"(2) solid waste management practices, including data on the different management methods and the cost, operation, and maintenance of such methods;

"(3) the amounts and percentages of resources (including energy) that can be recovered from solid waste by use of various   *solid waste* management practices and various technologies;

"(4) methods available to reduce the amount of solid waste that is generated;

"(5) existing and developing technologies for the recovery of energy or materials from solid waste and the costs, reliability, and risks associated with such technologies;

"(6) hazardous solid waste, including incidents of damage resulting from the disposal of hazardous solid wastes; inherently and potentially hazardous solid wastes; methods of neutralizing or properly disposing of hazardous solid wastes; facilities that properly dispose of hazardous wastes;

"(7) methods of financing resource recovery facilities or, sanitary landfills, or hazardous solid waste treatment facilities, whichever is appropriate for the entity developing such facility or landfill (taking into account the amount of solid waste reasonably expected to be available to such entity);

"(8) the availability of markets for the purchase of resources, either materials or energy, recovered from solid waste; and

"(9) research and development projects respecting solid waste management.

"(b) LIBRARY.—(1) The Administrator shall establish and maintain a central reference library for (A) the materials collected pursuant to subsection (a) of this section and (B) the actual performance and cost effectiveness records and other data and information with respect to—

"(i) the various methods of energy and resource recovery from solid waste,

"(ii) the various systems and means of resource conservation,

"(iii) the various systems and technologies for collection, transport, storage, treatment, and final disposition of solid waste, and

"(iv) other aspects of solid waste and hazardous solid waste management.

Such central reference library shall also contain, but not be limited to, the model codes and model accounting systems developed under this section, the information collected under subsection (d), and, subject to any applicable requirements of confidentiality, information respecting any aspect of solid waste provided by officers and employees of the Environmental Protection Agency which has been acquired by them in the conduct of their functions under this Act and which may be of value to Federal, State, and local authorities and other persons.

"(2) Information in the central reference library shall, to the extent practicable, be collated, analyzed, verified, and published and shall be made available to State and local governments and other persons at reasonable times and subject to such reasonable charges as may be necessary to defray expenses of making such information available.

"(c) MODEL ACCOUNTING SYSTEM.—In order to assist State and local governments in determining the cost and revenues associated with the collection and disposal of solid waste and with resource recovery operations, the Administrator shall develop and publish a recommended model cost and revenue accounting system applicable to the solid waste management functions of State and local governments. Such system shall be in accordance with generally accepted accounting principles. The Administrator shall periodically, but not less frequently than once every five years, review such accounting system and revise it as necessary.

"(d) MODEL CODES.—The Administrator is authorized, in cooperation with appropriate State and local agencies, to recommend model codes, ordinances, and statutes, providing for sound solid waste management.

"(e) INFORMATION PROGRAMS.—(1) The Administrator shall implement a program for the rapid dissemination of information on solid waste management, hazardous waste management, resource conservation, and methods of resource recovery from solid waste, including the results of any relevant research, investigations, experiments, surveys, studies, or other information which may be useful in the implementation of new or improved solid waste management practices and methods and information on any other technical, managerial, financial, or market aspect of resource conservation and recovery facilities.

"(2) The Administrator shall develop and implement educational programs to promote citizen understanding of the need for environmentally sound solid waste management practices.

"(f) COORDINATION.—In collecting and disseminating information under this section, the Administrator shall coordinate his actions and cooperate to the maximum extent possible with State and local authorities.

"(g) SPECIAL RESTRICTION.—Upon request, the full range of alternative technologies, programs or processes deemed feasible to meet the resource recovery or resource conservation needs of a jurisdiction shall be described in such a manner as to provide a sufficient evaluative basis from which the jurisdiction can make its decisions, but no officer or employee of the Environmental Protection Agency shall, in an official capacity, lobby for or otherwise represent an agency position in favor of resource recovery or resource conservation, as a policy alternative for adoption into ordinances, codes, regulations, or law by any State or political subdivision thereof.

"FULL-SCALE DEMONSTRATION FACILITIES

42 USC 6984.

"Sec. 8004. (a) AUTHORITY.—The Administrator may enter into contracts with public agencies or authorities or private persons for the construction and operation of a full-scale demonstration facility under this Act, or provide financial assistance in the form of grants to a full-scale demonstration facility under this Act only if the Administrator finds that—

"(1) such facility or proposed facility will demonstrate at full scale a new or significantly improved technology or process, a practical and significant improvement in solid waste management practice, or the technological feasibility and cost effectiveness of an existing, but unproven technology, process, or practice, and will not duplicate any other Federal, State, local, or commercial facility which has been constructed or with respect to which construction has begun (determined as of the date action is taken by the Administrator under this Act),

"(2) such contract or assistance meets the requirements of section 8001 and meets other applicable requirements of the Act,

"(3) such facility will be able to comply with the guidelines published under section 1008 and with other laws and regulations for the protection of health and the environment,

"(4) in the case of a contract for construction or operation, such facility is not likely to be constructed or operated by State, local, or private persons or in the case of an application for financial assistance, such facility is not likely to receive adequate financial assistance from other sources, and

"(5) any Federal interest in, or assistance to, such facility will be disposed of or terminated, with appropriate compensation, within such period of time as may be necessary to carry out the basic objectives of this Act.

"(b) TIME LIMITATION.—No obligation may be made by the Administrator for financial assistance under this subtitle for any full-scale demonstration facility after the date ten years after the enactment of this section. No expenditure of funds for any such full-scale demonstration facility under this subtitle may be made by the Administrator after the date fourteen years after such date of enactment.

"(c) COST SHARING.—Wherever practicable, in constructing, operating, or providing financial assistance under this subtitle to a full-scale demonstration facility, the Administrator shall endeavor to enter into agreements and make other arrangements for maximum practicable cost sharing with other Federal, State, and local agencies, private persons, or any combination thereof.

"(2) The Administrator shall enter into arrangements, wherever practicable and desirable, to provide monitoring of full-scale solid waste facilities (whether or not constructed or operated under this

PUBLIC LAW 94–580—OCT. 21, 1976   90 STAT. 2837

Act) for purposes of obtaining information concerning the performance, and other aspects, of such facilities. Where the Administrator provides only monitoring and evaluation instruments or personnel (or both) or funds for such instruments or personnel and provides no other financial assistance to a facility, notwithstanding section 8001(c)(3), title to any invention made or conceived of in the course of developing, constructing, or operating such facility shall not be required to vest in the United States and patents respecting such invention shall not be required to be issued to the United States.

"(d) PROHIBITION.—After the date of enactment of this section, the Administrator shall not construct or operate any full-scale facility (except by contract with public agencies or authorities or private persons).

"SPECIAL STUDY AND DEMONSTATION PROJECTS ON RECOVERY OF USEFUL ENERGY AND MATERIALS

"SEC. 8005. (a) STUDIES.—The Administrator shall conduct studies   42 USC 6985. and develop recommendations for administrative or legislative action on—

"(1) means of recovering materials and energy from solid waste, recommended uses of such materials and energy for national or international welfare, including identification of potential markets for such recovered resources, the impact of distribution of such resources on existing markets, and potentials for energy conservation through resource conservation and resource recovery;

"(2) actions to reduce waste generation which have been taken voluntarily or in response to governmental action, and those which practically could be taken in the future, and the economic, social, and environmental consequences of such actions;

"(3) methods of collection, separation, and containerization which will encourage efficient utilization of facilities and contribute to more effective programs of reduction, reuse, or disposal of wastes;

"(4) the use of Federal procurement to develop market demand for recovered resources;

"(5) recommended incentives (including Federal grants, loans, and other assistance) and disincentives to accelerate the reclamation or recycling of materials from solid wastes, with special emphasis on motor vehicle hulks;

"(6) the effect of existing public policies, including subsidies and economic incentives and disincentives, percentage depletion allowances, capital gains treatment and other tax incentives and disincentives, upon the recycling and reuse of materials, and the likely effect of the modification or elimination of such incentives and disincentives upon the reuse, recycling and conservation of such materials;

"(7) the necessity and method of imposing disposal or other charges on packaging, containers, vehicles, and other manufactured goods, which charges would reflect the cost of final disposal, the value of recoverable components of the item, and any social costs associated with nonrecycling or uncontrolled disposal of such items; and

90 STAT. 2838   PUBLIC LAW 94–580—OCT. 21, 1976

"(8) the legal constraints and institutional barriers to the acquisition of land needed for solid waste management, including land for facilities and disposal sites;

"(9) in consultation with the Secretary of Agriculture, agricultural waste management problems and practices, the extent of reuse and recovery of resources in such wastes, the prospects for improvement, Federal, State, and local regulations governing such practices, and the economic, social, and environmental consequences of such practices; and

"(10) in consultation with the Secretary of the Interior, mining waste management problems, and practices, including an assessment of existing authorities, technologies, and economics, and the environmental and public health consequences of such practices.

"(b) DEMONSTRATION.—The Administrator is also authorized to carry out demonstration projects to test and demonstrate methods and techniques developed pursuant to subsection (a).

"(c) APPLICATION OF OTHER SECTIONS.—Section 8001 (b) and (c) shall be applicable to investigations, studies, and projects carried out under this section.

"GRANTS FOR RESOURCE RECOVERY SYSTEMS AND IMPROVED SOLID WASTE DISPOSAL FACILITIES

42 USC 6986.

"SEC. 8006. (a) AUTHORITY.—The Administrator is authorized to make grants pursuant to this section to any State, municipal, or interstate or intermunicipal agency for the demonstration of resource recovery systems or for the construction of new or improved solid waste disposal facilities.

"(b) CONDITIONS.—(1) Any grant under this section for the demonstration of a resource recovery system may be made only if it (A) is consistent with any plans which meet the requirements of subtitle D of this Act; (B) is consistent with the guidelines recommended pursuant to section 1008 of this Act; (C) is designed to provide area-wide resource recovery systems consistent with the purposes of this Act, as determined by the Administrator, pursuant to regulations promulgated under subsection (d) of this section; and (D) provides an equitable system for distributing the costs associated with construction, operation, and maintenance of any resource recovery system among the users of such system.

"(2) The Federal share for any project to which paragraph (1) applies shall not be more than 75 percent.

"(c) LIMITATIONS.—(1) A grant under this section for the construction of a new or improved solid waste disposal facility may be made only if—

"(A) a State or interstate plan for solid waste disposal has been adopted which applies to the area involved, and the facility to be constructed (i) is consistent with such plan, (ii) is included in a comprehensive plan for the area involved which is satisfactory to the Administrator for the purposes of this Act, and (iii) is consistent with the guidelines recommended under section 1008, and

"(B) the project advances the state of the art by applying new and improved techniques in reducing the environmental impact of solid waste disposal, in achieving recovery of energy or resources, or in recycling useful materials.

"(2) The Federal share for any project to which paragraph (1) applies shall be not more than 50 percent in the case of a project serving an area which includes only one municipality, and not more than 75 percent in any other case.

"(d) REGULATIONS.—(1) The Administrator shall promulgate reg- *Regulations.* ulations establishing a procedure for awarding grants under this section which—

"(A) provides that projects will be carried out in communities of varying sizes, under such conditions as will assist in solving the community waste problems of urban-industrial centers, metropolitan regions, and rural areas, under representative geographic and environmental conditions; and

"(B) provides deadlines for submission of, and action on, grant requests.

(2) In taking action on applications for grants under this section, consideration shall be given by the Administrator (A) to the public benefits to be derived by the construction and the propriety of Federal aid in making such grant; (B) to the extent applicable, to the economic and commercial viability of the project (including contractual arrangements with the private sector to market any resources recovered); (C) to the potential of such project for general application to community solid waste disposal problems; and (D) to the use by the applicant of comprehensive regional or metropolitan area planning.

"(e) ADDITIONAL LIMITATIONS.—A grant under this section—

"(1) may be made only in the amount of the Federal share of (A) the estimated total design and construction costs, plus (B) in the case of a grant to which subsection (b)(1) applies, the first-year operation and maintenance costs;

"(2) may not be provided for land acquisition or (except as otherwise provided in paragraph (1)(B)) for operating or maintenance costs;

"(3) may not be made until the applicant has made provision satisfactory to the Administrator for proper and efficient operation and maintenance of the project (subject to paragraph (1)(B)); and

"(4) may be made subject to such conditions and requirements, in addition to those provided in this section, as the Administrator may require to properly carry out his functions pursuant to this Act.

For purposes of paragraph (1), the non-Federal share may be in any form, including, but not limited to, lands or interests therein needed for the project or personal property or services, the value of which shall be determined by the Administrator.

"(f) SINGLE STATE.—(1) Not more than 15 percent of the total of funds authorized to be appropriated for any fiscal year to carry out this section shall be granted under this section for projects in any one State.

"(2) The Administrator shall prescribe by regulation the manner *Regulation.* in which this subection shall apply to a grant under this section for a project in an area which includes all or part of more than one State.

"AUTHORIZATION OF APPROPRIATIONS

"SEC. 8007. There are authorized to be appropriated not to exceed *42 USC 6987.* $35,000,000 for the fiscal year 1978 to carry out the purposes of this subtitle (except for section 8002).".

90 STAT. 2840   PUBLIC LAW 94-580—OCT. 21, 1976

SOLID WASTE CLEANUP ON FEDERAL LANDS IN ALASKA

Study.
42 USC 6981
note.

SEC. 3. (a) The President shall direct such executive departments or agencies as he may deem appropriate to conduct a study, in consultation with representatives of the State of Alaska and the appropriate Native organizations, to determine the best overall procedures for removing existing solid waste on Federal lands in Alaska. Such study shall include, but shall not be limited to, a consideration of—

(1) alternative procedures for removing the solid waste in an environmentally safe manner, and

(2) the estimated costs of removing the solid waste.

Report to
congressional
committees.

(b) The President shall submit a report of the results together with appropriate supporting data and such recommendations as he deems desirable to the Committee on Public Works of the Senate and to the Committee on Interstate and Foreign Commerce of the House of Representatives not later than one year after the enactment of the Solid Waste Utilization Act of 1976. The President shall also submit, within six months after the study has been submitted to the committees, recommended administrative actions, procedures, and needed legislation to implement such procedures and the recommendations of the study.

Llangollen
Landfill, Del.,
leachate control
research
program.
42 USC 6981
note.
Cooperation with
EPA.

SEC. 4. (a) In order to demonstrate effective means of dealing with contamination of public water supplies by leachate from abandoned or other landfills, the Administrator of the Environmental Protection Agency is authorized to provide technical and financial assistance for a research program to control leachate from the Llangollen Landfill in New Castle County, Delaware.

(b) The research program authorized by this section shall be designed by the New Castle County areawide waste treatment management program, in cooperation with the Environmental Protection Agency, to develop methods for controlling leachate contamination from abandoned and other landfills that may be applied at the Llangollen Landfill and at other landfills throughout the Nation. Such research program shall investigate all alternative solutions or corrective actions, including—

(1) hydrogeologic isolation of the landfill combined with the collection and treatment of leachate;

(2) excavation of the refuse, followed by some type of incineration;

(3) excavation and transportation of the refuse to another landfill; and

(4) collection and treatment of contaminated leachate or ground water.

Such research program shall consider the economic, social, and environmental consequences of each such alternative.

(c) The Administrator of the Environmental Protection Agency shall make available personnel of the Agency, including those of the Solid and Hazardous Waste Research Laboratory (Cincinnati, Ohio), and shall arrange for other Federal personnel to be made available, to provide technical assistance and aid in such research. The Administrator may provide up to $250,000, of the sums appropriated under the Solid Waste Disposal Act, to the New Castle County areawide waste treatment management program to conduct such research, including obtaining consultant services.

42 USC 6901
note.

PUBLIC LAW 94–580—OCT. 21, 1976   90 STAT. 2841

(d) In order to prevent further damage to public water supplies during the period of this study, the Administrator of the Environmental Protection Agency shall provide up to $200,000 in each of fiscal years 1977 and 1978, of the sums appropriated under the Solid Waste Disposal Act for the operating costs of a counter-pumping program to contain the leachate from the Llangollen Landfill.

Counter-pumping program

42 USC 6901 note.

Approved October 21, 1976.

LEGISLATIVE HISTORY:

HOUSE REPORT No. 94–1491 accompanying H.R. 14496 (Comm. on Interstate and Foreign Commerce).
SENATE REPORT No. 94–869 (Comm. on Public Works).
CONGRESSIONAL RECORD, Vol. 122 (1976):
    June 30, considered and passed Senate.
    Sept. 27, considered and passed House, amended, in lieu of H.R. 14496.
    Sept. 30, Senate concurred in House amendments.
WEEKLY COMPILATION OF PRESIDENTIAL DOCUMENTS, Vol. 12, No. 43:
    Oct. 22, Presidential statement.

APPENDIX C
HAZARDOUS WASTE REGULATIONS

The following forms and associated instruction sheets, tables, and appendices are included in the chapters shown, and therefore are not duplicated in this Appendix.

(1) Chapter 4 (Standards Applicable to Generators of Hazardous Waste—Part 262)

- Fig. 4-1. EPA Forms 8700-13 and 8700-13A, Generators Annual Reporting of Off-Site Hazardous Waste Shipments.
- Associated instructions for using EPA forms.

(2) Chapter 6 (Standards Applicable to Owners and Operators of Hazardous Waste Treatment, Storage, and Disposal Facilities—Parts 264 and 265)

- Fig. 6-1. EPA Forms 8700-13 and 8700-13B, Facility Annual Report and Unmanifested Waste Report.
- Associated instructions for using EPA forms.
- Appendix I—Recordkeeping Instructions and associated Tables 1 and 2.

- Appendix III—EPA Interim Primary Drinking Water Standards.
- Appendix IV—Tests for Significance.
- Appendix V—Examples of Potentially Incompatible Waste.

(3) Chapter 7 (Permit Requirements and Procedures for Decisionmaking—Parts 122 and 124)

- Fig. 7-1. EPA Form 1 General (No. 3510-1), General Information for Consolidated Permit Program.
- Fig. 7-2. EPA Form 3 RCRA (No. 3510-3), Hazardous Waste Permit Application for Consolidated Permits Program.
- Associated instructions for using the two forms.

(4) Chapter 9 (Notification of Hazardous Waste Activity—Federal Register Public Notice)

- Fig. 9-1. EPA Form 8700-12, Notification of Hazardous Waste Activity.
- Associated instructions for using EPA form.

PART 260—HAZARDOUS WASTE MANAGEMENT SYSTEM: GENERAL

Subpart A—General

Sec.
260.1   Purpose, scope and applicability.
260.2   Availability of information, confidentiality of information.
260.3   Use of number and gender.

Subpart B—Definitions

260.10   Definitions.

Subpart C—Rulemaking Petitions

260.20   General.
260.21   Petitions for equivalent testing or analytical methods.
260.22   Petitions to amend Part 261 to exclude a waste produced at a particular facility.
Appendix I—Overview of Subtitle C Regulations

Authority: Secs. 1006, 2002(a), 3001 through 3007, 3010, and 7004, of the Solid Waste Disposal Act, as amended by the Resource Conservation and Recovery Act of 1976, as amended (42 U.S.C. 6905, 6912(a), 6921 through 6927, 6930, and 6974).

Subpart A—General

§ 260.1   Purpose, scope, and applicability.

(a) This part provides definitions of terms, general standards, and overview information applicable to Parts 260 through 265 of this Chapter.

(b) In this part: (1) Section 260.2 sets forth the rules that EPA will use in making information it receives available to the public and sets forth the requirements that generators, transporters, or owners or operators of treatment, storage, or disposal facilities must follow to assert claims of business confidentiality with respect to information that is submitted to EPA under Parts 260 through 265 of this Chapter.

(2) Section 260.3 establishes rules of grammatical construction for Parts 260 through 265 of this Chapter.

(3) Section 260.10 defines terms which are used in Parts 260 through 265 of this Chapter.

(4) Section 260.20 establishes procedures for petitioning EPA to

amend, modify, or revoke any provision of Parts 260 through 265 of this Chapter and establishes procedures governing EPA's action on such petitions.

(5) Section 260.21 establishes procedures for petitioning EPA to approve testing methods as equivalent to those prescribed in Parts 261, 264, or 265 of this Chapter.

(6) Section 260.22 establishes procedures for petitioning EPA to amend Subpart D of Part 261 to exclude a waste from a particular facility.

§ 260.2   Availability of information; confidentiality of information.

(a) Any information provided to EPA under Parts 260 through 265 of this Chapter will be made available to the public to the extent and in the manner authorized by the Freedom of Information Act, 5 U.S.C. section 552, section 3007(b) of RCRA and EPA regulations implementing the Freedom of Information Act and section 3007(b), Part 2 of this Chapter, as applicable.

(b) Any person who submits information to EPA in accordance with

Parts 260 through 265 of this Chapter may assert a claim of business confidentiality covering part or all of that information by following the procedures set forth in § 2.203(b) of this Chapter. Information covered by such a claim will be disclosed by EPA only to the extent, and by means of the procedures, set forth in Part 2, Subpart B of this Chapter. However, if no such claim accompanies the information when it is received by EPA, it may be made available to the public without further notice to the person submitting it.

§ 260.3 Use of number and gender.

As used in Parts 260 through 265 of this Chapter:

(a) Words in the masculine gender also include the feminine and neuter genders; and

(b) Words in the singular include the plural; and

(c) Words in the plural include the singular.

Subpart B—Definitions

§ 260.10 Definitions.

(a) When used in Parts 260 through 265 of this Chapter, the following terms have the meanings given below: (1) "Act" or "RCRA" means the Solid Waste Disposal Act, as amended by the Resource Conservation and Recovery Act of 1976, as amended, 42 U.S.C. section 6901 et seq.

(2) "Active portion" means that portion of a facility where treatment, storage, or disposal operations are being or have been conducted after the effective date of Part 261 of this Chapter and which is not a closed portion. (See also "closed portion" and "inactive portion".)

(3) "Administrator" means the Administrator of the Environmental Protection Agency, or his designee.

(4) "Aquifer" means a geologic formation, group of formations, or part of a formation capable of yielding a significant amount of ground water to wells or springs.

(5) "Authorized representative" means the person responsible for the overall operation of a facility or an operational unit (i.e., part of a facility), e.g., the plant manager, superintendent or person of equivalent responsibility.

(6) "Closed portion" means that portion of a facility which an owner or operator has closed in accordance with the approved facility closure plan and all applicable closure requirements. (See also "active portion" and "inactive portion".)

(7) "Confined aquifer" means an aquifer bounded above and below by impermeable beds or by beds of distinctly lower permeability than that of the aquifer itself; an aquifer containing confined ground water.

(8) "Constituent" or "hazardous waste constituent" means a constituent which caused the Administrator to list the hazardous waste in Part 261, Subpart D, of this Chapter, or a constituent listed in Table 1 of § 261.24 of this Chapter.

(9) "Container" means any portable device in which a material is stored, transported, treated, disposed of, or otherwise handled.

(10) "Contingency plan" means a document setting out an organized, planned, and coordinated course of action to be followed in case of a fire, explosion, or release of hazardous waste or hazardous waste constituents which could threaten human health or the environment.

(11) "Designated facility" means a hazardous waste treatment, storage, or disposal facility which has received an EPA permit (or a facility with interim status) in accordance with the requirements of 40 CFR Parts 122 and 124 of this Chapter, or a permit from a State authorized in accordance with Part 123 of this Chapter, that has been designated on the manifest by the generator pursuant to § 262.20.

(12) "Dike" means an embankment or ridge of either natural or man-made materials used to prevent the movement of liquids, sludges, solids, or other materials.

(13) "Discharge" or "hazardous waste discharge" means the accidental or intentional spilling, leaking, pumping, pouring, emitting, emptying, or dumping of hazardous waste into or on any land or water.

(14) "Disposal" means the discharge, deposit, injection, dumping, spilling, leaking, or placing of any solid waste or hazardous waste into or on any land or water so that such solid waste or hazardous waste or any constituent thereof may enter the environment or be emitted into the air or discharged into any waters, including ground waters.

(15) "Disposal facility" means a facility or part of a facility at which hazardous waste is intentionally placed into or on any land or water, and at which waste will remain after closure.

(16) "EPA hazardous waste number" means the number assigned by EPA to each hazardous waste listed in Part 261, Subpart D, of this Chapter and to each characteristic identified in Part 261, Subpart C, of this Chapter.

(17) "EPA identification number" means the number assigned by EPA to each generator, transporter, and treatment, storage, or disposal facility.

(18) "EPA region" means the states and territories found in any one of the following ten regions:

Region I—Maine, Vermont, New Hampshire, Massachusetts, Connecticut, and Rhode Island.

Region II—New York, New Jersey, Commonwealth of Puerto Rico, and the U.S. Virgin Islands.

Region III—Pennsylvania, Delaware, Maryland, West Virginia, Virginia, and the District of Columbia.

Region IV—Kentucky, Tennessee, North Carolina, Mississippi, Alabama, Georgia, South Carolina, and Florida.

Region V—Minnesota, Wisconsin, Illinois, Michigan, Indiana and Ohio.

Region VI—New Mexico, Oklahoma, Arkansas, Louisiana, and Texas.

Region VII—Nebraska, Kansas, Missouri, and Iowa.

Region VIII—Montana, Wyoming, North Dakota, South Dakota, Utah, and Colorado.

Region IX—California, Nevada, Arizona, Hawaii, Guam, American Samoa, Commonwealth of the Northern Mariana Islands.

Region X—Washington, Oregon, Idaho, and Alaska.

(19) "Equivalent method" means any testing or analytical method approved by the Administrator under §§ 260.20 and 260.21.

(20) "Existing hazardous waste management facility" or "existing facility" means a facility which was in operation, or for which construction had commenced, on or before October 21, 1976. Construction had commenced if:

(i) The owner or operator has obtained all necessary Federal, State, and local preconstruction approvals or permits; and either

(ii)(a) A continuous physical, on-site construction program has begun, or

(b) The owner or operator has entered into contractual obligations—which cannot be cancelled or modified without substantial loss—for construction of the facility to be completed within a reasonable time.

(21) "Facility" means all contiguous land, and structures, other appurtenances, and improvements on the land, used for treating, storing, or disposing of hazardous waste. A facility may consist of several treatment, storage, or disposal operational units (e.g., one or more landfills, surface impoundments, or combinations of them).

(22) "Federal agency" means any department, agency, or other instrumentality of the Federal Government, any independent agency or establishment of the Federal Government including any Government corporation, and the Government Printing Office.

(23) "Food-chain crops" means tobacco, crops grown for human

consumption, and crops grown for feed for animals whose products are consumed by humans.

(24) "Freeboard" means the vertical distance between the top of a tank or surface impoundment dike, and the surface of the waste contained therein.

(25) "Free liquids" means liquids which readily separate from the solid portion of a waste under ambient temperature and pressure.

(26) "Generator" means any person, by site, whose act or process produces hazardous waste identified or listed in Part 261 of this Chapter.

(27) "Ground water" means water below the land surface in a zone of saturation.

(28) "Hazardous waste" means a hazardous waste as defined in § 261.3 of this Chapter.

(29) "Inactive portion" means that portion of a facility which is not operated after the effective date of Part 261 of this Chapter. (See also "active portion" and "closed portion".)

(30) "Incinerator" means an enclosed device using controlled flame combustion, the primary purpose of which is to thermally break down hazardous waste. Examples of incinerators are rotary kiln, fluidized bed, and liquid injection incinerators.

(31) "Incompatible waste" means a hazardous waste which is unsuitable for:

(i) Placement in a particular device or facility because it may cause corrosion or decay of containment materials (e.g., container inner liners or tank walls); or

(ii) Commingling with another waste or material under uncontrolled conditions because the commingling might produce heat or pressure, fire or explosion, violent reaction, toxic dusts, mists, fumes, or gases, or flammable fumes or gases.
(See Part 265, Appendix V, of this Chapter for examples.)

(32) "Individual generation site" means the contiguous site at or on which one or more hazardous wastes are generated. An individual generation site, such as a large manufacturing plant, may have one or more sources of hazardous waste but is considered a single or individual generation site if the site or property is contiguous.

(33) "In operation" refers to a facility which is treating, storing, or disposing of hazardous waste.

(34) "Injection well" means a well into which fluids are injected. (See also "underground injection".)

(35) "Inner liner" means a continuous layer of material placed inside a tank or container which protects the construction materials of the tank or container from the contained waste or reagents used to treat the waste.

(36) "International shipment" means the transportation of hazardous waste into or out of the jurisdiction of the United States.

(37) "Landfill" means a disposal facility or part of a facility where hazardous waste is placed in or on land and which is not a land treatment facility, a surface impoundment, or an injection well.

(38) "Landfill cell" means a discrete volume of a hazardous waste landfill which uses a liner to provide isolation of wastes from adjacent cells or wastes. Examples of landfill cells are trenches and pits.

(39) "Land treatment facility" means a facility or part of a facility at which hazardous waste is applied onto or incorporated into the soil surface; such facilities are disposal facilities if the waste will remain after closure.

(40) "Leachate" means any liquid, including any suspended components in the liquid, that has percolated through or drained from hazardous waste.

(41) "Liner" means a continuous layer of natural or man-made materials, beneath or on the sides of a surface impoundment, landfill, or landfill cell, which restricts the downward or lateral escape of hazardous waste, hazardous waste constituents, or leachate.

(42) "Management" or "hazardous waste management" means the systematic control of the collection, source separation, storage, transportation, processing, treatment, recovery, and disposal of hazardous waste.

(43) "Manifest" means the shipping document originated and signed by the generator which contains the information required by Part 262, Subpart B, of this Chapter.

(44) "Manifest document number" means the serially increasing number assigned to the manifest by the generator for recording and reporting purposes.

(45) "Mining overburden returned to the mine site" means any material overlying an economic mineral deposit which is removed to gain access to that deposit and is then used for reclamation of a surface mine.

(46) "Movement" means that hazardous waste transported to a facility in an individual vehicle.

(47) "New hazardous waste management facility" or "new facility" means a facility which began operation, or for which construction commenced after October 21, 1976. (See also "Existing hazardous waste management facility".)

(48) "On-site" means the same or geographically contiguous property which may be divided by public or private right-of-way, provided the entrance and exit between the properties is at a cross-roads intersection, and access is by crossing as opposed to going along, the right-of-way. Non-contiguous properties owned by the same person but connected by a right-of-way which he controls and to which the public does not have access, is also considered on-site property.

(49) "Open burning" means the combustion of any material without the following characteristics:

(i) Control of combustion air to maintain adequate temperature for efficient combustion,

(ii) Containment of the combustion-reaction in an enclosed device to provide sufficient residence time and mixing for complete combustion, and

(iii) Control of emission of the gaseous combustion products.
(See also "incineration" and "thermal treatment".)

(50) "Operator" means the person responsible for the overall operation of a facility.

(51) "Owner" means the person who owns a facility or part of a facility.

(52) "Partial closure" means the closure of a discrete part of a facility in accordance with the applicable closure requirements of Parts 264 or 265 of this Chapter. For example, partial closure may include the closure of a trench, a unit operation, a landfill cell, or a pit, while other parts of the same facility continue in operation or will be placed in operation in the future.

(53) "Person" means an individual, trust, firm, joint stock company, Federal Agency, corporation (including a government corporation), partnership, association, State, municipality, commission, political subdivision of a State, or any interstate body.

(54) "Personnel" or "facility personnel" means all persons who work, at, or oversee the operations of, a hazardous waste facility, and whose actions or failure to act may result in noncompliance with the requirements of Parts 264 or 265 of this Chapter.

(55) "Pile" means any non-containerized accumulation of solid, nonflowing hazardous waste that is used for treatment or storage.

(56) "Point source" means any discernible, confined, and discrete conveyance, including, but not limited to any pipe, ditch, channel, tunnel, conduit, well, discrete fissure, container, rolling stock, concentrated animal feeding operation, or vessel or other floating craft, from which pollutants are or may

be discharged. This term does not include return flows from irrigated agriculture.

(57) "Publicly owned treatment works" or "POTW" means any device or system used in the treatment (including recycling and reclamation) of municipal sewage or industrial wastes of a liquid nature which is owned by a "State" or "municipality" (as defined by Section 502(4) of the CWA). This definition includes sewers, pipes, or other conveyances only if they convey wastewater to a POTW providing treatment.

(58) "Regional Administrator" means the Regional Administrator for the EPA Region in which the facility is located, or his designee.

(59) "Representative sample" means a sample of a universe or whole (e.g., waste pile, lagoon, ground water) which can be expected to exhibit the average properties of the universe or whole.

(60) "Run-off" means any rainwater, leachate, or other liquid that drains over land from any part of a facility.

(61) "Run-on" means any rainwater, leachate, or other liquid that drains over land onto any part of a facility.

(62) "Saturated zone" or "zone of saturation" means that part of the earth's crust in which all voids are filled with water.

(63) "Sludge" means any solid, semi-solid, or liquid waste generated from a municipal, commercial, or industrial wastewater treatment plant, water supply treatment plant, or air pollution control facility exclusive of the treated effluent from a wastewater treatment plant.

(64) "Solid waste" means a solid waste as defined in § 261.2 of this Chapter.

(65) "State" means any of the several States, the District of Columbia, the Commonwealth of Puerto Rico, the Virgin Islands, Guam, American Samoa, and the Commonwealth of the Northern Mariana Islands.

(66) "Storage" means the holding of hazardous waste for a temporary period, at the end of which the hazardous waste is treated, disposed of, or stored elsewhere.

(67) "Surface impoundment" or "impoundment" means a facility or part of a facility which is a natural topographic depression, man-made excavation, or diked area formed primarily of earthen materials (although it may be lined with man-made materials), which is designed to hold an accumulation of liquid wastes or wastes containing free liquids, and which is not an injection well. Examples of surface impoundments are holding, storage,

settling, and aeration pits, ponds, and lagoons.

(68) "Tank" means a stationary device, designed to contain an accumulation of hazardous waste which is constructed primarily of non-earthen materials (e.g., wood, concrete, steel, plastic) which provide structural support.

(69) "Thermal treatment" means the treatment of hazardous waste in a device which uses elevated temperatures as the primary means to change the chemical, physical, or biological character or composition of the hazardous waste. Examples of thermal treatment processes are incineration, molten salt, pyrolysis, calcination, wet air oxidation, and microwave discharge. (See also "incinerator" and "open burning".)

(70) "Totally enclosed treatment facility" means a facility for the treatment of hazardous waste which is directly connected to an industrial production process and which is constructed and operated in a manner which prevents the release of any hazardous waste or any constituent thereof into the environment during treatment. An example is a pipe in which waste acid is neutralized.

(71) "Transportation" means the movement of hazardous waste by air, rail, highway, or water.

(72) "Transporter" means a person engaged in the offsite transportation of hazardous waste by air, rail, highway, or water.

(73) "Treatment" means any method, technique, or process, including neutralization, designed to change the physical, chemical, or biological character or composition of any hazardous waste so as to neutralize such waste, or so as to recover energy or material resources from the waste, or so as to render such waste non-hazardous, or less hazardous; safer to transport, store, or dispose of; or amenable for recovery, amenable for storage, or reduced in volume.

(74) "Underground injection" means the subsurface emplacement of fluids through a bored, drilled or driven well; or through a dug well, where the depth of the dug well is greater than the largest surface dimension. (See also "injection well".)

(75) "Unsaturated zone" or "zone of aeration" means the zone between the land surface and the water table.

(76) "United States" means the 50 States, the District of Columbia, the Commonwealth of Puerto Rico, the U.S. Virgin Islands, Guam, American Samoa, and the Commonwealth of the Northern Mariana Islands.

(77) "Water (bulk shipment)" means the bulk transportation of hazardous waste which is loaded or carried on board a vessel without containers or labels.

(78) "Well" means any shaft or pit dug or bored into the earth, generally of a cylindrical form, and often walled with bricks or tubing to prevent the earth from caving in.

(79) "Well injection": (See "underground injection".)

Subpart C—Rulemaking Petitions

§ 260.20 General.

(a) Any person may petition the Administrator to modify or revoke any provision in Parts 260 through 265 of this Chapter. This section sets forth general requirements which apply to all such petitions. Section 260.21 sets forth additional requirements for petitions to add a testing or analytical method to Parts 261, 264 or 265. Section 260.22 sets forth additional requirements for petitions to exclude a waste at a particular facility from § 261.3 of this Chapter or the lists of hazardous wastes in Subpart D of Part 261.

(b) Each petition must be submitted to the Administrator by certified mail and must include:

(1) The petitioner's name and address;
(2) A statement of the petitioner's interest in the proposed action;
(3) A description of the proposed action, including (where appropriate) suggested regulatory language; and
(4) A statement of the need and justification for the proposed action, including any supporting tests, studies, or other information.

(c) The Administrator will make a tentative decision to grant or deny a petition and will publish notice of such tentative decision, either in the form of an advanced notice of proposed rulemaking, a proposed rule, or a tentative determination to deny the petition, in the **Federal Register** for written public comment.

(d) Upon the written request of any interested person, the Administrator may, at his discretion, hold an informal public hearing to consider oral comments on the tentative decision. A person requesting a hearing must state the issues to be raised and explain why written comments would not suffice to communicate the person's views. The Administrator may in any case decide on his own motion to hold an informal public hearing.

(e) After evaluating all public comments the Administrator will make a final decision by publishing in the **Federal Register** a regulatory amendment or a denial of the petition.

§ 260.21 Petitions for equivalent testing or analytical methods.

(a) Any person seeking to add a testing or analytical method to Parts 261, 264, or 265 of this Chapter may petition for a regulatory amendment under this section and § 260.20. To be successful, the person must demonstrate to the satisfaction of the Administrator that the proposed method is equal to or superior to the corresponding method prescribed in Parts 261, 264, or 265 of this Chapter, in terms of its sensitivity, accuracy, and precision (i.e., reproducibility).

(b) Each petition must include, in addition to the information required by § 260.20(b):

(1) A full description of the proposed method, including all procedural steps and equipment used in the method;

(2) A description of the types of wastes or waste matrices for which the proposed method may be used;

(3) Comparative results obtained from using the proposed method with those obtained from using the relevant or corresponding methods prescribed in Parts 261, 264, or 265 of this Chapter;

(4) An assessment of any factors which may interfere with, or limit the use of, the proposed method; and

(5) A description of the quality control procedures necessary to ensure the sensitivity, accuracy and precision of the proposed method.

(c) After receiving a petition for an equivalent method, the Administrator may request any additional information on the proposed method which he may reasonably require to evaluate the method.

(d) If the Administrator amends the regulations to permit use of a new testing method, the method will be incorporated in "Test Methods for the Evaluation of Solid Waste: Physical/Chemical Methods," SW–846, U.S. Environmental Protection Agency, Office of Solid Waste, Washington, D.C. 20460.

[Comment: This manual will be provided to any person on request, and will be available for inspection or copying at EPA headquarters or any EPA Regional Office.]

§ 260.22 Petitions to amend Part 261 to exclude a waste produced at a particular facility.

(a) Any person seeking to exclude a waste at a particular generating facility from the lists in Subpart D of Part 261 may petition for a regulatory amendment under this section and § 260.20. To be successful, the petitioner must demonstrate to the satisfaction of the Administrator that the waste produced by a particular generating facility does not meet any of the criteria under which the waste was listed as a hazardous waste and, in the case of an acutely hazardous waste listed under § 261.11(a)(2), that it also does not meet the criterion of § 261.11(a)(3). A waste which is so excluded may still, however, be a hazardous waste by operation of Subpart C of Part 261.

(b) The procedures in this section and § 260.20 may also be used to petition the Administrator for a regulatory amendment to exclude from § 261.3(a)(2)(ii) or (c), a waste which is described in those sections and is either a waste listed in Subpart D, contains a waste listed in Subpart D, or is derived from a waste listed in Subpart D. This exclusion may only be issued for a particular generating, storage, treatment, or disposal facility. The petitioner must make the same demonstration as required by paragraph (a) of this section, except that where the waste is a mixture of solid waste and one or more listed hazardous wastes or is derived from one or more hazardous wastes, his demonstration may be made with respect to each constituent listed waste or the waste mixture as a whole. A waste which is so excluded may still be a hazardous waste by operation of Subpart C of Part 261.

(c) If the waste is listed with codes "I", "C", "R", or "E" in Subpart D, the petitioner must show that demonstration samples of the waste do not exhibit the relevant characteristic defined in §§ 261.21, 261.22, 261.23, or 261.24 using any applicable test methods prescribed therein.

(d) If the waste is listed with code "T" in Subpart D, the petitioner must demonstrate that:

(1) Demonstration samples of the waste do not contain the constituent (as defined in Appendix VII) that caused the Administrator to list the waste, using the appropriate test methods prescribed in Appendix III; or

(2) The waste does not meet the criterion of § 261.11(a)(3) when considering the factors in § 261.11(a)(3) (i) through (xi).

(e) If the waste is listed with the code "H" in Subpart D, the petitioner must demonstrate that the waste does not meet both of the following criteria:

(1) The criterion of § 261.11(a)(2).

(2) The criterion of § 261.11(a)(3) when considering the factors listed in § 261.11(a)(3) (i) through (xi).

(f) [Reserved for listing radioactive wastes.]

(g) [Reserved for listed infectious wastes.]

(h) Demonstration samples must consist of enough representative samples, but in no case less than four samples, taken over a period of time sufficient to represent the variability or the uniformity of the waste.

(i) Each petition must include, in addition to the information required by § 260.20(b):

(1) The name and address of the laboratory facility performing the sampling or tests of the waste;

(2) The names and qualifications of the persons sampling and testing the waste;

(3) The dates of sampling and testing;

(4) The location of the generating facility;

(5) A description of the manufacturing processes or other operations and feed materials producing the waste and an assessment of whether such processes, operations, or feed materials can or might produce a waste that is not covered by the demonstration;

(6) A description of the waste and an estimate of the average and maximum monthly and annual quantities of waste covered by the demonstration;

(7) Pertinent data on and discussion of the factors delineated in the respective criterion for listing a hazardous waste, where the demonstration is based on the factors in § 261.11(a)(3);

(8) A description of the methodologies and equipment used to obtain the representative samples;

(9) A description of the sample handling and preparation techniques, including techniques used for extraction, containerization and preservation of the samples;

(10) A description of the tests performed (including results);

(11) The names and model numbers of the instruments used in performing the tests; and

(12) The following statement signed by the generator of the waste or his authorized representative:

I certify under penalty of law that I have personally examined and am familiar with the information submitted in this demonstration and all attached documents, and that, based on my inquiry of those individuals immediately responsible for obtaining the information, I believe that the submitted information is true, accurate, and complete. I am aware that there are significant penalties for submitting false information, including the possibility of fine and imprisonment.

(j) After receiving a petition for an exclusion, the Administrator may request any additional information which he may reasonably require to evaluate the petition.

(k) An exclusion will only apply to the waste generated at the individual facility covered by the demonstration and will not apply to waste from any other facility.

(l) The Administrator may exclude only part of the waste for which the demonstration is submitted where he has reason to believe that variability of the waste justifies a partial exclusion.

(m) The Administrator may (but shall not be required to) grant a temporary exclusion before making a final decision under § 260.20(d) whenever he finds that there is a substantial likelihood that an exclusion will be finally granted. The Administrator will publish notice of any such temporary exclusion in the **Federal Register**.

PART 261—IDENTIFICATION AND LISTING OF HAZARDOUS WASTE

Subpart A—General

Sec.

Authority: Secs. 1006, 2002(a), 3001, and 3002 of the Solid Waste Disposal Act, as amended by the Resource Conservation and Recovery Act of 1976, as amended (42 U.S.C. 6905, 6912, 6921 and 6922).

Subpart A—General

§ 261.1   Purpose and scope.

(a) This Part identifies those solid wastes which are subject to regulation as hazardous wastes under Parts 262 through 265 and Parts 122 through 124 of this Chapter and which are subject to the notification requirements of Section 3010 of RCRA. In this Part:

(1) Subpart A defines the terms "solid waste" and "hazardous waste," identifies those wastes which are excluded from regulation under Parts 262 through 265 and 122 through 124 and establishes special management requirements for hazardous waste produced by small quantity generators and hazardous waste which is used, re-used, recycled or reclaimed.

(2) Subpart B sets forth the criteria used by EPA to identify characteristics of hazardous waste and to list particular hazardous wastes.

(3) Subpart C identifies characteristics of hazardous waste.

(4) Subpart D lists particular hazardous wastes.

(b) This Part identifies only some of the materials which are hazardous wastes under Sections 3007 and 7003 of RCRA. A material which is not a hazardous waste identified in this part is still a hazardous waste for purposes of those sections if:

(1) In the case of Section 3007, EPA has reason to believe that the material may be a hazardous waste within the meaning of Section 1004(5) of RCRA.

(2) In the case of Section 7003, the statutory elements are established.

§ 261.2   Definition of solid waste.

(a) A solid waste is any garbage, refuse, sludge or any other waste material which is not excluded under § 261.4(a).

(b) An "other waste material" is any solid, liquid, semi-solid or contained gaseous material, resulting from industrial, commercial, mining or agricultural operations, or from community activities which:

(1) Is discarded or is being accumulated, stored or physically, chemically or biologically treated prior to being discarded; or

(2) Has served its original intended use and sometimes is discarded; or

(3) Is a manufacuring or mining by-product and sometimes is discarded.

(c) A material is "discarded" if it is abandoned (and not used, re-used, reclaimed or recycled) by being:

(1) Disposed of; or

(2) Burned or incinerated, except where the material is being burned as a fuel for the purpose of recovering usable energy; or

(3) Physically, chemically, or biologically treated (other than burned or incinerated) in lieu of or prior to being disposed of.

(d) A material is "disposed of" if it is discharged, deposited, injected, dumped, spilled, leaked or placed into or on any land or water so that such material or any constituent thereof may enter the environment or be emitted into the air or discharged into ground or surface waters.

(e) A "manufacturing or mining by-product" is a material that is not one of the primary products of a particular manufacturing or mining operation, is a secondary and incidental product of the particular operation and would not be solely and separately manufactured or mined by the particular manufacturing or mining operation. The term does not include an intermediate manufacturing or mining product which results from one of the steps in a manufacturing or mining process and is typically processed through the next step of the process within a short time.

§ 261.3   Definition of hazardous waste.

(a) A solid waste, as defined in § 261.2, is a hazardous waste if:

(1) It is not excluded from regulation as a hazardous waste under § 261.4(b); and

(2) It meets any of the following criteria:

(i) It is listed in Subpart D and has not been excluded from the lists in Subpart D under §§ 260.20 and 260.22 of this Chapter.

(ii) It is a mixture of solid waste and one or more hazardous wastes listed in Subpart D and has not been excluded from this paragraph under §§ 260.20 and 260.22 of this Chapter.

(iii) It exhibits any of the characteristics of hazardous waste identified in Subpart C.

(b) A solid waste which is not excluded from regulation under paragraph (a)(1) of this section becomes a hazardous waste when any of the following events occur:

(1) In the case of a waste listed in Subpart D, when the waste first meets the listing description set forth in Subpart D.

(2) In the case of a mixture of solid waste and one or more listed hazardous wastes, when a hazardous waste listed

in Subpart D is first added to the solid waste.

(3) In the case of any other waste (including a waste mixture), when the waste exhibits any of the characteristics identified in Subpart C.

(c) Unless and until it meets the criteria of paragraph (d):

(1) A hazardous waste will remain a hazardous waste.

(2) Any solid waste generated from the treatment, storage or disposal of a hazardous waste, including any sludge, spill residue, ash, emission control dust or leachate (but not including precipitation run-off), is a hazardous waste.

(d) Any solid waste described in paragraph (c) of this section is not a hazardous waste if it meets the following criteria:

(1) In the case of any solid waste, it does not exhibit any of the characteristics of hazardous waste identified in Subpart C.

(2) In the case of a waste which is a listed waste under Subpart D, contains a waste listed under Subpart D or is derived from a waste listed in Subpart D, it also has been excluded from paragraph (c) under §§ 260.20 and 260.22 of this Chapter.

§ 261.4 Exclusions.

(a) *Materials which are not solid wastes.* The following materials are not solid wastes for the purpose of this Part:

(1) (i) Domestic sewage; and

(ii) Any mixture of domestic sewage and other wastes that passes through a sewer system to a publicly-owned treatment works for treatment. "Domestic sewage" means untreated sanitary wastes that pass through a sewer system.

(2) Industrial wastewater discharges that are point source discharges subject to regulation under Section 402 of the Clean Water Act, as amended.

[Comment: This exclusion applies only to the actual point source discharge. It does not exclude industrial wastewaters while they are being collected, stored or treated before discharge, nor does it exclude sludges that are generated by industrial wastewater treatment.]

(3) Irrigation return flows.

(4) Source, special nuclear or by-product material as defined by the Atomic Energy Act of 1954, as amended, 42 U.S.C. 2011 *et seq.*

(5) Materials subjected to in-situ mining techniques which are not removed from the ground as part of the extraction process.

(b) *Solid wastes which are not hazardous wastes.* The following solid wastes are not hazardous wastes:

(1) Household waste, including household waste that has been collected, transported, stored, treated, disposed, recovered (e.g., refuse-derived fuel) or reused. "Household waste" means any waste material (including garbage, trash and sanitary wastes in septic tanks) derived from households (including single and multiple residences, hotels and motels.)

(2) Solid wastes generated by any of the following and which are returned to the soils as fertilizers:

(i) The growing and harvesting of agricultural crops.

(ii) The raising of animals, including animal manures.

(3) Mining overburden returned to the mine site.

(4) Fly ash waste, bottom ash waste, slag waste, and flue gas emission control waste generated primarily from the combustion of coal or other fossil fuels.

(5) Drilling fluids, produced waters, and other wastes associated with the exploration, development, or production of crude oil, natural gas or geothermal energy.

§ 261.5 Special requirements for hazardous waste generated by small quantity generators.

(a) Except as otherwise provided in this section, if a person generates, in a calendar month, a total of less than 1000 kilograms of hazardous wastes, those wastes are not subject to regulation under Parts 262 through 265 and Parts 122 through 124 of this Chapter, and the notification requirements of Section 3010 of RCRA.

(b) If a person whose waste has been excluded from regulation under paragraph (a) of this Section accumulates hazardous wastes in quantities greater than 1000 kilograms, those accumulated wastes are subject to regulation under Parts 262 through 265 and Parts 122 through 124 of this Chapter, and the notification requirements of Section 3010 of RCRA.

(c) If a person generates in a calendar month or accumulates at any time any of the following hazardous wastes in quantities greater than set forth below, those wastes are subject to regulation under Parts 262 through 265 and Parts 122 through 124 of this Chapter, and the notification requirements of Section 3010 of RCRA:

(1) One kilogram of any commercial product or manufacturing chemical intermediate having the generic name listed in § 261.33(e).

(2) One kilogram of any off-specification commercial chemical product or manufacturing chemical intermediate which, if it met

specifications, would have the generic name listed in § 261.33(e).

(3) Any containers identified in § 261.33(c) that are larger than 20 liters in capacity;

(4) 10 kilograms of inner liners from containers identified under § 261.33(c);

(5) 100 kilograms of any residue or contaminated soil, water or other debris resulting from the cleanup of a spill, into or on any land or water, of any commercial chemical product or manufacturing chemical intermediate having the generic name listed in § 261.33(e).

(d) In order for hazardous waste to be excluded from regulation under this section, the generator must comply with § 262.11 of this Chapter. He must also either treat or dispose of the waste in an on-site facility, or ensure delivery to an off-site treatment, storage or disposal facility, either of which is:

(1) Permitted by EPA under Part 122 of this Chapter, or by a State with a hazardous waste management program authorized under Part 123 of this Chapter;

(2) In interim status under Parts 122 and 265 of this Chapter; or,

(3) Permitted, licensed, or registered by a State to manage municipal or industrial solid waste.

(e) Hazardous waste subject to the reduced requirements of this section may be mixed with non-hazardous waste and remain subject to these reduced requirements even though the resultant mixture exceeds the quantity limitations identified in this section, unless the mixture meets any of the characteristics of hazardous waste identified in Subpart C.

§ 261.6 Special requirements for hazardous waste which is used, re-used, recycled or reclaimed.

(a) Except as otherwise provided in paragraph (b) of this section, a hazardous waste which meets either of the following criteria is not subject to regulation under Parts 262 through 265 or Parts 122 through 124 of this Chapter and is not subject to the notification requirements of Section 3010 of RCRA until such time as the Administrator promulgates regulations to the contrary:

(1) It is being beneficially used or re-used or legitimately recycled or reclaimed.

(2) It is being accumulated, stored or physically, chemically or biologically treated prior to beneficial use or re-use or legitimate recycling or reclamation.

(b) A hazardous waste which is a sludge, or which is listed in Subpart D, or which contains one or more hazardous wastes listed in Subpart D; and which is transported or stored prior to being used, re-used, recycled or reclaimed is subject to the following requirements with respect to such transportation or storage:

(1) Notification requirements under Section 3010 RCRA.

(2) Part 262 of this Chapter.

(3) Part 263 of this Chapter.

(4) Subparts A, B, C, D and E of Part 264 of this Chapter.

(5) Subparts A, B, C, D, E, G, H, I, J and L of Part 265 of this Chapter.

(6) Parts 122 and 124 of this Chapter, with respect to storage facilities.

Subpart B—Criteria for Identifying the Characteristics of Hazardous Waste and for Listing Hazardous Waste

§ 261.10   Criteria for identifying the characteristics of hazardous waste.

(a) The Administrator shall identify and define a characteristic of hazardous waste in Subpart C only upon determining that:

(1) A solid waste that exhibits the characteristic may:

(i) Cause, or significantly contribute to, an increase in mortality or an increase in serious irreversible, or incapacitating reversible, illness; or

(ii) Pose a substantial present or potential hazard to human health or the environment when it is improperly treated, stored, transported, disposed of or otherwise managed; and

(2) The characteristic can be:

(i) Measured by an available standardized test method which is reasonably within the capability of generators of solid waste or private sector laboratories that are available to serve generators of solid waste; or

(ii) Reasonably detected by generators of solid waste through their knowledge of their waste.

§ 261.11   Criteria for listing hazardous waste.

(a) The Administrator shall list a solid waste as a hazardous waste only upon determining that the solid waste meets one of the following criteria:

(1) It exhibits any of the characteristics of hazardous waste identified in Subpart C.

(2) It has been found to be fatal to humans in low doses or, in the absence of data on human toxicity, it has been shown in studies to have an oral LD 50 toxicity (rat) of less than 50 milligrams per kilogram, an inhalation LC 50 toxicity (rat) of less than 2 milligrams per liter, or a dermal LD 50 toxicity (rabbit) of less than 200 milligrams per kilogram or is otherwise capable of causing or significantly contributing to an increase in serious irreversible, or incapacitating reversible, illness. (Waste

listed in accordance with these criteria will be designated Acute Hazardous Waste.)

(3) It contains any of the toxic constituents listed in Appendix VIII unless, after considering any of the following factors, the Administrator concludes that the waste is not capable of posing a substantial present or potential hazard to human health or the environment when improperly treated, stored, transported or disposed of, or otherwise managed:

(i) The nature of the toxicity presented by the constituent.

(ii) The concentration of the constituent in the waste.

(iii) The potential of the constituent or any toxic degradation product of the constituent to migrate from the waste into the environment under the types of improper management considered in paragraph (a)(3)(vii) of this section.

(iv) The persistence of the constituent or any toxic degradation product of the constituent.

(v) The potential for the constituent or any toxic degradation product of the constituent to degrade into non-harmful constituents and the rate of degradation.

(vi) The degree to which the constituent or any degradation product of the constituent bioaccumulates in ecosystems.

(vii) The plausible types of improper management to which the waste could be subjected.

(viii) The quantities of the waste generated at individual generation sites or on a regional or national basis.

(ix) The nature and severity of the human health and environmental damage that has occurred as a result of the improper management of wastes containing the constituent.

(x) Action taken by other governmental agencies or regulatory programs based on the health or environmental hazard posed by the waste or waste constituent.

(xi) Such other factors as may be appropriate.

Substances will be listed on Appendix VIII only if they have been shown in scientific studies to have toxic, carcinogenic, mutagenic or teratogenic effects on humans or other life forms. (Wastes listed in accordance with these criteria will be designated Toxic wastes.)

(b) The Administrator may list classes or types of solid waste as hazardous waste if he has reason to believe that individual wastes, within the class or type of waste, typically or frequently are hazardous under the definition of hazardous waste found in Section 1004(5) of the Act.

(c) The Administrator will use the criteria for listing specified in this section to establish the exclusion limits referred to in § 261.5(c).

Subpart C—Characteristics of Hazardous Waste

§ 261.20   General.

(a) A solid waste, as defined in § 261.2, which is not excluded from regulation as a hazardous waste under § 261.4(b), is a hazardous waste if it exhibits any of the characteristics identified in this Subpart.

[Comment: § 262.11 of this Chapter sets forth the generator's responsibility to determine whether his waste exhibits one or more of the characteristics identified in this Subpart]

(b) A hazardous waste which is identified by a characteristic in this subpart, but is not listed as a hazardous waste in Subpart D, is assigned the EPA Hazardous Waste Number set forth in the respective characteristic in this Subpart. This number must be used in complying with the notification requirements of Section 3010 of the Act and certain recordkeeping and reporting requirements under Parts 262 through 265 and Part 122 of this Chapter.

(c) For purposes of this Subpart, the Administrator will consider a sample obtained using any of the applicable sampling methods specified in Appendix I to be a representative sample within the meaning of Part 260 of this Chapter.

[Comment: Since the Appendix I sampling methods are not being formally adopted by the Administrator, a person who desires to employ an alternative sampling method is not required to demonstrate the equivalency of his method under the procedures set forth in §§ 260.20 and 260.21.]

§ 261.21   Characteristic of ignitability.

(a) A solid waste exhibits the characteristic of ignitability if a representative sample of the waste has any of the following properties:

(1) It is a liquid, other than an aqueous solution containing less than 24 percent alcohol by volume, and has a flash point less than 60°C (140°F), as determined by a Pensky-Martens Closed Cup Tester, using the test method specified in ASTM Standard D–93–79, or a Setaflash Closed Cup Tester, using the test method specified in ASTM standard D–3278–78, or as determined by an equivalent test method approved by the Administrator under the procedures set forth in §§ 260.20 and 260.21.[1]

---

[1] ASTM Standards are available from ASTM, 1916 Race Street, Philadelphia, PA 19103.

(2) It is not a liquid and is capable, under standard temperature and pressure, of causing fire through friction, absorption of moisture or spontaneous chemical changes and, when ignited, burns so vigorously and persistently that is creates a hazard.

(3) It is an ignitable compressed gas as defined in 49 CFR 173.300 and as determined by the test methods described in that regulation or equivalent test methods approved by the Administrator under §§ 260.20 and 260.21.

(4) It is an oxidizer as defined in 49 CFR 173.151.

(b) A solid waste that exhibits the characteristic of ignitability, but is not listed as a hazardous waste in Subpart D, has the EPA Hazardous Waste Number of D001.

§ 261.22 Characteristic of corrosivity.

(a) A solid waste exhibits the characteristic of corrosivity if a representative sample of the waste has either of the following properties:

(1) It is aqueous and has a pH less than or equal to 2 or greater than or equal to 12.5, as determined by a pH meter using either the test method specified in the "Test Methods for the Evaluation of Solid Waste, Physical/Chemical Methods" [2] (also described in "Methods for Analysis of Water and Wastes" EPA 600/4–79–020, March 1979), or an equivalent test method approved by the Administrator under the procedures set forth in §§ 260.20 and 260.21.

(2) It is a liquid and corrodes steel (SAE 1020) at a rate greater than 6.35 mm (0.250 inch) per year at a test temperature of 55°C (130°F) as determined by the test method specified in NACE (National Association of Corrosion Engineers) Standard TM–01–69 [3] as standardized in "Test Methods for the Evaluation of Solid Waste, Physical/Chemical Methods," or an equivalent test method approved by the Administrator under the procedures set forth in §§ 260.20 and 260.21.

(b) A solid waste that exhibits the characteristic of corrosivity, but is not listed as a hazardous waste in Subpart D, has the EPA Hazardous Waste Number of D002.

---

[2] This document is available from Solid Waste Information, U.S. Environmental Protection Agency, 26 W. St. Clair Street, Cincinnati, Ohio 45268.

[3] The NACE Standard is available from the National Association of Corrosion Engineers, P.O. Box 986, Katy, Texas 77450.

§ 261.23 Characteristic of reactivity.

(a) A solid waste exhibits the characteristic of reactivity if a representative sample of the waste has *any* of the following properties:

(1) It is normally unstable and readily undergoes violent change without detonating.

(2) It reacts violently with water.

(3) It forms potentially explosive mixtures with water.

(4) When mixed with water, it generates toxic gases, vapors or fumes in a quantity sufficient to present a danger to human health or the environment.

(5) It is a cyanide or sulfide bearing waste which, when exposed to pH conditions between 2 and 12.5, can generate toxic gases, vapors or fumes in a quantity sufficient to present a danger to human health or the environment.

(6) It is capable of detonation or explosive reaction if it is subjected to a strong initiating source or if heated under confinement.

(7) It is readily capable of detonation or explosive decomposition or reaction at standard temperature and pressure.

(8) It is a forbidden explosive as defined in 49 CFR 173.51, or a Class A explosive as defined in 49 CFR 173.53 or a Class B explosive as defined in 49 CFR 173.88.

(b) A solid waste that exhibits the characteristic of reactivity, but is not listed as a hazardous waste in Subpart D, has the EPA Hazardous Waste Number of D003.

§ 261.24 Characteristic of EP Toxicity.

(a) A solid waste exhibits the characteristic of EP toxicity if, using the test methods described in Appendix II or equivalent methods approved by the Administrator under the procedures set forth in §§ 260.20 and 260.21, the extract from a representative sample of the waste contains any of the contaminants listed in Table I at a concentration equal to or greater than the respective value given in that Table. Where the waste contains less than 0.5 percent filterable solids, the waste itself, after filtering, is considered to be the extract for the purposes of this section.

(b) A solid waste that exhibits the characteristic of EP toxicity, but is not listed as a hazardous waste in Subpart D, has the EPA Hazardous Waste Number specified in Table I which corresponds to the toxic contaminant causing it to be hazardous.

Table I.—Maximum Concentration of Contaminants for Characteristic of EP Toxicity—Continued

EPA hazardous waste number	Contaminant	Maximum concentration (milligrams per liter)
D004	Arsenic	5.0
D005	Barium	100.0
D006	Cadmium	1.0
D007	Chromium	5.0
D008	Lead	5.0
D009	Mercury	0.2
D010	Selenium	1.0
D011	Silver	5.0
D012	Endrin (1,2,3,4,10,10-hexachloro-1,7-epoxy-1,4,4a,5,6,7,8,8a-octahydro-1,4-endo, endo-5,8-dimethano naphthalene.	0.02
D013	Lindane (1,2,3,4,5,6-hexachlorocyclohexane, gamma isomer.	0.4
D014	Methoxychlor (1,1,1-Trichloro-2,2-bis [p-methoxyphenyl]ethane).	10.0
D015	Toxaphene ($C_{10}H_{10}Cl_6$, Technical chlorinated camphene, 67–69 percent chlorine).	0.5
D016	2,4-D, (2,4-Dichlorophenoxyacetic acid).	10.0
D017	2,4,5-TP Silvex (2,4,5-Trichlorophenoxypropionic acid).	1.0

Subpart D—Lists of Hazardous Wastes

§ 261.30 General.

(a) A solid waste is a hazardous waste if it is listed in this Subpart, unless it has been excluded from this list under §§ 260.20 and 260.22.

(b) The Administrator will indicate his basis for listing the classes or types of wastes listed in this Subpart by employing one or more of the following Hazard Codes:

Ignitable Waste	(I)
Corrosive Waste	(C)
Reactive Waste	(R)
EP Toxic Waste	(E)
Acute Hazardous Waste	(H)
Toxic Waste	(T)

Appendix VII identifies the constituent which caused the Administrator to list the waste as an EP Toxic Waste (E) or Toxic Waste (T) in §§ 261.31 and 261.32.

(c) Each hazardous waste listed in this Subpart is assigned an EPA Hazardous Waste Number which precedes the name of the waste. This number must be used in complying with the notification requirements of Section 3010 of the Act and certain recordkeeping and reporting requirements under Parts 262 through 265 and Part 122 of this Chapter.

(d) Certain of the hazardous wastes listed in § 261.31 or § 261.32 have exclusion limits that refer to § 261.5(c)(5).

§ 261.31 Hazardous waste from nonspecific sources.

Industry and EPA hazardous waste No.	Hazardous waste	Hazard code
Generic:		
F001	The spent halogenated solvents used in degreasing, tetrachloroethylene, trichloroethylene, methylene chloride, 1,1,1-trichloroethane, carbon tetrachloride, and the chlorinated fluorocarbons; and sludges from the recovery of these solvents in degreasing operations.	(T)
F002	The spent halogenated solvents, tetrachloroethylene, methylene chloride, trichloroethylene, 1,1,1-trichloroethane, chlorobenzene, 1,1,2-trichloro-1,2,2-trifluoroethane, o-dichlorobenzene, trichlorofluoromethane and the still bottoms from the recovery of these solvents.	(T)
F003	The spent non-halogenated solvents, xylene, acetone, ethyl acetate, ethyl benzene, ethyl ether, n-butyl alcohol, cyclohexanone, and the still bottoms from the recovery of these solvents.	(I)
F004	The spent non-halogenated solvents, cresols and cresylic acid, nitrobenzene, and the still bottoms from the recovery of these solvents.	(T)
F005	The spent non-halogenated solvents, methanol, toluene, methyl ethyl ketone, methyl isobutyl ketone, carbon disulfide, isobutanol, pyridine and the still bottoms from the recovery of these solvents.	(I, T)
F006	Wastewater treatment sludges from electroplating operations.	(T)
F007	Spent plating bath solutions from electroplating operations.	(R, T)
F008	Plating bath sludges from the bottom of plating baths from electroplating operations.	(R, T)
F009	Spent stripping and cleaning bath solutions from electroplating operations.	(R, T)
F010	Quenching bath sludge from oil baths from metal heat treating operations.	(R, T)
F011	Spent solutions from salt bath pot cleaning from metal heat treating operations.	(R, T)
F012	Quenching wastewater treatment sludges from metal heat treating operations.	(T)
F013	Flotation tailings from selective flotation from mineral metals recovery operations.	(T)
F014	Cyanidation wastewater treatment tailing pond sediment from mineral metals recovery operations.	(T)
F015	Spent cyanide bath solutions from mineral metals recovery operations.	(R, T)
F016	Dewatered air pollution control scrubber sludges from coke ovens and blast furnaces.	(T)

§ 261.32 Hazardous waste from specific sources.

Industry and EPA hazardous waste No.	Hazardous waste	Hazard code
Wood Preservation: K001	Bottom sediment sludge from the treatment of wastewaters from wood preserving processes that use creosote and/or pentachlorophenol	(T)
Inorganic Pigments:		
K002	Wastewater treatment sludge from the production of chrome yellow and orange pigments	(T)
K003	Wastewater treatment sludge from the production of molybdate orange pigments	(T)
K004	Wastewater treatment sludge from the production of zinc yellow pigments	(T)
K005	Wastewater treatment sludge from the production of chrome green pigments	(T)
K006	Wastewater treatment sludge from the production of chrome oxide green pigments (anhydrous and hydrated)	(T)
K007	Wastewater treatment sludge from the production of iron blue pigments	(T)
K008	Oven residue from the production of chrome oxide green pigments.	(T)
Organic Chemicals:		
K009	Distillation bottoms from the production of acetaldehyde from ethylene	(T)
K010	Distillation side cuts from the production of acetaldehyde from ethylene	(T)
K011	Bottom stream from the wastewater stripper in the production of acrylonitrile	(R, T)
K012	Still bottoms from the final purification of acrylonitrile in the production of acrylonitrile	(T)
K013	Bottom stream from the acetonitrile column in the production of acrylonitrile	(R, T)
K014	Bottoms from the acetonitrile purification column in the production of acrylonitrile	(T)
K015	Still bottoms from the distillation of benzyl chloride	(T)
K016	Heavy ends or distillation residues from the production of carbon tetrachloride	(T)
K017	Heavy ends (still bottoms) from the purification column in the production of epichlorohydrin	(T)
K018	Heavy ends from fractionation in ethyl chloride production	(T)
K019	Heavy ends from the distillation of ethylene dichloride in ethylene dichloride production	(T)
K020	Heavy ends from the distillation of vinyl chloride in vinyl chloride monomer production	(T)
K021	Aqueous spent antimony catalyst waste from fluoromethanes production	(T)
K022	Distillation bottom tars from the production of phenol/acetone from cumene	(T)
K023	Distillation light ends from the production of phthalic anhydride from naphthalene	(T)
K024	Distillation bottoms from the production of phthalic anhydride from naphthalene	(T)
K025	Distillation bottoms from the production of nitrobenzene by the nitration of benzene	(T)
K026	Stripping still tails from the production of methyl ethyl pyridines	(T)
K027	Centrifuge residue from toluene diisocyanate production	(R, T)
K028	Spent catalyst from the hydrochlorinator reactor in the production of 1,1,1-trichloroethane	(T)
K029	Waste from the product stream stripper in the production of 1,1,1-trichloroethane	(T)
K030	Column bottoms or heavy ends from the combined production of trichloroethylene and perchloroethylene	(T)
Pesticides:		
K031	By-products salts generated in the production of MSMA and cacodylic acid	(T)
K032	Wastewater treatment sludge from the production of chlordane	(T)
K033	Wastewater and scrub water from the chlorination of cyclopentadiene in the production of chlordane	(T)
K034	Filter solids from the filtration of hexachlorocyclopentadiene in the production of chlordane	(T)
K035	Wastewater treatment sludges generated in the production of creosote	(T)
K036	Still bottoms from toluene reclamation distillation in the production of disulfoton	(T)
K037	Wastewater treatment sludges from the production of disulfoton	(T)
K038	Wastewater from the washing and stripping of phorate production	(T)
K039	Filter cake from the filtration of diethylphosphorodithoric acid in the production of phorate	(T)
K040	Wastewater treatment sludge from the production of phorate	(T)
K041	Wastewater treatment sludge from the production of toxaphene	(T)
K042	Heavy ends or distillation residues from the distillation of tetrachlorobenzene in the production of 2,4,5-T	(T)
K043	2,6-Dichlorophenol waste from the production of 2,4-D	(T)
Explosives:		
K044	Wastewater treatment sludges from the manufacturing and processing of explosives	(R)
K045	Spent carbon from the treatment of wastewater containing explosives	(R)
K046	Wastewater treatment sludges from the manufacturing, formulation and loading of lead-based initiating compounds	(T)
K047	Pink/red water from TNT operations	(R)
Petroleum Refining:		
K048	Dissolved air flotation (DAF) float from the petroleum refining industry	(T)
K049	Slop oil emulsion solids from the petroleum refining industry	(T)
K050	Heat exchanger bundle cleaning sludge from the petroleum refining industry	(T)
K051	API separator sludge from the petroleum refining industry	(T)
K052	Tank bottoms (leaded) from the petroleum refining industry	(T)
Leather Tanning Finishing:		
K053	Chrome (blue) trimmings generated by the following subcategories of the leather tanning and finishing industry: hair pulp/chrome tan/retan/wet finish; hair save/chrome tan/retan/wet finish; retan/wet finish; no beamhouse; through-the-blue; and shearling.	(T)

§ 261.32 Hazardous waste from specific sources. —Continued

Industry and EPA hazardous waste No.	Hazardous waste	Hazard code
K054	Chrome (blue) shavings generated by the following subcategories of the leather tanning and finishing industry: hair pulp/chrome tan/retan/wet finish; hair save/chrome tan/retan/wet finish; retan/wet finish; no beamhouse; through-the-blue; and shearling.	(T)
K055	Buffing dust generated by the following subcategories of the leather tanning and finishing industry: hair pulp/chrome tan/retan/wet finish; hair save/chrome tan/retan/wet finish; retan/wet finish; no beamhouse; and through-the-blue.	(T)
K056	Sewer screenings generated by the following subcategories of the leather tanning and finishing industry: hair pulp/chrome tan/retan/wet finish; hair save/chrome tan/retan/wet finish; retan/wet finish; no beamhouse; through-the-blue; and shearling.	(T)
K057	Wastewater treatment sludges generated by the following subcategories of the leather tanning and finishing industry: hair pulp/chrome tan/retan/wet finish; hair save/chrome tan/retan/wet finish; retan/wet finish; no beamhouse; through-the-blue and shearling.	(T)
K058	Wastewater treatment sludges generated by the following subcategories of the leather tanning and finishing industry: hair pulp/chrome tan/retan/wet finish; hair save/chrome tan/retan/wet finish; and through-the-blue.	(R, T)
K059	Wastewater treatment sludges generated by the following subcategory of the leather tanning and finishing industry: hair save/non-chrome tan/retan/wet finish.	(R)
Iron and Steel:		
K060	Ammonia still lime sludge from coking operations	(T)
K061	Emission control dust/sludge from the electric furnace production of steel	(T)
K062	Spent pickle liquor from steel finishing operations	(C, T)
K063	Sludge from lime treatment of spent pickle liquor from steel finishing operations	(T)
Primary Copper: K064	Acid plant blowdown slurry/sludge resulting from the thickening of blowdown slurry from primary copper production	(T)
Primary Lead: K065	Surface impoundment solids contained in and dredged from surface impoundments at primary lead smelting facilities	(T)
Primary Zinc:		
K066	Sludge from treatment of process wastewater and/or acid plant blowdown from primary zinc production	(T)
K067	Electrolytic anode slimes/sludges from primary zinc production	(T)
K068	Cadmium plant leach residue (iron oxide) from primary zinc production	(T)
Secondary Lead: K069	Emission control dust/sludge from secondary lead smelting	(T)

§ 261.33 Discarded Commercial Chemical Products, Off-Specification Species, Containers, and Spill Residues Thereof.

The following materials or items are hazardous wastes if and when they are discarded or intended to be discarded:

(a) Any commercial chemical product, or manufacturing chemical intermediate having the generic name listed in paragraphs (e) or (f) of this section.

(b) Any off-specification commercial chemical product or manufacturing chemical intermediate which, if it met specifications, would have the generic name listed in paragraphs (e) or (f) of this section.

(c) Any container or inner liner removed from a container that has been used to hold any commercial chemical product or manufacturing chemical intermediate having the generic name listed in paragraph (e) of this section, unless:

(1) The container or inner liner has been triple rinsed using a solvent capable of removing the commercial chemical product or manufacturing chemical intermediate;

(2) The container or inner liner has been cleaned by another method that has been shown in the scientific literature, or by tests conducted by the generator, to achieve equivalent removal; or

(3) In the case of a container, the inner liner that prevented contact of the commercial chemical product or manufacturing chemical intermediate with the container, has been removed.

(d) Any residue or contaminated soil, water or other debris resulting from the cleanup of a spill, into or on any land or water, of any commercial chemical product or manufacturing chemical intermediate having the generic name listed in paragraphs (e) or (f) of this Section.

[Comment: The phrase "commercial chemical product or manufacturing chemical intermediate having the generic name listed in . . ." refers to a chemical substance which is manufactured or formulated for commercial or manufacturing use. It does not refer to a material, such as a manufacturing process waste, that contains any of the substances listed in paragraphs (e) or (f). Where a manufacturing process waste is deemed to be a hazardous waste because it contains a substance listed in paragraphs (e) or (f), such waste will be listed in either §§ 261.31 or 261.32 or will be identified as a hazardous waste by the characteristics set forth in Subpart C of this Part.]

(e) The commercial chemical products or manufacturing chemical intermediates, referred to in paragraphs (a) through (d) of this section, are identified as acute hazardous wastes (H) and are subject to the small quantity exclusion defined in § 261.5(c). These wastes and their corresponding EPA Hazardous Waste Numbers are:

Hazardous waste No.	Substance [1]
	1080 see P058
	1081 see P057
	(Acetato)phenylmercury see P092
	Acetone cyanohydrin see P069
P001	3-(alpha-Acetonylbenzyl)-4-hydroxycoumarin and salts
P002	1-Acetyl-2-thiourea
P003	Acrolein
	Agarin see P007
	Agrosan GN 5 see P092
	Aldicarb see P069
	Aldifen see P048

Hazardous waste No.	Substance [1]
P004	Aldrin
	Algimycin see P092
P005	Allyl alcohol
P006	Aluminum phosphide (R)
	ALVIT see P037
	Aminoethylene see P054
P007	5-(Aminomethyl)-3-isoxazolol
P008	4-Aminopyridine
	Ammonium metavanadate see P119
P009	Ammonium picrate (R)
	ANTIMUCIN WDR see P092
	ANTURAT see P073
	AQUATHOL see P088
	ARETIT see P020
P010	Arsenic acid
P011	Arsenic pentoxide
P012	Arsenic trioxide
	Athrombin see P001
	AVITROL see P008
	Aziridene see P054
	AZOFOS see P061
	Azophos see P061
	BANTU see P072
P013	Barium cyanide
	BASENITE see P020
	BCME see P016
P014	Benzenethiol
	Benzoepin see P050
P015	Beryllium dust
P016	Bis(chloromethyl) ether
	BLADAN-M see P071
P017	Bromoacetone
P018	Brucine
P019	2-Butanone peroxide
	BUFEN see P092
	Butaphene see P020
P020	2-sec-Butyl-4,6-dinitrophenol
P021	Calcium cyanide
	CALDON see P020
P022	Carbon disulfide
	CERESAN see P092
	CERESAN UNIVERSAL see P092
	CHEMOX GENERAL see P020
	CHEMOX P.E. see P020
	CHEM-TOL see P090
P023	Chloroacetaldehyde
P024	p-Chloroaniline
P025	1-(p-Chlorobenzoyl)-5-methoxy-2-methylindole-3-acetic acid
P026	1-(o-Chlorophenyl)thiourea
P027	3-Chloropropionitrile
P028	alpha-Chlorotoluene
P029	Copper cyanide
	CRETOX see P108
	Coumadin see P001
	Coumafen see P001
P030	Cyanides

Hazardous waste No.	Substance
P031.........	Cyanogen
P032.........	Cyanogen bromide
P033.........	Cyanogen chloride
	Cyclodan see P050
P034.........	2-Cyclohexyl-4,6-dinitrophenol
	D-CON see P001
	DETHMOR see P001
	DETHNEL see P001
	DFP see P043
P035.........	2,4-Dichlorophenoxyacetic acid (2,4-D)
P036.........	Dichlorophenylarsine
	Dicyanogen see P031
P037.........	Dieldrin
	DIELDREX see P037
P038.........	Diethylarsine
P039.........	O,O-Diethyl-S-(2-(ethylthio)ethyl)ester of phosphorothioic acid
P040.........	O,O-Diethyl-O-(2-pyrazinyl)phosphorothioate
P041.........	O,O-Diethyl phosphoric acid, O-p-nitrophenyl ester
P042.........	3,4-Dihydroxy-alpha-(methylamino)-methyl benzyl alcohol
P043.........	Di-isopropylfluorophosphate
	DIMETATE see P044
	1,4:5,8-Dimethanonaphthalene, 1,2,3,4,10,10-hexachloro-1,4,4a,5,8,8a-hexahydro endo, endo see P060
P044.........	Dimethoate
P045.........	3,3-Dimethyl-1-(methylthio)-2-butanone-O-[(methylamino)carbonyl] oxime
P046.........	alpha,alpha-Dimethylphenethylamine
	Dinitrocyclohexylphenol see P034
P047.........	4,6-Dinitro-o-cresol and salts
P048.........	2,4-Dinitrophenol
	DINOSEB see P020
	DINOSEBE see P020
	Disulfoton see P039
P049.........	2,4-Dithiobiuret
	DNBP see P020
	DOLCO MOUSE CEREAL see P108
	DOW GENERAL see P020
	DOW GENERAL WEED KILLER see P020
	DOW SELECTIVE WEED KILLER see P020
	DOWICIDE G see P090
	DYANACIDE see P092
	EASTERN STATES DUOCIDE see P001
	ELGETOL see P020
P050.........	Endosulfan
P051.........	Endrin
	Epinephrine see P042
P052.........	Ethylcyanide
P053.........	Ethylenediamine
P054.........	Ethyleneimine
	FASCO FASCRAT POWDER see P001
	FEMMA see P091
P055.........	Ferric cyanide
P056.........	Fluorine
P057.........	2-Fluoroacetamide
P058.........	Fluoroacetic acid, sodium salt
	FOLODOL-80 see P071
	FOLODOL M see P071
	FOSFERNO M 50 see P071
	FRATOL see P058
	Fulminate of mercury see P065
	FUNGITOX OR see P092
	FUSSOF see P057
	GALLOTOX see P092
	GEARPHOS see P071
	GERUTOX see P020
P059.........	Heptachlor
P060.........	1,2,3,4,10,10-Hexachloro-1,4,4a,5,8,8a-hexahydro-1,4:5,8-endo, endo-dimethanonaphthalene
	1,4,5,6,7,7-Hexachloro-cyclic-5-norbornene-2,3-dimethanol sulfite see P050
P061.........	Hexachloropropene
P062.........	Hexaethyl tetraphosphate
	HOSTAQUICK see P092
	HOSTAQUIK see P092
	Hydrazomethane see P068
P063.........	Hydrocyanic acid
	ILLOXOL see P037
	INDOCI see P025
	Indomethacin see P025
	INSECTOPHENE see P050
	Isodrin see P060
P064.........	Isocyanic acid, methyl ester
	KILOSEB see P020
	KOP-THIODAN see P050
	KWIK-KIL see P108
	KWIKSAN see P092
	KUMADER see P001
	KYPFARIN see P001
	LEYTOSAN see P092
	LIQUIPHENE see P092

Hazardous waste No.	Substance [1]
	MALIK see P050
	MAREVAN see P001
	MAR-FRIN see P001
	MARTIN'D MAR-FRIN see P001
	MAVERAN see P001
	MEGATOX see P005
P065.........	Mercury fulminate
	MERSOLITE see P092
	METACID 50 see P071
	METAFOS see P071
	METAPHOR see P071
	METAPHOS see P071
	METASOL 30 see P092
P066.........	Methomyl
P067.........	2-Methylaziridine
	METHYL-E 605 see P071
P068.........	Methyl hydrazine
	Methyl isocyanate see P064
P069.........	2-Methyllactonitrile
P070.........	2-Methyl-2-(methylthio)propionaldehyde-o-(methylcarbonyl) oxime
	METHYL NIRON see P042
P071.........	Methyl parathion
	METRON see P071
	MOLE DEATH see P108
	MOUSE-NOTS see P108
	MOUSE-RID see P108
	MOUSE-TOX see P108
	MUSCIMOL see P007
P072.........	1-Naphthyl-2-thiourea
P073.........	Nickel carbonyl
P074.........	Nickel cyanide
P075.........	Nicotine and salts
P076.........	Nitric oxide
P077.........	p-Nitroaniline
P078.........	Nitrogen dioxide
P079.........	Nitrogen peroxide
P080.........	Nitrogen tetroxide
P081.........	Nitroglycerine (R)
P082.........	N-Nitrosodimethylamine
P083.........	N-Nitrosodiphenylamine
P084.........	N-Nitrosomethylvinylamine
	NYLMERATE see P092
	OCTALOX see P037
P085.........	Octamethylpyrophosphoramide
	OCTAN see P092
P086.........	Oleyl alcohol condensed with 2 moles ethylene oxide
	OMPA see P085
	OMPACIDE see P085
	OMPAX see P085
P087.........	Osmium tetroxide
P088.........	7-Oxabicyclo[2.2.1]heptane-2,3-dicarboxylic acid
	PANIVARFIN see P001
	PANORAM D-31 see P037
	PANTHERINE see P007
	PANWARFIN see P001
P089.........	Parathion
	PCP see P090
	PENNCAP-M see P071
	PENOXYL CARBON N see P048
P090.........	Pentachlorophenol
	Pentachlorophenate see P090
	PENTA-KILL see P090
	PENTASOL see P090
	PENWAR see P090
	PERMICIDE see P090
	PERMAGUARD see P090
	PERMATOX see P090
	PERMITE see P090
	PERTOX see P090
	PESTOX III see P085
	PHENMAD see P092
	PHENOTAN see P020
P091.........	Phenyl dichloroarsine
	Phenyl mercaptan see P014
P092.........	Phenylmercury acetate
P093.........	N-Phenylthiourea
	PHILIPS 1861 see P008
	PHIX see P092
P094.........	Phorate
P095.........	Phosgene
P096.........	Phosphine
P097.........	Phosphorothioic acid, 0,0-dimethyl ester, 0-ester with N,N-dimethyl benzene sulfonamide
	Phosphorothioic acid 0,0-dimethyl-0-(p-nitrophenyl) ester see P071
	PIED PIPER MOUSE SEED see P108
P098.........	Potassium cyanide
P099.........	Potassium silver cyanide
	PREMERGE see P020
P100.........	1,2-Propanediol
	Propargyl alcohol see P102
P101.........	Propionitrile

Hazardous waste No.	Substance [1]
P102.........	2-Propyn-1-o1
	PROTHROMADIN See P001
	QUICKSAM see P092
	QUINTOX see P037
	RAT AND MICE BAIT see P001
	RAT-A-WAY see P001
	RAT-B-GON see P001
	RAT-O-CIDE #2 see P001
	RAT-GUARD see P001
	RAT-KILL see P001
	RAT-MIX see P001
	RATS-NO-MORE see P001
	RAT-OLA see P001
	RATOREX see P001
	RATTUNAL see P001
	RAT-TROL see P001
	RO-DETH see P001
	RO-DEX see P108
	ROSEX see P001
	ROUGH & READY MOUSE MIX see P001
	SANASEED see P108
	SANTOBRITE see P090
	SANTOPHEN see P090
	SANTOPHEN 20 see P090
	SCHRADAN see P085
P103.........	Selenourea
P104.........	Silver Cyanide
	SMITE see P105
	SPARIC see P020
	SPOR-KIL see P092
	SPRAY-TROL BRAND RODEN-TROL see P001
	SPURGE see P020
P105.........	Sodium azide
	Sodium coumadin see P001
P106.........	Sodium cyanide
	Sodium fluoroacetate see P056
	SODIUM WARFARIN see P001
	SOLFARIN see P001
	SOLFOBLACK BB see P048
	SOLFOBLACK SB see P048
P107.........	Strontium sulfide
P108.........	Strychnine and salts
	SUBTEX see P020
	SYSTAM see P085
	TAG FUNGICIDE see P092
	TEKWAISA see P071
	TEMIC see P070
	TEMIK see P070
	TERM-I-TROL see P090
P109.........	Tetraethyldithiopyrophosphate
P110.........	Tetraethyl lead
P111.........	Tetraethylpyrophosphate
P112.........	Tetranitromethane
	Tetraphosphoric acid, hexaethyl ester see P062
	TETROSULFUR BLACK PB see P048
	TETROSULPHUR PBR see P048
P113.........	Thallic oxide
	Thallium peroxide see P113
P114.........	Thallium selenite
P115.........	Thallium (I) sulfate
	THIFOR see P092
	THIMUL see P092
	THIODAN see P050
	THIOFOR see P050
	THIOMUL see P050
	THIONEX see P050
	THIOPHENIT see P071
P116.........	Thiosemicarbazide
	Thiosulfan tionel see P050
P117.........	Thiuram
	THOMPSON'S WOOD FIX see P090
	TIOVEL see P050
P118.........	Trichloromethanethiol
	TWIN LIGHT RAT AWAY see P001
	USAF RH-8 see P069
	USAF EK-4890 see P002
P119.........	Vanadic acid, ammonium salt
P120.........	Vanadium pentoxide
	VOFATOX see P071
	WANADU see P120
	WARCOUMIN see P001
	WARFARIN SODIUM see P001
	WARFICIDE see P001
	WOFOTOX see P072
	YANOCK see P057
	YASOKNOCK see P058
	ZIARNIK see P092
P121.........	Zinc cyanide
P122.........	Zinc phosphide (R,T)
	ZOOCOUMARIN see P001

[1] The Agency included those trade names of which it was aware; an omission of a trade name does not imply that the omitted material is not hazardous. The material is hazardous if it is listed under its generic name.

(f) The commercial chemical products or manufacturing chemical intermediates, referred to in paragraphs (a), (b) and (d) of this section, are identified as toxic wastes (T) unless otherwise designated and are subject to the small quantity exclusion defined in § 261.5 (a) and (b). These wastes and their corresponding EPA Hazardous Waste Numbers are:

Hazardous Waste No.	Substance[1]
	AAF see U005
U001	Acetaldehyde
U002	Acetone (I)
U003	Acetonitrile (I,T)
U004	Acetophenone
U005	2-Acetylaminoflourene
U006	Acetyl chloride (C,T)
U007	Acrylamide
	Acetylene tetrachloride see U209
	Acetylene trichloride see U228
U008	Acrylic acid (I)
U009	Acrylonitrile
	AEROTHENE TT see U226
	3-Amino-5-(p-acetamidophenyl)-1H-1,2,4-triazole, hydrate see U011
U010	6-Amino-1,1a,2,8,8a,8b-hexahydro-8-(hydroxymethyl)8-methoxy-5-methylcarbamate azirino(2',3':3,4) pyrrolo(1,2-a) indole-4, 7-dione (ester)
U011	Amitrole
U012	Aniline (I)
U013	Asbestos
U014	Auramine
U015	Azaserine
U016	Benz[c]acridine
U017	Benzal chloride
U018	Benz[a]anthracene
U019	Benzene
U020	Benzenesulfonyl chloride (C,R)
U021	Benzidine
	1,2-Benzisothiazolin-3-one, 1,1-dioxide see U202
	Benzo[a]anthracene see U018
U022	Benzo[a]pyrene
U023	Benzotrichloride (C,R,T)
U024	Bis(2-chloroethoxy)methane
U025	Bis(2-chloroethyl) ether
U026	N,N-Bis(2-chloroethyl)-2-naphthylamine
U027	Bis(2-chloroisopropyl) ether
U028	Bis(2-ethylhexyl) phthalate
U029	Bromomethane
U030	4-Bromophenyl phenyl ether
U031	n-Butyl alcohol (I)
U032	Calcium chromate
	Carbolic acid see U188
	Carbon tetrachloride see U211
U033	Carbonyl fluoride
U034	Chloral
U035	Chlorambucil
U036	Chlordane
U037	Chlorobenzene
U038	Chlorobenzilate
U039	p-Chloro-m-cresol
U040	Chlorodibromomethane
U041	1-Chloro-2,3-epoxypropane
	CHLOROETHENE NU see U226
U042	Chloroethyl vinyl ether
U043	Chloroethene
U044	Chloroform (I,T)
U045	Chloromethane (I,T)
U046	Chloromethyl methyl ether
U047	2-Chloronaphthalene
U048	2-Chlorophenol
U049	4-Chloro-o-toluidine hydrochloride
U050	Chrysene
	C.I. 23060 see U073
U051	Cresote
U052	Cresols
U053	Crotonaldehyde
U054	Cresylic acid
U055	Cumene
	Cyanomethane see U003
U056	Cyclohexane (I)
U057	Cyclohexanone (I)
U058	Cyclophosphamide
U059	Daunomycin
U060	DDD

Hazardous Waste No.	Substance[1]
U061	DDT
U062	Diallate
U063	Dibenz[a,h]anthracene
	Dibenzo[a,h]anthracene see U063
U064	Dibenzo[a,i]pyrene
U065	Dibromochloromethane
U066	1,2-Dibromo-3-chloropropane
U067	1,2-Dibromoethane
U068	Dibromomethane
U069	Di-n-butyl phthalate
U070	1,2-Dichlorobenzene
U071	1,3-Dichlorobenzene
U072	1,4-Dichlorobenzene
U073	3,3'-Dichlorobenzidine
U074	1,4-Dichloro-2-butene
	3,3'-Dichloro-4,4'-diaminobiphenyl see U073
U075	Dichlorodifluoromethane
U076	1,1-Dichloroethane
U077	1,2-Dichloroethane
U078	1,1-Dichloroethylene
U079	1,2-trans-dichloroethylene
U080	Dichloromethane
	Dichloromethylbenzene see U017
U081	2,4-Dichlorophenol
U082	2,6-Dichlorophenol
U083	1,2-Dichloropropane
U084	1,3-Dichloropropene
U085	Diepoxybutane (I,T)
U086	1,2-Diethylhydrazine
U087	0,0-Diethyl-S-methyl ester of phosphorodithioic acid
U088	Diethyl phthalate
U089	Diethylstilbestrol
U090	Dihydrosafrole
U091	3,3'-Dimethoxybenzidine
U092	Dimethylamine (I)
U093	p-Dimethylaminoazobenzene
U094	7,12-Dimethylbenz[a]anthracene
U095	3,3'-Dimethylbenzidine
U096	alpha,alpha-Dimethylbenzylhydroperoxide (R)
U097	Dimethylcarbamoyl chloride
U098	1,1-Dimethylhydrazine
U099	1,2-Dimethylhydrazine
U100	Dimethylnitrosoamine
U101	2,4-Dimethylphenol
U102	Dimethyl phthalate
U103	Dimethyl sulfate
U104	2,4-Dinitrophenol
U105	2,4-Dinitrotoluene
U106	2,6-Dinitrotoluene
U107	Di-n-octyl phthalate
U108	1,4-Dioxane
U109	1,2-Diphenylhydrazine
U110	Dipropylamine (I)
U111	Di-n-propylnitrosamine
	EBDC see U114
	1,4-Epoxybutane see U213
U112	Ethyl acetate (I)
U113	Ethyl acrylate (I)
U114	Ethylenebisdithiocarbamate
U115	Ethylene oxide (I,T)
U116	Ethylene thiourea
U117	Ethyl ether (I)
U118	Ethylmethacrylate
U119	Ethyl methanesulfonate
	Ethylnitrile see U003
	Firemaster T23P see U235
U120	Fluoranthene
U121	Fluorotrichloromethane
U122	Formaldehyde
U123	Formic acid (C,T)
U124	Furan (I)
U125	Furfural (I)
U126	Glycidylaldehyde
U127	Hexachlorobenzene
U128	Hexachlorobutadiene
U129	Hexachlorocyclohexane
U130	Hexachlorocyclopentadiene
U131	Hexachloroethane
U132	Hexachlorophene
U133	Hydrazine (R,T)
U134	Hydrofluoric acid (C,T)
U135	Hydrogen sulfide
	Hydroxybenzene see U188
U136	Hydroxydimethyl arsine oxide
	4,4'-(Imidocarbonyl)bis(N,N-dimethyl)aniline see U014
U137	Indeno(1,2,3-cd)pyrene
U138	Iodomethane
U139	Iron Dextran
U140	Isobutyl alcohol

Hazardous Waste No.	Substance[1]
U141	Isosafrole
U142	Kepone
U143	Lasiocarpine
U144	Lead acetate
U145	Lead phosphate
U146	Lead subacetate
U147	Maleic anhydride
U148	Maleic hydrazide
U149	Malononitrile
	MEK Peroxide see U160
U150	Melphalan
U151	Mercury
U152	Methacrylonitrile
U153	Methanethiol
U154	Methanol
U155	Methapyrilene
	Methyl alcohol see U154
U156	Methyl chlorocarbonate
	Methyl chloroform see U226
U157	3-Methylcholanthrene
	Methyl chloroformate see U156
U158	4,4'-Methylene-bis-(2-chloroaniline)
U159	Methyl ethyl ketone (MEK) (I,T)
U160	Methyl ethyl ketone peroxide (R)
	Methyl iodide see U138
U161	Methyl isobutyl ketone
U162	Methyl methacrylate (R,T)
U163	N-Methyl-N'-nitro-N-nitrosoguanidine
U164	Methylthiouracil
	Mitomycin C see U010
U165	Naphthalene
U166	1,4-Naphthoquinone
U167	1-Naphthylamine
U168	2-Naphthylamine
U169	Nitrobenzene (I,T)
	Nitrobenzol see U169
U170	4-Nitrophenol
U171	2-Nitropropane (I)
U172	N-Nitrosodi-n-butylamine
U173	N-Nitrosodiethanolamine
U174	N-Nitrosodiethylamine
U175	N-Nitrosodi-n-propylamine
U176	N-Nitroso-n-ethylurea
U177	N-Nitroso-n-methylurea
U178	N-Nitroso-n-methylurethane
U179	N-Nitrosopiperidine
U180	N-Nitrosopyrrolidine
U181	5-Nitro-o-toluidine
U182	Paraldehyde
	PCNB see U185
U183	Pentachlorobenzene
U184	Pentachloroethane
U185	Pentachloronitrobenzene
U186	1,3-Pentadiene (I)
	Perc see U210
	Perchlorethylene see U210
U187	Phenacetin
U188	Phenol
U189	Phosphorous sulfide (R)
U190	Phthalic anhydride
U191	2-Picoline
U192	Pronamide
U193	1,3-Propane sultone
U194	n-Propylamine (I)
U196	Pyridine
U197	Quinones
U200	Reserpine
U201	Resorcinol
U202	Saccharin
U203	Safrole
U204	Selenious acid
U205	Selenium sulfide (R,T)
	Silvex see U233
U206	Streptozotocin
	2,4,5-T see U232
U207	1,2,4,5-Tetrachlorobenzene
U208	1,1,1,2-Tetrachloroethane
U209	1,1,2,2-Tetrachloroethane
U210	Tetrachloroethene
	Tetrachloroethylene see U210
U211	Tetrachloromethane
U212	2,3,4,6-Tetrachlorophenol
U213	Tetrahydrofuran (I)
U214	Thallium (I) acetate
U215	Thallium (I) carbonate
U216	Thallium (I) chloride
U217	Thallium (I) nitrate
U218	Thioacetamide
U219	Thiourea
U220	Toluene
U221	Toluenediamine
U222	o-Toluidine hydrochloride

Hazardous Waste No.	Substance[1]
U223.........	Toluene diisocyanate
U224.........	Toxaphene
	2,4,5-TP see U233
U225.........	Tribromomethane
U226.........	1,1,1-Trichloroethane
U227.........	1,1,2-Trichloroethane
U228.........	Trichloroethene
	Trichloroethylene see U228
U229.........	Trichlorofluoromethane
U230.........	2,4,5-Trichlorophenol
U231.........	2,4,6-Trichlorophenol
U232.........	2,4,5-Trichlorophenoxyacetic acid
U233.........	2,4,5-Trichlorophenoxypropionic acid alpha, alpha, alpha- Trichlorotoluene see U023
	TRI-CLENE see U228
U234.........	Trinitrobenzene (R,T)
U235.........	Tris(2,3-dibromopropyl) phosphate
U236.........	Trypan blue
U237.........	Uracil mustard
U238.........	Urethane
	Vinyl chloride see U043
	Vinylidene chloride see U078
U239.........	Xylene

[1] The Agency included those trade names of which it was aware; an omission of a trade name does not imply that it is not hazardous. The material is hazardous if it is listed under its generic name.

Appendix I—Representative Sampling Methods

The methods and equipment used for sampling waste materials will vary with the form and consistency of the waste materials to be sampled. Samples collected using the sampling protocols listed below, for sampling waste with properties similar to the indicated materials, will be considered by the Agency to be representative of the waste.

Extremely viscous liquid—ASTM Standard D140–70 Crushed or powdered material—ASTM Standard D346–75 Soil or rock-like material—ASTM Standard D420–69 Soil-like material—ASTM Standard D1452–65

Fly Ash-like material—ASTM Standard D2234–76 [ASTM Standards are available from ASTM, 1916 Race St., Philadelphia, PA 19103]

Containerized liquid wastes—"COLIWASA" described in "Test Methods for the Evaluation of Solid Waste, Physical/Chemical Methods," [1] U.S. Environmental Protection Agency, Office of Solid Waste, Washington, D.C. 20460. [Copies may be obtained from Solid Waste Information, U.S. Environmental Protection Agency, 26 W. St. Clair St., Cincinnati, Ohio 45268]

Liquid waste in pits, ponds, lagoons, and similar reservoirs.—"Pond Sampler" described in "Test Methods for the Evaluation of Solid Waste, Physical/Chemical Methods." [1]

This manual also contains additional information on application of these protocols.

[1] These methods are also described in "Samplers and Sampling Procedures for Hazardous Waste Streams," EPA 600/2–80–018, January 1980.

Appendix II— EP Toxicity Test Procedure

A. Extraction Procedure (EP)

1. A representative sample of the waste to be tested (minimum size 100 grams) should be obtained using the methods specified in Appendix I or any other methods capable of yielding a representative sample within the meaning of Part 260. [For detailed guidance on conducting the various aspects of the EP see "Test Methods for the Evaluation of Solid Waste, Physical/Chemical Methods," SW–846, U.S. Environmental Protection Agency Office of Solid Waste, Washington, D.C. 20460.[1]]

2. The sample should be separated into its component liquid and solid phases using the method described in "Separation Procedure" below. If the solid residue [2] obtained using this method totals less than 0.5% of the original weight of the waste, the residue can be discarded and the operator should treat the liquid phase as the extract and proceed immediately to Step 8.

3. The solid material obtained from the Separation Procedure should be evaluated for its particle size. If the solid material has a surface area per gram of material equal to, or greater than, 3.1 cm[2] or passes through a 9.5 mm (0.375 inch) standard sieve, the operator should proceed to Step 4. If the surface area is smaller or the particle size larger than specified above, the solid material should be prepared for extraction by crushing, cutting or grinding the material so that it passes through a 9.5 mm (0.375 inch) sieve or, if the material is in a single piece, by subjecting the material to the "Structural Integrity Procedure" described below.

4. The solid material obtained in Step 3 should be weighed and placed in an extractor with 16 times its weight of deionized water. Do not allow the material to dry prior to weighing. For purposes of this test, an acceptable extractor is one which will impart sufficient agitation to the mixture to not only prevent stratification of the sample and extraction fluid but also insure that all sample surfaces are continously

[1] Copies may be obtained from Solid Waste Information, U.S. Environmental Protection Agency, 26 W. St. Clair Street, Cincinnati, Ohio 45268.

[2] The percent solids is determined by drying the filter pad at 80° C until it reaches constant weight and then calculating the percent solids using the following equation:

$$\frac{(\text{weight of pad} + \text{solid}) - (\text{tare weight of pad})}{\text{initial weight of sample}} \times 100 = \% \text{ solids}$$

brought into contact with well mixed extraction fluid.

5. After the solid material and deionized water are placed in the extractor, the operator should begin agitation and measure the pH of the solution in the extractor. If the pH is greater than 5.0, the pH of the solution should be decreased to 5.0 ± 0.2 by adding 0.5 N acetic acid. If the pH is equal to or less than 5.0, no acetic acid should be added. The pH of the solution should be monitored, as described below, during the course of the extraction and if the pH rises above 5.2, 0.5N acetic acid should be added to bring the pH down to 5.0 ± 0.2. However, in no event shall the aggregate amount of acid added to the solution exceed 4 ml of acid per gram of solid. The mixture should be agitated for 24 hours and maintained at 20°–40° C (68°–104° F) during this time. It is recommended that the operator monitor and adjust the pH during the course of the extraction with a device such as the Type 45–A pH Controller manufactured by Chemtrix, Inc., Hillsboro, Oregon 97123 or its equivalent, in conjunction with a metering pump and reservoir of 0.5N acetic acid. If such a system is not available, the following manual procedure shall be employed:

(a) A pH meter should be calibrated in accordance with the manufacturer's specifications.

(b) The pH of the solution should be checked and, if necessary, 0.5N acetic acid should be manually added to the extractor until the pH reaches 5.0 ± 0.2. The pH of the solution should be adjusted at 15, 30 and 60 minute intervals, moving to the next longer interval if the pH does not have to be adjusted more than 0.5N pH units.

(c) The adjustment procedure should be continued for at least 6 hours.

(d) If at the end of the 24-hour extraction period, the pH of the solution is not below 5.2 and the maximum amount of acid (4 ml per gram of solids) has not been added, the pH should be adjusted to 5.0 ± 0.2 and the extraction continued for an additional four hours, during which the pH should be adjusted at one hour intervals.

6. At the end of the 24 hour extraction period, deionized water should be added to the extractor in an amount determined by the following equation:

V = (20)(W) − 16(W) − A
V = ml deionized water to be added
W = weight in grams of solid charged to extractor
A = ml of 0.5N acetic acid added during extraction

7. The material in the extractor should be separated into its component liquid and solid phases as described under "Separation Procedure."

8. The liquids resulting from Steps 2 and 7 should be combined: This

combined liquid (or the waste itself if it has less than ½ percent solids, as noted in Step 2) is the extract and should be analyzed for the presence of any of the contaminants specified in Table I of § 261.24 using the Analytical Procedures designated below.

Separation Procedure

Equipment: A filter holder, designed for filtration media having a nominal pore size of 0.45 micrometers and capable of applying a 5.3 kg/cm² (75 psi) hydrostatic pressure to the solution being filtered shall be used. For mixtures containing nonabsorptive solids, where separation can be affected without imposing a 5.3 kg/cm² pressure differential, vacuum filters employing a 0.45 micrometers filter media can be used. (For further guidance on filtration equipment or procedures see "Test Methods for Evaluating Solid Waste, Physical/Chemical Methods.")

Procedure: [3]

(i) Following manufacturer's directions, the filter unit should be assembled with a filter bed consisting of a 0.45 micrometer filter membrane. For difficult or slow to filter mixtures a prefilter bed consisting of the following prefilters in increasing pore size (0.65 micrometer membrane, fine glass fiber prefilter, and coarse glass fiber prefilter) can be used.

(ii) The waste should be poured into the filtration unit.

(iii) The reservoir should be slowly pressurized until liquid begins to flow from the filtrate outlet at which point the pressure in the filter should be immediately lowered to 10–15 psig. Filtration should be continued until liquid flow ceases.

(iv) The pressure should be increased stepwise in 10 psi increments to 75 psig and filtration continued until flow ceases or the pressurizing gas begins to exit from the filtrate outlet.

(v) The filter unit should be depressurized, the solid material removed and weighed and then transferred to the extraction apparatus, or, in the case of final filtration prior to analysis, discarded. Do not allow the

material retained on the filter pad to dry prior to weighing.

(vi) The liquid phase should be stored at 4°C for subsequent use in Step 8.

B. Structural Integrity Procedure

Equipment: A Structural Integrity Tester having a 3.18 cm (1.25 in.) diameter hammer weighing 0.33 kg (0.73 lbs.) and having a free fall of 15.24 cm (6 in.) shall be used. This device is available from Associated Design and Manufacturing Company, Alexandria, VA., 22314, as Part No. 125, or it may be fabricated to meet the specifications shown in Figure 1.

Procedure:

1. The sample holder should be filled with the material to be tested. If the sample of waste is a large monolithic block, a portion should be cut from the block having the dimensions of a 3.3 cm (1.3 in.) diameter x 7.1 cm (2.8 in.) cylinder. For a fixated waste, samples may be cast in the form of a 3.3 cm (1.3 in.) diameter x 7.1 cm (2.8 in.) cylinder for purposes of conducting this test. In such cases, the waste may be allowed to cure for 30 days prior to further testing.

2. The sample holder should be placed into the Structural Integrity Tester, then the hammer should be raised to its maximum height and dropped. This should be repeated fifteen times.

3. The material should be removed from the sample holder, weighed, and transferred to the extraction apparatus for extraction.

Analytical Procedures for Analyzing Extract Contaminants

The test methods for analyzing the extract are as follows:

(1) For arsenic, barium, cadmium, chromium, lead, mercury, selenium or silver: "Methods for Analysis of Water and Wastes," Environmental Monitoring and Support Laboratory, Office of Research and Development, U.S. Environmental Protection Agency, Cincinnati, Ohio 45268 (EPA–600/4–79–020, March 1979),

(2) For Endrin; Lindane; Methoxychlor; Toxaphene; 2,4-D; 2,4,5-TP Silver: in "Methods for Benzidine, Chlorinated Organic Compounds, Pentachlorophenol and Pesticides in Water and Wastewater," September 1978, U.S. Environmental Protection Agency, Environmental Monitoring and Support Laboratory, Cincinnati, Ohio 42568,

as standardized in "Test Methods for the Evaluation of Solid Waste, Physical/Chemical Methods."

For all analyses, the method of standard addition shall be used for the quantification of species concentration.

This method is described in "Test Methods for the Evaluation of Solid Waste." (It is also described in "Methods for Analysis of Water and Wastes.")

BILLING CODE 6560-01-M

---

[3] This procedure is intended to result in separation of the "free" liquid portion of the waste from any solid matter having a particle size ≥0.45um. If the sample will not filter, various other separation techniques can be used to aid in the filtration. As described above, pressure filtration is employed to speed up the filtration process. This does not alter the nature of the separation. If liquid does not separate during filtration, the waste can be centrifuged. If separation occurs during centrifugation the liquid portion (centrifugate) is filtered through the 0.45um filter prior to becoming mixed with the liquid portion of the waste obtained from the initial filtration. Any material that will not pass through the filter after centrifugation is considered a solid and is extracted.

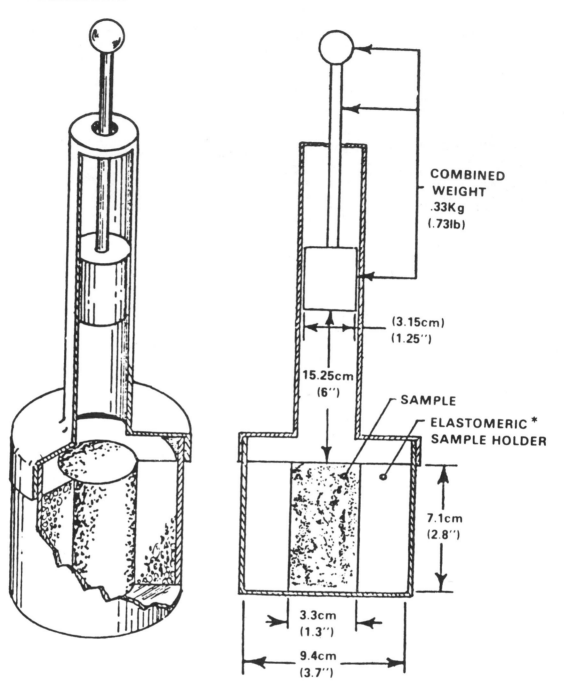

COMBINED
WEIGHT
.33Kg
(.73lb)

(3.15cm)
(1.25″)

15.25cm
(6″)

SAMPLE

ELASTOMERIC *
SAMPLE HOLDER

7.1cm
(2.8″)

3.3cm
(1.3″)

9.4cm
(3.7″)

*ELASTOMERIC SAMPLE HOLDER FABRICATED OF
MATERIAL FIRM ENOUGH TO SUPPORT THE SAMPLE

Figure 1

COMPACTION TESTER

BILLING CODE 6560-01-C

Appendix III—Chemical Analysis Test Methods

Tables 1, 2 and 3 specify the appropriate analytical procedures, described in "Test Methods for Evaluating Solid Waste" (SW–846), which should be used in determining whether the waste in question contains a given toxic constituent. Table 1 identifies the analytical class and the approved measurement techniques for each organic chemical listed in Appendix VII. Table 2 identifies the corresponding methods for the inorganic species. Table 3 identifies the specific sample preparation and measurement instrument introduction techniques which may be suitable for both the organic and inorganic species as well as the matrices of concern.

Prior to final selection of the analytical method the operator should consult the specific method descriptions in SW–846 for additional guidance on which of the approved methods should be employed for a specific waste analysis situation.

Table 1.—*Analytical Characteristics of Organic Chemicals*

Compound	Sample handling class/fraction	Non-GC methods	Measurement techniques		
				Conventional	
			GC/MS	GC	Detector
Acetonitrile	Volatile		8.24	8.03	NUU
Acrolein	Volatile		8.24	8.03	NSD
Acrylamide	Volatile		8.24	8.01	FID
Acrylonitrile	Volatile		8.24	8.03	NSD
Benzene	Volatile		8.24	8.02	PID
Benz(a)anthracene	Extractable/BN	8.10 (HPLC)	8.25	8.10	FID
Benzo(a)pyrene	Extractable/BN	8.10 (HPLC)	8.25	8.10	FID
Benzotrichloride	Extractable/BN		8.25	8.12	ECD
Benzyl chloride	Volatile or Extractable/BN		8.24	8.01	HSD
			8.25	8.12	ECD
Benz(b)fluoanthene	Extractable/BN	8.10 (HPLC)	8.25	8.10	FID
Bis(2-chloroethoxymethane)	Volatile		8.24	8.01	HSD
Bis(2-chloroethyl)ether	Volatile		8.24	8.01	HSD
Bis(2-chloroisopropyl)ether	Volatile		8.24	8.01	HSD
Carbon disulfide	Volatile		8.24	8.01	HSD
Carbon tetrachloride	Volatile		8.24	8.01	HSD
Chlordane	Extractable/BN		8.25	8.08	HSD
Chlorinated dibenzodioxins	Extractable/BN		8.25	8.08	ECD
Chlorinated biphenyls	Extractable/BN		8.25	8.08	HSD
Chloroacetaldehyde	Volatile		8.24	8.01	HSD
Chlorobenzene	Volatile		8.24	8.01	HSD
				8.02	PID
Chloroform	Volatile		8.24	8.01	HSD
Chloromethane	Volatile		8.24	8.01	HSD
2-Chlorophenol	Extractable/BN		8.25	8.04	FID, ECD
Chrysene	Extractable/BN	8.10 (HPLC)	8.25	8.10	FID
Creosote	Extractable/BN		[1]8.25	8.10	ECD
Cresol(s)	Extractable/A		8.25	8.04	FID, ECD
Cresylic acid(s)	Extractable/A		8.25	8.04	FID, ECD
Dichlorobenzene(s)	Extractable/BN		8.25	8.01	HSD
				8.02	PID
				8.12	ECD
Dichloroethane(s)	Volatile		8.24	8.01	HSD
Dichloromethane	Volatile		8.24	8.01	HSD
Dichlorophenoxy-acetic acid	Extractable/A		8.25	8.40	HSD
Dichloropropanol	Extractable/BN		8.25	8.12	ECD
2,4-Dimethylphenol	Extractable/A		8.25	8.04	FID, ECD
Dinitrobenzene	Extractable/BN		8.25	8.09	FID, ECD
4,6-Dinitro-o-cresol	Extractable/A		8.25	8.04	FID, ECD
2,4-Dinitrotoluene	Extractable/BN		8.25	8.09	FID, ECD
Endrin	Extractable/P		8.25	8.08	HSD
Ethyl ether	Volatile		8.24	8.01	FID
				8.02	FID
Formaldehyde	Volatile		8.24	8.01	FID
Formic acid	Extractable/BN		8.25	8.06	FID
Heptachlor	Extractable/P		8.25	8.06	HSD
Hexachlorobenzene	Extractable/BN		8.25	8.12	ECD
Hexachlorobutadiene	Extractable/BN		8.25	8.12	ECD
Hexachloroethane	Extractable/BN		8.25	8.12	ECD
Hexachlorocyclopentadiene	Extractable/BN		8.25	8.12	ECD
Lindane	Extractable/P		8.25	8.08	HSD
Maleic anhydride	Extractable/BN		8.25	8.06	ECD, FID
Methanol	Volatile		8.24	8.01	FID
Methomyl	Extractable/BN	8.32 (HPLC)			
Methyl ethyl ketone	Volatile		8.25	8.01	FID
				8.02	FID
Methyl isobutyl ketone	Volatile		8.25	8.01	FID
				8.02	FID
Naphthalene	Extractable/BN		8.25	8.10	FID
Napthoquinone	Extractable/BN		8.25	8.06	ECD, FID
				8.09	FID
Nitrobenzene	Extractable/BN		8.25	8.09	ECD, FID
4-Nitrophenol	Extractable/A		8.24	8.04	ECD, FID
Paraldehyde (trimer of acetaldehyde).	Volatile		8.24	8.01	FID

Table 1.—*Analytical Characteristics of Organic Chemicals—Continued*

Compound	Sample handling class/fraction	Non-GC methods	Measurement techniques		
			GC/MS	Conventional GC	Detector
Pentachlorophenol	Extractable/A		8.25	8.04	ECD
Phenol	Extractable/A		8.25	8.04	ECD, FID
Phorate	Extractable/BN			8.22	FPD
Phosphorodithioic acid esters	Extractable/BN			8.06	ECD, FID
				8.09	ECD, FID
				8.22	FPD
Phthalic anhydride	Extractable/BN		8.25	8.06	ECD, FID
				8.09	ECD, FID
2-Picoline	Extractable/BN		8.25	8.06	ECD, FID
				8.09	ECD, FID
Pyridine	Extractable/BN		8.25	8.06	ECD, FID
				8.09	ECD, FID
Tetrachlorobenzene(s)	Extractable/BN		8.25	8.12	ECD
Tetrachloroethane(s)	Volatile		8.24	8.01	HSD
Tetrachloroethene	Volatile		8.24	8.01	HSD
Tetrachlorophenol	Extractable/A		8.24	8.04	ECD
Toluene	Volatile		8.24	8.02	PID
Toluenediamine	Extractable/BN		8.25		
Toluene diisocyanate(s)	Extractable/nonaqueous		8.25	8.06	FID
Toxaphene	Extractable/P		8.25	8.08	HSD
Trichloroethane	Volatile		8.24	8.01	HSD
Trichloroethene(s)	Volatile		8.24	8.01	HSD
Trichlorofluoromethane	Volatile		8.24	8.01	HSD
Trichlorophenol(s)	Extractable/A		8.25	8.04	HSD
2,4,5-TP (Silvex)	Extractable/A		8.25	8.40	HSD
Trichloropropane	Volatile		8.24	8.01	HSD
Vinyl chloride	Volatile		8.24	8.01	HSD
Vinylidene chloride	Volatile		8.24	8.01	HSD
Xylene	Volatile		8.24	8.02	PID

[1] Analyze for phenanthrene and carbazole; if these are present in a ratio between 1.4:1 and 5:1, creosote should be considered present.

ECD = Electron capture detetor; FID = Flame ionization detector; FPD = Flame photometric detector; HSD = Halide specific detector; HPLC = High pressure liquid chromotography; NSD = Nitrogen-specific detector; PID = Photoionization detector.

Table 2—*Analytical Characteristics of Inorganic Species*

Species	Sample handling class	Measurement technique	Method number
Antimony	Digestion	Atomic absorbtion–furnace/flame	8.50
Arsenic	Hydride	Atomic absorbtion–flame	8.51
Barium	Digestion	Atomic absorbtion–furnace/flame	8.52
Cadmium	Digestion	Atomic absorbtion–furnace/flame	8.53
Chromium	Digestion	Atomic absorbtion–furnace/flame	8.54
Cyanides	Hydrolysis	Atomic absorbtion-spectroscopy	8.55
Lead	Digestion	Atomic absorbtion–furnace/flame	8.56
Mercury	Cold Vapor	Atomic absorbtion	8.57
Nickel	Digestion	Atomic absorbtion–furnace/flame	8.58
Selenium	Hydride digestion	Atomic absorbtion–furnace/flame	8.59
Silver	Digestion	Atomic absorbtion–furnace/flame	8.60

Table 3.—*Sample Prepartion/Sample Introduction Techniques*

Sample handling class	Physical characteristics of waste [1]		
	Fluid	Paste	Solid
Voltile	Purge and trap. Direct injection.	Purge and trap. Headspace	Headspace.
Semivolatile and nonvolatile.	Direct injection. Shake out	Shake out	Shake out. Soxhlet. Sonication.
Inorganic	Direct injection. Digestion. Hydride	Digestion. Hydride	Digestion. Hydride.

[1] For purposes of this Table, fluid refers to readily pourable liquids, which may or may not contain suspended particles. Paste-like materials, while fluid in the sense of flowability, can be thought of as being thixotropic or plastic in nature, e.g. paints. Solid materials are those wastes which can be handled without a container (i.e., can be piled up without appreciable sagging).

Procedure and Method Number(s)

Digestion—See appropriate procedure for element of interest.

Direct injection—8.80

Headspace—8.82

Hydride—See appropriate procedure for element of interest.

Purge & Trap—8.83

Shake out—8.84

Sonication—8.85

Soxhlet—8.86

Appendix VII.—*Basis for Listing Hazardous Wastes*

EPA hazardous waste No.	Hazardous constituents for which listed
F001	tetrachloroethylene, methylene chloride trichloroethyllene, 1,1,1-trichloroethane chlorinated fluorocarbons, carbon tetrachloride
F002	tetrachloroethylene, methylene chloride, trichloroethyllene, 1,1,1-trichloroethane, chlorobenzene, 1,1,2-trichloro-1,2,2-trifluoroethane, o-dichlorobenzene, trichlorofluoromethane
F003	N.A.
F004	cresols and cresylic acid, nitrobenzene
F005	methanol, toluene, methyl ethyl ketone, methyl isobutyl ketone, carbon disulfide, isobutanol, pyridine
F006	cadmium, chromium, nickel, cyanide (complexed)
F007	cyanide (salts)
F008	cyanide (salts)
F009	cyanide (salts)
F010	cyanide (salts)
F011	cyanide (salts)
F012	cyanide (complexed)
F013	cyanide (complexed)
F014	cyanide (complexed)
F015	cyanide (salts)
F016	cyanide (complexed)
K001	benzene, benz(a)anthracene, benzo(a)pyrene, chrysene, 4-nitrophenol, toluene, naphthalene phenol, 2-chlorophenol, 2,4-dimethyl phenol, 2,4,6-trichlorophenol, pentachlorophenol, 4,6-dinitro-o-cresol, tetrachlorophenol
K002	chromium, lead
K003	chromium, lead
K004	chromium
K005	chromium, lead
K006	chromium
K007	cyanide (complexed), chromium
K008	chromium
K009	chloroform, formaldehyde, methylene chloride, methyl chloride, paraldehyde, formic acid
K010	chloroform, formaldehyde, methylene chloride, methyl chloride, paraldehyde, formic acid, chloroacetaldehyde
K011	acrylonitrile, acetonitrile, hydrocyanic acid
K012	acrylonitrile, acetonitrile, acrolein, acrylamide
K013	hydrocyanic acid, acrylonitrile, acetonitrile
K014	acetonitrile, acrylamide
K015	benzyl chloride, chlorobenzene, toluene, benzotrichloride
K016	hexachlorobenzene, hexachlorobutadiene, carbon tetrachloride, hexachloroethane, perchloroethylene
K017	epichlorohydrin, chloroethers [bis(chloromethyl) ether and bis (2-chloroethyl) ethers], trichloropropane, dichloropropanols
K018	1,2-dichloroethane, trichloroethylene, hexachlorobutadiene, hexachlorobenzene
K019	ethylene dichloride, 1,1,1-trichloroethane, 1,1,2-trichloroethane, tetrachloroethanes (1,1,2,2-tetrachloroethane and 1,1,1,2-tetrachloroethane), trichloroethylene, tetrachloroethylene, carbon tetrachloride, chloroform, vinyl chloride, vinylidene chloride

Appendix VII.—Basis for Listing Hazardous Wastes—Continued

EPA hazardous waste No.	Hazardous constituents for which listed
K020.........	ethylene dichloride, 1,1,1-trichloroethane, 1,1,2-trichloroethanes, tetrachloroethanes (1,1,2,2-tetrachloroethane and 1,1,1,2-tetrachloroethane), trichloroethylene, tetrachloroethylene, carbon tetrachloride, chloroform, vinyl chloride, vinylidene chloride
K021.........	antimony, carbon tetrachloride, chloroform
K022.........	phenol, tars (polycyclic aromatic hydrocarbons)
K023.........	phthalic anahydride, maleic anhydride
K024.........	phthalic anhydride, polynuclear tar-like materials, naphthoquinone
K025.........	meta-dinitrobenzene, 2,4-dinitrotoluene
K026.........	paraldehyde, pyridines, 2-picoline
K027.........	toulene diisocyanate, toluene-2,4-diamine, tars (benzidimidazapone)
K028.........	1,1,1-trichloroethane, vinyl chloride
K029.........	1,2.-dichloroethane, 1,1,1-trichloroethane, vinyl chloride, vinlyidene chloride, chloroform
K030.........	hexachlorobenzene, hexachlorobutadiene, hexachloroethane, 1,1,1,2-tetrachloroethane, 1,1,2,2-tetrachloroethane, ethylene dichloride
K031.........	arsenic
K032.........	hexachlorocyclopentadiene
K033.........	hexachlorocyclopentadiene
K034.........	hexachlorocyclopentadiene
K035.........	cresote, benz(a)anthracene, benz(b)fluoroanthene, benzo(a)pyrene
K036.........	toulene, phosphorodithioic and phosphorothioic acid esters
K037.........	toulene, phosphorodithioic and phosphorothioic acid esters
K038.........	phorate, formaldehyde, phosphorodithioic and phosphorothioic acid esters
K039.........	phosphorodithioic and phosphorothioic acid esters
K040.........	phorate, formaldehyde, phosphorodithioic and phosphorothioic acid esters
K041.........	toxaphene
K042.........	hexachlorobenzene; ortho-dichlorobenzene
K043.........	2,4-dichlorophenol, 2,6-dichlorophenol, 2,4,6-trichlorophenol
K044.........	N.A.
K045.........	N.A.
K046.........	lead
K047.........	N.A.
K048.........	chromium, lead
K049.........	chromium, lead
K050.........	chromium
K051.........	chromium, lead
K052.........	lead
K053.........	chromium
K054.........	chromium
K055.........	chromium, lead
K056.........	chromium, lead
K057.........	chromium, lead
K058.........	chromium, lead
K059.........	N.A.
K060.........	cyanide, naphthalene, phenolic compounds, arsenic
K061.........	chromium, lead, cadmium
K062.........	chromium, lead
K063.........	chromium, lead
K064.........	lead, cadmium
K065.........	lead, cadmium
K066.........	lead, cadmium
K067.........	lead, cadmium
K068.........	lead, cadmium
K069.........	chromium, lead, cadmium

N.A.—Waste is hazardous because it meets either the ignitability, corrosivity or reactivity characteristic.

Appendix VIII—Hazardous Constituents

Acetaldehyde
(Acetato)phenylmercury
Acetonitrile
3-(alpha-Acetonylbenzyl)-4-hydroxycoumarin and salts
2-Acetylaminofluorene
Acetyl chloride
1-Acetyl-2-thiourea
Acrolein
Acrylamide
Acrylonitrile
Aflatoxins

Aldrin
Allyl alcohol
Aluminum phosphide
4-Aminobiphenyl
6-Amino-1,1a,2,8,8a,8b-hexahydro-8-(hydroxymethyl)-8a-methoxy-5-methylcarbamate azirino(2′,3′:3,4) pyrrolo(1,2-a)indole-4,7-dione (ester) (Mitomycin C)
5-(Aminomethyl)-3-isoxazolol
4-Aminopyridine
Amitrole
Antimony and compounds, N.O.S. [1]
Aramite
Arsenic and compounds, N.O.S.
Arsenic acid
Arsenic pentoxide
Arsenic trioxide
Auramine
Azaserine
Barium and compounds, N.O.S.
Barium cyanide
Benz[c]acridine
Benz[a]anthracene
Benzene
Benzenearsonic acid
Benzenethiol
Benzidine
Benzo[a]anthracene
Benzo[b]fluoranthene
Benzo[j]fluoranthene
Benzo[a]pyrene
Benzotrichloride
Benzyl chloride
Beryllium and compounds, N.O.S.
Bis(2-chloroethoxy)methane
Bis(2-chloroethyl) ether
N,N-Bis(2-chloroethyl)-2-naphthylamine
Bis(2-chloroisopropyl) ether
Bis(chloromethyl) ether
Bis(2-ethylhexyl) phthalate
Bromoacetone
Bromomethane
4-Bromophenyl phenyl ether
Brucine
2-Butanone peroxide
Butyl benzyl phthalate
2-sec-Butyl-4,6-dinitrophenol (DNBP)
Cadmium and compounds, N.O.S.
Calcium chromate
Calcium cyanide
Carbon disulfide
Chlorambucil
Chlordane (alpha and gamma isomers)
Chlorinated benzenes, N.O.S.
Chlorinated ethane, N.O.S.
Chlorinated naphthalene, N.O.S.
Chlorinated phenol, N.O.S.
Chloroacetaldehyde
Chloroalkyl ethers
p-Chloroaniline
Chlorobenzene
Chlorobenzilate
1-(p-Chlorobenzoyl)-5-methoxy-2-methylindole-3-acetic acid
p-Chloro-m-cresol
1-Chloro-2,3-epoxybutane
2-Chloroethyl vinyl ether
Chloroform
Chloromethane
Chloromethyl methyl ether
2-Chloronaphthalene

[1] The abbreviation N.O.S. signifies those members of the general class "not otherwise specified" by name in this listing.

2-Chlorophenol
1-(o-Chlorophenyl)thiourea
3-Chloropropionitrile
alpha-Chlorotoluene
Chlorotoluene, N.O.S.
Chromium and compounds, N.O.S.
Chrysene
Citrus red No. 2
Copper cyanide
Creosote
Crotonaldehyde
Cyanides (soluble salts and complexes), N.O.S.
Cyanogen
Cyanogen bromide
Cyanogen chloride
Cycasin
2-Cyclohexyl-4,6-dinitrophenol
Cyclophosphamide
Daunomycin
DDD
DDE
DDT
Diallate
Dibenz[a,h]acridine
Dibenz[a,j]acridine
Dibenz[a,h]anthracene(Dibenzo[a,h] anthracene)
7H-Dibenzo[c,g]carbazole
Dibenzo[a,e]pyrene
Dibenzo[a,h]pyrene
Dibenzo[a,i]pyrene
1,2-Dibromo-3-chloropropane
1,2-Dibromoethane
Dibromomethane
Di-n-butyl phthalate
Dichlorobenzene, N.O.S.
3,3′-Dichlorobenzidine
1,1-Dichloroethane
1,2-Dichloroethane
trans-1,2-Dichloroethane
Dichloroethylene, N.O.S.
1,1-Dichloroethylene
Dichloromethane
2,4-Dichlorophenol
2,6-Dichlorophenol
2,4-Dichlorophenoxyacetic acid (2,4-D)
Dichloropropane
Dichlorophenylarsine
1,2-Dichloropropane
Dichloropropanol, N.O.S.
Dichloropropene, N.O.S.
1,3-Dichloropropene
Dieldrin
Diepoxybutane
Diethylarsine
0,0-Diethyl-S-(2-ethylthio)ethyl ester of phosphorothioic acid
1,2-Diethylhydrazine
0,0-Diethyl-S-methylester phosphorodithioic acid
0,0-Diethylphosphoric acid, 0-p-nitrophenyl ester
Diethyl phthalate
0,0-Diethyl-0-(2-pyrazinyl)phosphorothioate
Diethylstilbestrol
Dihydrosafrole
3,4-Dihydroxy-alpha-(methylamino)-methyl benzyl alcohol
Di-isopropylfluorophosphate (DFP)
Dimethoate
3,3′-Dimethoxybenzidine
p-Dimethylaminoazobenzene
7,12-Dimethylbenz[a]anthracene
3,3′-Dimethylbenzidine
Dimethylcarbamoyl chloride

1,1-Dimethylhydrazine
1,2-Dimethylhydrazine
3,3-Dimethyl-1-(methylthio)-2-butanone-0-
((methylamino) carbonyl)oxime
Dimethylnitrosoamine
alpha,alpha-Dimethylphenethylamine
2,4-Dimethylphenol
Dimethyl phthalate
Dimethyl sulfate
Dinitrobenzene, N.O.S.
4,6-Dinitro-o-cresol and salts
2,4-Dinitrophenol
2,4-Dinitrotoluene
2,6-Dinitrotoluene Di-n-octyl phthalate
1,4-Dioxane
1,2-Diphenylhydrazine
Di-n-propylnitrosamine
Disulfoton
2,4-Dithiobiuret
Endosulfan
Endrin and metabolites
Epichlorohydrin
Ethyl cyanide
Ethylene diamine
Ethylenebisdithiocarbamate (EBDC)
Ethyleneimine
Ethylene oxide
Ethylenethiourea
Ethyl methanesulfonate
Fluoranthene
Fluorine
2-Fluoroacetamide
Fluoroacetic acid, sodium salt
Formaldehyde
Glycidylaldehyde
Halomethane, N.O.S.
Heptachlor
Heptachlor epoxide (alpha, beta, and gamma
isomers)
Hexachlorobenzene
Hexachlorobutadiene
Hexachlorocyclohexane (all isomers)
Hexachlorocyclopentadiene
Hexachloroethane
1,2,3,4,10,10-Hexachloro-1,4,4a,5,8,8a-
hexahydro-1,4:5,8-endo,endo-
dimethanonaphthalene
Hexachlorophene
Hexachloropropene
Hexaethyl tetraphosphate
Hydrazine
Hydrocyanic acid
Hydrogen sulfide
Indeno(1,2,3-c,d)pyrene
Iodomethane
Isocyanic acid, methyl ester
Isosafrole
Kepone
Lasiocarpine
Lead and compounds, N.O.S.
Lead acetate
Lead phosphate
Lead subacetate
Maleic anhydride
Malononitrile
Melphalan
Mercury and compounds, N.O.S.
Methapyrilene
Methomyl
2-Methylaziridine
3-Methylcholanthrene
4,4'-Methylene-bis-(2-chloroaniline)
Methyl ethyl ketone (MEK)
Methyl hydrazine
2-Methyllactonitrile
Methyl methacrylate

Methyl methanesulfonate
2-Methyl-2-(methylthio)propionaldehyde-o-
(methylcarbonyl) oxime
N-Methyl-N'-nitro-N-nitrosoguanidine
Methyl parathion
Methylthiouracil
Mustard gas
Naphthalene
1,4-Naphthoquinone
1-Naphthylamine
2-Naphthylamine
1-Naphthyl-2-thiourea
Nickel and compounds, N.O.S.
Nickel carbonyl
Nickel cyanide
Nicotine and salts
Nitric oxide
p-Nitroaniline
Nitrobenzene
Nitrogen dioxide
Nitrogen mustard and hydrochloride salt
Nitrogen mustard N-oxide and hydrochloride
salt
Nitrogen peroxide
Nitrogen tetroxide
Nitroglycerine
4-Nitrophenol
4-Nitroquinoline-1-oxide
Nitrosamine, N.O.S.
N-Nitrosodi-N-butylamine
N-Nitrosodiethanolamine
N-Nitrosodiethylamine
N-Nitrosodimethylamine
N-Nitrosodiphenylamine
N-Nitrosodi-N-propylamine
N-Nitroso-N-ethylurea
N-Nitrosomethylethylamine
N-Nitroso-N-methylurea
N-Nitroso-N-methylurethane
N-Nitrosomethylvinylamine
N-Nitrosomorpholine
N-Nitrosonornicotine
N-Nitrosopiperidine
N-Nitrosopyrrolidine
N-Nitrososarcosine
5-Nitro-o-toluidine
Octamethylpyrophosphoramide
Oleyl alcohol condensed with 2 moles
ethylene oxide
Osmium tetroxide
7-Oxabicyclo[2.2.1]heptane-2,3-dicarboxylic
acid
Parathion
Pentachlorobenzene
Pentachloroethane
Pentachloronitrobenzene (PCNB)
Pentacholorophenol
Phenacetin
Phenol
Phenyl dichloroarsine
Phenylmercury acetate
N-Phenylthiourea
Phosgene
Phosphine
Phosphorothioic acid, O,O-dimethyl ester, O-
ester with N,N-dimethyl benzene
sulfonamide
Phthalic acid esters, N.O.S.
Phthalic anhydride
Polychlorinated biphenyl, N.O.S.
Potassium cyanide
Potassium silver cyanide
Pronamide
1,2-Propanediol
1,3-Propane sultone
Propionitrile

Propylthiouracil
2-Propyn-1-ol
Pryidine
Reserpine
Saccharin
Safrole
Selenious acid
Selenium and compounds, N.O.S.
Selenium sulfide
Selenourea
Silver and compounds, N.O.S.
Silver cyanide
Sodium cyanide
Streptozotocin
Strontium sulfide
Strychnine and salts
1,2,4,5-Tetrachlorobenzene
2,3,7,8-Tetrachlorodibenzo-p-dioxin (TCDD)
Tetrachloroethane, N.O.S.
1,1,1,2-Tetrachloroethane
1,1,2,2-Tetrachloroethane
Tetrachloroethene (Tetrachloroethylene)
Tetrachloromethane
2,3,4,6-Tetrachlorophenol
Tetraethyldithiopyrophosphate
Tetraethyl lead
Tetraethylpyrophosphate
Thallium and compounds, N.O.S.
Thallic oxide
Thallium (I) acetate
Thallium (I) carbonate
Thallium (I) chloride
Thallium (I) nitrate
Thallium selenite
Thallium (I) sulfate
Thioacetamide
Thiosemicarbazide
Thiourea
Thiuram
Toluene
Toluene diamine
o-Toluidine hydrochloride
Tolylene diisocyanate
Toxaphene
Tribromomethane
1,2,4-Trichlorobenzene
1,1,1-Trichloroethane
1,1,2-Trichloroethane
Trichloroethene (Trichloroethylene)
Trichloromethanethiol
2,4,5-Trichlorophenol
2,4,6-Trichlorophenol
2,4,5-Trichlorophenoxyacetic acid (2,4,5-T)
2,4,5-Trichlorophenoxypropionic acid (2,4,5-
TP) (Silvex)
Trichloropropane, N.O.S.
1,2,3-Trichloropropane
0,0,0-Triethyl phosphorothioate
Trinitrobenzene
Tris(1-azridinyl)phosphine sulfide
Tris(2,3-dibromopropyl) phosphate
Trypan blue
Uracil mustard
Urethane
Vanadic acid, ammonium salt
Vanadium pentoxide (dust)
Vinyl chloride
Vinylidene chloride
Zinc cyanide
Zinc phosphide

[FR Doc. 80–14307 Filed 5–16–80; 8:45 am]
BILLING CODE 6560–01–M

Title 40 of the Code of Federal
Regulations is amended as follows:
  1. In § 261.31, add the following waste
streams:

§ 261.31 Hazardous waste from nonspecific sources.

Industry	EPA hazardous waste No.	Hazardous waste	Hazard code
Generic	F017	Paint residues or sludges from industrial painting in the mechanical and electrical products industry	(T)
	F018	Wastewater treatment sludge from industrial painting in the mechanical and electrical products industry.	(T)

  2. In § 261.32, add the following waste streams:

§ 261.32 Hazardous waste from specific sources.

Industry	EPA hazardous waste No.	Hazardous waste	Hazard code
Inorganic Chemicals	K071	Brine purification muds from the mercury cell process in chlorine production, where separately prepurified brine is not used	(T)
	K073	Chlorinated hydrocarbon wastes from the purification step of the diaphragm cell process using graphite anodes in chlorine production.	(T)
	K074	Wastewater treatment sludges from the production of TiO₂ pigment using chromium bearing ores by the chloride process.	(T)
Paint Manufacturing	K078	Solvent cleaning wastes from equipment and tank cleaning from paint manufacturing.	(I, T)
	K079	Water or caustic cleaning wastes from equipment and tank cleaning from paint manufacturing	(T)
	K081	Wastewater treatment sludges from paint manufacturing	(T)
	K082	Emission control dust or sludge from paint manufacturing	(T)
Organic Chemicals	K083	Distillation bottoms from aniline production	(T)
	K085	Distillation or fractionating column bottoms from the production of chlorobenzenes.	(T)
Ink Formulation	K086	Solvent washes and sludges, caustic washes and sludges, or water washes and sludges from cleaning tubs and equipment used in the formulation of ink from pigments, driers, soaps, and stabilizers containing chromium and lead.	(T)
Veterinary Pharmaceuticals	K084	Wastewater treatment sludges generated during the production of veterinary pharmaceuticals from arsenic or organo-arsenic compounds.	(T)
Coking	K087	Decanter tank tar sludge from coking operations	(T)
Primary Aluminum	K088	Spent potliners from primary aluminum reduction.	(T)
Ferroalloys	K090	Emission control dust or sludge from ferrochromium-silicon production.	(T)
	K091	Emission control dust or sludge from ferrochromium production	(T)
	K092	Emission control dust or sludge from ferromanganese production.	(T)

  3. In Appendix VII (Basis for Listing Hazardous Waste) add the following after F016:

EPA hazardous waste number	Hazardous constituents for which listed
F017	Cadmium, chromium, lead, cyanides, toluene, tetrachloroethylene.
F018	Cadmium, chromium, lead, cyanide, toluene, tetrachloroethylene.

  4. In Appendix VII (Basis for Listing Hazardous Waste) add the following after K069:

EPA hazardous waste number	Hazardous constituents for which listed
K071	Mercury.
K073	Chloroform, carbon tetrachloride, hexachloroethane, trichloroethane, tetrachloroethylene, dichloroethylene, 1,1,2,2-tetrachloroethane.
K074	Chromium.
K078	Chromium, lead.
K079	Lead, mercury, benzene, carbon tetrachloride, methylene chloride, tetrachlorethylene, naphthalene, di(2-ethylhexyl)phthalate, di-n-butylphthalate, toluene.
K081	Chromium, lead, mercury, nickel, methylene chloride, toluene.

EPA hazardous waste number	Hazardous constituents for which listed
K082	Antimony, cadmium, chromium, lead, nickel, silver, cyanides, phenol, mercury, pentachlorophenol, vinyl chloride, 3,3'-dichlorobenzidene, naphthalene, di(2-ethylhexyl)phthalate, di-n-butylphthalate, benzene, toluene, carbon tetrachloride, methylene chloride, trichloroethylene.
K083	Aniline, nitrobenzene, diphenylamine, phenylenediamine.
K084	Arsenic.
K085	Benzene, monochlorobenzene, dichlorobenzenes, trichlorobenzenes, tetrachlorobenzene, pentachlorobenzene, hexachlorobenzene, benzyl chloride.
K086	Chromium, lead.
K087	Phenol, naphthalene.
K088	Cyanide (complexes).
K090	Chromium.
K091	Chromium, lead.
K092	Chromium, lead.

5. In Appendix VIII (Hazardous Constituents) add the following constituents alphabetically:

— aniline
— diphenylamine
— phenylenediamine

6. In Appendix VIII (Hazardous Constituents) delete:

— chlorotoluene, N.O.S.

[FR Doc. 80-21186 Filed 7-15-80; 8:45 am]

BILLING CODE 6560-01-M

PART 262—STANDARDS APPLICABLE TO GENERATORS OF HAZARDOUS WASTE

Subpart A—General

Sec.
262.10 Purpose, scope, and applicability.
262.11 Hazardous waste determination.
262.12 EPA identification numbers.

Subpart B—The Manifest

262.20 General requirements.
262.21 Required information.
262.22 Number of copies.
262.23 Use of the manifest.

Subpart C—Pre-Transport Requirements

262.30 Packaging.
262.31 Labeling.
262.32 Marking.
262.33 Placarding.
262.34 Accumulation time.

Subpart D—Recordkeeping and Reporting

262.40 Recordkeeping.
262.41 Annual reporting.
262.42 Exception reporting.
262.43 Additional reporting.

Subpart E—Special Conditions

262.50 International shipments.
262.51 Farmers.

Appendix—Form

Annual Report (EPA Form 8700-13).

Authority: Secs. 2002(a), 3001, 3002, 3003, 3004, and 3005 of the Solid Waste Disposal Act, as amended by Resource Conservation and Recovery Act of 1976 and as amended by the Quiet Communities Act of 1978, (42 U.S.C. 6912(a), 6921, 6922, 6923, 6924, 6925)

Subpart A—General

§ 262.10 Purpose, scope, and applicability.

(a) These regulations establish standards for generators of hazardous waste.

(b) A generator who treats, stores, or disposes of hazardous waste on-site must only comply with the following sections of this Part with respect to that waste: Section 262.11 for determining whether or not he has a hazardous waste, § 262.12 for obtaining an EPA identification number, § 262.40(c) and (d) for Recordkeeping, § 262.43 for additional reporting and if applicable, § 262.51 for Farmers.

(c) Any person who imports hazardous waste into the United States must comply with the standards applicable to generators established in this Part.

(d) A farmer who generates waste pesticides which are hazardous waste and who complies with all of the requirements of § 262.51 is not required to comply with other standards in this Part or 40 CFR Parts 122, 264, or 265 with respect to such pesticides.

(e) A person who generates a hazardous waste as defined by 40 CFR Part 261 is subject to the compliance requirements and penalties prescribed in Section 3008 of the Act if he does not comply with the requirements of this Part.

Note.— A generator who treats, stores, or disposes of hazardous waste on-site must comply with the applicable standards and permit requirements set forth in 40 CFR Parts 264, 265, and 266 and Part 122.

§ 262.11 Hazardous waste determination.

A person who generates a solid waste, as defined in 40 CFR 261.2, must determine if that waste is a hazardous waste using the following method:

(a) He should first determine if the waste is excluded from regulation under 40 CFR 261.4 and 261.5.

(b) He must then determine if the waste is listed as a hazardous waste in Subpart D of 40 CFR Part 261.

Note.— Even if the waste is listed, the generator still has an opportunity under 40 CFR 260.22 to demonstrate to the Administrator that the waste from his particular facility or operation is not a hazardous waste.

(c) If the waste is not listed as a hazardous waste in Subpart D of 40 CFR Part 261, he must determine whether the waste is identified in Subpart C of 40 CFR Part 261 by either:

(1) Testing the waste according to the methods set forth in Subpart C of 40 CFR Part 261, or according to an equivalent method approved by the Administrator under 40 CFR 260.21; or

(2) Applying knowledge of the hazard characteristic of the waste in light of the materials or the processes used.

§ 262.12 EPA identification numbers.

(a) A generator must not treat, store, dispose of, transport, or offer for transportation, hazardous waste without having received an EPA identification number from the Administrator.

(b) A generator who has not received an EPA identification number may obtain one by applying to the Administrator using EPA form 8700-12. Upon receiving the request the Administrator will assign an EPA identification number to the generator.

(c) A generator must not offer his hazardous waste to transporters or to treatment, storage, or disposal facilities that have not received an EPA identification number.

Subpart B—The Manifest

§ 262.20 General requirements.

(a) A generator who transports, or offers for transportation, hazardous waste for off-site treatment, storage, or disposal must prepare a manifest before transporting the waste off-site.

(b) A generator must designate on the manifest one facility which is permitted to handle the waste described on the manifest.

(c) A generator may also designate on the manifest one alternate facility which is permitted to handle his waste in the event an emergency prevents delivery of the waste to the primary designated facility.

(d) If the transporter is unable to deliver the hazardous waste to the designated facility or the alternate facility, the generator must either designate another facility or instruct the transporter to return the waste.

§ 262.21 Required information.

(a) the manifest must contain all of the following information:

(1) A manifest document number;

(2) The generator's name, mailing address, telephone number, and EPA identification number;

(3) The name and EPA identification number of each transporter;

(4) The name, address and EPA identification number of the designated facility and an alternate facility, if any;

(5) The description of the waste(s) (e.g., proper shipping name, etc.) required by regulations of the U.S. Department of Transportation in 49 CFR 172.101, 172.202, and 172.203;

(6) The total quantity of each hazardous waste by units of weight or volume, and the type and number of containers as loaded into or onto the transport vehicle.

(b) The following certification must appear on the manifest: "This is to certify that the above named materials are properly classified, described, packaged, marked, and labeled and are in proper condition for transportation according to the applicable regulations of the Department of Transportation and the EPA."

§ 262.22 Number of copies.

The manifest consists of at least the number of copies which will provide the generator, each transporter, and the owner or operator of the designated facility with one copy each for their records and another copy to be returned to the generator.

§ 262.23 Use of the manifest.

(a) The generator must:

(1) Sign the manifest certification by hand; and

(2) Obtain the handwritten signature of the initial transporter and date of acceptance on the manifest; and

(3) Retain one copy, in accordance with § 262.40(a).

(b) The generator must give the transporter the remaining copies of the manifest.

(c) For shipment of hazardous waste within the United States solely by railroad or solely by water (bulk shipments only), the generator must send three copies of the manifest dated and signed in accordance with this section to the owner or operator of the designated facility. Copies of the manifest are not required for each transporter.

Note.—See § 263.20(e) for special provisions for rail or water (bulk shipment) transporters who deliver hazardous waste by rail or water to the designated facility.

Subpart C—Pre-Transport Requirements

§ 262.30 Packaging.

Before transporting hazardous waste or offering hazardous waste for transportation off-site, a generator must package the waste in accordance with the applicable Department of Transportation regulations on packaging under 49 CFR Parts 173, 178, and 179.

§ 262.31 Labeling.

Before transporting or offering hazardous waste for transportation off-site, a generator must label each package in accordance with the applicable Department of Transportation regulations on hazardous materials under 49 CFR Part 172.

§ 262.32 Marking.

(a) Before transporting or offering hazardous waste for transportation off-site, a generator must mark each package of hazardous waste in accordance with the applicable Department of Transportation regulations on hazardous materials under 49 CFR Part 172;

(b) Before transporting hazardous waste or offering hazardous waste for transportation off-site, a generator must mark each container of 110 gallons or less used in such transportation with the following words and information displayed in accordance with the requirements of 49 CFR 172.304:

HAZARDOUS WASTE—Federal Law Prohibits Improper Disposal. If found, contact the nearest police or public safety authority or the U.S. Environmental Protection Agency.
Generator's Name and Address ————.
Manifest Document Number ————.

§ 262.33 Placarding.

Before transporting hazardous waste or offering hazardous waste for transportation off-site, a generator must placard or offer the initial transporter the appropriate placards according to Department of Transportation regulations for hazardous materials under 49 CFR Part 172, Subpart F.

§ 262.34 Accumulation time.

(a) A generator may accumulate hazardous waste on-site without a permit for 90 days or less, provided that:

(1) All such waste is shipped off-site in 90 days or less;

(2) The waste is placed in containers which meet the standards of § 262.30 and are managed in accordance with 40 CFR 265.174 and 265.176 or in tanks, provided the generator complies with the requirements of Subpart J of 40 CFR Part 265 except § 265.193;

(3) The date upon which each period of accumulation begins is clearly marked and visible for inspection on each container;

(4) Each container is properly labeled and marked according to § 262.31 and § 262.32; and

(5) The generator complies with the requirements for owners or operators in Subparts C and D in 40 CFR Part 265 and with § 265.16.

(b) A generator who accumulates hazardous waste for more than 90 days is an operator of a storage facility and is subject to the requirements of 40 CFR Parts 264 and 265 and the permit requirements of 40 CFR Part 122.

Subpart D—Recordkeeping and Reporting

§ 262.40 Recordkeeping.

(a) A generator must keep a copy of each manifest signed in accordance with § 262.23(a) for three years or until he receives a signed copy from the designated facility which received the waste. This signed copy must be retained as a record for at least three years from the date the waste was accepted by the initial transporter.

(b) A generator must keep a copy of each Annual Report and Exception Report for a period of at least three years from the due date of the report (March 1).

(c) A generator must keep records of any test results, waste analyses, or other determinations made in accordance with § 262.11 for at least three years from the date that the waste was last sent to on-site or off-site treatment, storage, or disposal.

(d) The periods or retention referred to in this section are extended automatically during the course of any unresolved enforcement action regarding the regulated activity or as requested by the Administrator.

§ 262.41 Annual reporting.

(a) A generator who ships his hazardous waste off-site must submit Annual Reports:

(1) On EPA forms 8700–13 and 8700–13A according to the instructions on the form (See the Appendix to this Part);

(2) To the Regional Administrator for the Region in which the generator is located;

(3) No later than March 1 for the preceding calendar year.

(b) Any generator who treats, stores, or disposes of hazardous waste on-site must submit an Annual Report covering those wastes in accordance with the provisions of 40 CFR Parts 264, 265, and 266 and 40 CFR Part 122.

§ 262.42 Exception reporting.

(a) A generator who does not receive a copy of the manifest with the handwritten signature of the owner or operator of the designated facility within 35 days of the date the waste was accepted by the initial transporter must contact the transporter and/or the owner or operator of the designated facility to determine the status of the hazardous waste.

(b) A generator must submit an Exception Report to the EPA Regional Administrator for the Region in which the generator is located if he has not received a copy of the manifest with the handwritten signature of the owner or operator of the designated facility within 45 days of the date the waste was accepted by the initial transporter. The Exception Report must include:

(1) A legible copy of the manifest for which the generator does not have confirmation of delivery;

(2) A cover letter signed by the generator or his authorized representative explaining the efforts taken to locate the hazardous waste and the results of those efforts.

§ 262.43   Additional reporting.

The Administrator, as he deems necessary under section 2002(a) and section 3002(6) of the Act, may require generators to furnish additional reports concerning the quantities and disposition of wastes identified or listed in 40 CFR Part 261.

Subpart E—Special Conditions

§ 262.50   International shipments.

(a) Any person who exports hazardous waste to a foreign country or imports hazardous waste from a foreign country into the United States must comply with the requirements of this Part and with the special requirements of this section.

(b) When shipping hazardous waste outside the United States, the generator must:

(1) Notify the Administrator in writing four weeks before the initial shipment of hazardous waste to each country in each calendar year;

(i) The waste must be identified by its EPA hazardous waste identification number and its DOT shipping description;

(ii) The name and address of the foreign consignee must be included in this notice;

(iii) These notices must be sent to: Hazardous Waste Export, Division for Oceans and Regulatory Affairs (A–107), United States Environmental Protection Agency, Washington, D.C. 20460.

Note.—This requirement to notify will not be delegated to States authorized under 40 CFR Part 123. Therefore, all generators must notify the Administrator as required above.

(2) Require that the foreign consignee confirm the delivery of the waste in the foreign country. A copy of the manifest signed by the foreign consignee may be used for this purpose;

(3) Meet the requirements under § 262.21 for the manifest, except that:

(i) In place of the name, address, and EPA identification number of the designated facility, the name and address of the foreign consignee must be used;

(ii) The generator must identify the point of departure from the United States through which the waste must travel before entering a foreign country.

(c) A generator must file an Exception Report, if:

(1) He has not received a copy of the manifest signed by the transporter stating the date and place of departure from the United States within 45 days from the date it was accepted by the initial transporter; or

(2) Within 90 days from the date the waste was accepted by the initial transporter, the generator has not received written confirmation from the foreign consignee that the hazardous waste was received.

(d) When importing hazardous waste, a person must meet all requirements of § 262.21 for the manifest except that:

(1) In place of the generator's name, address and EPA identification number, the name and address of the foreign generator and the importer's name, address and EPA identification number must be used.

(2) In place of the generator's signature on the certification statement, the U.S. importer or his agent must sign and date the certification and obtain the signature of the initial transporter.

§ 262.51   Farmers.

A farmer disposing of waste pesticides from his own use which are hazardous wastes is not required to comply with the standards in this Part or other standards in 40 CFR Parts 122, 264 or 265 for those wastes provided he triple rinses each emptied pesticide container in accordance with § 261.33(c) and disposes of the pesticide residues on his own farm in a manner consistent with the disposal instructions on the pesticide label.

Appendix—Form—Annual Report (EPA Form 8700–13)

BILLING CODE 6560–01–M

PART 263—STANDARDS APPLICABLE TO TRANSPORTERS OF HAZARDOUS WASTE

Subpart A—General

Sec.
263.10   Scope.
263.11   EPA Identification Numbers.

Subpart B—Compliance With the Manifest System and Recordkeeping

263.20   The Manifest System.
263.21   Compliance with the Manifest.
263.22   Recordkeeping.

Subpart C—Hazardous Waste Discharges

263.30   Immediate Action.
263.31   Discharge Clean Up.

Authority: Sec. 2002(a), 3002, 3003, 3004 and 3005 of the Solid Waste Disposal Act as amended by the Resource Conservation and Recovery Act of 1976 and as amended by the Quiet Communities Act of 1978, (42 U.S.C. 6912, 6922, 6923, 6924, 6925).

Subpart A—General

§ 263.10   Scope.

(a) These regulations establish standards which apply to persons transporting hazardous waste within the United States if the transportation requires a manifest under 40 CFR Part 262.

Note.—The regulations set forth in Parts 262 and 263 establish the responsibilities of generators and transporters of hazardous waste in the handling, transportation, and management of that waste. In these regulations, EPA has expressly adopted certain regulations of the Department of Transportation (DOT) governing the transportation of hazardous materials. These regulations concern, among other things, labeling, marking, placarding, using proper containers, and reporting discharges. EPA has expressly adopted these regulations in order to satisfy its statutory obligation to promulgate regulations which are necessary to protect human health and the environment in the transportation of hazardous waste. EPA's adoption of these DOT regulations ensures consistency with the requirements of DOT and thus avoids the establishment of duplicative or conflicting requirements with respect to these matters. These EPA regulations which apply to both interstate and intrastate transportation of hazardous waste are enforceable by EPA.

DOT has revised its hazardous materials transportation regulations in order to encompass the transportation of hazardous waste and to regulate intrastate, as well as interstate, transportation of hazardous waste. Transporters of hazardous waste are cautioned that DOT's regulations are fully applicable to their activities and enforceable by DOT. These DOT regulations are codified in Title 49, Code of Federal Regulations, Subchapter C.

EPA and DOT worked together to develop standards for transporters of hazardous waste in order to avoid conflicting requirements. Except for transporters of bulk shipments of hazardous waste by water, a transporter who meets all applicable requirements of 49 CFR Parts 171 through 179 and the requirements of 40 CFR sections 263.11 and 263.31 will be deemed in compliance with this Part. Regardless of DOT's action, EPA retains its authority to enforce these regulations.

(b) These regulations do not apply to on-site transportation of hazardous waste by generators or by owners or operators of permitted hazardous waste management facilities.

(c) A transporter of hazardous waste must also comply with 40 CFR Part 262, Standards Applicable to Generators of Hazardous Waste, if he:

(1) Transports hazardous waste into the United States from abroad; or

(2) Mixes hazardous wastes of different DOT shipping descriptions by placing them into a single container.

Note.—Transporters who store hazardous waste are required to comply with the storage standards in 40 CFR Parts 264 and 265 and the permit requirements of 40 CFR Part 122.

§ 263.11   EPA identification number.

(a) A transporter must not transport hazardous wastes without having received an EPA identification number from the Administrator.

(b) A transporter who has not received an EPA identification number may obtain one by applying to the

Administrator using EPA Form 8700–12. Upon receiving the request, the Administrator will assign an EPA identification number to the transporter.

Subpart B—Compliance With the Manifest System and Recordkeeping

§ 263.20 The manifest system.

(a) A transporter may not accept hazardous waste from a generator unless it is accompanied by a manifest, signed by the generator in accordance with the provisions of 40 CFR Part 262.

(b) Before transporting the hazardous waste, the transporter must sign and date the manifest acknowledging acceptance of the hazardous waste from the generator. The transporter must return a signed copy to the generator before leaving the generator's property.

(c) The transporter must ensure that the manifest accompanies the hazardous waste.

(d) A transporter who delivers a hazardous waste to another transporter or to the designated facility must:

(1) Obtain the date of delivery and the handwritten signature of that transporter or of the owner or operator of the designated facility on the manifest; and

(2) Retain one copy of the manifest in accordance with § 263.22; and

(3) give the remaining copies of the manifest to the accepting transporter or designated facility.

(e) The requirements of paragraphs (c) and (d) of this section do not apply to rail or water (bulk shipment) transporters if:

(1) The hazardous waste is delivered by rail or water (bulk shipment) to the designated facility; and

(2) A shipping paper containing all the information required on the manifest (excluding the EPA identification numbers, generator certification, and signatures) accompanies the hazardous waste; and

(3) The delivering transporter obtains the date of delivery and handwritten signature of the owner or operator of the designated facility on either the manifest or the shipping paper; and

(4) The person delivering the hazardous waste to the initial rail or water (bulk shipment) transporter obtains the date of delivery and signature of the rail or water (bulk shipment) transporter on the manifest and forwards it to the designated facility; and

(5) A copy of the shipping paper or manifest is retained by each rail or water (bulk shipment) transporter in accordance with § 263.22.

(f) Transporters who transport hazardous waste out of the United States must:

(1) indicate on the manifest the date the hazardous waste left the United States; and

(2) sign the manifest and retain one copy in accordance with § 263.22(c); and

(3) return a signed copy of the manifest to the generator.

§ 263.21 Compliance with the manifest.

(a) The transporter must deliver the entire quantity of hazardous waste which he has accepted from a generator or a transporter to:

(1) The designated facility listed on the manifest; or

(2) The alternate designated facility, if the hazardous waste cannot be delivered to the designated facility because an emergency prevents delivery; or

(3) The next designated transporter; or

(4) The place outside the United States designated by the generator.

(b) If the hazardous waste cannot be delivered in accordance with paragraph (a) of this section, the transporter must contact the generator for further directions and must revise the manifest according to the generator's instructions.

§ 263.22 Recordkeeping.

(a) A transporter of hazardous waste must keep a copy of the manifest signed by the generator, himself, and the next designated transporter or the owner or operator of the designated facility for a period of three years from the date the hazardous waste was accepted by the initial transporter.

(b) For shipments delivered to the designated facility by rail or water (bulk shipment), each rail or water (bulk shipment) transporter must retain a copy of a shipping paper containing all the information required in § 263.20(e)(2) for a period of three years from the date the hazardous waste was accepted by the initial transporter.

(c) A transporter who transports hazardous waste out of the United States must keep a copy of the manifest indicating that the hazardous waste left the United States for a period of three years from the date the hazardous waste was accepted by the initial transporter.

(d) The periods of retention referred to in this Section are extended automatically during the course of any unresolved enforcement action regarding the regulated activity or as requested by the Administrator.

Subpart C—Hazardous Waste Discharges

§ 263.30 Immediate action.

(a) In the event of a discharge of hazardous waste during transportation, the transporter must take appropriate immediate action to protect human health and the environment (e.g., notify local authorities, dike the discharge area).

(b) If a discharge of hazardous waste occurs during transportation and an official (State or local government or a Federal Agency) acting within the scope of his official responsibilities determines that immediate removal of the waste is necessary to protect human health or the environment, that official may authorize the removal of the waste by transporters who do not have EPA identification numbers and without the preparation of a manifest.

(c) An air, rail, highway, or water transporter who has discharged hazardous waste must:

(1) Give notice, if required by 49 CFR 171.15, to the National Response Center (800–424–8802 or 202–426–2675); and

(2) Report in writing as required by 49 CFR 171.16 to the Director, Office of Hazardous Materials Regulations, Materials Transportation Bureau, Department of Transportation, Washington, D.C. 20590.

(d) A water (bulk shipment) transporter who has discharged hazardous waste must give the same notice as required by 33 CFR 153.203 for oil and hazardous substances.

§ 263.31 Discharge clean up.

A transporter must clean up any hazardous waste discharge that occurs during transportation or take such action as may be required or approved by Federal, State, or local officials so that the hazardous waste discharge no longer presents a hazard to human health or the environment.

[FR Doc. 80–14666 Filed 5–16–80; 8:45 am]

BILLING CODE 6560–01–M

PART 264—STANDARDS FOR OWNERS AND OPERATORS OF HAZARDOUS WASTE TREATMENT, STORAGE, AND DISPOSAL FACILITIES

Subpart A—General

Sec.
264.1 Purpose, scope and applicability.
264.2 [Reserved]
264.3 Relationship to interim status standards.
264.4 Imminent hazard action.
264.5–264.9 [Reserved]

Subpart B—General Facility Standards

264.10 Applicability.
264.11 Identification number.
264.12 Required notices.
264.13 General waste analysis.
264.14 Security.
264.15 General inspection requirements.
264.16 Personnel training.
264.17–264.29 [Reserved]

Authority: Secs. 1006, 2002(a), and 3004 of the Solid Waste Disposal Act, as amended by the Resource Conservation and Recovery Act of 1976, as amended (42 U.S.C. 6905, 6912(a), and 6924).

Subpart A—General

§ 264.1   Purpose, scope and applicability.

(a) The purpose of this Part is to establish minimum national standards which define the acceptable management of hazardous waste.

(b) The standards in this Part apply to owners and operators of all facilities which treat, store, or dispose of hazardous waste, except as specifically provided otherwise in this Part or Part 261 of this Chapter.

(c) The requirements of this Part apply to a person disposing of hazardous waste by means of ocean disposal subject to a permit issued under the Marine Protection, Research, and Sanctuaries Act only to the extent they are included in a RCRA permit by rule granted to such a person under Part 122 of this Chapter.

[Comment: These Part 264 regulations do apply to the treatment or storage of hazardous waste before it is loaded onto an ocean vessel for incineration or disposal at sea.]

(d) The requirements of this Part apply to a person disposing of hazardous waste by means of underground injection subject to a permit issued under an Underground Injection Control (UIC) program approved or promulgated under the Safe Drinking Water Act only to the extent they are required by § 122.45 of this Chapter.

[Comment: These Part 264 regulations do apply to the above-ground treatment or storage of hazardous waste before it is injected underground.]

(e) The requirements of this Part apply to the owner or operator of a POTW which treats, stores, or disposes of hazardous waste only to the extent they are included in a RCRA permit by rule granted to such a person under Part 122 of this Chapter.

(f) The requirements of this Part do not apply to a person who treats, stores, or disposes of hazardous waste in a State with a RCRA hazardous waste program authorized under Subparts A and B of Part 123 of this Chapter or with a RCRA Phase II hazardous waste program authorized under Subpart F of Part 123 of this Chapter, except that the requirements of this Part will continue to apply as stated in paragraph (d) of this Section, if the authorized State RCRA program does not cover disposal of hazardous waste by means of underground injection.

(g) The requirements of this Part do not apply to:

(1) The owner or operator of a facility permitted, licensed, or registered by a State to manage municipal or industrial solid waste, if the only hazardous waste the facility treats, stores, or disposes of is excluded from regulation under this Part by § 261.5 of this Chapter;

(2) The owner or operator of a facility which treats or stores hazardous waste, which treatment or storage meets the criteria in § 261.6(a) of this Chapter, except to the extent that § 261.6(b) of this Chapter provides otherwise;

(3) A generator accumulating waste on-site in compliance with § 262.34 of this Chapter;

(4) A farmer disposing of waste pesticides from his own use in compliance with § 262.51 of this Chapter; or

(5) The owner or operator of a totally enclosed treatment facility, as defined in § 260.10.

§ 264.2   [Reserved]

§ 264.3   Relationship to interim status standards.

A facility owner or operator who has fully complied with the requirements for interim status—as defined in Section

3005(e) of RCRA and regulations under § 122.23 of this Chapter—must comply with the regulations specified in Part 265 of this Chapter in lieu of the regulations in this Part, until final administrative disposition of his permit application is made.

[Comment: As stated in Section 3005(a) of RCRA, after the effective date of regulations under that Section, i.e., Parts 122 and 124 of this Chapter, the treatment, storage, or disposal of hazardous waste is prohibited except in accordance with a permit. Section 3005(e) of RCRA provides for the continued operation of an existing facility which meets certain conditions until final administrative disposition of the owner's or operator's permit application is made.]

§ 264.4   Imminent hazard action.

Notwithstanding any other provisions of these regulations, enforcement actions may be brought purusant to Section 7003 of RCRA.

§§ 264.5–264.9   [Reserved]

Subpart B—General Facility Standards

§ 264.10   Applicability.

The regulations in this Subpart apply to owners and operators of all hazardous waste facilities, except as § 264.1 provides otherwise.

§ 264.11   Identification number.

Every facility owner or operator must apply to EPA for an EPA identification number in accordance with the EPA notification procedures (45 FR 12746).

§ 264.12   Required notices.

(a) The owner or operator of a facility that has arranged to receive hazardous waste from a foreign source must notify the Regional Administrator in writing at least four weeks in advance of the date the waste is expected to arrive at the facility. Notice of subsequent shipments of the same waste from the same foreign source is not required.

(b) The owner or operator of a facility that receives hazardous waste from an off-site source (except where the owner or operator is also the generator) must inform the generator in writing that he has the appropriate permit(s) for, and will accept, the waste the generator is shipping. The owner or operator must keep a copy of this written notice as part of the operating record.

(c) Before transferring ownership or operation of a facility during its operating life, or of a disposal facility during the post-closure care period, the owner or operator must notify the new owner or operator in writing of the requirements of this Part and Part 122 of this Chapter.

[*Comment:* An owner's or operator's failure to notify the new owner or operator of the requirements of this Part in no way relieves the new owner or operator of his obligation to comply with all applicable requirements.]

§ 264.13 General waste analysis.

(a) (1) Before an owner or operator treats, stores, or disposes of any hazardous waste, he must obtain a detailed chemical and physical analysis of a representative sample of the waste. At a minimum, this analysis must contain all the information which must be known to treat, store, or dispose of the waste in accordance with the requirements of this Part or with the conditions of a permit issued under Part 122, Subparts A and B, and Part 124 of this Chapter.

(2) The analysis may include data developed under Part 261 of this Chapter, and existing published or documented data on the hazardous waste or on hazardous waste generated from similar processes.

[*Comment:* For example, the facility's records of analyses performed on the waste before the effective date of these regulations, or studies conducted on hazardous waste generated from processes similar to that which generated the waste to be managed at the facility, may be included in the data base required to comply with paragraph (a)(1) of this Section. The owner or operator of an off-site facility may arrange for the generator of the hazardous waste to supply part or all of the information required by paragraph (a)(1) of this Section. If the generator does not supply the information, and the owner or operator chooses to accept a hazardous waste, the owner or operator is responsible for obtaining the information required to comply with this Section.]

(3) The analysis must be repeated as necessary to ensure that it is accurate and up to date. At a minimum, the analysis must be repeated:

(i) When the owner or operator is notified, or has reason to believe, that the process or operation generating the hazardous waste has changed; and

(ii) For off-site facilities, when the results of the inspection required in paragraph (a)(4) of this Section indicate that the hazardous waste received at the facility does not match the waste designated on the accompanying manifest or shipping paper.

(4) The owner or operator of an off-site facility must inspect and, if necessary, analyze each hazardous waste movement received at the facility to determine whether it matches the identity of the waste specified on the

accompanying manifest or shipping paper.

(b) The owner or operator must develop and follow a written waste analysis plan which describes the procedures which he will carry out to comply with paragraph (a) of this Section. He must keep this plan at the facility. At a minimum, the plan must specify:

(1) The parameters for which each hazardous waste will be analyzed and the rationale for the selection of these parameters (i.e., how analysis for these parameters will provide sufficient information on the waste's properties to comply with paragraph (a) of this Section);

(2) The test methods which will be used to test for these parameters;

(3) The sampling method which will be used to obtain a representative sample of the waste to be analyzed. A representative sample may be obtained using either:

(i) One of the sampling methods described in Appendix I of Part 261 of this Chapter; or

(ii) An equivalent sampling method. [*Comment:* See § 261.20(c) of this Chapter for related discussion.]

(4) The frequency with which the initial analysis of the waste will be reviewed or repeated to ensure that the analysis is accurate and up to date; and

(5) For off-site facilities, the waste analyses that hazardous waste generators have agreed to supply.

(c) For off-site facilities, the waste analysis plan required in paragraph (b) of this Section must also specify the procedures which will be used to inspect and, if necessary, analyze each movement of hazardous waste received at the facility to ensure that it matches the identity of the waste designated on the accompanying manifest or shipping paper. At a minimum, the plan must describe:

(1) The procedures which will be used to determine the identity of each movement of waste managed at the facility; and

(2) The sampling method which will be used to obtain a representative sample of the waste to be identified, if the identification method includes sampling.

[*Comment:* Part 122, Subpart B, of this Chapter requires that the waste analysis plan be submitted with Part B of the permit application.]

§ 264.14 Security.

(a) The owner or operator must prevent the unknowing entry, and minimize the possibility for the unauthorized entry, of persons or livestock onto the active portion of his

facility, *unless* he can demonstrate to the Regional Administrator that:

(1) Physical contact with the waste, structures, or equipment within the active portion of the facility will not injure unknowing or unauthorized persons or livestock which may enter the active portion of a facility; and

(2) Disturbance of the waste or equipment, by the unknowing or unauthorized entry of persons or livestock onto the active portion of a facility, will not cause a violation of the requirements of this Part.

[*Comment:* Part 122, Subpart B, of this Chapter requires that an owner or operator who wishes to make the demonstration referred to above must do so with Part B of the permit application.]

(b) Unless the owner or operator has made a successful demonstration under paragraphs (a)(1) and (a)(2) of this Section, a facility must have:

(1) A 24-hour surveillance system (e.g., television monitoring or surveillance by guards or facility personnel) which continuously monitors and controls entry onto the active portion of the facility; or

(2) (i) An artificial or natural barrier (e.g., a fence in good repair or a fence combined with a cliff), which completely surrounds the active portion of the facility; and

(ii) A means to control entry, at all times, through the gates or other entrances to the active portion of the facility (e.g., an attendant, television monitors, locked entrance, or controlled roadway access to the facility).

[*Comment:* The requirements of paragraph (b) of this Section are satisfied if the facility or plant within which the active portion is located itself has a surveillance system, or a barrier and a means to control entry, which complies with the requirements of paragraph (b)(1) or (b)(2) of this Section.]

(c) Unless the owner or operator has made a successful demonstration under paragraphs (a)(1) and (a)(2) of this Section, a sign with the legend, "Danger—Unauthorized Personnel Keep Out", must be posted at each entrance to the active portion of a facility, and at other locations, in sufficient numbers to be seen from any approach to this active portion. The legend must be written in English and in any other language predominant in the area surrounding the facility (e.g., facilities in counties bordering the Canadian province of Quebec must post signs in French; facilities in counties bordering Mexico must post signs in Spanish), and must be legible from a distance of at least 25 feet. Existing signs with a legend other than "Danger—Unauthorized Personnel Keep Out" may be used if the legend on the sign indicates that only authorized personnel are allowed to enter the active portion, and that entry onto the active portion can be dangerous.

§ 264.15 General inspection requirements.

(a) The owner or operator must inspect his facility for malfunctions and deterioration, operator errors, and discharges which may be causing—or may lead to—(1) release of hazardous waste constituents to the environment or (2) a threat to human health. The owner or operator must conduct these inspections often enough to identify problems in time to correct them before they harm human health or the environment.

(b)(1) The owner or operator must develop and follow a written schedule for inspecting monitoring equipment, safety and emergency equipment, security devices, and operating and structural equipment (such as dikes and sump pumps) that are important to preventing, detecting, or responding to environmental or human health hazards.

(2) He must keep this schedule at the facility.

(3) The schedule must identify the types of problems (e.g., malfunctions or deterioration) which are to be looked for during the inspection (e.g., inoperative sump pump, leaking fitting, eroding dike, etc.).

(4) The frequency of inspection may vary for the items on the schedule. However, it should be based on the rate of possible deterioration of the equipment and the probability of an environmental or human health incident if the deterioration or malfunction or any operator error goes undetected between inspections. Areas subject to spills, such as loading and unloading areas, must be inspected daily when in use.

[Comment: Part 122, Subpart B, of this Chapter requires the inspection schedule to be submitted with Part B of the permit application. EPA will evaluate the schedule along with the rest of the application to ensure that it adequately protects human health and the environment. As part of this review, EPA may modify or amend the schedule as may be necessary.]

(c) The owner or operator must remedy any deterioration or malfunction of equipment or structures which the inspection reveals on a schedule which ensures that the problem does not lead to an environmental or human health hazard. Where a hazard is imminent or has already occurred, remedial action must be taken immediately.

(d) The owner or operator must record inspections in an inspection log or summary. He must keep these records for at least three years from the date of inspection. At a minimum, these records must include the date and time of the inspection, the name of the inspector, a notation of the observations made, and the date and nature of any repairs or other remedial actions.

§ 264.16 Personnel training.

(a)(1) Facility personnel must successfully complete a program of classroom instruction or on-the-job training that teaches them to perform their duties in a way that ensures the facility's compliance with the requirements of this Part. The owner or operator must ensure that this program includes all the elements described in the document required under paragraph (d)(3) of this Section.

(2) This program must be directed by a person trained in hazardous waste management procedures, and must include instruction which teaches facility personnel hazardous waste management procedures (including contingency plan implementation) relevant to the positions in which they are employed.

(3) At a minimum, the training program must be designed to ensure that facility personnel are able to respond effectively to emergencies by familiarizing them with emergency procedures, emergency equipment, and emergency systems, including, where applicable:

(i) Procedures for using, inspecting, repairing, and replacing facility emergency and monitoring equipment;

(ii) Key parameters for automatic waste feed cut-off systems;

(iii) Communications or alarm systems;

(iv) Response to fires or explosions;

(v) Response to ground-water contamination incidents; and

(vi) Shutdown of operations.

(b) Facility personnel must successfully complete the program required in paragraph (a) of this Section within six months after the effective date of these regulations or six months after the date of their employment or assignment to a facility, or to a new position at a facility, whichever is later. Employees hired after the effective date of these regulations must not work in unsupervised positions until they have completed the training requirements of paragraph (a) of this Section.

(c) Facility personnel must take part in an annual review of the initial training required in paragraph (a) of this Section.

(d) The owner or operator must maintain the following documents and records at the facility:

(1) The job title for each position at the facility related to hazardous waste management, and the name of the employee filling each job;

(2) A written job description for each position listed under paragraph (d)(1) of this Section. This description may be consistent in its degree of specificity with descriptions for other similar positions in the same company location or bargaining unit, but must include the requisite skill, education, or other qualifications, and duties of employees assigned to each position; (3) A written description of the type and amount of both introductory and continuing training that will be given to each person filling a position listed under paragraph (d)(1) of this Section;

(4) Records that document that the training or job experience required under paragraphs (a), (b), and (c) of this Section has been given to, and completed by, facility personnel.

(e) Training records on current personnel must be kept until closure of the facility; training records on former employees must be kept for at least three years from the date the employee last worked at the facility. Personnel training records may accompany personnel transferred within the same company.

§§ 264.17—264.29 [Reserved]

Subpart C—Preparedness and Prevention

§ 264.30 Applicability.

The regulations in this Subpart apply to owners and operators of all hazardous waste facilities, except as § 264.1 provides otherwise.

§ 264.31 Design and operation of facility.

Facilities must be designed, constructed, maintained, and operated to minimize the possibility of a fire, explosion, or any unplanned sudden or non-sudden release of hazardous waste or hazardous waste constituents to air, soil, or surface water which could threaten human health or the environment.

§ 264.32 Required equipment.

All facilities must be equipped with the following, unless it can be demonstrated to the Regional Administrator that none of the hazards posed by waste handled at the facility could require a particular kind of equipment specified below:

(a) An internal communications or alarm system capable of providing immediate emergency instruction (voice or signal) to facility personnel;

(b) A device, such as a telephone (immediately available at the scene of operations) or a hand-held two-way radio, capable of summoning emergency assistance from local police departments, fire departments, or State or local emergency response teams;

(c) Portable fire extinguishers, fire control equipment (including special extinguishing equipment, such as that using foam, inert gas, or dry chemicals), spill control equipment, and decontamination equipment; and

(d) Water at adequate volume and pressure to supply water hose streams, or foam producing equipment, or automatic sprinklers, or water spray systems.

[Comment: Part 122, Subpart B, of this Chapter requires that an owner or operator who wishes to make the demonstration referred to above must do so with Part B of the permit application.]

§ 264.33 Testing and maintenance of equipment.

All facility communications or alarm systems, fire protection equipment, spill control equipment, and decontamination equipment, where required, must be tested and maintained as necessary to assure its proper operation in time of emergency.

§ 264.34 Access to communications or alarm system.

(a) Whenever hazardous waste is being poured, mixed, spread, or otherwise handled, all personnel involved in the operation must have immediate access to an internal alarm or emergency communication device, either directly or through visual or voice contact with another employee, unless the Regional Administrator has ruled that such a device is not required under § 264.32.

(b) If there is ever just one employee on the premises while the facility is operating, he must have immediate access to a device, such as a telephone (immediately available at the scene of operation) or a hand-held two-way radio, capable of summoning external emergency assistance, unless the Regional Administrator has ruled that such a device is not required under § 264.32.

§ 264.35 Required aisle space.

The owner or operator must maintain aisle space to allow the unobstructed movement of personnel, fire protection equipment, spill control equipment, and decontamination equipment to any area of facility operation in an emergency, unless it can be demonstrated to the Regional Administrator that aisle space is not needed for any of these purposes.

[Comment: Part 122, Subpart B, of this Chapter requires that an owner or operator who wishes to make the demonstration referred to above must do so with Part B of the permit application.]

§ 264.36 Special handling for ignitable or reactive waste.

The owner or operator must take precautions to prevent accidental ignition or reaction of ignitable or reactive waste. This waste must be separated and protected from sources of ignition or reaction including but not limited to: open flames, smoking, cutting and welding, hot surfaces, frictional heat, sparks (static, electrical, or mechanical), spontaneous ignition (e.g., from heat-producing chemical reactions), and radiant heat. While ignitable or reactive waste is being handled, the owner or operator must confine smoking and open flame to specially designated locations. "No Smoking" signs must be conspicuously placed wherever there is a hazard from ignitable or reactive waste.

§ 264.37 Arrangements with local authorities.

(a) The owner or operator must attempt to make the following arrangements, as appropriate for the type of waste handled at his facility and the potential need for the services of these organizations:

(1) Arrangements to familiarize police, fire departments, and emergency response teams with the layout of the facility, properties of hazardous waste handled at the facility and associated hazards, places where facility personnel would normally be working, entrances to and roads inside the facility, and possible evacuation routes;

(2) Where more than one police and fire department might respond to an emergency, agreements designating primary emergency authority to a specific police and a specific fire department, and agreements with any others to provide support to the primary emergency authority;

(3) Agreements with State emergency response teams, emergency response contractors, and equipment suppliers; and

(4) Arrangements to familiarize local hospitals with the properties of hazardous waste handled at the facility and the types of injuries or illnesses which could result from fires, explosions, or releases at the facility.

(b) Where State or local authorities decline to enter into such arrangements, the owner or operator must document the refusal in the operating record.

§§ 264.38—264.49 [Reserved]

Subpart D—Contingency Plan and Emergency Procedures

§ 264.50 Applicability.

The regulations in this Subpart apply to owners and operators of all hazardous waste facilities, except as § 264.1 provides otherwise.

§ 264.51 Purpose and implementation of contingency plan.

(a) Each owner or operator must have a contingency plan for his facility. The contingency plan must be designed to minimize hazards to human health or the environment from fires, explosions, or any unplanned sudden or non-sudden release of hazardous waste or hazardous waste constituents to air, soil, or surface water.

(b) The provisions of the plan must be carried out immediately whenever there is a fire, explosion, or release of hazardous waste or hazardous waste constituents which could threaten human health or the environment.

§ 264.52 Content of contingency plan.

(a) The contingency plan must describe the actions facility personnel must take to comply with §§ 264.51 and 264.56 in response to fires, explosions, or any unplanned sudden or non-sudden release of hazardous waste or hazardous waste constituents to air, soil, or surface water at the facility.

(b) If the owner or operator has already prepared a Spill Prevention, Control, and Countermeasures (SPCC) Plan in accordance with Part 112 or Part 151 of this Chapter, or some other emergency or contingency plan, he need only amend that plan to incorporate hazardous waste management provisions that are sufficient to comply with the requirements of this Part.

(c) The plan must describe arrangements agreed to by local police departments, fire departments, hospitals, contractors, and State and local emergency response teams to coordinate emergency services, pursuant to § 264.37.

(d) The plan must list names, addresses, and phone numbers (office and home) of all persons qualified to act as emergency coordinator (see § 264.55), and this list must be kept up to date. Where more than one person is listed, one must be named as primary emergency coordinator and others must be listed in the order in which they will assume responsibility as alternates. For new facilities, this information must be supplied to the Regional Administrator at the time of certification, rather than at the time of permit application.

(e) The plan must include a list of all

emergency equipment at the facility (such as fire extinguishing systems, spill control equipment, communications and alarm systems (internal and external), and decontamination equipment), where this equipment is required. This list must be kept up to date. In addition, the plan must include the location and a physical description of each item on the list, and a brief outline of its capabilities.

(f) The plan must include an evacuation plan for facility personnel where there is a possibility that evacuation could be necessary. This plan must describe signal(s) to be used to begin evacuation, evacuation routes, and alternate evacuation routes (in cases where the primary routes could be blocked by releases of hazardous waste or fires).

§ 264.53   Copies of contingency plan.

A copy of the contingency plan and all revisions to the plan must be:

(a) Maintained at the facility; and

(b) Submitted to all local police departments, fire departments, hospitals, and State and local emergency response teams that may be called upon to provide emergency services.

[Comment: The contingency plan must be submitted to the Regional Administrator with Part B of the permit application under Part 122, Subparts A and B, of this Chapter and, after modification or approval, will become a condition of any permit issued.]

§ 264.54   Amendment of contingency plan.

The contingency plan must be reviewed, and immediately amended, if necessary, whenever:

(a) The facility permit is revised;

(b) The plan fails in an emergency;

(c) The facility changes—in its design, construction, operation, maintenance, or other circumstances—in a way that materially increases the potential for fires, explosions, or releases of hazardous waste or hazardous waste constituents, or changes the response necessary in an emergency;

(d) The list of emergency coordinators changes; or

(e) The list of emergency equipment changes.

[Comment: A change in the lists of facility emergency coordinators or equipment in the contingency plan constitutes a minor modification to the facility permit to which the plan is a condition.]

§ 264.55   Emergency coordinator.

At all times, there must be at least one employee either on the facility premises or on call (i.e., available to respond to an emergency by reaching the facility within a short period of time) with the responsibility for coordinating all emergency response measures. This emergency coordinator must be thoroughly familiar with all aspects of the facility's contingency plan, all operations and activities at the facility, the location and characteristics of waste handled, the location of all records within the facility, and the facility layout. In addition, this person must have the authority to commit the resources needed to carry out the contingency plan.

[Comment: The emergency coordinator's responsibilities are more fully spelled out in § 264.56. Applicable responsibilities for the emergency coordinator vary, depending on factors such as type and variety of waste(s) handled by the facility, and type and complexity of the facility.]

§ 264.56   Emergency Procedures.

(a) Whenever there is an imminent or actual emergency situation, the emergency coordinator (or his designee when the emergency coordinator is on call) must immediately:

(1) Activate internal facility alarms or communication systems, where applicable, to notify all facility personnel; and

(2) Notify appropriate State or local agencies with designated response roles if their help is needed.

(b) Whenever there is a release, fire, or explosion, the emergency coordinator must immediately identify the character, exact source, amount, and areal extent of any released materials. He may do this by observation or review of facility records or manifests, and, if necessary, by chemical analysis.

(c) Concurrently, the emergency coordinator must assess possible hazards to human health or the environment that may result from the release, fire, or explosion. This assessment must consider both direct and indirect effects of the release, fire, or explosion (e.g., the effects of any toxic, irritating, or asphyxiating gases that are generated, or the effects of any hazardous surface water run-off from water or chemical agents used to control fire and heat-induced explosions).

(d) If the emergency coordinator determines that the facility has had a release, fire, or explosion which could threaten human health, or the environment, outside the facility, he must report his findings as follows:

(1) If his assessment indicates that evacuation of local areas may be advisable, he must immediately notify appropriate local authorities. He must be available to help appropriate officials decide whether local areas should be evacuated; and

(2) He must immediately notify either the government official designated as the on-scene coordinator for that geographical area, (in the applicable regional contingency plan under Part 1510 of this Title) or the National Response Center (using their 24-hour toll free number 800/424-8802). The report must include:

(i) Name and telephone number of reporter;

(ii) Name and address of facility;

(iii) Time and type of incident (e.g., release, fire);

(iv) Name and quantity of material(s) involved, to the extent known;

(v) The extent of injuries, if any; and

(vi) The possible hazards to human health, or the environment, outside the facility.

(e) During an emergency, the emergency coordinator must take all reasonable measures necessary to ensure that fires, explosions, and releases do not occur, recur, or spread to other hazardous waste at the facility. These measures must include, where applicable, stopping processes and operations, collecting and containing release waste, and removing or isolating containers.

(f) If the facility stops operations in response to a fire, explosion, or release, the emergency coordinator must monitor for leaks, pressure buildup, gas generation, or ruptures in valves, pipes, or other equipment, wherever this is appropriate.

(g) Immediately after an emergency, the emergency coordinator must provide for treating, storing, or disposing of recovered waste, contaminated soil or surface water, or any other material that results from a release, fire, or explosion at the facility.

[Comment: Unless the owner or operator can demonstrate, in accordance with § 261.3(c) or (d) of this Chapter, that the recovered material is not a hazardous waste, the owner or operator becomes a generator of hazardous waste and must manage it in accordance with all applicable requirements of Parts 262, 263, and 264 of this Chapter.]

(h) The emergency coordinator must ensure that, in the affected area(s) of the facility:

(1) No waste that may be incompatible with the released material is treated, stored, or disposed of until cleanup procedures are completed; and

(2) All emergency equipment listed in the contingency plan is cleaned and fit for its intended use before operations are resumed.

(i) The owner or operator must notify the Regional Administrator, and appropriate State and local authorities, that the facility is in compliance with paragraph (h) of this Section before operations are resumed in the affected area(s) of the facility.

(j) The owner or operator must note in the operating record the time, date, and details of any incident that requires implementing the contingency plan. Within 15 days after the incident, he must submit a written report on the incident to the Regional Administrator. The report must include:

(1) Name, address, and telephone number of the owner or operator;

(2) Name, address, and telephone number of the facility;

(3) Date, time, and type of incident (e.g., fire, explosion);

(4) Name and quantity of material(s) involved;

(5) The extent of injuries, if any;

(6) An assessment of actual or potential hazards to human health or the environment, where this is applicable; and

(7) Estimated quantity and disposition of recovered material that resulted from the incident.

§§ 264.57–264.69 [Reserved]

Subpart E—Manifest System, Recordkeeping, and Reporting

§ 264.70 Applicability.

The regulations in this Subpart apply to owners and operators of both on-site and off-site facilities, except as § 264.1 provides otherwise. Sections 264.71, 264.72, and 264.76 do not apply to owners and operators of on-site facilities that do not receive any hazardous waste from off-site sources.

§ 264.71 Use of manifest system.

(a) If a facility receives hazardous waste accompanied by a manifest, the owner or operator, or his agent, must:

(1) Sign and date each copy of the manifest to certify that the hazardous waste covered by the manifest was received;

(2) Note any significant discrepancies in the manifest (as defined in § 264.72(a)) on each copy of the manifest;

[Comment: The Agency does not intend that the owner or operator of a facility whose procedures under § 264.13(c) include waste analysis must perform that analysis before signing the manifest and giving it to the transporter. Section 264.72(b), however, requires reporting an unreconciled discrepancy discovered during later analysis.]

(3) Immediately give the transporter at least one copy of the signed manifest;

(4) Within 30 days after the delivery, send a copy of the manifest to the generator; and

(5) Retain at the facility a copy of each manifest for at least three years from the date of delivery.

(b) If a facility receives, from a rail or water (bulk shipment) transporter, hazardous waste which is accompanied by a shipping paper containing all the information required on the manifest (excluding the EPA identification numbers, generator's certification, and signatures), the owner or operator, or his agent, must:

(1) Sign and date each copy of the shipping paper to certify that the hazardous waste covered by the shipping paper was received;

(2) Note any significant discrepancies in the shipping paper (as defined in § 264.72(a)) on each copy of the shipping paper;

[Comment: The Agency does not intend that the owner or operator of a facility whose procedures under § 264.13(c) include waste analysis must perform that analysis before signing the shipping paper and giving it to the transporter. Section 264.72(b), however, requires reporting an unreconciled discrepancy discovered during later analysis.]

(3) Immediately give the rail or water (bulk shipment) transporter at least one copy of the shipping paper;

(4) Within 30 days after the delivery, send a copy of the shipping paper to the generator; however, if the manifest is received within 30 days after the delivery, the owner or operator, or his agent, must sign and date the manifest and return it to the generator in lieu of the shipping paper; and

[Comment: Section 262.23(c) of this chapter requires the generator to send three copies of the manifest to the facility when hazardous waste is sent by rail or water (bulk shipment).]

(5) Retain at the facility a copy of each shipping paper and manifest for at least three years from the date of delivery.

§ 264.72 Manifest discrepancies.

(a) Manifest discrepancies are differences between the quantity or type of hazardous waste designated on the manifest or shipping paper, and the quantity or type of hazardous waste a facility actually receives. Significant discrepancies in quantity are: (1) For bulk waste, variations greater than 10 percent in weight, and (2) for batch waste, any variation in piece count, such as a discrepancy of one drum in a truckload. Significant discrepancies in type are obvious differences which can be discovered by inspection or waste analysis, such as waste solvent substituted for waste acid, or toxic constituents not reported on the manifest or shipping paper.

(b) Upon discovering a significant discrepancy, the owner or operator must attempt to reconcile the discrepancy with the waste generator or transporter

(e.g., with telephone conversations). If the discrepancy is not resolved within 15 days after receiving the waste, the owner or operator must immediately submit to the Regional Administrator a letter describing the discrepancy and attempts to reconcile it, and a copy of the manifest or shipping paper at issue.

§ 264.73 Operating record.

(a) The owner or operator must keep a written operating record at his facility.

(b) The following information must be recorded, as it becomes available, and maintained in the operating record until closure of the facility:

(1) A description and the quantity of each hazardous waste received, and the method(s) and date(s) of its treatment, storage, or disposal at the facility as required by Appendix I;

(2) The location of each hazardous waste within the facility and the quantity at each location. For disposal facilities, the location and quantity of each hazardous waste must be recorded on a map or diagram of each cell or disposal area. For all facilities, this information must include cross-references to specific manifest document numbers, if the waste was accompanied by a manifest;

(3) Records and results of waste analyses performed as specified in § 264.13;

(4) Summary reports and details of all incidents that require implementing the contingency plan as specified in § 264.56(j);

(5) Records and results of inspections as required by § 264.15(d) (except these data need be kept only three years); and

(6) For off-site facilities, notices to generators as specified in § 264.12(b).

§ 264.74 Availability, retention, and disposition of records.

(a) All records, including plans, required under this Part must be furnished upon request, and made available at all reasonable times for inspection, by any officer, employee, or representative of EPA who is duly designated by the Administrator.

(b) The retention period for all records required under this Part is extended automatically during the course of any unresolved enforcement action regarding the facility or as requested by the Administrator.

(c) A copy of records of waste disposal locations and quantities under § 264.73(b)(2) must be submitted to the Regional Administrator and local land authority upon closure of the facility.

§ 264.75 Annual report.

The owner or operator must prepare and submit a single copy of an annual report to the Regional Administrator by March 1 of each year. The report form and instructions in Appendix II must be used for this report. The annual report must cover facility activities during the previous calendar year and must include the following information:

(a) The EPA identification number, name, and address of the facility;

(b) The calendar year covered by the report;

(c) For off-site facilities, the EPA identification number of each hazardous waste generator from which the facility received a hazardous waste during the year; for imported shipments, the report must give the name and address of the foreign generator;

(d) A description and the quantity of each hazardous waste the facility received during the year. For off-site facilities, this information must be listed by EPA identification number of each generator;

(e) The method of treatment, storage, or disposal for each hazardous waste; and

(f) The certification signed by the owner or operator of the facility or his authorized representative.

§ 264.76 Unmanifested waste report.

If a facility accepts for treatment, storage, or disposal any hazardous waste from an off-site source without an accompanying manifest, or without an accompanying shipping paper as described in § 263.20(e)(2) of this Chapter, and if the waste is not excluded from the manifest requirement by § 261.5 of this Chapter, then the owner or operator must prepare and submit a single copy of a report to the Regional Administrator within 15 days after receiving the waste. The report form and instructions in Appendix II must be used for this report. The report must include the following information:

(a) The EPA identification number, name, and address of the facility;

(b) The date the facility received the waste;

(c) The EPA identification number, name, and address of the generator and the transporter, if available;

(d) A description and the quantity of each unmanifested hazardous waste and facility received;

(e) The method of treatment, storage, or disposal for each hazardous waste;

(f) The certification signed by the owner or operator of the facility or his authorized representative; and

(g) A brief explanation of why the waste was unmanifested, if known.

[Comment: Small quantities of hazardous waste are excluded from regulation under this Part and do not require a manifest. Where a facility receives unmanifested hazardous wastes, the Agency suggests that the owner or operator obtain from each generator a certification that the waste qualifies for exclusion. Otherwise, the Agency suggests that the owner or operator file an unmanifested waste report for the hazardous waste movement.]

§ 264.77 Additional reports.

In addition to submitting the annual report and unmanifested waste reports described in §§ 264.75 and 264.76, the owner or operator must also report to the Regional Administrator releases, fires, and explosions as specified in § 264.56(j).

PART 265—INTERIM STATUS STANDARDS FOR OWNERS AND OPERATORS OF HAZARDOUS WASTE TREATMENT, STORAGE, AND DISPOSAL FACILITIES

Subpart A—General

Subpart B—General Facility Standards

Subpart C—Preparedness and Prevention

Subpart D—Contingency Plan and Emergency Procedures

Subpart E—Manifest System, Recordkeeping, and Reporting

Subpart F—Ground-Water Monitoring

Subpart G—Closure and Post-Closure

Subpart H—Financial Requirements

Subpart I—Use and Management of Containers

Subpart J—Tanks

Authority: Secs. 1006, 2002(a) and 3004 of the Solid Waste Disposal Act, as amended by the Resource Conservation and Recovery Act of 1976, as amended (42 U.S.C. 6905, 6912(a), and 6924).

Subpart A—General

§ 265.1 Purpose, scope, and applicability.

(a) The purpose of this Part is to establish minimum national standards which define the acceptable management of hazardous waste during the period of interim status.

(b) The standards in this Part apply to owners and operators of facilities which treat, store, or dispose of hazardous waste who have fully complied with the requirements for interim status under Section 3005(e) of RCRA and § 122.22 of this Chapter, until final administrative disposition of their permit application is made. These standards apply to all treatment, storage, or disposal of hazardous waste at these facilities after the effective date of these regulations, except as specifically provided otherwise in this Part or Part 261 of this Chapter.

[*Comment:* As stated in Section 3005(a) of RCRA, after the effective date of regulations under that Section, i.e., Parts 122 and 124 of this Chapter, the treatment, storage, or disposal of hazardous waste is prohibited except in accordance with a permit. Section 3005(e) of RCRA provides for the continued operation of an existing facility which meets certain conditions until final administrative disposition of the owner's and operator's permit application is made.]

(c) The requirements of this Part do not apply to:

(1) A person disposing of hazardous waste by means of ocean disposal subject to a permit issued under the Marine Protection, Research, and Sanctuaries Act;

[*Comment:* These Part 265 regulations do apply to the treatment or storage of hazardous waste before it is loaded onto an ocean vessel for incineration or disposal at sea, as provided in paragraph (b) of this Section.]

(2) A person disposing of hazardous waste by means of underground injection subject to a permit issued under an Underground Injection Control (UIC) program approved or promulgated under the Safe Drinking Water Act;

[*Comment:* These Part 265 regulations do apply to the aboveground treatment or storage of hazardous waste before it is injected underground. These Part 265 regulations also apply to the disposal of hazardous waste by means of underground injection, as provided in paragraph (b) of this Section, until final administrative disposition of a person's permit application is made under RCRA or under an approved or promulgated UIC program.]

(3) The owner or operator of a POTW which treats, stores, or disposes of hazardous waste;

[*Comment:* The owner or operator of a facility under paragraphs (c)(1) through (c)(3) of this Section is subject to the requirements of Part 264 of this Chapter to the extent they are included in a permit by rule granted to such a person under Part 122 of this Chapter, or are required by § 122.45 of this Chapter.]

(4) A person who treats, stores, or disposes of hazardous waste in a State with a RCRA hazardous waste program authorized under Subparts A and B, or Subpart F, of Part 123 of this Chapter, except that the requirements of this Part will continue to apply as stated in paragraph (c)(2) of this Section, if the authorized State RCRA program does not cover disposal of hazardous waste by means of underground injection;

(5) The owner or operator of a facility permitted, licensed, or registered by a State to manage municipal or industrial solid waste, if the only hazardous waste the facility treats, stores, or disposes of is excluded from regulation under this Part by § 261.5 of this Chapter;

(6) The owner or operator of a facility which treats or stores hazardous waste, which treatment or storage meets the criteria in § 261.6(a) of this Chapter, except to the extent that § 261.6(b) of this Chapter provides otherwise;

(7) A generator accumulating waste on-site in compliance with § 262.34 of this Chapter, except to the extent the requirements are included in § 262.34 of this Chapter;

(8) A farmer disposing of waste pesticides from his own use in compliance with § 262.51 of this Chapter; or

(9) The owner or operator of a totally enclosed treatment facility, as defined in § 260.10.

§§ 265.2–265.3 [Reserved]

§ 265.4 Imminent hazard action.

Notwithstanding any other provisions of these regulations, enforcement actions may be brought pursuant to Section 7003 of RCRA.

§§ 265.5–265.9 [Reserved]

Subpart B—General Facility Standards

§ 265.10 Applicability

The regulations in this Subpart apply to owners and operators of all hazardous waste facilities, except as § 265.1 provides otherwise.

§ 265.11 Identification number.

Every facility owner or operator must apply to EPA for an EPA identification number in accordance with the EPA notification procedures (45 FR 12746).

§ 265.12 Required notices.

(a) The owner or operator of a facility that has arranged to receive hazardous waste from a foreign source must notify the Regional Administrator in writing at least four weeks in advance of the date of the waste is expected to arrive at the facility. Notice of subsequent shipments of the same waste from the same foreign source is not required.

(b) Before transferring ownership or operation of a facility during its operating life, or of a disposal facility during the post-closure care period, the owner or operator must notify the new owner or operator in writing of the requirements of this Part and Part 122 of this Chapter. (Also see § 122.23(c) of this Chapter.)

[Comment: An owner's or operator's failure to notify the new owner or operator of the requirements of this Part in no way relieves the new owner or operator of his obligation to comply with all applicable requirements.]

§ 265.13 General waste analysis.

(a)(1) Before an owner or operator treats, stores, or disposes of any hazardous waste, he must obtain a detailed chemical and physical analysis of a representative sample of the waste. At a minimum, this analysis must contain all the information which must be known to treat, store, or dispose of the waste in accordance with the requirements of this Part.

(2) The analysis may include data developed under Part 261 of this Chapter, and existing published or documented data on the hazardous waste or on waste generated from similar processes.

[Comment: For example, the facility's record of analyses performed on the waste before the effective date of these regulations, or studies conducted on hazardous waste generated from processes similar to that which generated the waste to be managed at the facility, may be included in the data base required to comply with paragraph (a)(1) of this Section. The owner or operator of an off-site facility may arrange for the generator of the hazardous waste to supply part or all of the information required by paragraph (a)(1) of this Section. If the generator does not supply the information, and the owner or operator chooses to accept a hazardous waste, the owner or operator is responsible for obtaining the information required to comply with this Section.]

(3) The analysis must be repeated as necessary to ensure that it is accurate and up to date. At a minimum, the analysis must be repeated:

(i) When the owner or operator is notified, or has reason to believe, that the process or operation generating the hazardous waste has changed; and

(ii) For off-site facilities, when the results of the inspection required in paragraph (a)(4) of this Section indicate that the hazardous waste received at the facility does not match the waste designated on the accompanying manifest or shipping paper.

(4) The owner or operator of an off-site facility must inspect and, if necessary, analyze each hazardous waste movement received at the facility to determine whether it matches the identity of the waste specified on the accompanying manifest or shipping paper.

(b) The owner or operator must develop and follow a written waste analysis plan which describes the procedures which he will carry out to comply with paragraph (a) of this Section. He must keep this plan at the facility. At a minimum, the plan must specify:

(1) The parameters for which each hazardous waste will be analyzed and the rationale for the selection of these parameters (i.e., how analysis for these parameters will provide sufficient information on the waste's properties to comply with paragraph (a) of this Section);

(2) The test methods which will be used to test for these parameters;

(3) The sampling method which will be used to obtain a representative sample of the waste to be analyzed. A representative sample may be obtained using either:

(i) One of the sampling methods described in Appendix I of Part 261 of this Chapter; or

(ii) An equivalent sampling method.

[Comment: See § 260.20(c) of this Chapter for related discussion.]

(4) The frequency with which the initial analysis of the waste will be reviewed or repeated to ensure that the analysis is accurate and up to date;

(5) For off-site facilities, the waste analyses that hazardous waste generators have agreed to supply; and

(6) Where applicable, the methods which will be used to meet the additional waste analysis requirements for specific waste management methods as specified in §§ 265.193, 265.225, 265.252, 265.273, 265.345, 265.375, and 265.402.

(c) For off-site facilities, the waste analysis plan required in paragraph (b) of this Section must also specify the procedures which will be used to inspect and, if necessary, analyze each movement of hazardous waste received at the facility to ensure that it matches the identity of the waste designated on the accompanying manifest or shipping paper. At a minimum, the plan must describe:

(1) The procedures which will be used to determine the identity of each movement of waste managed at the facility; and

(2) The sampling method which will be used to obtain a representative sample of the waste to be identified, if the identification method includes sampling.

§ 265.14 Security.

(a) The owner or operator must prevent the unknowing entry, and minimize the possibility for the unauthorized entry, of persons or livestock onto the active portion of his facility, unless:

(1) Physical contact with the waste, structures, or equipment with the active portion of the facility will not injure unknowing or unauthorized persons or livestock which may enter the active portion of a facility, and

(2) Disturbance of the waste or equipment, by the unknowing or unauthorized entry of persons or livestock onto the active portion of a facility, will not cause a violation of the requirements of this Part.

(b) Unless exempt under paragraphs (a)(1) and (a)(2) of this Section, a facility must have:

(1) A 24-hour surveillance system (e.g., television monitoring or surveillance by guards of facility personnel) which continuously monitors and controls entry onto the active portion of the facility; or

(2)(i) An artificial or natural barrier (e.g., a fence in good repair or a fence combined with a cliff), which completely surrounds the active portion of the facility; and

(ii) A means to control entry, at all times, through the gates or other entrances to the active portion of the facility (e.g., an attendant, television monitors, locked entrance, or controlled roadway access to the facility).

[Comment: The requirements of paragraph (b) of this Section are satisfied if the facility or plant within which the active portion is located itself has a surveillance system, or a barrier and a means to control entry, which complies with the requirements of paragraph (b)(1) or (b)(2) of this Section.]

(c) Unless exempt under paragraphs (a)(1) and (a)(2) of this Section, a sign with the legend, "Danger—Unauthorized Personnel Keep Out," must be posted at each entrance to the active portion of a facility, and at other locations, in sufficient numbers to be seen from any approach to this active portion. The legend must be written in English and in any other language predominant in the area surrounding the facility (e.g., facilities in counties bordering the Canadian province of Quebec must post signs in French; facilities in counties bordering Mexico must post signs in Spanish), and must be legible from a distance of at least 25 feet. Existing signs with a legend other than "Danger—Unauthorized Personnel Keep Out" may be used if the legend on the sign indicates that only authorized personnel are allowed to enter the active portion, and that entry onto the active portion can be dangerous.

[Comment: See § 265.117(b) for discussion of security requirements at disposal facilities during the post-closure care period.]

§ 265.15 General inspection requirements.

(a) The owner or operator must inspect his facility for malfunctions and deterioration, operator errors, and discharges which may be causing—or

may lead to—(1) release of hazardous waste constituents to the environment or (2) a threat to human health. The owner or operator must conduct these inspections often enough to identify problems in time to correct them before they harm human health or the environment.

(b)(1) The owner or operator must develop and follow a written schedule for inspecting all monitoring equipment, safety and emergency equipment, security devices, and operating and structural equipment (such as dikes and sump pumps) that are important to preventing, detecting, or responding to environmental or human health hazards.

(2) He must keep this schedule at the facility.

(3) The schedule must identify the types of problems (e.g., malfunctions or deterioration) which are to be looked for during the inspection (e.g., inoperative sump pump, leaking fitting, eroding dike, etc.).

(4) The frequency of inspection may vary for the items on the schedule. However, it should be based on the rate of possible deterioration of the equipment and the probability of an environmental or human health incident if the deterioration or malfunction or any operator error goes undetected between inspections. Areas subject to spills, such as loading and unloading areas, must be inspected daily when in use. At a minimum, the inspection schedule must include the items and frequencies called for in §§ 265.174, 265.194, 265.226, 265.347, 265.377, and 265.403.

(c) The owner or operator must remedy any deterioration or malfunction of equipment or structures which the inspection reveals on a schedule which ensures that the problem does not lead to an environmental or human health hazard. Where a hazard is imminent or has already occurred, remedial action must be taken immediately.

(d) The owner or operator must record inspections in an inspection log or summary. He must keep these records for at least three years from the date of inspection. At a minimum, these records must include the date and time of the inspection, the name of the inspector, a notation of the observations made, and the date and nature of any repairs or other remedial actions.

§ 265.16 Personnel training.

(a)(1) Facility personnel must successfully complete a program of classroom instruction or on-the-job training that teaches them to perform their duties in a way that ensures the facility's compliance with the requirements of this Part. The owner or operator must ensure that this program

includes all the elements described in the document required under paragraph (d)(3) of this Section.

(2) This program must be directed by a person trained in hazardous waste management procedures, and must include instruction which teaches facility personnel hazardous waste management procedures (including contingency plan implementation) relevant to the positions in which they are employed.

(3) At a minimum, the training program must be designed to ensure that facility personnel are able to respond effectively to emergencies by familiarizing them with emergency procedures, emergency equipment, and emergency systems, including where applicable:

(i) Procedures for using, inspecting, repairing, and replacing facility emergency and monitoring equipment;

(ii) Key parameters for automatic waste feed cut-off systems;

(iii) Communications or alarm systems;

(iv) Response to fires or explosions;

(v) Response to ground-water contamination incidents; and

(vi) Shutdown of operations.

(b) Facility personnel must successfuly complete the program required in paragraph (a) of this Section within six months after the effective date of these regulations or six months after the date of their employment or assignment to a facility, or to a new position at a facility, whichever is later. Employees hired after the effective date of these regulations must not work in unsupervised positions until they have completed the training requirements of paragraph (a) of this Section.

(c) Facility personnel must take part in an annual review of the initial training required in paragraph (a) of this Section.

(d) The owner or operator must maintain the following documents and records at the facility:

(1) The job title for each position at the facility related to hazardous waste management, and the name of the employee filling each job;

(2) A written job description for each position listed under paragraph (d)(1) of this Section. This description may be consistent in its degree of specificity with descriptions for other similar positions in the same company location or bargaining unit, but must include the requisite skill, education, or other qualifications, and duties of facility personnel assigned to each position;

(3) A written description of the type and amount of both introductory and continuing training that will be given to each person filling a position listed under paragraph (d)(1) of this Section;

(4) Records that document that the training or job experience required under paragraphs (a), (b), and (c) of this Section has been given to, and completed by, facility personnel.

(e) Training records on current personnel must be kept until closure of the facility. Training records on former employees must be kept for at least three years from the date the employee last worked at the facility. Personnel training records may accompany personnel transferred within the same company.

§ 265.17   General requirements for ignitable, reactive, or incompatible wastes.

(a) The owner or operator must take precautions to prevent accidental ignition or reaction of ignitable or reactive waste. This waste must be separated and protected from sources of ignition or reaction including but not limited to: open flames, smoking, cutting and welding, hot surfaces, frictional heat, sparks (static, electrical, or mechanical), spontaneous ignition (e.g., from heat-producing chemical reactions), and radiant heat. While ignitable or reactive waste is being handled, the owner or operator must confine smoking and open flame to specially designated locations. "No Smoking" signs must be conspicuously placed wherever there is a hazard from ignitable or reactive waste.

(b) Where specifically required by other Sections of this Part, the treatment, storage, or disposal of ignitable or reactive waste, and the mixture or commingling of incompatible wastes, or incompatible wastes and materials, must be conducted so that it does not:

(1) Generate extreme heat or pressure, fire or explosion, or violent reaction;

(2) Produce uncontrolled toxic mists, fumes, dusts, or gases in sufficient quantities to threaten human health;

(3) Produce uncontrolled flammable fumes or gases in sufficient quantities to pose a risk of fire or explosions;

(4) Damage the structural integrity of the device or facility containing the waste; or

(5) Through other like means threaten human health or the environment.

§§ 265.18–265.29   [Reserved]

Subpart C—Preparedness and Prevention

§ 265.30   Applicability.

The regulations in this Subpart apply to owners and operators of all hazardous waste facilities, except as § 265.1 provides otherwise.

§ 265.31   Maintenance and operation of facility.

Facilities must be maintained and operated to minimize the possibility of a fire, explosion, or any unplanned sudden or non-sudden release of hazardous waste or hazardous waste constituents to air, soil, or surface water which could threaten human health or the environment.

§ 265.32   Required equipment.

All facilities must be equipped with the following, *unless* none of the hazards posed by waste handled at the facility could require a particular kind of equipment specified below:

(a) An internal communications or alarm system capable of providing immediate emergency instruction (voice or signal) to facility personnel;

(b) A device, such as a telephone (immediately available at the scene of operations) or a hand-held two-way radio, capable of summoning emergency assistance from local police departments, fire departments, or State or local emergency response teams;

(c) Portable fire extinguishers, fire control equipment (including special extinguishing equipment, such as that using foam, inert gas, or dry chemicals), spill control equipment, and decontamination equipment; and

(d) Water at adequate volume and pressure to supply water hose streams, or foam producing equipment, or automatic sprinklers, or water spray systems.

§ 265.33   Testing and maintenance of equipment.

All facility communications or alarm systems, fire protection equipment, spill control equipment, and decontamination equipment, where required, must be tested and maintained as necessary to assure its proper operation in time of emergency.

§ 265.34   Access to communications or alarm system.

(a) Whenever hazardous waste is being poured, mixed, spread, or otherwise handled, all personnel involved in the operation must have immediate access to an internal alarm or emergency communication device, either directly or through visual or voice contact with another employee, *unless* such a device is not required under § 265.32.

(b) If there is ever just one employee on the premises while the facility is operating, he must have immediate access to a device, such as a telephone (immediately available at the scene of operation) or a hand-held two-way radio, capable of summoning external emergency assistance, *unless* such a device is not required under § 265.32.

§ 265.35   Required aisle space.

The owner or operator must maintain aisle space to allow the unobstructed movement of personnel, fire protection equipment, spill control equipment, and decontamination equipment to any area of facility operation in an emergency, *unless* aisle space is not needed for any of these purposes.

§ 265.36   [Reserved]

§ 265.37   Arrangements with local authorities.

(a) The owner or operator must attempt to make the following arrangements, as appropriate for the type of waste handled at his facility and the potential need for the services of these organizations:

(1) Arrangements to familiarize police, fire departments, and emergency response teams with the layout of the facility, properties of hazardous waste handled at the facility and associated hazards, places where facility personnel would normally be working, entrances to roads inside the facility, and possible evacuation routes;

(2) Where more than one police and fire department might respond to an emergency, agreements designating primary emergency authority to a specific police and a specific fire department, and agreements with any others to provide support to the primary emergency authority;

(3) Agreements with State emergency response teams, emergency response contractors, and equipment suppliers; and

(4) Arrangements to familiarize local hospitals with the properties of hazardous waste handled at the facility and the types of injuries or illnesses which could result from fires, explosions, or releases at the facility.

(b) Where State or local authorities decline to enter into such arrangements, the owner or operator must document the refusal in the operating record.

§ 265.38–265.49   [Reserved]

Subpart D—Contingency Plan and Emergency Procedures

§ 265.50   Applicability.

The regulations in this Subpart apply to owners and operators of all hazardous waste facilities, except as § 265.1 provides otherwise.

§ 265.51   Purpose and implementation of contingency plan.

(a) Each owner or operator must have a contingency plan for his facility. The contingency plan must be designed to minimize hazards to human health or the environment from fires, explosions, or any unplanned sudden or non-sudden release of hazardous waste or hazardous waste constituents to air, soil, or surface water.

(b) The provisions of the plan must be carried out immediately whenever there is a fire, explosion, or release of hazardous waste or hazardous waste constituents which could threaten human health or the environment.

§ 265.52 Content of contingency plan.

(a) The contingency plan must describe the actions facility personnel must take to comply with §§ 265.51 and 265.56 in response to fires, explosions, or any unplanned sudden or non-sudden release of hazardous waste or hazardous waste constituents to air, soil, or surface water at the facility.

(b) If the owner or operator has already prepared a Spill Prevention, Control, and Countermeasures (SPCC) Plan in accordance with Part 112 or Part 151 of this Chapter, or some other emergency or contingency plan, he need only amend that plan to incorporate hazardous waste management provisions that are sufficient to comply with the requirements of this Part.

(c) The plan must describe arrangements agreed to by local police departments, fire departments, hospitals, contractors, and State and local emergency response teams to coordinate emergency services, pursuant to § 265.37.

(d) The plan must list names, addresses, and phone numbers (office and home) of all persons qualified to act as emergency coordinator (see § 265.55), and this list must be kept up to date. Where more than one person is listed, one must be named as primary emergency coordinator and others must be listed in the order in which they will assume responsibility as alternates.

(e) The plan must include a list of all emergency equipment at the facility (such as fire extinguishing systems, spill control equipment, communications and alarm systems (internal and external), and decontamination equipment), where this equipment is required. This list must be kept up to date. In addition, the plan must include the location and a physical description of each item on the list, and a brief outline of its capabilities.

(f) The plan must include an evacuation plan for facility personnel where there is a possibility that evacuation could be necessary. This plan must describe signal(s) to be used to begin evacuation, evacuation routes, and alternate evacuation routes (in cases where the primary routes could be blocked by releases of hazardous waste or fires).

§ 265.53 Copies of contingency plan.

A copy of the contingency plan and all revisions to the plan must be:

(a) Maintained at the facility; and

(b) Submitted to all local police departments, fire departments, hospitals, and State and local emergency response teams that may be called upon to provide emergency services.

§ 265.54 Amendment of contingency plan.

The contingency plan must be reviewed, and immediately amended, if necessary, whenever:

(a) Applicable regulations are revised;

(b) The plan fails in an emergency;

(c) The facility changes—in its design, construction, operation, maintenance, or other circumstances—in a way that materially increases the potential for fires, explosions, or releases of hazardous waste or hazardous waste constituents, or changes the response necessary in an emergency;

(d) The list of emergency coordinators changes; or

(e) The list of emergency equipment changes.

§ 265.55 Emergency coordinator.

At all times, there must be at least one employee either on the facility premises or on call (i.e., available to respond to an emergency by reaching the facility within a short period of time) with the responsibility for coordinating all emergency response measures. This emergency coordinator must be thoroughly familiar with all aspects of the facility's contingency plan, all operations and activities at the facility, the location and characteristics of waste handled, the location of all records within the facility, and the facility layout. In addition, this person must have the authority to commit the resources needed to carry out the contingency plan.

[Comment: The emergency coordinator's responsibilities are more fully spelled out in § 265.56. Applicable responsibilities for the emergency coordinator vary, depending on factors such as type and variety of waste(s) handled by the facility, and type and complexity of the facility.]

§ 265.56 Emergency procedures.

(a) Whenever there is an imminent or actual emergency situation, the emergency coordinator (or his designee when the emergency coordinator is on call) must immediately:

(1) Activate internal facility alarms or communication systems, where applicable, to notify all facility personnel; and

(2) Notify appropriate State or local agencies with designated response roles if their help is needed.

(b) Whenever there is a release, fire, or explosion, the emergency coordinator must immediately identify the character, exact source, amount, and a real extent of any released materials. He may do this by observation or review of facility records or manifests and, if necessary, by chemical analysis.

(c) Concurrently, the emergency coordinator must assess possible hazards to human health or the environment that may result from the release, fire, or explosion. This assessment must consider both direct and indirect effects of the release, fire, or explosion (e.g., the effects of any toxic, irritating, or asphyxiating gases that are generated, or the effects of any hazardous surface water run-offs from water or chemical agents used to control fire and heat-induced explosions).

(d) If the emergency coordinator determines that the facility has had a release, fire, or explosion which could threaten human health, or the environment, outside the facility, he must report his findings as follows:

(1) If his assessment indicates that evacuation of local areas may be advisable, he must immediately notify appropriate local authorities. He must be available to help appropriate officials decide whether local areas should be evacuated; and

(2) He must immediately notify either the government official designated as the on-scene coordinator for that geographical area (in the applicable regional contingency plan under Part 1510 of this Title), or the National Response Center (using their 24-hour toll free number 800/424–8802). The report must include:

(i) Name and telephone number of reporter;

(ii) Name and address of facility;

(iii) Time and type of incident (e.g., release, fire);

(iv) Name and quantity of material(s) involved, to the extent known;

(v) The extent of injuries, if any; and

(vi) The possible hazards to human health, or the environment, outside the facility.

(e) During an emergency, the emergency coordinator must take all reasonable measures necessary to ensure that fires, explosions, and releases do not occur, recur, or spread to other hazardous waste at the facility. These measures must include, where applicable, stopping processes and operations, collecting and containing released waste, and removing or isolating containers.

(f) If the facility stops operations in response to a fire, explosion or release, the emergency coordinator must monitor for leaks, pressure buildup, gas generation, or ruptures in valves, pipes, or other equipment, wherever this is appropriate.

(g) Immediately after an emergency, the emergency coordinator must provide for treating, storing, or disposing of recovered waste, contaminated soil or surface water, or any other material that results from a release, fire, or explosion at the facility.

[*Comment:* Unless the owner or operator can demonstrate, in accordance with § 261.3(c) or (d) of this Chapter, that the recovered material is not a hazardous waste, the owner or operator becomes a generator of hazardous waste and must manage it in accordance with all applicable requirements of Parts 262, 263, and 265 of this Chapter.]

(h) The emergency coordinator must ensure that, in the affected area(s) of the facility:

(1) No waste that may be incompatible with the released material is treated, stored, or disposed of until cleanup procedures are completed; and

(2) All emergency equipment listed in the contingency plan is cleaned and fit for its intended use before operations are resumed.

(i) The owner or operator must notify the Regional Administrator, and appropriate State and local authorities, that the facility is in compliance with paragraph (h) of this Section before operations are resumed in the affected area(s) of the facility.

(j) The owner or operator must note in the operating record the time, date, and details of any incident that requires implementing the contingency plan. Within 15 days after the incident, he must submit a written report on the incident to the Regional Administrator. The report must include:

(1) Name, address, and telephone number of the owner or operator;

(2) Name, address, and telephone number of the facility;

(3) Date, time, and type of incident (e.g., fire, explosion);

(4) Name and quantity of material(s) involved;

(5) The extent of injuries, if any;

(6) An assessment of actual or potential hazards to human health or the environment, where this is applicable; and

(7) Estimated quantity and disposition of recovered material that resulted from the incident.

§§ 265.57–265.69 [Reserved]

Subpart E—Manifest System, Recordkeeping, and Reporting

§ 265.70 Applicability.

The regulations in this Subpart apply to owners and operators of both on-site and off-site facilities, except as § 265.1 provides otherwise. Sections 265.71, 265.72, and 265.76 do not apply to owners and operators of on-site facilities that do not receive any hazardous waste from off-site sources.

§ 265.71 Use of manifest system.

(a) If a facility receives hazardous waste accompanied by a manifest, the owner or operator, or his agent, must:

(1) Sign and date each copy of the manifest to certify that the hazardous waste covered by the manifest was received;

(2) Note any significant discrepancies in the manifest (as defined in § 265.72(a)) on each copy of the manifest;

[*Comment:* The Agency does not intend that the owner or operator of a facility whose procedures under § 265.13(c) include waste analysis must perform that analysis before signing the manifest and giving it to the transporter. Section 265.72(b), however, requires reporting an unreconciled discrepancy discovered during later analysis.]

(3) Immediately give the transporter at least one copy of the signed manifest;

(4) Within 30 days after the delivery, send a copy of the manifest to the generator; and

(5) Retain at the facility a copy of each manifest for at least three years from the date of delivery.

(b) If a facility receives, from a rail or water (bulk shipment) transporter, hazardous waste which is accompanied by a shipping paper containing all the information required on the manifest (excluding the EPA identification numbers, generator's certification, and signatures), the owner or operator, or his agent, must:

(1) Sign and date each copy of the shipping paper to certify that the hazardous waste covered by the shipping paper was received;

(2) Note any significant discrepancies in the shipping paper (as defined in § 265.72(a)) on each copy of the shipping paper;

[*Comment:* The Agency does not intend that the owner or operator of a facility whose procedures under § 265.13(c) include waste analysis must perform that analysis before signing the shipping paper and giving it to the transporter. Section 265.72(b), however, requires

reporting an unreconciled discrepancy discovered during later analysis.]

(3) Immediately give the rail or water (bulk shipment) transporter at least one copy of the shipping paper;

(4) Within 30 days after the delivery, send a copy of the shipping paper to the generator; however, if the manifest is received within 30 days after the delivery, the owner or operator, or his agent, must sign and date the manifest and return it to the generator in lieu of the shipping paper; and

[*Comment:* Section 262.23(c) of this Chapter requires the generator to send three copies of the manifest to the facility when hazardous waste is sent by rail or water (bulk shipment).]

(5) Retain at the facility a copy of each shipping paper and manifest for at least three years from the date of delivery.

§ 265.72 Manifest discrepancies.

(a) Manifest discrepancies are differences between the quantity or type of hazardous waste designated on the manifest or shipping paper, and the quantity or type of hazardous waste a facility actually receives. Significant discrepancies in quantity are: (1) for bulk waste, variations greater than 10 percent in weight, and (2) for batch waste, any variation in piece count, such as a discrepancy of one drum in a truckload. Significant discrepancies in type are obvious differences which can be discovered by inspection or waste analysis, such as waste solvent substituted for waste acid, or toxic constituents not reported on the manifest or shipping paper.

(b) Upon discovering a significant discrepancy, the owner or operator must attempt to reconcile the discrepancy with the waste generator or transporter (e.g., with telephone conversations). If the discrepancy is not resolved within 15 days after receiving the waste, the owner or operator must immediately submit to the Regional Administrator a letter describing the discrepancy and attempts to reconcile it, and a copy of the manifest or shipping paper at issue.

§ 265.73 Operating record.

(a) The owner or operator must keep a written operating record at his facility.

(b) The following information must be recorded, as it becomes available, and maintained in the operating record until closure of the facility:

(1) A description and the quantity of each hazardous waste received, and the method(s) and date(s) of its treatment, storage, or disposal at the facility as required by Appendix I;

(2) The location of each hazardous waste within the facility and the

quantity at each location. For disposal facilities, the location and quantity of each hazardous waste must be recorded on a map or diagram of each cell or disposal area. For all facilities, this information must include cross-references to specific manifest document numbers, if the waste was accompanied by a manifest;

[Comment: See §§ 265.119, 265.279, and 265.309 for related requirements.]

(3) Records and results of waste analyses and trial tests performed as specified in §§ 265.13, 265.193, 265.225, 265.252, 265.273, 265.345, 265.375, and 265.402;

(4) Summary reports and details of all incidents that require implementing the contingency plan as specified in § 265.56(j);

(5) Records and results of inspections as required by § 265.15(d) (except these data need be kept only three years);

(6) Monitoring, testing, or analytical data where required by §§ 265.90, 265.94, 265.276, 265.278, 265.280(d)(1), 265.347, and 265.377; and,

[Comment: As required by § 265.94, monitoring data at disposal facilities must be kept throughout the post-closure period.]

(7) All closure cost estimates under § 265.142 and, for disposal facilities, all post-closure cost estimates under § 265.144.

§ 265.74 Availability, retention, and disposition of records.

(a) All records, including plans, required under this Part must be furnished upon request, and made available at all reasonable times for inspection, by any officer, employee, or representative of EPA who is duly designated by the Administrator.

(b) The retention period for all records required under this Part is extended automatically during the course of any unresolved enforcement action regarding the facility or as requested by the Administrator.

(c) A copy of records of waste disposal locations and quantities under § 265.73(b)(2) must be submitted to the Regional Administrator and local land authority upon closure of the facility (see § 265.119).

§ 265.75 Annual report.

The owner or operator must prepare and submit a single copy of an annual report to the Regional Administrator by March 1 of each year. The report form and instructions in Appendix II must be used for this report. The annual report must cover facility activities during the previous calendar year and must include the following information:

(a) The EPA identification number, name, and address of the facility;

(b) The calendar year covered by the report;

(c) For off-site facilities, the EPA identification number of each hazardous waste generator from which the facility received a hazardous waste during the year; for imported shipments, the report must give the name and address of the foreign generator;

(d) A description and the quantity of each hazardous waste the facility received during the year. For off-site facilities, this information must be listed by EPA identification number of each generator;

(e) The method of treatment, storage, or disposal for each hazardous waste;

(f) Monitoring data under § 265.94(a)(2)(ii) and (iii), and (b)(2), where required;

(g) The most recent closure cost estimate under § 265.142, and, for disposal facilities, the most recent post-closure cost estimate under § 265.144; and

(h) The certification signed by the owner or operator of the facility or his authorized representative.

§ 265.76 Unmanifested waste report.

If a facility accepts for treatment, storage, or disposal any hazardous waste from an off-site source without an accompanying manifest, or without an accompanying shipping paper as described in § 263.20(e)(2) of this Chapter, and if the waste is not excluded from the manifest requirement by § 261.5 of this Chapter, then the owner or operator must prepare and submit a single copy of a report to the Regional Administrator within 15 days after receiving the waste. The report form and instructions in Appendix II must be used for this report. The report must include the following information:

(a) The EPA identification number, name, and address of the facility;

(b) The date the facility received the waste;

(c) The EPA identification number, name, and address of the generator and the transporter, if available;

(d) A description and the quantity of each unmanifested hazardous waste the facility received;

(e) The method of treatment, storage, or disposal for each hazardous waste;

(f) The certification signed by the owner or operator of the facility or his authorized representative; and

(g) A brief explanation of why the waste was unmanifested, if known.

[Comment: Small quantities of hazardous waste are excluded from regulation under this Part and do not

require a manifest. Where a facility receives unmanifested hazardous wastes, the Agency suggests that the owner or operator obtain from each generator a certification that the waste qualifies for exclusion. Otherwise, the Agency suggests that the owner or operator file an unmanifested waste report for the hazardous waste movement.]

§ 265.77 Additional reports.

In addition to submitting the annual report and unmanifested waste reports described in §§ 265.75 and 265.76, the owner or operator must also report to the Regional Administrator:

(a) Releases, fires, and explosions as specified in § 265.56(j);

(b) Ground-water contamination and monitoring data as specified in §§ 265.93 and 265.94; and

(c) Facility closure as specified in § 265.115.

§§ 265.78–265.89 [Reserved]

Subpart F—Ground-Water Monitoring

§ 265.90 Applicability.

(a) Within one year after the effective date of these regulations, the owner or operator of a surface impoundment, landfill, or land treatment facility which is used to manage hazardous waste must implement a ground-water monitoring program capable of determining the facility's impact on the quality of ground water in the uppermost aquifer underlying the facility, except as § 265.1 and paragraph (c) of this Section provide otherwise.

(b) Except as paragraphs (c) and (d) of this Section provide otherwise, the owner or operator must install, operate, and maintain a ground-water monitoring system which meets the requirements of § 265.91, and must comply with §§ 265.92–265.94. This ground-water monitoring program must be carried out during the active life of the facility, and for disposal facilities, during the post-closure care period as well.

(c) All or part of the ground-water monitoring requirements of this Subpart may be waived if the owner or operator can demonstrate that there is a low potential for migration of hazardous waste or hazardous waste constituents from the facility via the uppermost aquifer to water supply wells (domestic, industrial, or agricultural) or to surface water. This demonstration must be in writing, and must be kept at the facility. This demonstration must be certified by a qualified geologist or geotechnical engineer and must establish the following:

(1) The potential for migration of hazardous waste or hazardous waste constituents from the facility to the

uppermost aquifer, by an evaluation of:

(i) A water balance of precipitation, evapotranspiration, runoff, and infiltration; and

(ii) Unsaturated zone characteristics (i.e., geologic materials, physical properties, and depth to ground water); and

(2) The potential for hazardous waste or hazardous waste constituents which enter the uppermost aquifer to migrate to a water supply well or surface water, by an evaluation of:

(i) Saturated zone characteristics (i.e., geologic materials, physical properties, and rate of ground-water flow); and

(ii) The proximity of the facility to water supply wells or surface water.

(d) If an owner or operator assumes (or knows) that ground-water monitoring of indicator parameters in accordance with §§265.91 and 265.92 would show statistically significant increases (or decreases in the case of pH) when evaluated under § 265.93(b), he may, install, operate, and maintain an alternate ground-water monitoring system (other than the one described in §§ 265.91 and 265.92). If the owner or operator decides to use an alternate ground-water monitoring system he must:

(1) Within one year after the effective date of these regulations, submit to the Regional Administrator a specific plan, certified by a qualified geologist or geotechnical engineer, which satisfies the requirements of § 265.93(d)(3), for an alternate ground-water monitoring system;

(2) Not later than one year after the effective date of these regulations, initiate the determinations specified in § 265.93(d)(4);

(3) Prepare and submit a written report in accordance with § 265.93(d)(5);

(4) Continue to make the determinations specified in § 265.93(d)(4) on a quarterly basis until final closure of the facility; and

(5) Comply with the recordkeeping and reporting requirements in § 265.94(b).

§ 265.91 Ground-water monitoring system.

(a) A ground-water monitoring system must be capable of yielding ground-water samples for analysis and must consist of:

(1) Monitoring wells (at least one) installed hydraulically upgradient (i.e., in the direction of increasing static head) from the limit of the waste management area. Their number, locations, and depths must be sufficient to yield ground-water samples that are:

(i) Representative of background ground-water quality in the uppermost aquifer near the facility; and

(ii) Not affected by the facility; and

(2) Monitoring wells (at least three) installed hydraulically downgradient (i.e., in the direction of decreasing static head) at the limit of the waste management area. Their number, locations, and depths must ensure that they immediately detect any statistically significant amounts of hazardous waste or hazardous waste constituents that migrate from the waste management area to the uppermost aquifer.

(b) Separate monitoring systems for each waste management component of a facility are not required provided that provisions for sampling upgradient and downgradient water quality will detect any discharge from the waste management area.

(1) In the case of a facility consisting of only one surface impoundment, landfill, or land treatment area, the waste management area is described by the waste boundary (perimeter).

(2) In the case of a facility consisting of more than one surface impoundment, landfill, or land treatment area, the waste management area is described by an imaginary boundary line which circumscribes the several waste management components.

(c) All monitoring wells must be cased in a manner that maintains the integrity of the monitoring well bore hole. This casing must be screened or perforated, and packed with gravel or sand where necessary, to enable sample collection at depths where appropriate aquifer flow zones exist. The annular space (i.e., the space between the bore hole and well casing) above the sampling depth must be sealed with a suitable material (e.g., cement grout or bentonite slurry) to prevent contamination of samples and the ground water.

§ 265.92 Sampling and analysis.

(a) The owner or operator must obtain and analyze samples from the installed ground-water monitoring system. The owner or operator must develop and follow a ground-water sampling and analysis plan. He must keep this plan at the facility. The plan must include procedures and techniques for:

(1) Sample collection;

(2) Sample preservation and shipment;

(3) Analytical procedures; and

(4) Chain of custody control.

[Comment: See "Procedures Manual For Ground-water Monitoring At Solid Waste Disposal Facilities," EPA–530/SW–611, August 1977 and "Methods for Chemical Analysis of Water and Wastes," EPA–600/4–79–020, March 1979 for discussions of sampling and analysis procedures.]

(b) The owner or operator must determine the concentration or value of

the following parameters in ground-water samples in accordance with paragraphs (c) and (d) of this section:

(1) Parameters characterizing the suitability of the ground water as a drinking water supply, as specified in Appendix III.

(2) Parameters establishing ground-water quality:

(i) Chloride

(ii) Iron

(iii) Manganese

(iv) Phenols

(v) Sodium

(vi) Sulfate

[Comment: These parameters are to be used as a basis for comparison in the event a ground-water quality assessment is required under § 265.93(d).]

(3) Parameters used as indicators of ground-water contamination:

(i) pH

(ii) Specific Conductance

(iii) Total Organic Carbon

(iv) Total Organic Halogen

(c)(1) For all monitoring wells, the owner or operator must establish initial background concentrations or values of all parameters specified in paragraph (b) of this Section. He must do this quarterly for one year.

(2) For each of the indicator parameters specified in paragraph (b)(3) of this Section, at least four replicate measurements must be obtained for each sample and the initial background arithmetic mean and variance must be determined by pooling the replicate measurements for the respective parameter concentrations or values in samples obtained from upgradient wells during the first year.

(d) After the first year, all monitoring wells must be sampled and the samples analyzed with the following frequencies:

(1) Samples collected to establish ground-water quality must be obtained and analyzed for the parameters specified in paragraph (b)(2) of this Section at least annually.

(2) Samples collected to indicate ground-water contamination must be obtained and analyzed for the parameters specified in paragraph (b)(3) of this Section at least semi-annually.

(e) Elevation of the ground-water surface at each monitoring well must be determined each time a sample is obtained.

§ 265.93 Preparation, evaluation, and response.

(a) Within one year after the effective date of these regulations, the owner or operator must prepare an *outline* of a ground-water quality assessment program. The outline must describe a more comprehensive ground-water monitoring program (than that described in §§ 265.91 and 265.92) capable of determining:

(1) Whether hazardous waste or hazardous waste constituents have entered the ground water;

(2) The rate and extent of migration of hazardous waste or hazardous waste constituents in the ground water; and

(3) The concentrations of hazardous waste or hazardous waste constituents in the ground water.

(b) For each indicator parameter specified in § 265.92(b)(3), the owner or operator must calculate the arithmetic mean and variance, based on at least four replicate measurements on each sample, for each well monitored in accordance with § 265.92(d)(2), and compare these results with its initial background arithmetic mean. The comparison must consider individually each of the wells in the monitoring system, and must use the Student's t-test at the 0.01 level of significance (see Appendix IV) to determine statistically significant increases (and decreases, in the case of pH) over initial background.

(c)(1) If the comparisons for the *upgradient* wells made under paragraph (b) of this Section show a significant increase (or pH decrease), the owner or operator must submit this information in accordance with § 265.94(a)(2)(ii).

(2) If the comparisons for *downgradient* wells made under paragraph (b) of this Section show a significant increase (or pH decrease), the owner or operator must then immediately obtain additional ground-water samples from those downgradient wells where a significant difference was detected, split the samples in two, and obtain analyses of all additional samples to determine whether the significant difference was a result of laboratory error.

(d)(1) If the analyses performed under paragraph (c)(2) of this Section confirm the significant increase (or pH decrease), the owner or operator must provide written notice to the Regional Administrator—within seven days of the date of such confirmation—that the facility may be affecting ground-water quality.

(2) Within 15 days after the notification under paragraph (d)(1) of this Section, the owner or operator must develop and submit to the Regional Administrator a specific plan, based on the outline required under paragraph (a) of this Section and certified by a qualified geologist or geotechnical engineer, for a ground-water quality assessment program at the facility.

(3) The plan to be submitted under § 265.90(d)(1) or paragraph (d)(2) of this Section must specify:

(i) The number, location, and depth of wells;

(ii) Sampling and analytical methods for those hazardous wastes or hazardous waste constituents in the facility;

(iii) Evaluation procedures, including any use of previously-gathered ground-water quality information; and

(iv) A schedule of implementation.

(4) The owner or operator must implement the ground-water quality assessment plan which satisfies the requirements of paragraph (d)(3) of this Section, and, at a minimum, determine:

(i) The rate and extent of migration of the hazardous waste or hazardous waste constituents in the ground water; and

(ii) The concentrations of the hazardous waste or hazardous waste constituents in the ground water.

(5) The owner or operator must make his first determination under paragraph (d)(4) of this Section as soon as technically feasible, and, within 15 days after that determination, submit to the Regional Administrator a written report containing an assessment of the ground-water quality.

(6) If the owners or operator determines, based on the results of the first determination under paragraph (d)(4) of this Section, that no hazardous waste or hazardous waste constituents from the facility have entered the ground water, then he may reinstate the indicator evaluation program described in § 265.92 and paragraph (b) of this Section. If the owner or operator reinstates the indicator evaluation program, he must so notify the Regional Administrator in the report submitted under paragraph (d)(5) of this Section.

(7) If the owner or operator determines, based on the first determination under paragraph (d)(4) of this Section, that hazardous waste or hazardous waste constituents from the facility have entered the ground water, then he:

(i) Must continue to make the determinations required under paragraph (d)(4) of this Section on a quarterly basis until final closure of the facility, if the ground-water quality assessment plan was implemented prior to final closure of the facility; or

(ii) May cease to make the determinations required under paragraph (d)(4) of this Section, if the ground-water quality assessment plan was implemented during the post-closure care period.

(e) Notwithstanding any other provision of this Subpart, any ground-water quality assessment to satisfy the requirements of § 265.93(d)(4) which is initiated prior to final closure of the facility must be completed and reported in accordance with § 265.93(d)(5).

(f) Unless the ground water is monitored to satisfy the requirements of § 265.93(d)(4), at least annually the owner or operator must evaluate the data on ground-water surface elevations obtained under § 265.92(e) to determine whether the requirements under § 265.91(a) for locating the monitoring wells continues to be satisfied. If the evaluation shows that § 265.91(a) is no longer satisfied, the owner or operator must immediately modify the number, location, or depth of the monitoring wells to bring the ground-water monitoring system into compliance with this requirement.

§ 265.94 Recordkeeping and reporting.

(a) Unless the ground water is monitored to satisfy the requirements of § 265.93(d)(4), the owner or operator must:

(1) Keep records of the analyses required in § 265.92(c) and (d), the associated ground-water surface elevations required in § 265.92(e), and the evaluations required in § 265.93(b) throughout the active life of the facility, and, for disposal facilities, throughout the post-closure care period as well; and

(2) Report the following ground-water monitoring information to the Regional Administrator:

(i) During the first year when initial background concentrations are being established for the facility: concentrations or values of the parameters listed in § 265.92(b)(1) for each ground-water monitoring well within 15 days after completing each quarterly analysis. The owner or operator must separately identify for each monitoring well any parameters whose concentration or value has been found to exceed the maximum contaminant levels listed in Appendix III.

(ii) Annually: concentrations or values of the parameters listed in § 265.92(b)(3) for each ground-water monitoring well, along with the required evaluations for these parameters under § 265.93(b). The owner or operator must separately identify any significant differences from initial background found in the upgradient wells, in accordance with § 265.93(c)(1). During the active life of the facility, this information must be submitted as part of the annual report required under § 265.75.

(iii) As a part of the annual report required under § 265.75: results of the evaluation of ground-water surface elevations under § 265.93(f), and a description of the response to that evaluation, where applicable.

(b) If the ground water is monitored to satisfy the requirements of § 265.93(d)(4), the owner or operator must:

(1) Keep records of the analyses and evaluations specified in the plan, which satisfies the requirements of § 265.93(d)(3), throughout the active life of the facility, and, for disposal facilities, throughout the post-closure care period as well; and

(2) Annually, until final closure of the facility, submit to the Regional Administrator a report containing the results of his ground-water quality assessment program which includes, but is not limited to, the calculated (or measured) rate of migration of hazardous waste or hazardous waste constituents in the ground water during the reporting period. This report must be submitted as part of the annual report required under § 265.75.

§§ 265.95–265.109   [Reserved]

Subpart G—Closure and Post-Closure

§ 265.110   Applicability.

Except as § 265.1 provides otherwise:

(a) Sections 265.111–265.115 (which concern closure) apply to the owners and operators of all hazardous waste facilities; and

(b) Sections 265.117–265.120 (which concern post-closure care) apply to the owners and operators of all disposal facilities.

§ 265.111   Closure performance standard.

The owner or operator must close his facility in a manner that: (a) minimizes the need for further maintenance, and (b) controls, minimizes or eliminates, to the extent necessary to protect human health and the environment, post-closure escape of hazardous waste, hazardous waste constituents, leachate, contaminated rainfall, or waste decomposition products to the ground water, or surface waters, or to the atmosphere.

§ 265.112   Closure plan; amendment of plan.

(a) On the effective date of these regulations, the owner or operator must have a written closure plan. He must keep this plan at the facility. This plan must identify the steps necessary to completely close the facility at any point during its intended life and at the end of its intended life. The closure plan must include, at least:

(1) A description of how and when the facility will be partially closed, if applicable, and ultimately closed. The description must identify the maximum

extent of the operation which will be be unclosed during the life of the facility, and how the requirements of § 265.111 and the applicable closure requirements of §§ 265.197, 265.228, 265.280, 265.310, 265.351, 265.381, and 265.404 will be met;

(2) An estimate of the maximum inventory of wastes in storage or in treatment at any given time during the life of the facility;

(3) A description of the steps needed to decontaminate facility equipment during closure; and

(4) A schedule for final closure which must include, as a minimum, the anticipated date when wastes will no longer be received, the date when completion of final closure is anticipated, and intervening milestone dates which will allow tracking of the progress of closure. (For example, the expected date for completing treatment or disposal of waste inventory must be included, as must the planned date for removing any residual wastes from storage facilities and treatment processes.)

(b) The owner or operator may amend his closure plan at any time during the active life of the facility. (The active life of the facility is that period during which wastes are periodically received.) The owner or operator must amend his plan any time changes in operating plans or facility design affect the closure plan.

(c) The owner or operator must submit his closure plan to the Regional Administrator at least 180 days before the date he expects to begin closure. The Regional Administrator will modify, approve, or disapprove the plan within 90 days of receipt and after providing the owner or operator and the affected public (through a newspaper notice) the opportunity to submit written comments. If an owner or operator plans to begin closure within 180 days after the effective date of these regulations, he must submit the necessary plans on the effective date of these regulations.

§ 265.113   Time allowed for closure.

(a) Within 90 days after receiving the final volume of hazardous wastes, the owner or operator must treat all hazardous wastes in storage or in treatment, or remove them from the site, or dispose of them on-site, in accordance with the approved closure plan.

(b) The owner or operator must complete closure activities in accordance with the approved closure plan and within six months after receiving the final volume of wastes. The Regional Administrator may approve a longer closure period under § 265.112(c) if the owner or operator can demonstrate that: (1) the required or

planned closure activities will, of necessity, take him longer than six months to complete, and (2) that he has taken all steps to eliminate any significant threat to human health and the environment from the unclosed but inactive facility.

§ 265.114   Disposal or decontamination of equipment.

When closure is completed, all facility equipment and structures must have been properly disposed of, or decontaminated by removing all hazardous waste and residues.

§ 265.115   Certification of closure.

When closure is completed, the owner or operator must submit to the Regional Administrator certification both by the owner or operator and by an independent registered professional engineer that the facility has been closed in accordance with the specifications in the approved closure plan.

§ 265.116   [Reserved]

§ 265.117   Post-closure care and use of property; period of care.

(a) Post-closure care must consist of at least:

(1) Ground-water monitoring and reporting in accordance with the requirements of Subpart F; and

(2) Maintenance of monitoring and waste containment systems as specified in §§ 265.91, 265.223, 265.228, 265.280, and 265.310, where applicable.

(b) The Regional Administrator may require maintenance of any or all of the security requirements of § 265.14 during the post-closure period, when:

(1) Wastes may remain exposed after completion of closure; or

(2) Short term, incidental access by the public or domestic livestock may pose a hazard to human health.

(c) Post-closure use of property on or in which hazardous waste remains after closure must never be allowed to disturb the integrity of the final cover, liner(s), or any other components of any containment system, or the function of the facility's monitoring systems, unless the owner or operator can demonstrate to the Regional Administrator, either in the post-closure plan or by petition, that the disturbance:

(1) Is necessary to the proposed use of the property, and will not increase the potential hazard to human health or the environment; or

(2) Is necessary to reduce a threat to human health or the environment.

(d) The owner or operator of a disposal facility must provide post-closure care in accordance with the approved post-closure plan for at least

30 years after the date of completing closure. *However,* the owner or operator may petition the Regional Administrator to allow some or all of the requirements for post-closure care to be discontinued or altered before the end of the 30-year period. The petition must include evidence demonstrating the secure nature of the facility that makes continuing the specified post-closure requirement(s) unnecessary—e.g., no detected leaks and none likely to occur, characteristics of the waste, application of advanced technology, or alternative disposal, treatment, or re-use techniques. Alternately, the Regional Administrator may require the owner or operator to continue one or more of the post-closure care and maintenance requirements contained in the facility's post-closure plan for a specified period of time. The Regional Administrator may do this if he finds there has been noncompliance with any applicable standards or requirements, or that such continuation is necessary to protect human health or the environment. At the end of the specified period of time, the Regional Administrator will determine whether to continue or terminate post-closure care and maintenance at the facility. Anyone (a member of the public as well as the owner or operator) may petition the Regional Administrator for an extension or reduction of the post-closure care period based on cause. These petitions will be considered by the Regional Administrator at the time the post-closure plan is submitted and at five-year intervals after the completion of closure.

§ 265.118 Post-closure plan; amendment of plan.

(a) On the effective date of these regulations, the owner or operator of a disposal facility must have a written post-closure plan. He must keep this plan at the facility. This plan must identify the activities which will be carried on after final closure and the frequency of those activities. The post-closure plan must include at least:

(1) Ground-water monitoring activities and frequencies as specified in Subpart F for the post-closure period; and

(2) Maintenance activities and frequencies to ensure: (1) the integrity of the cap and final cover or other containment structures as specified in §§ 265.223, 265.228, 265.280, and 265.310, where applicable, and (2) the function of the facility's monitoring equipment as specified in § 265.91.

(b) The owner or operator may amend his post-closure plan at any time during the active life of the disposal facility or during the post-closure care period. The owner or operator must amend his plan any time changes in operating plans or facilities design affect his post-closure plan.

(c) The owner or operator of a disposal facility must submit his post-closure plan to the Regional Administrator at least 180 days before the date he expects to begin closure. The Regional Administrator will modify or approve the plan within 90 days of receipt and after providing the owner or operator and the affected public (through a newspaper notice) the opportunity to submit written comments. The plan may be modified to include security equipment maintenance under § 265.117(b). If an owner or operator of a disposal facility plans to begin closure within 180 days after the effective date of these regulations, he must submit the necessary plans on the effective date of these regulations. Any amendments to the plan under paragraph (b) of this Section which occur after approval of the plan must also be approved by the Regional Administrator before they may be implemented.

§ 265.119 Notice to local land authority.

Within 90 days after closure is completed, the owner or operator of a disposal facility must submit to the local land authority and to the Regional Administrator a survey plat indicating the location and dimensions of landfill cells or other disposal areas with respect to permanently surveyed benchmarks. This plat must be prepared and certified by a professional land surveyor. The plat filed with the local land authority must contain a note, prominently displayed, which states the owner's or operator's obligation to restrict disturbance of the site as specified in § 265.117(c). In addition, the owner or operator must submit to the Regional Administrator and to the local land authority a record of the type, location, and quantity of hazardous wastes disposed of within each cell or area of the facility. For wastes disposed of before these regulations were promulgated, the owner or operator must identify the type, location, and quantity of the wastes to the best of his knowledge and in accordance with any records he has kept.

§ 265.120 Notice in deed to property.

The owner of the property on which a disposal facility is located must record, in accordance with State law, a notation on the deed to the facility property—or on some other instrument which is normally examined during title search—that will in perpetuity notify any potential purchaser of the property that: (1) the land has been used to manage hazardous waste, and (2) its use is restricted under § 265.117(c).

§§ 265.121–265.139 [Reserved]

Subpart H—Financial Requirements

§ 265.140 Applicability.

(a) Section 265.142 applies to owners and operators of all hazardous waste facilities, except as this Section or § 265.1 provide otherwise.

(b) Section 265.144 applies only to owners and operators of disposal facilities.

(c) States and the Federal government are exempt from the requirements of this Subpart.

§ 265.141 [Reserved]

§ 265.142 Cost estimate for facility closure.

(a) On the effective date of these regulations, each facility owner or operator must have a written estimate of the cost of closing the facility in accordance with the requirements in §§ 265.111–265.115 and applicable closure requirements in §§ 265.197, 265.228, 265.280, 265.310, 265.351, 265.381, and 265.404. The owner or operator must keep this estimate, and all subsequent estimates required in this Section, at the facility. The estimate must equal the cost of closure at the point in the facility's operating life when the extent and manner of its operation would make closure the most expensive, as indicated by its closure plan (see § 265.112(a)). [*Comment:* For example, the closure cost estimate for a particular landfill may be for the cost of closure when its active disposal operations extend over 20 acres, if at all other times these operations extend over less than 20 acres. The estimate would not include costs of partial closures that the closure plan schedules before or after the time of maximum closure cost.]

(b) The owner or operator must prepare a new closure cost estimate whenever a change in the closure plan affects the cost of closure.

(c) On each anniversary of the effective date of these regulations, the owner or operator must adjust the latest closure cost estimate using an inflation factor derived from the annual Implicit Price Deflator for Gross National Product as published by the U.S. Department of Commerce in its *Survey of Current Business.* The inflation factor must be calculated by dividing the latest published annual Deflator by the Deflator for the previous year. The result is the inflation factor. The adjusted closure cost estimate must equal the latest closure cost estimate (see paragraph (b) of this Section) times the inflation factor. [*Comment:* The following is a sample calculation of the adjusted closure cost estimate: Assume that the latest closure cost estimate for a facility is $50,000, the latest published annual Deflator is 152.05, and the annual Deflator for the previous year is 141.70. The Deflators may be rounded to the nearest whole number. Dividing 152 by 142 gives the inflation factor, 1.07. Multiply $50,000 by 1.07 for a product of $53,500—the adjusted closure cost estimate.]

§ 265.143 [Reserved]

§ 265.144 Cost estimate for post-closure monitoring and maintenance.

(a) On the effective date of these regulations, the owner or operator of a disposal facility must have a written estimate of the annual cost of post-closure monitoring and maintenance of the facility in accordance with the applicable post-closure regulations in §§265.117–265.120, 265.228, 265.280, and 265.310. The owner or operator must keep this estimate, and all subsequent estimates required in this Section, at the facility.

(b) The owner or operator must prepare a new annual post-closure cost estimate whenever a change in the post-closure plan affects the cost of post-closure care (see § 265.118(b)). The latest post-closure cost estimate is calculated by multiplying the latest annual post-closure cost estimate by 30.

(c) On each anniversary of the effective date of these regulations, during the operating life of the facility, the owner or operator must adjust the latest post-closure cost estimate using the inflation factor calculated in accordance with § 265.142(c). The adjusted post-closure cost estimate must equal the latest post-closure cost estimate (see paragraph (b) of this Section) times the inflation factor.

§§ 265.145–265.169 [Reserved]

Subpart I—Use and Management of Containers

§ 265.170 Applicability.

The regulations in this Subpart apply to owners and operators of all hazardous waste facilities that store containers of hazardous waste, except as § 265.1 provides otherwise.

§ 265.171 Condition of containers.

If a container holding hazardous waste is not in good condition, or if it begins to leak, the owner or operator must transfer the hazardous waste from this container to a container that is in good condition, or manage the waste in some other way that complies with the requirements of this Part.

§ 265.172 Compatibility of waste with container.

The owner or operator must use a container made of or lined with

materials which will not react with, and are otherwise compatible with, the hazardous waste to be stored, so that the ability of the container to contain the waste is not impaired.

§ 265.173 Management of containers.

(a) A container holding hazardous waste must always be closed during storage, except when it is necessary to add or remove waste.

(b) A container holding hazardous waste must not be opened, handled, or stored in a manner which may rupture the container or cause it to leak.

[Comment: A container that is a hazardous waste listed in § 261.33 of this Chapter must be managed in compliance with the regulations of this Part. Re-use of containers in transportation is governed by U.S. Department of Transportation regulations, including those set forth in 49 CFR 173.28.]

§ 265.174 Inspections.

The owner or operator must inspect areas where containers are stored, at least weekly, looking for leaks and for deterioration caused by corrosion or other factors.

[Comment: See § 265.171 for remedial action required if deterioration or leaks are detected.]

§ 265.175 [Reserved]

§ 265.176 Special requirements for ignitable or reactive waste.

Containers holding ignitable or reactive waste must be located at least 15 meters (50 feet) from the facility's property line.

[Comment: See § 265.17(a) for additional requirements.]

§ 265.177 Special requirements for incompatible wastes.

(a) Incompatible wastes, or incompatible wastes and materials, (see Appendix V for examples) must not be placed in the same container, unless § 265.17(b) is complied with.

(b) Hazardous waste must not be placed in an unwashed container that previously held an incompatible waste or material (see Appendix V for examples), unless § 265.17(b) is complied with.

(c) A storage container holding a hazardous waste that is incompatible with any waste or other materials stored nearby in other containers, piles, open tanks, or surface impoundments must be separated from the other materials or protected from them by means of a dike, berm, wall, or other device.

[Comment: The purpose of this is to prevent fires, explosions, gaseous

emissions, leaching, or other discharge of hazardous waste or hazardous waste constituents which could result from the mixing of incompatible wastes or materials if containers break or leak.]

§ 265.178–265.189 [Reserved]

Subpart J—Tanks

§ 265.190 Applicability.

The regulations in this Subpart apply to owners and operators of facilities that use tanks to treat or store hazardous waste, except as § 265.1 provides otherwise.

§ 265.191 [Reserved]

§ 265.192 General operating requirements.

(a) Treatment or storage of hazardous waste in tanks must comply with § 265.17(b).

(b) Hazardous wastes or treatment reagents must not be placed in a tank if they could cause the tank or its inner liner to rupture, leak, corrode, or otherwise fail before the end of its intended life.

(c) Uncovered tanks must be operated to ensure at least 60 centimeters (2 feet) of freeboard, unless the tank is equipped with a containment structure (e.g., dike or trench), a drainage control system, or a diversion structure (e.g., standby tank) with a capacity that equals or exceeds the volume of the top 60 centimeters (2 feet) of the tank.

(d) Where hazardous waste is continuously fed into a tank, the tank must be equipped with a means to stop this inflow (e.g., a waste feed cutoff system or by-pass system to a stand-by tank).

[Comment: These systems are intended to be used in the event of a leak or overflow from the tank due to a system failure (e.g., a malfunction in the treatment process, a crack in the tank, etc.).]

§ 265.193 Waste analysis and trial tests.

(a) In addition to the waste analysis required by § 265.13, whenever a tank is to be used to:

(1) Chemically treat or store a hazardous waste which is substantially different from waste previously treated or stored in that tank; or

(2) Chemically treat hazardous waste with a substantially different process than any previously used in that tank; the owner or operator must, before treating or storing the different waste or using the different process:

(i) Conduct waste analyses and trial treatment or storage tests (e.g., bench scale or pilot plant scale tests); or

(ii) Obtain written, documented information on similar storage or treatment of similar waste under similar operating conditions;

to show that this proposed treatment or storage will meet all applicable requirements of § 265.192(a) and (b).

[Comment: As required by § 265.13, the waste analysis plan must include analyses needed to comply with §§ 265.198 and 265.199. As required by § 265.73, the owner or operator must place the results from each waste analysis and trial test, or the documented information, in the operating record of the facility.]

§ 265.194 Inspections.

(a) The owner or operator of a tank must inspect, where present:

(1) Discharge control equipment (e.g., waste feed cut-off systems, by-pass systems, and drainage systems), at least once each operating day, to ensure that it is in good working order;

(2) Data gathered from monitoring equipment (e.g., pressure and temperature gauges), at least once each operating day, to ensure that the tank is being operated according to its design;

(3) The level of waste in the tank, at least once each operating day, to ensure compliance with § 265.192(c);

(4) The construction materials of the tank, at least weekly, to detect corrosion or leaking of fixtures or seams; and

(5) The construction materials of, and the area immediately surrounding, discharge confinement structures (e.g., dikes), at least weekly, to detect erosion or obvious signs of leakage (e.g., wet spots or dead vegetation).

[Comment: As required by § 265.15(c), the owner or operator must remedy any deterioration or malfunction he finds.]

§§ 265.195–265.196 [Reserved]

§ 265.197 Closure.

At closure, all hazardous waste and hazardous waste residues must be removed from tanks, discharge control equipment, and discharge confinement structures.

[Comment: At closure, as throughout the operating period, unless the owner or operator can demonstrate, in accordance with § 261.3(c) or (d) of this Chapter, that any solid waste removed from his tank is not a hazardous waste, the owner or operator becomes a generator of hazardous waste and must manage it in accordance with all applicable requirements of Parts 262, 263, and 265 of this Chapter.]

§ 265.198 Special requirements for ignitable or reactive waste.

(a) Ignitable or reactive waste must not be placed in a tank, unless:

(1) The waste is treated, rendered, or mixed before or immediately after

placement in the tank so that (i) the resulting waste, mixture, or dissolution of material no longer meets the definition of ignitable or reactive waste under §§ 261.21 or 261.23 of this Chapter, and (ii) § 265.17(b) is complied with; or

(2) The waste is stored or treated in such a way that it is protected from any material or conditions which may cause the waste to ignite or react; or

(3) The tank is used solely for emergencies.

(b) The owner or operator of a facility which treats or stores ignitable or reactive waste in covered tanks must comply with the National Fire Protection Association's (NFPA's) buffer zone requirements for tanks, contained in Tables 2–1 through 2–6 of the "Flammable and Combustible Code—1977"

[Comment: See § 265.17(a) for additional requirements.]

§ 265.199 Special requirements for incompatible wastes.

(a) Incompatible wastes, or incompatible wastes and materials, (see Appendix V for examples) must not be placed in the same tank, unless § 265.17(b) is complied with.

(b) Hazardous waste must not be placed in an unwashed tank which previously held an incompatible waste or material, unless § 265.17(b) is complied with.

§§ 265.200–265.219 [Reserved]

Subpart K—Surface Impoundments

§ 265.220 Applicability.

The regulations in this Subpart apply to owners and operators of facilities that use surface impoundments to treat, store, or dispose of hazardous waste, except as § 265.1 provides otherwise.

§ 265.221 [Reserved]

§ 265.222 General operating requirements.

A surface impoundment must maintain enough freeboard to prevent any overtopping of the dike by overfilling, wave action, or a storm. There must be at least 60 centimeters (2 feet) of freeboard.

[Comment: Any point source discharge from a surface impoundment to waters of the United States is subject to the requirements of Section 402 of the Clean Water Act, as amended. Spills may be subject to Section 311 of that Act.]

§ 265.223 Containment system.

All earthen dikes must have a protective cover, such as grass, shale, or rock, to minimize wind and water erosion and to preserve their structural integrity.

§ 265.224 [Reserved]

§ 265.225 Waste analysis and trial tests.

(a) In addition to the waste analyses required by § 265.13, whenever a surface impoundment is to be used to:

(1) Chemically treat a hazardous waste which is substantially different from waste previously treated in that impoundment; or

(2) Chemically treat hazardous waste with a substantially different process than any previously used in that impoundment; the owner or operator must, before treating the different waste or using the different process:

(i) Conduct waste analyses and trial treatment tests (e.g., bench scale or pilot plant scale tests); or

(ii) Obtain written, documented information on similar treatment of similar waste under similar operating conditions; to show that this treatment will comply with § 265.17(b).

[Comment: As required by § 265.13, the waste analysis plan must include analyses needed to comply with §§ 265.229 and 265.230. As required by § 265.73, the owner or operator must place the results from each waste analysis and trial test, or the documented information, in the operating record of the facility.]

§ 265.226 Inspections.

(a) The owner or operator must inspect:

(1) The freeboard level at least once each operating day to ensure compliance with § 265.222, and

(2) The surface impoundment, including dikes and vegetation surrounding the dike, at least once a week to detect any leaks, deterioration, or failures in the impoundment.

[Comment: As required by § 265.15(c), the owner or operator must remedy any deterioration or malfunction he finds.]

§ 265.227 [Reserved]

§ 265.228 Closure and post-closure.

(a) At closure, the owner or operator may elect to remove from the impoundment:

(1) Standing liquids;

(2) Waste and waste residues;

(3) The liner, if any; and

(4) Underlying and surrounding contaminated soil.

(b) If the owner or operator removes all the impoundment materials in paragraph (a) of this Section, or can demonstrate under § 261.3(c) and (d) of this Chapter that none of the materials listed in paragraph (a) of this Section remaining at any stage of removal are hazardous wastes, the impoundment is not further subject to the requirements of this Part.

[Comment: At closure, as throughout the operating period, unless the owner or operator can demonstrate, in accordance with § 261.3 (c) or (d) of this Chapter, that any solid waste removed from the surface impoundment is not a hazardous waste, he becomes a generator of hazardous waste and must manage it in accordance with all applicable requirements of Parts 262, 263, and 265 of this Chapter. The surface impoundment may be subject to Part 257 of this Chapter even if it is not subject to this Part.]

(c) If the owner or operator does not remove all the impoundment materials in paragraph (a) of this Section, or does not make the demonstration in paragraph (b) of this Section, he must close the impoundment and provide post-closure care as for a landfill under Subpart G and § 265.310. If necessary to support the final cover specified in the approved closure plan, the owner or operator must treat remaining liquids, residues, and soils by removal of liquids, drying, or other means.

[Comment: The closure requirements under § 265.310 will vary with the amount and nature of the residue remaining, if any, and the degree of contamination of the underlying and surrounding soil. Section 265.117(d) allows the Regional Administrator to vary post-closure care requirements.]

§ 265.229 Special requirements for ignitable or reactive waste.

(a) Ignitable or reactive waste must not be placed in a surface impoundment, unless:

(1) The waste is treated, rendered, or mixed before or immediately after placement in the impoundment so that (i) the resulting waste, mixture, or dissolution of material no longer meets the definition of ignitable or reactive waste under §§ 261.21 or 261.23 of this Chapter, and (ii) § 265.17(b) is complied with; or

(2) The surface impoundment is used solely for emergencies.

§ 265.230 Special requirements for incompatible wastes.

Incompatible wastes, or incompatible wastes and materials, (see Appendix V for examples) must not be placed in the same surface impoundment, unless § 265.17(b) is complied with.

§§ 265.231–265.249 [Reserved]

Subpart L—Waste Piles

§ 265.250 Applicability.

The regulations in this Subpart apply to owners and operators of facilities that treat or store hazardous waste in piles, except as § 265.1 provides otherwise. Alternatively, a pile of hazardous waste may be managed as a landfill under Subpart N.

§ 265.251 Protection from wind.

The owner or operator of a pile containing hazardous waste which could be subject to dispersal by wind must cover or otherwise manage the pile so that wind dispersal is controlled.

§ 265.252 Waste analysis.

In addition to the waste analyses required by § 265.13, the owner or operator must analyze a representative sample of waste from each incoming movement before adding the waste to any existing pile, unless (1) the only wastes the facility receives which are amenable to piling are compatible with each other, or (2) the waste received is compatible with the waste in the pile to which it is to be added. The analysis conducted must be capable of differentiating between the types of hazardous waste the owner or operator places in piles, so that mixing of incompatible waste does not inadvertently occur. The analysis must include a visual comparison of color and texture.

[Comment: As required by § 265.13, the waste analysis plan must include analyses needed to comply with §§ 265.256 and 265.257. As required by § 265.73, the owner or operator must place the results of this analysis in the operating record of the facility.]

§ 265.253 Containment.

If leachate or run-off from a pile is a hazardous waste, then either:

(a) The pile must be placed on an impermeable base that is compatible with the waste under the conditions of treatment or storage, run-on must be diverted away from the pile, and any leachate and run-off from the pile must be collected and managed as a hazardous waste; or

(b)(1) The pile must be protected from precipitation and run-on by some other means; and

(2) No liquids or wastes containing free liquids may be placed in the pile.

[Comment: If collected leachate or run-off is discharged through a point source to waters of the United States, it is subject to the requirements of Section 402 of the Clean Water Act, as amended.]

(c) The date for compliance with paragraphs (a) and (b)(1) of this Section is 12 months after the effective date of this Part.

§§ 265.254–265.255 [Reserved]

§ 265.256 Special requirements for ignitable or reactive waste.

(a) Ignitable or reactive wastes must not be placed in a pile, unless:

(1) Addition of the waste to an existing pile (i) results in the waste or

mixture no longer meeting the definition of ignitable or reactive waste under §§ 261.21 or 261.23 of this Chapter, and (ii) complies with § 265.17(b); or

(2) The waste is managed in such a way that it is protected from any material or conditions which may cause it to ignite or react.

§ 265.257 Special requirements for incompatible wastes.

(a) Incompatible wastes, or incompatible wastes and materials, (see Appendix V for examples) must not be placed in the same pile, unless § 265.17(b) is complied with.

(b) A pile of hazardous waste that is incompatible with any waste or other material stored nearby in other containers, piles, open tanks, or surface impoundments must be separated from the other materials, or protected from them by means of a dike, berm, wall, or other device.

[Comment: The purpose of this is to prevent fires, explosions, gaseous emissions, leaching, or other discharge of hazardous waste or hazardous waste constituents which could result from the contact or mixing of incompatible wastes or materials.]

(c) Hazardous waste must not be piled on the same area where incompatible wastes or materials were previously piled, unless that area has been decontaminated sufficiently to ensure compliance with § 265.17(b).

§§ 265.258–265.269 [Reserved]

Subpart M—Land Treatment

§ 265.270 Applicability.

The regulations in this Subpart apply to owners and operators of hazardous waste land treatment facilities, except as §·265.1 provides otherwise.

§ 265.271 [Reserved]

§ 265.272 General operating requirements.

(a) Hazardous waste must not be placed in or on a land treatment facility unless the waste can be made less hazardous or non-hazardous by biological degradation or chemical reactions occurring in or on the soil.

(b) Run-on must be diverted away from the active portions of a land treatment facility.

(c) Run-off from active portions of a land treatment facility must be collected.

[Comment: If the collected run-off is a hazardous waste under Part 261 of this Chapter, it must be managed as a hazardous waste in accordance with all applicable requirements of Parts 262, 263, and 265 of this Chapter. If the collected run-off is discharged through a point source to waters of the United

States, it is subject to the requirements of Section 402 of the Clean Water Act, as amended.]

(d) The date for compliance with paragraphs (b) and (c) of this Section is 12 months after the effective date of this Part.

§ 265.273 Waste analysis.

In addition to the waste analyses required by § 265.13, before placing a hazardous waste in or on a land treatment facility, the owner or operator must:

(a) Determine the concentrations in the waste of any substances which exceed the maximum concentrations contained in Table I of § 261.24 of this Chapter that cause a waste to exhibit the EP toxicity characteristic;

(b) For any waste listed in Part 261, Subpart D, of this Chapter, determine the concentrations of any substances which caused the waste to be listed as a hazardous waste; and

(c) If food chain crops are grown, determine the concentrations in the waste of each of the following constituents: arsenic, cadmium, lead, and mercury, *unless* the owner or operator has written, documented data that show that the constituent is not present.

[*Comment:* Part 261 of this Chapter specifies the substances for which a waste is listed as a hazardous waste. As required by § 265.13, the waste analysis plan must include analyses needed to comply with §§ 265.281 and 265.282. As required by § 265.73, the owner or operator must place the results from each waste analysis, or the documented information, in the operating record of the facility.]

§§ 265.274–265.275 [Reserved]

§ 265.276 Food chain crops.

(a) An owner or operator of a hazardous waste land treatment facility on which food chain crops are being grown, or have been grown and will be grown in the future, must notify the Regional Administrator within 60 days after the effective date of this Part.

[*Comment:* The growth of food chain crops at a facility which has never before been used for this purpose is a significant change in process under § 122.23(c)(3) of this Chapter. Owners or operators of such land treatment facilities who propose to grow food chain crops after the effective date of this Part must comply with § 122.23(c)(3) of this Chapter.]

(b)(1) Food chain crops must not be grown on the treated area of a hazardous waste land treatment facility unless the owner or operator can

demonstrate, based on field testing, that any arsenic, lead, mercury, or other constituents identified under § 265.273(b):

(i) Will not be transferred to the food portion of the crop by plant uptake or direct contact, and will not otherwise be ingested by food chain animals (e.g., by grazing); or

(ii) Will not occur in greater concentrations in the crops grown on the land treatment facility than in the same crops grown on untreated soils under similar conditions in the same region.

(2) The information necessary to make the demonstration required by paragraph (b)(1) of this Section must be kept at the facility and must, at a minimum:

(i) Be based on tests for the specific waste and application rates being used at the facility; and

(ii) Include descriptions of crop and soil characteristics, sample selection criteria, sample size determination, analytical methods, and statistical procedures.

(c) Food chain crops must not be grown on a land treatment facility receiving waste that contains cadmium unless all requirements of paragraph (c)(1)(i) through (iii) of this Section or all requirements of paragraph (c)(2)(i) through (iv) of this Section are met.

(1) (i) The pH of the waste and soil mixture is 6.5 or greater at the time of each waste application, except for waste containing cadmium at concentrations of 2 mg/kg (dry weight) or less;

(ii) The annual application of cadmium from waste does not exceed 0.5 kilograms per hectare (kg/ha) on land used for production of tobacco, leafy vegetables, or root crops grown for human consumption. For other food chain crops, the annual cadmium application rate does not exceed:

Time period	Annual Cd application rate (kg/ha)
Present to June 30, 1984	2.0
July 1, 1984 to Dec. 31, 1986	1.25
Beginning Jan. 1, 1987	0.5

(iii) The cumulative application of cadmium from waste does not exceed the levels in either paragraph (c)(1)(iii)(A) of this Section or paragraph (c)(1)(iii)(B) of this Section.

(A)		
Soil cation exchange capacity (meq/100g)	Maximum cumulative application (kg/ha)	
	Background soil pH less than 6.5	Background soil pH greater than 6.5
Less than 5	5	5
5–15	5	10
Greater than 15	5	20

(B) For soils with a background pH of less than 6.5, the cumulative cadmium application rate does not exceed the levels below: *Provided,* that the pH of the waste and soil mixture is adjusted to and maintained at 6.5 or greater whenever food chain crops are grown.

Soil cation exchange capacity (meq/100g)	Maximum cumulative application (kg/ha)
Less than 5	5
5–15	10
Greater than 15	20

(2)(i) The only food chain crop produced is animal feed.

(ii) The pH of the waste and soil mixture is 6.5 or greater at the time of waste application or at the time the crop is planted, whichever occurs later, and this pH level is maintained whenever food chain crops are grown.

(iii) There is a facility operating plan which demonstrates how the animal feed will be distributed to preclude ingestion by humans. The facility operating plan describes the measures to be taken to safeguard against possible health hazards from cadmium entering the food chain, which may result from alternative land uses.

(iv) Future property owners are notified by a stipulation in the land record or property deed which states that the property has received waste at high cadmium application rates and that food chain crops should not be grown, due to a possible health hazard.

[*Comment:* As required by § 265.73, if an owner or operator grows food chain crops on his land treatment facility, he must place the information developed in this Section in the operating record of the facility.]

§ 265.277 [Reserved]

§ 265.278 Unsaturated zone (zone of aeration) monitoring.

(a) The owner or operator must have in writing, and must implement, an unsaturated zone monitoring plan which is designed to:

(1) Detect the vertical migration of hazardous waste and hazardous waste constituents under the active portion of the land treatment facility, and

(2) Provide information on the background concentrations of the

hazardous waste and hazardous waste constituents in similar but untreated soils nearby; this background monitoring must be conducted before or in conjunction with the monitoring required under paragraph (a)(1) of this Section.

(b) The unsaturated zone monitoring plan must include, at a minimum:

(1) Soil monitoring using soil cores, and

(2) Soil-pore water monitoring using devices such as lysimeters.

(c) To comply with paragraph (a)(1) of this Section, the owner or operator must demonstrate in his unsaturated zone monitoring plan that:

(1) The depth at which soil and soil-pore water samples are to be taken is below the depth to which the waste is incorporated into the soil;

(2) The number of soil and soil-pore water samples to be taken is based on the variability of:

(i) The hazardous waste constituents (as identified in § 265.273(a) and (b)) in the waste and in the soil; and

(ii) The soil type(s); and

(3) The frequency and timing of soil and soil-pore water sampling is based on the frequency, time, and rate of waste application, proximity to ground water, and soil permeability.

(d) The owner or operator must keep at the facility his unsaturated zone monitoring plan, and the rationale used in developing this plan.

(e) The owner or operator must analyze the soil and soil-pore water samples for the hazardous waste constituents that were found in the waste during the waste analysis under § 265.273 (a) and (b).

[Comment: As required by § 265.73, all data and information developed by the owner or operator under this Section must be placed in the operating record of the facility.]

§ 265.279   Recordkeeping.

The owner or operator of a land treatment facility must keep records of the application dates, application rates, quantities, and location of each hazardous waste placed in the facility, in the operating record required in § 265.73.

§ 265.280   Closure and post-closure.

(a) In the closure plan under § 265.112 and the post-closure plan under § 265.118, the owner or operator must address the following objectives and indicate how they will be achieved:

(1) Control of the migration of hazardous waste and hazardous waste constituents from the treated area into the ground water;

(2) Control of the release of contaminated run-off from the facility into surface water;

(3) Control of the release of airborne particulate contaminants caused by wind erosion; and

(4) Compliance with § 265.276 concerning the growth of food-chain crops.

(b) The owner or operator must consider at least the following factors in addressing the closure and post-closure care objectives of paragraph (a) of this Section:

(1) Type and amount of hazardous waste and hazardous waste constituents applied to the land treatment facility;

(2) The mobility and the expected rate of migration of the hazardous waste and hazardous waste constituents;

(3) Site location, topography, and surrounding land use, with respect to the potential effects of pollutant migration (e.g., proximity to ground water, surface water and drinking water sources);

(4) Climate, including amount, frequency, and pH of precipitation;

(5) Geological and soil profiles and surface and subsurface hydrology of the site, and soil characteristics, including cation exchange capacity, total organic carbon, and pH;

(6) Unsaturated zone monitoring information obtained under § 265.278; and

(7) Type, concentration, and depth of migration of hazardous waste constituents in the soil as compared to their background concentrations.

(c) The owner or operator must consider at least the following methods in addressing the closure and post-closure care objectives of paragraph (a) of this Section:

(1) Removal of contaminated soils;

(2) Placement of a final cover, considering: (i) Functions of the cover (e.g., infiltration control, erosion and run-off control, and wind erosion control), and (ii) Characteristics of the cover, including material, final surface contours, thickness, porosity and permeability, slope, length of run of slope, and type of vegetation on the cover;

(3) Collection and treatment of run-off

(4) Diversion structures to prevent surface water run-on from entering the treated area; and

(5) Monitoring of soil, soil-pore water, and ground water.

(d) In addition to the requirements of § 265.117, during the post-closure care period, the owner or operator of a land treatment facility must:

(1) Maintain any unsaturated zone monitoring system, and collect and analyze samples from this system in a manner and frequency specified in the post-closure plan;

(2) Restrict access to the facility as appropriate for its post-closure use; and

(3) Assure that growth of food chain crops complies with § 265.276.

§ 265.281   Special requirements for ignitable or reactive waste.

Ignitable or reactive wastes must not be land treated, unless the waste is immediately incorporated into the soil so that (1) the resulting waste, mixture, or dissolution of material no longer meets the definition of ignitable or reactive waste under §§ 261.21 or 261.23 of this Chapter, and (2) § 265.17(b) is complied with.

§ 265.282   Special requirements for incompatible wastes.

Incompatible wastes, or incompatible wastes and materials (see Appendix V for examples), must not be placed in the same land treatment area, unless § 265.17(b) is complied with.

§§ 265.283–265.299   [Reserved]

Subpart N—Landfills

§ 265.300   Applicability.

The regulations in this Subpart apply to owners and operators of facilities that dispose of hazardous waste in landfills, except as § 265.1 provides otherwise. A waste pile used as a disposal facility is a landfill and is governed by this Subpart.

§ 265.301   [Reserved]

§ 265.302   General operating requirements.

(a) Run-on must be diverted away from the active portions of a landfill.

(b) Run-off from active portions of a landfill must be collected.

[Comment: If the collected run-off is a hazardous waste under Part 261 of this Chapter, it must be managed as a hazardous waste in accordance with all applicable requirements of Parts 262, 263, and 265 of this Chapter. If the collected run-off is discharged through a point source to waters of the United States, it is subject to the requirements of Section 402 of the Clean Water Act, as amended.]

(c) The date for compliance with paragraphs (a) and (b) of this Section is 12 months after the effective date of this Part.

(d) The owner or operator of a landfill containing hazardous waste which is subject to dispersal by wind must cover or otherwise manage the landfill so that wind dispersal of the hazardous waste is controlled.

[*Comment:* As required by § 265.13, the waste analysis plan must include analyses needed to comply with §§ 265.312 and 265.313. As required by § 265.73, the owner or operator must place the results of these analyses in the operating record of the facility.]

§§ 265.303–265.308 [Reserved]

§ 265.309 Surveying and recordkeeping.

The owner or operator of a landfill must maintain the following items in the operating record required in § 265.73:

(a) On a map, the exact location and dimensions, including depth, of each cell with respect to permanently surveyed benchmarks; and

(b) The contents of each cell and the approximate location of each hazardous waste type within each cell.

§ 265.310 Closure and post-closure.

(a) The owner or operator must place a final cover over the landfill, and the closure plan under § 265.112 must specify the function and design of the cover. In the post-closure plan under § 265.118, the owner or operator must include the post-closure care requirements of paragraph (d) of this Section.

(b) In the closure and post-closure plans, the owner or operator must address the following objectives and indicate how they will be achieved:

(1) Control of pollutant migration from the facility via ground water, surface water, and air;

(2) Control of surface water infiltration, including prevention of pooling; and

(3) Prevention of erosion.

(c) The owner or operator must consider at least the following factors in addressing the closure and post-closure care objectives of paragraph (b) of this Section:

(1) Type and amount of hazardous waste and hazardous waste constituents in the landfill:

(2) The mobility and the expected rate of migration of the hazardous waste and hazardous waste constituents;

(3) Site location, topography, and surrounding land use, with respect to the potential effects of pollutant migration (e.g., proximity to ground water, surface water, and drinking water sources);

(4) Climate, including amount, frequency, and pH of precipitation;

(5) Characteristics of the cover including material, final surface contours, thickness, porosity and permeability, slope, length of run of slope, and type of vegetation on the cover; and

(6) Geological and soil profiles and surface and subsurface hydrology of the site.

(b) In addition to the requirements of § 265.117, during the post-closure care period, the owner or operator of a hazardous waste landfill must:

(1) Maintain the function and integrity of the final cover as specified in the approved closure plan;

(2) Maintain and monitor the leachate collection, removal, and treatment system (if there is one present in the landfill) to prevent excess accumulation of leachate in the system;

[*Comment:* If the collected leachate is a hazardous waste under Part 261 of this Chapter, it must be managed as a hazardous waste in accordance with all applicable requirements of Parts 262, 263, and 265 of this Chapter. If the collected leachate is discharged through a point source to waters of the United States, it is subject to the requirements of Section 402 of the Clean Water Act, as amended.]

(3) Maintain and monitor the gas collection and control system (if there is one present in the landfill) to control the vertical and horizontal escape of gases;

(4) Protect and maintain surveyed benchmarks; and

(5) Restrict access to the landfill as appropriate for its post-closure use.

§ 265.311 [Reserved]

§ 265.312 Special requirements for ignitable or reactive waste.

Ignitable or reactive waste must not be placed in a landfill, unless the waste is treated, rendered, or mixed before or immediately after placement in the landfill so that (1) the resulting waste, mixture, or dissolution of material no longer meets the definition of ignitable or reactive waste under §§ 261.21 or 261.23 of this Chapter, and (2) § 265.17(b) is complied with.

§ 265.313 Special requirements for incompatible wastes.

Incompatible wastes, or incompatible wastes and materials, (see Appendix V for examples) must not be placed in the same landfill cell, unless § 265.17(b) is complied with.

§ 265.314 Special requirements for liquid waste.

(a) Bulk or non-containerized liquid waste or waste containing free liquids must not be placed in a landfill, unless:

(1) The landfill has a liner which is chemically and physically resistant to the added liquid, and a functioning leachate collection and removal system with a capacity sufficient to remove all leachate produced; or

(2) Before disposal, the liquid waste or waste containing free liquids is treated or stabilized, chemically or physically (e.g., by mixing with an absorbent solid),

so that free liquids are no longer present.

(b) A container holding liquid waste or waste containing free liquids must not be placed in a landfill, unless:

(1) The container is designed to hold liquids or free liquids for a use other than storage, such as a battery or capacitor; or

(2) The container is very small, such as an ampule.

(c) The date for compliance with this Section is 12 months after the effective date of this Part.

§ 265.315 Special requirements for containers.

(a) An empty container must be crushed flat, shredded, or similarly reduced in volume before it is buried beneath the surface of a landfill.

(b) The date for compliance with this Section is 12 months after the effective date of this Part.

§§ 265.316–265.339 [Reserved]

Subpart O—Incinerators

§ 265.340 Applicability.

The regulations in this Subpart apply to owners and operators of facilities that treat hazardous waste in incinerators, except as § 265.1 provides otherwise.

§§ 265.341–265.342 [Reserved]

§ 265.343 General operating requirements.

Before adding hazardous waste, the owner or operator must bring his incinerator to steady state (normal) conditions of operation—including steady state operating temperature and air flow—using auxiliary fuel or other means.

§ 265.344 [Reserved]

§ 265.345 Waste analysis.

In addition to the waste analyses required by § 265.13, the owner or operator must sufficiently analyze any waste which he has not previously burned in his incinerator to enable him to establish steady state (normal) operating conditions (including waste and auxiliary fuel feed and air flow) and to determine the type of pollutants which might be emitted. At a minimum, the analysis must determine:

(a) Heating value of the waste;

(b) Halogen content and sulfur content in the waste; and

(c) Concentrations in the waste of lead and mercury, *unless* the owner or operator has written, documented data that show that the element is not present.

[*Comment:* As required by § 265.73, the owner or operator must place the results from each waste analysis, or the documented information, in the operating record of the facility.]

§ 265.346 [Reserved]

§ 265.347 Monitoring and inspections.

(a) The owner or operator must conduct, as a minimum, the following monotoring and inspections when incinerating hazardous wastes:

(1) Existing instruments which relate to combustion and emission control must be monitored at least every 15 minutes. Appropriate corrections to maintain steady state combustion conditions must be made immediately either automatically or by the operator. Instruments which relate to combustion and emission control would normally include those measuring waste feed, auxiliary fuel feed, air flow, incinerator temperature, scrubber flow, scrubber pH, and relevant level controls.

(2) The stack plume (emissions) must be observed visually at least hourly for normal appearance (color and opacity). The operator must immediately make any indicated operating corrections necessary to return visible emissions to their normal appearance.

(3) The complete incinerator and associated equipment (pumps, valves, conveyors, pipes, etc.) must be inspected at least daily for leaks, spills, and fugitive emissions, and all emergency shutdown controls and system alarms must be checked to assure proper operation.

§§ 265.348–265.350 [Reserved]

§ 265.351 Closure.

At closure, the owner or operator must remove all hazardous waste and hazardous waste residues (including but not limited to ash, scrubber waters, and scrubber sludges) from the incinerator.

[Comment: At closure, as throughout the operating period, unless the owner or operator can demonstrate, in accordance with § 261.3(c) or (d) of this Chapter, that any solid waste removed from his incinerator is not a hazardous waste, the owner or operator becomes a generator of hazardous waste and must manage it in accordance with all applicable requirements of Parts 262, 263, and 265 of this Chapter.]

§§ 265.352–265.369 [Reserved]

Subpart P—Thermal Treatment

§ 265.370 Applicability.

The regulations in this Subpart apply to owners and operators of facilities that thermally treat hazardous waste in devices other than incinerators, except as § 265.1 provides otherwise. Thermal treatment in incinerators is subject to the requirements of Subpart O.

§§ 265.371–265.372 [Reserved]

§ 265.373 General operating requirements.

Before adding hazardous waste, the owner or operator must bring his thermal treatment process to steady state (normal) conditions of operation—including steady state operating temperature—using auxiliary fuel or other means, unless the process is a non-continuous (batch) thermal treatment process which requires a complete thermal cycle to treat a discrete quantity of hazardous waste.

§ 265.374 [Reserved]

§ 265.375 Waste analysis.

In addition to the waste analyses required by § 265.13, the owner or operator must sufficiently analyze any waste which he has not previously treated in his thermal process to enable him to establish steady state (normal) or other appropriate (for a non-continuous process) operating conditions (including waste and auxiliary fuel feed) and to determine the type of pollutants which might be emitted. At a minimum, the analysis must determine:

(a) Heating value of the waste;

(b) Halogen content and sulfur content in the waste; and

(c) Concentrations in the waste of lead and mercury, unless the owner or operator has written, documented data that show that the element is not present.

[Comment: As required by § 265.73, the owner or operator must place the results from each waste analysis, or the documented information, in the operating record of the facility.]

§ 265.376 [Reserved]

§ 265.377 Monitoring and inspections.

(a) The owner or operator must conduct, as a minimum, the following monitoring and inspections when thermally treating hazardous waste:

(1) Existing instruments which relate to temperature and emission control (if an emission control device is present) must be monitored at least every 15 minutes. Appropriate corrections to maintain steady state or other appropriate thermal treatment conditions must be made immediately either automatically or by the operator. Instruments which relate to temperature and emission control would normally include those measuring waste feed, auxiliary fuel feed, treatment process temperature, and relevant process flow and level controls.

(2) The stack plume (emissions), where present, must be observed visually at least hourly for normal appearance (color and opacity). The operator must immediately make any indicated operating corrections necessary to return any visible emissions to their normal appearance.

(3) The complete thermal treatment process and associated equipment (pumps, valves, conveyors, pipes, etc.) must be inspected at least daily for leaks, spills, and fugitive emissions, and all emergency shutdown controls and system alarms must be checked to assure proper operation.

§§ 265.378–265.380 [Reserved]

§ 265.381 Closure.

At closure, the owner or operator must remove all hazardous waste and hazardous waste residues (including, but not limited to, ash) from the thermal treatment process or equipment.

[Comment: At closure, as throughout the operating period, unless the owner or operator can demonstrate, in accordance with § 261.3(c) or (d) of this Chapter, that any solid waste removed from his thermal treatment process or equipment is not a hazardous waste, the owner or operator becomes a generator of hazardous waste and must manage it in accordance with all applicable requirements of Parts 262, 263, and 265 of this Chapter.]

§ 265.382 Open burning; waste explosives.

Open burning of hazardous waste is prohibited except for the open burning and detonation of waste explosives. Waste explosives include waste which has the potential to detonate and bulk military propellants which cannot safely be disposed of through other modes of treatment. Detonation is an explosion in which chemical transformation passes through the material faster than the speed of sound (0.33 kilometers/second at sea level). Owners or operators choosing to open burn or detonate waste explosives must do so in accordance with the following table and in a manner that does not threaten human health or the environment.

Pounds of waste explosives or propellants	Minimum distance from open burning or detonation to the property of others
0 to 100	204 meters (670 feet).
101 to 1,000	380 meters (1,250 feet).
1,001 to 10,000	530 meters (1,730 feet).
10,001 to 30,000	690 meters (2,260 feet).

§§ 265.383–265.399 [Reserved]

Subpart Q—Chemical, Physical, and Biological Treatment

§ 265.400 Applicability.

The regulations in this Subpart apply to owners and operators of facilities which treat hazardous wastes by chemical, physical, or biological methods in other than tanks, surface impoundments, and land treatment facilities, except as § 265.1 provides otherwise. Chemical, physical, and biological treatment of hazardous waste in tanks, surface impoundments, and land treatment facilities must be conducted in accordance with Subparts J, K, and M, respectively.

§ 265.401 General operating requirements.

(a) Chemical, physical, or biological treatment of hazardous waste must comply with § 265.17(b).

(b) Hazardous wastes or treatment reagents must not be placed in the treatment process or equipment if they could cause the treatment process or equipment to rupture, leak, corrode, or otherwise fail before the end of its intended life.

(c) Where hazardous waste is continuously fed into a treatment process or equipment, the process or equipment must be equipped with a means to stop this inflow (e.g., a waste feed cut-off system or by-pass system to a standby containment device).

[Comment: These systems are intended to be used in the event of a malfunction in the treatment process or equipment.]

§ 265.402 Waste analysis and trial tests.

(a) In addition to the waste analysis required by § 265.13, whenever:

(1) A hazardous waste which is substantially different from waste previously treated in a treatment process or equipment at the facility is to be treated in that process or equipment, or

(2) A substantially different process than any previously used at the facility is to be used to chemically treat hazardous waste;

the owner or operator must, before treating the different waste or using the different process or equipment:

(i) Conduct waste analyses and trial treatment tests (e.g., bench scale or pilot plant scale tests); or

(ii) Obtain written, documented information on similar treatment of similar waste under similar operating conditions;

to show that this proposed treatment will meet all applicable requirements of § 265.401 (a) and (b).

[Comment: As required by § 265.13, the waste analysis plan must include analyses needed to comply with §§ 265.405 and 265.406. As required by § 265.73, the owner or operator must place the results from each waste analysis and trial test, or the documented information, in the operating record of the facility.]

§ 265.403 Inspections.

(a) The owner or operator of a treatment facility must inspect, where present:

(1) Discharge control and safety equipment (e.g., waste feed cut-off systems, by-pass systems, drainage systems, and pressure relief systems) at least once each operating day, to ensure that it is in good working order;

(2) Data gathered from monitoring equipment (e.g., pressure and temperature gauges), at least once each operating day, to ensure that the treatment process or equipment is being operated according to its design;

(3) The construction materials of the treatment process or equipment, at least weekly, to detect corrosion or leaking of fixtures or seams; and

(4) The construction materials of, and the area immediately surrounding, discharge confinement structures (e.g., dikes), at least weekly, to detect erosion or obvious signs of leakage (e.g., wet spots or dead vegetation).

[Comment: As required by § 265.15(c), the owner or operator must remedy any deterioration or malfunction he finds.]

§ 265.404 Closure.

At closure, all hazardous waste and hazardous waste residues must be removed from treatment processes or equipment, discharge control equipment, and discharge confinement structures.

[Comment: At closure, as throughout the operating period, unless the owner or operator can demonstrate, in accordance with § 261.3 (c) or (d) of this Chapter, that any solid waste removed from his treatment process or equipment is not a hazardous waste, the owner or operator becomes a generator of hazardous waste and must manage it in accordance with all applicable requirements of Parts 262, 263, and 265 of this Chapter.]

§ 265.405 Special requirements for ignitable or reactive waste.

(a) Ignitable or reactive waste must not be placed in a treatment process or equipment unless:

(1) The waste is treated, rendered, or mixed before or immediately after placement in the treatment process or equipment so that (i) the resulting waste, mixture, or dissolution of material no longer meets the definition of ignitable or reactive waste under § 261.21 or 261.23 or this Chapter, and (ii) § 265.17(b) is complied with; or (2) The waste is treated in such a way that it is protected from any material or conditions which may cause the waste to ignite or react.

§ 265.406 Special requirements for incompatible wastes.

(a) Incompatible wastes, or incompatible wastes and materials, (see Appendix V for examples) must not be placed in the same treatment process or equipment, unless § 265.17(b) is complied with.

(b) Hazardous waste must not be placed in unwashed treatment equipment which previously held an incompatible waste or material, unless § 265.17(b) is complied with.

§§ 265.407–265.429 [Reserved]

Subpart R—Underground Injection

§ 265.430 Applicability.

Except as § 265.1 provides otherwise:

(a) The owner or operator of a facility which disposes of hazardous waste by underground injection is excluded from the requirements of Subparts G and H of this Part.

(b) The requirements of this Subpart apply to owners and operators of wells used to dispose of hazardous waste which are classified as Class I under § 122.32(a) of this Chapter and which are classified as Class IV under § 122.32(d) of this Chapter.

[Comment: In addition to the requirements of Subparts A through E of this Part, the owner or operator of a facility which disposes of hazardous waste by underground injection ultimately must comply with the requirements of §§ 265.431–265.437. These Sections are reserved at this time. The Agency will propose regulations that would establish those requirements.]

§ 265.431–265.999 [Reserved]

PART 122—EPA ADMINISTERED PERMIT PROGRAMS: THE NATIONAL POLLUTANT DISCHARGE ELIMINATION SYSTEM; THE HAZARDOUS WASTE PERMIT PROGRAM; AND THE UNDERGROUND INJECTION CONTROL PROGRAM

Subpart A—Definitions and General Program Requirements

Sec.
122.1 What are the consolidated permit regulations?
122.2 Purpose and scope of Part 122.
122.3 Definitions.

Authority: Resource Conservation and Recovery Act, 42 U.S.C. § 6901 et seq.; Safe Drinking Water Act, 42 U.S.C. § 300f et seq.; and Clean Water Act, 33 U.S.C. § 1251 et seq.

Subpart A—Definitions and General Program Requirements

§ 122.1 What are the consolidated permit regulations?

(a) *Coverage.* (1) These consolidated permit regulations include provisions for five permit programs:

(i) The *Hazardous Waste Management* (HWM) Program under Subtitle C of the Solid Waste Disposal Act, as amended by the *Resource Conservation and Recovery Act of 1976* (RCRA) (Pub. L. 94–580, as amended by Pub. L. 95–609; 42 U.S.C. § 6901 et seq.);

(ii) The *Underground Injection Control* (UIC) Program under Part C of the *Safe Drinking Water Act* (SDWA) (Pub. L. 95–523, as amended by Pub. L. 95–190; 42 U.S.C. § 300f et seq.);

(iii) The *National Pollutant Discharge Elimination System* (NPDES) Program under sections 318, 402, and 405(a) of the *Clean Water Act* (CWA) (Pub. L. 92–500, as amended by Pub. L. 95–217 and Pub. L. 95–576; 33 U.S.C. § 1251 et seq.);

(iv) The *Dredge or Fill* (404) Program under section 404 of the *Clean Water Act;* and

(v) The *Prevention of Significant Deterioration* (PSD) Program under regulations implementing section 165 of the *Clean Air Act* (CAA), as amended, (Pub. L. 88–206 as amended; 42 U.S.C. § 7401 et seq.)

(2) For the RCRA, UIC, and NPDES programs, these regulations cover basic EPA permitting requirements (Part 122), what a State must do to obtain approval to operate its program in lieu of a Federal program and minimum requirements for administering the approved State program (Part 123), and procedures for EPA processing of permit applications and appeals (Part 124). For the 404 program, these regulations include only the requirements which must be met for a State to administer its own program in lieu of the U.S. Army Corps of Engineers in "State regulated waters," and provisions for EPA vetoes of State issued 404 permits. For the PSD program, these regulations cover only procedures for EPA processing of PSD permits in Part 124.

(b) *Structure.* (1) *Coverage of Parts.* These consolidated permit regulations are incorporated into three Parts of Title 40 of the Code of Federal Regulations:

(i) *Part 122.* This Part contains definitions for all of the programs except PSD. It also contains basic permitting requirements for EPA-administered RCRA, UIC, and NPDES programs, such as application requirements, standard permit conditions, and monitoring and reporting requirements.

(ii) *Part 123.* This Part describes what States must do to obtain EPA approval of their RCRA, UIC, NPDES, or 404 programs. It also sets forth the minimum requirements for administering these permit programs after approval.

(iii) *Part 124.* This Part establishes the procedures for EPA issuance of RCRA, UIC, NPDES, and PSD permits. It also establishes the procedures for administrative appeals of EPA permit decisions.

(2) *Subparts.* Parts 122, 123, and 124 are each organized into subparts. Each Part has a general Subpart A which contains requirements that apply to all the programs covered by that Part. Additional subparts supplement these general provisions with requirements which apply to one or more specified programs. In case of any inconsistency between Subpart A and any program-specific subpart, the program-specific subpart is controlling.

(3) Certain requirements set forth in Parts 122 and 124 are made applicable to approved State programs, including State 404 programs, by reference in Part 123. These references are set forth in § 123.7. If a section or paragraph of Parts 122 or 124 is applicable to States, through reference in § 123.7, that fact is signaled by the following words at the end of the section or paragraph heading: *(applicable to State programs, see § 123.7).* If these words are absent, the section (or paragraph) applies only to EPA-administered permits.

(4) The structure and coverage of these regulations by program is indicated in the following chart. A permit applicant or permittee that is interested in finding out about only one of the programs covered by these regulations can use this chart to determine which regulations to read. If a State is the permitting authority, the

Program	Coverage		
	Part 122	Part 123	Part 124
RCRA	Subparts A and B	Subparts A, B, and F	Subparts A, B, E, and F.
UIC	Subparts A and C	Subparts A and C	Subparts A and F.
NPDES	Subparts A and D	Subparts A and D	Subparts A, D, E, and F.
404	Subpart A	Subparts A and E	Subpart A.
PSD	None	None	Subparts A, C, and F.

applicant or permittee should read the State laws and program regulations which implement the requirements of Part 123 for the relevant program.

(c) *Relation to other requirements.* (1) *Consolidated permit application forms.* Applicants for EPA-issued RCRA Part A, UIC, NPDES, or PSD permits and persons seeking interim status under RCRA must submit their applications on EPA's consolidated permit application forms when available. (There will be no form for RCRA Part B applications and therefore no EPA application form is used. See § 122.25.) These forms, like these consolidated regulations, contain a general form covering all programs plus several program-specific forms. Although application forms have been consolidated, they, like permits, have been coordinated without losing their separate legal identities. There is no "consolidated permit." Each permit and application under a program is a separate document. Most of the information requested on these application forms (other than Form 5 for PSD) is required by these regulations. The essential information required in the general form (Form 1) is listed in § 122.4. The additional information required for RCRA Part A applications (Form 3) is listed in § 122.24, for UIC applications (Form 4) in § 122.37, and for NPDES applications (Forms 2a–d) in § 122.53. Applicants for State-issued permits must use State forms which must require at a minimum the information listed in these sections. All minimum information requirements for State 404 permit applications appear in § 123.94.

(2) *Technical regulations.* The five permit programs which are covered in these consolidated permit regulations each have separate additional regulations that contain technical requirements for those programs. These separate regulations are used by permit-issuing authorities to determine what requirements must be placed in permits if they are issued. These separate regulations are located as follows:

RCRA .... 40 CFR Parts 260–266.
UIC .... 40 CFR Part 146.
NPDES ..... 40 CFR Parts 125, 129, 133, 136.
              40 CFR Subchapter N (Parts 400–460).
404 .... 40 CFR Part 230.
PSD .... 40 CFR Part 52.

(d) *Authority.* The consolidation of these permit programs into one set of regulations is authorized by sections 101(f) and 501(a) of CWA, sections 1006 and 2002 of RCRA, section 1450 of the SDWA, and section 301 of the CAA.

(e) *Public participation.* This rule establishes the requirements for public participation in EPA and State permit issuance, enforcement, and related variance proceedings; and in the approval of State RCRA, UIC, NPDES, and 404 programs. These requirements carry out the purposes of the public participation requirements of 40 CFR Part 25 (Public Participation), and supersede the requirements of that Part as they apply to actions covered under Parts 122, 123, and 124.

(f) *State authorities.* Nothing in Parts 122, 123, or 124 precludes more stringent State regulation of any activity covered by these regulations, whether or not under an approved State program, except as provided for the RCRA program in § 123.33 (requirement that State RCRA programs under final authorization be consistent with the Federal program and other State programs).

§ 122.2 Purpose and scope of Part 122.

(a) *Subpart A* of Part 122 contains definitions (§ 122.3) and basic permitting requirements (§§ 122.4 through 122.19). Definitions are given for the RCRA, UIC, NPDES, and State 404 programs. Definitions for EPA processing of PSD permits are in Part 124, Subpart C. The permitting requirements apply to EPA administered RCRA, UIC, and NPDES programs. (Permit program requirements for the Federal 404 program administered by the Corps of Engineers do not appear in these regulations but are found in 33 CFR Parts 320–327.) In addition, the permitting requirements apply to State-administered RCRA, UIC, NPDES, and 404 programs to the extent specified by cross-reference in § 123.7.

(b) *Subparts B, C, and D* contain additional requirements for RCRA, UIC and NPDES permitting, respectively. They apply to EPA, and to approved States to the extent specified by cross-reference in § 123.7.

§ 122.3 Definitions.

The following definitions apply to Parts 122, 123, and 124, except Part 124 coverage of the PSD program (see § 124.2). Terms not defined in this section have the meaning given by the appropriate Act. When a defined term appears in a definition, the defined term is sometimes placed within quotation marks as an aid to readers. When a definition applies primarily to one or more programs, those programs appear in parentheses after the defined term.

Acidizing (UIC) means the injection of acid through the borehole or "well" into a "formation" to increase permeability and porosity by dissolving the acid-soluble portion of the rock constituents.

Administrator means the Administrator of the United States Environmental Protection Agency, or an authorized representative.

Applicable standards and limitations (NPDES) means all State, interstate, and Federal standards and limitations to which a "discharge" or a related activity is subject under the CWA, including "effluent limitations," water quality standards, standards of performance, toxic effluent standards or prohibitions, "best management practices," and pretreatment standards under sections 301, 302, 303, 304, 306, 307, 308, 403, and 405 of CWA.

Application means the EPA standard national forms for applying for a permit, including any additions, revisions or modifications to the forms; or forms approved by EPA for use in "approved States," including any approved modifications or revisions. For RCRA, application also includes the information required by the Director under § 122.25 (contents of Part B of the RCRA application).

Appropriate Act and regulations means the Clean Water Act (CWA); the Solid Waste Disposal Act, as amended by the Resource Conservation and Recovery Act (RCRA); or Safe Drinking Water Act (SDWA), whichever is applicable; and applicable regulations promulgated under those statutes. In the case of an "approved State program" appropriate Act and regulations includes State program requirements.

Approved program or *approved State* means a State or interstate program which has been approved or authorized by EPA under Part 123.

Aquifer (RCRA and UIC) means a geological "formation," group of formations, or part of a formation that is capable of yielding a significant amount of water to a well or spring.

Area of reviw (UIC) means the area surrounding an "injection well" described according to the criteria set forth in § 146.06.

Average monthly discharge limitation (NPDES) means the highest allowable average of "daily discharges" over a calendar month, calculated as the sum of all daily discharges measured during a calendar month divided by the number of daily discharges measured during that month.

Average weekly discharge limitation (NPDES) means the highest allowable average of "daily discharges" over a calendar week, calculated as the sum of all daily discharges measured during a calendar week divided by the number of daily discharges measured during that week.

Best management practices ("BMPs") (NPDES and 404) means schedules of activities, prohibitions of practices, maintenance procedures, and other management practices to prevent or reduce the pollution of "waters of the United States." For NPDES, BMPs also include treatment requirements, operating procedures, and practices to control plant site runoff, spillage or leaks, sludge or waste disposal, or drainage from raw material storage. For State 404 programs, BMPs also include methods, measures, practices, or design and performance standards, which facilitate compliance with section 404(b)(1) environmental guidelines (40 CFR Part 230), effluent limitations or prohibitions under section 307(a), and applicable water quality standards.

BMPs (NPDES and 404) means "best management practices."

Closure (RCRA) means the act of securing a "Hazardous Waste Management facility" pursuant to the requirements of 40 CFR Part 264.

Contaminant (UIC) means any physical, chemical, biological, or radiological substance or matter in water.

Contiguous zone (NPDES) means the entire zone established by the United States under Article 24 of the Convention on the Territorial Sea and the Contiguous Zone.

Continuous discharge (NPDES) means a "discharge" which occurs without interruption throughout the operating hours of the facility, except for infrequent shutdowns for maintenance, process changes, or other similar activities.

CWA means the Clean Water Act (formerly referred to as the Federal Water Pollution Control Act or Federal Water Pollution Control Act Amendments of 1972) Pub. L. 92–500, as amended by Pub. L. 95–217 and Pub. L. 95–576; 33 U.S.C. § 1251 *et seq.*

Daily discharge (NPDS) means the "discharge of a pollutant" meansured during a calendar day or any 24–hour period that reasonably represents the calendar day for purposes of sampling. For pollutants with limitations expressed in units of mass, the "daily discharge" is calculated as the total mass of the pollutant discharged over the day. For pollutants with limitations expressed in other units of measurement, the "daily discharge" is calculated as the average measurement of the pollutant over the day.

Direct discharge (NPDES) means the "discharge of a pollutant."

Director means the Regional Administrator or the State Director, as

the context requires, or an authorized representative. When there is no "approved State program," and there is an EPA administered program, "Director" means the Regional Administrator. When there is an approved State program, "Director" normally means the State Director. In some circumstances, however, EPA retains the authority to take certain actions even when there is an approved State program. (For example, when EPA has issued an NPDES permit prior to the approval of a State program, EPA may retain jurisdiction over that permit after program approval; see § 123.71.) In such cases, the term "Director" means the Regional Administrator and not the State Director.

Discharge (NPDES) when used without qualification means the "discharge of a pollutant."

Discharge of a pollutant (NPDES) means:

(a)(1) Any addition of any "pollutant" or combination of pollutants to "waters of the United States" from any "point source," or

(2) Any addition of any pollutant or combination of pollutants to the waters of the "contiguous zone" or the ocean from any point source other than a vessel or other floating craft which is being used as a means of transportation.

(b) This definition includes additions of pollutants into waters of the United States from: surface runoff which is collected or channelled by man; discharges through pipes, sewers, or other conveyances owned by a State, municipality, or other person which do not lead to a treatment works; and discharges through pipes, sewers, or other conveyances leading into privately owned treatment works.

This term does not include an addition of pollutants by any "indirect discharger."

Discharge Monitoring Report ("DMR") (NPDES) means the EPA uniform national form, including any subsequent additions, revisions, or modifications, for the reporting of self-monitoring results by permitees. DMRs must be used by "approved States" as well as by EPA. EPA will supply DMRs to any approved State upon request. The EPA national forms may be modified to substitute the State Agency name, address, logo, and other similar information, as appropriate, in place of EPA's.

Discharge of dredged material (404) means any addition from any "point source" of "dredged material" into "waters of the United States." The term includes the addition of dredged material into waters of the United States and the runoff or overflow from a contained land or water dredged material disposal area. Discharges of

pollutants into waters of the United States resulting from the subsequent onshore processing of dredged material are not included within this term and are subject to the NPDES program even though the extraction and deposit of such material may also require a permit from the Corps of Engineers or the State section 404 program.

Discharge of fill material (404) means the addition from any "point source" of "fill material" into "waters of the United States." The term includes the following activities in waters of the United States: placement of fill that is necessary for the construction of any structure; the building of any structure or impoundment requiring rock, sand, dirt, or other materials for its construction; site-development fills for recreational, industrial, commercial, residential, and other uses; causeways or road fills; dams and dikes; artificial islands; property protection and/or reclamation devices such as riprap, groins, seawalls, breakwaters, and revetments; beach nourishment; levees; fill for structures such as sewage treatment facilities, intake and outfall pipes associated with power plants and subaqueous utility lines; and artificial reefs.

Disposal (RCRA) means the discharge, deposit, injection, dumping, spilling, leaking, or placing of any "hazardous waste" into or on any land or water so that such hazardous waste or any constituent thereof may enter the environment or be emitted into the air or discharged into any waters, including ground water.

Disposal facility (RCRA) means a facility or part of a facility at which "hazardous waste" is intentionally placed into or on the land or water, and at which hazardous waste will remain after closure.

Disposal site (404) means that portion of the "waters of the United States" enclosed within fixed boundaries consisting of a bottom surface area and any overlaying volume of water. In the case of "wetland" on which water is not present, the disposal site consists of the wetland surface area. Fixed boundaries may consist of fixed geographic point(s) and associated dimensions, or of a discharge point and specific associated dimensions.

DMR (NPDES) means "Discharge Monitoring Report."

Draft permit means a document prepared under § 124.6 indicating the Director's tentative decision to issue or deny, modify, revoke and reissue, terminate, or reissue a "permit." A notice of intent to terminate a permit, and a notice of intent to deny a permit, as discussed in § 124.5, are types of "draft permits." A denial of a request for modification, revocation and reissuance, or termination, as discussed in § 124.5, is not a "draft permit." A "proposed permit" is not a "draft permit."

Drilling mud (UIC) means a heavy suspension used in drilling an "injection well," introduced down the drill pipe and through the drill bit.

Dredged material (404) means material that is excavated or dredged from "waters of the United States."

Effluent limitation (NPDES) means any restriction imposed by the Director on quantities, discharge rates, and concentrations of "pollutants" which are "discharged" from "point sources" into "waters of the United States," the waters of the "contiguous zone," or the ocean.

Effluent limitations guidelines (NPDES) means a regulation published by the Administrator under section 304(b) of CWA to adopt or revise "effluent limitations."

Effluents (404) means "dredged material" or "fill material," including return flow from confined sites.

Emergency permit means a RCRA, UIC, or State 404 "permit" issued in accordance with §§ 122.27, 122.40 or 123.96, respectively.

Environmental Protection Agency ("EPA") means the United States Environmental Protection Agency.

EPA means the United States "Environmental Protection Agency."

Exempted aquifer (UIC) means an "aquifer" or its portion that meets the criteria in the definition of "underground source of drinking water" but which has been exempted according to the procedures in § 122.35(b).

Existing HWM facility (RCRA) means a facility which was in operation or for which construction had commenced, on or before October 21, 1976. Construction had commenced if:

(a) The owner or operator had obtained all necessary Federal, State, and local preconstruction approvals or permits; and

(b)(1) A continuous physical, on-site construction program had begun, or

(2) The owner or operator had entered into contractual obligations—which cannot be cancelled or modified without substantial loss—for construction of the facility to be completed within a reasonable time.

[Note.—This definition reflects the literal language of the statute. However, EPA believes that amendments to RCRA now in conference will shortly be enacted and will change the date for determining when a facility is an "existing facility" to one no earlier than May of 1980; indications are that the conferees are considering October 30, 1980. Accordingly, EPA encourages every owner or operator of a facility which was built or under physical construction as of the promulgation date of these regulations to file Part A of its permit application so that it can be quickly processed for interim status when

the change in the law takes effect. When those amendments are enacted, EPA will amend this definition.]

Existing injection well (UIC) means an "injection well" other than a "new injection well."

Facility or activity means any "HWM facility," UIC "injection well," NPDES "point source," or State 404 dredge or fill activity, or any other facility or activity (including land or appurtenances thereto) that is subject to regulation under the RCRA, UIC, NPDES, or 404 programs.

Fill material (404) means any "pollutant" which replaces portions of the "waters of the United States" with dry land or which changes the bottom elevation of a water body for any purpose.

Final authorization (RCRA) means approval by EPA of a State program which has met the requirements of § 3006(b) of RCRA and the applicable requirements of Part 123, Subparts A and B.

Fluid (UIC) means any material or substance which flows or moves whether in a semisolid, liquid, sludge, gas, or any other form or state.

Formation (UIC) means a body of rock characterized by a degree of lithologic homogeneity which is prevailingly, but not necessarily, tabular and is mappable on the earth's surface or traceable in the subsurface.

Formation fluid (UIC) means "fluid" present in a "formation" under natural conditions as opposed to introduced fluids, such as "drilling mud."

General permit (NPDES and 404) means an NPDES or 404 "permit" authorizing a category of discharges under the CWA within a geographical area. For NPDES, a general permit means a permit issued under § 122.59. For 404, a general permit means a permit issued under § 123.95.

Generator (RCRA) means any person, by site location, whose act or process produces "hazardous waste" identified or listed in 40 CFR Part 261.

Ground water (RCRA and UIC) means water below the land surface in a zone of saturation.

Hazardous substance (NPDES) means any substance designated under 40 CFR Part 116 pursuant to section 311 of CWA.

Hazardous waste (RCRA and UIC) means a hazardous waste as defined in 40 CFR § 261.3.

Hazardous Waste Management facility ("HWM facility") means all contiguous land, and structures, other appurtenances, and improvements on the land, used for treating, storing, or disposing of "hazardous waste." A facility may consist of several

"treatment," "storage," or "disposal" operational units (for example, one or more landfills, surface impoundments, or combinations of them).

HWM facility (RCRA) means "Hazardous Waste Management facility."

Indirect discharger (NPDES) means a nondomestic discharger introducing "pollutants" to a "publicly owned treatment works."

Injection well (RCRA and UIC) means a "well" into which "fluids" are being injected.

Injection zone (UIC) means a geological "formation," group of formations, or part of a formation receiving fluids through a "well."

In operation (RCRA) means a facility which is treating, storing, or disposing of "hazardous waste."

Interim authorization (RCRA) means approval by EPA of a State hazardous waste program which has met the requirements of § 3006(c) of RCRA and applicable requirements of Part 123, Subpart F.

Interstate agency means an agency of two or more States established by or under an agreement or compact approved by the Congress, or any other agency of two or more States having substantial powers or duties pertaining to the control of pollution as determined and approved by the Administrator under the "appropriate Act and regulations."

Major facility means any RCRA, UIC, NPDES, or 404 "facility or activity" classified as such by the Regional Administrator, or, in the case of "approved State programs," the Regional Administrator in conjunction with the State Director.

Manifest (RCRA and UIC) means the shipping document originated and signed by the "generator" which contains the information required by Subpart B of 40 CFR Part 262.

Maximum daily discharge limitation (NPDES) means the highest allowable "daily discharge."

Municipality (NPDES) means a city, town, borough, county, parish, district, association, or other public body created by or under State law and having jurisdiction over disposal or sewage, industrial wastes, or other wastes, or an Indian tribe or an authorized Indian tribal organization, or a designated and approved management agency under section 208 of CWA.

National Pollutant Discharge Elimination System means the national program for issuing, modifying, revoking and reissuing, terminating, monitoring and enforcing permits, and imposing and enforcing pretreatment requirements, under sections 307, 402, 318, and 405 of CWA. The term includes an "approved program."

New discharger (NPDES) means any building, structure, facility, or installation:

(a)(1) From which there is or may be a new or additional "discharge of pollutants" at a "site" at which on October 18, 1972 it had never discharged pollutants; and

(2) Which has never received a finally effective NPDES "permit" for discharges at that site; and

(3) Which is not a "new source."

(b) This definition includes an "indirect discharger" which commences discharging into "waters of the United States." It also includes any existing mobile point source, such as an offshore oil drilling rig, seafood processing rig, seafood processing vessel, or aggregate plant, that begins discharging at a location for which it does not have an existing permit.

New HWM facility (RCRA) means a "Hazardous Waste Management facility" which began operation or for which construction commenced after October 21, 1976.

New injection well (UIC) means a "well" which began injection after a UIC program for the State applicable to the well is approved.

New source (NPDES) means any building, structure, facility, or installation from which there is or may be a "discharge of pollutants," the construction of which commenced:

(a) After promulgation of standards of performance under section 306 of CWA which are applicable to such source; or

(b) After proposal of standards of performance in accordance with section 306 of CWA which are applicable to such source, but only if the standards are promulgated in accordance with section 306 within 120 days of their proposal.

NPDES means "National Pollutant Discharge Elimination System."

Off-site (RCRA) means any site which is not "on-site."

On-site (RCRA) means on the same or geographically contiguous property which may be divided by public or private right(s)-of-way, provided the entrance and exit between the properties is at a cross-roads intersection, and access is by crossing as opposed to going along, the right(s)-of-way. Non-contiguous properties owned by the same person but connected by a right-of-way which the person controls and to which the public does not have access, is also considered on-site property.

Owner or operator means the owner or operator of any "facility or activity" subject to regulation under the RCRA, UIC, NPDES, or 404 programs.

Permit means an authorization, license, or equivalent control document issued by EPA or an "approved State" to implement the requirements of this Part and Parts 123 and 124. "Permit" includes RCRA "permit by rule" (§ 122.26), UIC area permit (§ 122.39), NPDES or 404 "general permit" (§§ 122.59 and 123.95), and RCRA, UIC, or 404 "emergency permit" (§§ 122.27, 122.40, and 123.96). Permit does not include RCRA interim status (§ 122.23), UIC authorization by rule (§ 122.37), or any permit which has not yet been the subject of final agency action, such as a "draft permit" or a "proposed permit."

Permit by rule (RCRA) means a provision of these regulations stating that a "facility or activity" is deemed to have a RCRA permit if it meets the requirements of the provision.

Person means an individual, association, partnership, corporation, municipality, State or Federal agency, or an agent or employee thereof.

Phase I (RCRA) means that phase of the Federal hazardous waste management program commencing on the effective date of the last of the following to be initially promulgated: 40 CFR Parts 122, 123, 260, 261, 262, 263, and 265. Promulgation of Phase I refers to promulgation of the regulations necessary for Phase I to begin.

Phase II (RCRA) means that phase of Federal hazardous waste management program commencing on the effective date of the first Subpart of 40 CFR Part 264, Subparts F through R to be initially promulgated. Promulgation of Phase II refers to promulgation of the regulations necessary for Phase II to begin.

Physical construction (RCRA) means excavation, movement of earth, erection of forms or structures, or similar activity to prepare an "HWM facility" to accept "hazardous waste."

Plugging (UIC) means the act or process of stopping the flow of water, oil, or gas in "formations" penetrated by a borehole or "well."

Point source (NPDES and 404) means any discernible, confined, and discrete conveyance, including but not limited to any pipe, ditch, channel, tunnel, conduit, well, discrete fissure, container, rolling stock, concentrated animal feeding operation, vessel, or other floating craft, from which pollutants are or may be discharged. This term does not include return flows from irrigated agriculture.

Pollutant (NPDES and 404) means dredged spoil, solid waste, incinerator residue, filter backwash, sewage, garbage, sewage sludge, munitions, chemical wastes, biological materials, radioactive materials (except those regulated under the Atomic Energy Act of 1954, as amended (42 U.S.C. § 2011 *et seq.*)), heat, wrecked or discarded equipment, rock, sand, cellar dirt and industrial, municipal, and agricultural waste discharged into water. It does not mean:

(a) Sewage from vessels; or

(b) Water, gas, or other material which is injected into a well to facilitate production of oil or gas, or water derived in association with oil and gas production and disposed of in a well, if the well used either to facilitate production or for disposal purposes is approved by authority of the State in which the well is located, and if the State determines that the injection or disposal will not result in the degradation of ground or surface water resources.

[Note.—Radioactive materials covered by the Atomic Energy Act are those encompassed in its definition of source, byproduct, or special nuclear materials. Examples of materials not covered include radium and accelerator-produced isotopes. See *Train* v. *Colorado Public Interest Research Group, Inc.,* 426 U.S. 1 (1976).]

POTW means "publicly owned treatment works."

Primary industry category (NPDES) means any industry category listed in the NRDC settlement agreement (*Natural Resources Defense Council et al.* v. *Train,* 8 E.R.C. 2120 (D.D.C. 1976), modified 12 E.R.C. 1833 (D.D.C. 1979); also listed in Appendix A of Part 122.

Privately owned treatment works (NPDES) means any device or system which is (a) used to treat wastes from any facility whose operator is not the operator of the treatment works and (b) not a "POTW."

Process wastewater (NPDES) means any water which, during manufacturing or processing, comes into direct contact with or results from the production or use of any raw material, intermediate product, finished product, byproduct, or waste product.

Proposed permit (NPDES) means a State NPDES "permit" prepared after the close of the public comment period (and, when applicable, any public hearing and administrative appeals) which is sent to EPA for review before final issuance by the State. A "proposed permit" is not a "draft permit."

Publicly owned treatment works ("POTW") means any device or system used in the treatment (including recycling and reclamation) of municipal sewage or industrial wastes of a liquid nature which is owned by a "State" or "municipality." This definition includes sewers, pipes, or other conveyances only if they convey wastewater to a POTW providing treatment.

Radioactive waste (UIC) means any waste which contains radioactive material in concentrations which exceed those listed in 10 CFR Part 20, Appendix B, Table II, Column 2, or exceed the "Criteria for Identifying and Applying Characteristics of Hazardous Waste and for Listing Hazardous Waste" in 40 CFR Part 261, whichever is applicable.

RCRA means the Solid Waste Disposal Act as amended by the Resource Conservation and Recovery Act of 1976 (Pub. L. 94–580, as amended by Pub. L. 95–609, 42 U.S.C. § 6901 *et seq.*).

Recommencing discharger (NPDES) means a source which recommences discharge after terminating operations.

Regional Administrator means the Regional Administrator of the appropriate Regional Office of the Environmental Protection Agency or the authorized representative of the Regional Administrator.

Schedule of compliance means a schedule of remedial measures included in a "permit," including an enforceable sequence of interim requirements (for example, actions, operations, or milestone events) leading to compliance with the "appropriate Act and regulations."

SDWA means the Safe Drinking Water Act (Pub. L. 95–523, as amended by Pub. L. 95–1900; 42 U.S.C. § 300f *et seq.*).

Secondary industry category (NPDES) means any industry category which is not a "primary industry category."

Secretary (NPDES and 404) means the Secretary of the Army, acting through the Chief of Engineers.

Section 404 program or *State 404 program* or *404* means an "approved State program" to regulate the "discharge of dredged material" and the "discharge of fill material" under section 404 of the Clean Water Act in "State regulated waters."

Sewage from vessels (NPDES) means human body wastes and the wastes from toilets and other receptacles intended to receive or retain body wastes that are discharged from vessels and regulated under section 312 of CWA, except that with respect to commercial vessels on the Great Lakes this term includes graywater. For the purposes of this definition, "graywater" means galley, bath, and shower water.

Sewage sludge (NPDES) means the solids, residues, and precipitate separated from or created in sewage by the unit processes of a "publicly owned treatment works." "Sewage" as used in this definition means any wastes, including wastes from humans, households, commercial establishments, industries, and storm water runoff, that are discharged to or otherwise enter a publicly owned treatment works.

Site means the land or water area where any "facility or activity" is physically located or conducted, including adjacent land used in connection with the facility or activity.

State means any of the 50 States, the District of Columbia, Guam, the Commonwealth of Puerto Rico, the Virgin Islands, American Samoa, the Trust Territory of the Pacific Islands (except in the case of RCRA), and the Commonwealth Northern Mariana Islands (except in the case of CWA).

State Director means the chief administrative officer of any State or interstate agency operating an "approved program," or the delegated representative of the State Director. If responsibility is divided among two or more State or interstate agencies, "State Director" means the chief administrative officer of the State or interstate agency authorized to perform the particular procedure or function to which reference is made.

State/EPA Agreement means an agreement between the Regional Administrator and the State which coordinates EPA and State activities, responsibilities and programs including those under the RCRA, SDWA, and CWA programs.

State regulated waters (404) means those "waters of the United States" in which the Corps of Engineers suspends the issuance of section 404 permits upon approval of a State's section 404 permit program by the Administrator under section 404(h). These waters shall be identified in the program description as required by § 123.4(h)(1). The Secretary shall retain jurisdiction over the following waters (see CWA section 404(g)(1)):

(a) Waters which are subject to the ebb and flow of the tide;

(b) Waters which are presently used, or are susceptible to use in their natural condition or by reasonable improvement as a means to transport interstate or foreign commerce shoreward to their ordinary high water mark; and

(c) "Wetlands" adjacent to waters in (a) and (b).

Storage (RCRA) means the holding of "hazardous waste" for a temporary period, at the end of which the hazardous waste is treated, disposed, or stored elsewhere.

Stratum (plural *strata*) (UIC) means a single sedimentary bed or layer, regardless of thickness, that consists of generally the same kind of rock material.

Total dissolved solids (UIC and NPDES) means the total dissolved (filterable) solids as determined by use of the method specified in 40 CFR Part 136.

Toxic pollutant (NPDES and 404) means any pollutant listed as toxic under section 307(a)(1) of CWA.

Transporter (RCRA) means a person engaged in the off-site transportation of "hazardous waste" by air, rail, highway or water.

Treatment (RCRA) means any method, technique, or process, including neutralization, designed to change the physical, chemical, or biological character or composition of any "hazardous waste" so as to neutralize such wastes, or so as to recover energy or material resources from the waste, or so as to render such waste non-hazardous, or less hazardous; safer to transport, store or dispose of; or amenable for recovery, amenable for storage, or reduced in volume.

UIC means the Underground Injection Control program under Part C of the Safe Drinking Water Act, including an "approved program."

Underground injection (UIC) means a "well injection."

Underground source of drinking water ("USDW") (RCRA and UIC) means an "aquifer" or its portion:

(a)(1) Which supplies drinking water for human consumption; or

(2) In which the ground water contains fewer than 10,000 mg/l "total dissolved solids;" and

(b) Which is not an "exempted aquifer."

USDW (RCRA and UIC) means "underground source of drinking water."

Variance (NPDES) means any mechanism or provision under sections 301 or 316 of CWA or under 40 CFR Part 125, or in the applicable "effluent limitations guidelines" which allows modification to or waiver of the generally applicable effluent limitation requirements or time deadlines of CWA. This includes provisions which allow the establishment of alternative limitations based on fundamentally different factors or on sections 301(c), 301(g), 301(h), 301(i), or 316(a) of CWA.

Waters of the United States or *Waters of the U.S.* means:

(a) All waters which are currently used, were used in the past, or may be susceptible to use in interstate or foreign commerce, including all waters which are subject to the ebb and flow of the tide;

(b) All interstate waters, including interstate "wetlands;"

(c) All other waters such as intrastate lakes, rivers, streams (including intermittent streams), mudflats, sandflats, "wetlands," sloughs, prairie potholes, wet meadows, playa lakes, or natural ponds the use, degradation, or destruction of which would affect or could affect interstate or foreign commerce including any such waters:

(1) Which are or could be used by interstate or foreign travelers for recreational or other purposes;

(2) From which fish or shellfish are or could be taken and sold in interstate or foreign commerce; or

(3) Which are used or could be used for industrial purposes by industries in interstate commerce;

(d) All impoundments of waters

otherwise defined as waters of the United States under this definition;

(e) Tributaries of waters identified in paragraphs (1)–(4) of this definition;

(f) The territorial sea; and

(g) "Wetlands" adjacent to waters (other than waters that are themselves wetlands) identified in paragraphs (a)–(f) of this definition.

Waste treatment systems, including treatment ponds or lagoons designed to meet the requirements of CWA (other than cooling ponds as defined in 40 CFR § 423.11(m) which also meet the criteria of this definition) are not waters of the United States. This exclusion applies only to manmade bodies of water which neither were originally created in waters of the United States (such as a disposal area in wetlands) nor resulted from the impoundment of waters of the United States.

Well (UIC) means a bored, drilled or driven shaft, or a dug hole, whose depth is greater than the largest surface dimension.

Well injection (UIC) means the subsurface emplacement of "fluids" through a bored, drilled, or driven "well;" or through a dug well, where the depth of the dug well is greater than the largest surface dimension.

Wetlands means those areas that are inundated or saturated by surface or ground water at a frequency and duration sufficient to support, and that under normal circumstances do support, a prevalence of vegetation typically adapted for life in saturated soil conditions. Wetlands generally include swamps, marshes, bogs, and similar areas.

§ 122.4 Application for a permit.

(Applicable to State programs, see § 123.7.)

(a) *Permit application.* Any person who is required to have a permit (including new applicants and permittees with expiring permits) shall complete, sign, and submit an application to the Director as described in this section and in §§ 122.23 (RCRA), 122.38 (UIC), 122.53 (NPDES), and 123.94 (404). Persons currently authorized with interim status under RCRA (§ 122.23) or UIC authorization by rule (§ 122.37) shall apply for permits when required by the Director. Persons covered by RCRA permits by rule (§ 122.26), and NPDES or 404 dischargers covered by general permits under § 122.59 or 123.97, respectively, need not apply. Procedures for applications, issuance and administration of emergency permits are found exclusively in §§ 122.27 (RCRA), 122.40 (UIC), and 123.96 (404).

(b) *Who applies?* When a facility or activity is owned by one person but is operated by another person, it is the operator's duty to obtain a permit,

except that for RCRA only, the owner must also sign the permit application.

(c) *Completeness.* The Director shall not issue a permit under a program before receiving a complete application for a permit under that program except for NPDES and 404 general permits, RCRA permits by rule, or emergency permits. An application for a permit under a program is complete when the Director receives an application form and any supplemental information which are completed to his or her satisfaction. The completeness of any application for a permit shall be judged independently of the status of any other permit application or permit for the same facility or activity. For EPA—administered RCRA, UIC, and NPDES programs, an application which is reviewed under § 124.3 is complete when the Director receives either a complete application or the information listed in a notice of deficiency.

(d) *Information requirements.* All applicants for RCRA, UIC, or NPDES permits (for State 404 permits see § 123.94) shall provide the following information to the Director, using the application form provided by the Director (additional information required of applicants is set forth in §§ 122.24 and 122.25 (RCRA), 122.38 (UIC), and 122.53 (NPDES)).

(1) The activities conducted by the applicant which require it to obtain permits under RCRA, UIC, NPDES, or PSD.

(2) Name, mailing address, and location of the facility for which the application is submitted.

(3) Up to four SIC codes which best reflect the principal products or services provided by the facility.

(4) The operator's name, address, telephone number, ownership status, and status as Federal, State, private, public, or other entity.

(5) Whether the facility is located on Indian lands.

(6) A listing of all permits or construction approvals received or applied for under any of the following programs:

(i) Hazardous Waste Management program under RCRA.

(ii) UIC program under SDWA.

(iii) NPDES program under CWA.

(iv) Prevention of Significant Deterioration (PSD) program under the Clean Air Act.

(v) Nonattainment program under the Clean Air Act.

(vi) National Emission Standards for Hazardous Pollutants (NESHAPS) preconstruction approval under the Clean Air Act.

(vii) Ocean dumping permits under the Marine Protection Research and Sanctuaries Act.

(viii) Dredge or fill permits under section 404 of CWA.

(ix) Other relevant environmental permits, including State permits.

(7) A topographic map (or other map if a topographic map is unavailable) extending one mile beyond the property boundaries of the source, depicting the facility and each of its intake and discharge structures; each of its hazardous waste treatment, storage, or disposal facilities; each well where fluids from the facility are injected underground; and those wells, springs, other surface water bodies, and drinking water wells listed in public records or otherwise known to the applicant in the map area.

(8) A brief description of the nature of the business.

(e) *Recordkeeping.* Applicants shall keep records of all data used to complete permit applications and any supplemental information submitted under §§ 122.4(d), 122.24, and 122.25 (RCRA); 122.38 (UIC); 122.53 (NPDES); and 123.94 (404) for a period of at least 3 years from the date the application is signed.

§ 122.5 Continuation of expiring permits.

(a) *EPA permits.* When EPA is the permit-issuing authority, the conditions of an expired permit continue in force under 5 U.S.C. § 558(c) until the effective date of a new permit (see § 124.15) if:

(1) The permittee has submitted a timely application under §§ 122.25 (RCRA), 122.38 (UIC), or 122.53 (NPDES) which is a complete (under § 122.4(c)) application for a new permit; and

(2) The Regional Administrator, through no fault of the permittee, does not issue a new permit with an effective date under § 124.15 on or before the expiration date of the previous permit (for example, when issuance is impracticable due to time or resource constraints).

(b) *Effect.* Permits continued under this section remain fully effective and enforceable.

(c) *Enforcement.* When the permittee is not in compliance with the conditions of the expiring or expired permit the Regional Administrator may choose to do any or all of the following:

(1) Initiate enforcement action based upon the permit which has been continued;

(2) Issue a notice of intent to deny the new permit under § 124.6. If the permit is denied, the owner or operator would then be required to cease the activities authorized by the continued permit or be subject to enforcement action for operating without a permit;

(3) Issue a new permit under Part 124 with appropriate conditions; or

(4) Take other actions authorized by these regulations.

(d) *State continuation.*

(1) An EPA 1 (or, in the case of 404, Corps of Engineers) issued permit does not continue in force beyond its expiration date under Federal law if at that time a State is the permitting authority. States authorized to administer the RCRA, UIC, NPDES or 404 programs may continue either EPA (or Corps of Engineers) or State-issued permits until the effective date of the new permits, if State law allows. Otherwise, the facility or activity is operating without a permit from the time of expiration of the old permit to the effective date of the State-issued new permit.

§ 122.6 Signatories to permit applications and reports.

(Applicable to State programs, see § 123.7.)

(1) *Applications.* All permit applications, except those submitted for Class II wells under the UIC program (see paragraph (b) of this section), shall be signed as follows:

(1) *For a corporation:* by a principal executive officer of at least the level of vice-president;

(2) *For a partnership or sole proprietorship:* by a general partner or the proprietor, respectively; or

(3) *For a municipality, State, Federal, or other public agency:* by either a principal executive officer or ranking elected official.

(b) *Reports.* All reports required by permits, other information requested by the Director, and all permit applications submitted for Class II wells under § 122.38 for the UIC program shall be signed by a person described in paragraph (a) of this section, or by a duly authorized representative of that person. A person is a duly authorized representative only if:

(1) The authorization is made in writing by a person described in paragraph (a) of this section;

(2) The authorization specifies either an individual or a position having responsibility for the overall operation of the regulated facility or activity, such as the position of plant manager, operator of a well or a well field, superintendent, or position of equivalent responsibility. (A duly authorized representative may thus be either a named individual or any individual occupying a named position.); and

(3) The written authorization is submitted to the Director.

(c) *Changes to authorization.* If an authorization under paragraph (b) of this section is no longer accurate because a different individual or position has responsibility for the overall operation of the facility, a new authorization satisfying the requirements of paragraph (b) of this section must be submitted to the Director prior to or together with any reports, information, or applications to be signed by an authorized representative.

(d) *Certification.* Any person signing a document under paragraphs (a) or (b) of this section shall make the following certification:

"I certify under penalty of law that I have personally examined and am familiar with the information submitted in this document and all attachments and that, based on my inquiry of those individuals immediately responsible for obtaining the information, I believe that the information is true, accurate, and complete. I am aware that there are significant penalties for submitting false information, including the possibility of fine and imprisonment."

§ 122.7 Conditions applicable to all permits.

(Applicable to State programs, see § 123.7.)

The following conditions apply to all RCRA, UIC, NPDES, and 404 permits. For additional conditions applicable to all permits for each of the programs individually, see sections 122.28 (RCRA), 122.41 (UIC), 122.60 and 122.61 (NPDES) and 123.97 (404). All conditions applicable to all permits, and all additional conditions applicable to all permits for individual programs, shall be incorporated into the permits either expressly or by reference. If incorporated by reference, a specific citation to these regulations (or the corresponding approved State regulations) must be given in the permit.

(a) *Duty to comply.* The permittee must comply with all conditions of this permit. Any permit noncompliance constitutes a violation of the appropriate Act and is grounds for enforcement action; for permit termination, revocation and reissuance, or modification; or for denial of a permit renewal application.

(b) *Duty to reapply.* If the permittee wishes to continue an activity regulated by this permit after the expiration date of this permit, the permittee must apply for and obtain a new permit.

(c) *Duty to halt or reduce activity.* It shall not be a defense for a permittee in an enforcement action that it would have been necessary to halt or reduce the permitted activity in order to maintain compliance with the conditions of this permit.

(d) *Duty to mitigate.* The permittee shall take all reasonable steps to minimize or correct any adverse impact on the environment resulting from noncompliance with this permit.

(e) *Proper operation and maintenance.* The permittee shall at all times properly operate and maintain all facilities and systems of treatment and control (and related appurtenances) which are installed or used by the permittee to achieve compliance with the conditions of this permit. Proper operation and maintenance includes effective performance, adequate funding, adequate operator staffing and training, and adequate laboratory and process controls, including appropriate quality assurance procedures. This provision requires the operation of back-up or auxiliary facilities or similar systems *only* when necessary to achieve compliance with the conditions of the permit.

(f) *Permit actions.* This permit may be modified, revoked and reissued, or terminated for cause. The filing of a request by the permittee for a permit modification, revocation and reissuance, or termination, or a notification of planned changes or anticipated noncompliance, does not stay any permit condition.

(g) *Property rights.* This permit does not convey any property rights of any sort, or any exclusive privilege.

(h) *Duty to provide information.* The permittee shall furnish to the Director, within a reasonable time, any information which the Director may request to determine whether cause exists for modifying, revoking and reissuing, or terminating this permit, or to determine compliance with this permit. The permittee shall also furnish to the Director, upon request, copies of records required to be kept by this permit.

(i) *Inspection and entry.* The permittee shall allow the Director, or an authorized representative, upon the presentation of credentials and other documents as may be required by law, to:

(1) Enter upon the permittee's premises where a regulated facility or activity is located or conducted, or where records must be kept under the conditions of this permit;

(2) Have access to and copy, at reasonable times, any records that must be kept under the conditions of this permit;

(3) Inspect at reasonable times any facilities, equipment (including monitoring and control equipment), practices, or operations regulated or required under this permit; and

(4) Sample or monitor at reasonable times, for the purposes of assuring permit compliance or as otherwise authorized by the appropriate Act, any

substances or parameters at any location.

(j) *Monitoring and records.*

(1) Samples and measurements taken for the purpose of monitoring shall be representative of the monitored activity.

(2) The permittee shall retain records of all monitoring information, including all calibration and maintenance records and all original strip chart recordings for continuous monitoring instrumentation, copies of all reports required by this permit, and records of all data used to complete the application for this permit, for a period of at least 3 years from the date of the sample, measurement, report or application. This period may be extended by request of the Director at any time.

(3) Records of monitoring information shall include:

(i) The date, exact place, and time of sampling or measurements;

(ii) The individual(s) who performed the sampling or measurements;

(iii) The date(s) analyses were performed;

(iv) The individual(s) who performed the analyses;

(v) The analytical techniques or methods used; and

(vi) The results of such analyses.

(k) *Signatory requirement.* All applications, reports, or information submitted to the Director shall be signed and certified. (See § 122.6.)

(l) *Reporting requirements.* (1) *Planned changes.* The permittee shall give notice to the Director as soon as possible of any planned physical alternations or additions to the permitted facility.

(2) *Anticipated noncompliance.* The permittee shall give advance notice to the Director of any planned changes in the permitted facility or activity which may result in noncompliance with permit requirements.

(3) *Transfers.* This permit is not transferable to any person except after notice to the Director. The Director may require modification or revocation and reissuance of the permit to change the name of the permittee and incorporate such other requirements as may be necessary under the appropriate Act. (See § 122.14; in some cases, modification or revocation and reissuance is mandatory.)

(4) *Monitoring reports.* Monitoring results shall be reported at the intervals specified elsewhere in this permit.

(5) *Compliance schedules.* Reports of compliance or noncompliance with, or any progress reports on, interim and final requirements contained in any compliance schedule of this permit shall be submitted no later than 14 days following each schedule date.

(6) *Twenty-four hour reporting.* The permittee shall report any noncompliance which may endanger health or the environment. Any information shall be provided orally within 24 hours from the time the permittee becomes aware of the circumstances. A written submission shall also be provided within 5 days of the time the permittee becomes aware of the circumstances. The written submission shall contain a description of the noncompliance and its cause; the period of noncompliance, including exact dates and times, and if the noncompliance has not been corrected, the anticipated time it is expected to continue; and steps taken or planned to reduce, eliminate, and prevent reoccurrence of the noncompliance.

(7) *Other noncompliance.* The permittee shall report all instances of noncompliance not reported under paragraphs (1)(4), (5), and (6) of this section, at the time monitoring reports are submitted. The reports shall contain the information listed in paragraph (l)(6) of this section.

(8) *Other information.* Where the permittee becomes aware that it failed to submit any relevant facts in a permit application, or submitted incorrect information in a permit application or in any report to the Director, it shall promptly submit such facts or information.

§ 122.8 Establishing permit conditions.

(Applicable to State programs, see § 122.7.)

(a) *All programs.* In addition to conditions required in all permits for all programs (§ 122.7), the Director shall establish conditions, as required on a case-by-case basis, in permits for all programs under §§ 122.9 (duration of permits), 122.10(a) (schedules of compliance), 122.11 (monitoring), and for EPA permits only 122.10(b) (alternate schedules of compliance) and 122.12 (considerations under Federal law).

(b) *Individual programs.*

(1) In addition to conditions required in all permits for a particular program (§§ 122.28 for RCRA, 122.41 for UIC, 122.60 and 122.61 for NPDES, and 123.97 for 404), the Director shall establish conditions in permits for the individual programs, as required on a case-by-case basis, to provide for and assure compliance with all applicable requirements of the appropriate Act and regulations.

(2) For a State issued permit, an applicable requirement is a State statutory or regulatory requirement which takes effect prior to final administrative disposition of a permit. For a permit issued by EPA, an applicable requirement is a statutory or regulatory requirement (including any interim final regulation) which takes effect prior to the issuance of the permit (except as provided in § 124.86(c) for RCRA, UIC and NPDES permits being processed under Subparts E or F of Part 124). Section 124.14 (reopening of comment period) provides a means for reopening EPA permit proceedings at the discretion of the Director where new requirements become effective during the permitting process and are of sufficient magnitude to make additional preceedings desirable. For State and EPA administered programs, an applicable requirement is also any requirement which takes effect prior to the modification or revocation and reissuance of a permit, to the extent allowed in § 122.15.

(3) New or reissued permits, and to the extent allowed under § 122.15 modified or revoked and reissued permits, shall incorporate each of the applicable requirements referenced in §§ 122.29 (RCRA), 122.42 (UIC), 122.62 and 122.63 (NPDES); and 123.98 (404).

(c) *Incorporation.* All permit conditions shall be incorporated either expressly or by reference. If incorporated by reference, a specific citation to the applicable regulations or requirements must be given in the permit.

§ 122.9 Duration of permits.

(Applicable to State programs, see § 123.7.)

(a) *NPDES and section 404.* NPDES and section 404 permits shall be effective for a fixed term not to exceed 5 years.

(b) *RCRA.* RCRA permits shall be effective for a fixed term not to exceed 10 years. (See also § 122.30 (interim permits for UIC wells)).

(c) *UIC.* UIC permits for Class I and Class V wells shall be effective for a fixed term not to exceed 10 years. UIC permits for Class II and III wells shall be issued for a period up to the operating life of the facility. The Director shall review each issued Class II or III well UIC permit at least once every 5 years to determine whether it should be modified, revoked and reissued, terminated, or a minor modification made as provided in §§ 122.15, 122.16, and 122.17.

(d) Except as provided in § 122.5, the term of a permit shall not be extended by modification beyond the maximum duration specified in this section.

(e) The Director may issue any permit for a duration that is less than the full allowable term under this section.

§ 122.10 Schedules of compliance.

(a) *General (applicable to State programs, see § 123.7).* The permit may, when appropriate, specify a schedule of compliance leading to compliance with the appropriate Act and regulations.

(1) *Time for compliance.* Any schedules of compliance under this section shall require compliance as soon as possible.

(i) *For NPDES,* in addition, schedules of compliance shall require compliance not later than the applicable statutory deadline under the CWA.

(ii) *For UIC,* in addition, schedules of compliance shall require compliance not later than 3 years after the effective date of the permit.

(2) *For NPDES only.* The first NPDES permit issued to a new source, a new discharger which commenced discharge after August 13, 1979, or a recommencing discharger shall not contain a schedule of compliance under this section. See also § 122.66(d)(4).

(3) *Interim dates.* Except as provided in paragraph (b)(1)(ii) of this section, if a permit establishes a schedule of compliance which exceeds 1 year from the date of permit issuance, the schedule shall set forth interim requirements and the dates for their achievement.

(i) The time between interim dates shall not exceed 1 year.

(ii) If the time necessary for completion of any interim requirement (such as the construction of a control facility) is more than 1 year and is not readily divisible into stages for completion, the permit shall specify interim dates for the submission of reports of progress toward completion of the interim requirements and indicate a projected completion date.

[Note.—Examples of interim requirements include: (1) submit a complete Step 1 construction grant (for POTWs); (2) let a contract for construction of required facilities; (3) commence construction of required facilities; (4) complete construction of required facilities.]

(4) *Reporting.* The permit shall be written to require that no later than 14 days following each interim date and the final date of compliance, the permittee shall notify the director in writing of its compliance or noncompliance with the interim or final requirements, or submit progress reports if paragraph (a)(1)(ii) of this section is applicable.

(b) *Alternative schedules of compliance.* A RCRA, UIC, or NPDES permit applicant or permittee may cease conducting regulated activities (by receiving a terminal volume of hazardous waste for HWM facilities, plugging and abandonment for UIC wells, or termination of direct discharge for NPDES sources) rather than continue to operate and meet permit requirements as follows:

(1) If the permittee decides to cease conducting regulated activities at a given time within the term of a permit which has already been issued:

(i) The permit may be modified to contain a new or additional schedule leading to timely cessation of activities; or

(ii) The permittee shall cease conducting permitted activities before noncompliance with any interim or final compliance schedule requirement already specified in the permit.

(2) If the decision to cease conducting regulated activities is made before issuance of a permit whose term will include the termination date, the permit shall contain a schedule leading to termination which will ensure timely compliance with applicable requirements, or for NPDES, compliance no later than the statutory deadline.

(3) If the permittee is undecided whether to cease conducting regulated activities, the Director may issue or modify a permit to contain two schedules as follows:

(i) Both schedules shall contain an identical interim deadline requiring a final decision on whether to cease conducting regulated activities no later than a date which ensures sufficient time to comply with applicable requirements in a timely manner if the decision is to continue conducting regulated activities;

(ii) One schedule shall lead to timely compliance with applicable requirements, and for NPDES, compliance no later than the statutory deadline;

(iii) The second schedule shall lead to cessation of regulated activities by a date which will ensure timely compliance with applicable requirements, or for NPDES, compliance no later than the statutory deadline.

(iv) Each permit containing two schedules shall include a requirement that after the permittee has made a final decision under paragraph (b)(3)(i) of this section it shall follow the schedule leading to compliance if the decision is to continue conducting regulated activities, and follow the schedule leading to termination if the decision is to cease conducting regulated activities.

(4) The applicant's or permittee's decision to cease conducting regulated activities shall be evidenced by a firm public commitment satisfactory to the Director, such as a resolution of the board of directors of a corporation.

§ 122.11 Requirements for recording and reporting of monitoring results.

(Applicable to State programs, see § 123.7.)

All permits shall specify:

(a) Requirements concerning the proper use, maintenance, and installation, when appropriate, of monitoring equipment or methods (including biological monitoring methods when appropriate);

(b) Required monitoring including type, intervals, and frequency sufficient to yield data which are representative of the monitored activity including, when appropriate, continuous monitoring;

(c) Applicable reporting requirements based upon the impact of the regulated activity and as specified in Parts 264 and 266 (RCRA), Part 146 (UIC), § 122.62 (NPDES), and, when applicable, 40 CFR Part 230 (404). Reporting shall be no less frequent than specified in the above regulations.

§ 122.12 Considerations under Federal law.

Permits shall be issued in a manner and shall contain conditions consistent with requirements of applicable Federal laws. These laws may include:

(a) The *Wild and Scenic Rivers Act,* 16 U.S.C. 1273 *et seq.* Section 7 of the Act prohibits the Regional Administrator from assisting by license or otherwise the construction of any water resources project that would have a direct, adverse effect on the values for which a national wild and scenic river was established.

(b) The *National Historic Preservation Act of 1966,* 16 U.S.C. 470 *et seq.* Section 106 of the Act and implementing regulations (36 CFR Part 800) require the Regional Administrator, before issuing a license, to adopt measures when feasible to mitigate potential adverse effects of the licensed activity and properties listed or eligible for listing in the National Register of Historic Places. The Act's requirements are to be implemented in cooperation with State Historic Preservation Officers and upon notice to, and when appropriate, in consultation with the Advisory Council on Historic Preservation.

(c) The *Endangered Species Act,* 16 U.S.C. 1531 *et seq.* Section 7 of the Act and implementing regulations (50 CFR Part 402) require the Regional Administrator to ensure, in consultation with the Secretary of the Interior or Commerce, that any action authorized by EPA is not likely to jeopardize the continued existence of any endangered or threatened species or adversely affect its critical habitat.

(d) The *Coastal Zone Management Act,* 16 U.S.C. 1451 *et seq.* Section 307(c) of the Act and implementing regulations (15 CFR Part 930) prohibit EPA from issuing a permit for an activity affecting land or water use in the coastal zone until the applicant certifies that the proposed activity complies with the State Coastal Zone Management program, and the State or its designated agency concurs with the certification (or the Secretary of Commerce overrides the State's nonconcurrence).

(e) The *Fish and Wildlife Coordination Act,* 16 U.S.C. 661 *et seq.,* requires that the Regional Administrator, before issuing a permit proposing or authorizing the impoundment (with certain exemptions), diversion, or other control or modification of any body of water, consult with the appropriate State agency exercising jurisdiction over wildlife resources to conserve those resources.

(f) *Executive orders.* (Reserved.)

(g) *For NPDES only,* the National Environmental Policy Act, 33 U.S.C. 4321 *et seq.,* may require preparation of an Environmental Impact Statement and the inclusion of EIS-related permit conditions, as provided in § 122.67(c).

§ 122.13 Effect of a permit.

(a) *(Applicable to State programs, see § 123.7(a)].* Except for Class II and III wells under UIC, and except for any toxic effluent standards and prohibitions imposed under section 307 of the CWA for NPDES, compliance with a permit during its term constitutes compliance, for purposes of enforcement, with Subtitle C of RCRA, Part C of SWDA, sections 301, 302, 306, 307, 318, 403, and 405 of CWA for NPDES, and sections 301, 307, and 403 of CWA for 404. However, a permit may be modified, revoked and reissued, or terminated during its term for cause as set forth in §§ 122.15 and 122.16.

(b) *(Applicable to State programs, see § 123.7(a).)* The issuance of a permit does not convey any property rights of any sort, or any exclusive privilege.

(c) The issuance of a permit does not authorize any injury to persons or property or invasion of other private rights, or any infringement of State or local law or regulations.

§ 122.14 Transfer of permits.

(Applicable to State programs, see § 122.7.)

(a) *Transfers by modification.* Except as provided in paragraph (b) of this section, a permit may be transferred by the permittee to a new owner or operator only if the permit has been modified or revoked and reissued (under § 122.15(b)(2)), or a minor modification made (under § 122.17(d)), to identify the new permittee and incorporate such other requirements as may be necessary under the appropriate Act.

(b) *Automatic transfers.* As an alternative to transfers under paragraph (a) of this section, any NPDES permit or UIC permit for a well not injecting hazardous waste may be automatically transferred to a new permittee if:

(1) The current permittee notifies the Director at least 30 days in advance of the proposed transfer date in paragraph (b)(2) of this section;

(2) The notice includes a written agreement between the existing and new permittees containing a specific date for transfer of permit responsibility, coverage, and liability between them and, in the case of UIC permits, the notice demonstrates that the financial responsibility requirements of § 122.42(g) will be met by the new permittee; and

(3) The Director does not notify the existing permittee and the proposed new permittee of his or her intent to modify or revoke and reissue the permit. A modification under this subparagraph may also be a minor modification under § 122.17. If this notice is not received, the transfer is effective on the date specified in the agreement mentioned in paragraph (b)(2) of this section.

§ 122.15 Modification or revocation and reissuance of permits.

(Applicable to State programs, see § 123.7).

When the Director receives any information (for example, inspects the facility, receives information submitted by the permittee as required in the permit (see § 122.7), receives a request for modification or revocation and reissuance under § 124.5, or conducts a review of the permit file) he or she may determine whether or not one or more of the causes listed in paragraphs (a) and (b) of this section for modification or revocation and reissuance or both exist. If cause exists, the Director may modify or revoke and reissue the permit accordingly, subject to the limitations of paragraph (c) of this section, and may request an updated application if necessary. When a permit is modified, only the conditions subject to modification are reopened. If a permit is revoked and reissued, the entire permit is reopened and subject to revision and the permit is reissued for a new term. See § 124.5(c)(2). If cause does not exist under this section or § 122.17, the Director shall not modify or revoke and reissue the permit. If a permit modification satisfies the criteria in § 122.17 for "minor modifications" the permit may be modified without a draft permit or public review. Otherwise, a draft permit must be prepared and other procedures in Part 124 (or procedures of an approved State program) followed.

(a) *Causes for modification.* The following are causes for modification but not revocation and reissuance of permits. However, for Class II or III wells under UIC, the following may be causes for revocation and reissuance as well as modification; and the following may be causes for revocation and reissuance as well as modification under any program when the permittee requests or agrees.

(1) *Alterations.* There are material and substantial alterations or additions to the permitted facility or activity which occurred after permit issuance which justify the application of permit conditions that are different or absent in the existing permit.

[Note.—For NPDES, certain reconstruction activities may cause the new source provisions of § 122.67 to be applicable.]

(2) *Information.* The Director has received information. Permits other than for UIC Class II and III wells may be modified during their terms for this cause only if the information was not available at the time of permit issuance (other than revised regulations, guidance, or test methods) and would have justified the application of different permit conditions at the time of issuance. For UIC area permits (§ 122.39), NPDES general permits (§ 122.59) and 404 general permits (§ 123.95) this cause shall include any information indicating that cumulative effects on the environment are unacceptable.

(3) *New regulations.* The standards or regulations on which the permit was based have been changed by promulgation of amended standards or regulations or by judicial decision after the permit was issued. Permits other than for UIC Class II or III wells may be modified during their terms for this cause only as follows:

(i) For promulgation of amended standards or regulations, when:

. (A) The permit condition requested to be modified was based on a promulgated Part 260–266 (RCRA) or Part 146 (UIC) regulation, or a promulgated effluent limitation guideline or EPA approved or promulgated water quality standard (NPDES); and

(B) EPA has revised, withdrawn, or modified that portion of the regulation or effluent limitation guideline on which the permit condition was based, or has approved a State action with regard to a water quality standard on which the permit condition was based; and

(C) A permittee requests modification in accordance with § 124.5 within ninety (90) days after **Federal Register** notice of the action on which the request is based.

(ii) For judicial decisions, a court of competent jurisdiction has remanded and stayed EPA promulgated regulations or effluent limitation guidelines, if the remand and stay concern that portion of the regulations or guidelines on which the permit condition was based and a request is filed by the permittee in accordance with § 124.5 within ninety (90) days of judicial remand.

(iii) For changes based upon modified State certifications of NPDES permits, see § 124.55(b).

(4) *Compliance schedules.* The Director determines good cause exists for modification of a compliance schedule, such as an act of God, strike, flood, or materials shortage or other events over which the permittee has little or no control and for which there is no reasonably available remedy. However, in no case shall an NPDES compliance schedule be modified to extend beyond an applicable CWA statutory deadline. See also § 122.17(c) (minor modifications) and paragraph (a)(5)(xi) of this section (NPDES innovative technology).

(5) *For NPDES only,* the Director may modify a permit:

(i) When the permittee has filed a request for a variance under CWA sections 301(c), 301(g), 301(h), 301(i), 301(k), or 316(a), or for "fundamentally different factors" within the time specified in § 122.53, and the Director processes the request under the applicable provisions of §§ 124.61, 124.62, and 124.64.

(ii) When required to incorporate an applicable 307(a) toxic effluent standard or prohibition (see § 122.62(b)).

(iii) When required by the "reopener" conditions in a permit, which are established in the permit under § 122.62(b) (for CWA toxic effluent limitations) or 40 CFR § 403.10(e) (pretreatment program):

(iv) Upon request of a permittee who qualifies for effluent limitations on a net basis under § 122.63(h).

(v) When a discharger is no longer eligible for net limitations, as provided in § 122.63(h)(1)(ii)(B).

(vi) As necessary under 40 CFR § 403.8(e) (compliance schedule for development of pretreatment program).

(vii) Upon failure of an approved State to notify, as required by section 402(b)(3), another State whose waters may be affected by a discharge from the approved State.

(viii) When the level of discharge of any pollutant which is not limited in the permit exceeds the level which can be achieved by the technology-based treatment requirements appropriate to the permittee under § 125.3(c).

(ix) When the permittee begins or expects to begin to use or manufacture as an intermediate or final product or byproduct any toxic pollutant which was not reported in the permit application under § 122.53(d)(9).

(x) To establish a "notification level" as provided in § 122.62(f).

(xi) To modify a schedule of compliance to reflect the time lost during construction of an innovative or alternative facility, in the case of a POTW which has received a grant under section 202(a)(3) of CWA for 100% of the costs to modify or replace facilities constructed with a grant for innovative and alternative wastewater technology under section 202(a)(2). In no case shall the compliance schedule be modified to extend beyond an applicable CWA statutory deadline for compliance.

(6) *For 404 only,* the Director shall modify a permit to reflect toxic effluent standards or prohibitions or water quality standards, under the "reopener" condition of § 123.97(g).

(b) *Causes for modification or revocation and reissuance.* The following are causes to modify or, alternatively, revoke and reissue a permit:

(1) Cause exists for termination under § 122.16, and the Director determines that modification or revocation and reissuance is appropriate.

(2) The Director has received notification (as required in the permit, see § 122.17(1)(3)) of a proposed transfer of the permit. A permit also may be modified to reflect a transfer after the effective date of an automatic transfer (§ 122.14(b)) but will not be revoked and reissued after the effective date of the transfer except upon the request of the new permittee.

(c) *Facility siting.* For RCRA and UIC, suitability of the facility location will not be considered at the time of permit modification or revocation and reissuance unless new information or standards indicate that a threat to human health or the environment exists which was unknown at the time of permit issuance.

§ 122.16 Termination of permits.

(Applicable to State programs, see § 122.7.)

(a) The following are causes for terminating a permit during its term, or for denying a permit renewal application:

(1) Noncompliance by the permittee with any condition of the permit;

(2) The permittee's failure in the application or during the permit issuance process to disclose fully all relevant facts, or the permittee's misrepresentation of any relevant facts at any time; or

(3) A determination that the permitted activity endangers human health or the environment and can only be regulated to acceptable levels by permit modification or termination.

(4) *For NPDES and 404 only,* permits may be modified or terminated when there is a change in any condition that requires either a temporary or a permanent reduction or elimination of any discharge controlled by the permit (for example, plant closure or termination of discharge by connection to a POTW).

(b) The Director shall follow the applicable procedures in Part 124 or State procedures in terminating any RCRA, UIC, NPDES, or 404 permit under this section.

§ 122.17 Minor modifications of permits.

Upon the consent of the permittee, the Director may modify a permit to make the corrections or allowances for changes in the permitted activity listed in this section, without following the procedures of Part 124. Any permit modification not processed as a minor modification under this section must be made for cause and with Part 124 draft permit and public notice as required in § 122.15. Minor modifications may only:

(a) Correct typographical errors;

(b) Require more frequent monitoring or reporting by the permittee;

(c) Change an interim compliance date in a schedule of compliance, provided the new date is not more than 120 days after the date specified in the existing permit and does not interfere with attainment of the final compliance date requirement; or

(d) Allow for a change in ownership or operational control of a facility where the Director determines that no other change in the permit is necessary, provided that a written agreement containing a specific date for transfer of permit responsibility, coverage, and liability between the current and new permittees has been submitted to the Director.

(e) *For RCRA only,* change the lists of facility emergency coordinators or equipment in the permit's contingency plan.

(f) *For UIC only,*

(1) Change quantities or types of fluids injected which are within the capacity of the facility as permitted and, in the judgment of the Director, after reviewing information required under §§ 146.16, 146.26 and 146.36, would not

interfere with the operation of the facility or its ability to meet conditions prescribed in the permit, and would not change its classification.

(2) Change construction requirements approved by the Director pursuant to § 122.42(a) (establishing UIC permit conditions), provided that any such alteration shall comply with the requirements of this Part and Part 146.

(g) *For NPDES only,*

(1) Change the construction schedule for a discharger which is a new source. No such change shall affect a discharger's obligation to have all pollution control equipment installed and in operation prior to discharge under § 122.66.

(2) Delete a point source outfall when the discharge from that outfall is terminated and does not result in discharge of pollutants from other outfalls except in accordance with permit limits.

(h) *For 404 only,* extend the term of a State section 404 permit, so long as the modification does not extend the term of the permit beyond 5 years from its original effective date.

§ 122.18 Noncompliance and program reporting by the Director.

(Applicable to State programs, see § 123.7.)

The Director shall prepare quarterly and annual reports as detailed below. When the State is the permit-issuing authority, the State Director shall submit any reports required under this section to the Regional Administrator. When EPA is the permit-issuing authority, the Regional Administrator shall submit any report required under this section to EPA Headquarters. For purposes of this section only, RCRA permittees shall include RCRA interim status facilities, when appropriate.

(a) *Quarterly reports for RCRA, UIC, and NPDES.* The Director shall submit quarterly narrative reports for major facilities as follows:

(1) *Format.* The report shall use the following format:

(i) Provide separate lists for RCRA, UIC, and NPDES permittees; the NPDES permittees shall be further subcategorized as non-POTWs, POTWs, and Federal permittes;

(ii) For facilities or activities with permits under more than one program, provide an additional list combining information on noncompliance for each such facility;

(iii) Alphabetize each list by permittee name. When two or more permittees have the same name, the lowest permit number shall be entered first.

(iv) For each entry on a list, include the following information in the following order:

(A) Name, location, and permit number of the noncomplying permittee.

(B) A brief description and date of each instance of noncompliance for that permittee. Instances of noncompliance may include one or more of the kinds set forth in paragraph (a)(2) of this section. When a permittee has noncompliance of more than one kind under a single program, combine the information into a single entry for each such permittee.

(C) The date(s) and a brief description of the action(s) taken by the Director to ensure compliance.

(D) Status of the instance(s) of noncompliance with the date of the review of the status or the date of resolution.

(E) Any details which tend to explain or mitigate the instance(s) of noncompliance.

(2) *Instances of noncompliance to be reported.* Any instances of noncompliance within the following categories shall be reported in successive reports until the noncompliance is reported as resolved. Once noncompliance is reported as resolved it need not appear in subsequent reports.

(i) *Failure to complete construction elements.* When the permittee has failed to complete, by the date specified in the permit, an element of a compliance schedule involving either planning for construction (for example, award of a contract, preliminary plans), or a construction step (for example, begin construction, attain operation level); and the permittee has not returned to compliance by accomplishing the required element of the schedule within 30 days from the date a compliance schedule report is due under the permit.

(ii) *Modifications to schedules of compliance.* When a schedule of compliance in the permit has been modified under §§ 122.15 or 122.17 because of the permittee's noncompliance.

(iii) *Failure to complete or provide compliance schedule or monitoring reports.* When the permittee has failed to complete or provide a report required in a permit compliance schedule (for example, progress report or notice of noncompliance or compliance) or a monitoring report; and the permittee has not submitted the complete report within 30 days from the date it is due under the permit for compliance schedules, or from the date specified in the permit for monitoring reports.

(iv) *Deficient reports.* When the required reports provided by the permittee are so deficient as to cause

misunderstanding by the Director and thus impede the review of the status of compliance.

(v) *Noncompliance with other permit requirements.* Noncompliance shall be reported in the following circumstances:

(A) Whenever the permittee has violated a permit requirement (other than reported under paragraphs (a)(2) (i) or (ii) of this section), and has not returned to compliance within 45 days from the date reporting of noncompliance was due under the permit; or

(B) When the Director determines that a pattern of noncompliance exists for a major facility permittee over the most recent four consecutive reporting periods. (*For NPDES only,* this pattern of noncompliance is based on violations of monthly averages and excludes parameters where there is continuous monitoring.) This pattern includes any violation of the same requirement in two consecutive reporting periods, and any violation of one or more requirements in each of four consecutive reporting periods; or

(C) When the Director determines significant permit noncompliance or other significant event has occurred, such as a discharge of a toxic or hazardous substance by an NPDES facility, a fire or explosion at an RCRA facility, or migration of fluids into a USDW.

(vi) *All other.* Statistical information shall be reported quarterly on all other instances of noncompliance by major facilities with permit requirements not otherwise reported under paragraph (a) of this section.

(3) *For RCRA only,* the Director shall submit, in a manner and form prescribed by the Administrator, quarterly reports concerning noncompliance by transporters (for example, recordkeeping requirements), and by generators that send their wastes to off-site treatment, storage, or disposal facilities.

(b) *Quarterly reports for State 404 programs.* The Director shall submit noncompliance reports for section 404 discharges specified under § 123.6(f)(1)(i) (A)–(E) containing the following information:

(1) Name, location, and permit number of each noncomplying permittee;

(2) A brief description and date of each instance of noncompliance, which should include the following:

(i) Any unauthorized discharges of dredged or fill material subject to the State's jurisdiction or any noncompliance with permit conditions; and

(ii) A description of investigations conducted and of any enforcement actions taken or contemplated.

(c) *Annual reports for RCRA, UIC, and NPDES.*

(1) *Annual noncompliance report.* Statistical reports shall be submitted by the Director on nonmajor RCRA, UIC, and NPDES permittees indicating the total number reviewed, the number of noncomplying nonmajor permittees, the number of enforcement actions, and number of permit modifications extending compliance deadlines. The statistical information shall be organized to follow the types of noncompliance listed in paragraph (a) of this section.

(2) *For NPDES only,* a separate list of nonmajor discharges which are one or more years behind in construction phases of the compliance schedule shall also be submitted in alphabetical order by name and permit number.

(3) *For RCRA only,* in addition to the annual noncompliance report, the Director shall prepare a "program report" which contains information (in a manner and form prescribed by the Administrator) on generators and transporters; the permit status of regulated facilities; and summary information on the quantities and types of hazardous wastes generated, transported, stored, treated, and disposed during the preceding year. This summary information shall be reported according to EPA characteristics and lists of hazardous wastes at 40 CFR Part 261.

(4) *For State-administered UIC programs only,* in addition to the annual noncompliance report, the State Director shall:

(i) Submit each year a program report to the Administrator (in a manner and form prescribed by the Administrator) consisting of:

(A) A detailed description of the State's implementation of its program;

(B) Suggested changes if any to the program description (see § 123.4(f)) which are necessary to more accurately reflect the State's progress in issuing permits;

(C) An updated inventory of active underground injection operations in the State.

(ii) In addition to complying with the requirements of paragraph (c)(4)(i) of this section the State Director shall provide the Administrator within 3 months of the completion of the second full year of State operation of the UIC program a supplemental report containing the information required in 40 CFR Part 146 on corrective actions taken by operators of new Class II wells based upon these regulations.

(d) *Annual reports for State 404 programs.* The State Director shall submit to the Regional Administrator an annual report assessing the cumulative impacts of the State's permit program on the integrity of State regulated waters. This report shall include:

(1) The number and nature of individual permits issued by the State during the year. This should include the locations and types of water bodies where permitted activities are sited (for example, wetlands, rivers, lakes, and other categories which the Director and Regional Administrator may establish);

(2) The number of acres of each of the categories of waters in paragraph (d)(1) of this section which were filled or which received any discharge or dredged material during the year (either by authorized or known unauthorized activities);

(3) The number and nature of permit applications denied; and permits modified, revoked and reissued, or terminated during the year.

(4) The number and nature of permits issued under emergency conditions, as provided in § 123.96;

(5) The approximate number of persons in the State discharging dredged or fill material under general permits and an estimate of the cumulative impacts of these activities.

(e) *Schedule.*

(1) *For all quarterly reports.* On the last working day of May, August, November, and February, the State Director shall submit to the Regional Administrator information concerning noncompliance with RCRA, UIC, NPDES, and State 404 permit requirements by major dischargers (or for 404, other dischargers specified under § 123.6(f)(1)(i)(A)–(E)) in the State in accordance with the following schedule. The Regional Administrator shall prepare and submit information for EPA-issued permits to EPA Headquarters in accordance with the same schedule:

Quarters Covered by Reports on Noncompliance by Major Dischargers

[Date for completion of reports]

January, February, and March	May 31 [1]
April, May, and June	Aug. 31 [1]
July, August, and September	Nov. 30 [1]
October, November, and December	Feb. 28 [1]

[1] Reports must be made available to the public for inspection and copying on this date.

(2) *For all annual reports.* The period for annual reports shall be for the calendar year ending December 31, with reports completed and available to the public no more than 60 days later.

§ 122.19 Confidentiality of information.

(a) In accordance with 40 CFR Part 2, any information submitted to EPA pursuant to these regulations may be claimed as confidential by the submitter. Any such claim must be asserted at the time of submission in the manner prescribed on the application form or instructions or, in the case of other submissions, by stamping the words "confidential business information" on each page containing such information. If no claim is made at the time of submission, EPA may make the information available to the public without further notice. If a claim is asserted, the information will be treated in accordance with the procedures in 40 CFR Part 2 (Public Information).

(b) (*Applicable to State programs, see § 123.7.*) Claims of confidentiality for the following information will be denied:

(1) The name and address of any permit applicant or permittee;

(2) *For UIC permits,* information which deals with the existence, absence, or level of contaminants in drinking water;

(3) *For NPDES permits,* permit applications and permits; and

(4) *For NPDES and 404 permits,* effluent data.

(c) (*Applicable to State programs, see § 123.7.*) *For NPDES only,* information required by NPDES application forms provided by the Director under §§ 122.4 and 122.53 may not be claimed confidential. This includes information submitted on the forms themselves and any attachments used to supply information required by the forms.

(d) (*Applicable to State programs, see § 122.7.*) *For RCRA only,*

(1) Claims or confidentiality for permit application information must be substantiated at the time the application is submitted and in the manner prescribed in the application instructions.

(2) If a submitter does not provide substantiation, the Director will notify it by certified mail of the requirement to do so. If the Director does not receive the substantiation within 10 days after the submitter receives the notice, the Director shall place the unsubstantiated information in the public file.

Subpart B—Additional Requirements for Hazardous Waste Programs Under the Resource Conservation and Recovery Act

§ 122.21 Purpose and scope of Subpart B.

(a) *Content of Subpart B.* The regulations in this Subpart set forth the specific requirements for the RCRA permit program. They apply to EPA, and to approved States to the extent set forth in Part 123. Sections of this Subpart which are applicable to States are indicated at the section headings as follows: (Applicable to State RCRA programs, see § 123.7). The regulations in this Subpart supplement the requirements in Part 122, Subpart A, which contains requirements for all programs.

(b) *Authority for this Subpart and other RCRA Subtitle C Regulations.*

(1) Section 3001 of RCRA requires EPA (i) to establish criteria for identifying the characteristics of hazardous waste and for listing hazardous waste, and (ii) using those criteria to identify the characteristics of hazardous waste and list particular wastes considered to be hazardous.

(2) Section 3002 of RCRA requires EPA to establish standards applicable to generators of hazardous waste. Section 3002 also requires establishment of a manifest system to assure that hazardous waste which is transported off-site goes to a permitted treatment, storage, or disposal facility.

(3) Section 3003 of RCRA requires EPA to establish standards applicable to transporters of hazardous waste.

(4) Section 3004 of RCRA requires EPA to establish standards for the location, design, construction, monitoring, and operation of hazardous waste treatment, storage, and disposal facilities.

(5) Section 3005 of RCRA requires EPA to publish regulations requiring each person owning or operating a hazardous waste treatment, storage, or disposal facility to obtain a RCRA permit.

(6) Section 3006 of RCRA requires EPA to publish guidelines to assist States in developing hazardous waste management programs.

(7) Section 3010 of RCRA requires any person who generates or transports hazardous waste, or who owns or operates a facility for the treatment, storage, or disposal of hazardous waste, to notify EPA (or States having approved hazardous waste programs under section 3006 of RCRA) of such activity within 90 days of the promulgation or revision of regulations under section 3001 of RCRA. Section 3010 provides that no hazardous waste subject to regulations under Subtitle C of RCRA may be transported, treated, stored, or disposed of unless the required notification has been given.

(8) The following chart indicates where the regulations for sections 3001 through 3006 and the public notice for section 3010 appear in the **Federal Register.**

Section of RCRA	Coverage	Final regulation	Location
Subtitle C	Overview and definitions	40 CFR Part 260	45 FR 12724; Feb. 26, 1980; and [45 FR ——].
3001	Identification and listing of hazardous waste.	40 CFR Part 261	[—— FR ——]
3002	Generators of hazardous waste.	40 CFR Part 262	45 FR 12724, Feb. 26, 1980.
3003	Transporters of hazardous waste.	40 CFR Part 263	45 FR 12737, Feb. 26, 1980.
3004	Standards for HWM facilities.	40 CFR Parts 264, 265, and 266.	[—— FR ——]
3005	Permit requirements for HWM facilities.	40 CFR Parts 122 and 124	These regulations.
3006	Guidelines for State programs.	40 CFR Part 123	These regulations.
3010	Preliminary notification of HW activity.	(Public Notice)	45 FR 12746, Feb. 26, 1980.

(c) *Overview of the RCRA Permit Program.* Not later than 90 days after the promulgation or revision of regulations in 40 CFR Part 261 (identifying and listing hazardous wastes) all generators and transporters of hazardous waste, and all owners or operators of hazardous waste treatment, storage, or disposal facilities must file a notification of that activity under section 3010. Six months after the initial promulgation of the Part 261 regulations, treatment, storage, or disposal of hazardous waste by any person who has not applied for or received a RCRA permit is prohibited. A RCRA permit application consists of two parts, Part A (see § 122.24) and Part B (see § 122.25). For "existing HWM facilities," the requirement to submit an application is satisfied by submitting only Part A of the permit application

until the date the Director sets for submitting Part B of the application. (Part A consists of Forms 1 and 3 of the Consolidated Permit Application Forms.) Timely submission of both notification under section 3010 and Part A qualifies owners and operators of existing HWM facilities for *interim status* under section 3005(e) of RCRA. Facility owners and operators with interim status are treated as having been issued a permit until EPA or a State with interim authorization for Phase II or final authorization under Part 123 makes a final determination on the permit application. Facility owners and operators with interim status must comply with *interim status standards* set forth at 40 CFR Part 265 or with the equivalent provisions of a State program which has received interim or final

authorization under Part 123. Facility owners and operators with interim status are not relieved from complying with other State requirements. For existing HWM facilities the Director shall set a date, giving at least six months notice, for submission of Part B of the application. There is no form for Part B of the application; rather, Part B must be submitted in narrative form and contain the information set forth at § 122.25. Owners or operators of new HWM facilities must submit Part A and Part B of the permit application at least 180 days before physical construction is expected to commence.

(d) *Scope of the RCRA permit requirement.* RCRA requires a permit for the "treatment," "storage," or "disposal" of any "hazardous waste" as identified or listed in 40 CFR Part 261. The terms "treatment," "storage," "disposal," and "hazardous waste" are defined in § 122.3.

(1) *Specific inclusions (applicable to State RCRA programs, see § 123.7).* Owners and operators of certain facilities require RCRA permits as well as permits under other programs for certain aspects of the facility operation. RCRA permits are required for:

(i) Injection wells that dispose of hazardous waste, and associated surface facilities that treat, store, or dispose of hazardous waste. (See § 122.30.) However, the owner and operator with a UIC permit in a State with an approved or promulgated UIC program, will be deemed to have a RCRA permit for the injection well itself if they comply with the requirements of § 122.26(b) (permit by rule for injection wells).

(ii) Treatment, storage, or disposal of hazardous waste at facilities requiring an NPDES permit. However, the owner and operator of a publicly owned treatment works receiving hazardous waste will be deemed to have a RCRA permit for that waste if they comply with the requirements of § 122.26(c) (permit by rule for POTWs).

(iii) Barges or vessels that dispose of hazardous waste by ocean disposal and onshore hazardous waste treatment or storage facilities associated with an ocean disposal operation. However, the owner and operator will be deemed to have a RCRA permit for ocean disposal from the barge or vessel itself if they comply with the requirements of § 122.26(a) (permit by rule for ocean disposal barges and vessels).

(2) *Specific exclusions.* The following persons are among those who are not required to obtain a RCRA permit:

(i) Generators who accumulate hazardous waste on-site for less than 90 days, as provided in 40 CFR § 262.34.

(ii) Farmers who dispose of hazardous waste pesticides from their own use as provided in 40 CFR § 262.51.

(iii) Persons who own or operate facilities solely for the treatment, storage, or disposal of hazardous waste excluded from regulations under this Part by 40 CFR § 261.4 or § 261.5 (small generator exemption).

(iv) Owners or operators of totally enclosed treatment facilities as defined in 40 CFR § 260.10.

(v) Owners or operators of totally enclosed treatment facilities as defined in 40 CFR § 260.10.

§ 122.22 Application for a permit.

(Applicable to State RCRA programs, see § 123.7.)

(a) *Existing HWM facilities.* (1) Not later than six months after the first promulgation of regulations in 40 CFR Part 261 listing and identifying hazardous wastes, all owners and operators of existing hazardous waste treatment, storage, or disposal facilities must submit Part A of their permit application with the Regional Administrator.

(2) At any time after promulgation of Phase II the owner and operator of an existing HWM facility may be required to submit Part B of their permit application. The State Director may require submission of Part B (or equivalent completion of the State RCRA application process) if the State in which the facility is located has received interim authorization for Phase II or final authorization; if not, the Regional Administrator may require submission of Part B. Any owner or operator shall be allowed at least six months from the date of request to submit Part B of the application. Any owner or operator of an existing HWM facility may voluntarily submit Part B of the application at any time.

(3) Failure to furnish a requested Part B application on time, or to furnish in full the information required by the Part B application, is grounds for termination of interim status under Part 124.

(b) *New HWM Facilities.* (1) No person shall begin physical construction on a new HWM facility without having submitted Part A and Part B of its permit application and received a finally effective RCRA permit.

(2) An application for a permit for a new HWM facility (including both Part A and Part B) may be filed any time after promulgation of Phase II. The application shall be filed with the Regional Administrator if at the time of application the State in which the new HWM facility is proposed to be located has not received interim authorization for Phase II or final authorization;

otherwise it shall be filed with the State Director. All applications must be submitted at least 180 days before physical construction is expected to commence.

(c) *Updating permit applications.* (1) If any owner or operator of a HWM facility has filed Part A of a permit application and has not yet filed Part B, the owner or operator shall file an amended Part A application:

(i) With the Regional Administrator, if the facility is located in a State which has not obtained interim authorization for Phase II or final authorization, within six months after the promulgation of revised regulations under Part 261 listing or identifying additional hazardous wastes, if the facility is treating, storing, or disposing of any of those newly listed or identified wastes.

[Note.— EPA intends to promulgate regulations in June of 1980 listing or designating additional wastes beyond those listed or designated in its initial promulgation of Part 261. The wastes to be listed or designated in June are set forth in an Appendix to the initial promulgation. EPA encourages facilities applying for interim status before that second set of wastes is actually published to list or designate any of the wastes in that set which they are treating, storing, or disposing of. That will avoid the need to extensively update the Part A application when the June 1980 promulgation occurs.]

(ii) With the State Director, if the facility is located in a State which has obtained Phase II interim authorization or final authorization, no later than the effective date of regulatory provisions listing or designating wastes as hazardous in that State in addition to those listed or designated under the previously approved State program, if the facility is treating, storing, or disposing of any of those newly listed or designated wastes; or

(iii) As necessary to comply with provisions of § 122.23 for changes during interim status or the analogous provisions of a State program approved for final authorization or interim authorizaton for Phase II. Revised Part A applications necessary to comply with the provisions of § 122.23 shall be filed with the Regional Administrator if the State in which the facility in question is located does not have Phase II interim authorization or final authorization; otherwise it shall be filed with the State Director.

(2) The owner or operator of a facility who fails to comply with the updating requirements of paragraph (c)(1) of this section does not receive interim status as to the wastes not covered by duly filed Part A applications.

(d) *Reapplications.* Any HWM facility with an effective permit shall submit a new application at least 180 days before the expiration date of the effective permit, unless permission for a later date has been granted by the Director. (The Director shall not grant permission for applications to be submitted later than the expiration date of the existing permit.)

§ 122.23 Interim status.

(a) *Qualifying for interim status.* Any person who owns or operates an "existing HWM facility" shall have interim status and shall be treated as having been issued a permit to the extent he or she has:

(1) Notified the Administrator within 90 days from the promulgation or revision of Part 261 as required in Section 3010 of RCRA (this may be done by completing EPA form 8700–12); and

(2) Complied with the requirements of § 122.22 (a) and (c) governing submission of Part A applications;

(3) When EPA determines on examination or reexamination of a Part A application that it fails to meet the standards of these regulations, it may notify the owner or operator that the application is deficient and that the owner or operator is therefore not entitled to interim status. The owner or operator will then be subject to EPA enforcement for operating without a permit.

(b) *Coverage.* During the interim status period the facility shall not:

(1) Treat, store, or dispose of hazardous waste not specified in Part A of the permit application;

(2) Employ processes not specified in Part A of the permit application; or

(3) Exceed the design capacities specified in Part A of the permit application.

(c) *Changes during interim status.* (1) New hazardous wastes not previously identified in Part A of the permit application may be treated, stored, or disposed of at a facility if the owner or operator submits a revised Part A permit application prior to such a change;

(2) Increases in the design capacity of processes used at a facility may be made if the owner or operator submits a revised Part A permit application prior to such a change (along with a justification explaining the need for the change) and the Director approves the change because of a lack of available treatment, storage, or disposal capacity at other hazardous waste management facilities;

(3) Changes in the processes for the treatment, storage, or disposal of hazardous waste may be made at a

facility or additional processes may be added if the owner or operator submits a revised Part A permit application prior to such a change (along with a justification explaining the need for the change) and the Director approves the change because:

(i) It is necessary to prevent a threat to human health or the environment because of an emergency situation, or

(ii) It is necessary to comply with Federal regulations (including the interim status standards at 40 CFR Part 265) or State or local laws.

(4) Changes in the ownership or operational control of a facility may be made if the new owner or operator submits a revised Part A permit application no later than 90 days prior to the scheduled change. When a transfer of ownership or operational control of a facility occurs, the old owner or operator shall comply with the requirements of 40 CFR Part 265, Subpart H (financial requirements), until the new owner or operator has demonstrated to the Director that it is complying with that Subpart. All other interim status duties are transferred effective immediately upon the date of the change of ownership or operational control of the facility. Upon demonstration to the Director by the new owner or operator of compliance with that Subpart, the Director shall notify the old owner or operator in writing that it no longer needs to comply with that Part as of the date of demonstration.

(5) In no event shall changes be made to an HWM facility during interim status which amount to reconstruction of the facility. Reconstruction occurs when the capital investment in the changes to the facility exceeds fifty percent of the capital cost of a comparable entirely new HWM facility.

(d) *Interim status standards.* During interim status, owners or operators shall comply with the interim status standards at 40 CFR Part 265.

(e) *Grounds for termination of interim status.* Interim status terminates when:

(1) Final administrative disposition of a permit application is made; or

(2) Interim status is terminated as provided in § 122.22(a)(3).

§ 122.24 Contents of Part A.

(Applicable to State RCRA programs, see § 123.7.)

In addition to the information in § 122.4(d), Part A of the RCRA application shall include the following information:

(a) The latitude and longitude of the facility.

(b) The name, address, and telephone number of the owner of the facility.

(c) An indication of whether the facility is new or existing and whether it is a first or revised application.

(d) For existing facilities, a scale drawing of the facility showing the location of all past, present, and future treatment, storage, and disposal areas.

(e) For existing facilities, photographs of the facility clearly delineating all existing structures; existing treatment, storage, and disposal areas; and sites of future treatment, storage, and disposal areas.

(f) A description of the processes to be used for treating, storing, and disposing of hazardous waste, and the design capacity of these items.

(g) A specification of the hazardous wastes listed or designated under 40 CFR Part 261 to be treated, stored, or disposed at the facility, an estimate of the quantity of such wastes to be treated, stored, or disposed annually. and a general description of the processes to be used for such wastes.

§ 122.25 Contents of Part B.

(Applicable to State RCRA programs, see § 123.7.)

Part B of the RCRA application includes the following:

(a) *General information requirements.* The following information is required for all facilities:

(1) A general description of the facility.

(2) Chemical and physical analyses of the hazardous wastes to be handled at the facility. At a minimum, these analyses shall contain all the information which must be known to treat, store, or dispose of the wastes in accordance with Part 264.

(3) A copy of the waste analysis plan required by § 264.13(b) and, if applicable, § 264.13(c).

(4) A description of the security procedures and equipment required by § 264.14, or a justification demonstrating the reasons for requesting a waiver of this requirement.

(5) A copy of the general inspection schedule required by § 264.15(b).

(6) A justification of any request for a waiver(s) of the preparedness and prevention requirements of § 264.30.

(7) A copy of the contingency plan required by Part 264, Subpart D.

(8) A description of procedures, structures, or equipment used at the facility to,

(i) Prevent uncontrolled reaction of incompatible wastes (for example, procedures to avoid fires, explosions, or toxic gases).

(ii) Prevent hazards in unloading operations (for example, ramps, special forklifts).

(iii) Prevent runoff from hazardous waste handling areas to other areas of the facility or environment, or to prevent flooding (for example, berms, dikes, trenches).

(iv) Prevent contamination of water supplies.

(v) Mitigate effects of equipment failure and power outages.

(vi) Prevent undue exposure of personnel to hazardous waste (for example, protective clothing).

(9) Traffic pattern, volume and control (for example, show turns across traffic lanes, and stacking lanes (if appropriate); provide access road surfacing and load bearing capacity; show traffic control signals; provide estimates of traffic volume (number, types of vehicles)).

b. [*Reserved.*]

[Note.—The requirements set forth in § 122.25(a) reflect those permit application requirements related to the initial promulgation of Part 264. Additional permit application requirements including specific design and operating data, financial plans, and site engineering information will be promulgated when the remaining portions of Part 264 are promulgated.]

§ 122.26 Permits by rule.

(Applicable to State RCRA programs, see § 123.7.)

Notwithstanding any other provision of this Part or Part 124, the following shall be deemed to have a RCRA permit if the conditions listed are met:

(a) *Ocean disposal barges or vessels.* The owner or operator of a barge or other vessel which accepts hazardous waste for ocean disposal, if the owner or operator:

(1) Has a permit for ocean dumping issued under 40 CFR Part 220 (Ocean Dumping, authorized by the Marine Protection, Research, and Sanctuaries Act, as amended, 33 U.S.C. § 1420 *et seq.*);

(2) Complies with the conditions of that permit; and

(3) Complies with the following hazardous waste regulations:

(i) 40 CFR § 264.11, Identification number;

(ii) 40 CFR § 264.71, Use of manifest system;

(iii) 40 CFR § 264.72, Manifest discrepancies;

(iv) 40 CFR § 264.73(a) and (b)(1), Operating record;

(v) 40 CFR § 264.75, Annual report; and

(vi) 40 CFR § 264.76, Unmanifested waste report.

(b) *Injection wells.* The owner or operator of an injection well disposing of hazardous waste, if the owner or operator:

(1) Has a permit for underground injection issued under Part 122, Subpart C or Part 123, Subpart C; and

(2) Complies with the conditions of that permit and the requirements of § 122.45 (requirements for wells managing hazardous waste).

(c) *Publicly owned treatment works.* The owner or operator of a POTW which accepts for treatment hazardous waste, if the owner or operator:

(1) Has an NPDES permit;

(2) Complies with the conditions of that permit; and

(3) Complies with the following regulations:

(i) 40 CFR § 264.11, Identification number;

(ii) 40 CFR § 264.71, Use of manifest system;

(iii) 40 CFR § 264.72, Manifest discrepancies;

(iv) 40 CFR § 264.73 (a) and (b)(1), Operating record;

(v) 40 CFR § 264.75, Annual report;

(vi) 40 CFR § 264.76, Unmanifested waste report; and

(4) If the waste meets all Federal, State, and local pretreatment requirements which would be applicable to the waste if it were being discharged into the POTW through a sewer, pipe, or similar conveyance.

§ 122.27   Emergency permits.

(Applicable to State RCRA programs, see § 123.7.)

Notwithstanding any other provision of this Part or Part 124, in the event the Director finds an imminent and substantial endangerment to human health or the environment the Director may issue a temporary emergency permit to a facility to allow treatment, storage, or disposal of hazardous waste for a non-permitted facility or not covered by the permit for a facility with an effective permit. This emergency permit:

(a) May be oral or written. If oral, it shall be followed within five days by a written emergency permit;

(b) Shall not exceed 90 days in duration;

(c) Shall clearly specify the hazardous wastes to be received, and the manner and location of their treatment, storage, or disposal;

(d) May be terminated by the Director at any time without process if he or she determines that termination is appropriate to protect human health and the environment;

(e) Shall be accompanied by a public notice published under § 124.11(b) including:

(1) Name and address of the office granting the emergency authorization;

(2) Name and location of the permitted HWM facility;

(3) A brief description of the wastes involved;

(4) A brief description of the action authorized and reasons for authorizing it; and

(5) Duration of the emergency permit; and

(f) Shall incorporate, to the extent possible and not inconsistent with the emergency situation, all applicable requirements of this Part and 40 CFR Parts 264 and 266.

§ 122.28   Additional conditions applicable to all RCRA permits.

(Applicable to State RCRA programs, see § 122.7.)

The following conditions, in addition to those set forth in § 122.7, apply to all RCRA permits:

(a) In addition to § 122.7(a) (duty to comply): the permittee need not comply with the conditions of this permit to the extent and for the duration such noncompliance is authorized in an emergency permit. (See § 122.27.)

(b) In addition to § 122.7(j) (monitoring): the permittee shall maintain records from all ground monitoring wells and associated groundwater surface elevations, for the active life of the facility, and for disposal facilities for the post-closure care period as well.

(c) In addition to § 122.7(l)(1) (notice of planned changes): for a new HWM facility, the permittee may not commence treatment, storage, or disposal of hazardous waste; and for a facility being modified the permittee may not treat, store, or dispose of hazardous waste in the modified portion of the facility, until:

(1) The permittee has submitted to the Director by certified mail or hand delivery a letter signed by the permittee and a registered professional engineer stating that the facility has been constructed or modified in compliance with the permit; and

(2)(i) The Director has inspected the modified or newly constructed facility and finds it is in compliance with the conditions of the permit; or

(ii) Within 15 days of the date of submission of the letter in paragraph (c)(1) of this section, the permittee has not received notice from the Director of his or her intent to inspect, prior inspection is waived and the permittee may commence treatment, storage, or disposal of hazardous waste.

(d) The following shall be included as information which must be reported orally within 24 hours under § 122.7(l)(6):

(1) Information concerning release of any hazardous waste that may cause an endangerment to public drinking water supplies.

(2) Any information of a release or discharge of hazardous waste, or of a fire or explosion from a HWM facility, which could threaten the environment or human health outside the facility. The description of the occurrence and its cause shall include:

(i) Name, address, and telephone number of the owner or operator;

(ii) Name, address, and telephone number of the facility;

(iii) Date, time, and type of incident;

(iv) Name and quantity of material(s) involved;

(v) The extent of injuries, if any;

(vi) An assessment of actual or potential hazards to the environment and human health outside the facility, where this is applicable; and

(vii) Estimated quantity and disposition of recovered material that resulted from the incident.

The Director may waive the five day written notice requirement in favor of a written report within fifteen days.

(e) The following reports required by Part 264 shall be submitted in addition to those required by § 122.7(l) (reporting requirements):

(1) Manifest discrepancy report: if a significant discrepancy in a manifest is discovered, the permittee must attempt to reconcile the discrepancy. If not resolved within fifteen days, the permittee must submit a letter report including a copy of the manifest to the Director. (See 40 CFR § 264.72.)

(2) Unmanifested waste report: must be submitted to the Director within 15 days of receipt of unmanifested waste. (See § 264.76.)

(3) Annual report: an annual report must be submitted covering facility activities during the previous calendar year. (See 40 CFR § 264.75.)

(4) [Reserved.]

[Note.—The above reports are required in Part 264 as initially promulgated. Additional reports will be required and added to this section when remaining portions of Part 264 are promulgated.]

§ 122.29   Establishing RCRA permit conditions.

(Applicable to State RCRA programs, see § 123.7.)

In addition to the conditions established under § 122.8(a), each RCRA permit shall include each of the applicable requirements specified in 40 CFR Parts 264 and 266.

§ 122.30   Interim permits for UIC wells.

(Applicable to State programs, see § 123.7.)

The Director may issue a permit under this Part to any Class I UIC well (see

§ 122.32) injecting hazardous wastes within a State in which no UIC program has been approved or promulgated. Any such permit shall apply and insure compliance with all applicable requirements of 40 CFR Part 264, Subpart R (RCRA standards for wells), and shall be for a term not to exceed two years. No such permit shall be issued after approval or promulgation of a UIC program in the State. Any permit under this section shall contain a condition providing that it will terminate upon final action by the Director under a UIC program to issue or deny a UIC permit for the facility.

§ 122.45   Requirements for wells injecting hazardous waste.

(Applicable to State UIC programs, see § 123.7.)

(a) *Applicability.* The regulations in this section apply to all generators of hazardous waste, and to the owners or operators of all hazardous waste management facilities, using any class of well to inject hazardous wastes accompanied by a manifest. (See also § 122.36.)

(b) *Authorization.* The owner or operator of any well that is used to inject hazardous wastes accompanied by a manifest or delivery document shall apply for authorization to inject as specified in § 122.38 within 6 months after the approval of an applicable State program.

(c) *Requirements.* In addition to requiring compliance with the applicable requirements of this Part and 40 CFR Part 146, Subparts B–F, the Director shall, for each facility meeting the requirements of paragraph (b) of this section, require that the owner or operator comply with the following:

(1) *Notification.* The owner or operator shall comply with the notification requirements of Section 3010 of Pub. L. 94–580.

(2) *Identification number.* The owner or operator shall comply with the requirements of 40 CFR § 264.11.

(3) *Manifest system.* The owner or operator shall comply with the applicable recordkeeping and reporting requirements for manifested wastes in 40 CFR § 264.71.

(4) *Manifest discrepancies.* The owner or operator shall comply with 40 CFR § 264.72.

(5) *Operating record.* The owner or operator shall comply with 40 CFR § 264.73(a), (b)(1), and (b)(2).

(6) *Annual report.* The owner or operator shall comply with 40 CFR § 264.75.

(7) *Unmanifested waste report.* The owner or operator shall comply with 40 CFR § 264.75.

(8) *Personnel training.* The owner or operator shall comply with the applicable personnel training requirements of 40 CFR § 264.16.

(9) *Certification of closure.* When abandonment is completed, the owner or operator must submit to the Director certification by the owner or operator and certification by an independent registered professional engineer that the facility has been closed in accordance with the specifications in § 122.42(f).

(d) *Additional requirements for Class IV wells.* [Reserved].

PART 123—STATE PROGRAM REQUIREMENTS

Subpart A—General Program Requirements

Authority: Resource Conservation and Recovery Act, 42 U.S.C. 6901 *et seq.*; Safe Drinking Water Act, 42 U.S.C. 300(f) *et seq.*; Clean Water Act, 33 U.S.C. 1251 *et seq.*

Subpart A—General Program Requirements

§ 123.1   Purpose and scope.

(a) This part specifies the procedures EPA will follow in approving, revising, and withdrawing State programs under the following statutes and the requirements State programs must meet to be approved by the Administrator under:

(1) Section 3006(b) (hazardous waste-final authorization) and section 3006(c) (hazardous waste-interim authorization) of RCRA;

(2) Section 1422 (underground injection control—UIC) of SDWA;

(3) Sections 318, 402, and 405 (National Pollutant Discharge

Elimination System—NPDES) of CWA; and

(4) Section 404 (dredged or fill material) of CWA.

(b) Subpart A contains requirements applicable to all programs listed in paragraph (a) except hazardous waste programs operating under interim authorization. All requirements applicable to hazardous waste programs operating under interim authorization are contained in Subpart F. (References in this subpart to "programs under this Part" do not refer to hazardous waste programs operating under interim authorization.) Subpart A includes the elements which must be part of submissions to EPA for program approval, the substantive provisions which must be present in State programs for them to be approved, and the procedures EPA will follow in approving, revising, and withdrawing State programs. Subpart B contains additional requirements for States seeking final authorization under RCRA. Subpart C contains additional requirements for State UIC programs. Subpart D specifies additional requirements for State NPDES programs. Subpart E specifies additional requirements for State section 404 programs.

(c) State submissions for program approval must be made in accordance with the procedures set out in Subpart A and, in the case of State 404 programs with the procedures set out in Subpart E. (Submissions for interim authorization shall be made in accordance with Subpart F.) This includes developing and submitting to EPA a program description (§ 123.4), an Attorney General's statement (§ 123.5), a Memorandum of Agreement with the Regional Administrator (§ 123.6) and with the Secretary in the case of section 404 programs (§ 123.99).

(d) The substantive provisions which must be included in State programs for them to be approved include requirements for permitting, compliance evaluation, enforcement, public participation, and sharing of information. The requirements are found both in Subpart A (§§ 123.7 to 123.11) and in the program specific subparts. Many of the requirements for State programs are made applicable to States by cross-referencing other EPA regulations. In particular, many of the provisions of Parts 122 and 124 are made applicable to States by the references contained in § 123.7.

(e) Upon submission of a complete program, EPA will conduct a public hearing, if interest is shown, and determine whether to approve or disapprove the program taking into consideration the requirements of this Part, the appropriate Act and any comments received.

(f) The Administrator shall approve State programs which conform to the applicable requirements of this Part.

(g) Upon approval of a State program, the Administrator (or the Secretary in the case of section 404 programs) shall suspend the issuance of Federal permits for those activities subject to the approved State program.

(h) Any State program approved by the Administrator shall at all times be conducted in accordance with the requirements of this Part.

(i) States are encouraged to consolidate their permitting activities. While approval of State programs under this Part will facilitate such consolidation, these regulations do not require consolidation. Each of the four programs under this Part may be applied for and approved separately.

(j) Partial State programs are not allowed under NPDES, 404, or RCRA (for programs operating under final authorization). However, in many cases States will lack authority to regulate activities on Indian lands. This lack of authority does not impair a State's ability to obtain full program approval in accordance with this Part, i.e., inability of a State to regulate activities on Indian lands does not constitute a partial program. Similarly, a State can assume primary enforcement responsibility for the UIC program, notwithstanding § 123.51(e), when the State program is unable to regulate activities on Indian lands within the State. EPA, or in the case of section 404 programs the Secretary, will administer the program on Indian lands if the State does not seek this authority.

[Note.—States are advised to contact the United States Department of the Interior, Bureau of Indian Affairs, concerning authority over Indian lands.]

(k) Except as provided in § 123.32, nothing in this Part precludes a State from:

(1) Adopting or enforcing requirements which are more stringent or more extensive than those required under this Part;

(2) Operating a program with a greater scope of coverage than that required under this Part. Where an approved State program has a greater scope of coverage than required by Federal law the additional coverage is not part of the Federally approved program.

[Note.—For example, when a State requires permits for discharges into publicly owned treatment works, these permits are not NPDES permits. Also, State assumption of the section 404 program is limited to certain waters, as provided in § 123.91(c). The Federal program operated by the Corps of Engineers continues to apply to the remaining waters in the State even after program approval. However, this does not restrict States from regulating discharges of dredged or fill materials into those waters over which the Secretary retains section 404 jurisdiction.]

§ 123.2 Definitions.

The definitions in Part 122 apply to all subparts of this Part, including Subpart F.

§ 123.3 Elements of a program submission.

(a) Any State that seeks to administer a program under this Part shall submit to the Administrator at least three copies of a program submission. The submission shall contain the following:

(1) A letter from the Governor of the State requesting program approval;

(2) A complete program description, as required by § 123.4, describing how the State intends to carry out its responsibilities under this Part;

(3) An Attorney General's statement as required by § 123.5;

(4) A Memorandum of Agreement with the Regional Administrator as required by § 123.6, and, in the case of State section 404 programs, a Memorandum of Agreement with the Secretary as required by § 123.99;

(5) Copies of all applicable State statutes and regulations, including those governing State administrative procedures;

(6) The showing required by § 123.39(c) (RCRA programs only) and § 123.54(b) (UIC programs only) of the State's public participation activities prior to program submission.

(b) Within 30 days of receipt by EPA of a State program submission, EPA will notify the State whether its submission is complete. If EPA finds that a State's submission is complete, the statutory review period (i.e., the period of time allotted for formal EPA review of a proposed State program under the appropriate Act) shall be deemed to have begun on the date of receipt of the State's submission. If EPA finds that a State's submission is incomplete, the statutory review period shall not begin until all the necessary information is received by EPA.

(c) If the State's submission is materially changed during the statutory review period, the statutory review period shall begin again upon receipt of the revised submission.

(d) The State and EPA may extend the statutory review period by agreement.

§ 123.4 Program description.

Any State that seeks to administer a program under this part shall submit a description of the program it proposes to administer in lieu of the Federal program under State law or under an interstate compact. The program description shall include:

(a) A description in narrative form of the scope, structure, coverage and processes of the State program.

(b) A description (including organization charts) of the organization and structure of the State agency or agencies which will have responsibility for administering the program, including the information listed below. If more than one agency is responsible for administration of a program, each agency must have statewide jurisdiction over a class of activities. The responsibilities of each agency must be delineated, their procedures for coordination set forth, and an agency may be designated as a "lead agency" to facilitate communications between EPA and the State agencies having program responsibility. In the case of State RCRA programs, such a designation is mandatory (see paragraph (f)(4) of this section). When the State proposes to administer a program of greater scope of coverage than is required by Federal law, the information provided under this paragraph shall indicate the resources dedicated to administering the Federally required portion of the program.

(1) A description of the State agency staff who will carry out the State program, including the number, occupations, and general duties of the employees. The State need not submit complete job descriptions for every employee carrying out the State program.

(2) An itemization of the estimated costs of establishing and administering the program for the first two years after approval, including cost of the personnel listed in paragraph (b)(1) of this section, cost of administrative support, and cost of technical support.

(3) An itemization of the sources and amounts of funding, including an estimate of Federal grant money, available to the State Director for the first two years after approval to meet the costs listed in paragraph (b)(2) of this section, identifying any restrictions or limitations upon this funding.

(c) A description of applicable State procedures, including permitting procedures and any State administrative or judicial review procedures.

(d) Copies of the permit form(s), application form(s), reporting form(s), and manifest format the State intends to employ in its program. Forms used by States need not be identical to the forms used by EPA but should require the same basic information, except that State NPDES programs are required to use standard Discharge Monitoring Reports (DMR). The State need not provide copies of uniform national forms

it intends to use but should note its intention to use such forms. State section 404 application forms must include the information required by § 123.94 and State section 404 permit forms must include the information and conditions required by § 123.97.

[Note.—States are encouraged to use uniform national forms established by the Administrator. If uniform national forms are used, they may be modified to include the State Agency's name, address, logo, and other similar information, as appropriate, in place of EPA's.]

(e) A complete description of the State's compliance tracking and enforcement program.

(f) *State RCRA programs only.* In the case of State RCRA programs, the program description shall also include:

(1) A description of the State manifest tracking system, and of the procedures the State will use to coordinate information with other approved State programs and the Federal program regarding interstate and international shipments.

(2) An estimate of the number of the following:

(i) Generators;

(ii) Transporters: and

(iii) On- and off-site storage, treatment and disposal facilities, and a brief description of the types of facilities and an indication of the permit status of these facilities.

(3) If available, an estimate of the annual quantities of hazardous wastes:

(i) Generated within the State;

(ii) Transporters; and State; and

(iii) Stored, treated, or disposed of within the State:

(A) on-site; and

(B) off-site.

(4) When more than one agency within a State has responsibility for administering the State program, an identification of a "lead agency" and a description of how the State agencies will coordinate their activities.

(g) *State UIC programs only.* In the case of a submission for approval of a State UIC program the State's program description shall also include:

(1) A schedule for issuing permits within five years after program approval to all injection wells within the State which are required to have permits under this Part and Part 122;

(2) The priorities (according to criteria set forth in 40 CFR § 146.09) for issuing permits, including the number of permits in each class of injection well which will be issued each year during the first five years of program operation;

(3) A description of how the Director will implement the mechanical integrity testing requirements of 40 CFR § 146.08, including the frequency of testing that

will be required and the number of tests that will be reviewed by the Director each year;

(4) A description of the procedure whereby the Director will notify owners and operators of injection wells of the requirement that they apply for and obtain a permit. The notification required by this paragraph shall require applications to be filed as soon as possible, but not later than four years after program approval for all injection wells requiring a permit;

(5) A description of any rule under which the Director proposes to authorize injections, including the text of the rule;

(6) For any existing enhanced recovery and hydrocarbon storage wells which the Director proposes to authorize by rule, a description of the procedure for reviewing the wells for compliance with applicable monitoring, reporting, construction, and financial responsibility requirements of §§ 122.41 and 122.42, and 40 CFR Part 146;

(7) A description of and schedule for the State's program to establish and maintain a current inventory of injection wells which must be permitted under State law;

(8) Where the Director has designated underground sources of drinking water in accordance with § 122.35(a), a description and identification of all such designated sources in the State;

(9) A description of aquifers, or parts thereof, which the Director has identified under § 122.35(b) as exempted aquifers, and a summary of supporting data;

(10) A description of and schedule for the State's program to ban Class IV wells prohibited under § 122.36; and

(11) A description of and schedule for the State's program to establish an inventory of Class V wells and to assess the need for a program to regulate Class V wells.

(h) *State 404 programs only.* In the case of a submission for approval of a section 404 program the State's program description shall also include:

(1) A description of State regulated waters.

[Note.—States should obtain from the Secretary an identification of those waters of the U.S. within the State over which the Corps of Engineers retains authority under section 404(g) of CWA.]

(2) A categorization, by type and quantity, of discharges within the State, and an estimate of the number of discharges within each category for which the discharger must file for a permit.

(3) An estimate of the number and percent of activities within each category for which the State has already issued a State permit regulating the discharge.

(4) In accordance with § 123.92(a)(6), a description of the specific best management practices requirements proposed to be used to satisfy the exemption provisions of section 404(f)(1)(E) of CWA for construction or maintenance of farm roads, forest roads, or temporary roads for moving mining equipment.

(5) A description of how the State section 404 agency(ies) will interact with other State and local agencies.

(6) A description of how the State will coordinate its enforcement strategy with that of the Corps of Engineers and EPA.

(7) Where more than one agency within a State has responsibility for administering the State program:

(i) A memorandum of understanding among all the responsible State agencies which establishes:

(A) Procedures for obtaining and exchanging information necessary for each agency to determine and assess the cumulative impacts of all activities authorized under the State program;

(B) Common reporting requirements; and

(C) Any other appropriate procedures not inconsistent with section 404 of CWA or these regulations;

(ii) A description of procedures for coordinating compliance monitoring and enforcement, distributing among the responsible agencies information received from applicants and permittees, and issuing reports required by section 404 of CWA or these regulations.

(8) Where several State 404 permits are required for a single project, a description of procedures for:

(i) Ensuring that all the necessary State 404 permits are issued before any of the permits go into effect; and

(ii) Concurrent processing and, where appropriate, joint processing of all of the necessary State 404 permits.

§ 123.5   Attorney General's statement.

(a) Any State that seeks to administer a program under this Part shall submit a statement from the State Attorney General (or the attorney for those State or interstate agencies which have independent legal counsel) that the laws of the State, or an interstate compact, provide adequate authority to carry out the program described under § 123.4 and to meet the requirements of this Part. This statement shall include citations to the specific statutes, administrative regulations, and, where appropriate, judicial decisions which demonstrate adequate authority. State statutes and regulations cited by the State Attorney General or independent legal counsel shall be in the form of lawfully adopted State statutes and regulations at the time the statement is signed and shall be fully effective by the time the program is approved. To qualify as "independent

legal counsel" the attorney signing the statement required by this section must have full authority to independently represent the State agency in court on all matters pertaining to the State program.

[Note.—EPA will supply States with an Attorney General's statement format on request.]

(b) When a State seeks authority over activities on Indian lands, the statement shall contain an appropriate analysis of the State's authority.

(c) *State NPDES programs only.* In the case of State NPDES programs, the Attorney General's statement shall certify that the State has adequate legal authority to issue and enforce general permits if the State seeks to implement the general permit program under § 122.59.

(d) *State section 404 programs only.*

(1) In the case of State section 404 programs the State Attorney General's statement shall contain an analysis of State law regarding the prohibition on taking private property without just compensation, including any applicable judicial interpretations, and an assessment of the effect such law will have on the successful implementation of the State's regulation of the discharge of dredged or fill material.

(2) In the case of State section 404 programs, where more than one agency has responsibility for administering the State program, the Attorney General's Statement shall include certification that each agency has full authority to administer the program within its category of jurisdiction and that the State as a whole has full authority to administer a complete State section 404 program.

§ 123.6   Memorandum of Agreement with the Regional Administrator.

(a) Any State that seeks to administer a program under this Part shall submit a Memorandum of Agreement. The Memorandum of Agreement shall be executed by the State Director and the Regional Administrator and shall become effective when approved by the Administrator. In addition to meeting the requirements of paragraph (b) of this section, the Memorandum of Agreement may include other terms, conditions, or agreements consistent with this Part and relevant to the administration and enforcement of the State's regulatory program. The Administrator shall not approve any Memorandum of Agreement which contains provisions which restrict EPA's statutory oversight responsibility.

(b) The Memorandum of Agreement shall include the following:

(1) Provisions for the prompt transfer from EPA to the State of pending permit

applications and any other information relevant to program operation not already in the possession of the State Director (e.g., support files for permit issuance, compliance reports, etc.). When existing permits are transferred from EPA to the State for administration, the Memorandum of Agreement shall contain provisions specifying a procedure for transferring the administration of these permits. If a State lacks the authority to directly administer permits issued by the Federal government, a procedure may be established to transfer responsibility for these permits.

[Note.—For example, EPA and the State and the permittee could agree that the State would issue a permit(s) identical to the outstanding Federal permit which would simultaneously be terminated.]

(2) Provisions specifying classes and categories of permit applications, draft permits, and proposed permits that the State will send to the Regional Administrator for review, comment and, where applicable, objection.

[Note.—The nature and basis of EPA review of State permits and permit applications differs among the programs governed by this Part. See §§ 123.38 (RCRA), 123.75 (NPDES) and 123.101 (404).]

(3) Provisions specifying the frequency and content of reports, documents and other information which the State is required to submit to EPA. The State shall allow EPA to routinely review State records, reports, and files relevant to the administration and enforcement of the approved program. State reports may be combined with grant reports where appropriate. These procedures shall implement the requirements of § 123.74 (NPDES programs only) and § 123.100 (404 programs only).

(4) Provisions on the State's compliance monitoring and enforcement program, including:

(i) Provisions for coordination of compliance monitoring activities by the State and by EPA. These may specify the basis on which the Regional Administrator will select facilities or activities within the State for EPA inspection. The Regional Administrator will normally notify the State at least 7 days before any such inspection; and

(ii) Procedures to assure coordination of enforcement activities.

(5) When appropriate, provisions for joint processing of permits by the State and EPA, for facilities or activities which require permits from both EPA and the State under different programs. See § 124.4.

[Note.—To promote efficiency and to avoid duplication and inconsistency, States are encouraged to enter into joint processing agreements with EPA for permit issuance. Likewise, States are encouraged (but not required) to consider steps to coordinate or consolidate their own permit programs and activities.]

(6) Provisions for modification of the Memorandum of Agreement in accordance with this Part.

(c) The Memorandum of Agreement, the annual program grant and the State/EPA Agreement should be consistent. If the State/EPA Agreement indicates that a change is needed in the Memorandum of Agreement, the Memorandum of Agreement may be amended through the procedures set forth in this part. The State/EPA Agreement may not override the Memorandum of Agreement.

[Note.—Detailed program priorities and specific arrangements for EPA support of the State program will change and are therefore more appropriately negotiated in the context of annual agreements rather than in the MOA. However, it may still be appropriate to specify in the MOA the basis for such detailed agreements, e.g., a provision in the MOA specifying that EPA will select facilities in the State for inspection annually as part of the State/EPA agreement.]

(d) *State RCRA prorgrams only.* In the case of State RCRA programs the Memorandum of Agreement shall also provide that:

(1) EPA may conduct compliance inspections of all generators, transporters, and HWM facilities in each year for which the State is operating under final authorization. The Regional Administrator and the State Director may agree to limitations on compliance inspections of generators, transporters, and non-major HWM facilities.

(2) No limitations on EPA compliance inspections of generators, transporters, or non-major HWM facilities under paragraph (d)(1) of this section shall restrict EPA's right to inspect any generator, transporter, or HWM facility which it has cause to believe is not in compliance with RCRA; however, before conducting such an inspection, EPA will normally allow the State a reasonable opportunity to conduct a compliance evaluation inspection.

(3) The State Director shall promptly forward to EPA copies of draft permits and permit applications for all major HWM facilities for review and comment. The Regional Administrator and the State Director may agree to limitations regarding review of and comment on draft permits and/or permit applications for non-major HWM facilities. The State Director shall supply EPA copies of final permits for all major HWM facilities.

(4) The Regional Administrator shall promptly forward to the State Director information obtained prior to program approval in notifications provided under

section 3010(a) of RCRA. The Regional Administrator and the State Director shall agree on procedures for the assignment of EPA identification numbers for new generators, transporters, treatment, storage, and disposal facilities.

(5) The State Director shall review all permits issued under State law prior to the date of program approval and modify or revoke and reissue them to require compliance with the requirements of this Part. The Regional Administrator and the State Director shall establish a time within which this review must take place.

(e) *State NPDES programs only.* In the case of State NPDES programs the Memorandum of Agreement shall also specify the extent to which EPA will waive its right to review, object to, or comment upon State-issued permits under sections 402(d)(3), (e) or (f) of CWA. While the Regional Administrator and the State may agree to waive EPA review of certain "classes or categories" of permits, no waiver of review may be granted for the following discharges:

(1) Discharges into the territorial sea;

(2) Discharges which may affect the waters of a State other than the one in which the discharge originates;

(3) Discharges proposed to be regulated by general permits (see § 122.59);

(4) Discharges from publicly owned treatment works with a daily average discharge exceeding 1 million gallons per day;

(5) Discharges of uncontaminated cooling water with a daily average discharge exceeding 500 million gallons per day;

(6) Discharges from any major discharger or from any discharger within any of the 21 industrial categories listed in Appendix A to Part 122;

(7) Discharges from other sources with a daily average discharge exceeding 0.5 (one-half) million gallons per day, except that EPA review of permits for discharges of non-process wastewater may be waived regardless of flow.

(f) *State section 404 programs only.* (1) In the case of State section 404 programs, the Memorandum of Agreement with the Regional Administrator shall also specify:

(i) The categories (including any class, type, or size within such categories) of discharges for which EPA will waive review of State-issued permit applications, draft permits, and draft general permits. While the Regional Administrators and the State, after consultation with the Corps of Engineers, the U.S. Fish and Wildlife Service, and the National Marine Fisheries Service, may agree to waive

Federal review of certain "classes or categories" of permits, no waiver may be granted for the following activities:

(A) Discharges which may affect the waters of a State other than the one in which the discharge originates;

(B) Major discharges;

(C) Discharges into critical areas established under State or Federal law including fish and wildlife sanctuaries or refuges, National and historical monuments, wilderness areas and preserves, National and State parks, components of the National Wild and Scenic Rivers system, the designated critical habitat of threatened or endangered species, and sites identified or proposed under the National Historic Preservation Act;

(D) Discharges proposed to be regulated by general permits; or

(E) Discharges known or suspected to contain toxic pollutants in toxic amounts under section 307(a)(1) of CWA or hazardous substances in reportable quantities under section 311 of CWA.

(ii) A definition of major discharges.

(2) In the case of State section 404 programs, where more than one agency within a State has responsibility for administering the program, all of the responsible agencies shall be parties to the Memorandum of Agreement.

(g) *State NPDES and Section 404 programs only.* Whenever a waiver is granted under paragraphs (e) or (f)(1) of this section, the Memorandum of Agreement shall contain:

(1) A statement that the Regional Administrator retains the right to terminate the waiver as to future permit actions, in whole or in part, at any time by sending the State Director written notice of termination; and

(2) A statement that the State shall supply EPA and, in the case of State section 404 programs, the Corps of Engineers, the U.S. Fish and Wildlife Service, and the National Marine Fisheries Servcie (unless receipt is waived in writing), with copies of final permits.

§ 123.7 Requirements for permitting.

(a) All State programs under this Part must have legal authority to implement each of the following provisions and must be administered in conformance with each; except that States are not precluded from omitting or modifying any provisions to impose more stringent requirements:

(1) § 122.4—(Application for a permit), except in the case of § 122.4(d) for State section 404 programs;

(2) § 122.6—(Signatories);

(3) § 122.7—(Applicable permit conditions);

(4) § 122.8—(Establishing permit conditions);

(5) § 122.9—(Duration);

(6) § 122.10(a)—(Schedules of compliance);

(7) § 122.11—(Monitoring requirements);

(8) § 122.13 (a) and (b)—(Effect of permit);

(0) § 122.14—(Permit transfer);

(10) § 122.15—(Permit modification);

(11) § 122.16—(Permit termination);

(12) § 122.18—(Noncompliance reporting);

(13) § 122.19 (b)–(d)—(Confidential information);

(14) § 124.3(a)—(Application for a permit);

(15) § 124.5 (a), (c), (d), and (f)—(Modification of permits), except as provided in § 123.100(b)(2) for State section 404 programs;

(16) § 124.6 (a), (c), (d), and (e)—(Draft permit), except as provided in § 123.100(b)(2) for State section 404 programs;

(17) § 124.8—(Fact sheets), except as provided in § 123.100(b)(2) for State section 404 programs;

(18) § 124.10 (a)(1)(ii), (a)(1)(iii), (a)(1)(v), (b), (c), (d), and (e)—(Public notice);

(19) § 124.11—(Public comments and requests for hearings);

(20) § 124.12(a)—(Public hearings); and

(21) § 124.17 (a) and (c)—(Response to comments).

[Note.—States need not implement provisions identical to the above listed provisions or the provisions listed in §§ 123.7 (b)–(d). Implemented provisions must, however, establish requirements at least as stringent as the corresponding listed provisions. While States may impose more stringent requirements, they may not make one requirement more lenient as a tradeoff for making another requirement more stringent; for example, by requiring that public hearings be held prior to issuing any permit while reducing the amount of advance notice of such a hearing.

State programs may, if they have adequate legal authority, implement any of the provisions of Parts 122 and 124. See, for example, § 122.5(d) (continuation of permits) and § 124.4 (consolidation of permit processing).

(b) *State RCRA programs only.* Any State hazardous waste program shall have legal authority to implement each of the following provisions and must be administered in conformance with each, except that States are not precluded from omitting or modifying any provisions to impose more stringent requirements:

(1) § 122.21(d)(2)—(Specific inclusions);

(2) § 122.22—(Application for a permit);

(3) § 122.24—(Contents of Part A);

(4) § 122.25—(Contents of Part B);

[Note.—States need not use a two part permit application process. The State application process must, however, require information in sufficient detail to satisfy the requirements of §§ 122.24 and 122.25.]

(5) § 122.26—(Permit by rule);

(6) § 122.27—(Emergency permits);

(7) § 122.28—(Additional permit conditions);

(8) § 122.29—(Establishing permit conditions); and

(9) § 122.30—(Interim permits for UIC wells).

(c) *State UIC programs only.* State UIC programs shall have legal authority to implement each of the following provisions and must be administered in conformance with each; except that States are not precluded from omitting or modifying any provisions to impose more stringent requirements:

(1) § 122.32—(Classification of injection wells);

(2) § 122.33—(Prohibition of unauthorized injection);

(3) § 122.34—(Prohibition of movement of fluids into underground sources of drinking water);

(4) § 122.35—(Identification of underground sources of drinking water and exempted aquifers);

(5) § 122.36—(Elimination of Class IV wells);

(6) § 122.37—(Authorization by rule);

(7) § 122.38—(Authorization by permit);

(8) § 122.39—(Area permits);

(9) § 122.41—(Additional permit conditions);

(10) § 122.42—(Establishing permit conditions);

(11) § 122.44—(Corrective action); and

(12) § 122.45—(Requirements for wells managing hazardous wastes).

(d) *State NPDES programs only.* State NPDES programs shall have legal authority to implement each of the following provisions and must be administered in conformance with each; except that States are not precluded from omitting or modifying any provisions to impose more stringent requirements:

(1) § 122.52—(Prohibitions);

(2) § 122.53 (a), (d)–(g) and (i)–(k)—(Application for a permit);

(3) § 122.54—(Concentrated animal feeding operations);

(4) § 122.55—(Concentrated aquatic animal production facilities);

(5) § 122.56—(Aquaculture projects);

(6) § 122.57—(Separate storm sewers);

(7) § 122.58—(Silviculture);

(8) § 122.59—(General permits), *provided that* States which do not seek

to implement the general permit program under § 122.59 need not do so;

(9) § 122.60—(Conditions applicable to all permits);

(10) § 122.61—(Conditions applicable to specified categories of permits);

(11) § 122.62—(Establishing permit conditions);

(12) § 122.63—(Calculating NPDES conditions);

(13) § 122.64—(Duration of permit);

(14) § 122.65—(Disposal into wells);

(15) § 124.56—(Fact sheets);

(16) § 124.57(a)—(Public notice);

(17) § 124.59—(Comments from government agencies);

(18) Subparts A, B, C, D, H, I, J, K and L of Part 125; and

(19) 40 CFR Parts 129, 133, and Subchapter N.

[Note.—For example, a State may impose more stringent requirements in an NPDES program by omitting the upset provision of § 122.60 or by requiring more prompt notice of an upset.]

(e) *State NPDES and 404 programs only.* (1) State NPDES and 404 permit programs shall have an approved continuing planning process under 40 CFR § 35.1500 and shall assure that the approved planning process is at all times consistent with CWA.

(2) State NPDES and 404 programs shall ensure that any board or body which approves all or portions of permits shall not include as a member any person who receives, or has during the previous 2 years received, a significant portion of income directly or indirectly from permit holders or applicants for a permit.

(i) For the purposes of this paragraph:

(A) "Board or body" includes any individual, including the Director, who has or shares authority to approve all or portions of permits either in the first instance, as modified or reissued, or on appeal.

(B) "Significant portion of income" means 10 percent or more of gross personal income for a calendar year, except that it means 50 percent or more of gross personal income for a calendar year if the recipient is over 60 years of age and is receiving that portion under retirement, pension, or similar arrangement.

(C) "Permit holders or applicants for a permit" does not include any department or agency of a State government, such as a Department of Parks or a Department of Fish and Wildlife.

(D) "Income" includes retirement benefits, consultant fees, and stock dividends.

(ii) For the purposes of this subparagraph, income is not received "directly or indirectly from permit holders or applicants for a permit" when it is derived from mutual fund payments, or from other diversified investments for which the recipient does not know the identity of the primary sources of income.

§ 123.8   Requirements for compliance evaluation programs.

(a) State programs shall have procedures for receipt, evaluation, retention and investigation for possible enforcement of all notices and reports required of permittees and other regulated persons (and for investigation for possible enforcement of failure to submit these notices and reports).

(b) State programs shall have inspection and surveillance procedures to determine, independent of information supplied by regulated persons, compliance or noncompliance with applicable program requirements. The State shall maintain:

(1) A program which is capable of making comprehensive surveys of all facilities and activities subject to the State Director's authority to identify persons subject to regulation who have failed to comply with permit application or other program requirements. Any compilation, index, or inventory of such facilities and activities shall be made available to the Regional Administrator upon request;

(2) A program for periodic inspections of the facilities and activities subject to regulation. These inspections shall be conducted in a manner designed to:

(i) Determine compliance or noncompliance with issued permit conditions and other program requirements;

(ii) Verify the accuracy of information submitted by permittees and other regulated persons in reporting forms and other forms supplying monitoring data; and

(iii) Verify the adequacy of sampling, monitoring, and other methods used by permittees and other regulated persons to develop that information;

(3) A program for investigating information obtained regarding violations of applicable program and permit requirements; and

(4) Procedures for receiving and ensuring proper consideration of information submitted by the public about violations. Public effort in reporting violations shall be encouraged, and the State Director shall make available information on reporting procedures.

(c) The State Director and State officers engaged in compliance evaluation shall have authority to enter any site or premises subject to

regulation or in which records relevant to program operation are kept in order to copy any records, inspect, monitor or otherwise investigate compliance with the State program including compliance with permit conditions and other program requirements. States whose law requires a search warrant before entry conform with this requirement.

(d) Investigatory inspections shall be conducted, samples shall be taken and other information shall be gathered in a manner (e.g., using proper "chain of custody" procedures) that will produce evidence admissible in an enforcement proceeding or in court.

(e) *State NPDES programs only.* State NPDES compliance evaluation programs shall have procedures and ability for:

(1) Maintaining a comprehensive inventory of all sources covered by NPDES permits and a schedule of reports required to be submitted by permittees to the State agency;

(2) Initial screening (i.e., pre-enforcement evaluation) of all permit or grant-related compliance information to identify violations and to establish priorities for further substantive technical evaluation;

(3) When warranted, conducting a substantive technical evaluation following the initial screening of all permit or grant-related compliance information to determine the appropriate agency response;

(4) Maintaining a management information system which supports the compliance evaluation activities of this Part; and

(5) Inspecting the facilities of all major dischargers at least annually.

§ 123.9   Requirements for enforcement authority.

(a) Any State agency administering a program shall have available the following remedies for violations of State program requirements:

(1) To restrain immediately and effectively any person by order or by suit in State court from engaging in any unauthorized activity which is endangering or causing damage to public health or the environment;

[Note.—This paragraph requires that States have a mechanism (e.g., an administrative cease and desist order or the ability to seek a temporary restraining order) to stop any unauthorized activity endangering public health or the environment.]

(2) To sue in courts of competent jurisdiction to enjoin any threatened or continuing violation of any program requirement, including permit conditions, without the necessity of a prior revocation of the permit;

(3) To assess or sue to recover in court civil penalties and to seek criminal remedies, including fines, as follows:

(i) *State RCRA programs only.* (A) Civil penalties shall be recoverable for any program violation in at least the amount of $10,000 per day.

(B) Criminal remedies shall be obtainable against any person who knowingly transports any hazardous waste to an unpermitted facility; who treats, stores, or disposes of hazardous waste without a permit; or who makes any false statement or representation in any application, label, manifest, record, report, permit or other document filed, maintained, or used for purposes of program compliance. Criminal fines shall be recoverable in at least the amount of $10,000 per day for each violation, and imprisonment for at least six months shall be available.

(ii) *State UIC programs only.* (A) For all wells except Class II wells, civil penalties shall be recoverable for any program violation in at least the amount of $2,500 per day. For Class II wells, civil penalties shall be recoverable for any program violation in at least the amount of $1,000 per day.

(B) Criminal fines shall be recoverable in at least the amount of $5,000 per day against any person who willfully violates any program requirement, or, for Class II wells, pipeline (production) severance shall be imposable against any person who willfully violates any program requirement.

(iii) *State NPDES and section 404 programs only.* (A) Civil penalties shall be recoverable for the violation of any NPDES or section 404 permit condition; any NPDES or section 404 filing requirement; any duty to allow or carry out inspection, entry or monitoring activities; or any regulation or orders issued by the State Director. Such penalties shall be assessable in at least the amount of $5,000 per day for each violation.

(B) Criminal fines shall be recoverable against any person who willfully or negligently violates any applicable standards or limitations; any NPDES or section 404 permit condition; or any NPDES or section 404 filing requirement. Such fines shall be assessable in at least the amount of $10,000 per day for each violation.

[Note.—States which provide criminal remedies based on "criminal negligence," "gross negligence" or strict liability satisfy the requirement of paragraph (a)(3)(iii)(B) of this section.]

(C) Criminal fines shall be recoverable against any person who knowingly makes any false statement, representation or certification in any

NPDES or section 404 form, in any notice or report required by an NPDES or section 404 permit, or who knowingly renders inaccurate any moitoring device or method required to be maintained by the Director. Such fines shall be recoverable in at least the amount of $5,000 for each instance of violation.

[Note.—In many States the State Director will be represented in State courts by the State Attorney General or other appropriate legal officer. Although the State Director need not appear in court actions he or she should have power to request that any of the above actions be brought.]

(b)(1) The maximum civil penalty or criminal fine (as provided in paragraph (a)(3) of this section) shall be assessable for each instance of violation and, if the violation is continuous, shall be assessable up to the maximum amount for each day of violation.

(2) The burden of proof and degree of knowledge or intent required under State law for establishing violations under paragraph (a)(3) of this section, shall be no greater than the burden of proof or degree of knowledge or intent EPA must provide when it brings an action under the appropriate Act.

[Note.—For example, this requirement is not met if State law includes mental state as an element of proof for civil violations.]

(c) Any civil penalty assessed, sought or agreed upon by the State Director under paragraph (a)(3) of this section shall be appropriate to the violation. A civil penalty agreed upon by the State Director in settlement of administrative or judicial litigation may be adjusted by a percentage which represents the likelihood of success in establishing the underlying violation(s) in such litigation. If such civil penalty, together with the costs of expeditious compliance, would be so severely disproportionate to the resources of the violator as to jeopardize continuance in business, the payment of the penalty may be deferred or the penalty may be forgiven in whole or part, as circumstances warrant. In the case of a penalty for a failure to meet a statutory or final permit compliance deadline, "appropriate to the violation," as used in this paragraph, means a penalty which is equal to:

(1) An amount appropriate to redress the harm or risk to public health or the environment; plus

(2) An amount appropriate to remove the economic benefit gained or to be gained from delayed compliance; plus

(3) An amount appropriate as a penalty for the violator's degree of recalcitrance, defiance, or indifference to requirements of the law; plus

(4) An amount appropriate to recover unusual or extraordinary enforcement costs thrust upon the public; minus

(5) An amount, if any, appropriate to reflect any part of the noncompliance attributable to the government itself; and minus

(6) An amount appropriate to reflect any part of the noncompliance caused by factors completely beyond the violator's control (e.g., floods, fires).

[Note.—In addition to the requirements of this paragraph, the State may have other enforcement remedies. The following enforcement options, while not mandatory, are highly recommended:

Procedures for assessment by the State of the costs of investigations, inspections, or monitoring surveys which lead to the establishment of violations;

Procedures which enable the State to assess or to sue any persons responsible for unauthorized activities for any expenses incurred by the State in removing, correcting, or terminating any adverse effects upon human health and the environment resulting from the unauthorized activity, whether or not accidental;

Procedures which enable the State to sue for compensation for any loss or destruction of wildlife, fish or aquatic life, or their habitat, and for any other damages caused by unauthorized activity, either to the State or to any residents of the State who are directly aggrieved by the unauthorized activity, or both; and

Procedures for the administrative assessment of penalties by the Director.]

(d) Any State administering a program shall provide for public participation in the State enforcement process by providing either:

(1) Authority which allows intervention as of right in any civil or administrative action to obtain remedies specified in paragraphs (a) (1), (2) or (3) of this section by any citizen having an interest which is or may be adversely affected; or

(2) Assurance that the State agency or enforcement authority will:

(i) Investigate and provide written responses to all citizen complaints submitted pursuant to the procedures specified in § 123.8(b)(4);

(ii) Not oppose intervention by any citizen when permissive intervention may be authorized by statute, rule, or regulation; and

(iii) Publish notice of and provide at least 30 days for public comment on any proposed settlement of a State enforcement action.

§ 123.10   Sharing of information.

(a) Any information obtained or used in the administration of a State program shall be available to EPA upon request without restriction. If the information has been submitted to the State under a claim of confidentiality, the State must submit that claim to EPA when providing information under this section. Any information obtained from a State and subject to a claim of confidentiality will be treated in accordance with the regulations in 40 CFR Part 2. If EPA obtains from a State information that is not claimed to be confidential, EPA may make that information available to the public without further notice.

(b) EPA shall furnish to States with approved programs the information in its files not submitted under a claim of confidentiality which the State needs to implement its approved program. EPA shall furnish to States with approved programs information submitted to EPA under a claim of confidentiality, which the State needs to implement its approved program, subject to the conditions in 40 CFR Part 2.

§ 123.11   Coordination with other programs.

(a) Issuance of State permits under this Part may be coordinated with issuance of RCRA, UIC, NPDES, and 404 permits whether they are controlled by the State, EPA, or the Corps of Engineers. See § 124.4.

(b) The State Director of any approved program which may affect the planning for and development of hazardous waste management facilities and practices shall consult and coordinate with agencies designated under section 4006(b) of RCRA (40 CFR Part 255) as responsible for the development and implementation of State solid waste management plans under section 4002(b) of RCRA (40 CFR Part 256).

§ 123.12   Approval process.

The process for EPA approval of State programs is set out in §§ 123.39 (RCRA), 123.54 (UIC), 123.77 (NPDES), and 123.104 (404).

§ 123.13   Procedures for revision of State programs.

(a) Either EPA or the approved State may initiate program revision. Program revision may be necessary when the controlling Federal or State statutory or regulatory authority is modified or supplemented. The State shall keep EPA fully informed of any proposed modifications to its basic statutory or regulatory authority, its forms, procedures, or priorities.

(b) Revision of a State program shall be accomplished as follows:

(1) The State shall submit a modified program description, Attorney General's statement, Memorandum of Agreement, or such other documents as EPA determines to be necessary under the circumstances.

(2) Whenever EPA determines that the proposed program revision is substantial, EPA shall issue public notice and provide an opportunity to comment for a period of at least 30 days. The public notice shall be mailed to interested persons and shall be published in the **Federal Register** and in enough of the largest newspapers in the State to provide Statewide coverage. The public notice shall summarize the proposed revisions and provide for the opportunity to request a public hearing. Such a hearing will be held if there is significant public interest based on requests received.

(3) The Administrator shall approve or disapprove program revisions based on the requirements of this Part and of the appropriate Act.

(4) A program revision shall become effective upon the approval of the Administrator. Notice of approval of any substantial revision shall be published in the **Federal Register**. Notice of approval of non-substantial program revisions may be given by a letter from the Administrator to the State Governor or his designee.

(c) States with approved programs shall notify EPA whenever they propose to transfer all or part of any program from the approved State agency to any other State agency, and shall identify any new division of responsibilities among the agencies involved. The new agency is not authorized to administer the program until approved by the Administrator under paragraph (b) of this section. Organizational charts required under § 123.4(b) shall be revised and resubmitted.

(d) Whenever the Administrator has reason to believe that circumstances have changed with respect to a State program, he may request, and the State shall provide, a supplemental Attorney General's statement, program description, or such other documents or information as are necessary.

(e) *State RCRA programs only.* All new programs must comply with these regulations immediately upon approval. Any approved program which requires revision because of a modification to this Part or to 40 CFR Parts 122, 124, 260, 261, 262, 263, 264, 265 or 266 shall be so revised within one year of the date of promulgation of such regulation, unless a State must amend or enact a statute in order to make the required revision in which case such revision shall take place within two years.

(f) *State UIC programs only.* The State shall submit the information required under paragraph (b)(1) of this section within 270 days of any amendment to this Part or 40 CFR Parts 122, 124, or 146 which revises or adds any requirement

respecting an approved State UIC program.

(g) *State NPDES programs only.* All new programs must comply with these regulations immediately upon approval. Any approved State section 402 permit program which requires revision to conform to this Part shall be so revised within one year of the date of promulgation of these regulations, unless a State must amend or enact a statute in order to make the required revision in which case such revision shall take place within 2 years, except that revision of State programs to implement the requirements of 40 CFR Part 403 (pretreatment) shall be accomplished as provided in 40 CFR § 403.10. In addition, approved States shall submit, within 6 months, copies of their permit forms for EPA review and approval. Approved States shall also assure that permit applicants, other than POTWs, either (1) whose permits expire after November 30, 1980 or (2) whose permits expire before November 30, 1980 and who have not reapplied for a permit prior to April 30, 1980, submit, as part of their application, the information required under §§ 122.4(d) and 122.53 (d) or (e), as appropriate.

(h) *State section 404 programs only.* The Regional Administrator shall consult with the Corps of Engineers, the U.S. Fish and Wildlife Service, and the National Marine Fisheries Service regarding any substantial program revision, and shall consider their recommendations prior to approval of any such revision.

§ 123.14   Criteria for withdrawal of State programs.

(a) The Administrator may withdraw program approval when a State program no longer complies with the requirements of this Part, and the State fails to take corrective action. Such circumstances include the following:

(1) When the State's legal authority no longer meets the requirements of this Part, including:

(i) Failure of the State to promulgate or enact new authorities when necessary; or

(ii) Action by a State legislature or court striking down or limiting State authorities.

(2) When the operation of the State program fails to comply with the requirements of this Part, including:

(i) Failure to exercise control over activities required to be regulated under this Part, including failure to issue permits;

(ii) Repeated issuance of permits which do not conform to the requirements of this Part; or

(iii) Failure to comply with the public participation requirements of this Part.

(3) When the State's enforcement program fails to comply with the requirements of this Part, including:

(i) Failure to act on violations of permits or other program requirements;

(ii) Failure to seek adequate enforcement penalties or to collect administrative fines when imposed; or

(iii) Failure to inspect and monitor activities subject to regulation.

(4) When the State program fails to comply with the terms of the Memorandum of Agreement required under § 123.6.

§ 123.15   Procedures for withdrawal of State programs.

(a) A State with a program approved under this Part may voluntarily transfer program responsibilities required by Federal law to EPA (or to the Secretary in the case of 404 programs) by taking the following actions, or in such other manner as may be agreed upon with the Administrator.

(1) The State shall give the Administrator (and the Secretary in the case of section 404 programs) 180 days notice of the proposed transfer and shall submit a plan for the orderly transfer of all relevant program information not in the possession of EPA (or the Secretary in the case of section 404 programs) (such as permits, permit files, compliance files, reports, permit applications) which are necessary for EPA (or the Secretary in the case of section 404 programs) to administer the program.

(2) Within 60 days of receiving the notice and transfer plan, the Administrator (and the Secretary in the case of section 404 programs) shall evaluate the State's transfer plan and shall identify any additional information needed by the Federal government for program administration and/or identify any other deficiencies in the plan.

(3) At least 30 days before the transfer is to occur the Administrator shall publish notice of the transfer in the **Federal Register** and in enough of the largest newspapers in the State to provide Statewide coverage, and shall mail notice to all permit holders, permit applicants, other regulated persons and other interested persons on appropriate EPA and State mailing lists.

(b) The following procedures apply when the Administrator orders the commencement of proceedings to determine whether to withdraw approval of a State program, other than a UIC program. The process for withdrawing approval of State UIC programs is set out in § 123.55.

(1) *Order.* The Administrator may order the commencement of withdrawal proceedings on his or her own initiative or in response to a petition from an interested person alleging failure of the State to comply with the requirements of this Part as set forth in § 123.14. The Administrator shall respond in writing to any petition to commence withdrawal proceedings. He may conduct an informal investigation of the allegations in the petition to determine whether cause exists to commence proceedings under this paragraph. The Administrator's order commencing proceedings under this paragraph shall fix a time and place for the commencement of the hearing and shall specify the allegations against the State which are to be considered at the hearing. Within 30 days the State shall admit or deny these allegations in a written answer. The party seeking withdrawal of the State's program shall have the burden of coming forward with the evidence in a hearing under this paragraph.

(2) *Definitions.* For purposes of this paragraph the definitions of "Act," "Administrative Law Judge," "Hearing," "Hearing Clerk," and "Presiding Officer" in 40 CFR § 22.03 apply in addition to the following:

(i) "Party" means the petitioner, the State, the Agency, and any other person whose request to participate as a party is granted.

(ii) "Person" means the Agency, the State and any individual or organization having an interest in the subject matter of the proceeding.

(iii) "Petitioner" means any person whose petition for commencement of withdrawal proceedings has been granted by the Administrator.

(3) *Procedures.* The following provisions of 40 CFR Part 22 (Consolidated Rules of Practice) are applicable to proceedings under this paragraph:

(i) § 22.02—(use of number/gender);

(ii) § 22.04(c)—(authorities of Presiding Officer);

(iii) § 22.06—(filing/service of rulings and orders);

(iv) § 22.07(a) and (b)—*except that,* the time for commencement of the hearing shall not be extended beyond the date set in the Administrator's order without approval of the Administrator— (computation/extension of time);

(v) § 22.08—*however,* substitute "order commencing proceedings" for "complaint"—(Ex Parte contacts);

(vi) § 22.09—(examination of filed documents);

(vii) § 22.11(a), (c) and (d), *however,* motions to intervene must be filed within 15 days from the date the notice of the Administrator's order is first published—(intervention);

(viii) § 22.16 *except that,* service shall be in accordance with paragraph (b)(4) of this section, the first sentence in § 22.16(c) shall be deleted, and, the word "recommended" shall be substituted for the word "initial" in § 22.16(c)— (motions);

(ix) § 22.19(a), (b) and (c)—(prehearing conference);

(x) § 22.22—(evidence);

(xi) § 22.23—(objections/offers of proof);

(xii) § 22.25—(filing the transcript); and

(xiii) § 22.26—(findings/conclusions).

(4) *Record of proceedings.* (i) The hearing shall be either stenographically reported verbatim or tape recorded, and thereupon transcribed by an official reporter designated by the Presiding Officer;

(ii) All orders issued by the Presiding Officer, transcripts of testimony, written statements of position, stipulations, exhibits, motions, briefs, and other written material of any kind submitted in the hearing shall be a part of the record and shall be available for inspection or copying in the Office of the Hearing Clerk, upon payment of costs. Inquiries may be made at the Office of the Administrative Law Judges, Hearing Clerk, 401 M Street, S.W., Washington, D.C. 20460;

(iii) Upon notice to all parties the Presiding Officer may authorize corrections to the transcript which involve matters of substance;

(iv) An original and two (2) copies of all written submissions to the hearing shall be filed with the Hearing Clerk;

(v) A copy of each such submission shall be served by the person making the submission upon the Presiding Officer and each party of record. Service under this paragraph shall take place by mail or personal delivery;

(vi) Every submission shall be accompanied by an acknowledgement of service by the person served or proof of service in the form of a statement of the date, time, and manner of service and the names of the persons served, certified by the person who made service; and

(vii) The Hearing Clerk shall maintain and furnish to any person upon request, a list containing the name, service address, and telephone number of all parties and their attorneys or duly authorized representatives.

(5) *Participation by a person not a party.* A person who is not a party may, at the discretion of the Presiding Officer, be permitted to make a limited appearance by making an oral or written statement of his/her position on the issues within such limits and on such conditions as may be fixed by the Presiding Officer, but he/she may not otherwise participate in the proceeding.

(6) *Rights of parties.* All parties to the proceeding may:

(i) Appear by counsel or other representative in all hearing and pre-hearing proceedings;

(ii) Agree to stipulations of facts which shall be made a part of the record.

(7) *Recommended decision.* (i) Within 30 days after the filing of proposed findings and conclusions, and reply briefs, the Presiding Officer shall evaluate the record before him/her, the proposed findings and conclusions and any briefs filed by the parties and shall prepare a recommended decision, and shall certify the entire record, including the recommended decision, to the Administrator.

(ii) Copies of the recommended decision shall be served upon all parties.

(iii) Within 20 days after the certification and filing of the record and recommended decision, all parties may file with the Administrator exceptions to the recommended decision and a supporting brief.

(8) *Decision by Administrator.* (i) Within 60 days after the certification of the record and filing of the Presiding Officer's recommended decision, the Administrator shall review the record before him and issue his own decision.

(ii) If the Administrator concludes that the State has administered the program in conformity with the appropriate Act and regulations his decision shall constitute "final agency action" within the meaning of 5 U.S.C. §704.

(iii) If the Administrator concludes that the State has not administered the program in conformity with the appropriate Act and regulations he shall list the deficiencies in the program and provide the State a reasonable time, not to exceed 90 days, to take such appropriate corrective action as the Administrator determines necessary.

(iv) Within the time prescribed by the Administrator the State shall take such appropriate corrective action as required by the Administrator and shall file with the Administrator and all parties a statement certified by the State Director that appropriate corrective action has been taken.

(v) The Administrator may require a further showing in addition to the certified statement that corrective action has been taken.

(vi) If the State fails to take appropriate corrective action and file a certified statement thereof within the time prescribed by the Administrator, the Administrator shall issue a supplementary order withdrawing approval of the State program. If the State takes appropriate corrective action, the Administrator shall issue a supplementary order stating that approval of authority is not withdrawn.

(vii) The Administrator's supplementary order shall constitute final Agency action within the meaning of 5 U.S.C. § 704.

(c) Withdrawal of authorization under this section and the appropriate Act does not relieve any person from complying with the requirements of State law, nor does it affect the validity of actions by the State prior to withdrawal.

Subpart B—Additional Requirements for State Hazardous Waste Programs

§ 123.31   Purpose and scope.

(a) This Subpart specifies additional requirements a State program must meet in order to obtain final authorization under section 3006(b) of RCRA. All of the requirements a State program must meet in order to obtain interim authorization under section 3006(c) of RCRA are specified in Subpart F.

(b) States approved under this Subpart are authorized to administer and enforce their hazardous waste program in lieu of the Federal program.

(c) States may apply for final authorization at any time after the initial promulgation of Phase II. State programs under final authorization may not take effect until the effective date of Phase II.

(d) States operating under interim authorization may apply for and receive final authorization as specified in paragraph (c) of this section. Notwithstanding approval under Subpart F, such States must meet all the requirements of Subpart A and this subpart in order to qualify for final authorization.

(e) States need not have been approved under Subpart F in order to qualify for final authorization.

§ 123.32   Consistency.

To obtain approval, a State program must be consistent with the Federal program and State programs applicable in other States and in particular must comply with the provisions below. For purposes of this section the phrase "State programs applicable in other States" refers only to those State hazardous waste programs which have received final authorization under this Part.

(a) Any aspect of the State program which unreasonably restricts, impedes, or operates as a ban on the free movement across the State border of hazardous wastes from or to other States for treatment, storage, or disposal at facilities authorized to operate under the Federal or an approved State program shall be deemed inconsistent.

(b) Any aspect of State law or of the State program which has no basis in human health or environmental protection and which acts as a prohibition on the treatment, storage or disposal of hazardous waste in the State may be deemed inconsistent.

(c) If the State manifest system does not meet the requirements of this Part, the State program shall be deemed inconsistent.

§ 123.33   Requirements for identification and listing of hazardous wastes.

The State program must control all the hazardous wastes controlled under 40 CFR Part 261 and must adopt a list of hazardous wastes and a set of characteristics for identifying hazardous wastes equivalent to those under 40 CFR Part 261.

§ 123.34   Requirements for generators of hazardous wastes.

(a) The State program must cover all generators covered by 40 CFR Part 262. States must require new generators to contact the State and obtain an EPA identification number before they perform any activity subject to regulation under the approved State hazardous waste program.

(b) The State shall have authority to require and shall require all generators to comply with reporting and recordkeeping requirements equivalent to those under 40 CFR §§ 262.40 and 262.41. States must require that generators keep these records at least 3 years.

(c) The State program must require that generators who accumulate hazardous wastes for short periods of time prior to shipment off-site do so in containers meeting DOT shipping requirements under 49 CFR Parts 173, 178 and 179 or accumulate such wastes in tanks in accordance with State storage standards authorized by EPA under the approved State program.

(d) The State program must require that generators comply with requirements that are equivalent to the requirements for the packaging, labeling, marking, and placarding of hazardous waste under 40 CFR §§ 262.30 to 262.33, and are consistent with relevant DOT regulations under 49 CFR Parts 172, 173, 178 and 179.

(e) The State program shall provide requirements respecting international shipments which are equivalent to those at 40 CFR § 262.50, except that advance notification of international shipments, as required by 40 CFR § 262.50(b)(1),

shall be filed with the Administrator. The State may require that a copy of such advance notice be filed with the State Director, or may require equivalent reporting procedures.

[Note.—Such notices shall be mailed to Hazardous Waste Export, Division for Oceans and Regulatory Affairs (A-107), U.S. Environmental Protection Agency, Washington, D.C. 20460.]

(f) The State must require that all generators of hazardous waste who transport (or offer for transport) such hazardous waste off-site:

(1) Use a manifest system that ensures that interstate and intrastate shipments of hazardous waste are designated for delivery, and, in the case of intrastate shipments, are delivered to facilities that are authorized to operate under an approved State program or the Federal program;

(2) Initiate the manifest and designate on the manifest the storage, treatment, or disposal facility to which the waste is to be shipped;

(3) Ensure that all wastes offered for transport are accompanied by the manifest, except in the case of shipments by rail or water specified in 40 CFR §§ 262.23(c) and 263.20(e). The State program shall provide requirements for shipments by rail or water equivalent to those under 40 CFR §§ 262.23(c) and 263.20(e).

(4) Investigate instances where manifests have not been returned by the owner or operator of the designated facility and report such instances to the State in which the shipment originated.

(g) In the case of interstate shipments for which the manifest has not been returned, the State program must provide for notification to the State in which the facility designated on the manifest is located and to the State in which the shipment may have been delivered (or to EPA in the case of unauthorized States).

(h) The State must follow the Federal manifest format (40 CFR § 262.21) and may supplement the format to a limited extent subject to the consistency requirements of the Hazardous Materials Transportation Act (49 U.S.C. 1801 et seq.).

§ 123.35   Requirements for transporters of hazardous wastes.

(a) The State program must cover all transporters covered by 40 CFR Part 263. New transporters must be required to contact the State and obtain an EPA identification number from the State before they accept hazardous waste for transport.

(b) The State shall have the authority to require and shall require all transporters to comply with recordkeeping requirements equivalent to those found at 40 CFR § 263.22. States must require that records be kept at least 3 years.

(c) The State must require the transporter to carry the manifest during transport, except in the case of shipments by rail or water specified in 40 CFR § 263.20(e), and to deliver wastes only to the facility designated on the manifest. The State program shall provide requirements for shipments by rail or water equivalent to those under 40 CFR § 263.20(e).

(d) For hazardous wastes that are discharged in transit, the State program must require that transporters notify appropriate State, local, and Federal agencies of such discharges, and clean up such wastes, or take action so that such wastes do not present a hazard to human health or the environment. These requirements shall be equivalent to those found at 40 CFR §§ 263.30 and 263.31.

§ 123.36 Requirements for hazardous waste management facilities.

The State shall have standards for hazardous waste management facilities which are equivalent to 40 CFR Parts 264 and 266. These standards shall include:

(a) Technical standards for tanks, containers, waste piles, incineration, chemical, physical and biological treatment facilities, surface impoundments, landfills, and land treatment facilities;

(b) Financial responsibility during facility operation;

(c) Preparedness for and prevention of discharges or releases of hazardous waste; contingency plans and emergency procedures to be followed in the event of a discharge or release of hazardous waste;

(d) Closure and post-closure requirements including financial requirements to ensure that money will be available for closure and post-closure monitoring and maintenance;

(e) Groundwater monitoring;

(f) Security to prevent unauthorized access to the facility;

(g) Facility personnel training;

(h) Inspections, monitoring, recordkeeping, and reporting;

(i) Compliance with the manifest system, including the requirement that facility owners or operators return a signed copy of the manifest to the generator to certify delivery of the hazardous waste shipment;

(j) Other requirements to the extent that they are included in 40 CFR Parts 264 and 266.

§ 123.37 Requirements with respect to permits and permit applications.

(a) State law must require permits for owners and operators of all hazardous waste management facilities required to obtain a permit under 40 CFR Part 122 and prohibit the operation of any hazardous waste management facility without such a permit, *except that* States may, if adequate legal authority exists, authorize owners and operators of any facility which would qualify for interim status under the Federal program to remain in operation until a final decision is made on the permit application. Where State law authorizes such continued operation it shall require compliance by owners and operators of such facilities with standards at least as stringent as EPA's interim status standards at 40 CFR Part 265.

(b) The State must require all new HWM facilities to contact the State and obtain an EPA identification number before commencing treatment, storage, or disposal of hazardous waste.

(c) All permits issued by the State shall require compliance with the standards adopted by the State under § 123.36.

(d) All permits issued under State law prior to the date of approval of final authorization shall be reviewed by the State Director and modified or revoked and reissued to require compliance with the requirements of this Part.

§ 123.38 EPA review of State permits.

(a) The Regional Administrator may comment on permit applications and draft permits as provided in the Memorandum of Agreement under § 123.6.

(b) Where EPA indicates, in a comment, that issuance of the permit would be inconsistent with the approved State program, EPA shall include in the comment:

(1) A statement of the reasons for the comment (including the section of RCRA or regulations promulgated thereunder that support the comment); and

(2) The actions that should be taken by the State Director in order to address the comments (including the conditions which the permit would include if it were issued by the Regional Administrator).

(c) A copy of any comment shall be sent to the permit applicant by the Regional Administrator.

(d) The Regional Administrator shall withdraw such a comment when satisfied that the State has met or refuted his or her concerns.

(e) Under section 3008(a)(3) of RCRA, EPA may terminate a State-issued permit in accordance with the procedures of Part 124, Subpart E, or

bring an enforcement action in accordance with the procedures of 40 CFR Part 22 in the case of a violation of a State program requirement. In exercising these authorities, EPA will observe the following conditions:

(1) The Regional Administrator may take action under section 3008(a)(3) of RCRA against a holder of a State-issued permit at any time on the ground that the permittee is not complying with a condition of that permit.

(2) The Regional Administrator may take action under section 3008(a)(3) of RCRA against a holder of a State-issued permit at any time on the grounds that the permittee is not complying with a condition that the Regional Administrator in commenting on the permit application or draft permit stated was necessary to implement approved State program requirements, whether or not that condition was included in the final permit.

(3) The Regional Administrator may not take action under section 3008(a)(3) of RCRA against a holder of a State-issued permit on the ground that the permittee is not complying with a condition necessary to implement approved State program requirements unless the Regional Administrator stated in commenting on the permit application or draft permit that that condition was necessary.

(4) The Regional Administrator may take action under section 7003 of RCRA against a permit holder at any time whether or not the permit holder is complying with permit conditions.

§ 123.39 Approval process.

(a) Prior to submitting an application to EPA for approval of a State program, the State shall issue public notice of its intent to seek program approval from EPA. This public notice shall:

(1) Be circulated in a manner calculated to attract the attention of interested persons including:

(i) Publication in enough of the largest newspapers in the State to attract statewide attention; and

(ii) Mailing to persons on the State agency mailing list and to any other persons whom the agency has reason to believe are interested;

(2) Indicate when and where the State's proposed submission may be reviewed by the public;

(3) Indicate the cost of obtaining a copy of the submission;

(4) Provide for a comment period of not less than 30 days during which time interested members of the public may express their views on the proposed program;

(5) Provide that a public hearing will be held by the State or EPA if sufficient

public interest is shown or, alternatively, schedule such a public hearing. Any public hearing to be held by the State on its application for authorization shall be scheduled no earlier than 30 days after the notice of hearing is published;

(6) Briefly outline the fundamental aspects of the State program; and

(7) Identify a person that an interested member of the public may contact with any questions.

(b) If the proposed State program is substantially modified after the public comment period provided in paragraph (a)(4) of this section, the State shall, prior to submitting its program to the Administrator, provide an opportunity for further public comment in accordance with the procedures of paragraph (a) of this section, *provided* that the opportunity for further public comment may be limited to those portions of the State's application which have been changed since the prior public notice.

(c) After complying with the requirements of paragraphs (a) and (b) of this section the State may submit, in accordance with § 123.3, a proposed program to EPA for approval. Such formal submission may only be made after the date of promulgation of Phase II. The program submission shall include copies of all written comments received by the State, a transcript, recording, or summary of any public hearing which was held by the State, and a responsiveness summary which identifies the public participation activities conducted, describes the matters presented to the public, summarizes significant comments received and responds to these comments.

(d) Within 90 days from the date of receipt of a complete program submission for final authorization, the Administrator shall make a tentative determination as to whether or not he expects to grant authorization to the State program. If the Administrator indicates that he may not approve the State program he shall include a general statement of his areas of concern. The Administrator shall give notice of this tentative determination in the **Federal Register** and in accordance with paragraph (a)(1) of this section. Notice of the tentative determination of authorization shall also:

(1) Indicate that a public hearing will be held by EPA no earlier than 30 days after notice of the tentative determination of authorization. The notice may require persons wishing to present testimony to file a request with the Regional Administrator, who may cancel the public hearing if sufficient public interest in a hearing is not expressed;

(2) Afford the public 30 days after the notice to comment on the State's submission and the tentative determination; and

(3) Note the availability of the State submission for inspection and copying by the public.

(e) Within 90 days of the notice given pursuant to paragraph (d) of this section, the Administrator shall make a final determination whether or not to approve the State's program, taking into account any comments submitted. The Administrator will grant final authorization only after the effective date of Phase II. The Administrator shall give notice of this final determination in the **Federal Register** and in acccordance with paragraph (a)(1) of this section. The notification shall include a concise statement of the reasons for this determination, and a response to significant comments received.

Subpart F—Requirements for Interim Authorization of State Hazardous Waste Programs

§ 123.121 Purpose and scope.

(a) This subpart specifies all of the requirements a State program must meet in order to obtain interim authorization under section 3006(c) of RCRA. The requirements a State program must meet in order to obtain final authorization under section 3006(b) of RCRA are specified in Subparts A and B.

(b) Interim authorization of State programs under this Subpart may occur in two phases. The first phase (Phase I) allows States to administer a hazardous waste program in lieu of and corresponding to that portion of the Federal program which covers identification and listing of hazardous waste (40 CFR Part 261), generators (40 CFR Part 262) and transporters (40 CFR Part 263) of hazardous wastes, and establishes preliminary (interim status) standards for hazardous waste treatment, storage and disposal facilities (40 CFR Part 265). The second phase (Phase II) allows States with interim authority for Phase I to establish a permit program for hazardous waste treatment, storage and disposal facilities in lieu of and corresponding to the Federal hazardous waste permit program (40 CFR Parts 264 and 266). States may apply for interim authorization either sequentially (application for interim authorization for Phase I followed by an amendment of that application for Phase II) or all at once (application for interim authorization for both Phases I and II at the same time) as long as they adhere to the schedule in § 123.122.

(c) The Administrator shall approve a State program which meets the applicable requirements of this Subpart.

(d) Upon approval of a State program for Phase II, the Administrator shall suspend the issuance of Federal permits for those activities subject to the approved State program.

(e) Any State program approved by the Administrator under this Subpart shall at all times be conducted in accordance with this Subpart.

(f) Lack of authority to regulate activities on Indian lands does not impair a State's ability to obtain interim authorization under this Subpart. EPA will administer the program on Indian lands if the State does not seek this authority.

[Note.—States are advised to contact the United States Department of Interior, Bureau of Indian Affairs, concerning authority over Indian lands.]

(g) Nothing is this Subpart precludes a State from:

(1) Adopting or enforcing requirements which are more stringent or more extensive than those required under this Subpart.

(2) Operating a program with a greater scope of coverage than that required under this Subpart. Where an approved program has a greater scope of coverage than required by Federal law the additional coverage is not part of the Federally approved program.

§ 123.122 Schedule.

(a) Interim authorization for Phase I shall not take effect until Phase I commences. Interim authorization for Phase II shall not take effect until Phase II commences.

(b) Interim authorization may extend for a 24-month period from the commencement of Phase II. At the end of this period all interim authorizations automatically expire and EPA shall administer the Federal program in any State which has not received final authorization.

(c) A State may apply for interim authorization at any time prior to expiration of the 6th month of the 24-month period beginning with the commencement of Phase II.

(1) States applying for interim authorization prior to the promulgation of Phase II shall apply only for interim authorization for Phase I.

(2) States applying for interim authorization after the promulgation of Phase II but before the commencement of Phase II may apply either for interim authorization for both Phase I and Phase II or only for interim authorization for Phase I.

(3) States applying for interim authorization after the commencement of Phase II shall apply for interim authorization for both Phase I and Phase II, unless they have already applied for interim authorization for Phase I.

(4) States which have received interim authorization for Phase I shall amend their original submission to meet the requirements for interim authorization for Phase II not later than 6 months after the effective date of Phase II.

(d) No State may apply for interim authorization for Phase II unless it has received interim authorization for Phase I or is simultaneously applying for interim authorization for both Phase I and Phase II.

§ 123.123 Elements of a program submission.

(a) States applying for interim authorization shall submit at least three copies of a program submission to EPA containing the following:

(1) A letter from the Governor of the State requesting State program approval;

(2) A complete program description, as required by § 123.124, describing how the State intends to carry out its responsibilities under this subpart;

(3) An Attorney General's statement as required by § 123.125;

(4) A Memorandum of Agreement with the Regional Administrator as required by § 123.126;

(5) An authorization plan as required by § 123.127;

(6) Copies of all applicable State statutes and regulations, including those governing State administrative procedures.

(b) Within 30 days of receipt by EPA of a State program submission, EPA will notify the State whether its submission is complete. If a State's submission is found to be complete, EPA's formal review of the proposed State program shall be deemed to have begun on the date of receipt of the State's submission. See § 123.135. If a State's submission is found to be incomplete, formal review shall not begin until all the necessary information is received by EPA.

(c) If the State's submission is materially changed during the formal review period, the formal review period shall recommence upon receipt of the revised submission.

(d) States simultaneously applying for interim authorization for both Phase I and Phase II shall prepare a single submission.

(e) States applying for interim authorization for Phase II shall amend their submission for interim authorization for Phase I as specified in §§ 123.124 to 123.127.

§ 123.124 Program description.

Any State that wishes to administer a program under this Subpart shall submit to the Regional Administrator a complete description of the program it proposes to administer in lieu of the Federal program under State law. A State applying only for interim authorization for Phase II shall amend its program description for interim authorization for Phase I as necessary to reflect the program it proposes to administer to meet the requirements for interim authorization for Phase II. The program description shall include:

(a) A description in narrative form of the scope, structure, coverage, and processes of the State program.

(b) A description (including organization charts) of the organization and structure of the State agency or agencies which will have responsibility for administering the program including the information listed below. If more than one agency is responsible for administration of the program, each agency must have Statewide jurisdiction over a class of activities. The responsibilities of each agency must be delineated, their procedures for coordination set forth, and one of the agencies must be designated a "lead agency" to facilitate communications between EPA and the State agencies having program responsibility. Where the State proposes to administer a program of greater scope of coverage than is required by Federal law, the information provided under this section shall indicate the resources dedicated to administering the Federally required portion of the program.

(1) A description of the State agency staff who will be engaged in carrying out the State program, including the number, occupations, and general duties of the employees. The State need not submit complete job descriptions for every employee engaged in carrying out the State program.

(2) An itemization of the proposed or actual costs of establishing and administering the program, including cost of the personnel listed in paragraph (b)(1) of this section, cost of administrative support and cost of technical support.

(3) An itemization of the sources and amounts of funding, including an estimate of Federal grant money, available to the State Director to meet the costs listed in paragraph (b)(2) of this section identifying any restrictions or limitations upon this funding.

(c) A description of applicable State procedures, including permitting procedures, and any State appellate review procedures.

[Note.—States applying only for interim authorization for Phase I need describe permitting procedures only to the extent they will be utilized to assure compliance with standards substantially equivalent to 40 CFR Part 265.]

(d) Copies of the forms and the manifest format the State intends to use in its program. Forms used by the State need not be identical to the forms used by EPA, but should require the same basic information. If the State chooses to use uniform national forms it should so note.

(e) A complete description of the State's compliance monitoring and enforcement program.

(f) A description of the State manifest system if the State has such a system and of the procedures the State will use to coordinate information with other approved State programs and the Federal program regarding interstate and international shipments.

(g) An estimate of the number of the following:

(1) Generators;

(2) Transporters; and

(3) On- and off-site treatment, storage and disposal facilities including a brief description of the types of facilities and an indication, if applicable, of the permit status of these facilities.

§ 123.125 Attorney General's statement.

(a) Any State seeking to administer a program under this Subpart shall submit a statement from the State Attorney General (or the attorney for those State or interstate agencies which have independent legal counsel), that the laws of the State, or the interstate compact, provide adequate authority to carry out the program described under § 123.124 and to meet the applicable requirements of this Subpart. This statement shall include citations to the specific statutes, administrative regulations, and, where appropriate, judicial decisions which demonstrate adequate authority. Except as provided in § 123.128(d), the State Attorney General or independent legal counsel must certify that the enabling legislation for the program for Phase I was in existence within 90 days of the promulgation of Phase I. In the case of a State applying for interim authorization for Phase II, the State Attorney General or independent legal counsel must certify that the enabling legislation for the program for Phase II was in existence within 90 days of the promulgation of Phase II. State statutes and regulations cited by the State Attorney General or independent legal counsel shall be lawfully adopted at the time the statement is signed and shall be fully effective by the time the program is approved. To qualify as "independent legal counsel" the attorney signing the statement required by this section must have full authority to independently represent the State agency in court on all matters pertaining to the State program. In the case of a State applying only for interim authorization for Phase II, the Attorney General's statement submitted for interim authorization for Phase I shall be amended and recertified to demonstrate adequate authority to carry out all the requirements of this Subpart.

(b)(1) In the case of a State applying only for interim authorization for Phase I, the Attorney General's statement shall certify that the authorization plan under § 123.127(a), if carried out, would provide the State with enabling authority and regulations adequate to meet the requirements for final authorization contained in Phase I.

(2) In the case of a State applying for interim authorization for Phase II, the Attorney General's statement shall certify that the authorization plan under § 123.127(b), if carried out, would provide the State with enabling authority and regulations adequate to meet all the requirements for final authorization.

(c) Where a State seeks authority over activities on Indian lands, the statement shall contain an appropriate analysis of the State's authority.

§ 123.126   Memorandum of Agreement.

(a) The State Director and the Regional Administrator shall execute a Memorandum of Agreement (MOA). In addition to meeting the requirements of paragraph (b) of this section, and, if applicable, paragraph (c) of this section, the Memorandum of Agreement may include other terms, conditions, or agreements relevant to the administration and enforcement of the State's regulatory program which are not inconsistent with this subpart. No Memorandum of Agreement shall be approved which contains provisions which restrict EPA's statutory oversight responsibility. In the case of a State applying for interim authorization for Phase II, the Memorandum of Agreement shall be amended and re-executed to include the requirements of paragraph (c) of this section and any revisions to the requirements of paragraph (b) of this section.

(b) The Memorandum of Agreement shall include the following:

(1) Provisions for the prompt transfer from EPA to the State of information obtained in notifications made pursuant to section 3010 of RCRA and received by EPA prior to the approval of the State program, EPA identification numbers for new generators, transporters, and treatment, storage, and disposal facilities, and any other information relevant to effective program operation not already in the possession of the State Director (e.g., pending permit applications, compliance reports, etc.).

(2) Provisions specifying the frequency and content of reports, documents, and other information which the State is required to submit to EPA. The State shall allow EPA to routinely review State records, reports, and files relevant to the administration and enforcement of the approved program. State reports may be combined with grant reports when appropriate.

(3) Provisions on the State's compliance monitoring and enforcement program, including:

(i) Provisions for coordination of compliance monitoring activities by the State and EPA. These may specify the basis on which the Regional Administrator will select facilities or activities within the State for EPA inspection. The Regional Administrator will normally notify the State at least 7 days before any such inspection; and

(ii) Procedures to assure coordination of enforcement activities.

(4) Provisions for modification of the Memorandum of Agreement in accordance with this Part.

(5) A provision allowing EPA to conduct compliance inspections of all generators, transporters, and HWM facilities during interim authorization. The Regional Administrator and the State Director may agree to limitations regarding compliance inspections of generators, transporters, and non-major HWM facilities.

(6) A provision that no limitations on EPA compliance inspections of generators, transporters, and non-major HWM facilities under paragraph (b)(5) of this section shall restrict EPA's right to inspect any HWM facility, generator, or transporter which it has cause to believe is not in compliance with RCRA; however, before conducting such an inspection, EPA will normally allow the State a reasonable opportunity to conduct a compliance evaluation inspection.

(7) A provision delineating respective State and EPA responsibilities during the interim authorization period.

(c) In the case of a State applying for interim authorization for Phase II, the Memorandum of Agreement shall also include the following:

(1) Provisions for prompt transfer from EPA to the State of pending permit applications and support files for permit issuance. Where existing permits are transferred to the State for administration, the Memorandum of Agreement shall contain provisions specifying a procedure for transferring responsibility for these permits. If a State lacks the authority to directly administer permits issued by the Federal government, a procedure may be established to transfer responsibility for these permits.

(2) Provisions specifying classes and categories of permit applications and draft permits that the State Director will send to the Regional Administrator for review and comment. The State Director shall promptly forward to EPA copies of permit applications and draft permits for all major HWM facilities. The Regional Administrator and the State Director may agree to limitations regarding review of and comment on permit applications and draft permits for non-major HWM facilities. The State Director shall supply EPA copies of final permits for all major HWM facilities.

(3) Where appropriate, provisions for joint processing of permits by the State and EPA for facilities or activities which require permits under different programs, from both EPA and the State.

§ 123.127   Authorization plan.

The State must submit an "authorization plan" which shall describe the additions and modifications necessary for the State program to qualify for final authorization as soon as practicable, but no later than the end of the interim authorization period. This plan shall include the nature of and schedules for any changes in State legislation and regulations; resource levels; actions the State must take to control the complete universe of hazardous waste listed or designated under section 3001 of RCRA as soon as possible; the manifest and permit systems; and the surveillance and enforcement program which will be necessary in order for the State to become eligible for final authorization.

(a) In the case of a State applying only for interim authorization for Phase I, the authorization plan shall describe the additions and modifications necessary for the State program to meet the requirements for final authorization contained in Phase I.

(b) In the case of a State applying for interim authorization for Phase II, the authorization plan under paragraph (a) of this section shall be amended to describe the further additions and modifications necessary for the State program to meet all the requirements for final authorization.

§ 123.128 **Program requirements for interim authorization for Phase I.**

The following requirements are applicable to States applying for interim authorization for Phase I. If a State does not have legislative authority or regulatory control over certain activities that do not occur in the State, the State may be granted interim authorization for Phase I provided the State authorization plan under § 123.127 provides for the development of a complete program as soon as practicable after receiving interim authorization.

(a) *Requirements for identification and listing of hazardous waste.* The State program must control a universe of hazardous wastes generated, transported, treated, stored, and disposed of in the State which is nearly identical to that which would be controlled by the Federal program under 40 CFR Part 261.

(b) *Requirements for generators of hazardous waste.* (1) This paragraph applies unless the State comes within the exceptions described under paragraph (d) of this section.

(2) The State program must cover all generators of hazardous wastes controlled by the State.

(3) The State shall have the authority to require and shall require all generators covered by the State program to comply with reporting and recordkeeping requirements substantially equivalent to those found at 40 CFR §§ 262.40 and 262.41.

(4) The State program must require that generators who accumulate hazardous wastes for short periods of time prior to shipment do so in a manner that does not present a hazard to human health or the environment.

(5) The State program shall provide requirements respecting international shipments which are substantially equivalent to those at 40 CFR § 262.50, except that advance notification of international shipment, as required by 40 CFR § 262.50(b)(1), shall be filed with the Administrator. The State may require that a copy of such advance notice be filed with the State Director, or may require equivalent reporting procedures.

[Note.—Such notices shall be mailed to Hazardous Waste Export, Division for Oceans and Regulatory Affairs (A–107), U.S. Environmental Protection Agency, Washington, D.C. 20460.]

(6) The State program must require that such generators of hazardous waste who transport (or offer for transport) such hazardous waste off-site use a manifest system that ensures that inter- and intrastate shipments of hazardous waste are designated for delivery, and,

in the case of intrastate shipments, are delivered only to facilities that are authorized to operate under an approved State program or the Federal program.

(7) The State manifest system must require that:

(i) The manifest itself identify the generator, transporter, designated facility to which the hazardous waste will be transported, and the hazardous waste being transported;

(ii) The manifest accompany all wastes offered for transport, except in the case of shipments by rail or water specified in §§ 262.23(c) and 263.20(e); and

(iii) Shipments of hazardous waste that are not delivered to a designated facility are either identified and reported by the generator to the State in which the shipment originated or are independently identified by the State in which the shipment originated.

(8) In the case of interstate shipments for which the manifest has not been returned, the State program must provide for notification to the State in which the facility designated on the manifest is located and to the State in which the shipment may have been delivered (or to EPA in the case of unauthorized States).

(c) *Requirements for transporters of hazardous wastes.* (1) This paragraph applies unless the State comes within the exceptions described under paragraph (d) of this section.

(2) The State program must cover all transporters of hazardous waste controlled by the State.

(3) The State shall have the authority to require and shall require all transporters covered by the State program to comply with recordkeeping requirements substantially equivalent to those found at 40 CFR § 263.22.

(4) The State program must require such transporters of hazardous waste to use a manifest system that ensures that inter- and intrastate shipments of hazardous waste are delivered only to facilities that are authorized under an approved State program or the Federal program.

(5) The State program must require that transporters carry the manifest with all shipments, except in the case of shipments by rail or water specified in 40 CFR § 263.20(e).

(6) For hazardous wastes that are discharged in transit, the State program must require that transporters notify appropriate State, local, and Federal agencies of the discharges, and clean up the wastes or take action so that the wastes do not present a hazard to human health or the environment. These

requirements shall be substantially equivalent to those found at 40 CFR §§ 263.30 and 263.31.

(d) *Limited exceptions from generator, transporter, and related manifest requirements.* A State applying for interim authorization for Phase I which meets all the requirements for such interim authorization except that it does not have statutory or regulatory authority for the manifest system or other generator or transporter requirements discussed in paragraphs (b) and (c) of this section may be granted interim authorization, if the State authorization plan under § 123.127 delineates the necessary steps for obtaining this authority no later than the end of the interim authorization period under § 123.122(b). A State may apply for interim authorization to implement the manifest system and other generator and transporter requirements if the enabling legislation for that part of the program was in existence within 90 days of the promulgation of Phase I. If such application is made, it shall be made as part of the State's submission for interim authorization for Phase II. Until the State manifest system and other generator and transporter requirements are approved by EPA, all Federal requirements for generators and transporters (including use of the Federal manifest system) shall apply in such States and enforcement responsibility for that part of the program shall remain with the Federal Government. The universe of wastes for which these Federal requirements apply shall be the universe of wastes controlled by the State under paragraph (a) of this section.

(e) *Requirements for hazardous waste treatment, storage, and disposal facilities.* States must have standards applicable to HWM facilities which are substantially equivalent to 40 CFR Part 265. State law shall prohibit the operation of facilities not in compliance with such standards. These standards shall include:

(1) Preparedness for and prevention of releases of hazardous waste controlled by the State under paragraph (a) of this section and contingency plans and emergency procedures to be followed in the event of a release of such hazardous waste;

(2) Closure and post-closure requirements;

(3) Groundwater monitoring;

(4) Security to prevent unknowing and unauthorized access to the facility;

(5) Facility personnel training;

(6) Inspection, monitoring, recordkeeping, and reporting;

(7) Compliance with the manifest system including the requirement that

the facility owner or operator or the State in which the facility is located must return a copy of the manifest to the generator or to the State in which the generator is located indicating delivery of the waste shipment; and

(8) Other facility standards to the extent that they are included in 40 CFR Part 265, except that Subpart R (standards for injection wells) may be included in the State standards at the State's option.

(f) *Requirements for enforcement authority.* (1) Any State agency administering a program under this Subpart shall have the following authority to remedy violations of State program requirements:

(i) Authority to restrain immediately by order or by suit in State court any person from engaging in any unauthorized activity which is endangering or causing damage to public health or the environment;

(ii) To sue in courts of competent jurisdiction to enjoin any threatened or continuing violation of any program requirement, including, where appropriate, permit conditions, without the necessity of a prior revocation of the permit; and

(iii) For any program violation, to assess or sue to recover in court civil penalties in at least the amount of $1000 per day or to seek criminal fines in at least the amount of $1000 per day.

(2) Any State agency administering a program under this Subpart shall provide for public participation in the State enforcement process by providing either:

(i) Authority which allows intervention as of right in any civil or administrative action to obtain remedies specified in paragraph (f)(1) of this section by any citizen having an interest which is or may be adversely affected; or

(ii) Assurance that the State agency or enforcement authority will:

(A) Investigate and provide written responses to all citizen complaints submitted pursuant to the procedures specified in paragraph (g)(2)(iv) of this section;

(B) Not oppose intervention by any citizen where permissive intervention may be authorized by statute, rule, or regulation; and

(C) Publish and provide at least 30 days for public comment on any proposed settlement of a State enforcement action.

(g) *Requirements for compliance evaluation programs.* (1) A State program under this Subpart shall have procedures for receipt, evaluation, recordkeeping, and investigation for possible enforcement of all required notices and reports.

(2) A State program shall have independent inspection and surveillance authority and procedures to determine compliance or noncompliance with applicable program requirements. This shall include:

(i) The capability to make comprehensive surveys of any activities subject to the State Director's authority in order to identify persons subject to regulation who have failed to comply with program requirements;

(ii) A program for periodic inspections of the activities subject to regulation;

(iii) The capability to investigate evidence of violations of applicable program and permit requirements; and

(iv) Procedures for receiving and ensuring proper consideration of information submitted by the public about violations. Public effort in reporting violations shall be encouraged, and the State Director shall make available information on reporting procedures.

(3) The State officers engaged in compliance evaluation activities shall have authority to enter any conveyance, vehicle, facility, or premises subject to regulation or in which records relevant to program operation are kept in order to inspect, monitor, or otherwise investigate compliance with the State program. States whose law requires a search warrant prior to entry conform with this requirement.

(4) Investigatory inspections shall be conducted, samples shall be taken, and other information shall be gathered in a manner (e.g., using proper "chain of custody" procedures) that will produce evidence admissible in an enforcement proceeding or in court.

§ 123.129   Additional program requirements for interim authorization for Phase II.

In addition to the requirements of § 123.128, the following requirements are applicable to States applying for interim authorization for Phase II.

(a) State programs must have standards applicable to hazardous waste management facilities that provide substantially the same degree of human health and environmental protection as the standards promulgated under 40 CFR Parts 264 and 266.

(b) State programs shall require a permit for owners and operators of those hazardous waste treatment, storage, and disposal facilities which handle any waste controlled by the State under § 123.128(a) and for which a permit is required under 40 CFR Part 122. The State program shall prohibit the operation of such facilities without a permit, provided States may authorize owners and operators of facilities which would qualify for interim status under the Federal program (if State law so authorizes) to remain in operation pending permit action. Where State law authorizes such continued operation it shall require compliance by owners and operators of such facilities with standards substantially equivalent to EPA's interim status standards under 40 CFR Part 265.

(c) All permits issued by the State under this section shall require compliance with the standards adopted by the State in accordance with paragraph (a) of this section.

(d) State programs shall have requirements for permitting which are substantially equivalent to the provisions listed in §§ 123.7(a) and (b).

(e) No permit may be issued by a State with interim authorization for Phase II with a term greater than ten years.

§ 123.130   Interstate movement of hazardous waste.

(a) If a waste is transported from a State where it is listed or designated as hazardous under the program applicable in that State, whether that is the Federal program or an approved State program, into a State with interim authorization where it is not listed or designated, the waste must be manifested in accordance with the laws of the State where the waste was generated and must be treated, stored, or disposed of as required by the laws of the State into which it has been transported.

(b) If a waste is transported from a State with interim authorization where it is not listed or designated as hazardous into a State where it is listed or designated as hazardous under the program applicable in that State, whether that is the Federal program or an approved State program, the waste must be treated, stored, or disposed of in accordance with the law applicable in the State into which it has been transported.

(c) In all cases of interstate movement of hazardous waste, as defined by 40 CFR Part 261, generators and transporters must meet DOT requirements in 49 CFR Parts 172, 173, 178, and 179 (e.g., for shipping paper, packaging, labeling, marking, and placarding). `

§ 123.131   Progress reports.

The State Director shall submit a semi-annual progress report to the EPA Regional Administrator within 4 weeks of the date 6 months after Phase I commences, and at 6-month intervals thereafter until the expiration of interim authorization. The reports shall briefly summarize, in a manner and form prescribed by the Regional Administrator, the State's compliance in meeting the requirements of the authorization plan, the reasons and proposed remedies for any delay in meeting milestones, and the anticipated problems and solutions for the next reporting period.

§ 123.132 Sharing of information.

(a) Any information obtained or used in the administration of a State program shall be available to EPA upon request without restriction. If the information has been submitted to the State under a claim of confidentiality, the State must submit that claim to EPA when providing information under this Subpart. Any information obtained from a State and subject to a claim of confidentiality will be treated in accordance with the regulations in 40 CFR Part 2. If EPA obtains from a State information that is not claimed to be confidential, EPA may make that information available to the public without further notice.

(b) EPA shall furnish to States with approved programs the information in its files not submitted under a claim of confidentiality which the State needs in order to implement its approved program. EPA shall furnish to States with approved programs information submitted to EPA under a claim of confidentiality, which the State needs in order to implement its approved program, subject to the conditions in 40 CFR Part 2.

§ 123.133 Coordination with other programs.

(a) Issuance of State permits under this Part may be coordinated, as provided in Part 124, with issuance of NPDES, 404, and UIC permits whether they are controlled by the State, EPA, or the Corps of Engineers.

(b) The State Director of any approved program which may affect the planning for and development of hazardous waste management facilities and practices shall consult and coordinate with agencies designated under section 4006(b) of RCRA (40 CFR Part 255) as responsible for the development and implementation of State solid waste management plans under section 4002(b) of RCRA (40 CFR Part 256).

§ 123.134 EPA review of State permits.

(a) The Regional Administrator may comment on permit applications and draft permits as provided in the Memorandum of Agreement under § 123.126.

(b) Where EPA indicates, in a comment, that issuance of the permit would be inconsistent with the approved State program, EPA shall include in the comment:

(1) A statement of the reasons for the comment (including the section of RCRA or regulations promulgated thereunder that support the comment); and

(2) The actions that should be taken by the State Director in order to address the comments (including the conditions which the permit would include if it were issued by the Regional Administrator).

(c) A copy of any comment shall be sent to the permit applicant by the Regional Administrator.

(d) The Regional Administrator shall withdraw such comment when satisfied that the State has met or refuted his or her concerns.

(e) Under section 3008(a)(3) of RCRA, EPA may terminate a State-issued permit in accordance with the procedures of Part 124, Subpart E or bring an enforcement action in accordance with the procedures of 40 CFR Part 22 in the case of a violation of a State program requirement. In exercising these authorities, EPA will observe the following conditions:

(1) The Regional Administrator may take action under section 3008(a)(3) of RCRA against a holder of a State-issued permit at any time on the ground that the permittee is not complying with a condition of that permit.

(2) The Regional Administrator may take action under section 3008(a)(3) of RCRA against a holder of a State-issued permit at any time on the ground that the permittee is not complying with a condition that the Regional Administrator in commenting on the permit application or draft permit stated was necessary to implement approved State program requirements, whether or not that condition was included in the final permit.

(3) The Regional Administrator may not take action under section 3008(a)(3) of RCRA against a holder of a State-issued permit on the ground that the permittee is not complying with a condition necessary to implement approved State program requirements unless the Regional Administrator stated in commenting on the permit application or draft permit that that condition was necessary.

(4) The Regional Administrator may take action under section 7003 of RCRA against a permit holder at any time whether or not the permit holder is complying with the permit conditions.

§ 123.135 Approval process.

(a) Within 30 days of receipt of a complete program submission for interim authorization, the Regional Administrator shall:

(1) Issue notice in the **Federal Register** and in accordance with § 123.39(a)(1) of a public hearing on the State's application for interim authorization. Such public hearing will be held by EPA no earlier than 30 days after notice of the hearing, provided that if significant public interest in a hearing is not expressed, the hearing may be cancelled if a statement to this effect is included in the public notice. The State shall participate in any public hearing held by EPA.

(2) Afford the public 30 days after the notice to comment on the State's submission; and

(3) Note the availability of the State's submission for inspection and copying by the public. The State submission shall, at a minimum, be available in the main office of the lead State agency and in the EPA Regional Office.

(b) Within 90 days of the notice in the **Federal Register** required by paragraph (a)(1) of this section, the Administrator shall make a final determination whether or not to approve the State's program taking into account any comments submitted. The Administrator will give notice of this final determination in the **Federal Register** and in accordance with § 123.39(a)(1). The notification shall include a concise statement of the reasons for this determination, and a response to significant comments received.

(c) Where a State has received interim authorization for Phase I the same procedures required in paragraphs (a) and (b) of this section shall be used in determining whether this amended program submission meets the requirements of the Federal program.

§ 123.136 Withdrawal of State programs.

(a) The criteria and procedures for withdrawal set forth in §§ 123.14 and 15 apply to this section.

(b) In addition to the criteria in § 123.14, a State program may be withdrawn if a State which has obtained interim authorization fails to meet the schedule for or accomplish the additions or revisions of its program set forth in its authorization plan.

§ 123.137 Reversion of State programs.

(a) A State program approved for interim authorization for Phase I shall terminate on the last day of the 6th month after the effective date of Phase II and EPA shall administer and enforce the Federal program in the State commencing on that date if the State has failed to submit by that date an amended submission pursuant to § 123.122(c)(4).

(b) A State program approved for interim authorization for Phase I shall terminate and EPA shall administer and enforce the Federal program in the State if the Regional Administrator determines pursuant to § 123.135(c) that a program submission amended pursuant to § 123.122(c)(4) does not meet the requirements of the Federal program.

PART 124—PROCEDURES FOR DECISIONMAKING

Subpart A—General Program Requirements

Authority: Resource Conservation and Recovery Act, 42 U.S.C. § 6901 *et seq;* Safe Drinking Water Act, 42 U.S.C. § 300(f) *et seq;* Clean Water Act, 33 U.S.C. § 1251 *et seq;* and Clean Air Act, 42 U.S.C. § 1857 *et seq.*

Hearings Available Under This Part

Programs	Subpart (A) Public hearing	Subpart (E) Evidentiary hearing	Subpart (F) Panel hearing
RCRA	On draft permit, at Director's discretion or on request (§ 124.12).	(1) Permit termination (RCRA section 3008). (2) With NPDES evidentiary hearing (§ 124.74(b)(2))	(1) At RA's discretion in lieu of public hearing (§§ 124.12 and 124.111(a)(3)). (2) When consolidated with NPDES draft permit processed under Subpart F (§ 124.111(a)(1)i).
UIC	On draft permit, at Director's discretion or on request (§ 124.12).	With NPDES evidentiary hearing (§ 124.74(b)(2))	(1) At RA's discretion in lieu of public hearing (§§ 124.12 and 124.111(a)(3)). (2) When consolidated with NPDES draft permit processed under Subpart F (§ 124.111(a)(1)(i).
PSD	On draft permit, at Director's discretion or on request (§ 124.12).	Not available (§ 124.71(c))	When consolidated with NPDES draft permit processed under Subpart F if RA determines that CAA one year deadline will not be violated.
NPDES (other than general permit).	On draft permit, at Director's discretion or on request (§ 124.12).	(1) On request to challenge any permit condition or variance (§ 124.74). (2) At RA's discretion for any 301(h) request. (§ 124.64(b)).	(1) At RA's discretion when first decision on permit or variance request (§ 124.111). (2) At RA's discretion when request for evidentiary hearing is granted under § 124.75(a)(2). (§§ 124.74(c)(8) and 124.111(a)(2)). (3) At RA's discretion for any 301(h) request (§ 124.64(b)).
NPDES (general permit)	On draft permit, at Director's discretion or on request (§ 124.12).	Not available (§ 124.71)(a)	At RA's discretion in lieu of public hearing (§ 124.111(a)(3)).
404	On draft permit or on application when no draft permit, at Director's discretion or on request (§ 124.12).	Not available (§ 124.71)	Not available (§ 124.111).

Subpart A—General Program Requirements

§ 124.1   Purpose and scope.

(a) This Part contains EPA procedures for issuing, modifying, revoking and reissuing, or terminating all RCRA, UIC, PSD and NPDES "permits" other than RCRA and UIC "emergency permits" (see §§ 122.27 and 122.40) and RCRA "permits by rule" (§ 122.26). The latter kinds of permits are governed by Subpart A of Part 122. RCRA interim status and UIC authorization by rule are not "permits" and are covered by specific provisions in Subpart A of Part 122. This Part also does not apply to permits issued, modified, revoked and reissued or terminated by the Corps of Engineers. Those procedures are specified in 33 CFR Parts 320–327.

(b) Part 124 is organized into six subparts. Subpart A contains general procedural requirements applicable to all permit programs covered by these regulations. Subparts B through F supplement these general provisions with requirements that apply to only one or more of the programs. Subpart A describes the steps EPA will follow in receiving permit applications, preparing draft permits, issuing public notice, inviting public comment and holding public hearings on draft permits. Subpart A also covers assembling an administrative record, responding to comments, issuing a final permit decision, and allowing for administrative appeal of the final permit decision. Subpart B is reserved for specific procedural requirements for RCRA permits. There are none of these at present but they may be added in the future. Subpart C contains definitions and specific procedural requirements for PSD permits. Subpart D applies to NPDES permits until an evidentiary hearing begins, when Subpart E procedures take over for EPA-issued NPDES permits and EPA-terminated RCRA permits. Subpart F, which is based on the "initial licensing" provisions of the Administrative Procedure Act (APA), can be used instead of Subparts A through E in appropriate cases.

(c) Part 124 offers an opportunity for three kinds of hearings: a public hearing under Subpart A, an evidentiary hearing under Subpart E, and a panel hearing under Subpart F. This chart describes when these hearings are available for each of the five permit programs.

(d) This Part is designed to allow permits for a given facility under two or more of the listed programs to be processed separately or together at the choice of the Regional Administrator. This allows EPA to combine the processing of permits only when appropriate, and not necessarily in all cases. The Regional Administrator may consolidate permit processing when the permit applications are submitted, when draft permits are prepared, or when final permit decisions are issued. This Part also allows consolidated permits to be subject to a single public hearing under § 124.12, a single evidentiary hearing under § 124.75, or a single non-adversary panel hearing under § 124.120. Permit applicants may recommend whether or not their applications should be consolidated in any given case.

(e) Certain procedural requirements set forth in Part 124 must be adopted by States in order to gain EPA approval to operate RCRA, UIC, NPDES, and 404 permit programs. These requirements are listed in § 123.7 and signaled by the following words at the end of the appropriate Part 124 section or paragraph heading: (applicable to State programs, see § 123.7). Part 124 does not apply to PSD permits issued by an approved State.

(f) To coordinate decisionmaking when different permits will be issued by EPA and approved State programs, this Part allows applications to be jointly processed, joint comment periods and hearings to be held, and final permits to be issued on a cooperative basis whenever EPA and a State agree to take such steps in general or in individual cases. These joint processing agreements may be provided in the Memorandum of Agreement developed under § 123.6.

§ 124.2   Definitions.

(a) The definitions in Part 122 apply to this Part except for PSD permits which are governed by the definitions in § 124.41.

(b) For the purposes of Part 124, the term "Director" means the State Director or Regional Administrator and is used when the accompanying provision is required of EPA administered programs and of State programs under § 123.7. The term "Regional Administrator" is used when the accompanying provision applies exclusively to EPA-issued permits and is not applicable to State programs under § 123.7. While States are not required to implement these latter provisions, they are not precluded from doing so, notwithstanding use of the term "Regional Administrator."

(c) The term "formal hearing" means any evidentiary hearing under Subpart E or any panel hearing under Subpart F but does not mean a public hearing conducted under § 124.12.

§ 124.3   Application for a permit.

(a) (Applicable to State programs, see § 123.7). (1) Any person who requires a permit under the RCRA, UIC, NPDES, or PSD programs shall complete, sign, and submit to the Director an application for each permit required under §§ 122.21 (RCRA), 122.31 (UIC), 40 CFR 52.21 (PSD), and 122.51 (NPDES). Applications are not required for RCRA permits by rule (§ 122.26), underground injections authorized by rule (§ 122.37), NPDES general permits (§ 122.59) and 404 general permits (§ 123.95).

(2) The Director shall not begin the processing of a permit until the applicant has fully complied with the application requirements for that permit. See §§ 122.4, 122.22 (RCRA), 122.38 (UIC), 40 CFR 52.21 (PSD), and 122.53 (NPDES).

(3) Permit applications (except for PSD permits) must comply with the signature and certification requirements of § 122.6.

(b) In the case of a PSD permit issued to a facility or activity which 40 CFR § 52.21(k) exempts from the requirements of § 52.21 (l), (n), and (p), no proceedings under this Part shall be held to the extent that the Regional Administrator determines that proceedings providing the public with at least as much participation as this Part in the material determinations involved have already been held in the process of granting construction approval under the applicable State implementation plan. The Regional Administrator shall briefly document that finding and make it available to any member of the public upon request. The Regional Administrator shall prepare a draft permit under § 124.6 and follow the applicable procedures under this Part to the extent he or she is unable to make a finding under this subparagraph.

(c) The Regional Administrator shall review for completeness every application for an EPA-issued permit. Each application for an EPA-issued permit submitted by a new HWM facility, a new UIC injection well, a major PSD stationary source or major PSD modification, or an NPDES new source or NPDES new discharger should be reviewed for completeness by the Regional Administrator within 30 days of its receipt. Each application for an EPA-issued permit submitted by an existing HWM facility (both Parts A and B of the application), existing injection well or existing NPDES source should be reviewed for completeness within 60 days of receipt. Upon completing the review, the Regional Administrator shall notify the applicant in writing whether the application is complete. If the application is incomplete, the Regional Administrator shall list the information necessary to make the application complete. When the application is for an existing HWM facility, an existing UIC injection well or an existing NPDES source, the Regional Administrator shall specify in the notice of deficiency a date for submitting the necessary information. The Regional Administrator shall notify the applicant that the application is complete upon receiving this information. After the application is completed, the Regional Administrator may request additional information from an applicant but only when necessary to clarify, modify, or supplement previously submitted material. Requests for such additional information will not render an application incomplete.

(d) If an applicant fails or refuses to correct deficiencies in the application, the permit may be denied and appropriate enforcement actions may be taken under the applicable statutory provision including RCRA section 3008, SDWA sections 1423 and 1424, CAA section 167, and CWA sections 308, 309, 402(h), and 402(k).

(e) If the Regional Administrator decides that a site visit is necessary for any reason in conjunction with the processing of an application, he or she shall notify the applicant and a date shall be scheduled.

(f) The effective date of an application is the date on which the Regional Administrator notifies the applicant that the application is complete as provided in paragraph (c) of this section.

(g) For each application from a major new HWM facility, major new UIC injection well, major NPDES new source, or major NPDES new discharger, the Regional Administrator shall, no later than the effective date of the application, prepare and mail to the applicant a project decision schedule. (This paragraph does not apply to PSD permits.) The schedule shall specify target dates by which the Regional Administrator intends to:

(1) Prepare a draft permit;

(2) Give public notice;

(3) Complete the public comment period, including any public hearing;

(4) Issue a final permit; and

(5) In the case of an NPDES permit, complete any formal proceedings under Subparts E or F.

§ 124.4 Consolidation of permit processing.

(a)(1) Whenever a facility or activity requires a permit under more than one statute covered by these regulations, processing of two or more applications for those permits may be consolidated. The first step in consolidation is to prepare each draft permit at the same time.

(2) Whenever draft permits are prepared at the same time, the statements of basis (required under

§ 124.7 for EPA-issued permits only) or fact sheets (§ 124.8), administrative records (required under § 124.9 for EPA-issued permits only), public comment periods (§ 124.10), and any public hearings (§ 124.12) on those permits should also be consolidated. The final permits may be issued together. They need not be issued together if in the judgment of the Regional Administrator or State Director(s), joint processing would result in unreasonable delay in the issuance of one or more permits.

(b) Whenever an exisiting facility or activity requires additional permits under one or more of the statutes covered by these regulations, the permitting authority may coordinate the expiration date(s) of the new permit(s) with the expiration date(s) of the existing permit(s) so that all permits expire simultaneously. Processing of the subsequent applications for renewal permits may then be consolidated.

(c) Processing of permit applications under paragraphs (a) or (b) of this section may be consolidated as follows:

(1) The Director may consolidate permit processing at his or her discretion whenever a facility or activity requires all permits either from EPA or from an approved State.

(2) The Regional Administrator and the State Director(s) may agree to consolidate draft permits whenever a facility or activity requires permits from both EPA and an approved State.

(3) Permit applicants may recommend whether or not the processing of their applications should be consolidated.

(d) Whenever permit processing is consolidated and the Regional Administrator invokes the "initial licensing" provisions of Subpart F for an NPDES, RCRA, or UIC permit, any permit(s) with which that NPDES, RCRA or UIC permit was consolidated shall likewise be processed under Subpart F.

(e) Except with the written consent of the permit applicant, the Regional Administrator shall not consolidate processing a PSD permit with any other permit under paragraphs (a) or (b) of this section or process a PSD permit under Subpart F as provided in paragraph (d) of this section when to do so would delay issuance of the PSD permit more than one year from the effective date of the application under § 124.3(f).

§ 124.5 Modification, revocation and reissuance, or termination of permits.

(a) (*Applicable to State programs, see § 123.7*). Permits (other than PSD permits) may be modified, revoked and reissued, or terminated either at the request of any interested person (including the permittee) or upon the Director's initiative. However, permits

may only be modified, revoked and reissued, or terminated for the reasons specified in §§ 122.15 or 122.16. All requests shall be in writing and shall contain facts or reasons supporting the request.

(b) If the Director decides the request is not justified, he or she shall send the requester a brief written response giving a reason for the decision. Denials of requests for modification, revocation and reissuance, or termination are not subject to public notice, comment, or hearings. Denials by the Regional Administrator may be informally appealed to the Administrator by a letter briefly setting forth the relevant facts. The Administrator may direct the Regional Administrator to begin modification, revocation and reissuance, or termination proceedings under paragraph (c) of this section. The appeal shall be considered denied if the Administrator takes no action on the letter within 60 days after receiving it. This informal appeal is, under 5 U.S.C. § 704, a prerequisite to seeking judicial review of EPA action in denying a request for modification, revocation and reissuance, or termination.

(c) (*Applicable to State programs, see § 123.7*). (1) If the Director tentatively decides to modify or revoke and reissue a permit under § 122.15, he or she shall prepare a draft permit under § 124.6 incorporating the proposed changes. The Director may request additional information and, in the case of a modified permit, may require the submission of an updated permit application. In the case of revoked and reissued permits, the Director shall require the submission of a new application.

(2) In a permit modification under this section, only those conditions to be modified shall be reopened when a new draft permit is prepared. All other aspects of the existing permit shall remain in effect for the duration of the unmodified permit. When a permit is revoked and reissued under this section, the entire permit is reopened just as if the permit had expired and was being reissued. During any revocation and reissuance proceeding the permittee shall comply with all conditions of the existing permit until a new final permit is reissued.

(3) "Minor modifications" as defined in § 122.17 are not subject to the requirements of this section.

(d) (*Applicable to State programs, see § 123.7*). If the Director tentatively decides to terminate a permit under § 122.16, he or she shall issue a notice of intent to terminate. A notice of intent to terminate is a type of draft permit which follows the same procedures as any draft permit prepared under § 124.6. In the case of EPA-issued permits, a notice of intent to terminate shall not be issued if the Regional Administrator and the permittee agree to termination in the course of transferring permit responsibility to an approved State under § 123.6(b)(1).

(e) When EPA is the permitting authority, all draft permits (including notices of intent to terminate) prepared under this section shall be based on the administrative record as defined in § 124.9.

(f) (*Applicable to State programs, see § 123.7*). Any request by the permittee for modification to an existing 404 permit (other than a request for a minor modification as defined in § 122.17) shall be treated as a permit application and shall be processed in accordance with all requirements of § 124.3.

(g)(1) [Reserved for PSD Modification Provisions]

(2) PSD permits may be terminated only by rescission under § 52.21(w) or by automatic expiration under § 52.21(s). Applications for rescission shall be processed under § 52.21(w) and are not subject to this Part.

§ 124.6 Draft permits.

(a) (*Applicable to State programs, see § 123.7*). Once an application is complete, the Director shall tentatively decide whether to prepare a draft permit (except in the case of State section 404 permits for which no draft permit is required under § 123.100) or to deny the application.

(b) If the Director tentatively decides to deny the permit application, he or she shall issue a notice of intent to deny. A notice of intent to deny the permit application is a type of draft permit which follows the same procedures as any draft permit prepared under this section. See § 124.6(e). If the Director's final decision (§ 124.15) is that the tentative decision to deny the permit application was incorrect, he or she shall withdraw the notice of intent to deny and proceed to prepare a draft permit under paragraph (d) of this section.

(c) (*Applicable to State programs, see § 123.7*). If the Director tentatively decides to issue an NPDES or 404 general permit, he or she shall prepare a draft general permit under paragraph (d) of this section.

(d) (*Applicable to State programs, see § 123.7*). If the Director decides to prepare a draft permit, he or she shall prepare a draft permit that contains the following information:

(1) All conditions under §§ 122.7 and 122.8 (except for PSD permits);

(2) All compliance schedules under § 122.10 (except for PSD permits);

(3) All monitoring requirements under § 122.11 (except for PSD permits); and

(4) For:

(i) RCRA permits, standards for treatment, storage, and/or disposal and other permit conditions under § 122.28;

(ii) UIC permits, permit conditions under § 122.42;

(iii) PSD permits, permit conditions under 40 CFR § 52.21;

(iv) 404 permits, permit conditions under §§ 123.97 and 123.98;

(v) NPDES permits, effluent limitations, standards, prohibitions and conditions under §§ 122.60 and 122.61, including when applicable any conditions certified by a State agency under § 124.55, and all variances that are to be included under § 124.63.

(e) (*Applicable to State programs, see § 123.7*). All draft permits prepared by EPA under this section shall be accompanied by a statement of basis (§ 124.7) or fact sheet (§ 124.8), and shall be based on the administrative record (§ 124.9), publicly noticed (§ 124.10) and made available for public comment (§ 124.11). The Regional Administrator shall give notice of opportunity for a public hearing (§ 124.12), issue a final decision (§ 124.15) and respond to comments (§ 124.17). For RCRA, UIC or PSD permits, an appeal may be taken under § 124.19 and, for NPDES permits, an appeal may be taken under § 124.74. Draft permits prepared by a State shall be accompanied by a fact sheet if required under § 124.8.

§ 124.7 Statement of basis.

EPA shall prepare a statement of basis for every draft permit for which a fact sheet under § 124.8 is not prepared. The statement of basis shall briefly describe the derivation of the conditions of the draft permit and the reasons for them or, in the case of notices of intent to deny or terminate, reasons supporting the tentative decision. The statement of basis shall be sent to the applicant and, on request, to any other person.

§ 124.8 Fact sheet.

(*Applicable to State programs, see § 123.7.*)

(a) A fact sheet shall be prepared for every draft permit for a major HWM, UIC, 404, or NPDES facility or activity, for every 404 and NPDES general permit (§§ 123.95 and 122.59), for every NPDES draft permit that incorporates a variance or requires an explanation under § 124.56(b), and for every draft permit which the Director finds is the subject of widespread public interest or raises major issues. The fact sheet shall briefly set forth the principal facts and the significant factual, legal, methodological

and policy questions considered in preparing the draft permit. The Director shall send this fact sheet to the applicant and, on request, to any other person.

(b) The fact sheet shall include, when applicable:

(1) A brief description of the type of facility or activity which is the subject of the draft permit;

(2) The type and quantity of wastes, fluids, or pollutants which are proposed to be or are being treated, stored, disposed of, injected, emitted, or discharged.

(3) For a PSD permit, the degree of increment consumption expected to result from operation of the facility or activity.

(4) A brief summary of the basis for the draft permit conditions including references to applicable statutory or regulatory provisions and appropriate supporting references to the administrative record required by § 124.9 (for EPA-issued permits);

(5) Reasons why any requested variances or alternatives to required standards do or do not appear justified;

(6) A description of the procedures for reaching a final decision on the draft permit including:

(i) The beginning and ending dates of the comment period under § 124.10 and the address where comments will be received;

(ii) Procedures for requesting a hearing and the nature of that hearing; and

(iii) Any other procedures by which the public may participate in the final decision.

(7) Name and telephone number of a person to contact for additional information.

(8) For NPDES permits, provisions satisfying the requirements of § 124.56.

§ 124.9 Administrative record for draft permits when EPA is the permitting authority.

(a) The provisions of a draft permit prepared by EPA under § 124.6 shall be based on the administrative record defined in this section.

(b) For preparing a draft permit under § 124.6, the record shall consist of:

(1) The application, if required, and any supporting data furnished by the applicant;

(2) The draft permit or notice of intent to deny the application or to terminate the permit;

(3) The statement of basis (§ 124.7) or fact sheet (§ 124.8);

(4) All documents cited in the statement of basis or fact sheet; and

(5) Other documents contained in the supporting file for the draft permit.

(6) For NPDES new source draft permits only, any environmental assessment, environmental impact statement (EIS), finding of no significant impact, or environmental information document and any supplement to an EIS that may have been prepared. NPDES permits other than permits to new sources as well as all RCRA, UIC and PSD permits are not subject to the environmental impact statement provisions of section 102(2)(C) of the National Environmental Policy Act, 42 U.S.C. 4321.

(c) Material readily available at the issuing Regional Office or published material that is generally available, and that is included in the administrative record under paragraphs (b) and (c) of this section, need not be physically included with the rest of the record as long as it is specifically referred to in the statement of basis or the fact sheet.

(d) This section applies to all draft permits when public notice was given after the effective date of these regulations.

§ 124.10 Public notice of permit actions and public comment period.

(a) *Scope.*

(1) The Director shall give public notice that the following actions have occurred:

(i) A permit application has been tentatively denied under § 124.6(b);

(ii) (*Applicable to State programs, see § 123.7*). A draft permit has been prepared under § 124.6(d);

(iii) (*Applicable to State programs, see § 123.7*). A hearing has been scheduled under § 124.12, Subpart E, or Subpart F;

(iv) An appeal has been granted under § 124.19(c);

(v) (*Applicable to State programs, see § 123.7*). A State section 404 application has been received in cases when no draft permit will be prepared (see § 123.100); or

(vi) An NPDES new source determination has been made under § 122.66.

(2) No public notice is required when a request for permit modification, revocation and reissuance, or termination is denied under § 124.5(b). Written notice of that denial shall be given to the requester and to the permittee.

(3) Public notices may describe more than one permit or permit action.

(b) *Timing (applicable to State programs, see § 123.7).* (1) Public notice of the preparation of a draft permit (including a notice of intent to deny a permit application) required under paragraph (a) of this section shall allow at least 30 days for public comment. For EPA-issued permits, if the Regional Administrator determines under 40 CFR Part 6, Subpart F that an Environmental Impact Statement (EIS) shall be prepared for an NPDES new source, public notice of the draft permit shall not be given until after a draft EIS is issued.

(2) Public notice of a public hearing shall be given at least 30 days before the hearing. (Public notice of the hearing may be given at the same time as public notice of the draft permit and the two notices may be combined.)

(c) *Methods (applicable to State programs, see § 123.7).* Public notice of activities described in paragraph (a)(1) of this section shall be given by the following methods:

(1) By mailing a copy of a notice to the following persons (any person otherwise entitled to receive notice under this paragraph may waive his or her rights to receive notice for any classes and categories of permits):

(i) The applicant (except for NPDES and 404 general permits when there is no applicant);

(ii) Any other agency which the Director knows has issued or is required to issue a RCRA, UIC, PSD, NPDES or 404 permit for the same facility or activity (including EPA when the draft permit is prepared by the State);

(iii) Federal and State agencies with jurisdiction over fish, shellfish, and wildlife resources and over coastal zone management plans, the Advisory Council on Historic Preservation, State Historic Preservation Officers, and other appropriate government authorities, including any affected States;

(iv) For NPDES and 404 permits only, any State agency responsible for plan development under CWA section 208(b)(2), 208(b)(4) or 303(e) and the U.S. Army Corps of Engineers, the U.S. Fish and Wildlife Service and the National Marine Fisheries Service;

(v) For NPDES permits only, any user identified in the permit application of a privately owned treatment works;

(vi) For 404 permits only, any reasonably ascertainable owner of property adjacent to the regulated facility or activity and the Regional Director of the Federal Aviation Administration if the discharge involves the construction of structures which may affect aircraft operations or for purposes associated with seaplane operations;

(vii) For PSD permits only, affected State and local air pollution control agencies, the chief executives of the city and county where the major stationary source or major modification would be located, any comprehensive regional land use planning agency and any State, Federal Land Manager, or Indian Governing Body whose lands may be affected by emissions from the regulated activity;

(viii) Persons on a mailing list developed by:

(A) Including those who request in writing to be on the list;

(B) Soliciting persons for "area lists" from participants in past permit proceedings in that area; and

(C) Notifying the public of the opportunity to be put on the mailing list through periodic publication in the public press and in such publications as Regional and State funded newsletters, environmental bulletins, or State law journals. (The Director may update the mailing list from time to time by requesting written indication of continued interest from those listed. The Director may delete from the list the name of any person who fails to respond to such a request.)

(2) For major permits and NPDES and 404 general permits, publication of a notice in a daily or weekly newspaper within the area affected by the facility or activity; and for EPA-issued NPDES general permits, in the Federal Register;

[Note.—The Director is encouraged to provide as much notice as possible of the NPDES or 404 draft general permit to the facilities or activities to be covered by the general permit.]

(3) When the program is being administered by an approved State, in a manner constituting legal notice to the public under State law; and

(4) Any other method reasonably calculated to give actual notice of the action in question to the persons potentially affected by it, including press releases or any other forum or medium to elicit public participation.

(d) *Contents (applicable to State programs, see § 123.7).* (1) *All public notices.* All public notices issued under this Part shall contain the following minimum information:

(i) Name and address of the office processing the permit action for which notice is being given;

(ii) Name and address of the permittee or permit applicant and, if different, of the facility or activity regulated by the permit, except in the case of NPDES and 404 draft general permits under §§ 122.59 and 123.95;

(iii) A brief description of the business conducted at the facility or activity described in the permit application or the draft permit, for NPDES or 404 general permits when there is no application.

(iv) Name, address and telephone number of a person from whom interested persons may obtain further information, including copies of the draft permit or draft general permit, as the case may be, statement of basis or fact sheet, and the application; and

(v) A brief description of the comment procedures required by §§ 124.11 and 124.12 and the time and place of any hearing that will be held, including a statement of procedures to request a hearing (unless a hearing has already been scheduled) and other procedures by which the public may participate in the final permit decision.

(vi) For EPA-issued permits, the location of the administrative record required by § 124.9, the times at which the record will be open for public inspection, and a statement that all data submitted by the applicant is available as part of the administrative record.

(vii) For NPDES permits only, a general description of the location of each existing or proposed discharge point and the name of the receiving water. For draft general permits, this requirement will be satisfied by a map or description of the permit area. For EPA-issued NPDES permits only, if the discharge is from a new source, a statement as to whether an environmental impact statement will be or has been prepared.

(viii) For 404 permits only,

(A) The purpose of the proposed activity (including, in the case of fill material, activities intended to be conducted on the fill), a description of the type, composition, and quantity of materials to be discharged and means of conveyance; and any proposed conditions and limitations on the discharge;

(B) The name and water quality standards classification, if applicable, of the receiving waters into which the discharge is proposed, and a general description of the site of each proposed discharge and the portions of the site and the discharges which are within State regulated waters;

(C) A description of the anticipated environmental effects of activities conducted under the permit;

(D) References to applicable statutory or regulatory authority; and

(E) Any other available information which may assist the public in evaluating the likely impact of the proposed activity upon the integrity of the receiving water.

(ix) Any additional information considered necessary or proper.

(2) *Public notices for hearings.* In addition to the general public notice described in paragraph (d)(1) of this section, the public notice of a hearing under § 124.12, Subpart E, or Subpart F shall contain the following information:

(i) Reference to the date of previous public notices relating to the permit;

(ii) Date, time, and place of the hearing;

(iii) A brief description of the nature and purpose of the hearing, including the applicable rules and procedures; and

(iv) For 404 permits only, a summary of major issues raised to date during the public comment period.

(e) *(Applicable to State programs, see § 123.7).* In addition to the general public notice described in paragraph (d)(1) of this section, all persons identified in paragraphs (c)(1) (i), (ii), (iii), and (iv) of this section shall be mailed a copy of the fact sheet or statement of basis (for EPA-issued permits), the permit application (if any) and the draft permit (if any).

§ 124.11   Public comments and requests for public hearings.

(Applicable to State programs, see § 123.7.)

During the public comment period provided under § 124.10, any interested person may submit written comments on the draft permit or the permit application for 404 permits when no draft permit is required (see § 123.100) and may request a public hearing, if no hearing has already been scheduled. A request for a public hearing shall be in writing and shall state the nature of the issues proposed to be raised in the hearing. All comments shall be considered in making the final decision and shall be answered as provided in § 124.17.

§ 124.12   Public hearings.

(a) *(Applicable to State programs, see § 123.7.)* The Director shall hold a public hearing whenever he or she finds, on the basis of requests, a significant degree of public interest in a draft permit(s). The Director also may hold a public hearing at his or her discretion, whenever, for instance, such a hearing might clarify one or more issues involved in the permit decision. Public notice of the hearing shall be given as specified in § 124.10.

(b) Whenever a public hearing will be held and EPA is the permitting authority, the Regional Administrator shall designate a Presiding Officer for the hearing who shall be responsible for its scheduling and orderly conduct.

(c) Any person may submit oral or written statements and data concerning the draft permit. Reasonable limits may be set upon the time allowed for oral statements, and the submission of statements in writing may be required. The public comment period under § 124.10 shall automatically be extended to the close of any public hearing under this section. The hearing officer may also extend the comment period by so stating at the hearing.

(d) A tape recording or written transcript of the hearing shall be made available to the public.

(e) At his or her discretion, the Regional Administrator may specify that RCRA and UIC permits be processed under the procedures in Subpart F.

§ 124.13   Obligation to raise issues and provide information during the public comment period.

All persons, including applicants, who believe any condition of a draft permit is inappropriate or that the Director's tentative decision to deny an application, terminate a permit, or prepare a draft permit is inappropriate, must raise all reasonably ascertainable issues and submit all reasonably available arguments and factual grounds supporting their position, including all supporting material, by the close of the public comment period (including any public hearing) under § 124.10. All supporting materials shall be included in full and may not be incorporated by reference, unless they are already part of the administrative record in the same proceeding, or consist of State or Federal statutes and regulations, EPA documents of general applicability, or other generally available reference materials. Commenters shall make supporting material not already included in the administrative record available to EPA as directed by the Regional Administrator. (A comment period longer than 30 days will often be necessary in complicated proceedings to give commenters a reasonble opportunity to comply with the requirements of this section. Commenters may request longer comment periods and they should be freely established under § 124.10 to the extent they appear necessary.)

§ 124.14   Reopening of the public comment period.

(a) If any data information or arguments submitted during the public comment period, including information or arguments required under § 124.13, appear to raise substantial new questions concerning a permit, the Regional Administrator may take one or more of the following actions:

(1) Prepare a new draft permit, appropriately modified, under § 124.6;

(2) Prepare a revised statement of basis under § 124.7, a fact sheet or revised fact sheet under § 124.8 and reopen the comment period under § 124.14; or

(3) Reopen or extend the comment period under § 124.10 to give interested persons an opportunity to comment on the information or arguments submitted.

(b) Comments filed during the reopened comment period shall be limited to the substantial new questions that caused its reopening. The public notice under § 124.10 shall define the scope of the reopening.

(c) For RCRA, UIC, or NPDES permits, the Regional Administrator may also, in the circumstances described above, elect to hold further proceedings under Subpart F. This decision may be combined with any of the actions enumerated in paragraph (a) of this section.

(d) Public notice of any of the above actions shall be issued under § 124.10.

§ 124.15 Issuance and effective date of permit.

(a) After the close of the public comment period under § 124.10 on a draft permit, the Regional Administrator shall issue a final permit decision. The Regional Administrator shall notify the applicant and each person who has submitted written comments or requested notice of the final permit decision. This notice shall include reference to the procedures for appealing a decision on a RCRA, UIC, or PSD permit or for contesting a decision on an NPDES permit or a decision to terminate a RCRA permit. For the purposes of this section, a final permit decision means a final decision to issue, deny, modify, revoke and reissue, or terminate a permit.

(b) A final permit decision shall become effective 30 days after the service of notice of the decision under paragraph (a) of this section, unless:

(1) A later effective date is specified in the decision; or

(2) Review is requested under § 124.19 (RCRA, UIC, and PSD permits) or an evidentiary hearing is requested under § 124.74 (NPDES permit and RCRA permit terminations); or

(3) No comments requested a change in the draft permit, in which case the permit shall become effective immediately upon issuance.

§ 124.16 Stays of contested permits conditions.

(a) *Stays.* (1) If a request for review of a RCRA or UIC permit under § 124.19 or an NPDES permit under § 124.74 or § 124.114 is granted or if conditions of a RCRA or UIC permit are consolidated for reconsideration in an evidentiary hearing on an NPDES permit under §§ 124.74, 124.82 or 124.114, the effect of the contested permit conditions shall be stayed and shall not be subject to judicial review pending final agency action. (No stay of a PSD permit is available under this section.) If the permit involves a new facility or new injection well, new source, new discharger or a recommencing discharger, the applicant shall be without a permit for the proposed new facility, injection well, source or discharger pending final agency action. See also § 124.60.

(2) Uncontested conditions which are not severable from those contested shall be stayed together with the contested conditions. Stayed provisions of permits for existing facilities, injection wells, and sources shall be identified by the Regional Administrator. All other provisions of the permit for the existing facility, injection well, or source shall remain fully effective and enforceable.

(b) *Stays based on cross effects.* (1) A stay may be granted based on the grounds that an appeal to the Administrator under § 124.19 of one permit may result in changes to another EPA-issued permit only when each of the permits involved has been appealed to the Administrator and he or she has accepted each appeal.

(2) No stay of an EPA-issued RCRA, UIC, or NPDES permit shall be granted based on the staying of any State-issued permit except at the discretion of the Regional Administrator and only upon written request from the State Director.

(c) Any facility or activity holding an existing permit must:

(1) Comply with the conditions of that permit during any modification or revocation and reissuance proceeding under § 124.5; and

(2) To the extent conditions of any new permit are stayed under this section, comply with the conditions of the existing permit which correspond to the stayed conditions, unless compliance with the existing conditions would be technologically incompatible with compliance with other conditions of the new permit which have not been stayed.

§ 124.17 Response to comments.

(a) *(Applicable to State programs, see § 123.7).* At the time that any final permit decision is issued under § 124.15, the Director shall issue a response to comments. States are only required to issue a response to comments when a final permit is issued. This response shall:

(1) Specify which provisions, if any, of the draft permit have been changed in the final permit decision, and the reasons for the change; and

(2) Briefly describe and respond to all significant comments on the draft permit or the permit application (for section 404 permits only) raised during the public comment period, or during any hearing.

(b) For EPA-issued permits, any documents cited in the response to comments shall be included in the administrative record for the final permit decision as defined in § 124.18. If new points are raised or new material supplied during the public comment period, EPA may document its response to those matters by adding new materials to the administrative record.

(c) *(Applicable to State programs, see § 123.7).* The response to comments shall be available to the public.

§ 124.18 Administrative record for final permit when EPA is the permitting authority.

(a) The Regional Administrator shall base final permit decisions under § 124.15 on the administrative record defined in this section.

(b) The administrative record for any final permit shall consist of the administrative record for the draft permit and:

(1) All comments received during the public comment period provided under § 124.10 (including any extension or reopening under § 124.14);

(2) The tape or transcript of any hearing(s) held under § 124.12;

(3) Any written materials submitted at such a hearing;

(4) The response to comments required by § 124.17 and any new material placed in the record under that section;

(5) For NPDES new source permits only, any final environmental impact statement and any supplement to the final EIS;

(6) Other documents contained in the supporting file for the permit; and

(7) The final permit.

(c) The additional documents required under paragraph (b) of this section should be added to the record as soon as possible after their receipt or publication by the Agency. The record shall be complete on the date the final permit is issued.

(d) This section applies to all final RCRA, UIC, PSD, and NPDES permits when the draft permit was subject to the administrative record requirements of § 124.9 and to all NPDES permits when the draft permit was included in a public notice after October 12, 1979.

(e) Material readily available at the issuing Regional Office, or published materials which are generally available and which are included in the administrative record under the standards of this section or of § 124.17 ("Response to comments"), need not be physically included in the same file as the rest of the record as long as it is specifically referred to in the statement of basis or fact sheet or in the response to comments.

§ 124.19 Appeal of RCRA, UIC, and PSD permits.

(a) Within 30 days after a RCRA, UIC, or PSD final permit decision has been issued under § 124.15, any person who filed comments on that draft permit or participated in the public hearing may petition the Administrator to review any condition of the permit decision. Any person who failed to file comments or failed to participate in the public hearing on the draft permit may petition for administrative review only to the extent of the changes from the draft to the final permit decision. The 30-day period within which a person may request review under this section begins with the service of notice of the Regional Administrator's action unless a later date is specified in that notice. The petition shall include a statement of the reasons supporting that review, including a demonstration that any issues being raised were raised during the public comment period (including any public hearing) to the extent required by these regulations and when appropriate, a showing that the condition in question is based on:

(1) A finding of fact or conclusion of law which is clearly erroneous, or

(2) An exercise of discretion or an important policy consideration which the Administrator should, in his or her discretion, review.

(b) The Administrator may also decide on his or her initiative to review any condition of any RCRA, UIC, or PSD permit issued under this Part. The Administrator must act under this paragraph within 30 days of the service date of notice of the Regional Administrator's action.

(c) Within a reasonable time following the filing of the petition for review, the Administrator shall issue an order either granting or denying the petition for review. To the extent review is denied, the conditions of the final permit decision become final agency action. Public notice of any grant of review by the Administrator under paragraph (a) or (b) of this section shall be given as provided in §124.10. Public notice shall set forth a briefing schedule for the appeal and shall state that any interested person may file an amicus brief. Notice of denial of review shall be sent only to the person(s) requesting review.

(d) The Administrator may defer consideration of an appeal of a RCRA or UIC permit under this section until the completion of formal proceedings under Subpart E or F relating to an NPDES permit issued to the same facility or activity upon concluding that:

(1) The NPDES permit is likely to raise issues relevant to a decision of the RCRA or UIC appeals;

(2) The NPDES permit is likely to be appealed; and

(3) Either: (i) The interests of both the facility or activity and the public are not likely to be materially adversely affected by the deferral; or

(ii) Any adverse effect is outweighed by the benefits likely to result from a consolidated decision on appeal.

(e) A petition to the Administrator under paragraph (a) of this section is, under 5 U.S.C. § 704, a prerequisite to the seeking of judicial review of the final agency action.

(f)(1) For purposes of judicial review under the appropriate Act, final agency action occurs when a final RCRA, UIC, or PSD permit is issued or denied by EPA and agency review procedures are exhausted. A final permit decision shall be issued by the Regional Administrator: (i) When the Administrator issues notice to the parties that review has been denied; (ii) when the Administrator issues a decision on the merits of the appeal and the decision does not include a remand of the proceedings; or (iii) upon the completion of remand proceedings if the proceedings are remanded, unless the Administrator's remand order specifically provides that appeal of the remand decision will be required to exhaust administrative remedies.

(2) Notice of any final agency action regarding a PSD permit shall promptly be published in the Federal Register.

§ 124.20 Computation of time.

(a) Any time period scheduled to begin on the occurrence of an act or event shall begin on the day after the act or event.

(b) Any time period scheduled to begin before the occurrence of an act or event shall be computed so that the period ends on the day before the act or event.

(c) If the final day of any time period falls on a weekend or legal holiday, the time period shall be extended to the next working day.

(d) Whenever a party or interested person has the right or is required to act within a prescribed period after the service of notice or other paper upon him or her by mail, 3 days shall be added to the prescribed time.

§ 124.21 Effective date of Part 124.

(a) Except for paragraphs (b) and (c) of this section, Part 124 will become effective July 18, 1980. Because this effective date will precede the processing of any RCRA or UIC permits, Part 124 will apply in its entirety to all RCRA and UIC permits.

(b) All provisions of Part 124 pertaining to the RCRA program will become effective on November 19, 1980.

(c) All provisions of Part 124 pertaining to the UIC program will become effective July 18, 1980, but shall not be implemented until the effective date of 40 CFR Part 146.

(d) This Part does not significantly change the way in which NPDES permits are processed. Since October 12, 1979,

NPDES permits have been the subject to almost identical requirements in the revised NPDES regulations which were promulgated on June 7, 1979. See 44 FR 32948. To the extent this Part changes the revised NPDES permit regulations, those changes will take effect as to all permit proceedings in progress on July 3, 1980.

(e) This Part also does not significantly change the way in which PSD permits are processed. For the most part, these regulations will also apply to PSD proceedings in progress on July 18, 1980. However, because it would be disruptive to require retroactively a formal administrative record for PSD permits issued without one, § § 124.9 and 124.18 will apply to PSD permits for which draft permits were prepared after the effective date of these regulations.

Subpart B—Specific Procedures Applicable to RCRA Permits [Reserved]

Subpart E—Evidentiary Hearings for EPA-Issued NPDES Permits and EPA-Terminated RCRA Permits

§ 124.71 Applicability.

(a) The regulations in this Subpart govern all formal hearings conducted by EPA under CWA section 402, except for those conducted under Subpart F. They also govern all evidentiary hearings conducted under RCRA section 3008 in connection with the termination of a RCRA permit. This includes termination of interim status for failure to furnish information needed to made a final decision. A formal hearing is available to challenge any NPDES permit issued under § 124.15 except for a general permit. Persons affected by a general permit may not challenge the conditions of a general permit as of right in further agency proceedings. They may instead either challenge the general permit in court, or apply for an individual NPDES permit under § 122.53 as authorized in § 122.59 and then request a formal hearing on the issuance or denial of an individual permit. (The Regional Administrator also has the discretion to use the procedures of Subpart F for general permits. See § 124.111.)

(b) In certain cases, evidentiary hearings under this Subpart may also be held on the conditions of UIC permits, or of RCRA permits which are being issued, modified, or revoked and reissued, rather than terminated or suspended. This will occur when the conditions of the UIC or RCRA permit in question are closely linked with the conditions of an NPDES permit as to which an evidentiary hearing has been granted. See § 124.74(b)(2). Any interested person may challenge the Regional Administrator's initial new source determination by requesting an evidentiary hearing under this Part. See § 122.66.

(c) PSD permits may never be subject to an evidentiary hearing under this Subpart. Section 124.74(b)(2)(iv) provides only for consolidation of PSD permits with other permits subject to a panel hearing under Subpart F.

§ 124.72 Definitions.

For the purpose of this Subpart, the following definitions are applicable:

"Hearing Clerk" means The Hearing Clerk, U.S. Environmental Protection Agency, 401 M Street, S.W., Washington, D.C. 20460.

"Judicial Officer" means a permanent or temporary employee of the Agency appointed as a Judicial Officer by the Administrator under these regulations and subject to the following conditions:

(a) A Judicial Officer shall be a licensed attorney. A Judicial Officer shall not be employed in the Office of Enforcement or the Office of Water and Waste Management, and shall not participate in the consideration or decision of any case in which he or she performed investigative or prosecutorial functions, or which is factually related to such a case.

(b) The Administrator may delegate any authority to act in an appeal of a given case under this Subpart to a Judicial Officer who, in addition, may perform other duties for EPA, provided that the delegation shall not preclude a Judicial Officer from referring any motion or case to the Administrator when the Judicial·Officer decides such action would be appropriate. The Administrator, in deciding a case, may consult with and assign the drafting of preliminary findings of fact and conclusions and/or a preliminary decision to any Judicial Officer.

"Party" means the EPA trial staff under § 124.78 and any person whose request for a hearing under § 124.74 or whose request to be admitted as a party or to intervene under § 124.79 or § 124.117 has been granted.

"Presiding Officer" for the purposes of this Subpart means an Administrative Law Judge appointed under 5 U.S.C. 3105 and designated to preside at the hearing. Under Subpart F other persons may also serve as hearing officers. See § 124.119.

"Regional Hearing Clerk" means an employee of the Agency designated by a Regional Administrator to establish a repository for all books, records, documents, and other materials relating to hearings under this Subpart.

§ 124.73 Filing and submission of documents.

(a) All submissions authorized or required to be filed with the Agency under this Subpart shall be filed with the Regional Hearing Clerk, unless otherwise provided by regulation. Submissions shall be considered filed on the date on which they are mailed or delivered in person to the Regional Hearing Clerk.

(b) All submissions shall be signed by the person making the submission, or by an attorney or other authorized agent or representative.

(c)(1) All data and information referred to or in any way relied upon in any submission shall be included in full and may not be incorporated by reference, unless previously submitted as part of the administrative record in the same proceeding. This requirement does not apply to State or Federal statutes and regulations, judicial decisions published in a national reporter system, officially issued EPA documents of general applicability, and any other generally available reference material which may be incorporated by reference. Any party incorporating materials by reference shall provide copies upon request by the Regional Administrator or the Presiding Officer.

(2) If any part of the material submitted is in a foreign language, it shall be accompanied by an English translation verified under oath to be complete and accurate, together with the name, address, and a brief statement of the qualifications of the person making the translation. Translations of literature or other material in a foreign language shall be accompanied by copies of the original publication.

(3) Where relevant data or information is contained in a document also containing irrelevant matter, either the irrelevant matter shall be deleted or the relevant portions shall be indicated.

(4) Failure to comply with the requirements of this section or any other requirement in this Subpart may result in the noncomplying portions of the submission being excluded from consideration. If the Regional Administrator or the Presiding Officer, on motion by any party or *sua sponte,* determines that a submission fails to meet any requirement of this Subpart, the Regional Administrator or Presiding Officer shall direct the Regional Hearing Clerk to return the submission, together with a reference to the applicable regulations. A party whose materials have been rejected has 14 days to correct the errors and resubmit, unless the Regional Administrator or the Presiding Officer finds good cause to allow a longer time.

(d) The filing of a submission shall not mean or imply that it in fact meets all applicable requirements or that it contains reasonable grounds for the action requested or that the action requested is in accordance with law.

(e) The original of all statements and documents containing factual material, data, or other information shall be signed in ink and shall state the name, address, and the representative capacity of the person making the submission.

§ 124.74 Requests for evidentiary hearing.

(a) Within 30 days following the service of notice of the Regional Administrator's final permit decision under § 124.15, any interested person may submit a request to the Regional Administrator under paragraph (b) of this section for an evidentiary hearing to reconsider or contest that decision. If such a request is submitted by a person other than the permittee, the person shall simultaneously serve a copy of the request on the permittee.

(b)(1) In accordance with § 124.76, such requests shall state each legal or factual question alleged to be at issue, and their relevance to the permit decision, together with a designation of the specific factual areas to be adjudicated and the hearing time estimated to be necessary for adjudication. Information supporting the request or other written documents relied upon to support the request shall be submitted as required by § 124.73 unless they are already part of the administrative record required by § 124.18.

[Note.—This paragraph allows the submission of requests for evidentiary hearings even though both legal and factual issues may be raised, or only legal issues may be raised. In the latter case, because no factual issues were raised, the Regional Administrator would be required to deny the request. However, on review of the denial the Administrator is authorized by § 124.91(a)(1) to review policy or legal conclusions of the Regional Administrator. EPA is requiring an appeal to the Administrator even of purely legal issues involved in a permit decision to ensure that the Administrator will have an opportunity to review any permit before it will be final and subject to judicial review.]

(2) Persons requesting an evidentiary hearing on an NPDES permit under this section may also request an evidentiary hearing on a RCRA or UIC permit. PSD permits may never be made part of an evidentiary hearing under Subpart E. This request is subject to all the requirements of paragraph (b)(1) of this section and in addition will be granted only if:

(i) Processing of the RCRA or UIC permit at issue was consolidated with the processing of the NPDES permit as provided in § 124.4;

(ii) The standards for granting a hearing on the NPDES permit are met;

(iii) The resolution of the NPDES permit issues is likely to make necessary or appropriate modification of the RCRA or UIC permit; and

(iv) If a PSD permit is involved, a permittee who is eligible for an evidentiary hearing under Subpart E on his or her NPDES permit requests that the formal hearing be conducted under the procedures of Subpart F and the Regional Administrator finds that consolidation is unlikely to delay final permit issuance beyond the PSD one-year statutory deadline.

(c) These requests shall also contain:

(1) The name, mailing address, and telephone number of the person making such request;

(2) A clear and concise factual statement of the nature and scope of the interest of the requester;

(3) The names and addresses of all persons whom the requester represents; and

(4) A statement by the requester that, upon motion of any party granted by the Presiding Officer, or upon order of the Presiding Officer *sua sponte* without cost or expense to any other party, the requester shall make available to appear and testify, the following:

(i) The requester;

(ii) All persons represented by the requester; and

(iii) All officers, directors, employees, consultants, and agents of the requester and the persons represented by the requester.

(5) Specific references to the contested permit conditions, as well as suggested revised or alternative permit conditions (including permit denials) which, in the judgment of the requester, would be required to implement the purposes and policies of the CWA.

(6) In the case of challenges to the application of control or treatment technologies identified in the statement of basis or fact sheet, identification of the basis for the objection, and the alternative technologies or combination of technologies which the requester believes are necessary to meet the requirements of the CWA.

(7) Identification of the permit obligations that are contested or are inseverable from contested conditions and should be stayed if the request is granted by reference to the particular contested conditions warranting the stay.

(8) Hearing requests also may ask that a formal hearing be held under the procedures set forth in Subpart F. An applicant may make such a request even if the proceeding does not constitute "initial licensing" as defined in § 124.111.

(d) If the Regional Administrator grants an evidentiary hearing request, in whole or in part, the Regional Administrator shall identify the permit conditions which have been contested by the requester and for which the evidentiary hearing has been granted. Permit conditions which are not contested or for which the Regional Administrator has denied the hearing request shall not be affected by, or considered at, the evidentiary hearing. The Regional Administrator shall specify these conditions in writing in accordance with § 124.60(c).

(e) The Regional Administrator must grant or deny all requests for an evidentiary hearing on a particular permit. All requests that are granted for a particular permit shall be combined in a single evidentiary hearing.

(f) The Regional Administrator (upon notice to all persons who have already submitted hearing requests) may extend the time allowed for submitting hearing requests under this section for good cause.

§ 124.75   Decision on request for a hearing.

(a)(1) Within 30 days following the expiration of the time allowed by § 124.74 for submitting an evidentiary hearing request, the Regional Administrator shall decide the extent to which, if at all, the request shall be granted, provided that the request conforms to the requirements of § 124.74, and sets forth material issues of fact relevant to the issuance of the permit.

(2) When an NPDES permit for which a hearing request has been granted constitutes "initial licensing" under § 124.111, the Regional Administrator may elect to hold a formal hearing under the procedures of Subpart F rather than under the procedures of this Subpart even if no person has requested that Subpart F be applied. If the Regional Administrator makes such a decision, he or she shall issue a notice of hearing under § 124.116. All subsequent proceedings shall then be governed by §§ 124.117 through 124.121, except that any reference to a draft permit shall mean the final permit.

(3) Whenever the Regional Administrator grants a request made under § 124.74(c)(8) for a formal hearing under Subpart F on an NPDES permit that does not constitute an initial license under § 124.111, the Regional Administrator shall issue a notice of hearing under § 124.116 including a statement that the permit will be processed under the procedures of Subpart F unless a written objection is received within 30 days. If no valid

objection is received, the application shall be processed in accordance with §§ 124.117 through 124.121, except that any reference to a draft permit shall mean the final permit. If a valid objection is received, this Subpart shall be applied instead.

(b) If a request for a hearing is denied in whole or in part, the Regional Administrator shall briefly state the reasons. That denial is subject to review by the Administrator under § 124.91.

§ 124.76   Obligation to submit evidence and raise issues before a final permit is issued.

No evidence shall be submitted by any party to a hearing under this Subpart that was not submitted to the administrative record required by § 124.18 as part of the preparation of and comment on a draft permit, unless good cause is shown for the failure to submit it. No issues shall be raised by any party that were not submitted to the administrative record required by § 124.18 as part of the preparation of and comment on a draft permit unless good cause is shown for the failure to submit them. Good cause includes the case where the party seeking to raise the new issues or introduce new information shows that it could not reasonably have ascertained the issues or made the information available within the time required by § 124.15; or that it could not have reasonably anticipated the relevance or materiality of the information sought to be introduced. Good cause exists for the introduction of data available on operation authorized under § 124.60(a)(2).

§ 124.77   Notice of hearing.

Public notice of the grant of an evidentiary hearing regarding a permit shall be given as provided in § 124.57(b) and by mailing a copy to all persons who commented on the draft permit, testified at the public hearing, or submitted a request for a hearing. Before the issuance of the notice, the Regional Administrator shall designate the Agency trial staff and the members of the decisional body (as defined in § 124.78).

§ 124.78   Ex parte communications.

(a) For purposes of this section, the following definitions shall apply:

(1) "Agency trial staff" means those Agency employees, whether temporary or permanent, who have been designated by the Agency under § 124.77 or § 124.116 as available to investigate, litigate, and present the evidence, arguments, and position of the Agency in the evidentiary hearing or nonadversary panel hearing.

Appearance as a witness does not necessarily require a person to be designated as a member of the Agency trial staff;

(2) "Decisional body" means any Agency employee who is or may reasonably be expected to be involved in the decisional process of the proceeding including the Administrator, Judicial Officer, Presiding Officer, the Regional Administrator (if he or she does not designate himself or herself as a member of the Agency trial staff), and any of their staff participating in the decisional process. In the case of a non-adversary panel hearing, the decisional body shall also include the panel members, whether or not permanently employed by the Agency;

(3) *Ex parte* communication" means any communication, written or oral, relating to the merits of the proceeding between the decisional body and an interested person outside the Agency or the Agency trial staff which was not originally filed or stated in the administrative record or in the hearing. *Ex parte* communications do not include:

(i) Communications between Agency employees other than between the Agency trial staff and the members of the decisional body;

(ii) Discussions between the decisional body and either:

(A) Interested persons outside the Agency, or

(B) The Agency trial staff, *if* all parties have received prior written notice of the proposed communications and have been given the opportunity to be present and participate therein.

(4) "Interested person outside the Agency" includes the permit applicant, any person who filed written comments in the proceeding, any person who requested the hearing, any person who requested to participate or intervene in the hearing, any participant in the hearing and any other interested person not employed by the Agency at the time of the communications, and any attorney of record for those persons.

(b)(1) No interested person outside the Agency or member of the Agency trial staff shall make or knowingly cause to be made to any members of the decisional body, an *ex parte* communication on the merits of the proceedings.

(2) No member of the decisional body shall make or knowingly cause to be made to any interested person outside the Agency or member of the Agency trial staff, an *ex parte* communication on the merits of the proceedings.

(3) A member of the decisional body who receives or who makes or who knowingly causes to be made a communication prohibited by this subsection shall file with the Regional Hearing Clerk all written communications or memoranda stating the substance of all oral communications together with all written responses and memoranda stating the substance of all oral responses.

(c) Whenever any member of the decisionmaking body receives an *ex parte* communication knowingly made or knowingly caused to be made by a party or representative of a party in violation of this section, the person presiding at the stage of the hearing then in progress may, to the extent consistent with justice and the policy of the CWA, require the party to show cause why its claim or interest in the proceedings should not be dismissed, denied, disregarded, or otherwise adversely affected on account of such violation.

(d) The prohibitions of this section begin to apply upon issuance of the notice of the grant of a hearing under § 124.77 or § 124.116. This prohibition terminates at the date of final agency action.

§ 124.79 Additional parties and issues.

(a) Any person may submit a request to be admitted as a party within 15 days after the date of mailing, publication, or posting of notice of the grant of an evidentiary hearing, whichever occurs last. The Presiding Officer shall grant requests that meet the requirements of §§ 124.74 and 124.76.

(b) After the expiration of the time prescribed in paragraph (a) of this section any person may file a motion for leave to intervene as a party. This motion must meet the requirements of §§ 124.74 and 124.76 and set forth the grounds for the proposed intervention. No factual or legal issues, besides those raised by timely hearing requests, may be proposed except for good cause. A motion for leave to intervene must also contain a verified statement showing good cause for the failure to file a timely request to be admitted as a party. The Presiding Officer shall grant the motion only upon an express finding on the record that:

(1) Extraordinary circumstances justify granting the motion;

(2) The intervener has consented to be bound by:

(i) Prior written agreements and stipulations by and between the existing parties; and

(ii) All orders previously entered in the proceedings; and

(3) Intervention will not cause undue delay or prejudice the rights of the existing parties.

§ 124.80 Filing and service.

(a) An original and one (1) copy of all written submissions relating to an evidentiary hearing filed after the notice is published shall be filed with the Regional Hearing Clerk.

(b) The party filing any submission shall also serve a copy of each submission upon the Presiding Officer and each party of record. Service shall be by mail or personal delivery.

(c) Every submission shall be accompanied by an acknowledgement of service by the person served or a certificate of service citing the date, place, time, and manner of service and the names of the persons served.

(d) The Regional Hearing Clerk shall maintain and furnish a list containing the name, service address, and telephone number of all parties and their attorneys or duly authorized representatives to any person upon request.

§ 124.81 Assignment of Administrative Law Judge.

No later than the date of mailing, publication, or posting of the notice of a grant of an evidentiary hearing, whichever occurs last, the Regional Administrator shall refer the proceeding to the Chief Administrative Law Judge who shall assign an Administrative Law Judge to serve as Presiding Officer for the hearing.

§ 124.82 Consolidation and severance.

(a) The Administrator, Regional Administrator, or Presiding Officer has the discretion to consolidate, in whole or in part, two or more proceedings to be held under this Subpart, whenever it appears that a joint hearing on any or all of the matters in issue would expedite or simplify consideration of the issues and that no party would be prejudiced thereby. Consolidation shall not affect the right of any party to raise issues that might have been raised had there been no consolidation.

(b) If the Presiding Officer determines consolidation is not conducive to an expeditious, full, and fair hearing, any party or issues may be severed and heard in a separate proceeding.

§ 124.83 Prehearing conferences.

(a) The Presiding Officer, *sua sponte*, or at the request of any party, may direct the parties or their attorneys or duly authorized representatives to appear at a specified time and place for one or more conferences before or during a hearing, or to submit written proposals or correspond for the purpose of considering any of the matters set forth in paragraph (c) of this section.

(b) The Presiding Officer shall allow a reasonable period before the hearing begins for the orderly completion of all prehearing procedures and for the submission and disposition of all prehearing motions. Where the circumstances warrant, the Presiding Officer may call a prehearing conference to inquire into the use of available procedures contemplated by the parties and the time required for their completion, to establish a schedule for their completion, and to set a tentative date for beginning the hearing.

(c) In conferences held, or in suggestions submitted, under paragraph (a) of this section, the following matters may be considered:

(1) Simplification, clarification, amplification, or limitation of the issues.

(2) Admission of facts and of the genuiness of documents, and stipulations of facts.

(3) Objections to the introduction into evidence at the hearing of any written testimony, documents, papers, exhibits, or other submissions proposed by a party, except that the administrative record required by § 124.19 shall be received in evidence subject to the provisions of § 124.85(d)(2). At any time before the end of the hearing any party may make, and the Presiding Officer shall consider and rule upon, motions to strike testimony or other evidence other than the administrative record on the grounds of relevance, competency, or materiality.

(4) Matters subject to official notice may be taken.

(5) Scheduling as many of the following as are deemed necessary and proper by the Presiding Officer:

(i) Submission of narrative statements of position on each factual issue in controversy;

(ii) Submission of written testimony and documentary evidence (e.g., affidavits, data, studies, reports, and any other type of written material) in support of those statements; or

(iii) Requests by any party for the production of additional documentation, data, or other information relevant and material to the facts in issue.

(6) Grouping participants with substantially similar interests to eliminate redundant evidence, motions, and objections.

(7) Such other matters that may expedite the hearing or aid in the disposition of the matter.

(d) At a prehearing conference or at some other reasonable time set by the Presiding Officer, each party shall make available to all other parties the names of the expert and other witnesses it expects to call. At its discretion or at the request of the Presiding Officer, a party may include a brief narrative summary of any witness's anticipated testimony. Copies of any written testimony, documents, papers, exhibits, or materials which a party expects to introduce into evidence, and the administrative record required by § 124.18 shall be marked for identification as ordered by the Presiding Officer. Witnesses, proposed written testimony, and other evidence may be added or amended upon order of the Presiding Officer for good cause shown. Agency employees and consultants shall be made available as witnesses by the Agency to the same extent that production of such witnesses is required of other parties under § 124.74(c)(4). (See also § 124.85(b)(16).)

(e) The Presiding Officer shall prepare a written prehearing order reciting the actions taken at each prehearing conference and setting forth the schedule for the hearing, unless a transcript has been taken and accurately reflects these matters. The order shall include a written statement of the areas of factual agreement and disagreement and of the methods and procedures to be used in developing the evidence and the respective duties of the parties in connection therewith. This order shall control the subsequent course of the hearing unless modified by the Presiding Officer for good cause shown.

§ 124.84 Summary determination.

(a) Any party to an evidentiary hearing may move with or without supporting affidavits and briefs for a summary determination in its favor upon any of the issues being adjudicated on the basis that there is no genuine issue of material fact for determination. This motion shall be filed at least 45 days before the date set for the hearing, except that upon good cause shown the motion may be filed at any time before the close of the hearing.

(b) Any other party may, within 30 days after service of the motion, file and serve a response to it or a countermotion for summary determination. When a motion for summary determination is made and supported, a party opposing the motion may not rest upon mere allegations or denials but must show, by affidavit or by other materials subject to consideration by the Presiding Officer, that there is a genuine issue of material fact for determination at the hearing.

(c) Affidavits shall be made on personal knowledge, shall set forth facts that would be admissible in evidence, and shall show affirmatively that the affiant is competent to testify to the matters stated therein.

(d) The Presiding Officer may set the matter for oral argument and call for the submission of proposed findings, conclusions, briefs, or memoranda of law. The Presiding Officer shall rule on the motion not more than 30 days after the date responses to the motion are filed under paragraph (b) of this section.

(e) If all factual issues are decided by summary determination, no hearing will be held and the Presiding Officer shall prepare an initial decision under § 124.89. If summary determination is denied or if partial summary determination is granted, the Presiding Officer shall issue a memorandum opinion and order, interlocutory in character, and the hearing will proceed on the remaining issues. Appeals from interlocutory rulings are governed by § 124.90.

(f) Should it appear from the affidavits of a party opposing a motion for summary determination that he or she cannot for reasons stated present, by affidavit or otherwise, facts essential to justify his or her opposition, the Presiding Officer may deny the motion or order a continuance to allow additional affidavits or other information to be obtained or may make such other order as is just and proper.

§ 124.85 Hearing procedure.

(a)(1) The permit applicant always bears the burden of persuading the Agency that a permit authorizing pollutants to be discharged should be issued and not denied. This burden does not shift.

[Note.—In many cases the documents contained in the administrative record, in particular the fact sheet or statement of basis and the response to comments, should adequately discharge this burden.]

(2) The Agency has the burden of going forward to present an affirmative case in support of any challenged condition of a final permit.

(3) Any hearing participant who, by raising material issues of fact, contends:

(i) That particular conditions or requirements in the permit are improper or invalid, and who desires either:

(A) The inclusion of new or different conditions or requirements; or

(B) The deletion of those conditions or requirements; or

(ii) That the denial or issuance of a permit is otherwise improper or invalid, shall have the burden of going forward to present an affirmative case at the conclusion of the Agency case on the challenged requirement.

(b) The Presiding Officer shall conduct a fair and impartial hearing, take action to avoid unnecessary delay in the disposition of the proceedings, and maintain order. For these purposes, the Presiding Officer may:

(1) Arrange and issue notice of the date, time, and place of hearings and conferences;

(2) Establish the methods and procedures to be used in the development of the evidence;

(3) Prepare, after considering the views of the participants, written statements of areas of factual disagreement among the participants;

(4) Hold conferences to settle, simplify, determine, or strike any of the issues in a hearing, or to consider other matters that may facilitate the expeditious disposition of the hearing;

(5) Administer oaths and affirmations;

(6) Regulate the course of the hearing and govern the conduct of participants;

(7) Examine witnesses;

(8) Identify and refer issues for interlocutory decision under § 124.90;

(9) Rule on, admit, exclude, or limit evidence;

(10) Establish the time for filing motions, testimony, and other written evidence, briefs, findings, and other submissions;

(11) Rule on motions and other procedural matters pending before him, including but not limited to motions for summary determination in accordance with § 124.84;

(12) Order that the hearing be conducted in stages whenever the number of parties is large or the issues are numerous and complex;

(13) Take any action not inconsistent with the provisions of this Subpart for the maintenance of order at the hearing and for the expeditious, fair, and impartial conduct of the proceeding;

(14) Provide for the testimony of opposing witnesses to be heard simultaneously or for such witnesses to meet outside the hearing to resolve or isolate issues or conflicts;

(15) Order that trade secrets be treated as confidential business information in accordance with § 122.19 and 40 CFR Part 2; and

(16) Allow such cross-examination as may be required for a full and true disclosure of the facts. No cross-examination shall be allowed on questions of policy except to the extent required to disclose the factual basis for permit requirements, or on questions of law, or regarding matters (such as the validity of effluent limitations guidelines) that are not subject to challenge in an evidentiary hearing. No Agency witnesses shall be required to testify or be made available for cross-examination on such matters. In deciding whether or not to allow cross-examination, the Presiding Officer shall consider the likelihood of clarifying or resolving a disputed issue of material fact compared to other available methods. The party seeking cross-examination has the burden of demonstrating that this standard has been met.

(c) All direct and rebuttal evidence at an evidentiary hearing shall be submitted in written form, unless, upon

motion and good cause shown, the Presiding Officer determines that oral presentation of the evidence on any particular fact will materially assist in the efficient identification and clarification of the issues. Written testimony shall be prepared in narrative form.

(d)(1) The Presiding Officer shall admit all relevant, competent, and material evidence, except evidence that is unduly repetitious. Evidence may be received at any hearing even though inadmissible under the rules of evidence applicable to judicial proceedings. The weight to be given evidence shall be determined by its reliability and probative value.

(2) The administrative record required by § 124.18 shall be admitted and received in evidence. Upon motion by any party the Presiding Officer may direct that a witness be provided to sponsor a portion or portions of the administrative record. The Presiding Officer, upon finding that the standards in § 124.85(b)(3) have been met, shall direct the appropriate party to produce the witness for cross-examination. If a sponsoring witness cannot be provided, the Presiding Officer may reduce the weight accorded the appropriate portion of the record.

[Note.—Receiving the administrative record into evidence automatically serves several purposes: (1) it documents the prior course of the proceeding; (2) it provides a record of the views of affected persons for consideration by the agency decisionmaker; and (3) it provides factual material for use by the decisionmaker.]

(3) Whenever any evidence or testimony is excluded by the Presiding Officer as inadmissible, all such evidence or testimony existing in written form shall remain a part of the record as an offer of proof. The party seeking the admission of oral testimony may make an offer of proof, by means of a brief statement on the record describing the testimony excluded.

(4) When two or more parties have substantially similar interests and positions, the Presiding Officer may limit the number of attorneys or other party representatives who will be permitted to cross-examine and to make and argue motions and objections on behalf of those parties. Attorneys may, however, engage in cross-examination relevant to matters not adequately covered by previous cross-examination.

(5) Rulings of the Presiding Officer on the admissibility of evidence or testimony, the propriety of cross-examination, and other procedural matters shall appear in the record and shall control further proceedings, unless reversed as a result of an interlocutory appeal taken under § 124.90.

(6) All objections shall be made promptly or be deemed waived. Parties shall be presumed to have taken exception to an adverse ruling. No objection shall be deemed waived by further participation in the hearing.

§ 124.86 Motions.

(a) Any party may file a motion (including a motion to dismiss a particular claim on a contested issue), with the Presiding Officer on any matter relating to the proceeding. All motions shall be in writing and served as provided in § 124.80 except those made on the record during an oral hearing before the Presiding Officer.

(b) Within 10 days after service of any written motion, any part to the proceeding may file a response to the motion. The time for response may be shortened to 3 days or extended for an additional 10 days by the Presiding Officer for good cause shown.

(c) Notwithstanding § 122.52, any party may file with the Presiding Officer a motion seeking to apply to the permit any regulatory or statutory provision issued or made available after the issuance of the permit under § 124.15. The Presiding Officer shall grant any motion to apply a new statutory provision unless he or she finds it contrary to legislative intent. The Presiding Officer may grant a motion to apply a new regulatory requirement when appropriate to carry out the purpose of CWA, and when no party would be unduly prejudiced thereby.

§ 124.87 Record of hearings.

(a) All orders issued by the Presiding Officer, transcripts of oral hearings or arguments, written statements of position, written direct and rebuttal testimony, and any other data, studies, reports, documentation, information and other written material of any kind submitted in the proceeding shall be a part of the hearing record and shall be available to the public except as provided in § 122.19, in the Office of the Regional Hearing Clerk, as soon as it is received in that office.

(b) Evidentiary hearings shall be either stenographically reported verbatim or tape recorded, and thereupon transcribed. After the hearing, the reporter shall certify and file with the Regional Hearing Clerk:

(1) The original of the transcript, and

(2) The exhibits received or offered into evidence at the hearing.

(c) The Regional Hearing Clerk shall promptly notify each of the parties of the filing of the certified transcript of proceedings. Any party who desires a copy of the transcript of the hearing may obtain a copy of the hearing transcript from the Regional Hearing Clerk upon payment of costs.

(d) The Presiding Officer shall allow witnesses, parties, and their counsel an opportunity to submit such written proposed corrections of the transcript of any oral testimony taken at the hearing, pointing out errors that may have been made in transcribing the testimony, as are required to make the transcript conform to the testimony. Except in unusual cases, no more than 30 days shall be allowed for submitting such corrections from the day a complete transcript of the hearing becomes available.

§ 124.88 Proposed findings of fact and conclusions; brief.

Within 45 days after the certified transcript is filed, any party may file with the Regional Hearing Clerk proposed findings of fact and conclusions of law and a brief in support thereof. Briefs shall contain appropriate references to the record. A copy of these findings, conclusions, and brief shall be served upon all the other parties and the Presiding Officer. The Presiding Officer, for good cause shown, may extend the time for filing the proposed findings and conclusions and/or the brief. The Presiding Officer may allow reply briefs.

§ 124.89 Decisions.

(a) The Presiding Officer shall review and evaluate the record, including the proposed findings and conclusions, any briefs filed by the parties, and any interlocutory decisions under § 124.90 and shall issue and file his initial decision with the Regional Hearing Clerk. The Regional Hearing Clerk shall immediately serve copies of the initial decision upon all parties (or their counsel of record) and the Administrator.

(b) The initial decision of the Presiding Officer shall automatically become the final decision 30 days after its service unless within that time:

(1) A party files a petition for review by the Administrator pursuant to § 124.91; or

(2) The Administrator *sua sponte* files a notice that he or she will review the decision pursuant to § 124.91.

§ 124.90 Interlocutory appeal.

(a) Except as provided in this section, appeals to the Administrator may be taken only under § 124.91. Appeals from orders or rulings may be taken under this section only if the Presiding Officer, upon motion of a party, certifies those orders or rulings to the Administrator for appeal on the record. Requests to the Presiding Officer for certification must be filed in writing within 10 days of service of notice of the order, ruling, or decision and shall state briefly the grounds relied on.

(b) The Presiding Officer may certify an order or ruling for appeal to the Administrator if:

(1) The order or ruling involves an important question on which there is substantial ground for difference of opinion, and

(2) Either:

(i) An immediate appeal of the order or ruling will materially advance the ultimate completion of the proceeding; or

(ii) A review after the final order is issued will be inadequate or ineffective.

(c) If the Administrator decides that certification was improperly granted, he or she shall decline to hear the appeal. The Administrator shall accept or decline all interlocutory appeals within 30 days of their submission; if the Administrator takes no action within that time, the appeal shall be automatically dismissed. When the Presiding Officer declines to certify an order or ruling to the Administrator for an interlocutory appeal, it may be reviewed by the Administrator only upon appeal from the initial decision of the Presiding Officer, except when the Administrator determines, upon motion of a party and in exceptional circumstances, that to delay review would not be in the public interest. Such motion shall be made within 5 days after receipt of notification that the Presiding Officer has refused to certify an order or ruling for interlocutory appeal to the Administrator. Ordinarily, the interlocutory appeal will be decided on the basis of the submissions made to the Presiding Officer. The Administrator may, however, allow briefs and oral argument.

(d) In exceptional circumstances, the Presiding Officer may stay the proceeding pending a decision by the Administrator upon an order or ruling certified by the Presiding Officer for an interlocutory appeal, or upon the denial of such certification by the Presiding Officer.

(e) The failure to request an interlocutory appeal shall not prevent taking exception to an order or ruling in an appeal under § 124.91.

§ 124.91 Appeal to the Administrator.

(a)(1) Within 30 days after service of an initial decision, or a denial in whole or in part of a request for an evidentiary hearing, any party or requester, as the case may be, may appeal any matter set forth in the initial decision or denial, or any adverse order or ruling to which the party objected during the hearing, by filing with the Administrator notice of appeal and petition for review. The petition shall include a statement of the supporting reasons and, when appropriate, a showing that the initial decision contains:

(i) A finding of fact or conclusion of law which is clearly erroneous, or

(ii) An exercise of discretion or policy which is important and which the Administrator should review.

(2) Within 15 days after service of a petition for review under paragraph (a)(1) of this section, any other party to the proceeding may file a responsive petition.

(3) Policy decisions made or legal conclusions drawn in the course of denying a request for an evidentiary hearing may be reviewed and changed by the Administrator in an appeal under this section.

(b) Within 30 days of an initial decision or denial of a request for an evidentiary hearing the Administrator may, *sua sponte*, review such decision. Within 7 days after the Administrator has decided under this section to review an initial decision or the denial of a request for an evidentiary hearing, notice of that decision shall be served by mail upon all affected parties and the Regional Administrator.

(c)(1) Within a reasonable time following the filing of the petition for review, the Administrator shall issue an order either granting or denying the petition for review. When the Administrator grants a petition for review or determines under paragraph (b) of this section to review a decision, the Administrator may notify the parties that only certain issues shall be briefed.

(2) Upon granting a petition for review, the Regional Hearing Clerk shall promptly forward a copy of the record to the Judicial Officer and shall retain a complete duplicate copy of the record in the Regional Office.

(d) Notwithstanding the grant of a petition for review or a determination under paragraph (b) of this section to review a decision, the Administrator may summarily affirm without opinion an initial decision or the denial of a request for an evidentiary hearing.

(e) A petition to the Administrator under paragraph (a) of this section for review of any initial decision or the denial of an evidentiary hearing is, under 5 U.S.C. § 704, a prerequisite to the seeking of judicial review of the final decision of the Agency.

(f) If a party timely files a petition for review or if the Administrator *sua sponte* orders review, then, for purposes of judicial review, final Agency action on an issue occurs as follows:

(1) If the Administrator denies review or summarily affirms without opinion as provided in § 124.91(d), then the initial decision or denial becomes the final Agency action and occurs upon the service of notice of the Administrator's action.

(2) If the Administrator issues a decision without remanding the proceeding then the final permit, redrafted as required by the Administrator's original decision, shall be reissued and served upon all parties to the appeal.

(3) If the Administrator issues a decision remanding the proceeding, then final Agency action occurs upon completion of the remanded proceeding, including any appeals to the Administrator from the results of the remanded proceeding.

(g) The petitioner may file a brief in support of the petition within 21 days after the Administrator has granted a petition for review. Any other party may file a responsive brief within 21 days of service of the petitioner's brief. The petitioner then may file a reply brief within 14 days of service of the responsive brief. Any person may file an *amicus brief* for the consideration of the Administrator within the same time periods that govern reply briefs. If the Administrator determines, *sua sponte*, to review an initial Regional Administrator's decision or the denial of a request for an evidentiary hearing, the Administrator shall notify the parties of the schedule for filing briefs.

(h) Review by the Administrator of an initial decision or the denial of an evidentiary hearing shall be limited to the issues specified under paragraph (a) of this section, except that after notice to all parties, the Administrator may raise and decide other matters which he or she considers material on the basis of the record.

Subpart F—Non-Adversary Panel Procedures

§ 124.111 Applicability.

(a) Except as set forth in this Subpart, this Subpart applies in lieu of, and to complete exclusion of, Subparts A through E in the following cases:

(1)(i) In any proceedings for the issuance of any NPDES permit which constitutes "initial licensing" under the Administrative Procedure Act, when the Regional Administrator elects to apply this Subpart and explicitly so states in the public notice of the draft permit under § 124.10 or in a supplemental notice under § 124.14. If an NPDES draft permit is processed under this Subpart, any other draft permits which have been consolidated with the NPDES draft permit under § 124.4 shall likewise be processed under this Subpart, except for PSD permits when the Regional Administrator makes a finding under § 124.4(e) that consolidation would be likely to result in missing the one year

statutory deadline for issuing a final PSD permit under the CAA.

(ii) "Initial licensing" includes both the first decision on an NPDES permit applied for by a discharger that has not previously held one and the first decision on any variance requested by a discharger.

(iii) To the extent this Subpart is used to process a request for a variance under CWA section 301(h), the term "Adminstrator or a person designated by the Administrator" shall be substituted for the term "Regional Administrator".

(2) In any proceeding for which a hearing under this Subpart was granted under § 124.75 following a request for a formal hearing under § 124.74. See §§ 124.74(c)(8) and 124.75(a)(2).

(3) Whenever the Regional Administrator determines as a matter of discretion that the more formalized mechanisms of this Subpart should be used to process draft NPDES general permits (for which evidentiary hearings are unavailable under § 124.71), or draft RCRA or draft UIC permits.

(b) EPA shall not apply these procedures to a decision on a variance where Subpart E proceedings are simultaneously pending on the other conditions of the permit. See § 124.64(b).

§ 124.112 Relation to other subparts.

The following provisions of Subparts A through E apply to proceedings under this Subpart:

(a)(1) §§ 124.1 through 124.10.

(2) § 124.14 "Reopening of comment period."

(3) § 124.16 "Stays of contested permit conditions."

(4) § 124.20 "Computation of time."

(b)(1) § 124.41 "Definitions applicable to PSD permits."

(2) § 124.42 "Additional procedures for PSD permits affecting Class I Areas."

(c)(1) §§ 124.51 through 124.56.

(2) § 124.57 (c) "Public notice."

(3) §§ 124.58 through 124.66.

(d)(1) § 124.72 "Definitions," except for the definition of "Presiding Officer," see § 124.119.

(2) § 124.73 "Filing."

(3) § 124.78 *"Ex parte* communications."

(4) § 124.80 "Filing and service."

(5) § 124.85(a) (Burden of proof).

(6) § 124.86 "Motions."

(7) § 124.87 "Record of hearings."

(8) § 124.90 "Interlocutory appeal."

(e) In the case of permits to which this Subpart is made applicable after a final permit has been issued under § 124.15, either by the grant under § 124.75 of a hearing request under § 124.74, or by notice of supplemental proceedings

under § 124.14, §§ 124.13 and 124.76 shall also apply.

§ 124.113 Public notice of draft permits and public comment period.

Public notice of a draft permit under this Subpart shall be given as provided in §§ 124.10 and 124.57. At the discretion of the Regional Administrator, the public comment period specified in this notice may include an opportunity for a public hearing under § 124.12.

§ 124.114 Request for hearing.

(a) By the close of the comment period under § 124.113, any person may request the Regional Administrator to hold a panel hearing on the draft permit by submitting a written request containing the following:

(1) A brief statement of the interest of the person requesting the hearing;

(2) A statement of any objections to the draft permit;

(3) A statement of the issues which such person proposes to raise for consideration at the hearing; and

(4) Statements meeting the requirements of § 124.74(c)(1)–(5).

(b) Whenever (1) a written request satisfying the requirements of paragraph (a) of this section has been received and presents genuine issues of material fact, or (2) the Regional Administrator determines *sua sponte* that a hearing under this Subpart is necessary or appropriate, the Regional Administrator shall notify each person requesting the hearing and the applicant, and shall provide public notice under § 124.57(c). If the Regional Administrator determines that a request does not meet the requirements of paragraph (a) of this section or does not present genuine issues of fact, the Regional Administrator may deny the request for the hearing and shall serve written notice of that determination on all persons requesting the hearing.

(c) The Regional Administrator may also decide before a draft permit is prepared under § 124.6 that a hearing should be held under this section. In such cases, the public notice of the draft permit shall explicitly so state and shall contain the information required by § 124.57(c). This notice may also provide for a hearing under § 124.12 before a hearing is conducted under this section.

§ 124.115 Effect of denial of or absence of request for hearing.

If no request for a hearing is made under § 124.114, or if all such request are denied under that section, the Regional Administrator shall then prepare a recommended decision under § 124.124. Any person whose hearing request has been denied may then appeal that recommended decision to the Administrator as provided in § 124.91.

§ 124.116   Notice of hearing.

(a) Upon granting a request for a hearing under § 124.114 the Regional Administrator shall promptly publish a notice of the hearing as required under § 124.57(c). The mailed notice shall include a statement which indicates whether the Presiding Officer or the Regional Administrator will issue the Recommended decision. The mailed notice shall also allow the participants at least 30 days to submit written comments as provided under § 124.118.

(b) The Regional Administrator may also give notice of a hearing under this section at the same time as notice of a draft permit under § 124.113. In that case the comment periods under §§ 124.113 and 124.118 shall be merged and held as a single public comment period.

(c) The Regional Administrator may also give notice of hearing under this section in response to a hearing request under § 124.74 as provided in § 124.75.

§ 124.117   Request to participate in hearing.

(a) Persons desiring to participate in any hearing noticed under this section, shall file a request to participate with the Regional Hearing Clerk before the deadline set forth in the notice of the grant of the hearing. Any person filing such a request becomes a party to the proceedings within the meaning of the Administrative Procedure Act. The request shall include:

(1) A brief statement of the interest of the person in the proceeding;

(2) A brief outline of the points to be addressed;

(3) An estimate of the time required; and

(4) The requirements of § 124.74(c)(1)–(5).

(5) If the request is submitted by an organization, a nonbinding list of the persons to take part in the presentation.

(b) As soon as practicable, but in no event later than 2 weeks before the scheduled date of the hearing, the Presiding Officer shall make a hearing schedule available to the public and shall mail it to each person who requested to participate in the hearing.

§ 124.118   Submission of written comments on draft permit.

(a) No later than 30 days before the scheduled start of the hearing (or such other date as may be set forth in the notice of hearing), each party shall file all of its comments on the draft permit, based on information in the administrative record and any other information which is or reasonably

could have been available to that party. All comments shall include any affidavits, studies, data, tests, or other materials relied upon for making any factual statements in the comments.

(b)(1) Written comments filed under paragraph (a) of this section shall constitute the bulk of the evidence submitted at the hearing. Oral statements at the hearing should be brief and in the nature of argument. They shall be restricted either to points that could not have been made in written comments, or to emphasize points which are made in the comments, but which the party believes can more effectively be argued in the hearing context.

(2) Notwithstanding the foregoing, within two weeks prior to the deadline specified in paragraph (a) of this section for the filing of comments, any party may move to submit all or part of its comments orally at the hearing in lieu of submitting written comments and the Presiding Officer shall, within one week, grant such motion if the Presiding Officer finds that the party will be prejudiced if required to submit the comments in written form.

(c) Parties to any hearing may submit written material in response to the comments filed by other parties under paragraph (a) of this section at the time they appear at the panel stage of the hearing under § 124.120.

§ 124.119   Presiding Officer.

(a)(1)(i) Before giving notice of a hearing under this Subpart in a proceeding involving an NPDES permit, the Regional Administrator shall request that the Chief Administrative Law Judge assign an Administrative Law Judge as the Presiding Officer. The Chief Administrative Law Judge shall then make the assignment.

(ii) If all parties to such a hearing waive in writing their statutory right to have an Administrative Law Judge named as the Presiding Officer in a hearing subject to this subparagraph the Regional Administrator may name a Presiding Officer under paragraph (a)(2)(ii) of this section.

(2) Before giving notice of a hearing under this Subpart in a proceeding which does not involve an NPDES permit or a RCRA permit termination, the Regional Administrator shall either:

(i) Request that the Chief Administrative Law Judge assign an Administrative Law Judge as the Presiding Officer. The Chief Administrative Law Judge may thereupon make such an assignment if he concludes that the other duties of his office allow, or

(ii) Name a lawyer permanently or temporarily employed by the Agency and without prior connection with the proceeding to serve as Presiding Officer.

(iii) If the Chief Administrative Law Judge declines to name an Administrative Law Judge as Presiding Officer upon receiving a request under subparagraph (2)(i) of this section, the Regional Administrator shall name a Presiding Officer under paragraph (a)(2)(ii) of this section.

(b) It shall be the duty of the Presiding Officer to conduct a fair and impartial hearing. The Presiding Officer shall have the authority:

(1) Conferred by § 124.85(b)(1)–(15), § 124.83(b) and (c), and;

(2) To receive relevant evidence, provided that all comments under §§ 124.113 and 124.118, the record of the panel hearing under § 124.120, and the administrative record, as defined in § 124.9 or in § 124.18 as the case may be shall be received in evidence, and

(3) Either upon motion or *sua sponte*, to change the date of the hearing under § 124.120, or to recess such a hearing until a future date. In any such case the notice required by § 124.10 shall be given.

§ 124.120   Panel hearing.

(a) A Presiding Officer shall preside at each hearing held under this Subpart. An EPA panel shall also take part in the hearing. The panel shall consist of three or more EPA temporary or permanent employees having special expertise or responsibility in areas related to the hearing issue, at least two or whom shall not have taken part in writing the draft permit. If appropriate for the evaluation of new or different issues presented at the hearing, the panel membership, at the discretion of the Regional Administrator, may change or may include persons not employed by EPA.

(b) At the time of the hearing notice under § 124.116, the Regional Administrator shall designate the persons who shall serve as panel members for the hearing and the Regional Administrator shall file with the Regional Hearing Clerk the name and address of each person so designated. The Regional Administrator may also designate EPA employees who will provide staff support to the panel but who may or may not serve as panel members. The designated persons shall be subject to the *ex parte* rules in § 124.78. The Regional Administrator may also designate Agency trial staff as defined in § 124.78 for the hearing.

(c) At any time before the close of the hearing the Presiding Officer, after consultation with the panel, may request that any person having knowledge concerning the issues raised in the hearing and not then scheduled to participate therein appear and testify at the hearing.

(d) The panel members may question any person participating in the panel hearing. Cross-examination by persons other than panel members shall not be permitted at this stage of the proceeding except when the Presiding Officer determines, after consultation with the panel, that the cross-examination would expedite consideration of the issues. However, the parties may submit written questions to the Presiding Officer for the Presiding Officer to ask the participants, and the Presiding Officer may, after consultation with the panel, and at his or her sole discretion, ask these questions.

(e) At any time before the close of the hearing, any party may submit to the Presiding Officer written questions specifically directed to any person appearing or testifying in the hearing. The Presiding Officer, after consultation with the panel may, at his sole discretion, ask the written question so submitted.

(f) Within 10 days after the close of the hearing, any party shall submit such additional written testimony, affidavits, information, or material as they consider relevant or which the panel may request. These additional submissions shall be filed with the Regional Hearing Clerk and shall be a part of the hearing record.

§ 124.121 Opportunity for cross-examination.

(a) Any party to a panel hearing may submit a written request to cross-examine any issue of material fact. The motion shall be submitted to the Presiding Officer within 15 days after a full transcript of the panel hearing is filed with the Regional Hearing Clerk and shall specify:

(1) The disputed issue(s) of material fact. This shall include an explanation of why the questions at issue are factual rather than of an analytical or policy nature, the extent to which they are in dispute in light of the then-existing record, and the extent to which they are material to the decision on the application; and

(2) The person(s) to be cross-examined, and an estimate of the time necessary to conduct the cross-examination. This shall include a statement explaining how the cross-examination will resolve the disputed issues of material fact.

(b) After receipt of all motions for cross-examination under paragraph (a) of this section, the Presiding Officer,

after consultation with the hearing panel, shall promptly issue an order either granting or denying each request Orders granting requests for cross-examination shall be served on all parties and shall specify:

(1) The issues on which cross-examination is granted;

(2) The persons to be cross-examined on each issue;

(3) The persons allowed to conduct cross-examination;

(4) Time limits for the examination of witnesses by each cross-examiner; and

(5) The date, time, and place of the supplementary hearing at which cross-examination shall take place.

(c) In issuing this order, the Presiding Officer may determine that two or more parties have the same or similar interests and that to prevent unduly repetitious cross-examination, they should be required to choose a single representative for purposes of cross-examination. In that case, the order shall simply assign time for cross-examination without further identifying the representative. If the designated parties fail to choose a single representative, the Presiding Officer may divide the assigned time among the representatives or issue any other order which justice may require.

(d) The Presiding Officer and, to the extent possible, the members of the hearing panel shall be present at the supplementary hearing. During the course of the hearing, the Presiding Officer shall have authority to modify any order issued under paragraph (b) of this section. A record will be made under § 124.87.

(e)(1) No later than the time set for requesting cross-examination, a party may request that alternative methods of clarifying the record (such as the submission of additional written information) be used in lieu of or in addition to cross-examination. The Presiding Officer shall issue an order granting or denying this request at the time he or she issues (or would have issued) an order granting or denying a request for cross-examination, under paragraph (b) of this section. If the request for an alternative method is granted, the order shall specify the alternative and any other relevant information (such as the due date for submitting written information).

(2) In passing on any request for cross-examination submitted under paragraph (a) of this section, the Presiding Officer may, as a precondition to ruling on the merits of the request, require alternative means of clarifying the record to be used whether or not a request to do so has been made. The party requesting cross-examination shall

have one week to comment on the results of using the alternative method. After considering these comments the Presiding Officer shall issue an order granting or denying the request for cross-examination.

(f) The provisions of § 124.85(d)(2) apply to proceedings under this Subpart.

§ 124.122 Record for final permit.

The record on which the final permit shall be based in any proceeding under this Subpart consists of:

(a) The administrative record compiled under §§ 124.9 or 124.18 as the case may be;

(b) Any material submitted under § 124.78 relating to *ex parte* contacts;

(c) All notices issued under § 124.113;

(d) All requests for hearings, and rulings on those requests, received or issued under § 124.114;

(e) Any notice of hearing issued under § 24.116;

(f) Any request to participate in the hearing received under § 124.117;

(g) All comments submitted under § 124.118, any motions made under that section and the rulings on them, and any comments filed under § 124.113;

(h) The full transcript and other material received into the record of the panel hearing under § 124.120;

(i) Any motions for, or rulings on, cross-examination filed or issued under § 124.121;

(j) Any motions for, orders for, and the results of, any alternatives to cross-examination under § 124.121; and

(k) The full transcript of any cross-examination held.

§ 124.123 Filing of brief, proposed findings of fact and conclusions of law and proposed modified permit.

Unless otherwise ordered by the Presiding Officer, each party may, within 20 days after all requests for cross-examination are denied or after a transcript of the full hearing including any cross-examination becomes available, submit proposed findings of fact; conclusions regarding material issues of law, fact, or discretion; a proposed modified permit (if such person is urging that the draft or final permit be modified); and a brief in support thereof; together with references to relevant pages of transcript and to relevant exhibits. Within 10 days thereafter each party may file a reply brief concerning matters contained in opposing briefs and containing alternative findings of fact; conclusions regarding material issues of law, fact, or discretion; and a proposed modified permit where appropriate. Oral argument may be held at the discretion of the Presiding Officer on motion of any party or *sua sponte.*

§ 124.124   Recommended decision.

The person named to prepare the decision shall, as soon as practicable after the conclusion of the hearing, evaluate the record of the hearing and prepare and file a recommended decision with the Regional Hearing Clerk. That person may consult with, and receive assistance from, any member of the hearing panel in drafting the recommended decision, and may delegate the preparation of the recommended decision to the panel or to any member or members of it. This decision shall contain findings of fact, conclusions regarding all material issues of law, and a recommendation as to whether and in what respect the draft or final permit should be modified. After the recommended decision has been filed, the Regional Hearing Clerk shall serve a copy of that decision on each party and upon the Administrator.

§ 124.125   Appeal from or review of recommended decision.

(a)(1) Within 30 days after service of the recommended decision, any party may take exception to any matter set forth in that decision or to any adverse order or ruling of the Presiding Officer to which that party objected, and may appeal those exceptions to the Administrator as provided in § 124.91, except that references to "initial decision" will mean recommended decision under § 124.124.

§ 124.126   Final decision.

As soon as practicable after all appeal proceedings have been completed, the Administrator shall issue a final decision. That final decision shall include findings of fact; conclusions regarding material issue of law, fact, or discretion, as well as reasons therefore; and a modified permit to the extent appropriate. It may accept or reject all or part of the recommended decision. The Administrator may delegate some or all of the work of preparing this decision to a person or persons without substantial prior connection with the matter. The Administrator or his or her designee may consult with the Presiding Officer, members of the hearing panel, or any other EPA employee other than members of the Agency Trial Staff under § 124.78 in preparing the final decision. The Hearing Clerk shall file a copy of the decision on all parties.

§ 124.127   Final decision if there is no review.

If no party appeals a recommended decision to the Administrator, and if the Administrator does not elect to review

it, the recommended decision becomes the final decision of the Agency upon the expiration of the time for filing any appeals.

§ 124.128   Delegation of authority; time limitations.

(a) The Administrator may delegate to a Judicial Officer any or all of his or her authority under this Subpart.

(b) The failure of the Administrator, Regional Administrator, or Presiding Officer to do any act within the time periods specified under this Part shall not waive or diminish any right, power, or authority of the United States Environmental Protection Agency.

(c) Upon a showing by any party that it has been prejudiced by a failure of the Administrator, Regional Administrator, or Presiding Officer to do any act within the time periods specified under this Part the Administrator, Regional Administrator, or Presiding Officer, as the case may be, may grant that party such relief of a procedural nature (including extension of any time for compliance or other action) as may be appropriate.

Appendix A to Part 124—Guide to Decisionmaking Under Part 124

This Appendix is designed to assist in reading the procedural requirements set out in Part 124. It consists of two flow charts.

Figure 1 diagrams the more conventional sequence of procedures EPA expects to follow in processing permits under this Part. It outlines how a permit will be applied for, how a draft permit will be prepared and publicly noticed for comment, and how a final permit will be issued under the procedures in Subpart A.

This permit may then be appealed to the Administrator, as specified both in Subpart A (for RCRA, UIC, or PSD permits), or Subpart E or F (for NPDES permits). The first flow chart also briefly outlines which permit decisions are eligible for which types of appeal.

Part 124 also contains special "non-adversary panel hearing" procedures based on the "initial licensing" provisions of the Administrative Procedure Act. These procedures are set forth in Subpart F. In some cases, EPA may only decide to make those procedures applicable after it has gone through the normal Subpart A procedures on a draft permit. This process is also diagrammed in Figure 1.

Figure 2 sets forth the general procedure to be followed where these Subpart F procedures have been made applicable to a permit from the beginning.

Both flow charts outline a sequence of events directed by arrows. The boxes set forth elements of the permit process; and the diamonds indicate key decisionmaking points in the permit process.

The charts are discussed in more detail below.

Figure 1—Conventional EPA Permitting Procedures

This chart outlines the procedures for issuing permits whenever EPA does not make use of the special "panel hearing" procedures in Subpart F. The major steps depicted on this chart are as follows:

1. The permit process can begin in any one of the following ways:

a. Normally, the process will begin when a person applies for a permit under §§ 122.4 and 124.3.

b. In other cases, EPA may decide to take action on its own initiative to change a permit or to issue a general permit. This leads directly to preparation of a draft permit under § 124.6.

c. In addition, the permittee or any interested person (other than for PSD permits) may request modification, revocation and reissuance or termination of a permit under §§ 122.15, 122.16 and 124.5.

Those requests can be handled in either of two ways:

i. EPA may tentatively decide to grant the request and issue a new draft permit for public comment, either with or without requiring a new application.

ii. If the request is denied, an informal appeal to the Administrator is available.

2. The next major step in the permit process is the preparation of a draft permit. As the chart indicates, preparing a draft permit also requires preparation of either a statement of basis (§ 124.7), a fact sheet (§ 124.8) or, compilation of an "administrative record" (§ 124.9), and public notice (§ 124.10).

3. The next stage is the public comment period (§ 124.11). A public hearing under § 124.12 may be requested before the close of the public comment period.

EPA has the discretion to hold a public hearing, even if there were no requests during the public comment period. If EPA decides to schedule one, the public comment period will be extended through the close of the hearing. EPA also has the discretion to conduct the public hearing under Subpart F panel procedures. (See Figure 2.)

The regulations provide that all arguments and factual materials that a person wishes EPA to consider in connection with a particular permit must be placed in the record by the close of the public comment period (§ 124.13).

4. Section 124.14 states that EPA, at any time before issuing a final permit decision may decide to either reopen or extend the comment period, prepare a new draft permit and begin the process again from that point, or for RCRA and UIC permits, or for NPDES permits that constitute "initial licensing", to begin "panel hearing" proceedings under Subpart F. These various results are shown schematically.

5. The public comment period and any public hearing will be followed by issuance of a final permit decision (§ 124.15). As the chart shows, the final permit must be accompanied by a response to comments (§ 124.17) and be based on the administrative record (§ 124.18).

6. After the final permit is issued, it may be appealed to higher agency authority. The exact form of the appeal depends on the type of permit involved.

a. RCRA, UIC or PSD permits standing alone will be appealed directly to the Administrator under § 124.19.

b. NPDES permits which do not involve "initial licensing" may be appealed in an evidentiary hearing under Subpart E. The regulations provide (§ 124.74) that if such a hearing is granted for an NPDES permit and if RCRA or UIC permits have been consolidated with that permit under § 124.4 then closely related conditions of those RCRA or UIC permits may be reexamined in an evidentiary hearing. PSD permits, however, may never be reexamined in a Subpart E hearing.

c. NPDES permits which do involve "initial licensing" may be appealed in a panel hearing under Subpart F. The regulations provide that if such a hearing is granted for an NPDES permit, consolidated RCRA, UIC, or PSD permits may also be reexamined in the same proceeding.

As discussed below, this is only one of several ways the panel hearing procedures may be used under these regulations.

7. This chart does not show EPA appeal procedures in detail. Procedures for appeal to the Administrator under § 124.19 are self-explanatory; Subpart F procedures are diagrammed in Figure 2; and Subpart E procedures are basically the same that would apply in any evidentiary hearing.

However, the chart at this stage does reflect the provisions of § 124.60(b), which allows EPA, even after a formal hearing has begun, to "recycle" a permit back to the draft permit stage at any time before that hearing has resulted in an initial decision.

Figure 2—Non-Adversary Panel Procedures

This chart outlines the procedures for processing permits under the special "panel hearing" procedures of Subpart F. These procedures were designed for making decisions that involve "initial licensing" NPDES permits. Those permits include the first decisions on an NPDES permit applied for by any discharger that has not previously held one, and the first decision on any statutory variance. In addition, these procedures will be used for any RCRA, UIC, or PSD permit which has been consolidated with such an NPDES permit, and may be used, if the Regional Administrator so chooses, for the issuance of individual RCRA or UIC permits. The steps depicted on this chart are as follows:

1. Application for a permit. These proceedings will generally begin with an application, since NPDES initial licensing always will begin with an application.

2. Preparation of a draft permit. This is identical to the similar step in Figure 1.

3. Public comment period. This again is identical to the similar step in Figure 1. The Regional Administrator has the opportunity to schedule an informal public hearing under § 124.12 during this period.

4. Requests for a panel hearing must be received by the end of the public comment period under § 124.113. See § 124.114.

If a hearing request is denied, or if no hearing requests are received, a recommended decision will be issued based on the comments received. The recommended decision may then be appealed to the Administrator. See § 124.115.

5. If a hearing is granted, notice of the hearing will be published in accordance with § 124.116 and will be followed by a second comment period during which requests to participate and the bulk of the remaining evidence for the final decision will be received (§§ 124.117 and 124.118).

The regulations also allow EPA to move directly to this stage by scheduling a hearing when the draft permit is prepared. In such cases the comment period on the draft permit under § 124.113 and the prehearing comment period under § 124.118 would occur at the same time. EPA anticipates that this will be the more frequent practice when permits are processed under panel procedures.

This is also a stage at which EPA can switch from the conventional procedures diagrammed in Figure 1 to the panel hearing procedures. As the chart indicates, EPA would do this by scheduling a panel hearing either through use of the "recycle" provision in § 124.14 or in response to a request for a formal hearing under § 124.74.

6. After the close of the comment period, a panel hearing will be held under § 124.120, followed by any cross-examination granted under § 124.121. The recommended decision will then be prepared (§ 124.124) and an opportunity for appeal provided under § 124.125. A final decision will be issued after appeal proceedings, if any, are concluded.

BILLING CODE 6560-01-M

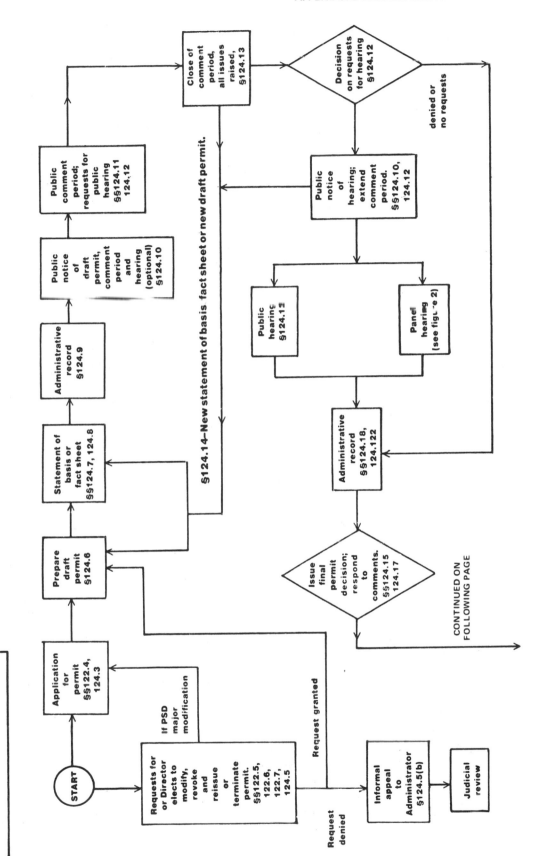

Figure 1-Conventional
EPA Permitting Procedures

CONTINUED ON
FOLLOWING PAGE

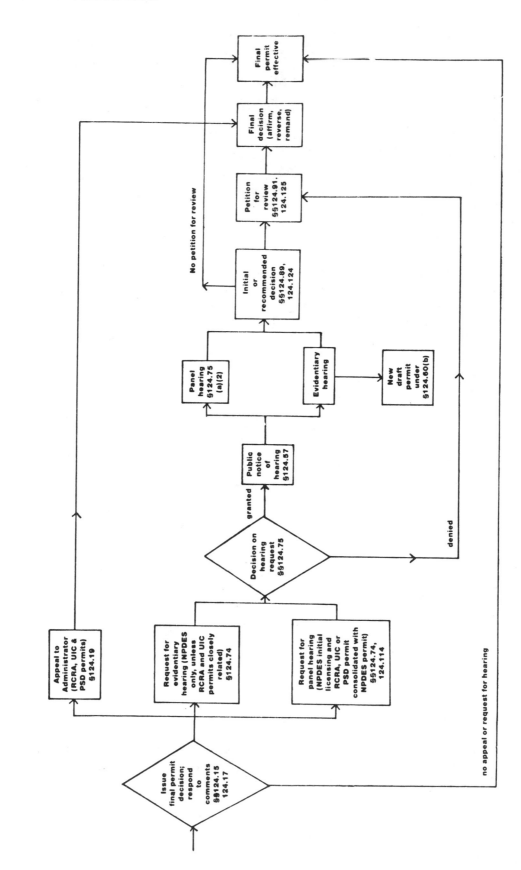

EPA Appeal Procedures

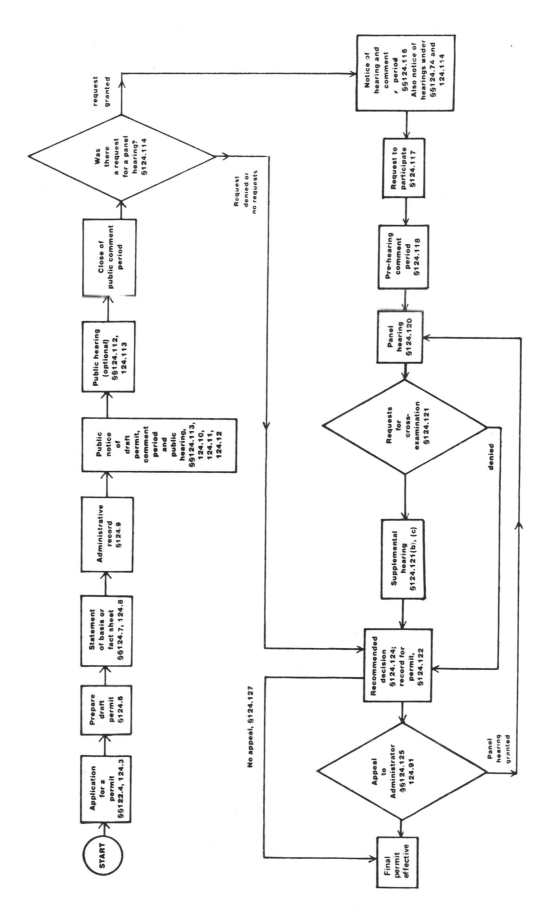

Figure 2-Non-Adversary Panel Procedures

BILLING CODE 6560-01-C

ENVIRONMENTAL PROTECTION AGENCY

[Docket No. 3010; FRL 1419-4]

Preliminary Notification of Hazardous Waste Activity

AGENCY: Environmental Protection Agency.

ACTION: Publication of notification form.

SUMMARY: Section 3010 of the Resource Conservation and Recovery Act (RCRA), 42 U.S.C. 6930 requires any person who generates or transports hazardous waste or who owns or operates a facility for the treatment, storage, or disposal of hazardous waste to notify the Environmental Protection Agency (or States having authorized hazardous waste permit programs) of the hazardous waste activity within 90 days of the promulgation or revision of RCRA Section 3001 regulations. This notification will give EPA and the public a "snapshot" of the hazardous waste activity regulated under RCRA. In accordance with Section 6001 of RCRA, Federal Agencies must comply with the notification requirement of Section 3010. Section 3010 states that unless notification has been given, "no identified or listed hazardous waste subject to this subtitle may be transported, treated, stored, or disposed of." In addition, existing treatment, storage, and disposal facilities that do not provide notification pursuant to this section become ineligible for Interim Status under Section 3005(e). This notice sets forth procedures and a form which should be used when filing a notification of hazardous waste activity.

ADDRESS: The official docket for this notice is located in Room 2503, U.S. Environmental Protection Agency, 401 M Street, S.W., Washington, D.C. and is available for viewing from 9:00 am to 4:00 pm, Monday through Friday, excluding holidays.

FOR FURTHER INFORMATION ON THIS NOTICE CONTACT: Mr. Terrence Kafara, State Programs and Resource Recovery Division, Office of Solid Waste, WH–563, U.S. Environmental Protection Agency, Washington, D.C. 20460 (202) 755–9150.

SUPPLEMENTARY INFORMATION:

I. Authority

This notice is issued under authority of Sections 2002(a) and 3010 of the Solid Waste Disposal Act, as amended by the Resource Conservation and Recovery Act of 1976 and as amended by the Quiet Communities Act of 1978 ("RCRA" or "the Act"), 42 U.S.C. 6912(a) and 6930.

II. Background

The Resource Conservation and Recovery Act provides for the development and implementation of a comprehensive program to protect human health and the environment from the improper management of hazardous waste. A fundamental premise of the statute is that human health and the environment will best be protected by careful management of the transportation, treatment, storage, and disposal of hazardous waste, in accordance with standards developed under the Act.

Elsewhere in today's Federal Register, standards for generators (Section 3002) and transporters (Section 3003) of hazardous waste are published. Within the next several months, the Environmental Protection Agency ("EPA" or "the Agency") will promulgate regulations for determining which wastes are hazardous (Section 3001); requirements for the treatment, storage, and disposal of hazardous waste (Section 3004); procedures for obtaining a hazardous waste permit (Section 3005); and procedures for the delegation of program responsibility from EPA to the States (Section 3006). The publication of these regulations sets in motion a series of events which will culminate in full implementation of the hazardous waste control program. One of these events is the notification of hazardous waste activity required by Section 3010 of RCRA.

Section 3010 requires all persons engaging in hazardous waste management activities to notify EPA or States having authorized hazardous waste permit programs. In developing the Notification Form included in this notice, EPA considered two distinct approaches. The first was to expand the form to include the gathering of additional information which could be used in establishing enforcement priorities. The second approach consisted of a form whose requests were limited to that information necessary to fulfill the statutory requirement for notification.

EPA chose the latter approach for two reasons. First, the successful establishment of a computer data base depends upon extensive use of a Notification Form which is designed for direct key punching and entry into the computer. The inclusion of information beyond that amount, which is legally required by Section 3010, would likely result in decreased use of the form by persons filing notifications. Second, the information required for the establishment of surveillance and enforcement priorities can be obtained

in other, more appropriate ways, such as annual reporting requirements.

III. Implementation

EPA has identified approximately 400,000 persons, businesses, and Federal Agencies which may be required to file notifications. EPA will mail to each of these persons a notification package including this Public Notice, an explanatory letter from the EPA Administrator, a copy of the EPA regulations on identification and listing of hazardous waste (Title 40, Code of Federal Regulations, Part 261), information from other EPA regulations which may be needed to assist persons in filing a notification, and a postcard which may be used by owners and operators of treatment, storage, and disposal facilities to request Part A of the permit application. The notification must be filed with EPA within 90 days of promulgation of 40 CFR Part 261. Part A of the permit application must be filed within six months of promulgation of the 40 CFR Part 261 for owners and operators of those facilities who wish to continue their operations under the Interim Status provisions of Section 3005(e) of the Act.

Receipt of a notification package from EPA does not mean that a person must file a notification, but rather that the person has been identified as a member of an industry or other activity likely to handle hazardous waste. For example, many persons in the wood and woodworking industry will receive notification packages. While many businesses in that industry generate hazardous waste, some will receive packages although they do not ordinarily produce hazardous waste.

Conversely, non-receipt of a notification package does not mean that a person is not required to file a notification. Each person must make a determination of whether he/she handles hazardous waste. EPA's mass mailing is intended to assist people in determining whether they are required to file a notification, and failure of the Agency to mail a notification package to any affected person will not relieve that person of the obligation to file a notification according to the requirements of the Act.

Each notification package will also include several pre-printed labels for use in filing the notification. The twelve-digit (12) number on the upper left hand corner of each label is that person's EPA Identification Number. It must be used on hazardous waste manifests and annual and other reports. Persons who do not receive notification packages will be assigned an EPA Identification

Number upon receipt by EPA of their notifications.

EPA will return to each notifier an acknowledgement of receipt of the notification, which will include the notifier's EPA Identification Number. Persons who received notification packages will already know their EPA Identification Numbers and may begin to use them as required, that is, six months after promulgation of 40 CFR Part 261, even if the EPA acknowledgement is delayed. This acknowledgement in no way constitutes an endorsement by EPA of the adequacy of the notification or of the notifier's business practices, rather, it serves as a confirmation that EPA received the notification.

IV. Who Must File

In order to transport, offer for transportation, treat, store, or dispose of hazardous waste, after the effective date of 40 CFR 261, a person must have filed a notification and received an EPA Identification Number. Regulations governing the notification process were proposed on July 11, 1978 (43 Fed. Reg. 29908 et seq.). The proposed regulations stated that persons conducting hazardous waste activities "at the time of promulgation or revision of Section 3001 regulations" were required to file a notification. Many commenters requested that the Agency clarify the question of who must file. EPA, accordingly, has been more specific in the instructions to the form.

It should be emphasized that the notification process applies in general to persons handling hazardous waste at the time of promulgation or amendment of the Section 3001 regulations. There are certain persons who do not have to file a notification though they handle hazardous waste, for example, persons generating small quantities of hazardous waste. The Section 3001 regulations define which persons who handle hazardous waste are exempt from notifying.

Hazardous waste management facilities which are no longer in operation are not required to notify because it is EPA's view that the intent of Congress was that the Notification process was to be a snapshot of current hazardous waste management practices for the benefit of EPA and the public. Further, it was not intended, nor is it an appropriate vehicle for finding abandoned hazardous waste disposal sites in part, because the RCRA Subtitle C Regulations do not cover those abandoned sites. New legislation to find and clean-up abandoned sites is being considered by EPA and the Congress. It is noted that a notification is required

for a facility storing hazardous waste at the time of promulgation or amendment to the Section 3001 regulations even though no new wastes are being added. These facilities are, in effect, continuing to store hazardous waste and therefore they are considered to be in operation.

Generators of hazardous waste who begin operation after the initial notification period must, prior to shipping hazardous waste, apply for an EPA Identification Number using the Notification Form in accordance with the regulations published under Section 3002 (40 CFR Part 262). Similarly, new hazardous waste transporters must also apply for an EPA Identification Number on the Notification Form in accordance with the regulations published under Section 3003 (40 CFR Part 263) prior to moving any hazardous waste. The owners/operators of new treatment, storage, and disposal facilities and facilities which failed to meet the requirements for Interim Status (Notification within 90 days and submission of a Part A permit application within 180 days) may not operate until they receive a permit. This requirement includes generators who treat, store, or dispose of hazardous waste on-site.

Persons who have provided proper notification of hazardous waste activity may later begin to handle additional hazardous wastes not included in the original notification. In the administration of this program, EPA will not require these persons to file a new notification under Section 3010 with respect to those wastes. Such a requirement would be costly to both EPA and the regulated community with no corresponding benefit.

The Section 3010 notification will also provide information on the location of Class IV injection wells. Regulations promulgated under the Safe Drinking Water Act provide for EPA to conduct a survey of these wells. Under the Section 3010 notification, hazardous waste generators and owners and operators of facilities for treating, storing, or disposing of hazardous waste must indicate if an injection well is located at their installation. Requiring this information under the Notification process will greatly assist EPA in preparing for the survey under the Safe Drinking Water Act.

V. Information Required

Section 3010 requires a person who notifies EPA of his hazardous waste activity to state "the location and general description of such activity and the identified or listed hazardous wastes handled." The proposed regulations required submission of that information,

with the optional addition of the amount of hazardous waste handled annually.

Several changes were made from the proposed regulations to the final Public Notice. The "amount" item was deleted, because it would be extremely difficult for many persons to make an accurate estimate of the amount of hazardous waste handled in the past. The information thus obtained might be inaccurate or misleading, and therefore useless for informational purposes. A number of commenters suggested deletion of this item.

The "description of activity" requirement has been simplified. The proposed rule required a designation of which type of hazardous waste activity was being conducted (generation, transportation, treatment, storage, or disposal) and a general description of the business activity which produced the waste, through use of SIC (Standard Industrial Classification) code numbers. The SIC code requirement was deleted because, in many cases, the code number would not reveal the information necessary for EPA to make use of the data. For instance, a manufacturer's SIC code would not necessarily reveal that the company produces electroplating wastes, because the SIC code is keyed to the product (e.g., automobiles) and not to the waste. More accurate information as to the amount and type of waste will be obtained through annual and other reports, permit applications, and EPA access to records.

The proposed form listed six characteristics of non-listed hazardous waste to be checked by persons handling such waste: ignitable, reactive, infectious, radioactive, corrosive, and toxic. The Section 3001 regulations, however, will not include characteristics for identifying waste as infectious or radioavtive, but will use those properties as a basis for listing specific hazardous wastes. The form has been modified accordingly.

The proposal also allowed 180 days after promulgation of the Section 3001 regulations for a final notification with respect to toxic waste. Persons who had reason to believe they handled toxic waste were required to file a notification within 90 days of promulgation of the Section 3001 regulations, but were permitted to indicate that it was "undetermined" whether they handled toxic waste. Then, at the expiration of the 180-day period, such persons were to be considered as handling toxic waste unless they had submitted statements to the effect that the waste was not toxic. The reason for this variation in procedure was to allow for anticipated delays in obtaining a laboratory analysis establishing whether each waste met the proposed standard for toxicity. The Agency decided, however, not to require notifiers to report on whether each waste is toxic. Rather, notifiers are simply required to determine whether at least one waste handled meets the standard for toxicity. This change greatly reduces the initial analytical burden on notifiers, and it is not unreasonable to expect completion of the notification within 90 days of promulgation of the Section 3001 regulations. Therefore the 180-day provision has been deleted. It should be noted that following the effective date of the hazardous waste regulatory program (6 months after promulgation of Section 3001) the determination as to toxicity must be completed by generators for each waste as required by 40 CFR Part 262, Standards Applicable to Generators of Hazardous Waste. Commenters stated that many transporters particularly those in the rail industry, would be unable to identify the waste they transport and therefore should be exempt from this notification provision. The Act requires all persons who handle hazardous waste to notify. The purpose of Section 3010 is to provide EPA information on the indentity and hazardous waste handled by persons involved in hazardous waste activities. This information is essential for EPA's implementation of the Act. If necessary, transporters will be able to obtain the information on hazardous wastes from the shippers.

Finally, the proposed regulations were silent on the question of determining which wastes should be included in the notification. The proposal required persons conducting hazardous waste activities to notify with respect to those wastes handled "at the time of promulgation or revision of Section 3001 regulations." The instructions in this final Public Notice are more specific regarding the time period to be used. Any hazardous wastes handled during the three-month period immediately prior to the date of filing the notification must be included. Notifiers may also include other wastes which they anticipate they will be handling. For example, manufacturers may, from time to time, need to dispose of some of the chemical substances listed in 40 CFR Part 261. Notifiers who did not handle these wastes during the previous three months may include them in their notifications.

The form included in this notice offers a standardized approach for persons required to file notifications to provide the necessary information. Use of the form will assist EPA in the orderly initiation of its data management system, which is designed to facilitate key punching for entry into the computer. Some commenters suggested that EPA accept data processing printouts in lieu of a Notification Form. EPA will accept printouts. However, EPA believes it will be simpler for persons to use the form rather than printouts when notifying. If a printout is used, the notifier will have to provide for each waste on the printout the appropriate EPA Hazardous Waste Number or, for non-listed hazardous wastes, the general characteristics of the wastes. Notifiers who do not use the form must include all required information, including the certification in Item X of the form, signed by the person notifying or his authorized representative.

VI. Claims of Confidentiality

In the proposed regulation, EPA requested public comment on two alternatives for handling claims of condifentiality. The first would have permitted unsubstantiated claims of confidentiality at the time of notification. If a request for the information under the Freedom of Information Act were to be received, or in anticipation of such a request, EPA would have required substantiation of the claim. The second would require the necessary substantiation to be submitted with the notification.

Although many comments received on the question favored the first option, EPA chose the second based on three primary consideration. First, if EPA obtains the substantiation of the claim of confidentiality with the notification, the Regional Counsel can determine the business confidentiality before the information is entered into the Agency's computer, thus simplifying the recordkeeping procedure. Second, since EPA expects to receive many requests for the information, it would place an unnecessary burden on the Agency to require it to make a separate request for substantiation of each claim of confidentiality. No additional burden will be placed on notifiers because EPA would request substantiation of the claim in any event. In fact, notifiers will have 90 days to prepare the substantiation, instead of the 15 business days permitted by EPA's Freedom of Information Act Regulations. Finally, EPA believes that, in light of the very general nature of the information requested, very few notifiers will be able to justify a claim that information in the notification is entitled to confidential treatment.

The final notice, therefore, requires that all claims of confidentiality be accompanied by a written substantiation of the claim, in accordance with 40 CFR Part 2, Subpart B. The substantiation must address such questions as measures taken to guard against undesired disclosure of the information, the extent to which the information has been disclosed to others, and the harm which will result to the claimant's competitive position if the information is disclosed. The specific questions to be answered are listed in the instructions to the Notification Form.

Failure to include substantiation of a claim of confidentiality at the time of submission of notification will constitute a waiver of the claim, and the information in the notification will be available to the public.

VII. EPA Identification Number

The proposed regulations listed several types of identification numbers already assigned to many businesses and Federal Agencies, and instructed the notifier to use an appropriate number as his EPA hazardous waste identification number. A number of comments on this matter were received, some suggesting still more types of numbers to be used and others suggesting that, in the interest of conformity with the identification numbers to be assigned to treatment, storage and disposal facilities, the same numbering system should be used. EPA selected the latter option, and will use the Dun and Bradstreet Data Universal Numbering (DUN) system. The DUN system is the most nearly complete listing of U.S. businesses. Federal Agencies, which are not included in the DUN system, will be assigned their General Service Administration Real Property Number.

VIII. Use of Public Notice in Lieu of Final Promulgation of Regulations

The proposed rule included requirements as to who must file notifications, where and when to file, and what information notifications should contain, as well as a sample form. Also included was a provision for the temporary authorization of States for the sole purpose of receiving notifications.

EPA's primary reason for proposing regulations rather than a Public Notice was to establish rules covering the authorization of States to receive the notification (Limited Interim Authorization). However, EPA decided, for the reasons discussed below, to abandon the concept of Limited Interim Authorization. EPA has therefore decided not to promulgate notification regulations, but to issue this Public Notice instead.

The effect of this notice is, first, to provide a mechanism for implementation of Section 3010; second, to establish a certification statement which must be signed by anyone submitting a notification, and third, to establish a procedure for submission of claims of confidentiality.

IX. Limited Interim Authorization of States

The proposed regulations included procedures for granting "Limited Interim Authorization" to States in order to enable them to conduct the notification program. This concept was developed because it is very unlikely that any State hazardous waste program will be authorized during the 90-day notification period. It was intended that "Limited Interim Authorization" be granted separate and apart from the interim and full authorizations granted under section 3006 of RCRA.

Some commenters supported "Limited Interim Authorization" as a means for involving States in hazardous waste management as early as possible, while others objected on the grounds that "Limited Interim Authorization" is illegal and potentially burdensome. While EPA recognizes the benefits of involving the States during the initial implementation stages of RCRA, it has concluded that the use of "Limited Interim Authorization" is not authorized by RCRA. Section 3010 is explicit in stating that only "the Administrator" or "States having authorized hazardous waste permit programs under Section 3006" may receive notifications. The creation of a nonstatutory class of State authorizations in light of such an express Congressional directive raises serious legal questions. Therefore, since procedures involved in authorizing State programs under section 3006 are not likely to be completed during the 90-day notification period, EPA will carry out the notification process in full.

X. OMB Review

Under the Federal Reports Act of 1942, the Office of Management and Budget (OMB) reviews reporting requirements in proposed forms and regulations in order to minimize the reporting burden on respondents and the cost to the government. EPA submitted Form 8700–12 implementing section 3010 notification requirements to OMB on December 6, 1979. Reporting requirements implementing sections 3002 and 3003 were submitted on February 4, 1980.

In the course of OMB's preliminary review, OMB staff raised no objection to the form itself. OMB has, however, stated that it cannot complete its review of these requirements until the Administrator of EPA has made decisions regarding the coverage of the RCRA program under Subtitle C and OMB has had an opportunity to review the reporting burden of the entire regulatory program.

XI. Regulatory Analysis

In accordance with Executive Orders 12044 and 11821 as amended by Executive Order 11949, OMB Circular A–107 and EPA policy as stipulated in 39 Fed. Reg. 37419 (October 21, 1974), draft analyses of economic and environmental impacts have been prepared for the proposed Subtitle C, Hazardous Waste Management Regulations. The final Environmental Impact Statement will be available following promulgation of 40 CFR Part 261.

Dated: February 19, 1980.

Douglas M. Costle,
Administrator.

BILLING CODE 6560-01-M

APPENDIX D
PROPOSED HAZARDOUS WASTE REGULATIONS

§ 261.32 [Amended]

1. In § 261.32, add the following waste streams:

Industry	EPA hazardous waste No.*	Hazardous waste	Haz-ardous code
Wood preservation.		Wastewater from wood preserving processes that use creosote or pentachlorophenol.	(T).
Organic chemicals.		Distillation bottoms from the production of 1,1,1-trichloroethane.	(T).
		Heavy ends from the heavy ends column from the production of 1,1,1-trichloroethane.	(T).
		Vacuum stripper discharge from chlordene chlorinator in the production of chlordane.	
		Distillation light ends from the production of phthalic anhydride from ortho-xylene.	(T).
		Distillation bottoms from the production of phthalic anhydride from ortho-xylene.	(T).
Pesticides........		Untreated process wastewater from the production of toxaphene.	(T).
		Untreated wastewater from the production of 2,4-D.	(T).
		Wastewater from the production of methomyl.	(T).
		Process wastewater from creosote production.	(T).
Secondary lead.		Waste leaching solution from acid leaching of emission control dust/sludge from secondary lead smelting.	(T).

*The EPA Hazardous Waste Number will not be assigned until the listed waste is promulgated.

§ 261.32 [Amended]

1. In § 261.32, add the following waste streams:

Industry	EPA hazardous waste No.¹	Hazardous waste	Hazard code
Veterinary Pharmaceuticals............	....	Distillation tar residues from the distillation of aniline-based compounds in the production of veterinary pharmaceuticals from arsenic or organo-arsenic compounds.	(T)
		Residue from the use of activated carbon for decolorization in the production of veterinary pharmaceuticals from arsenic or organo-arsenic compounds.	(T)
Organic Chemicals	....	Process residues from aniline extraction from the production of aniline.	(T)
		Combined wastewater streams generated from nitrobenzene/aniline production.	(T)
		Separated aqueous stream from the reactor product washing step in the batch production of chlorobenzenes.	(T)
Gray Iron Foundries and Ductile Iron Foundries.	....	Emission control dust from gray and ductile iron foundry cupola furnaces.	(T)
Inorganic Chemicals.......	....	Wastewater treatment sludge from the mercury cell process in chlorine production.	(T)

Subpart H—Financial Requirements

§ 265.140 Applicability.

(a) The requirements of §§ 265.142, 265.143, 265.146, 265.147, and 265.149 apply to owners and operators of all hazardous waste facilities, except as provided otherwise in this section or in § 265.1.

(b) The requirements of §§ 265.144 and 265.145 apply only to owners and operators of disposal facilities.

*   *   *   *   *

§ 265.141 Definitions.

When used in Part 265, the following terms have the meanings given below:

(a) "Assets" means debit balances carried forward upon a closing of books of account representing property values or rights acquired that are recognized and measured in conformity with generally accepted accounting principles.

(b) "Current assets" means cash and other assets that are reasonably expected to be realized in cash or sold or consumed during the normal operating cycle of a business or within one year, if the operating cycle is shorter than one year.

(c) "Current liabilities" means liabilities expected to be satisfied by either the use of assets classified as current in the same balance sheet or the creation of other current liabilities; or those expected to be satisfied within a relatively short period of time, usually one year.

(d) "Liabilities" means obligations carried forward upon a closing of books of account that are recognized and measured in conformity with generally accepted accounting principles.

(e) "Marketable securities" means securities that are traded on recognized established securities markets where there are independent bona fide offers to buy and sell and where payment will be received in settlement of a sale within a relatively short time conforming to trade custom.

(f) "Net working capital" means the excess of current assets over current liabilities.

(g) "Net worth" means the excess of total assets over total liabilities and is equivalent to owner's equity.

(h) "Standby letter of credit" means an irrevocable engagement by an issuing bank, at the request of an owner or operator, that it will honor demands for payment made by the U.S. Environmental Protection Agency for the period of the letter of credit and under terms specified for letters of credit in these regulations.

(i) "Surety bond" means a contract by which a surety company engages to be answerable for the default or debts by an owner or operator on responsibilities relating to closure or post-closure care, and agrees to satisfy these responsibilities if the owner or operator does not, in accordance with the terms specified for surety bonds in these regulations.

(j) "Total-liabilities-to-net-worth ratio" means the value of total liabilities, which includes the sum of short-term and long-term obligations, divided by the value of net worth.

(k) "Trust fund" means a fund established by an owner or operator and held by a financial institution as the trustee with a fiduciary responsibility to carry out the terms of the trust as specified in these regulations for the benefit of the U.S. Environmental Protection Agency.

*   *   *   *   *

§ 265.143 Financial assurance for facility closure.

By the effective date of these regulations, an owner or operator of each facility must establish financial assurance for closure of the facility. He must choose from among the following options:

(a) *Closure trust fund.*

(1) The owner or operator may establish a closure trust fund. The trustee must be a bank or other financial institution. The beneficiary of the trust fund must be the U.S. Environmental Protection Agency.

(2) The trust agreement must be executed on EPA Form 8700–15 (see Appendix II). The owner or operator must send the properly executed trust agreement to the Regional Administrator by certified mail within 10 days after the effective date of the agreement.

(3) Replacement of a trust fund with another form or forms of financial assurance allowed in this section must be preceded by the written consent of the Regional Administrator. The owner or operator must report any change of trustee to the Regional Administrator within 10 days after such a change becomes effective.

(4) Payments to the trust fund must be in cash or marketable securities. The value of each security must be determined in accordance with the Internal Revenue Service method for valuing securities for estate tax purposes (26 CFR 20.2031–2). In all valuations of the trust fund for purposes of these regulations, securities must be valued by this IRS method.

(5) Payments to the closure trust fund must be made annually over the operating life of the facility as estimated in the closure plan (§ 265.112(a)) or 20 years, whichever period is shorter; this period is hereafter referred to as the "pay-in" period. The first payment must be equal to the adjusted closure cost estimate (see § 265.142) divided by the pay-in period in years. The first payment must be made by the effective date of these regulations. Subsequent payments must be made no later than 30 days after each anniversary date of the first payment. The trust agreement must require the trustee to notify the Regional Administrator by certified mail within 5 days after the end of the 30-day period if he does not receive payment within such period. Upon receiving such notification, the Regional Administrator may order the facility to begin closure unless the owner or operator has established other financial assurance as allowed in this section.

(6) The owner or operator must adjust the amount of each annual payment after the first one by multiplying the amount of the previous year's payment by the inflation factor calculated in accordance with § 265.142(c).

(7) If a new closure cost estimate is prepared in accordance with § 265.142(b), the next annual payment must be calculated as follows:

Step 1—Divide the adjusted closure cost estimate by the number of years in the pay-in period as of the effective date of these regulations.

Step 2—Multiply the result by the number of payments made to the fund.

Step 3—From the result of step 2 subtract the current value of the fund. The result is the amount which needs to be distributed over the remaining pay-in period.

Step 4—Divide the result of step 3 by the remaining years in the pay-in period.

Step 5—Add the result of step 4 to the result of step 1 to obtain the new payment.

(For an example of this calculation, see Appendix I.)

(8) The owner or operator must determine the value of the trust fund each year within 30 days prior to the date each annual payment is due to be made. If the total value of the fund has decreased since the previous year's valuation, the next payment must be calculated using the steps in paragraph (a)(7) of this section. The owner or operator may also use the calculation in paragraph (a)(7) to determine his next payment if the value of the fund has increased. If the value of the fund exceeds the *total* amount of the adjusted closure cost estimate, the owner or operator may submit a written request to the Regional Administrator for release of the amount in excess of the adjusted closure cost estimate. This request must be accompanied by a written statement from the trustee confirming the value of the fund.

(9) An owner or operator may accelerate payments into the trust fund or he may deposit the full amount of the closure cost estimate at the time the fund is established, but the trust fund must be valued annually and its value must be maintained at no less than the value that the fund would have had if annual payments had been made as specified in paragraphs (a)(5)–(8) of this section.

(10) If an owner or operator establishes a closure trust fund after the effective date of these regulations, having initially used one of the other mechanisms specified in this section, his first payment must be in the amount that the trust fund would have contained if it had been established on the effective date of these regulations in accordance with the requirements of this section.

(11) If the operating life of a facility extends beyond the maximum 20-year pay-in period, the owner or operator must determine the value of the trust fund every year after the 20th year until closure begins. Whenever the closure cost estimate changes during this period in accordance with § 265.142 (b) or (c), the owner or operator must compare the new estimate with the latest annual value of the fund. If the value of the fund is less than the amount of the adjusted closure cost estimate, the owner or operator must deposit cash or marketable securities into the fund so that its value equals the amount of the estimate. Such payment must be made within 60 days of the change in the closure cost estimate. If the value of the fund is greater than the total amount of the adjusted closure cost estimate, the owner or operator may submit a written request to the Regional Administrator for release of funds in excess of the estimate. This request must be accompanied by a written statement from the trustee confirming the value of the fund.

(12) Within 30 days after receiving a request from the owner or operator for release of excess funds as specified in paragraphs (a) (8) and (11), the Regional Administrator must direct the trustee in writing to release such excess funds to the owner or operator unless the Regional Administrator finds that the closure cost estimate was not prepared and adjusted in accordance with § 265.142.

(13) An owner or operator may request reimbursement for closure expenditures by submitting itemized bills to the Regional Administrator. Within 30 days after receiving bills for closure activities, the Regional Administrator must direct the trustee in writing to pay those bills which the Regional Administrator determines to be in accordance with the closure plan or are otherwise justified. Such payments must be made so long as the value of the fund after payment is at least 20 percent of the value that the fund had before any closure bills were paid.

(14) If an owner or operator substitutes another form or forms of financial assurance specified in this section for all or part of the trust fund, he may apply to the Regional Administrator for release of funds from the trust fund. Within 30 days after receiving such request, the Regional Administrator must direct the trustee in writing to release the excess funds to the owner or operator.

(15) The terms of the trust must require the trustee to make disbursements as specified in this paragraph. The trustee will disburse monies from the trust fund to parties designated by the Regional

Administrator upon written notification from the Regional Administrator that:

(i) The value of the trust fund exceeds the amount of the adjusted closure cost estimate; or

(ii) The itemized bills are in accordance with the approved closure plan or are otherwise justified, and they must be paid if the value of the trust fund after such payment is at least 20 percent of the value that the fund had before any closure bills were paid; or

(iii) The owner or operator has established other financial assurance for closure as allowed in this section for part or all of the trust fund; or

(iv) There has been a legal determination, a copy of which is attached to this notification, of a violation of the closure requirements of these regulations rendered in a proceeding brought pursuant to Section 3008 of RCRA.

(16) The trust agreement must require the trustee to release all funds remaining in the trust fund to the owner or operator upon receipt from him of the original or an authenticated copy of the Regional Administrator's letter, specified in paragraph (h) of this section, notifying the owner or operator that he is no longer required to comply with the requirements of this section for financially assuring closure of the facility.

(b) *Surety bond guaranteeing performance of closure.* (1) An owner or operator may meet the requirements of this section by obtaining a surety bond guaranteeing performance of closure. A surety company issuing a bond in accordance with these regulations must, at a minimum, be authorized to do business in the United States and be certified by the U.S. Treasury Department, in Circular 570, to write bonds in the penal sum of the bond to be issued. The obligee of the bond must be the U.S. Environmental Protection Agency.

(2) The bond must be executed on EPA Form 8700–16 (see Appendix III). The terms of the bond must provide that the surety will send the properly executed bond to the Regional Administrator by certified mail within 10 days after the effective date of the bond.

(3) The surety bond must guarantee that the owner or operator will perform facility closure in accordance with the closure plan. The surety bond must be written in an amount equal to or greater than the adjusted closure cost estimate (see § 265.142). The surety bond must be written so that whenever closure activities begin or are ordered to begin by the Regional Administrator during the term of the bond, the bond coverage includes completion of closure in accordance with the closure plan.

(4) If the closure cost estimate increases beyond the amount of the penal sum of the bond, the owner or operator must, within 30 days of such increase in the estimate, cause the penal sum of the bond to be increased or obtain other financial assurance, as specified in this section, to cover the increase. If the closure cost estimate decreases, the penal sum of the bond may be reduced to the amount of the adjusted closure cost estimate. At the request of the owner or operator, the Regional Administrator must send written notice to the surety of any reduction in the required penal sum within 30 days after receiving the request.

(5) The terms of the surety bond must provide that the surety company may cancel the bond by sending notice to the owner or operator and to the Regional Administrator by certified mail. Cancellation must not be effective for at least 90 days after the Regional Administrator receives the notice. The owner or operator, within 5 days of receiving a notice of cancellation from the surety, must notify the Regional Administrator by certified mail that he has received such a notice. The owner or operator may cancel the bond by providing 30 days' notice to the surety company if the Regional Administrator has given prior written consent based on his having received evidence of other financial assurance as specified in this section.

(6) Thirty days after receiving a notice of cancellation from the surety the Regional Administrator may order the owner or operator to begin closure unless the Regional Administrator has received evidence of other financial assurance as specified in this section.

(7) A surety becomes liable on a bond obligation only when a proceeding brought pursuant to the provisions of Section 3008 of RCRA has determined that the owner or operator has violated the closure requirements of these regulations. The terms of the bond must require that, following such a determination, the surety must:

(i) Complete closure of the facility in accordance with the closure plan; or

(ii) Pay the amount of the penal sum into an escrow account as directed by the Regional Administrator.

(8) The Regional Administrator must direct the depositary of an escrow account established under paragraph (b)(7)(ii) of this section to disburse funds to designated parties for the purpose of completing closure.

(c) *Standby letter of credit assuring funds for closure.* (1) An owner or operator may meet the requirements of this section by obtaining an irrevocable standby letter of credit. The letter must be written in favor of the Regional Administrator of the U.S. Environmental Protection Agency and must be for a period of at least one year. The letter of credit may be issued by any bank which is a member of the Federal Reserve System.

(2) The letter of credit must be executed on EPA Form 8700–17 (see Appendix IV). The terms of the letter must provide that the issuing bank will send the properly executed letter of credit to the Regional Administrator by certified mail within 10 days after the effective date of the letter.

(3) The credit must be issued for at least the amount of the adjusted closure cost estimate (see § 265.142).

(4) If the closure cost estimate increases beyond the amount of the credit, the owner or operator must, within 30 days of such increase in the estimate, cause the amount of the credit to be increased or obtain other financial assurance, as specified in this section, to cover the increase. If the closure cost estimate decreases, the credit may be reduced to the amount of the adjusted closure cost estimate. At the request of the owner or operator, the Regional Administrator must send written notice to the issuing bank of any reduction in the required credit within 30 days after receiving the request.

(5) The letter of credit must contain a clause providing for automatic annual extensions of the credit, subject to 60 days' written notice by the issuing bank to both the owner or operator and the Regional Administrator, by certified mail, of the bank's intention not to renew the credit. The owner or operator, within 5 days of receiving notice of nonrenewal from the bank, must notify the Regional Administrator by certified mail that he has received such a notice. The owner or operator may cancel the letter of credit by providing 30 days' notice to the issuing bank if the Regional Administrator has given prior written consent based on his having received evidence of other financial assurance as specified in this section.

(6) Thirty days after receiving a notice of nonrenewal from the bank the Regional Administrator may draw upon the credit up to the full amount of the credit unless he has received evidence that the owner or operator has established other financial assurance as specified in this section. If the Regional Administrator draws upon the letter of credit following a notice of nonrenewal, the issuing bank must, under the terms of the letter, deposit the amount of the draft immediately and directly into an interest-bearing escrow account. Disbursements from the escrow account must be made in the same manner as

specified for trust funds in paragraphs (a)(12)–(16) of this section.

(7) If the closure cost estimate increases beyond the amount of the funds in the escrow account, the owner or operator must, within 30 days of such increase, add to the account or establish other financial assurance as specified in this section to cover the increase. If the owner or operator fails to do so, the Regional Administrator may order him to begin closure.

(8) The Regional Administrator may otherwise draw upon the letter of credit only upon a legal determination of a violation of the closure requirements of these regulations rendered in a proceeding brought pursuant to the provisions of Section 3008 of RCRA. The terms of the letter must provide that, if the Regional Administrator draws upon the letter of credit following such a determination, the issuing bank will immediately and directly deposit the amount of the draft into an interest-bearing escrow account. The letter must require the escrow depositary to disburse monies from the escrow account to persons designated by the Regional Administrator to complete closure of the facility.

(d) *Use of more than one type of financial instrument.* An owner or operator may meet the requirements of this section by establishing more than one type of financial instrument. These instruments are limited to a trust fund, surety bond, or letter of credit as specified in paragraphs (a), (b), and (c), respectively, of this section (e.g., a letter of credit may assure half the closure cost and a trust fund the remaining half).

(e) *Financial test and guaranty for closure.* (1) An owner or operator may meet the requirements of this section by having all of the following financial characteristics:

(i) At least $10 million in net worth in the United States.

(ii) A total-liabilities-to-net-worth ratio of not more than three.

(iii) Net working capital in the United States of at least twice the adjusted closure cost estimate (see § 265.142).

(2) These characteristics must be demonstrated in a financial statement which has been audited by an independent certified public accountant and which contains unconsolidated balance sheets dated no more than 140 days prior to the current date. The owner or operator who intends to use a financial test to meet both closure and post-closure requirements for a single facility or to meet closure and/or post-closure requirements for more than one facility must indicate in the statement which requirements for which facilities are to be met through the financial test and must demonstrate that his net working capital in the United States is at least twice the sum of all the adjusted estimates of closure and post-closure costs to be covered by the financial test. The owner or operator must have the financial statement available at the facility and must provide data from the statement if requested as part of annual reports to the Regional Administrator under § 265.75.

(3) If at any time during the operating life of the facility the owner or operator fails to meet the requirements of paragraph (e)(1) of this section, he must notify the Regional Administrator by certified mail within 5 days of learning of failure to meet the requirements. Evidence of other financial assurance as specified in this section must be sent to the Regional Administrator by certified mail within 30 days from the time that the owner or operator learns of failure to meet the requirements; otherwise the Regional Administrator may order him to begin closure.

(4) An owner or operator may meet the requirements of this section by obtaining another entity's written guaranty providing financial assurance, in an amount equal to the adjusted closure cost estimate, for the owner's or operator's compliance with the closure requirements of these regulations. The guarantor must meet the requirements for owners or operators in paragraphs (e) (1) and (2) of this section.

(5) The guaranty must be executed on EPA Form 8700–18 (see Appendix V). The owner or operator must send the properly executed guaranty to the Regional Administrator by certified mail within 10 days after the effective date of the guaranty.

(6) Under the terms of the guaranty, the guarantor must notify the Regional Administrator and the owner or operator by certified mail if he at any time fails to meet the requirements of paragraph (e)(1) of this section. The guarantor must send such notice within 5 days after learning of failure to meet the requirements.

(7) The owner or operator must, within 30 days of receiving such notification, establish other financial assurance as specified in this section and provide evidence of such assurance to the Regional Administrator. If he fails to do so, the Regional Administrator may order him to begin closure.

(8) The guarantor may cancel the guaranty with 90 days' notice to the Regional Administrator and the owner or operator by certified mail, except that the guaranty must remain in effect if closure begins or is ordered to begin by the Regional Administrator before the end of the 90 days. Evidence of other financial assurance as specified in this section must be provided to the Regional Administrator within 30 days after a notice of cancellation is received by the Regional Administrator; otherwise, he may order the owner or operator to begin closure.

(9) The guaranty may be cancelled at any time following the mutual written consent of the owner or operator, the Regional Administrator, and the guarantor.

(10) Under the terms of the guaranty, in the event of a legal determination of a violation of the closure requirements rendered in a proceeding brought pursuant to Section 3008 of RCRA, the guarantor must pay parties designated by the Regional Administrator to complete closure in accordance with the closure plan.

(f) *Revenue test for municipalities.* (1) If the owner or operator is a municipality (as defined by RCRA), it may meet the requirements of this section by having annual revenues from property, sales, and/or income taxes equal to 10 times the adjusted closure cost estimate (see § 265.142). To be acceptable, these tax revenues must be legally available to cover closure responsibilities, i.e., they must not be dedicated to other purposes or otherwise precluded from use in meeting closure responsibilities.

(2) The owner or operator must send a letter signed by the chief financial officer of the municipality to the Regional Administrator stating that the municipality meets the requirements of paragraph (f)(1) of this section. The letter must be sent by certified mail within 10 days after the owner or operator begins use of the revenue test to meet the requirements of this section.

(3) If at any time during the operating life of the facility the annual tax revenues fail to meet the minimum multiple specified in paragraph (f)(1), the owner or operator must notify the Regional Administrator by certified mail within 5 days of learning of failure to meet the requirement. The owner or operator must send evidence of other financial assurance as specified in this section to the Regional Administrator by certified mail within 30 days from the time that the owner or operator learns of failure to meet the minimum multiple; otherwise the Regional Administrator may order the owner or operator to begin closure.

(g) *Use of a single financial mechanism for multiple facilities.* An owner or operator may use a single financial mechanism, as specified in paragraphs (a) through (f) of this section, to meet the requirements of this section for more than one facility of which he is the owner or operator. The amount of funds available through the mechanism must be no less than the sum of funds that would be available if a separate mechanism had been established for each facility.

(h) *Release of the owner or operator from the requirements of this section.* Within 60 days of receiving certifications from the owner or operator and an independent registered professional engineer that closure has been accomplished in accordance with the closure plan (see § 265.115), the Regional Administrator must, unless he has reason to believe that closure has not been in accordance with the closure plan, send a letter to the owner or operator notifying him that he no longer has to comply with the requirements of this section for the facility in question.

[Comment: It should be noted that this letter from the Regional Administrator to the owner or operator releases him only from requirements for financial assurance for closure of the facility; it does not release him from legal responsibility for meeting the closure standards.]

§ 265.145   Financial assurance for post-closure monitoring and maintenance.

By the effective date of these regulations, an owner or operator of each disposal facility must establish financial assurance for 30 years of post-closure care of the facility. He must choose from among the following options:

(a) *Post-closure trust fund.* (1) The owner or operator may establish a post-closure trust fund. The trustee must be a bank or other financial institution. The beneficiary of the trust fund must be the U.S. Environmental Protection Agency.

(2) The trust agreement must be executed on EPA Form 8700–19 (see Appendix VI). The owner or operator must send the properly executed trust agreement to the Regional Administrator by certified mail within 10 days after the effective date of the agreement.

(3) Replacement of a trust fund with another form or forms of financial assurance allowed in this section must be preceded by written consent of the Regional Administrator. The owner or operator must report any change of trustee to the Regional Administrator within 10 days after such a change becomes effective.

(4) Payments to the trust fund must be in cash or marketable securities. The value of each security must be determined in accordance with the Internal Revenue Service method for valuing securities for estate tax purposes (26 CFR 20.2031–2). In all valuations of the trust fund for purposes of these regulations, securities must be valued by this IRS method.

(5) Payments to the post-closure trust fund must be made annually over the operating life of the facility as estimated in the closure plan (§ 265.112(a)) or 20 years, whichever period is shorter; this period is hereafter referred to as the "pay-in" period. The first payment must be equal to the adjusted post-closure cost estimate (see § 265.144) divided by the pay-in-period in years. The first payment must be made by the effective date of these regulations. Subsequent payments must be made no later than 30 days after each anniversary date of the first payment. The trust agreement must require the trustee to notify the Regional Administrator by certified mail within 5 days after the 30-day period if he does not receive payment within such period. Upon receiving such notification, the Regional Administrator may order the facility to begin closure unless the owner or operator has established other financial assurance as allowed in this section.

(6) The owner or operator must adjust the amount of each annual payment after the first one by multiplying the amount of the previous year's payment by the inflation factor calculated in accordance with § 265.142(c).

(7) If a new post-closure cost estimate is prepared in accordance with § 265.144(b), the next annual payment must be calculated as follows:

Step 1—Divide the adjusted post-closure cost estimate by the number of years in the pay-in period as of the effective date of these regulations.

Step 2—Multiply the result by the number of payments made to the fund.

Step 3—From the result of step 2 subtract the current value of the fund. The result is the amount which needs to be distributed over the remaining pay-in period.

Step 4—Divide the result of step 3 by the remaining years in the pay-in period.

Step 5—Add the result of step 4 to the result of step 1 to obtain the new payment.

(Appendix I provides an example of a calculation of a new closure trust fund payment using these same steps.)

(8) The owner or operator must determine the value of the trust fund each year during the operating life of the facility within 30 days prior to the date each annual payment is due to be made. If the total value of the fund has decreased since the previous year's valuation, the next payment must be calculated using the steps in paragraph (a)(7) of this section. The owner or operator may also use the calculation in paragraph (a)(7) to determine his next payment if the value of the fund has increased. If the value of the fund exceeds the *total* amount of the adjusted post-closure cost estimate, the owner or operator may submit a written request to the Regional Administrator for release of the amount in excess of the adjusted post-closure cost estimate. This request must be accompanied by a written statement from the trustee confirming the value of the fund.

(9) An owner or operator may accelerate payments into the trust fund or he may deposit the full amount of the post-closure cost estimate at the time the fund is established, but the trust fund must be valued annually and its value must be maintained at no less than the value that the fund would have had if payments and valuations had been made as specified in paragraphs (a)(5)–(8) of this section.

(10) If an owner or operator establishes a post-closure trust fund after the effective date of these regulations, having initially used one of the other mechanisms specified in this section, his first payment must be in the amount that the trust fund would have contained if it had been established on the effective date of these regulations in accordance with the requirements of this section.

(11) If the operating life of a facility extends beyond the maximum 20-year pay-in period, the owner or operator must determine the value of the trust fund every year after the 20th year until closure begins. Whenever the post-closure cost estimate changes during this period in accordance with § 265.144 (b) or (c), the owner or operator must compare the new estimate with the latest annual value of the fund. If the value of the fund is less than the amount of the adjusted post-closure cost estimate, the owner or operator must deposit cash or marketable securities into the fund so that its value equals the amount of the estimate. Such payment must be made within 60 days of the change in the post-closure cost estimate. If the value of the fund is greater than the total amount of the adjusted post-closure estimate, the owner or operator may submit a written request to the Regional Administrator for release of funds in excess of the estimate. This request must be accompanied by a written statement from the trustee confirming the value of the fund.

(12) Within 30 days after receiving a request from the owner or operator for release of excess funds as specified in paragraphs (a)(8) and (11), the Regional Administrator must direct the trustee in writing to release such excess funds to the owner or operator unless the Regional Administrator finds that the post-closure cost estimate was not prepared and adjusted in accordance

with § 265.144.

(13) An owner or operator may request reimbursement for post-closure expenditures by submitting itemized bills to the Regional Administrator. Within 30 days after receiving the bills for post-closure activities, the Regional Administrator must direct the trustee in writing to pay those bills which the Regional Administrator determines to be in accordance with the post-closure plan or are otherwise justified.

(14) If an owner or operator substitutes another form of financial assurance specified in this section for all or part of the trust fund, he may apply to the Regional Administrator for release of funds from the trust fund. Within 30 days after receiving such a request, the Regional Administrator must direct the trustee in writing to release the excess funds to the owner or operator.

(15) Reversion of excess funds after closure.

(i) If, under the provisions of § 265.117(d), the Regional Administrator follows termination or reduction of some or all of the requirements of a post-closure plan before the end of the 30-year period, the excess portion of the trust fund must be released by the Regional Administrator.

(ii) At the end of the post-closure care period or the end of 30 years of post-closure care, whichever comes earlier, the Regional Administrator must direct the trustee to release any funds remaining in the trust to the owner or operator.

(16) The terms of the trust must require the trustee to make disbursements as specified in this paragraph. The trustee will disburse monies from the trust fund to parties designated by the Regional Administrator upon written notification from the Regional Administrator that:

(i) The value of the trust fund during the operating life of the facility exceeds the amount of the adjusted post-closure cost estimate; or

(ii) The itemized bills are in accordance with the approved post-closure plan or are otherwise justified; or

(iii) The owner or operator has established other financial assurance for post-closure care as allowed in this section for part or all of the trust fund; or

(iv) There has been a legal determination, a copy of which is attached to this notification, of a violation of the post-closure requirements of these regulations rendered in a proceeding brought pursuant to Section 3008 of RCRA; or

(v) The post-closure care period has ended or the requirements for post-closure care have been reduced.

(b) *Surety bond guaranteeing a lump-sum payment for post-closure care.* (1) An owner or operator may meet the requirements of this section by obtaining a surety bond guaranteeing a lump-sum payment into a post-closure trust fund. A surety company issuing a bond in accordance with these regulations must, at a minimum, be authorized to do business in the United States and be certified by the U.S. Treasury Department, in Circular 570, to write bonds in the penal sum of the bond to be issued. The obligee of the bond must be the U.S. Environmental Protection Agency.

(2) The bond must be executed on EPA Form 8700–20 (see Appendix VII). The terms of the bond must provide that the surety will send the properly executed bond to the Regional Administrator by certified mail within 10 days after the effective date of the bond.

(3) Such surety bond must guarantee that the owner or operator will, within 30 days after the beginning of closure of the facility, pay a lump sum equal to the final post-closure cost estimate prepared in accordance with § 265.144 into a trust fund that complies with the provisions of paragraph (a) of this section. The surety bond must be written so that whenever closure activities begin or are ordered to begin by the Regional Administrator during the term of the bond, the bond coverage includes completion of the payment obligation guaranteed by the bond.

(4) If the post-closure cost estimate increases beyond the amount of the penal sum of the bond, the owner or operator must, within 30 days of such increase in the estimate, cause the penal sum of the bond to be increased or obtain other financial assurance, as specified in this section, to cover the increase. If the post-closure cost estimate decreases, the penal sum of the bond may be reduced to the amount of the adjusted post-closure cost estimate. At the request of the owner or operator, the Regional Administrator must send written notice to the surety of any reduction in the required penal sum within 30 days after receiving the request.

(5) The terms of the surety bond must provide that the surety company may cancel the bond by sending notice to the owner or operator and to the Regional Administrator by certified mail. Cancellation must not be effective for at least 90 days after the Regional Administrator receives the notice. The owner or operator, within 5 days of receiving a notice of cancellation from the surety, must notify the Regional Administrator by certified mail that he

has received such a notice. The owner or operator may cancel the bond by providing 30 days' notice to the surety company if the Regional Administrator has given prior written consent based on his having received evidence of other financial assurance as specified in this section.

(6) Thirty days after receiving a notice of cancellation from the surety, the Regional Administrator may order the owner or operator to begin closure unless the Regional Administrator has received evidence of other financial assurance as specified in this section.

(7) A surety becomes liable on a bond obligation only when the owner or operator fails to perform as guaranteed by the bond and fails to provide other financial assurance of post-closure care as specified in this section.

(8) The Regional Administrator must notify the surety in writing within 60 days after the beginning of closure that the owner or operator has:

(i) Established financial assurance for post-closure care that satisfies the requirements of this section; or

(ii) Failed to fulfill the payment obligation guaranteed by the bond. The Regional Administrator will then direct the surety in the placement of funds in a trust fund meeting the specifications of paragraph (a) of this section.

(c) *Standby letter of credit assuring a lump-sum payment at the time of closure for post-closure care.* (1) An owner or operator may meet the requirements of this section by obtaining an irrevocable standby letter of credit assuring a lump-sum payment at the time of closure to provide for post-closure care. The letter must be written in favor of the Regional Administrator of the U.S. Environmental Protection Agency and must be for a period of at least one year. The letter of credit may be issued by any bank which is a member of the Federal Reserve System.

(2) The letter of credit must be executed on EPA Form 8700–17 (see Appendix IV). The terms of the letter must provide that the issuing bank will send the properly executed letter of credit to the Regional Administrator by certified mail within 10 days after the effective date of the letter.

(3) The credit must be issued for an amount equal to the adjusted post-closure cost estimate (see § 265.144).

(4) If the post-closure cost estimate increases beyond the amount of the credit, the owner or operator must, within 30 days of such increase in the estimate, cause the credit to be increased or obtain other financial assurance, as specified in this section, to cover the increase. If the post-closure cost estimate decreases, the credit may

be reduced to the amount of the adjusted post-closure cost estimate. At the request of the owner or operator, the Regional Administrator must send written notice to the issuing bank of any reduction in the required credit within 30 days after receiving the request.

(5) The letter of credit must contain a clause providing for automatic annual extensions of the credit subject to 60 days' written notice by the issuing bank to both the owner or operator and the Regional Administrator, by certified mail, of the bank's intention not to renew the credit. The owner or operator, within 5 days of receiving a notice of nonrenewal from the bank, must notify the Regional Administrator by certified mail that he has received such a notice. The owner or operator may cancel the letter of credit by providing 30 days' notice to the issuing bank if the Regional Administrator has given prior written consent based on his having received evidence of other financial assurance as specified in this section.

(6) Thirty days after receiving a notice of nonrenewal from the bank, the Regional Administrator may draw upon the credit up to the full amount of the credit unless he has evidence that the owner or operator has established other financial assurance as specified in this section. The terms of the letter must provide that if the Regional Administrator draws upon the letter of credit following a notice of nonrenewal the issuing bank will deposit the amount of the draft immediately and directly into an interest-bearing escrow account. Disbursements from the escrow account must be made in the same manner as specified for trust funds in paragraphs (a)(12)–(16) of this section.

(7) If the post-closure cost estimate increases beyond the amount of the funds in the escrow account, the owner or operator must, within 30 days of such increase, add to the account or establish other financial assurance as specified in this section to cover the increase. If the owner or operator fails to do so, the Regional Administrator may order him to begin closure.

(8) The Regional Administrator may otherwise draw on the credit only if the owner or operator fails to establish, within 30 days after the beginning of closure, other financial assurance for post-closure care as specified in this section. The issuing bank must, under the terms of the letter, deposit the amount of such a draft immediately and directly into an interest-bearing escrow account. Disbursements from the escrow account must be made in the same manner as specified for trust funds in paragraphs (a)(13)–(16) of this section.

(d) *Surety bond guaranteeing*

performance of post-closure duties. (1) An owner or operator may meet the requirements of this section by obtaining a surety bond guaranteeing performance of post-closure care. A surety company issuing a bond in accordance with these regulations must, at a minimum, be authorized to do business in the United States and be certified by the U.S. Treasury Department, in Circular 570, to write bonds in the penal sum of the bond to be issued. The obligee of the bond must be the U.S. Environmental Protection Agency.

(2) The bond must be executed on EPA Form 8700–21 (see Appendix VIII). The terms of the bond must provide that the surety will send the properly executed bond to the Regional Administrator by certified mail within 10 days after the effective date of the bond.

(3) The surety bond must guarantee that the owner or operator will satisfy the post-closure care requirements of these regulations for 30 years or for the post-closure care period, whichever period is shorter. The surety bond must be written in the amount of the adjusted post-closure cost estimate (see § 265.144).

(4) If the post-closure cost estimate increases beyond the amount of the penal sum of the bond, the owner or operator must, within 30 days of such increase in the estimate, cause the penal sum of the bond to be increased or obtain other financial assurance, as specified in this section, to cover the increase. If the post-closure cost estimate decreases, the penal sum of the bond may be reduced to the amount of the adjusted post-closure cost estimate. At the request of the owner or operator, the Regional Administrator must send written notice to the surety of any reduction in the required penal sum within 30 days after receiving the request.

(5) Under the terms of the bond, the surety company may cancel the bond during the operating life of the facility by sending notice to the Regional Administrator and to the owner or operator by certified mail. Cancellation must not be effective for at least 90 days after the Regional Administrator receives the notice. The owner or operator, within 5 days of receiving notice of cancellation from the surety, must notify the Regional Administrator by certified mail that he has received such a notice. The owner or operator may cancel the bond at any time by providing 30 days' notice to the surety company if the Regional Administrator has given prior written consent based on his having received evidence of other

financial assurance as specified in this section.

(6) Thirty days after receiving a cancellation notice from the surety, the Regional Administrator may order the owner or operator to begin closure unless the Regional Administrator has received evidence of other financial assurance as specified in this section.

(7) The surety bond must be written so that whenever closure activities begin or the Regional Administrator orders them to begin during the term of the bond, the bond coverage extends to the end of 30 years of post-closure care or to the end of the post-closure care period, whichever is shorter. The owner or operator, as the principal of the bond, must notify the surety of the date on which post-closure care begins in accordance with the post-closure plan for the facility.

(8) As post-closure obligations are completed, the penal sum of the bond may be reduced commensurately, so that the balance of the penal sum of the bond will equal the remaining cost obligations of the owner or operator for post-closure care. At the request of the owner or operator, the Regional Administrator must send written notice to the surety of any reduction in the required penal sum within 30 days after receiving the request.

(9) A surety becomes liable on a bond obligation only when a proceeding brought pursuant to the provisions of Section 3008 of RCRA has determined that the owner or operator has violated the post-closure requirements of these regulations. Following such a determination the surety must:

(i) Complete post-closure care of the facility in accordance with the post-closure plan; or

(ii) Pay the amount of the penal sum of the bond into a trust fund meeting the specifications of paragraph (a) of this section as directed by the Regional Administrator.

(e) *Standby letter of credit assuring funds during the post-closure period.* (1) An owner or operator may meet the requirements of this section by obtaining an irrevocable standby letter of credit assuring availability of funds during the post-closure period. The letter must be written in favor of the Regional Administrator of the U.S. Environmental Protection Agency and must be for a period of at least one year. The letter of credit may be issued by any bank which is a member of the Federal Reserve System.

(2) The letter of credit must be executed on EPA Form 8700–17 (see Appendix IV). The terms of the letter must provide that the issuing bank will send the properly executed letter of

credit to the Regional Administrator by certified mail within 10 days after the effective date of the letter.

(3) The credit must be issued for the amount of the adjusted post-closure cost estimate (see § 265.144).

(4) If the post-closure cost estimate increases beyond the amount of the credit, the owner or operator must, within 30 days of such increase in the estimate, cause the amount of the credit to be increased or obtain other financial assurance, as specified in this section, to cover the increase. If the post-closure cost estimate decreases, the amount of the credit may be reduced to the amount of the adjusted post-closure cost estimate. At the request of the owner or operator, the Regional Administrator must send written notice to the surety of any reduction in the required credit within 30 days after receiving the request.

(5) As post-closure obligations are completed, the credit guarantee may be reduced commensurately, so that the remaining credit will equal the remaining cost obligations of the owner or operator for post-closure care. At the request of the owner or operator, the Regional Administrator must send written notice to the bank of any reduction in the required credit guarantee within 30 days after receiving the request.

(6) The letter of credit must contain a clause providing for automatic annual extensions of the credit subject to 60 days' written notice by the issuing bank to both the owner or operator and the Regional Administrator, by certified mail, of the bank's intention not to renew the credit. The owner or operator, within 5 days of receiving a notice of nonrenewal from the bank, must notify the Regional Administrator by certified mail that he has received such a notice. The owner or operator may cancel the letter of credit by providing 30 days' notice to the issuing bank if the Regional Administrator has given prior written consent based on his having received evidence of other financial assurance as specified in this section.

(7) Thirty days after receiving a notice of nonrenewal from the bank, the Regional Administrator may draw upon the credit up to the full amount of the credit unless he has received evidence that the owner or operator has established other financial assurance as specified in this section. The terms of the letter must provide that if the Regional Administrator draws upon the letter of credit following a notice of nonrenewal, the issuing bank will deposit the amount of the draft immediately and directly into an interest-bearing escrow account.

Disbursements from the escrow account must be made in the same manner as specified for trust funds in paragraphs (a)(12)–(16) of this section.

(8) If the escrow account specified in paragraph (e)(7) of this section is established during operating life, and if the post-closure cost estimate increases beyond the amount of the funds in the escrow account, the owner or operator must, within 30 days of such increase, add to the account or establish other financial assurance as specified in this section to cover the increase. If the owner or operator fails to do so, the Regional Administrator may order him to begin closure.

(9) The Regional Administrator may otherwise draw upon the letter of credit only upon a legal determination of a violation of the post-closure requirements of these regulations rendered in a proceeding brought pursuant to the provisions of Section 3008 of RCRA. The terms of the letter must provide that if the Regional Administrator draws upon the letter of credit following such a determination, the issuing bank will immediately and directly deposit the amount of the draft into an interest-bearing escrow account. The letter of credit must require the escrow depositary to disburse monies from the escrow account to persons designated by the Regional Administrator to carry out post-closure care of the facility.

(f) *Use of more than one type of financial instrument.* An owner or operator may meet the requirements of this section by establishing more than one type of financial instrument. These instruments are limited to a trust fund, surety bonds, or letters of credit as specified in paragraphs (a) through (e) of this section (e.g., a letter of credit may assure half the post-closure cost and a trust fund the remaining half).

(g) *Financial test and guaranty for post-closure care.* (1) An owner or operator may meet the requirements of this section by having all of the following financial characteristics:

(i) At least $10 million in net worth in the United States.

(ii) A total-liabilities-to-net-worth ratio of not more than three.

(iii) Net working capital in the United States of at least twice the adjusted post-closure cost estimate (see § 265.144).

(2) These characteristics must be demonstrated in a financial statement which has been audited by an independent certified public accountant and which contains unconsolidated balance sheets dated no more than 140 days prior to the current date. The owner or operator who intends to use a financial test to meet both closure and

post-closure requirements for a single facility or to meet closure and/or post-closure requirements for more than one facility must indicate in the statement which requirements are to be met for which facilities through the financial test and must demonstrate that his net working capital in the United States is at least twice the sum of all the adjusted estimates of closure and post-closure costs to be covered by the financial test. The owner or operator must have the financial statement available at the facility and must provide data from the statement if requested as part of annual reports to the Regional Administrator under § 265.75.

(3) If the owner or operator fails to meet the requirements of paragraph (g)(1) of this section at any time before the end of the post-closure care period or 30 years of post-closure care, whichever comes earlier, he must notify the Regional Administrator by certified mail within 5 days of learning of failure to meet the requirements. Evidence of other financial assurance as specified in this section must be sent to the Regional Administrator by certified mail within 30 days from the time that the owner or operator learns of failure to meet the requirements of paragraph (g)(1). If he does not establish other financial assurance, and this lapse in financial assurance occurs during operating life, the Regional Administrator may order the owner or operator to begin closure.

(4) An owner or operator may meet the requirements of this section by obtaining another entity's written guaranty providing financial assurance, in an amount equal to the adjusted post-closure cost estimate, for compliance by the owner or operator with the post-closure requirements of these regulations. The guarantor must meet the requirements for owners or operators in paragraphs (g) (1) and (2) of this section.

(5) The guaranty must be executed on EPA Form 8700–18 (see Appendix V). The owner or operator must send the properly executed guaranty to the Regional Administrator by certified mail within 10 days after the effective date of the guaranty.

(6) Under the terms of the guaranty, the guarantor must notify the Regional Administrator and the owner or operator by certified mail if he fails to meet the requirements of paragraph (g)(1) of this section at any time before the end of the post-closure period or the end of 30 years of post-closure care, whichever comes earlier. The guarantor must send such notice within 5 days after learning of failure to meet the requirements.

(7) The owner or operator must, within 30 days of such notification,

establish other financial assurance as specified in this section and provide evidence of such assurance to the Regional Administrator. If he fails to do so, and such failure occurs during operating life, the Regional Administrator may order him to begin closure.

(8) The guarantor may cancel the guaranty during the operating life of the facility with 90 days' notice to the Regional Administrator and the owner or operator by certified mail, except that the guaranty must remain in effect if closure begins or is ordered to begin by the Regional Administrator before the end of the 90 days. Evidence of other financial assurance as specified in this section must be provided to the Regional Administrator within 30 days after a notice of cancellation is received by the Regional Administrator; otherwise, he may order the owner or operator to begin closure.

(9) The guaranty may be cancelled at any time following the mutual written consent of the owner or operator, the Regional Administrator, and the guarantor.

(10) Under the guaranty, in the event of a legal determination of a violation of the post-closure requirements rendered in a proceeding brought pursuant to Section 3008 of RCRA, the guarantor must pay parties designated by the Regional Administrator to complete post-closure care for 30 years or the post-closure care period, whichever period is shorter.

(h) *Revenue test for municipalities.* (1) If the owner or operator is a municipality (as defined by RCRA), it may meet the requirements of this section by having annual revenues from property, sales, and/or income taxes equal to 10 times the adjusted post-closure cost estimate (see § 265.144). To be acceptable, these tax revenues must be legally available to cover post-closure responsibilities, i.e., they must not be dedicated to other purposes or otherwise precluded from use for post-closure care.

(2) The owner or operator must send a letter signed by the chief financial officer of the municipality to the Regional Administrator stating that the municipality meets the requirements of paragraph (h)(1) of this section. The letter must be sent by certified mail within 10 days after the owner or operator begins use of the revenue test to meet the requirements of this section.

(3) If the annual tax revenues fail to meet the minimum multiple specified in paragraph (h)(1) at any time before the end of the post-closure care period or 30 years of post-closure care, whichever comes earlier, the owner or operator must notify the Regional Administrator

by certified mail within 5 days of learning of failure to meet the requirements. The owner or operator must send evidence of other financial assurance as specified in this section to the Regional Administrator by certified mail within 30 days from the time that the owner or operator learns of failure to meet the minimum multiple. If he does not establish other financial assurance, and this lapse in financial assurance occurs during operating life, the Regional Administrator may order the owner or operator to begin closure.

(i) *Use of a single financial mechanism for multiple facilities.* An owner or operator may use a single financial mechanism, as specified in paragraphs (a) through (h) of this section, to meet the requirements of this section for more than one facility of which he is the owner or operator. The amount of funds available through the mechanism must be no less than the sum of funds that would be available if a separate mechanism had been established for each facility.

§ 265.146 Use of a single mechanism for financial assurance of both closure and post-closure care.

An owner or operator may use a single mechanism to provide financial assurance for both closure and post-closure care of one or more facilities of which he is the owner or operator. Such a mechanism must be one of the following:

(a) A trust fund that meets the specifications of both § 265.143(a) and § 265.145(a).

(b) A surety bond that meets the specifications of both § 265.143(b) and § 265.145 (b) or (d).

(c) A letter of credit that meets the specifications of both § 265.143(c) and § 265.145 (c) or (e).

(d) A guaranty that meets the specifications of both § 265.143(e) and § 265.145(g).

(e) The financial test as specified under both § 265.143(e) and § 265.145(g).

(f) The revenue test as specified under both § 265.143(f) and § 265.145(h).

The amount of funds available under the mechanism must be no less than the sum of funds that would be available if a separate mechanism had been established for financial assurance of closure and of post-closure care of each facility.

§ 265.147 Liability requirement.

An owner or operator of a hazardous waste treatment, storage, or disposal facility or group of facilities must have and maintain liability insurance from an

insurer licensed or eligible to insure facilities in the jurisdiction where any one facility is located, for sudden and accidental occurrences in the amount of $1 million per occurrence with an annual aggregate per firm of $2 million, exclusive of legal defense costs, for claims arising out of injury to persons or property from the operations of each such hazardous waste facility or group of facilities. The deductible written into the insurance policy must not exceed 5 percent of the per incident limit of liability of the policy.

§ 265.148 [Reserved]

§ 265.149 Applicability of State financial requirements.

(a) A facility may be located in a State in which existing hazardous waste regulations include liability requirements and requirements for financial assurance for closure and post-closure care. If so, the owner or operator may use existing State-authorized financial mechanisms in meeting the requirements of §§ 265.143, 265.145, and 265.147 provided that:

(1) The State-authorized mechanism is a mechanism allowed in §§ 265.143, 265.145, or 265.147; or

(2) The State mechanism provides substantially equivalent assurance (e.g., escrow account) or liability coverage as the mechanisms of §§ 265.143, 265.145, and 265.147.

The owner or operator must obtain an additional financial assurance mechanism for closure or for post-closure care, chosen from § 265.143 for closure and from § 265.145 for post-closure care, or additional liability insurance as specified in § 265.147, if the amount of funds available from the State mechanisms is less than that required by this Subpart. The total amount of funds available through the combination of the State and Federal mechanisms must equal at least the amount required in §§ 265.143, 265.145, and 265.147.

(b) If a State assumes legal responsibility for an owner's or operator's compliance with the closure or post-closure requirements or liability requirements of these regulations *or* assures that funds will be available from State sources to cover such requirements, the owner or operator will be in compliance with such requirements of this Subpart to the extent the State's assurances are substantially equivalent to meeting the requirements of this Subpart. The owner or operator must send a letter to the Regional Administrator describing the nature of the State's responsibility regarding his facility's closure, post-closure care, and/or his liability, and citing the State regulation providing for such assumption of responsibility. The letter must be sent by certified mail within 10 days after the effective date of these EPA regulations or the date on which State assumption of responsibility for the facility becomes effective. A copy of the letter must be sent to the responsible State agency(ies).

Appendix I to Part 265

The following is an example of the calculation in § 265.143(a)(7) using these assumptions: The closure cost estimate at the time the closure trust fund was established was $70,000. Five annual payments have been made. The current value of the fund is $25,000 (including earnings of the fund and yearly increases in the payments as a result of the adjustment for inflation required by paragraph (a)(6)). The total pay-in period is 20 years. Now the owner or operator has changed the estimate to $120,000 because of a change in the closure plan and therefore needs to recalculate his next payment.

Step 1—The adjusted estimate, $120,000, divided by the pay-in period, 20 years, is $6,000.

Step 2—$6,000 multiplied by the number of payments made, 5, is $30,000.

Step 3—$30,000 minus the current value of the fund, $25,000, is $5,000.

Step 4—$5,000 divided by the remaining years in the pay-in period, 15, is $333.

Step 5—Adding $333 to the $6,000 from Step 1 gives the new payment, $6,333.

Appendix II to Part 265

EPA Form 8700–15

U.S. Environmental Protection Agency

Closure Trust Agreement

As provided for in 40 CFR 265.143(a) under authority of the Resource Conservation and Recovery Act of 1976, as amended (42 USC 6901)

[1] EPA Facility Identification No. ————— Adjusted closure cost estimate, in accordance with 40 CFR 265.142: $ —————

On this —— day of —————, 19——, I (owner or operator) —————, am placing property described below in trust for the U.S. Environmental Protection Agency (EPA) to be held by (name of financial institution) ————— as trustee under the terms set forth below. The trust shall be named the "Closure Trust" for the following hazardous waste management facilities:

—————————————————————

(name and address of facility, or write in "see attached Schedule A" [1] if more than one facility).

1. Purpose Clause

Pursuant to the financial assurance requirements of 40 CFR 265.143, the purpose of this trust is to pay for the costs of closing the above-named facility(ies) in accordance

—————

[1] If closure of more than one facility is covered by the trust, list on a separate sheet the EPA Facility Identification Numbers, names, and addresses, and adjusted closure cost estimates for all the facilities, clearly label this list "Schedule A," and attach it to this agreement. Show total of cost estimates.

[2] List property included in initial transfer on separate sheet, clearly label this list "Schedule B," and attach it to this agreement.

3. Period Clause

This trust shall continue until terminated upon the happening of one of the following conditions:

(a) When (owner or operator) ————— presents to the trustee the original or an with the closure requirements of 40 CFR Part 265.

2. Property Clause

It is agreed to by (owner or operator) ————— as grantor of this trust that the trust will be funded in accordance with the requirements of § 265.143(a) of the regulations. The initial transfer of property to the trust shall consist of the property listed in Schedule B, attached hereto. [2] authenticated copy of the letter(s) signed by the EPA Regional Administrator(s) stating that he is no longer required to provide financial assurance for closure of the above-named facility(ies). In such an event, all remaining trust property, less final trust administration expenses, shall be delivered to (owner or operator) —————.

(b) By the mutual written consent of the grantor of this trust, the EPA Regional Administrator(s) of the Region(s) in which the facility(ies) is (are) located, the trustee of this trust at any time.

4. Operation of the Trust, Duties of the Trustee

(name of financial institution acting as trustee) ————— acknowledges below its receipt of the trust property listed in Schedule B and its acceptance of the obligations and duties of the trustee as defined below.

(a) The trustee agrees to notify the EPA Regional Administrator(s) by certified mail within five days following the expiration of the thirty-day period after the anniversary of the establishment of the trust, as specified in § 265.143(a)(5).

(b) The trustee may resign from its obligations as trustee by submitting a written notice of its intent to the grantor and to the EPA Regional Administrator(s).

(c) The trustee is to make payments out of the trust only under the conditions specified in 40 CFR 265.143(a)(15).

—————————————————————

(date)   (signature of grantor)

—————————————————————

(address of grantor) —————————

(authorized signature for trustee) —————

(name of trustee) —————————————

(address of trustee) ————————————

(signature of notary) ————————————

Mail original to the EPA Regional Administrator within 10 days of the effective date by certified mail. If more than one facility is covered and the facilities are in more than one Region, send original to Regional Administrator of Region in which the largest number of facilities are located and copies to the other Regional Administrator(s), by certified mail.

Appendix III to Part 265

EPA Form 8700–16

U.S. Environmental Protection Agency

Closure Performance Bond

As provided for in 40 CFR 265.143(b) under authority of the Resource Conservation and Recovery Act of 1976, as amended (42 USC 6901)

[1] EPA Facility Identification No. ————— Adjusted closure cost estimate, in accordance

with 40 CFR 265.142: $ —————

Know all men by these presents, that we, (owner of operator) ————— of (address) —————, as Principal and (name of surety company) —————, a company created and existing under the laws of (State) —————, as Surety, are held and firmly bound unto the U.S. Environmental Protection Agency (EPA) in the penal sum of ————— U.S. dollars ($—————) for payment of which, well and truly to be made, we bind ourselves, our heirs, executors, administrators, successors and assigns, jointly and severally, and firmly by these presents.

Whereas, the Principal intends to obtain interim status, as defined by Section 3005 of the Resource Conservation and Recovery Act of 1976, as amended, for one or more hazardous waste management facilities, and such status depends upon compliance with the standards of 40 CFR Part 265, which includes the requirement, specified in § 265.143, that the owner or operator of each such facility must establish financial assurance that the applicable closure requirements of Part 265 will be met, and

Whereas, this bond is written to assure compliance with the closure requirements of Part 265 for the following hazardous waste management facilities: (name and address of facility or write in "see attached Schedule A" [1] if more than one facility) —————, and shall inure to the benefit of EPA in accordance with Part 265,

Now, therefore, the condition of this obligation is such that, if the Principal shall faithfully fulfill the closure requirements of 40 CFR Part 265 at each of the facilities guaranteed by this bond, pursuant to all applicable statutes, rules and regulations, and shall close each such facility in accordance with the closure plan required by the said Part 265, then, and only then, the above obligation shall be void; otherwise to be and to remain in full force and effect.

The Surety shall become liable on this bond obligation only upon legal determination rendered in a proceeding brought pursuant to Section 3008 of the Resource Conservation and Recovery Act, as amended, that the Principal has violated the closure requirements of 40 CFR Part 265. Following such a determination, the Surety must either complete closure of the facility in accordance with the approved closure plan for the facility or pay the amount of the penal sum into an escrow account as directed by an EPA Regional Administrator.

The liability of the Surety shall not be discharged by any payment or succession of payments hereunder, unless and until such payment or payments shall amount in the aggregate to the penal sum of the bond, but in no event shall the Surety's obligation hereunder exceed the amount of said penal sum. The insolvency or bankruptcy of the Principal shall not constitute a defense to the Surety with regard to claims of liability on the bond obligations, and in the event of said

—————

[1] If closure of more than one facility is covered by the bond, list on a separate sheet the EPA Facility Identification Numbers, names, addresses, and adjusted closure cost estimates for all the facilities, clearly label this list "Schedule A," and attach it to this bond. Show total of cost estimates.

insolvency or bankruptcy, the Surety must pay any unsatisfied final judgments obtained on such claims. The Surety agrees to furnish written notice forthwith to the Regional Administrator(s) of the EPA Region(s) in which the facility(ies) is (are) located of all suits filed, judgments rendered, and payments made by the Surety under this bond.

This bond is effective the ———— day of ————, 19——, at the address of the Principal as stated herein and shall continue in force until terminated as hereinafter provided. The Surety may terminate this bond by written notice sent by certified mail to the Principal and to the EPA Regional Administrator(s) of the Region(s) in which the facility(ies) is (are) located, such termination to become effective ninety (90) days after actual receipt of said notice by EPA; provided, however, no such termination shall become effective with respect to any facility closure guaranteed by this bond if closure of said facility has begun or has been ordered to begin by an EPA Regional Administrator. The Principal may terminate this bond by sending written notice to the Surety, such termination to become effective thirty (30) days after receipt of such notice by the Surety; provided, however, that such notice is accompanied by written authorization for termination of the bond by the Regional Administrator(s) of the EPA Region(s) in which the bonded facility(ies) is (are) located.

If more than one surety company joins in executing this bond, such action shall constitute joint and several liability on the part of the sureties.

In witness whereof, the Principal and Surety have executed this instrument on the ———— day of ————, 19——.
(Seal) ————————————————
(Surety)
(Seal) ————————————————
(Principal)
(Seal) ————————————————
(attorney-in-fact)   (address of
Principal) ——————————————

Surety Bond No. ————

Mail original to the EPA Regional Administrator within 10 days of the effective date by certified mail. If more than one facility is covered and the facilities are in more than one Region, send original to Regional Administrator of Region in which the largest number of facilities are located and copies to the other Regional Administrator(s), by certified mail.

Appendix IV to Part 265

EPA Form 8700–17

U.S. Environmental Protection Agency

Standby Letter of Credit

As provided for in 40 CFR 265.143(c), 265.145(c), and 265.145(e) under authority of the Resource Conservation and Recovery Act of 1976, as amended (42 USC 6901)
[1]EPA Facility Identification No. ————
Adjusted cost estimate(s) for the facility, for

[1]If more than one facility is covered by this Letter of Credit, list on a separate sheet the EPA Facility Identification Numbers, names, addresses, and adjusted closure and/or post-closure cost estimates for all the facilities, clearly label this list "Schedule A," and attach it to this Letter of Credit. Show total(s) of cost estimates.

closure and/or post-closure care to be covered by this Letter of Credit, in accordance with 40 CFR 265.142 and 265.144: $———— (closure) $———— (post-closure)

Administrator(s) for Region(s)————
U.S. Environmental Protection Agency
Address(es) ————
(Address to EPA Regional Administrator(s) of Region(s) in which the facility(ies) is (are) located.)

Dear Sir or Madam: We hereby establish our Irrevocable Standby Letter of Credit No. ————, in favor of the Regional Administrator(s) for Region(s) —— of the U.S. Environmental Protection Agency for the account of (owner or operator) ———— up to the aggregate amount of ———— U.S. dollars ($———) available by your drafts as specified below.

This Letter of Credit is effective as of today's date and will expire on the ———— day of ———— 19——, subject to the operation of the renewal clause below.

The purpose of this Letter of Credit is to provide financial assurance to the U.S. Environmental Protection Agency of compliance with the ("closure," "post-closure," or "closure and post-closure") ———— requirements of 40 CFR Part 265 as they apply to (name and address of facility, or write in "see attached Schedule A" [1] if more than one facility) ————. Such assurance is required for closure by 40 CFR 265.143 and for post-closure care by 40 CFR 265.145. This Letter of Credit provides assurance for (check those that apply):
—Closure in accordance with the letter-of-credit specifications of 40 CFR 265.143(c)
—A lump-sum payment at closure for the purpose of assuring post-closure care in accordance with letter-of-credit specifications of 40 CFR 265.145(c)
—Funds for the performance of post-closure care in accordance with letter-of-credit specifications of 40 CFR 265.145(e)

All drafts on this Letter of Credit submitted in writing and accompanied by your signature will be promptly paid and deposited in an interest-bearing escrow account in this Bank. If a draft on the escrow account is accompanied by a copy of an order from a Federal Administrative Law Judge or a Federal District Court Judge setting forth a determination of a violation of the above-mentioned closure and/or post-closure requirements, we will pay the party or parties designated by the court or the EPA Regional Administrator(s).

Alternatively, payments may be made out of any amount in escrow following a draft upon this Letter of Credit by the mutual written consent of (owner or operator) ———— and the EPA Regional Administrator(s), pursuant to 40 CFR 265.143(c)(6), 265.145(c)(6) and (8), or 265.145(e)(7), as applicable.

It is a condition of this Letter of Credit that it will be automatically extended for one-year periods from the expiration date set forth above, unless sixty (60) days before that date we notify you by certified mail of our intent not to renew the credit. In that case, for the remainder of the period of the Letter of Credit, you may draw upon the credit up to the aggregate amount of the credit remaining, such draft to be deposited in escrow as

described above. This Letter of Credit may be terminated by (owner or operator) ———— by sending written notice to this Bank, such termination to become effective thirty (30) days after receipt of such notice by this Bank; provided, however, that such notice is accompanied by your written authorization for termination of the Letter of Credit.

This Letter of Credit is subject to Article Five of the Uniform Commercial Code and the "Uniform Customs and Practices for Documentary Credits" (1974 Revision) described in International Chamber of Commerce Brochure No. 290.

All communications concerning this Letter of Credit are to be addressed to: (name and address of responsible officer of the issuing bank) ————.

————————————————————
(date)   (authorized signature)
(print or type name of person signing) ————
(title of person signing) ————————
(name of bank)————————————
Mail to the EPA Regional Administrator(s) within 10 days of the effective date by certified mail.

Appendix V to Part 265

EPA Form 8700–18

U.S. Environmental Protection Agency

Guaranty

As provided for in 40 CFR 265.143(e) and 265.145(g), under authority of the Resource Conservation and Recovery Act of 1976, as amended (42 USC 6901)
[1]EPA Facility Identification No.———— —
Adjusted cost estimates(s) for the facility, for closure and/or post-closure care to be covered by this guaranty, in accordance with 40 CFR 265.142 and 265.144: $————(closure) $———— (post-closure)

Guaranty made this ———— day of ————, 19——, by (name of guaranteeing entity) ————, a business entity organized under the laws of the State of ————, with its principal office at ————, herein referred to as guarantor, to the U.S. Environmental Protection Agency (EPA) as obligee on behalf of (owner or operator) ———— of (business address) ————.

Recitals

1. Guarantor meets or exceeds the financial test requirements of 40 CFR 265.143(e) and/or 265.145(g). Guarantor agrees to notify the EPA Regional Administrator(s) for the Region(s) in which the facility(ies) listed below is (are) located and (owner or operator) ———— within five days after the guarantor learns of its failure to meet any of the test requirements at any time during the life of this guaranty.

2. (Owner or operator) ———— operates or owns a hazardous waste facility at (address of facility, or write in "see attached Schedule A" [1] if more than one facility is covered) ————.

[1]If more than one facility is covered by this guaranty, list on a separate sheet the EPA Facility Identification Numbers, names, addresses, and the adjusted closure and/or post-closure estimates for all the facilities, clearly label this list "Schedule A," and attach it to this guaranty. Show total(s) of cost estimates.

Statement of Guaranty

For value received from (owner or operator) ————, the guarantor guarantees to the U.S. Environmental Protection Agency (EPA) that in the event that (owner or operator) ————, fails to comply with the ("closure," "post-closure," or "closure and post-closure") ———— requirements of 40 CFR part 265 applicable to (name and address of facility or write in "see attached Schedule A") ————, the guarantor agrees to pay the persons(s) designated by EPA or to pay EPA itself, following a legal determination of a violatio₁ of the regulations, an amount sufficient to bring the above-mentioned facility(ies) into compliance with the applicable regulations, but not to exceed the adjusted cost estimate(s) as prepared in accordance with 40 CFR 265.142 and 265.144.

This guaranty is good for so long as (owner or operator) ———— must comply with the applicable financial assurance requirements of 40 CFR 265.143 and 265.145 for the above-named facility(ies).

The guarantor may terminate this guaranty by sending notice by certified mail to the EPA Administrator(s) for the Region(s) in which the facility(ies) is (are) located and to (owner or operator) ————, such termination to become effective ninety (90) days after actual receipt of the notice by EPA; provided, however, that no such termination shall become effective if closure begins or is ordered to begin by an EPA Regional Administrator before the end of the 90 days. Furthermore, if compliance with post-closure requirements is guaranteed, no such termination may become effective if closure has taken place.

This guaranty may be terminated at any time subject to the mutual, prior written consent of the guarantor, the EPA Regional Administrator(s) of the Region(s) in which the facility(ies) is (are) located, and (owner or operator) ————.

(effective date)————(name of guarantor)

(authorized signature for guarantor) ————
(print or type name of person signing) ————
(title of person signing) ————
(signature of witness or notary) ————
Mail original to the EPA Regional Administrator within 10 days of the effectivᵢ date by certified mail. If more than one facility is covered and the facilities are in more than one Region, send original to Regional Administrator of Region in which the largest number of facilities are located and copies to the other Regional Administrator(s), by certified mail.

Appendix VI to Part 265

EPA Form 8700–19

U.S. Environmental Protection Agency

Post-Closure Trust Agreement

As provided for in 40 CFR 265.145(a), undeɪ authority of the Resource Conservation and Recovery Act of 1976, as amended (42 USC 6901)
[1] EPA Facility Identification No. ————
Adjusted post-closure cost estimate, in accordance with 40 CFR 265.144: $ ————

On this ———— day of ————, 19 ——, I (owner or operator) ————, am placing property described below in trust for the U.S. Environmental Protection Agency (EPA) to be held by (name of financial institution) ———— as trustee under the terms set forth below. The trust shall be named the "Post-Closure Trust" for the following hazardous waste management facility(ies):

———————————————
(name and address of facility, or write in "see attached Schedule A" [1] if more than one facility).

1. Purpose Clause

Pursuant to the financial assurance requirements of 40 CFR 265.145, the purpose of this trust is to pay for the costs of post-closure care of the above-named facility(ies) in accordance with the post-closure requirements of 40 CFR Part 265.

2. Property Clause

It is agreed to by (owner or operator) ———— as grantor of this trust that the trust will be funded in accordance with the requirements of § 265.145(a) of the regulations. The initial transfer of property to the trust shall consist of the property listed in Schedule B, attached hereto.[2]

3. Period Clause

This trust shall continue until terminated upon the happening of one of the following conditions:

(a) Upon written notice(s) from the EPA Regional Administrator(s) that (owner or operator) ———— is no longer required to maintain financial assurance for post-closure care of the above-named facility(ies). In such an event, all remaining trust property, less final trust administration expenses, shall be delivered to (owner or operator) ————.

(b) By the mutual written consent of the grantor of this trust, the EPA Regional Administrator(s) of the Region(s) in which the facility(ies) is (are) located, the trustee of this trust at any time.

4. Operation of the Trust, Duties of the Trustee

(name of financial institution acting as trustee) ———— acknowledges below its receipt of the trust property listed in Schedule B and its acceptance of the obligations and duties of the trustee as defined below.

(a) The trustee agrees to notify the EPA Regional Administrator(s) by certified mail within five days following the expiration of the thirty-day period after the anniversary of the establishment of the trust, as specified in § 265.145(a)(5).

(b) The trustee may resign from its obligations as trustee by submitting written notice of its intent to the grantor and to the EPA Regional Administrator(s).

(c) The trustee is to make payments out of the trust only under the conditions specified in 40 CFR 265.145(a)(16).

———————————————
[1] If post-closure care of more than one facility is covered by the trust, list on a separate sheet the EPA Facility Identification Numbers, names, and addresses, and adjusted post-closure cost estimates for the facilities, clearly label this list "Schedule A," and attach it to this agreement. Show total of cost estimates.

[2] List property included in initial transfer on separate sheet, clearly label this list "Schedule B." and attach it to this agreement.

———————————————
(date)   (signature of grantor)

(address of grantor) ————
(authorized signature for trustee) ————
(name of trustee) ————
(address of trustee) ————
(signature of notary) ————
Mail original to the EPA Regional Administrator within 10 days of the effective date by certified mail. If more than one facility is covered and the facilities are in more then one Region, send original to Regional Administrator of Region in which the largest number of facilities are located and copies to the other Regional Administrator(s), by certified mail.

Appendix VII to Part 265

EPA Form 8700–20

U.S. Environmental Protection Agency

Bond for Payment to Post-Closure Care Trust Fund

As provided for in 40 CFR 265.145(b) under authority of the Resource Conservation and Recovery Act of 1976, as amended (42 USC 6901)
[1] EPA Facility Identification No. ————
Adjusted post-closure cost estimate, in accordance with 40 CFR 265.144: $ ————

Know all men by these presents, that we, (owner of operator) ———— of (address) ————, as Principal and (name of surety company) ————, a company created and existing under the laws of (State) ————, as Surety, are held and firmly bound unto the U.S. Environmental Protection Agency (EPA) in the penal sum of ———— U.S. dollars ($————) for payment of which, well and truly to be made, we bind ourselves, our heirs, executors, administrators, successors and assigns, jointly and severally, and firmly by these presents.

Whereas, the Principal intends to obtain interim status, as defined by Section 3005 of the Resource Conservation and Recovery Act of 1976, as amended, for one or more hazardous waste disposal facilities, and such status depends upon compliance with the standards of 40 CFR Part 265, which includes the requirement, specified in § 265.145, that the owner or operator of each such facility must establish financial assurance that the applicable requirements of Part 265 for post-closure care will be met, and

Whereas, this bond is written to assure that the Principal will establish a trust fund in accordance with § 265.145 for the purpose of providing for post-closure care of the following hazardous waste disposal facilities: (name and address of facility or write in "see attached Schedule A" [1] if more than one facility) ————, and shall inure to the benefit of EPA in accordance with said Part 265,

Now, therefore, the condition of this obligation is such that, if the Principal shall faithfully, for each of the facilities guaranteed

———————————————
[1] If provision for post-closure care of more than one facility is covered by the bond, list on a separate sheet the EPA Facility Identification Numbers, names, addresses, and adjusted post-closure cost estimates for all the facilities, clearly lable this list "Schedule A," and attach it to this bond. Show total of cost estimates.

by this bond, within 30 days after beginning closure, make full payment in the amount of the final adjusted post-closure cost estimate calculated in accordance with § 265.144 into a trust fund meeting the requirements of § 265.145(a) to assure the costs of 30 years of post-closure care, pursuant to all applicable statutes, rules and regulations, then and only then, the above obligation shall be void; otherwise to be and to remain in full force and effect.

The Surety shall become liable on this bond obligation only when the Principal fails to make payment in accordance with § 265.145(b)(3). Upon notification by an EPA Regional Administrator that the Principal has failed to fulfill the payment obligation, the Surety will place funds in the amount of the payment obligation into a trust fund as directed by an EPA Regional Administrator.

The liability of the Surety shall not be discharged by any payment or succession of payments hereunder, unless and until such payment or payments shall amount in the aggregate to the penal sum of the bond, but in no event shall the Surety's obligation hereunder exceed the amount of said penal sum. The insolvency or bankruptcy of the Principal shall not constitute a defense to the Surety with regard to claims of liability on the bond obligations, and in the event of said insolvency or bankruptcy, the Surety must pay any unsatisfied final judgments obtained on such claims. The Surety agrees to furnish written notice forthwith to the Regional Administrator(s) of the EPA Region(s) in which the facility(ies) is (are) located of all suits filed, judgments rendered, and payments made by the Surety under this bond.

This bond is effective the ———— day of ————, 19——, at the address of the Principal as stated herein and shall continue in force for each facility guaranteed by this bond until ninety (90) days following the beginning of closure of that facility or until receipt of written notice sent by EPA to the Surety of satisfactory completion of the financial assurance obligation of the Principal with regard to post-closure care of that facility, the sooner, or until otherwise terminated as hereinafter provided. The Surety may terminate this bond by written notice sent by certified mail to the Principal and to the EPA Regional Administrator(s) for the Region(s) in which the facility(ies) is (are) located, such termination to become effective ninety (90) days after actual receipt of said notice by EPA; provided, however, that no such termination shall become effective if closure of said facility has begun, or has been ordered to begin by an EPA Regional Administrator. The Principal may terminate this bond by sending written notice to the Surety, such termination to become effective thirty (30) days after receipt of such notice by the Surety; provided, however, that such notice is accompanied by written authorization for termination of the bond by the Regional Administrator(s) of the EPA Region(s) in which the bonded facility(ies) is (are) located.

If more than one surety company joins in executing this bond, such action shall constitute joint and several liability on the part of the sureties.

In witness whereof, the Principal and Surety have executed this instrument on the

———— day of ————, 19——.
(Seal) ————————————————
(Surety)
(Seal) ————————————————
(Principal)
(Seal) ————————————————
(attorney-in-fact)   (address of Principal) ————————————————
Surety Bond No. ————

Mail original to the EPA Regional Administrator within 10 days of the effective date by certified mail. If more than one facility is covered and the facilities are in more than one Region, send original to Regional Administrator of Region in which the largest number of facilities are located and copies to the other Regional Administrator(s), by certified mail.

Appendix VIII to Part 265

EPA Form 8700–21

U.S. Environmental Protection Agency

Post-Closure Performance Bond

As provided for in 40 CFR 265.145(d), under authority of the Resource Conservation and Recovery Act of 1976, as amended (42 USC 6901)

[1] EPA Facility Identification No. ————
Adjusted post-closure cost estimate, in accordance with 40 CFR 265.144: $ ————

Know all men by these presents, that we, (owner or operator) ———— of (address) ————, as Principal and (name of surety company) ————, a company created and existing under the laws of (State) ————, as Surety, are held and firmly bound unto the U.S. Environmental Protection Agency (EPA) in the penal sum of ———— U.S. dollars ($——) for payment of which, well and truly to be made, we bind ourselves, our heirs, executors, administrators, successors and assigns, jointly and severally, and firmly by these presents.

Whereas, the Principal intends to obtain interim status, as defined by Section 3005 of the Resource Conservation and Recovery Act of 1976, as amended, for one or more hazardous waste disposal facilities, and such status depends upon compliance with the standards of 40 CFR Part 265, which includes the requirement, specified in § 265.145, that the owner or operator of each such facility must establish financial assurance that the applicable requirements of Part 265 for post-closure care will be met, and

Whereas, this bond is written to assure compliance with the post-closure requirements of 40 CFR Part 265 for the following hazardous waste disposal facilities: (name and address of facility or write in "see attached Schedule A" [1] if more than one facility) ————, and shall inure to the benefit of EPA in accordance with said Part 265,

Now, therefore, the condition of this obligation is such that, if the Principal shall faithfully fulfill the applicable post-closure requirements set forth in 40 CFR Part 265 for each of the facilities guaranteed by this bond,

[1] If post-closure care of more than one facility is covered by the bond, list on a separate sheet the EPA Facility Identification Numbers, names, and addresses, and adjusted post-closure cost estimates for all the facilities, clearly label this list "Schedule A," and attach it to this bond. Show total of cost estimates.

pursuant to all applicable statutes, rules and regulations, and shall carry out the post-closure plan required by Part 265, then, and only then, the above obligation shall be void; otherwise to be and to remain in full force and effect.

The Surety shall become liable on this bond obligation only upon a legal determination rendered in a proceeding pursuant to Section 3008 of the Resource Conservation and Recovery Act, as amended, that the Principal has violated the post-closure requirements of 40 CFR Part 265. Following such a determination, the Surety must either complete post-closure care of the facility in accordance with the approved post-closure plan for the facility or pay the amount of the penal sum into a trust fund as directed by an EPA Regional Administrator.

The liability of the Surety shall not be discharged by any payment or succession of payments hereunder, unless and until such payment or payments shall amount in the aggregate to the penal sum of the bond, but in no event shall the Surety's obligation hereunder exceed the amount of said penal sum. The insolvency or bankruptcy of the Principal shall not constitute a defense to the Surety with regard to claims of liability on the bond obligations, and in the event of said insolvency or bankruptcy, the Surety must pay any unsatisfied final judgments obtained on such claims. The Surety agrees to furnish written notice forthwith to the Regional Administrator(s) of the EPA Region(s) in which the facility(ies) is (are) located of all suits filed, judgments rendered, and payments made by said Surety under this bond.

This bond is effective the ———— day of ————, 19——, at the address of the Principal as stated herein and shall continue in force until the end of 30 years of post-closure care unless prior notice is received by the Surety from EPA, or until terminated as hereinafter provided. The Surety may terminate this bond by written notice sent by certified mail to the Principal and to the EPA Regional Administrator(s) of the Region(s) in which the facility(ies) is (are) located, such termination to become effective ninety (90) days after actual receipt of such notice by the Agency; provided, however, that no such termination shall become effective if closure of any said facility has taken place, has begun, or has been ordered to begin by an EPA Regional Administrator. The Principal may terminate this bond by sending written notice to the Surety, such termination to become effective thirty (30) days after receipt of such notice by the Surety; provided, however, that such notice is accompanied by written authorization for termination of the bond by the Regional Administrator(s) of the EPA Region(s) in which the bonded facility(ies) is (are) located.

If more than one surety company joins in executing this bond, such action shall constitute joint and several liability on the part of the sureties.

In witness whereof, the Principal and Surety have executed this instrument on the ———— day of ————, 19——.
(Seal) ————————————————
(Surety)
(Seal) ————————————————
(Principal)

(attorney-in-fact)   (address of
Principal) ——————————————
Surety Bond No. —————

Mail original to the EPA Regional
Administrator within 10 days of the effective
date by certified mail. If more than one
facility is covered and the facilities are in
more then one Region, send original to
Regional Administrator of Region in which
the largest number of facilities are located
and copies to the other Regional
Administrator(s), by certified mail.

[FR Doc. 80–14310 Filed 5–16–80; 8:45 am]

BILLING CODE 6560–01–M

It is proposed to further amend Title
40 CFR, Part 265, by adding §§ 265.431—
265.437 to Subpart R, which has been
promulgated in today's **Federal Register**
as follows:

§ 265.431   Definitions.

The following definitions promulgated
in § 122.3 of this Chapter apply:

Formation means a body of rock
characterized by a degree of lithologic
homogeneity; which is prevailingly, but
not necessarily, tabular and mappable
on the earth's surface or traceable in the
subsurface.

Formation fluid means "fluid" present
in a " formation" under natural
conditions as opposed to introduced
fluids, such as drilling mud.

Injection well means a "well" into
which "fluids" are being injected.

Injection zone means a geological
"formation", group of formations, or part
of a formation receiving fluids through a
well.

Plugging means the act or process of
stopping the flow of water, oil, or gas in
formations penetrated by a borehole or
well.

Underground source of drinking water
("USDW") means an aquifer or its
portion: (a) which supplies drinking
water for human consumption; or (b) in
which the ground water contains fewer
than 10,000 mg/l "total dissolved
solids".

§ 265.432   General operating requirements.

The owner or operator of a Class I
well for disposal of hazardous waste
must prevent migration of hazardous
waste or hazardous waste constituents
into or between underground sources of
drinking water as follows:

(a) Wells must be cased and cemented
between the well bore and casing;

(b) Hazardous waste must be injected
through tubing, with a packer set
immediately above the injection zone
and with the annulus between the
tubing and the long string of casings
filled with fluid, or by another equally
effective technique for which the owner
or operator has a written demonstration,
available for review by the Regional

Administrator, indicating that it
provides a comparable level of
protection to underground sources of
drinking water.

(c) Injection of hazardous waste
between the outermost casing and the
well bore is prohibited; and

(d) Injection pressure at the well head
must not exceed a maximum pressure
which must be calculated so as to
assure that the pressure in the injection
zone during injection does not initiate
new fractures or propagate existing
fractures in the injection zone, initiate
fractures in the confining zone or
otherwise cause the migration of
hazardous waste, hazardous waste
constituents, or formation fluids into an
underground source of drinking water.

§ 265.433   Waste analysis.

For disposal of hazardous waste by
underground injection the owner or
operator must, in addition to the waste
analyses required by § 265.13:

(a) Conduct waste analyses and trial
tests; or

(b) Present written, documented
information from his or similar disposal
operations to show that this disposal
will comply with § 265.17(b) and for
Class I wells, that the waste is
compatible with fluids in the injection
zone and minerals in both the injection
zone and the confining zone and will not
damage the mechanical integrity of the
well.

§ 265.434   Monitoring and response.

(a) The owner or operator of a facility
which disposes of hazardous waste by
underground injection into a Class I well
must:

(1) On the effective date of these
regulations develop and submit to the
Regional Administrator a plan for a
monitoring program capable of
determining compliance with § 265.432,
by:

(i) Demonstrating the mechanical
integrity of the injection well to satisfy
§ 265.432(a) and (b); and

(ii) Demonstrating that the pressure of
the injected fluids remains within
allowable limits to satisfy § 265.432(d).

(2) The plan to be submitted under
paragraph (a) of this section must
specify:

(i) For demonstrating mechanical
integrity:

(A) The annual pressure range to be
maintained and basis for determining it
for the specific well tubing, packer and
casing characteristics and for the
anticipated injection fluid temperatures;

(B) The devices and procedures for
continuous monitoring and recording of
the annular pressure, and evaluation of
that information; and

(C) Procedures for immediate

response to changes in the annular
pressure outside the allowable range,
and for restoration of mechanical
integrity;

(ii) For demonstrating that injection
fluid pressure remains within allowable
limits:

(A) The calculated fracture pressure
and the basis for determining it for the
specific formation and zone of injection;

(B) The calculated allowable injection
pressure to be measured at the well
head and the basis for determining it for
specific injection fluid characteristics
(i.e., specific gravity, viscosity and
temperature);

(C) The techniques and procedures for
continuous monitoring and recording of
the injection pressure at the well head,
for evaluation of that information; and

(D) Procedures for immediate
response to an increase in the well head
pressure above the allowable limit, to
restore pressure to within allowable
limits.

(3) On the effective date of these
regulations the owner or operator must
implement the monitoring plan which
satisfies paragraph (a)(2) of this section
and determine the mechanical integrity
of the well and the injection zone
pressure.

(4) The owner or operator must keep
records of the monitoring data and
evaluations specified in paragraphs
(a)(2) (i) and (ii) of this section
throughout the active life of the facility.

(5) The owner or operator must submit
an annual report to the Regional
Administrator which assures
compliance with § 265.432. He must
separately identify in the annual report
those corrective actions, specified in
paragraphs (a)(2)(i) (C) and (a)(2)(ii)(D)
of this section which were implemented
during the reporting period, and an
explanation of the circumstances which
required corrective action.

(b) The owner or operator of a facility
which disposes of hazardous waste by
underground injection into a Class IV
well which discharges above an
underground source of drinking water
must monitor the ground water in
accordance with the requirements of
Subpart F of this Part.

§ 265.435   Closure and post-closure.

(a) The owner or operator must close
his facility in a manner that:

(1) Will prevent the migration of
hazardous waste or hazardous waste
constituents into or between
underground sources of drinking water
via the well structure; and

(2) Will minimize the need for further
maintenance to protect human health
and the environment.

(b) On the effective date of these

regulations, the owner or operator must have a written closure plan. He must keep this plan at the facility. This plan must identify the steps necessary to completely close the facility. The closure plan must:

(1) Identify the techniques to be used to close the well in accordance with paragraphs (c) and (d) of this Section;

(2) Describe the steps which are necessary to decontaminate facility equipment during closure; and

(3) Include a schedule for final closure which specifies the anticipated date when wastes will no longer be received, the anticipated date when final closure will be completed, and intervening milestone dates for tracking the progress of closure.

(c) The owner or operator may amend his closure plan at any time during the active life of the facility. The owner or operator must amend his plan any time changes in operating plans or facility design affect the closure plan.

(d) The owner or operator must submit his closure plan to the Regional Administrator at least 180 days before the date he expects to begin closure. The Regional Administrator will modify, approve, or disapprove the plan within 90 days of receipt and after providing the owner or operator and the affected public (through a newspaper notice) the opportunity to submit written comments. If an owner or operator plans to begin closure within 180 days after the effective date of these regulations, he must submit the necessary plans on the effective date of these regulations.

(e) Within 90 days after receiving the final volume of hazardous wastes, the owner or operator must treat all hazardous wastes in storage or in treatment, or remove them from the site, or dispose of them on-site, in accordance with the approved closure plan.

(f) The owner or operator must complete closure activities in accordance with the approved closure plan and within six months after receiving the final volume of wastes. The Regional Administrator may approve a longer closure period under paragraph (d) of this section if the owner or operator can demonstrate that:

(1) The required or planned closure activities will, of necessity, take him longer than six months to complete, and

(2) That he has taken all steps to eliminate any significant threat to human health and the environment from the unclosed but inactive facility.

(g) The owner or operator of a Class I well must close by plugging to satisfy paragraph (a) of this section.

(h) At closure, the owner or operator of a Class IV well which discharges above an underground source of

drinking water must;

(1) Remove the hazardous waste remaining in the well; and

(2) Close the well in a manner which satisfies paragraph (a) of this section.

[*Comment:* At closure, as throughout the operating period, unless the owner or operator can demonstrate, in accordance with § 261.3 (c) or (d) of this Chapter, that any solid waste removed from the injection well is not a hazardous waste, he becomes a generator of hazardous waste and must manage it in accordance with all applicable requirements of Parts 262, 263, and 265 of this Chapter.]

(i) When closure is completed, the owner or operator must submit to the Regional Administrator certification both by the owner or operator and by an independent registered professional engineer that the facility has been closed in accordance with the specifications in the approved closure plan.

(j) The owner or operator of a Class IV well which discharges above an underground source of drinking water must provide post-closure care in accordance with the applicable requirements of §§ 265.117–265.120 (see Subpart G of this Part).

§ 265.436   Financial requirements.

(a) On the effective date of these regulations, the owner or operator of a facility which disposes of hazardous waste by underground injection must have a written estimate of the cost of closing the facility in accordance with the requirements in § 265.435. The owner or operator must keep this estimate, and all subsequent estimates required in this Section, at the facility.

(b) The owner or operator must prepare a new closure cost estimate whenever a change in the closure plan affects the cost of closure.

(c) The owner or operator must maintain financial responsibility in the form of performance bonds or other equivalent form of financial assurance to close a facility which disposes of hazardous waste by underground injection. In lieu of individual performance bonds, owners or operators may furnish a bond or other equivalent form of financial guarantee covering all facilities which dispose of hazardous waste by underground injection in any one State.

(d) On the effective date of these regulations an owner or operator of a facility which disposes of hazardous waste by underground injection in a Class IV well which discharges above underground sources of drinking water must have a written estimate of the annual cost of post-closure monitoring

and maintenance in accordance with the applicable post-closure requirements in §§ 265.117–265.120. This estimate, and all subsequent estimates, must be kept at the facility.

(e) The cost estimate required in paragraph (d) of this section must be revised whenever a change in the post-closure care plan affects the cost of post-closure care (see § 265.118(b)). The latest post-closure cost estimate is calculated by multiplying the latest annual post-closure cost estimate by 30.

(f) On each anniversary of the effective date of these regulations, the owner or operator must adjust the latest post-closure cost estimate using an inflation factor derived from the annual Implicit Price Deflator for Gross National Product as published by the U.S. Department of Commerce in its *Survey of Current Business.* The inflation factor must be calculated by dividing the latest annual published Deflator by the Deflator for the previous year. The result is the inflation factor. The adjusted post-closure cost estimate must equal the latest post-closure cost estimate times the inflation factor.

§ 265.437   Special requirements for ignitable, reactive or incompatible wastes.

Ignitable, reactive or incompatible wastes (see Appendix V for examples) must not be disposed by underground injection unless § 265.17(b) is satisfied.

[FR Doc. 80–14310 Filed 5–16–80; 8:45 am]

BILLING CODE 6560–01–M

APPENDIX E
DOT REGULATIONS OF INTEREST (CONTENTS ONLY)

173.182 Nitrates.
173.183 Potassium nitrate mixed (fused) with sodium nitrite.
173.184 Nitrocellulose or collodion cotton, wet; or nitrocellulose, colloided, granular, or flake, wet, nitrostarch, wet, or nitroguanidine, wet.
173.185 Paper stock, wet.
173.186 Paper waste, wet.
173.187 Peroxide of sodium.
173.188 Phosphoric anhydride.
173.189 Phosphorus, amorphous, red.
173.190 Phosphorus, white or yellow.
173.191 Phosphorus pentachloride.
173.192 Picrate of ammonia (ammonium picrate), picric acid, trinitrobenzoic acid, and urea nitrate wet.
173.193 Picric acid, trinitrobenzoic acid, or urea nitrate, wet.
173.194 Potassium permanganate.
173.195 Pyroxylin plastic scrap.
173.196 [Reserved]
173.197 Pyroxylin plastics, in sheets, rolls, rods, or tubes.
173.197a Smokeless powder for small arms.
173.198 Sodium hydride.
173.199 Rags, oily.
173.200 Rags, wet.
173.201 Rubber scrap, rubber buffings, reclaimed rubber, and regenerated rubber.
173.202 Sodium and potassium, metallic liquid alloy.
173.203 Tetranitromethane.
173.204 Sodium hydrosulfite.
173.205 Sodium picramate, wet.
173.206 Sodium or potassium, metallic; sodium amide; sodium potassium alloys; sodium aluminum hydride; lithium metal; lithium silicon; lithium ferro silicon; lithium hydride; lithium borohydride; lithium aluminum hydride; lithium acetylide-ethylene diamine complex; aluminum hydride; cesium metal; rubidium metal; zirconium hydride, powdered.
173.207 Sulfide of sodium or sulfide of potassium, fused or concentrated, when ground.
173.208 Titanium metal powder, wet or dry.
173.209 Tankage, garbage, and tankage fertilizers.
173.210 Tankages, rough ammoniate.
173.211 Textile waste, wet.
173.212 Trinitrobenzene and trinitrotoluene, wet.
173.213 Wool waste, wet.
173.214 Hafnium metal or zirconium metal, wet, minimum 25 percent water by weight, mechanically produced, finer than 270 mesh particle size; hafnium metal or zirconium metal, dry, in an atmosphere of inert gas, mechanically produced, finer than 270 mesh particle size; hafnium metal or zirconium metal, wet, minimum 25 percent water by weight, chemically produced (See Note 1), finer than 20 mesh particle size; hafnium metal or zirconium metal, dry, in an atmosphere of inert gas, chemically produced (See Note 1), finer than 20 mesh particle size.
173.216 Zirconium picramate, wet.
173.217 Calcium hypochlorite mixture, dry; lithium hypochlorite mixture, dry; mono-(trichloro) tetra-(monopotassium dichloro)-penta-s-triazinetrione, dry; potassium dichloro-s-triazinetrione, dry; sodium dichloro-s-triazinetrione, dry; trichloro-s-triazinetrione, dry.
173.218 Isopropyl percarbonate, unstabilized.

173.219 Potassium perchlorate.
173.220 Magnesium or zirconium scrap consisting of borings, clippings, shavings, sheets, turnings, or scalpings, and magnesium metallic (other than scrap), powdered, pellets, turnings, or ribbon.
173.221 Liquid organic peroxides, n.o.s., and liquid organic peroxide solutions, n.o.s.
173.222 Acetyl peroxide and acetyl benzoyl peroxide, solution.
173.223 Peracetic acid.
173.224 Cumene hydroperoxide, dicumyl peroxide, diisopropylbenzene hydroperoxide, paramenthane hydroperoxide, and tertiary butylisopropyl benzene hydroperoxide.
173.225 Phosphorus trisulfide, phosphorus sesquisulfide, phosphorus heptasulfide, and phosphorus pentasulfide.
173.226 Thorium metal, powdered.
172.227 Urea peroxide.
173.228 Zinc ammonium nitrite.
173.229 Chlorate and borate mixtures or chlorate and magnesium chloride mixtures.
173.230 Sodium, metallic, dispersion in organic solvent.
173.231 Calcium, metallic, crystalline.
173.232 Aluminum, metallic powder.
173.233 Nickel catalyst, finely divided, activated or spent.
173.234 Sodium nitrite and sodium nitrite mixtures.
173.235 Ammonium bichromate (ammonium dichromate).
173.236 Decaborane.
173.237 Chlorine dioxide hydrate, frozen; chloric acid.
173.238 Aircraft rocket engines (commercial) and/or aircraft rocket engine igniters (commercial).
173.239 Barium azide—50 percent or more water wet.
173.239a Ammonium perchlorate.

Subpart F—Corrosive Materials: Definition and Preparation

Sec.
173.240 Corrosive material; definition.
173.241 Outage.
173.242 Bottles containing corrosive liquids.
173.243 Closing and cushioning.
173.244 Limited quantities of corrosive materials.
173.245 Corrosive liquids not specifically provided for.
173.245a Corrosive liquids, n.o.s. shipped in bulk.
173.245b Corrosive solids not specifically provided for.
173.246 Antimony pentafluoride, bromide pentafluoride, iodine pentafluoride, bromine trifluoride, and chlorine trifluoride.
173.247 Acetyl bromide, acetyl chloride, acetyl iodide, antimony pentachloride, benzoyl chloride, boron trifluoride-acetic acid complex, chromyl chloride, dichloroacetyl chloride, diphenylmethyl bromide solution, pyro sulfuryl chloride, silicon chloride, sulfur chloride (mono and di), sulfuryl chloride, thionyl chloride, tin tetrachloride (anhydrous), titanium tetrachloride, and trimethyl acetyl chloride.
173.247a Vanadium tetrachloride and vanadium oxytrichloride.

173.248 Acid sludge, sludge acid, spent sulfuric acid, or spent mixed acid.
173.249 Alkaline corrosive liquids, n.o.s.; Alkaline liquids, n.o.s.; Alkaline corrosive battery fluid; Potassium fluoride solution; Potassium hydrogen fluoride solution; Sodium aluminate, liquid; Sodium hydroxide solution; Potassium hydroxide solution; Boiler compound, liquid, solution.
173.249a Cleaning compound, liquid; Coal tar dye, liquid; Dye intermediate, liquid; Mining reagent, liquid; and Textile treating compound mixture, liquid.
173.250 Automobiles, other self-propelled vehicles, engines or other mechanical apparatus.
173.250a Benzene phosphorus dichloride and benzene phosphorus thiodichloride.
173.251 Boron trichloride and boron tribromide.
173.252 Bromine.
173.253 Chloracetyl chloride.
173.254 Chlorosulfonic acid and mixtures of chlorosulfonic acid-sulfur trioxide.
173.255 Dimethyl sulfate.
173.256 Compounds, cleaning, liquid.
173.257 Electrolyte (acid) and alkaline corrosive battery fluid.
173.258 Electrolyte, acid, or alkaline corrosive battery fluid, packed with storage batteries.
173.259 Electrolyte, acid, or alkaline corrosive battery fluid, packed with battery charger, radio current supply device, or electronic equipment and actuating devices.
173.260 Electric storage batteries, wet.
173.261 Fire-extinguisher charges.
173.262 Hydrobromic acid.
173.263 Hydrochloric (muriatic) acid, hydrochloric (muriatic) acid mixtures, hydrochloric (muriatic) acid solution, inhibited, sodium chlorite solution (not exceeding 42 percent sodium chlorite), and cleaning compounds, liquids, containing hydrochloric (muriatic) acid.
173.264 Hydrofluoric acid; White acid.
173.265 Hydrofluosilicic acid.
173.266 Hydrogen peroxide solution in water.
173.267 Mixed acid (nitric and sulfuric acid) (nitrating acid).
173.268 Nitric acid.
173.269 Perchloric acid.
173.270 Phosphorus tribromide.
173.271 Phosphorus oxybromide, phosphorus oxychloride, phosphorus trichloride, and thiophosphoryl chloride.
173.272 Sulfuric acid.
173.273 Sulfur trioxide, stabilized.
173.274 Fluosulfonic acid.
173.275 Difluorophosphoric acid, anhydrous, monofluorophosphoric acid, anhydrous, hexafluorophosphoric acid, and mixtures thereof.
173.276 Anhydrous hydrazine and hydrazine solution.
173.277 Hypochlorite solutions.
173.278 Nitrohydrochloric acid.
173.279 Anisoyl chloride.
173.280 Trichlorosilanes.
173.281 Benzyl bromide (bromotoluene, alpha).
173.282 Isopropyl percarbonate, stabilized.
173.283 Fluoboric acid.
173.284—173.285 [Reserved]
173.286 Chemical kits.
173.287 Chromic acid solution.
173.288 Chloroformates.

pentachloride, potassium hydrogen sulfate, sodium aluminate, sodium hydrogen sulfate, and/or sodium hydrogen sulfite (each in solid form).
173.850 Lime, unslaked; quicklime; and calcium oxide.
173.860 Mercury, metallic.
173.861 Gallium metal, liquid.
173.862 Gallium metal, solid.

Subpart M—Other Regulated Material; ORM-C

173.910 Ammonium sulfate nitrate.
173.915 Battery parts.
173.920 Bleaching powder.
173.925 Box toe board.
173.930 Burlap bags, used and unwashed or not cleaned.
173.931 Burlap cloth, burlap bags, new, used, and washed, or vacuum cleaned, wheel cleaned, or otherwise mechanically cleaned.
173.945 Calcium cyanamide, not hydrated.
173.952 Castor beans and castor pomace.
173.955 Coconut meal pellets.
173.960 Copra.
173.965 Cotton and other fibers.
173.970 Cotton batting, batting dross, wadding, seed hull fiber, shavings, pulp, and cut linters.
173.975 Cotton sweepings; and textile, cotton, felt, or wool waste.
173.980 Excelsior.
173.985 Exothermic ferrochrome, ferromanganese, and silicon-chrome.
173.990 Feed, wet, mixed.
173.995 Fish scrap and fish meal.
173.1000 Garbage tankage, rough ammoniate tankage, or tankage fertilizer.
173.1005 Hay or straw.
173.1010 Lead dross or scrap.
173.1020 Magnetized material.
173.1025 Metal borings, shavings, turnings or cuttings.
173.1030 Oakum or twisted jute packing.
173.1035 Oiled material.
173.1040 Pesticide, water-reactive.
173.1045 Petroleum coke, uncalcined.
173.1060 Rosin.
173.1065 Rubber curing compound, solid.
173.1070 Sawdust or wood shavings.
173.1075 Scrap paper or waste.
173.1080 Sulfur.
173.1085 Yeast, active (in liquid or compressed form).

Subpart N—Other Regulated Material; ORM-D

173.1200 Consumer Commodity.

PART 178—SHIPPING CONTAINER SPECIFICATIONS

Sec.
178.0 Purpose, scope, and applicability.

Subpart A—Specifications for Carboys, Jugs in Tubs, and Rubber Drums

178.1 Specification 1A; boxed carboys.
178.2 Specification 1B; boxed lead carboys.
178.3 Specification 1C; carboys in kegs.
178.4 Specification 1D; boxed glass carboys.
178.5 Specification 1X; boxed carboys, 5 to 6½ gallons, for export only.
178.6 Specification 1EX; glass carboys in plywood drums.
178.7 Specification 1E; glass carboys in plywood drums.
178.8 Specification 28; metal-jacketed lead carboys.

178.9 Specification 28A; metal-jacketed lead carboys.
178.12 Specification 34B; aluminum carboys.
178.13 Specification 1H; polyethylene carboys in low carbon steel or other equally efficient metal crates.
178.14 Specification 1K; glass carboys cushioned with expandable polystyrene in wooden wirebound box outside containers.
178.15 Specification 31; jugs in tubs.
178.16 Specification 35; non-reusable molded polyethylene drum for use without overpack; removable head required.
178.18 Specification 43A; rubber drums.
178.19 Specification 34; reusable molded polyethylene container for use without overpack. Removable head not authorized.

Subpart B—Specifications for Inside Containers, and Linings

178.20 Specification 2A; inside containers, metal cans pails and kits.
178.21 Specification 2T; polyethylene container.
178.22 Specification 2C; inside containers, corrugated fiberboard cartons.
178.23 Specification 2D; inside containers, duplex paper bags.
178.24 Specification 2U; molded or thermoformed polyethylene containers having rated capacity of over one gallon. Removable head containers or containers fabricated from film not authorized.
178.24a Specification 2E; inside polyethylene bottle.
178.25 Specification 2F; inside metal containers and liners.
178.26 Specification 2G; inside containers, fiber cans and boxes.
178.27 Specification 2TL; polyethylene container.
178.28 Specification 2J; inside containers, waterproof paper bags for linings.
178.29 Specification 2K; inside containers, paper bags for linings.
178.30 Specification 2L; lining for boxes.
178.31 Specification 2M; waterproofed paper lining.
178.32 Specification 2N; inside containers, metal cans.
178.33 Specification 2P; inside nonrefillable metal containers.
178.33a Specification 2Q; inside nonrefillable metal containers.
178.34 Specification 2R; inside containment vessel.
178.35 Specification 2S; polyethylene container.
178.35a Specification 2SL; molded or thermoformed polyethylene container.

Subpart C—Specifications for Cylinders

178.36 Specification 3A; seamless steel cylinders or 3AX; seamless steel cylinders of capacity over 1,000 pounds water volume.
178.37 Specification 3AA; seamless steel cylinders made of definitely prescribed steels or 3AAX; seamless steel cylinders made of definitely prescribed steels of capacity over 1,000 pounds water volume.
178.38 Specification 3B; seamless steel cylinders.
178.39 Specification 3BN; seamless nickel cylinders.

178.40 Specification 3C; seamless steel cylinders.
178.41 Specification 3D; seamless steel cylinders.
178.42 Specification 3E; seamless steel cylinders.
178.43 Specification 3A480X; seamless steel cylinders.
178.44 Specification 3HT; inside containers, seamless steel cylinders for aircraft use made of definitely prescribed steel.
178.45 Specification 3T; seamless steel cylinder.
178.47 Specification 4DS; inside containers, welded stainless steel for aircraft use.
178.48 Specification 4; forge welded steel cylinders.
178.49 Specification 4A; forge welded steel cylinders.
178.50 Specification 4B; welded and brazed steel cylinders.
178.51 Specification 4BA; welded or brazed steel cylinders made of definitely prescribed steels.
178.52 Specification 4C; welded and brazed steel cylinders.
178.53 Specification 4D; inside containers, welded steel for aircraft use.
178.54 Specification 4B240-FLW; welded or welded and brazed cylinders with fusion-welded longitudinal seam.
178.55 Specification 4B240ET; welded and brazed cylinders made from electric resistance welded tubing.
178.56 Specification 4AA480; welded steel cylinders made of definitely prescribed steels.
178.57 Specification 4L; welded cylinders insulated.
178.58 Specification 4DA; inside containers, welded steel for aircraft use.
178.59 Specification 8; steel cylinders with approved porous filling for acetylene.
178.60 Specification 8AL; steel cylinders with approved porous filling for acetylene.
178.61 Specification 4BW; welded steel cylinders made of definitely prescribed steels with electric-arc welded longitudinal seam.
178.63 [Reserved]
178.65 Specification 39; non-reusable (non-refillable) cylinder.
178.66—178.67 [Reserved]
178.68 Specification 4E; welded aluminum cylinders.

Subpart D—Specifications for Metal Barrels, Drums, Kegs, Cases, Trunks, and Boxes

178.80 Specification 5; steel barrels or drums.
178.81 Specification 5A; steel barrels or drums.
178.82 Specification 5B; steel barrels or drums.
178.83 Specification 5C; steel barrels or drums.
178.84 Specification 5D; steel barrels or drums, lined.
178.85 Specification 5F; steel drums.
178.87 Specification 5H; steel barrels or drums, lead lined.
178.88 Specification 5K; nickel barrels or drums.
178.89 Specification 5L; steel barrels or drums.
178.90 Specification 5M; monel drums.
178.91 Specification 5X; steel drums, aluminum lined.
178.92 Specification 5P; lagged steel drums.
178.97 Specification 6A; steel barrels or drums.

178.98 Specification 6B; steel barrels or drums.
178.99 Specification 6C; steel barrels or drums.
178.100 Specification 6J; steel barrels and drums.
178.101 Specification 6K; steel barrels or drums.
178.102 Specification 6D; cylindrical steel overpack, straight sided, for inside plastic container.
178.103 Specification 6L; metal packaging.
178.104 Specification 6M; metal packaging.
178.107 Specification 42B; aluminum drums.
178.108 Specification 42C; aluminum barrels or drums.
178.109 Specification 42D; aluminum drums.
178.110 Specification 42F; aluminum barrels or drums.
178.111 Specification 42G; aluminum drums.
178.112 Specification 42H; aluminum drums; removable head containers not authorized.
178.115 Specification 17C; steel drums.
178.116 Specification 17E; steel drums.
178.117 Specification 17F; steel drums.
178.118 Specification 17H; steel drums.
178.119 Specification 17X; steel barrels or drums.
178.120 Specification 20PF phenolic-foam insulated, metal overpack.
178.121 Specification 21PF fire and shock resistant, phenolic-foam insulated, metal overpack.
178.130 Specification 37K; steel drums.
178.131 Specification 37A; steel drums.
178.132 Specification 37B; steel drums.
178.133 Specification 37P; steel drums with polyethylene liner.
178.134 Specification 37M; cylindrical steel overpack, straight sided for inside plastic container; nonreusable containers.
178.135 Specification 37C; steel drums.
178.136 Specification 42E; aluminum drums.
178.137 Specification 37D; steel drum. Non-reusable container. Open-head not authorized.
178.140 Specification 13; metal kegs.
178.141 Specification 13A; metal drums.
178.146 Specification 32A; metal cases, riveted or lock seamed.
178.147 Specification 32B; metal cases, welded or riveted.
178.148 Specification 32C; metal trunks.
178.149 Specification 32D; metal boxes for old and worn-out motion-picture film no longer exhibitable.
178.150 Specification 33A; polystyrene cases. Nonreusable containers.

Subpart E—Specifications for Wooden Barrels, Kegs, Boxes, Kits, and Drums

178.156 Specification 10B; wooden barrels and kegs (tight).
178.165 Specification 14; wooden boxes, nailed.
178.168 Specification 15A; wooden boxes, nailed.
178.169 Specification 15B; wooden boxes, nailed.
178.170 Specification 15C; wooden boxes, nailed.
178.171 Specification 15D; wooden boxes, nailed.
178.172 Specification 15E; wooden boxes, fiberboard lined.
178.176 Specification 15L; wooden boxes with inside containers for desensitized liquid explosives.

178.177 Specification 15M; wooden boxes, metal lined, with inside containers for desensitized liquid explosives.
178.181 Specification 15X; wooden boxes for two five-gallon cans.
178.182 Specification 15P; glued plywood, or wooden box for inside containers.
178.185 Specification 16A; plywood or wooden boxes, wirebound.
178.186 Specification 16B; wooden boxes, wirebound.
178.187 Specification 16D; wooden wirebound overwrap for inside containers.
178.190 Specification 19A; wooden boxes, glued plywood cleated.
178.191 Specification 19B; wooden boxes, glued plywood, nailed.
178.193 Specification 18B; wooden kits.
178.194 Specification 20WC wooden protective jacket.
178.195 Specification 21WC wooden-steel protective overpack.
178.196 Specification 22A; wooden drums, glued plywood.
178.197 Specification 22B; wooden drums, glued plywood.
178.198 Specification 22C; plywood drum for plastic inside container.

Subpart F—Specifications for Fiberboard Boxes, Drums, and Mailing Tubes

178.205 Specification 12B; fiberboard boxes.
178.206 Specification 12C; fiberboard boxes.
178.207 Specification 12D; fiberboard boxes.
178.208 Specification 12E; fiberboard boxes.
178.209 Specification 12H; fiberboard boxes.
178.210 Specification 12A; fiberboard boxes.
178.211 Specification 12P; fiberboard boxes. Nonreusable containers for one inside plastic container greater than 1-gallon capacity, as prescribed in Part 173 of this chapter.
178.212 Specification 12R; paper-faced expanded polystyrene board boxes. Nonreusable containers.
178.214 Specification 23F; fiberboard boxes.
178.218 Specification 23G; special cylindrical fiberboard box for high explosives.
178.219 Specification 23H; fiberboard boxes.
178.224 Specification 21C; fiber drum.
178.225 Specification 21P; fiber drum overpack for inside plastic container.
178.226 Specification 29; mailing tubes.

Subpart G—Specifications for Bags, Cloth, Burlap, Paper or Plastic

Sec.
178.230 Specification 36A; lined cloth bags (triplex).
178.233 Specification 36B; burlap bags, lined.
178.234 Specification 36C; burlap bags, paper lined.
178.236 Specification 44B; multiwall paper bags.
178.237 Specification 44C; multiwall paper bags.
178.238 Specification 44D; multiwall paper bags.
178.239 Specification 44E; multiwall paper bags.
178.240 Specification 45B; bags, cloth and paper, lined.
178.241 Specification 44P; all-plastic bags.

Subpart H—Specifications for Portable Tanks

178.245 Specification 51; steel portable tanks.
178.251 General design and construction requirements applicable to specifications 56 (§ 178.252) and 57 portable tanks (§ 178.253).
178.252 Specification 56; metal portable tank.
178.253 Specification 57; metal portable tank.
178.255 Specification 60; steel portable tanks.

Subpart I—[Reserved]

Subpart J—Specifications for Containers for Motor Vehicle Transportation

178.315 Specification MC 200; containers for liquid nitroglycerin, desensitized liquid nitroglycerin or diethylene glycol dinitrate.
178.318 Specification MC 201; container for blasting caps, electric blasting caps and percussion caps.
178.337 Specification MC 331; cargo tanks constructed of steel, primarily for transportation of compressed gases as defined in the Compressed Gas Section.
178.340 General design and construction requirements applicable to specifications MC 306 (§ 178.341), MC 307 (§ 178.342), and MC 312 (§ 178.343) cargo tanks.
178.341 Specification MC 306; cargo tanks.
178.342 Specification MC 307; cargo tanks.
178.343 Specification MC 312; cargo tanks.

Subpart K—Specifications for General Packagings

178.350 Specification 7A; general packaging, Type A.

PART 179—SPECIFICATIONS FOR TANK CARS

Subpart A—Introduction, Approvals and Reports

Sec.
179.1 General.
179.2 Definitions and abbreviations.
179.3 Procedure for securing approval.
179.4 Changes in specifications for tank cars.
179.5 Certificate of construction.
179.6 Repairs and alterations.

Subpart B—General Design Requirements

179.10 Tank mounting.
179.11 Welding certification.
179.12 Interior heater systems.
179.13 Tank car capacity and gross weight limitation.
179.14 Tank car couplers.

Subpart C—Specifications for Pressure Tank Car Tanks (Classes DOT-105, 109, 112, and 114)

179.100 General specification applicable to pressure tank car tanks.
179.101 Individual specification requirements applicable to pressure tank car tanks.
179.102 Special commodity requirements for pressure tank car tanks.
179.103 Special requirements for class 114A * * * tank car tanks.
179.104 Special requirements for spec. 105-A200-F tank car tanks.

179.105 Special requirements for Specifications 112 and 114 tank cars.

Subpart D—Specifications for Nonpressure Tank Car Tanks (Classes DOT-103, 104, 111AF, 111AW, and 115AW)

179.200 General specifications applicable to non-pressure tank cars tanks (Classes DOT-103, 104, and 111).

179.201 Individual specification requirements applicable to non-pressure tank car tanks.

179.202 Special commodity requirements for non-pressure tank car tanks.

179.220 General specifications applicable to nonpressure tank car tanks consisting of an inner container supported within an outer shell (class DOT-115).

179.221 Individual specification requirements applicable to tank car tanks consisting of an inner container supported within an outer shell.

Subpart E—Specifications for Multi-Unit Tank Car Tanks (Classes DOT-106A and 110AW)

Sec.

179.300 General specifications applicable to multi-unit tank car tanks designed to be removed from car structure for filling and emptying (Classes DOT-106A and 110AW).

179.301 Individual specification requirements for multi-unit tank car tanks.

179.302 Special commodity requirements for multi-unit tank car tanks.

Subpart F—Specifications for Liquefied Hydrogen Tank Car Tanks and Seamless Steel Tanks (Classes DOT-113A-W and 107A)

179.400 General specifications applicable to liquefied hydrogen tank car tanks.

179.401 Individual specification requirements for liquefied hydrogen tank car tanks.

179.500 Specification DOT-107A * * * *, seamless steel tank car tanks.

Index

Date Due